ORGANIC PHOSPHORUS COMPOUNDS

ORGANIC PHOSPHORUS COMPOUNDS

Volume 1

G. M. KOSOLAPOFF
Auburn University

and

L. MAIER
Monsanto Research S. A.

WILEY-INTERSCIENCE, a Division of John Wiley & Sons, Inc.
New York • London • Sydney • Toronto

Library of Congress Cataloging in Publication Data

Kosolapoff, Gennady M

Organic phosphorous compounds.

1950 ed. published under title: Organophosphorus compounds.

Includes bibliographies.

1. Organophosphorus compounds. I. Maier, L., joint author. II. Title.

QD412.P1K55 1972 547'.07 72-1359

ISBN 0-471-50440-8 (v. 1)

Printed in the United States of America

10 9 8 7 6 5 4 3 2

Preface

In the first edition, published in 1950, an attempt was made to give a comprehensive coverage of the organic chemistry of phosphorus compounds. This objective was achieved in 376 pages of text and tables. The subsequent vast expansion of knowledge about organic phosphorus compounds made it impossible for one person to encompass the entire field. It was therefore decided to divide the subject into 21 chapters and seek the help of experts for the various chapters. We consider ourselves very fortunate to have obtained the cooperation of many distinguished experts in the field.

The unexpected quantity of the material made necessary the publishing of several volumes. The unusually large number of organic phosphorus compounds is explained by the ability of phosphorus to occur in different valent and oxidation states and to bind up to six organic groups directly or through heteroatoms. For the purpose of clarity the subject is divided systematically into different classes of compounds. Unfortunately, analogous reaction types could not be dealt with under one point and are therefore discussed in several chapters insofar as they lead to different reaction products.

As in the first edition, the chapters are planned to make it easy to find the methods of preparing various compound types. The chapters also list known substances in each type (including the compounds of the first edition), the principal physical properties of individual substances, and proper references to the original literature. We encouraged our coauthors to list all the compounds that have been identified.

Compounds that contain two or more phosphorus atoms in the molecule with different valent or oxidation states are listed under the first heading in the list of contents. For example, $R_2PCH{=}PR_3$ is listed under phosphines in Chapter 1 and not under phosphine alkylenes in Chapter 5A; $[R_2PCH_2PR_3]^+X^-$ is listed under phosphines in Chapter 1 and not under phosphonium salts in Chapter 4; $R_2P(O)CH_2P(S)R_2$ is listed under phosphine oxides in Chapter 6 and not under phosphine sulfides in Chapter 7; and $[(EtO)_2(O)PCH_2]_3P{=}O$ is listed under phosphine oxides in Chapter 6 and not under phosphonic acids and derivatives in Chapter 18.

Each chapter has a brief introductory statement followed by a discussion of the methods of synthesis that are applicable to the particular class of compounds. Each type of synthesis is labeled with a number that appears immediately after the structural formula of each compound in the compound list. The synthetic section of each chapter is followed by a summary of the general chemical reactions and physical properties of the particular class.

The nomenclature used in these volumes follows, in general, the system adopted by the American Chemical Society's Committee on the Nomenclature of Organic Phosphorus Compounds (Ann: Chem. Eng. News, 30, 4515 (1952), and J. Chem. Soc., 1952, 5122).

The following abbreviations are used: Me = CH_3, Et = C_2H_5, i-Pr = iso-C_3H_7, Pr = n-C_3H_7, Bu = n-C_4H_9, i-Bu = iso-C_4H_9, t-Bu = t-C_4H_9, Am = n-C_5H_{11}, Ph = C_6H_5, M = metal, m. or M. = melting point, b. = boiling point, b_{11} = boiling point at 11 torr, $\bar{D}$(P-C) = mean dissociation energy of P-C, elect. diff. = electron diffraction, microw. spect. = microwave spectrum, μ = dipole moment, ^{31}P = ^{31}P-chemical shift, ^{1}H = ^{1}H-NMR, L = ligand, n_D^{20} = refractive index at 20° referring to the D-line of sodium, d_4^{20} = density at 20° in reference to water at 4°, and coupl. const. = coupling constant.

In compiling these volumes use was made of the chemical literature published before 1970. In several chapters the literature appearing during 1970 was included. The patent literature was included whenever abstracts are available in Chemical Abstracts.

We have encouraged our coauthors to exhaustively review the literature on organic phosphorus compounds. The task was a formidable one, and it was in particular very time-consuming to prepare the extensive list of compounds. We wish to thank all our coauthors for cooperating in such an excellent way and for making their chapters as complete as possible.

One of us (L. Maier) greatly appreciates the continued encouragement and interest of Dr. H. H. Zeiss, President of Monsanto Research S. A.

L. Maier
G. M. Kosolapoff

Zürich, Switzerland
Auburn, Alabama
April 1972

Contents

Introduction

LUDWIG MAIER

Monsanto Research S. A., Zürich

Although the Arabic alchemists and Paracelsus (1493 to 1541) may have had phosphorus available to them, the discovery of elemental phosphorus is generally ascribed to Hennig Brand between 1667 and 1677.[1]

Organic compounds of phosphorus have been in existence at least since the beginning of life on this planet. Such substances were probably prepared artificially for the first time in the Middle Ages, for example, by oxidation of phosphorus in the presence of turpentine (Robert Boyle, 1681[2]).

The scientifically planned study of these compounds may be regarded as having begun in the early part of the nineteenth century. In the succeeding years of that century, the chemistry of organophosphorus compounds developed at a rather rapid pace. This development continued with increased vigor in the opening decades of the present century and has not lost any of its impetus, as may be witnessed by the voluminous literature on the subject in its various aspects.

The literature contains evidence of the work of many thousands of chemists who made great and small contributions to our present knowledge of the organic chemistry of phosphorus. The esterification of dihydrated phosphoric acids with alcohols represents, in all probability, the first research in this field (Lassaigne, 1820). In 1847 Thénard (1818 to 1884)[3] prepared several phosphines, and in 1872 A. W. von Hofmann (1813 to 1892)[4] synthesized the first alkylphosphonic acids. A systematic investigation of organic phosphorus compounds, however, began only in 1874 with the work of A. Michaelis (1847 to 1916). During his illustrious lifetime he was unquestionably the outstanding leader in this subject area not only in Germany but in all the world. Over a period of about four decades, he and his students were responsible for a great amount of synthetic work and for the development of at least the master-stroke outlines of almost all the methods of preparation in use at present. The work of Michaelis gave the foundation for the formation of P-C, P-S, P-Se, P-N, and P-C bonds. The Michaelis-Becker reaction[5] is used widely for the synthesis of alkylphosphonates.

The period that overlaps the later stages of activity of the Michaelis school and extends to the present time developed a number of pioneering investigators. Among them is the founder of the present Russian school of phosphorus chemistry, A. E. Arbuzov. With extraordinary persistence and energy, A. E. Arbuzov determined the structure of phosphorous acid and its derivatives. One of the most versatile methods for the formation of carbon-phosphorus bonds involves the reaction of an ester of trivalent phosphorus with an alkyl halide. This reaction, discovered by Michaelis and Kaehne[6] has been explored in great detail by A. E. Arbuzov.[7] It is widely employed for the synthesis of phosphonate and phosphinate esters and phosphine oxides and is known as the Michaelis-Arbuzov reaction. Today, frequent contributors to the field of organic phosphorus chemistry from Russia are the schools of Kazan (B. A. Arbuzov, A. N. Pudovik), Moscow (M. I. Kabachnik), Kiev (A. V. Kirsanov), and Leningrad (A. A. Petrov).

After the Michaelis and Arbuzov period, a third period of organic phosphorus chemistry began around 1930. Organic phosphorus compounds were used as plasticizers for synthetics, as extraction agents, as oxidation inhibitors for lubricants, as flotation agents, and as complexing agents for transition metals. Of particular interest is the discovery of the toxic and insecticidal properties of organic phosphorus compounds by W. Lange and G. v. Krueger,[8] and G. Schrader[9] and B. C. Saunders[10] (for summaries see ref. 11). After World War II these discoveries created a whole new industry.

The development of several new methods of phosphorylation culminated in the authentic preparation of adenosine triphosphate, mononucleotides, and oligonucleotides (H. R. Todd, H. G. Khorana, F. Cramer).

The Wittig reaction, that is, the conversion of an aldehyde or ketone into an olefin by reaction with a phosphine-ylid,[12] opened up a new field of organophosphorus chemistry. And the preparation of optically active phosphines by Horner and his group[13] greatly stimulated the entire field of phosphine chemistry.

The stereochemistry of the double salts of phosphines with metal halides has attracted chemists for many years (e.g., F. G. Mann, J. Chatt). However, added motivation for research in this field came from a report by Reppe et al.[14] describing the catalytic activity of phosphine-nickel complexes in the trimerization and tetramerization of acetylenes and their derivatives.[14] As a result of these studies, phosphine-stabilized hydrides, alkyls, and aryls of transition metals have been prepared.

Recently, the direct synthesis of organic phosphorus compounds from elemental phosphorus has attracted great interest, and it is now possible to prepare a wide variety of organic phosphorus compounds by this method (for a recent review see ref. 15).

Similar to the first edition of Organophosphorus Compounds,[16] this second edition has been designed to fit the requirements of organic and inorganic chemists and also those engaged in coordination chemistry. It should enable them to locate suitable methods of preparation for the various types of organic compounds of phosphorus and to find the chemical and physical properties of individual compounds. Only synthetically available substances of rather firmly established constitution have been considered for this compilation, and the numerous biological materials that cannot be assigned permanent structures have been omitted. The purely biochemical aspects are multiplying and changing so rapidly that a summary of them has not been attempted at this time.

The listings of known compounds that appear at the end of each of the 21 chapters consist of structural formulas which will not become obsolete. The arrangement of these compounds in order of complexity--and in a few chapters in which too many compounds were known, strictly according to the number of carbon atoms (Chemical Abstracts system)--eliminates the need for an extensive index of compound names, as the individuals can be readily found by scanning the listings. This feature is also an aid in keeping the sizes of the books reasonable. The extraordinary variety of organic phosphorus compounds requires a strict subdivision according to the type of compounds. A great help in dividing the subject into the 21 chapters listed below has been K. Sasse's Organische Phosphorverbindungen[17].

Chapter 1:	Primary, Secondary, and Tertiary Phosphines
Chapter 2:	Organophosphorous-Metal Compounds, Biphosphines, Triphosphines, Tetraphosphines, Cyclopolyphosphines, and Corresponding Oxides, Sulfides, and Selenides
Chapter 3A:	Phosphine Complexes with Metals
Chapter 3B:	Phosphite, Phosphonite, and Aminophosphine Complexes
Chapter 4:	Quarternary Phosphonium Compounds
Chapter 5A:	Phosphine Alkylenes and Other Phosphorus Ylids
Chapter 5B:	Penta- and Hexaorganophosphorous Compounds
Chapter 6:	Tertiary Phosphine Oxides
Chapter 7:	Tertiary Phosphine Sulfides, Selenides, and Tellurides
Chapter 8:	Halo- and Pseudohalophosphines
Chapter 9:	Phosphonyl- (Thiono-, Seleno-) and Phosphinyl- (Thiono-, Seleno-) Halides and Pseudohalides

Chapter 10: Phosphonous Acids (Thio-, Seleno Analogs) and Derivatives
Chapter 11: Phosphinous Acids and Derivatives
Chapter 12: Organic Derivatives of Hypophosphorus, Hypodiphosphorous, and Hypophosphoric Acid
Chapter 13: Organic Derivatives of Phosphorous Acid and Thiophosphorous Acid
Chapter 14: Phosphinic Acids and Derivatives
Chapter 15: Organic Derivatives of Phosphoric Acid
Chapter 16: Phosphorous (V)-Nitrogen Compounds with Phosphorous in Coordination Number Four
Chapter 17: Cyclophosphazenes
Chapter 18: Phosphonic Acids
Chapter 19: Organic Derivatives of Thio (Seleno-, Telluro-) Phosphoric Acid

Generally, the literature is covered completely up to 1969, inclusive. In several cases literature from 1970 and 1971 has been included. However, for a complete coverage of the literature from 1970 to the present, the Specialist Periodical Reports on Organophosphorus Chemistry from the Chemical Society[18] should be consulted.

REFERENCES

1. Gmelins Handbuch der Anorganischen Chemie, 8th ed., Part 1.A, Phosphor, 1965.
2. Robert Boyle, New Experiments and Observations Made upon the Icy Noctiluca, London, 1681 and 1682.
3. P. E. Thénard, C. R. Acad. Sci., Paris, 25, 892 (1847).
4. A. W. von Hofmann, Chem. Ber., 5, 104 (1872).
5. A. Michaelis and T. L. Becker, Chem. Ber., 30, 1003 (1897).
6. A. Michaelis and R. Kaehne, Chem. Ber., 31, 1048 (1898).
7. A. E. Arbuzov, J. Russ. Phys. Chem. Soc., 38, 687 (1906).
8. W. Lange and G. Krueger, Chem. Ber., 65, 1598 (1932).
9. G. Schrader, D.B.P. 767,511 (1937); D.B.P. 767,830 (1939).
10. H. Mc Combie, B. C. Saunders, and G. J. Stacey, J. Chem. Soc., 1945, 921.
11. G. Schrader, Die Entwicklung Neuer Insektizider Phosphorsäure-Ester, Verlag Chemie, 1963; R. D. O'Brien, Toxic Phosphorus Esters, Academic, 1960; D. F. Heath, Organophosphorus Poisons, Pergamon, 1961; B. C. Saunders, Phosphorus and Fluorine, Cambridge University Press, 1957.
12. G. Wittig and G. Geissler, Ann. Chem., 580, 44 (1953).

13. L. Horner, H. Winkler, A. Rapp, A. Mentrup, H. Hoffmann, and P. Beck, Tetrahedron Lett., 1961, 161.
14. W. Reppe and W. J. Schweckendiek, Ann. Chem., 560, 104 (1948).
15. L. Maier, Fortschr. Chem. Forsch., 19, 1-60 (1971).
16. G. M. Kosolapoff, Organophosphorus Compounds, Wiley, 1950.
17. K. Sasse, in Houben-Weyl, Organische Phosphorverbindungen, Vol. XII/1 and 2, Thieme, 1963.
18. S. Trippett, R. S. Davidson, D. W. Hutchinson, R. Keat, R. A. Shaw, and J. C. Tebby, Organophosphorus Chemistry, Specialist Periodical Reports, the Chemical Society, London, 1970. (One volume per year.)

Chapter 1. Primary, Secondary, and Tertiary Phosphines

LUDWIG MAIER

Monsanto Research S. A., Zürich, Switzerland

The subject of primary, secondary, and tertiary phosphines was last reviewed in 1963 by L. Maier and by K. Sasse (see Bibliography). Since then several new methods have become available to prepare this class of compound, and interesting properties and new reactions have been uncovered.

Most fruitful in this respect was the introduction of silanes as reducing agents.

The preparative procedures for the synthesis of optically active tertiary phosphines, first reported in 1961,[635] have been greatly improved. This in turn has stimulated study of the stereochemistry of the reactions of tertiary phosphines. As a result, the stereochemistry of alkylation, oxidation, sulfurization, phosphine imine formation, deoxygenation, desulfurization, and many other reactions of tertiary phosphines is now known.

Furthermore, the more than 100 suggested applications attest to the wide interest this class of compound has received recently. It is the purpose of this chapter to summarize the methods, to point out their advantages and their limitations for the preparation of phosphines, and to discuss the reactions and properties of this class of compounds. Roman numerals are given to the different methods of preparation and are used in the list of compounds to indicate the method of preparation for a specific phosphine.

Although emphasis has been placed upon the literature appearing since 1962, earlier literature has been included in order to cover the whole subject completely. Great efforts have been made to include all known compounds in the list of compounds, together with all known properties.

The literature concerning the subject of this review is covered up to May 1970, including patent literature so far as abstracts are available in *Chemical Abstracts*. The nomenclature follows in general the system adopted by the American Chemical Society's Committee on the Nomenclature of Organic Phosphorus Compounds.

A. PRIMARY PHOSPHINES

A.1. Preparation of Primary Phosphines

I. BY REDUCTION OF DERIVATIVES OF THE PHOSPHONOUS AND PHOSPHONIC TYPES

Prior to 1950 there was no information available on the reduction of halophosphines by methods that involve the use of solvents inert to the halophosphines. However, a rather indirect reduction was achieved in one case. When diphenylphosphinous chloride was caused to react with zinc at an elevated temperature and the reaction product treated with water, diphenylphosphine was obtained.[327] The reaction probably proceeded by the way of zinc derivatives--Ph_2PZnCl or $(Ph_2P)_2Zn$--which yielded the product upon hydrolytic attack.

The introduction of $LiAlH_4$ as a reducing agent proved most fruitful for the preparation of phosphines. Since its first use by Horvat and Furst to prepare phenylphosphine from phenylphosphonous dichloride,[636] a wide variety of organic phosphorus compounds have been reduced to the corresponding phosphines, as illustrated in the following scheme:

$$2RPX_2 + LiAlH_4 \longrightarrow 2RPH_2 + LiAlX_4$$

$$4RPOCl_2 + 3LiAlH_4 \longrightarrow 4RPH_2 + 2LiAlCl_4 + LiAl(OH)_4$$

$$2RP(OH)_2 + LiAlH_4 \longrightarrow 2RPH_2 + LiAl(OH)_4$$

$$4RPO(OR')_2 + 3LiAlH_4 \longrightarrow 4RPH_2 + LiAl(OH)_4 + 2LiAl(OR')_4$$

$$2RP(OR')_2 + LiAlH_4 \longrightarrow 2RPH_2 + LiAl(OR')_4$$

$$4RP(S)Cl_2 + 3LiAlH_4 \longrightarrow 4RPH_2 + 2LiAlCl_4 + LiAl(SH)_4$$

The reduction of phosphonous dihalides with $LiAlH_4$ is now one of the most widely used methods for the preparation of primary phosphines.[100,362,619,639,819,821,906,1083,1252,1418] The reduction probably proceeds through the formation of complexes of the type $Li[H_{4-x}Al(PHR)_x]$ (value of x not known), since work-up of a reaction mixture from $LiAlH_4$ and $PhPCl_2$ with D_2O gave deuterated phenylphosphine,[411] while reduction with $LiAlD_4$ and work-up with H_2O gave nondeuterated phenylphosphine.[411] To avoid formation of by-products, such as cyclopolyphosphines formed by the reaction of RPH_2 with $RPCl_2$ [911] (see Chapter 2), $LiAlH_4$ should be used in excess. As an exception, reduction of β-bromoethylphosphonous dibromide with $LiAlH_4$ gave not the expected β-bromoethylphosphine. Instead, phosphirane (1), the first known three-membered phosphorus ring compound, was obtained.[1402]

$$\text{BrCH}_2\text{CH}_2\text{PBr}_2 + \text{LiAlH}_4 \longrightarrow \underset{\displaystyle \underset{H}{|}}{\overset{\displaystyle \overset{\text{H}_2\text{C} \longrightarrow \text{CH}_2}{\diagdown \quad \diagup}}{P}} + \text{CH}_3\text{CH}_2\text{PH}_2 + \text{C}_2\text{H}_4 + \text{PH}_3$$

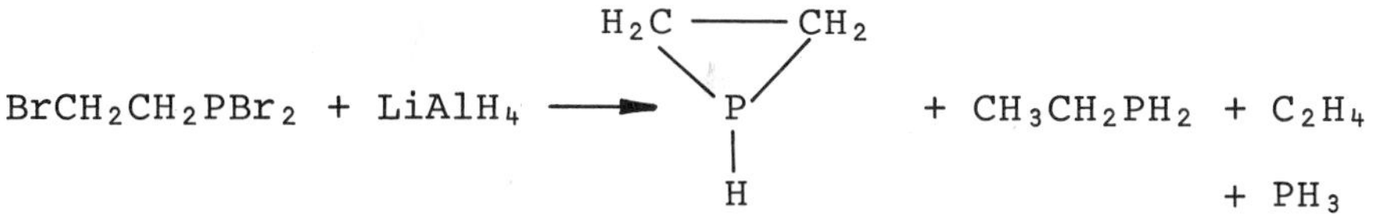

(1)

Silanes of the type $SiHCl_3$ and Ph_2SiH_2 reduced phenylphosphonous dichloride to phenylphosphine in high yield (∿80%).[410] As an intermediate, a compound with a Si-P bond was postulated. Protic solvents, for example, alcohols, water, cleave this bond and liberate the phosphines, which can then be directly distilled.[410]

Other procedures for the formation of primary phosphines involve the reduction of alkylphosphonous dichlorides with hydrogen sulfide[491] and the electrolytic reduction of phenylphosphonous dichloride in glycol ether solution.[311]

Phenylphosphonous dichloride could also be reduced to phenylphosphine by lithium hydride,[619,1418] by lithium in tetrahydrofuran at -40°,[128] or by a finely divided dispersion of sodium in toluene[1083] followed by treatment with water.[609,619,1083] Diisobutylaluminum hydride,[1426] potassium borohydride, and sodium hydride were said to be unreactive.[619] One patent, however, claimed that sodium hydride could be used for this purpose in the presence of catalysts such as BR_3, AlR_3, or GaR_3.[756] Lithium borohydride reduced phenylphosphonous dichloride to a crystalline borine adduct, $C_6H_5PH_2 \cdot BH_3$, m. 38.5°. A deficiency of the reducing agent in the latter reaction yielded an unstable liquid, $C_6H_5PHCl \cdot BH_3$, which gradually lost hydrogen chloride and hydrogen and formed a high-molecular-weight polymer, $(C_6H_5PBH)_x$.[1431] Alkyl- and arylphosphonic dichlorides[400,619,919,1083] and alkylenebis(phosphonic dichlordies)[537] were also reduced by $LiAlH_4$ and $LiBH_4$[400] to the corresponding phosphines in yields of 50 to 90%. $LiBH_4$, which is a milder reducing agent than $LiAlH_4$, has the advantage that it does not attack nitro, nitrile, and amide groups;[1223] thus the preparation of nitro-, cyano, and carbamidosubstituted phenylphosphines by this method seems possible. Cyclohexylphosphine was also successfully prepared from $C_6H_{11}P(O)Cl_2$ using a sodium dispersion in toluene at 90° as the reducing agent.[609] In an attempt to reduce phenylphosphonous acid with $LiAlH_4$, a small yield of $C_6H_5PH_2$ (13%) was obtained,[1419] but $CH_3C_6H_4P(O)(OH)_2$ and $C_6H_5P(O)(OH)_2$[400] were not reduced by this reagent. The preparation of phosphines from phosphonates is of preparative importance because the starting material is easily accessible from the Michaelis-Arbuzov[801] (p. 12, and Chapter 18) reaction. Phosphonates[609,919,1083] and diphosphonates[921] gave, on reduction with $LiAlH_4$, the corresponding phosphines in yields varying from 40 to 90%. Again, reduction of phosphonates with silanes (Ph_2SiH_2 or methylpolysiloxane) seems to produce, in general, higher yields

$$2(RO)_2P(O)(CH_2)_n(O)P(OR)_2 + 3LiAlH_4 \longrightarrow 2H_2P(CH_2)_nPH_2 + LiAl(OH)_4 + 2LiAl(OR)_4$$

of primary phosphines.[410] Although only one phosphonite, namely, $C_6H_5P(OC_2H_5)_2$, was reduced with $LiAlH_4$ to a primary phosphine,[1211] there is no doubt that other phosphonites can also be reduced. Similarly, alkylenebis(phosphonites) yielded with $LiAlH_4$ primary diphosphines of the type $H_2P(CH_2)_nPH_2$ (n = 4,6).[1212] Recently, it was found that thiophosphonic dihalides also yielded primary phosphines upon reduction with $LiAlH_4$.[932,1009] This reduction is free of by-products, and the yields of primary phosphines are high.[932] Finally, the reduction of quasiphosphonium compounds of the type $[(C_6H_5O)_3PR]^+I^-$ and $[(R_2N)_3PR]^+I^-$ to primary phosphines was also effected with sodium suspensions in boiling toluene.[609] And a patent claims the catalytic reduction of alkylphosphonous dichlorides with hydrogen (1000 psi) at 250° in the presence of Et_3N, carbon, and palladium.[502]

II. FROM METAL PHOSPHIDES AND ALKYLATING AGENTS

Although the preparation of tertiary phosphines from metal phosphides and alkyl halides has been known for over a century, this method became important only recently for the preparation of primary phosphines, when procedures were found that restrict the degree of substitution to the number of phosphorus-to-metal bonds present in the starting material. Thus a metal phosphide of the type MPH_2 yields, with an alkyl halide, primarily a primary phosphine. The metals used in this reaction are usually lithium, sodium, potassium, magnesium, or calcium. The sodium, potassium,[736,1401,1412] or calcium[848,1401] intermediates may be prepared from the metals and phosphine, PH_3, in liquid

$$MPH_2 + RX \longrightarrow RPH_2 + MX$$

ammonia with appropriate precautions to repress substitutions beyond the desired limit (for other methods see Chapter 2).

Treatment of metal phosphides, MPH_2, with alkyl halides,[213,361,756,1401,1412] sodium salts of halocarboxylic acids,[680,682] alkylene dihalides,[847,1399,1400] β-chloroethylamines,[681] epoxides,[782,1094] or episulfides[428,1271] in an inert organic solvent, liquid ammonia, or hexamethylphosphoramide resulted in the formation of primary phosphines. Bromo- and iodobenzene reacted only slowly with KPH_2 at -33.5° in liquid ammonia,[1412] and 2-bromo-p-cymene gave with $NaPH_2$ or $LiPH_2$ in ether a mixture of 2- and 3-p-cymylphosphines,[1343] (<u>2a</u> and <u>2b</u>). 1,2-Dichloroethane does

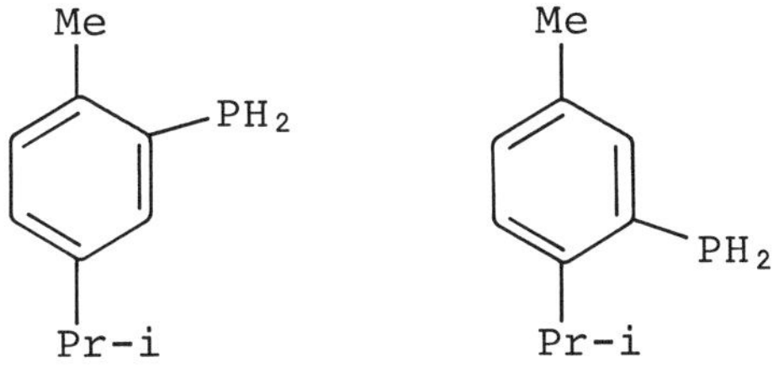

(2a) (2b)

not give, with $NaPH_2$ in liquid ammonia, ethylene diphosphine, $H_2PCH_2CH_2PH_2$, as reported previously.[847] Instead, a phosphirane (1) is produced.[1399,1400,1402] The interaction of α,ω-dichloroalkanes, $Cl(CH_2)_xCl$ (x = 3 and higher), and $NaPH_2$ in liquid ammonia produced, however, alkylene diphosphines[847,1399,1400] in addition to cyclic compounds.[1399,1400]

The highest yield of methylphosphine was obtained when $Ca(PH_2)_2$ was made to react with methyl chloride.[1401] In order to obtain primary phosphines free of nitrogen-containing impurities in reactions conducted in liquid ammonia, special techniques are required in the isolation process.[1401] A more clean-cut reaction is secured when triphenylmethysodium,[35] phenyllithium[35,707,818] or butyllithium[1082] is made the source of the metal. In this case phosphine may be converted to the monosodium or monolithium derivative if the reaction is carried out in ether[35,818] or in a 4:1 ether-pentane mixture.[707] High yields of primary phosphines were isolated when the alkylation of $NaPH_2$ was carried out in hexamethylphosphoramide.[213]

III. FROM ELEMENTAL PHOSPHORUS AND VARIOUS REAGENTS

Addition of alkyl iodides to white phosphorus in aqueous sodium hydroxide results in the formation of a spectrum of phosphine derivatives. The yields of the individual members are usually below 20 to 30%.[65] The reaction probably proceeds through the formation of alkali salts of the lower acids of phosphorus as, for example[801]:

$$P_4 + 6NaOH + 2RI \longrightarrow 2NaI + 2Na_2HPO_3 + 2RPH_2$$

Also, a series of phosphine derivatives is produced when white phosphorus is heated with an aliphatic alcohol in a sealed vessel for several hours at 250°.[116] In addition, appreciable amounts of their oxidation products, such as the phosphonic acids, are also isolated.[116] Ethylphosphine has been obtained in 20% yield, which appears to be the maximum to be expected.[116] Phenol and sodium phenolate react with red phosphorus in the presence of small amounts of water (without water no reaction occurs)[1388] when heated for 4 hr to 250 to 265° at 100 atm, yielding hydrogen and small amounts of phenylphosphine and diphenylphosphine. In addition to the last-mentioned compounds, phenylphosphonic acid is also produced. Apparently, the reaction of red phosphorus with phenol or sodium phenolate proceeds in the presence of water through dismutation, analogous to that of phosphorus with water alone, so that, instead of PH_3 and inorganic acids, their phenylated analogs are formed.

When phosphorus vapor, with argon as a carrier gas, and methane was swept through a discharge tube, reaction occurred, forming PH_3 as the main product and small amounts of CH_3PH_2.[1457] Direct addition of elemental phosphorus to olefins was achieved when the mixture was heated to 250 to 400° under a hydrogen pressure of 700 to 1000 atm. Small amounts of iodine or alkyl iodides lower the reaction temperatures.[1067] Cyclohexylphosphine has been obtained in 10% yield by this procedure. Somewhat higher yields of phenylphosphine (25 to 40%) were obtained in the reaction of white phosphorus with phenylmagnesium bromide or with phenyllithium.[1176] By using the same conditions, butylphosphine was isolated in only 7 to 9% yields after hydrolysis of the reaction mixture. Secondary phosphines were also formed but in lower yields.

$$6PhLi + P_4 \longrightarrow Li_3P + Li_2PPh + LiPPh_2 + PPh_3$$

In thise reactions much of the phosphorus was converted to nonmelting, insoluble yellow solids, believed to be organopolyphosphines.[1176] The latter were the only products in reactions of phenylsodium or tri-n-butylaluminum with phosphorus. In addition to other products, very small amounts of butylphosphine were produced (1.9%) when white phosphorus was electrolyzed in dimethylformamide in the presence of butyl bromide.[745]

IV. BY HYDROLYSIS OF ALKYL- AND ARYLPHOSPHONOUS DIHALIDES

When a phosphonous acid, prepared by hydrolysis of a

phosphonous dihalide, is subjected to heat, it disproportionates to yield a primary phosphine and a phosphonic acid. This reaction was the most widely used one before

$$3RP(OH)_2 \longrightarrow RPH_2 + 2RP(O)(OH)_2$$

the introduction of reducing agents, such as $LiAlH_4$ or sodium[785,984,801] (p. 12). Although this method has been used in the past decade,[398,436,949,1362,1418] it is not likely to play a great role in future preparations of primary phosphines, with the exception of trifluoromethylphosphine, CF_3PH_2, which is best prepared by hydrolysis of CF_3PI_2[100] with H_2O (24% CF_3PH_2) or by hydrolysis of $[CF_3P]_4$ (21% CF_3PH_2) and $(CF_3P)_5$ at 140°.[905] High yields of CF_3PH_2 are also obtained from the reaction of $(CF_3P)_{4/5}$ with HI in the presence of mercury.[208]

V. UNDER FRIEDEL-CRAFTS CONDITIONS

When a mixture of PCl_3, benzene, and $AlCl_3$ was heated for 2 hr at 75° in the presence of methylcyclohexane and then hydrolyzed with ice water, phenyl and diphenylphosphine were obtained.[1254] The yields and the scope of this reaction are not known. However, it seems to be among the most economical reactions suitable for the synthesis of phosphine derivatives.

Treatment of the complex $H_3P{\cdot}AlCl_3$ (formed from $AlCl_3$ and PH_3 at 70 to 80° in a 1:1 ratio) with alkyl chlorides or bromides, followed by hydrolysis with ice water containing hydrogen chloride, also produced primary phosphines in yields ranging from 15 to 83%. Aromatic halides did not react with the phosphine-aluminum trichloride complex, $H_3P{\cdot}AlCl_3$, and alkyl iodides favored side reactions. The yield of primary phosphines increased with increasing chain length of the alkyl halide. Whereas higher alkyl halides produced only primary phosphines in this reaction, lower alkyl halides gave small amounts of secondary phosphines in addition to the primary phosphines.[1085] A recent report indicates that the reaction of n-dodecyl bromide with $PH_3{\cdot}AlCl_3$ under the conditions described above does not produce n-dodecyl phosphine to any appreciable extent. Instead, a mixture of primary phosphines believed to be structural isomers of $n\text{-}C_{12}H_{25}PH_2$ was obtained.[535] It seems reasonable that this reaction proceeds by a carbonium ion mechanism analogously to other Friedel-Crafts reactions and is thus unsuitable for the preparation of n-alkyl phosphines.

VI. FROM PH_3 OR PH_4I AND ALKYLATING AGENTS

Among the oldest methods known for the preparation of phosphines are the reactions of phosphonium iodide with alkyl halides in the presence of zinc oxide or powdered zinc,[588,589,591,801] and of phosphine with alkyl iodide. Phosphonium bromide and alkyl bromide with zinc oxide may be used instead of the corresponding iodide.[36] The reactions are usually carried out at temperatures of about 150°. Thereby a complex mixture of primary, secondary, and tertiary phosphines, as well as phosphonium salts is formed. Because of the different basicity, the individuals can be separated fairly readily. Addition of water to the cooled reaction mixture liberates some by-product phosphine and the primary phosphine, both of which are very weak

$$2RI + 2PH_4I + ZnO \longrightarrow 2RPH_2{\cdot}HI + ZnI_2 + H_2O$$

$$2RI + PH_4I + ZnO \longrightarrow R_2PH{\cdot}HI + ZnI_2 + H_2O$$

bases. The secondary and the tertiary phosphine are liberated by addition of alkali and are then separated by fractional distillation[801] (p. 10).

Under conditions analogous to those cited above, polyhalides and PH_4I also yield primary phosphines; for example, chloroform yields methylphosphine, 1,2-dibromoethane gives ethylphosphine, and benzotrichloride produces benzylphosphine.[592] The yields are poor and the usual product mixtures are formed.

Alkylation of PH_3 with methyl iodide in dimethyl sulfoxide in the presence of potassium hydroxide gave a high yield of methylphosphine (75%).[740] Similar high yields of primary and secondary phosphines were claimed to be obtained in the reaction of PH_3 with alcohols over dehydrated aluminosilicate catalysts at 200°.[509]

VII. BY PYROLYSIS OF BIPHOSPHINES AND TRIPHOSPHINES

1,2-Bis(trifluoromethyl)biphosphine, $CF_3(H)P{-}P(H)CF_3$, when heated for 18 hr at 200° and then 6 hr at 225°, yielded CF_3PH_2 in a 74% yield.[905] Analogously, 1,2,3-tris(trifluoromethyl)triphosphine, $CH_3(H)P{-}PCF_3{-}P(H)CF_3$, when left in contact with an active nickel surface at 25° for 24 hr, gave CF_3PH_2 in nearly quantitative yield.[905] A small yield of CF_3PH_2 was also obtained when trifluoroiodomethane and phosphine were heated at 240° for 4 hr.[532] Pyrolysis of $MeHPCH_2PHMe$[1400] and $PhHPCH_2CH_2PHPh$[702] gave also primary phosphines among other products.

VIII. BY MISCELLANEOUS METHODS

Fast-neutron irradiation of PH_3 in the presence of ethylene yields $H_3{}^{31}SiCH_2CH_2PH_2$ in addition to other products.[426] One of the most economical methods for the preparation of primary, secondary, and tertiary phosphines is the reaction of phosphine, PH_3, with olefins and aldehydes. Since, however, primary, secondary, and tertiary phosphines are formed in most cases simultaneously, these reactions are discussed in Section C.1.XII and XIII.

A.2. General Chemistry of Primary Phosphines

A.2.1. Oxidation

Primary phosphines are easily oxidized and, in particular, those containing the lower aliphatic radicals ignite spontaneously in air. Uncontrolled oxidation yields phosphonous and phosphonic acid[801] (p. 23; ref. 401, p. 394). Recently, however, the lowest oxidation product of primary phosphines, the primary phosphine oxides having the formula $RP(O)H_2$ have been isolated.[172,174] They were initially obtained in the reaction of PH_3 with ketones (see Section C.1.XIII) and then, more directly, in the controlled oxidation of primary phosphines with stoichiometric amounts of N_2O_4[47] or hydrogen peroxide in ethanol at 0°.[172,174] Pri-

$$RPH_2 + H_2O_2 \xrightarrow{EtOH} RP(=O)H_2 + H_2O$$

mary phosphine oxides are, in general, rather unstable (see Chapter 10). They are easily oxidized further to phosphonic acids.

$$RP(=O)H_2 \xrightarrow{[O]} RP(=O)(OH)_2$$

The anilinium salts of the latter and the adducts of primary phosphine oxides with benzaldehyde have been used for their characterization.[172,174]

$$RP(=O)H_2 + 2C_6H_5CHO \xrightarrow{H^+} RP(=O)[CH(OH)C_6H_5]_2$$

A.2.2. Sulfurization and Selenation

Primary phosphines react readily with sulfur and give a host of products, depending upon the ratio of $RPH_2:S$ used.[786,912,919] When phenylphosphine is heated with one equivalent of sulfur to 50°, phenylphosphine sulfide, $PhP(S)H_2$, is apparently produced.[912] This compound could, however, not be isolated pure, since it readily splits off H_2S with the formation of $[PhP]_4S$, for which a four- (3) or five-membered (4) ring structure was proposed.[912,929]

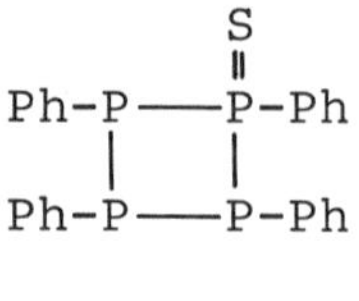
(3)

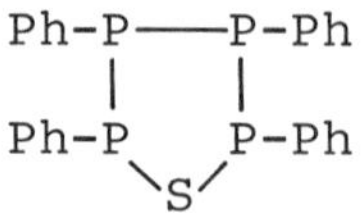
(4)

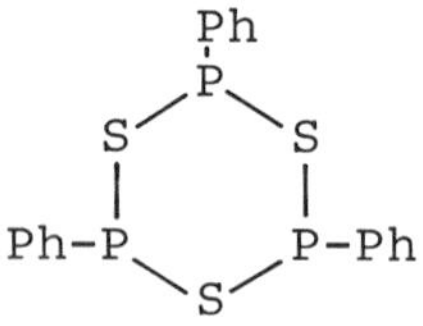
(5)

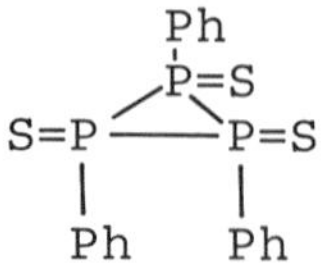
(6)

Reaction of phenylphosphine with 2 equivalents of sulfur leads to the formation of $(PhPS)_3$.[912] The structure of this product is not yet known with certainty. Out of the six possible structures (excluding S-S bonds), a six-membered ring structure (5) has been proposed.[278,923,929] Mass spectroscopic and IR studies seem to indicate a three-membered ring structure (6).[89] However, we found in our product in the IR an absorption band at 578 cm^{-1}, which indicates the presence of PSP bonds.[933]

Treatment of phenylphosphine with three equivalents of sulfur leads to the formation of phenyldithiophosphonic anhydride, which has a four-membered ring structure with trans configuration (7).[912]

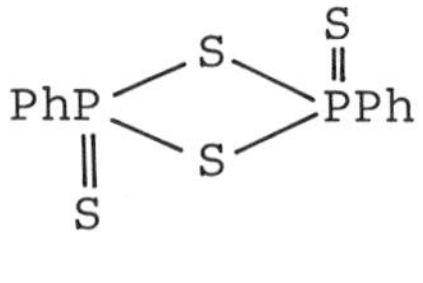
(7)

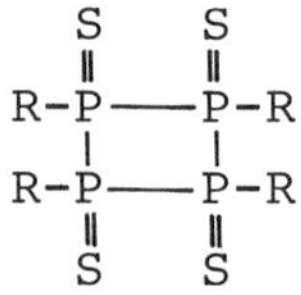
(8)

In contrast to the reaction of phenylphosphine with sulfur, primary aliphatic substituted phosphines yield $(RPS)_4$, when treated with sulfur in any ratio.[919] A four-membered ring structure (8) has been proposed for these products.[919] Phenylphosphine also reacts with selenium to give in one step $(PhPSe)_x$.[919]

A.2.3. Metalation

The metalation of primary phosphines can be effected by the free alkali metals[669,1084,1406,1407] or by the organoalkali compounds[707] in inert solvents. The degree of substitution depends on the conditions, and the monosubstituted metal phosphides, RHPM (M = Na, K), as well as the disubstituted metal phosphides RPM_2 (M = Li, Na, K), have been prepared (see Chapter 2, Section A). Phenylphosphine also reacts with Grignard reagents with substitution of the hydrogen atoms by the MgX group.[737] This reaction proceeds more smoothly when phenylmagnesium bromide is used[949] instead of ethylmagnesium bromide.[737]

$$C_6H_5PH_2 + 2RMgX \longrightarrow C_6H_5P(MgX)_2 + 2RH$$

A.2.4. Addition to Unsaturated Compounds

Thionylaniline adds phenylphosphine in benzene solution to give thionylanilidenphenylphosphine which is not stable and rearranges to phenylthiophosphonic anilide, m. 78 to 81°.[55] The addition of primary phosphines to carbon-

$$PhN{=}S{=}O + H_2PPh \longrightarrow (PhNHSOPHPh) \longrightarrow PhP(=S)(OH){-}NH{-}Ph \text{ or } PhP(=O)(SH){-}NHPh$$

carbon double bonds (Section C.1.XII) and carbonyl-containing compounds (Section C.1.XIII) is discussed in later sections.

A.2.5. Reaction with Acid Halides

Phenylphosphine, when treated with phosgene, yields phenylphosphonous dichloride.[988] The course of the reaction is not known, but it may well proceed through the formation of tetraphenylcyclotetraphosphine which is then cleaved by phosgene to give the final product.

$$C_6H_5PH_2 + 2COCl_2 \longrightarrow C_6H_5PCl_2 + 2CO + 2HCl$$

$$4C_6H_5PH_2 + 4COCl_2 \longrightarrow [C_6H_5P]_4 + 4CO + 8HCl$$

$$[C_6H_5P]_4 \xrightarrow{+\ 4COCl_2} 4C_6H_5PCl_2 + 4CO$$

Other primary phosphines as well as secondary phosphines react with phosgene in inert solvents in the same way to give the corresponding chlorophosphines in excellent yields.[553] Thionylchloride has also been reported to react with phenylphosphine in benzene solution with the formation of phenylthiophosphonic dichloride, phenylthiophosphonic anhydride, $(C_6H_5PSO)_3$ (9), m. 143°, and higher polymeric materials.[55]

$$C_6H_5PH_2 + SOCl_2 \longrightarrow C_6H_5PSCl_2 + H_2O$$

$$C_6H_5PH_2 + SOCl_2 \longrightarrow (C_6H_5PSO)_x + 2HCl \quad (x = \text{preferentially } 3)$$

$(C_6H_5PSO)_3$ (9) is cleaved by phosphorus pentachloride at room temperature to phenylthiophosphonic dichloride in nearly quantitative yield and is hydrolyzed with hot water to phenylphosphonic acid and hydrogen sulfide. Furthermore, compound (9) is stable toward oxidizing agents. From this evidence it was concluded that compound (9) had either structure (9a) or (9b). An x-ray analysis proved structure 9b (L. Maier et al.).

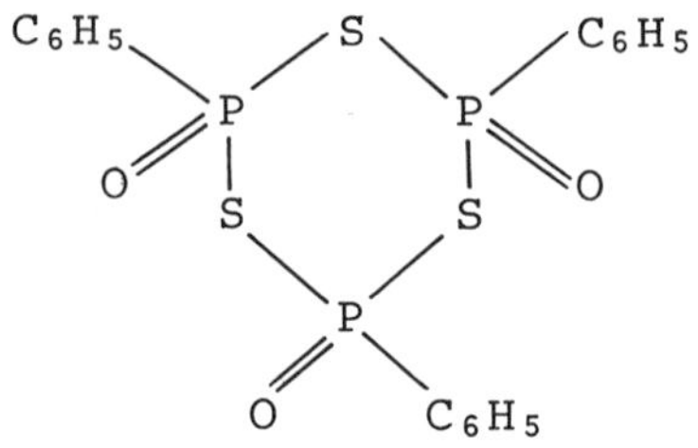

(9a)

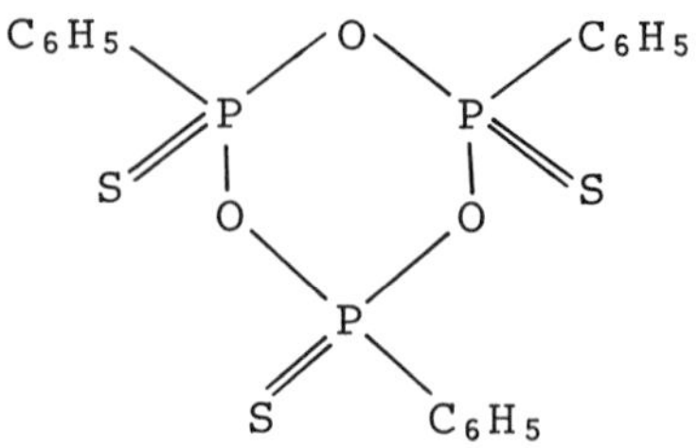

(9b)

The analogous reaction of phenylphosphine with SO_2Cl_2 leads to (PhPS) and $(PhPO_2)_x$.[55] It does not seem profitable to speculate about the structure of PhPS ($b_{0.3}$ 117°), since no details concerning this reaction yet have been published. Treatment of $PhPH_2$ with BuSCl gave high yields

$$2PhPH_2 + 2SO_2Cl_2 \longrightarrow PhPS + PhPO_2 + 4HCl + SO_2$$

of dibutyl phenyldithiophosphonite, $PhP(SBu)_2$.[1112] The same products were obtained when primary phosphines were allowed to react with disulfides in the presence of radical

$$PhPH_2 + 2BuSCl \longrightarrow PhP(SBu)_2 + 2HCl$$

$$RPH_2 + R'SSR' \xrightarrow[\text{initiator}]{\text{radical}} RP(SR')_2 + 2R'SH$$

initiators.[465] 3-Hydroxypropylphosphine, however, reacted differently with diphenyldisulfide and gave a 1,2-oxaphospholane derivative, $PhS\overline{POCH_2CH_2CH_2}$.[463] Chloramine and dimethylchloramine produced with phenylphosphine a cyclopolyphosphine, $(PhP)_5$, and ammonium chloride.[565] Interaction of $PhPH_2$ and S_2Cl_2 leads to the formation of phenyldithiophosphonic anhydride (7).[414,912] Carbonic acid chlorides yield with primary phosphines in the presence of a hydrogen chloride acceptor acyl phosphides.[678] Phosphino-

substituted mixed acid anhydrides react only very slowly
with the weakly basic phosphino group and can therefore

$$RPH_2 + R'COCl \xrightarrow{K_2CO_3} RPH(COR')$$

be converted directly, without protection of the H_2P func-
tion, into phosphinosubstituted carbonic acid amides.[680]
Similarly, "dipeptides" and "tripeptides," which contain
one PH_2 group in place of a NH_2 group, have been obtained
by this procedure.[680] Primary phosphines also react easily
with phosphonous and phosphinous halides, and with metal
and transition metal halides.

$H_2PCH_2COOH + ClCO_2Et$

$\xrightarrow{Et_3N} H_2PCH_2C\overset{O\ O}{\overset{\|\ \|}{C}}OCOEt$

$\xrightarrow{NH_3} H_2PCH_2\overset{O}{\overset{\|}{C}}NH_2$

$\xrightarrow[HCl]{H_2N-\bigcirc-CO_2M} H_2PCH_2\overset{O}{\overset{\|}{C}}NH-\bigcirc-CO_2H$

$H_2NCH_2CONH-\bigcirc-CO_2M$

$H_2PCH_2\overset{O}{\overset{\|}{C}}NHCH_2-\overset{O}{\overset{\|}{C}}NH-\bigcirc-CO_2H$

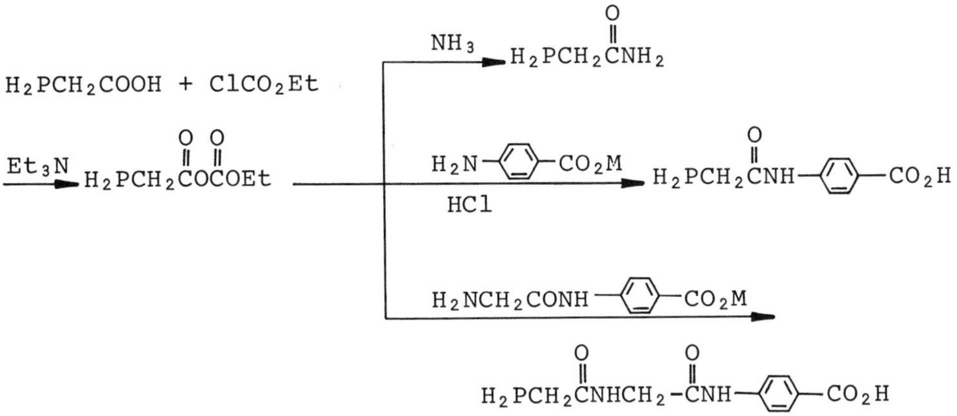

These reactions are discussed in Chapter 2, while complexes of primary phosphines with transition metals are summarized in Chapter 3A.

A.2.6. Reaction with Halogens

Primary phosphines react vigorously with halogens and
give, in the uncontrolled reaction, alkyltetrahalophosphor-
anes (RPX_4). If, however, the reaction is carried out with
stoichiometric amounts of a halogen in an inert solvent at
25°, phosphonous dihalides are formed.[385,1407,1428] Thus

$$RPH_2 + 2X_2 \longrightarrow RPX_2 + 2HX$$

butylphosphonous dibromide was obtained in 44% yield in
the reaction of butylphosphine with bromine in chloroform.
[1407] In general, however, this reaction is not a good
method for preparing phosphonous dihalides (see Chapter 8).

A.2.7. Reaction with Lewis Acids

The best known examples of this type of reaction are

the addition compounds of primary phosphines with trimethylborine[158,159] and diborane.[924,1341] These and other complexes are discussed in Chapter 3. Phenylphosphine also gave a 1:1 adduct with SO_3 in Freon solution which was, however, only stable up to -78°.[1256]

A.2.8. Phosphonium Salt Formation

Primary phosphines as a class are only weak bases, and the salts formed with hydrogen iodide are decomposed by water. Use has been made of this reaction in the separation of primary phosphines from secondary and tertiary phosphines, since the hydrogen iodide adducts of these are decomposed only by alkali[801] (p. 10). The addition products of primary phosphines with alkyl halides are decomposed by alkali.

$$RPH_2 + R'X \longrightarrow [RR'PH_2]^+X^- \xrightarrow{OH^-} RR'PH + H_2O + X^-$$

This reaction has been used for the preparation of secondary phosphines (see Section B). A more detailed discussion of the basic and nucleophilic properties is given in Section D.

A.2.9. Hydrolysis

The bond between carbon and phosphorus is generally regarded as stable to hydrolysis, but in perfluoroalkyl derivatives of phosphorus the strong electron-withdrawing organic group makes the phosphorus atom susceptible to nucleophilic attack, and easy carbon-phosphorus bond fission results.[99] Thus CF_3PH_2 is readily decomposed by aqueous alkali to give 65 to 70% of the theoretical amount of fluoride ion.[100] Carbon-phosphorus bond fission has also been found to occur when polyfluoroalkylphosphines are subjected to aqueous base hydrolysis, but only if the α-carbon of the alkyl group contains fluorine.[179,385] A phosphaalkene intermediate has been suggested.

$$CHFClCF_2PH_2 \xrightarrow{\text{aq. NaOH}} [CHFClCF{=}PH] \xrightarrow{H_2O} CHFClCHFPH\cdot OH \xrightarrow{OH^-}$$

$$H_3PO_2 + [CHFClCHF^- \longrightarrow Cl^- + CHF{=}CHF]$$

This intermediate seems also to be formed in the reaction of $CHF_2CF_2PH_2$ with liquid ammonia. As the final product, a P-N polymer of composition $(CHF_2CH_2PNH)_x$ is obtained.[448] Aniline gave a similar polymer of the type $(CHF_2CH_2PNPh)_x$,[473] whereas dimethylamine afforded the amides $CHF_2CH_2PF(NMe_2)$ and $CHF_2CH_2P(NMe_2)_2$.[473] In every case both α-fluorines have been lost from the original tetrafluoroethyl group in the phosphine, and a mechanism involving phosphaalkene intermediates can account for this.

B. SECONDARY PHOSPHINES

B.1. Preparation of Secondary Phosphines

I. BY REDUCTION OF DERIVATIVES OF THE PHOSPHINOUS AND PHOSPHINIC TYPES

The methods described for the preparation of primary phosphines (Section A.1) by the reduction process are essentially applicable to the preparation of secondary phosphines. Naturally, the starting material must contain two P-C bonds since no such linkages are formed during the reduction process.

The types of compounds that have been reduced to secondary phosphines are illustrated in the equations below. The reactions are usually carried out in ether with excess $LiAlH_4$. After refluxing, the mixture is hydrolyzed, the organic layer separated, and the phosphine isolated in the usual way. As an exception, $(CF_3)_2PI$ could not be reduced to the secondary phosphine with $LiAlH_4$,[100] although catalytic reduction was successful (see Section B.1.VI).

	Ref.
$4R_2PX + LiAlH_4 \longrightarrow 4R_2PH + LiAlX_4$	619,822
$4R_2P(O)X + 2LiAlH_4 \longrightarrow 4R_2PH + LiAl(OH)_4 + LiAlX_4$	619,637, 706,1002
$4R_2P(S)X + 2LiAlH_4 \longrightarrow 4R_2PH + LiAlX_4 + LiAl(SH)_4$	820
$4R_2P(O)OR' + 2LiAlH_4 \longrightarrow 4R_2PH + LiAl(OH)_4 + LiAl(OR')_4$	619,716, 1330
$4R_2PX_3 + 3LiAlH_4 \longrightarrow 4R_2PH + 3LiAlX_4 + 4H_2$	619
$4R_2P(O)H + LiAlH_4 \longrightarrow 4R_2PH + LiAl(OH)_4$	619,940
$Ph_2P(S)SH + LiAlH_4 \xrightarrow{Et_2O}$ no reduction	706

$$2R_2P(O)OH + LiAlH_4 \xrightarrow[Et_2O]{THF} 2R_2PH + LiAl(OH)_4$$ 543,567, 940

The reduction of phosphinous halides probably also proceeds through the formation of complexes of the type $Li[H_{4-x}Al(PR_2)_x]$ (value of x unknown), since work-up of a reduction mixture from $LiAlH_4$ and Ph_2PCl with D_2O gave deuterated diphenylphosphine,[411] while reduction with $LiAlD_4$ and work-up with H_2O gave nondeuterated diphenylphosphine.[411] The reduction of diphenylphosphinic acid, its aluminum salt, and its ethyl ester by $LiAlH_4$ in tetrahydrofuran goes through diphenylphosphine oxide to the secondary phosphine.[940] The main driving force of the reduction has been ascribed to the intermediate formation of P-O-Al groups,[940] for example,

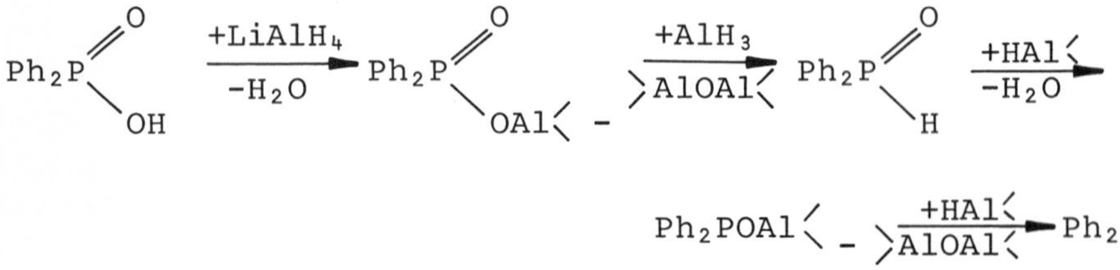

As a side reaction, cleavage of the solvent tetrahydrofuran by the diphenylphosphide ion, Ph_2P^-, was observed and 4-hydroxybutyldiphenylphosphine was isolated as a byproduct.[425,937,940] Other phosphinic acids were similarly reduced to the secondary phosphines, but o-chlorophenylphenylphosphinic acid could only be reduced to the secondary phosphine oxide.[419,940] A secondary phosphine oxide was also obtained in the $LiAlH_4$ reduction of 2,8-dimethylphenoxaphosphinic chloride and the corresponding phosphinate.[471] o-, m-, and p-Methoxyphenylphenylphosphinic acids on reduction with $LiAlH_4$ in tetrahydrofuran gave a mixture of the corresponding secondary methoxyphenylphenylphosphine and the tertiary methoxyphenylmethylphenylphosphine.[941,956]

$$Ph(MeOC_6H_4)P(O)OH \xrightarrow{LiAlH_4} Ph(MeOC_6H_4)PH + PhMe(MeOC_6H_4)P$$

The product of the tertiary phosphine was attributed to the activity of the phosphide ion, $MeOC_6H_4PhP^-$, to demethylate part of the secondary phosphine, undergoing P-methylation itself to the tertiary phosphine.[941,954] This demethylation mechanism is supported by the observation that diphenylphosphide ion when boiled with anisole for 4 hr in tetrahydrofuran gives phenol and methyldiphenylphosphine in high yield.[938,941,954,956] Certain other alkyl aryl ethers, cyclic ethers, and alkyl aryl sulfides were similarly dealkylated by the diphenylphosphide ion.[954]

Although reduction of Ph_2PCl with Ph_3SnH gave only a 54% yield of Ph_2PH, this reducing agent has the advantage that functional groups can be present.[1281] Functional groups may also be present when silanes are used as reducing agents.[410] In addition, silanes of the type $SiHCl_3$, R_3SiH, R_2SiH_2, or $RSiH_3$ reduced not only phosphinous halides to secondary phosphines in high yield, but also phosphinic acids, phosphinic chlorides, and phosphinates.[410]

$$R_2P(O)X \xrightarrow[\text{or } 3SiHCl_3/NaOH/H_2O]{3R_3SiH} R_2PH + \text{siloxane}$$

The reaction of $NaBH_4$ with phosphinous,[194,501] phosphinic,[194,195] and thiophosphinic chlorides[907] has been used successfully for the preparation of phosphinoborines, $(R_2PBH_2)_x$. Small amounts of the secondary phosphines were obtained as by-products in this reaction.[194,195,907] The thermal decomposition of the monoborine adduct of dimethylaminodimethylphosphine at 200° for 30 hr produced dimethylphosphine in 26% yield.[194] Analogously, the first cyclic secondary phosphine was obtained in 30% yield by thermal decomposition of the monoborine adduct of dimethylaminophospholane at 210° for 21 hr.[193]

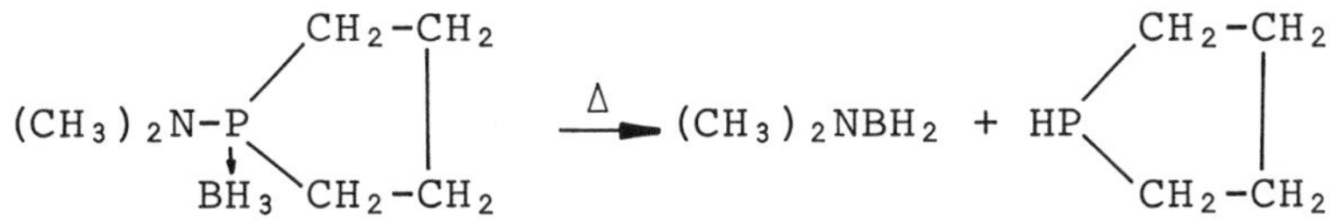

Before the introduction of silanes as reducing agents, alkali metals were used extensively to reduce compounds of the phosphinous and phosphinic type to secondary phosphines. R_2PX,[561,609,756,823,1049,1050,1350] R_2PNR_2,[618] (p. 140) R_2PS_2H,[609] and $R_2P(O)Cl$[609] have been successfully reduced with sodium, potassium, lithium, or sodium hydride in boiling dibutyl ether, tetrahydrofuran, or a hydrocarbon, and the yields of the secondary phosphines obtained upon hydrolysis are comparable to those obtained with $LiAlH_4$ as the reducing agent. The reduction proceeds stepwise, and diphenylphosphinous chloride with sodium yields first tetraphenylbiphosphine which reacts with excess sodium to give sodium diphenylphosphide.[823] Dialkylphosphinous chlorides with sodium in boiling dipropyl or dibutyl ether,[701] or with lithium in tetrahydrofuran,[561] similarly yield tetraalkylbiphosphines. Further reduction occurs only slowly in dipropyl or dibutyl ether. In decalin[1050] or tetrahydrofuran,[561] however, the P-P bond is cleaved easily by sodium, potassium,[1050] or lithium,[651] and alkali dialkylphosphides are formed. Hydrolysis or alcoholysis of

$$2R_2PCl + 2Na \longrightarrow R_2P\text{-}PR_2 + 2NaCl$$

$$R_2P\text{-}PR_2 + 2Na \longrightarrow 2R_2PNa$$

these products gives secondary phosphines. A similar stepwise reduction has been observed in the reaction of diphenylphosphinic chloride with sodium. A 1:2 ratio of the two components gave the sodium derivative of diphenylphosphine oxide, which then underwent further reaction with excess sodium to yield sodium diphenylphosphide.[609]

$$Ph_2POCl + 2Na \longrightarrow Ph_2P(O)Na + NaCl$$

$$Ph_2P(O)Na + 2Na \longrightarrow Ph_2PNa + Na_2O$$

Electrolytic reduction of Ph_2PCl in diglyme gave Ph_2PH in quantitative yield.[311] A high yield of $(C_6F_5)_2PH$ was also obtained from the reduction of $(C_6F_5)_2PCl$ with PH_3 in a sealed glass tube without a solvent.[246]

II. FROM METAL PHOSPHIDES AND ALKYLATING AGENTS

In much the same way as in phosphine, PH_3, the hydrogen atoms in primary phosphines can be selectively replaced by an alkali metal to give metal phosphides of the type RPHM. The sodium and potassium intermediates may be prepared from the metals and a primary phosphine in liquid ammonia,[569,1401,1407] or more conveniently by refluxing a primary phosphine with finely divided sodium metal[1084,1252] or potassium metal[656,669,1406,1407] in an inert solvent at temperatures from 50 to 80°. Only at above 90° is the second hydrogen atom also replaced by sodium.

In order to increase the rate of reaction, utilization has been made of the disubstituted product by carrying out the reaction above 90°, since it has been found that it interacts with a primary phosphine to give the monosubstituted product[1084] (for a detailed discussion see Chapter 2, Section A).

$$RPH_2 + 2Na \xrightarrow[140^\circ]{\text{petrol ether}} RPNa_2 + H_2$$

$$RPNa_2 + RPH_2 \longrightarrow 2RPHNa$$

Treatments of these monoalkali phosphides with alkylating agents such as alkyl halides,[422,569,1084,1401,1406,1407] sodium salts of halocarboxylic acids,[682,717] episulfides,[428,712,1271] epoxides,[162] and chloroalkylamines[681,694] produces secondary phosphines, usually in high yield.

$$RPHM + R'X \longrightarrow RR'PH + MX$$

As an exception, interaction of sodium phosphide and 1,2-dichloroalkanes led to the formation of substituted phosphiranes.[213]

$$RCHClCH_2Cl + 2NaPHR' \longrightarrow RCH\underset{\underset{\overset{|}{R'}}{P}}{\diagdown \diagup} CH_2 + 2NaCl + R'PH_2$$

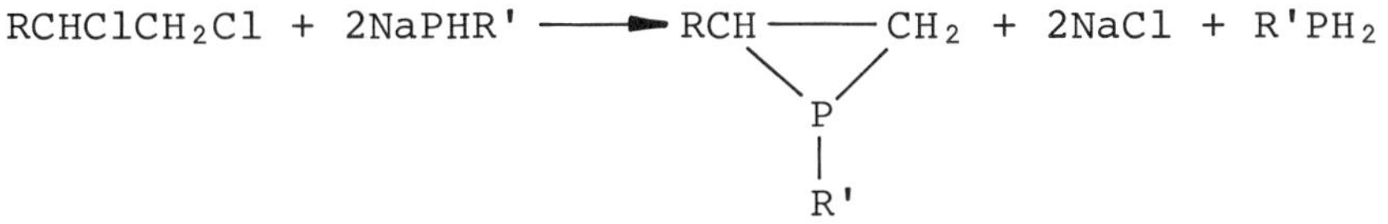

Secondary phosphines are also produced from a disubstituted metal derivative of PH_3 and an alkyl halide,[1401] for example, in liquid ammonia CaPH, prepared according to the equation below, and methyl chloride gave dimethylphosphine

$$Ca(PH_2)_2 \cdot 6NH_3 \xrightarrow[\text{vacuo}]{50°} CaPH + PH_3 + 6NH_3$$

in 33% yield. In general, however, the preparation of the disubstituted metal derivatives of PH_3 is troublesome, and the yields of the secondary phosphines are low. Furthermore, this reaction yields only symmetrical secondary phosphines, while from the monosubstituted metal derivative of primary phosphines unsymmetrical secondary phosphines can also be prepared.

Interaction of alkali phosphides and alkylene dihalides produced disecondary alkylene diphosphines, but only when the chain between the two phosphorus atoms contained three or more carbon atoms.[596,656,669,672,800]

$$2RHPM + X(CH_2)_nX \longrightarrow RHP(CH_2)_nPHR + 2MX$$

$$n \geqq 3$$

Alkylene dihalides should be used in stoichiometric amounts, otherwise phosphonium salt formation takes place. In some cases a side reaction has been observed with the formation of cyclic phosphines.

III. FROM PH_3 OR RPH_2 AND ALKYLATING AGENTS

Phosphine itself reacts rather unsatisfactorily with alkyl iodides on being heated in sealed tubes, with or without zinc iodide,[332,587] and forms a variety of products (see Section A.1.VI). Somewhat better results are obtained when primary or secondary phosphines are used as starting materials.[591,801] Simple dialkylphosphines, for example, Me_2PH and $MeC_{12}H_{25}PH$, have been obtained in about

70% yield by moderately heating ($\sim 50°$) for 2.5 hr methyl
or n-alkyl iodides and primary phosphines without solvent,
[162,535] or by carrying out the reaction at room temperature
in methanol.[260] The products are obtained in the form of
salts, HI, from which they are liberated by alkali.[260,535,591] Essentially higher yields of secondary phosphines
were isolated ($\sim 90\%$) when the $AlCl_3$ complex of a primary
phosphine was treated with excess methyl iodide at reflux.
The phosphine was liberated by the addition of powdered
NaOH in methanol.[535] Recent improvements in the alkylation of PH_3 itself, that is, carrying out the alkylation
with methyl iodide in dimethyl sulfoxide in the presence
of KOH, allows the isolation of Me_2PH in 65% yield.[741]
Acyl halides react with primary phosphines in the presence
of K_2CO_3 or tertiary amines as HCl acceptors to form alkyl-
acylphosphines in good yield.[678] Methyl- and ethylphosphine give with 2,3-dichlorotetrafluoropropene the secondary phosphines, 2-chlorotetrafluoroallylmethyl- or ethylphosphine, by an SN'_2 mechanism,[124] directly without the
addition of base.

$$RPH_2 + CF_2\text{=}CCl\text{-}CF_2\text{-}Cl \longrightarrow [RPH_2CF_2 \cdot CCl\text{=}CF_2Cl^-] \longrightarrow$$

$$RHPCF_2CCl\text{=}CF_2 + HCl$$

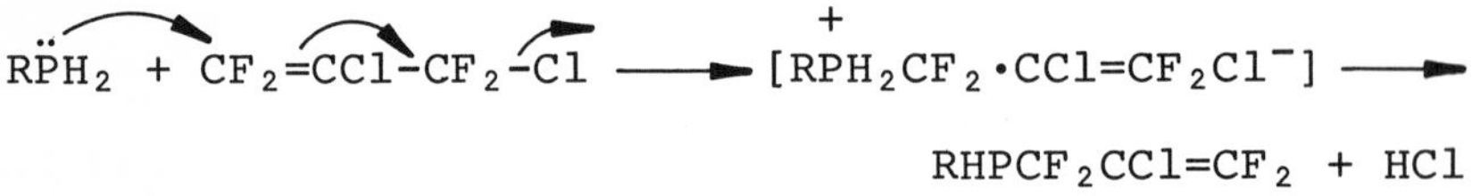

In a patent it is claimed[1467] that tris(sec-butyl)-
phosphine, when heated with methanesulfonic acid and PH_3
in equivalent amounts in an autoclave for 1 hr at 85°,
gave di-2-butylphosphine and small amounts of tetra(i-
butyl)biphosphine. The scope of this reaction is not known,
but a more detailed study appears worthwhile.

IV. BY REDUCTION OF ORGANOSUBSTITUTED BIPHOSPHINES
     AND BIPHOSPHINE DISULFIDES

One of the most convenient methods (easy accessibility
of starting material, see Chapter 2) for the preparation
of secondary phosphines is the reduction of organosubstituted biphosphine disulfides with $LiAlH_4$ in dioxan, diglyme,
or ether.[706,909,965,1050,1079,1386] Since tetraalkyl- and
tetraarylbiphosphines are not cleaved with $LiAlH_4$ in diox-
ane or ether but are cleaved at higher temperatures,[711] it
has been proposed that biphosphine disulfides form first
an addition compound with $LiAlH_4$, in which the P-P bond is
polarized, thus making the reductive cleavage of this bond
possible.[706] Homolytic cleavage of P-P bonds in biphosphines[561,709,823,1046,1050,1350] and bisphosphine disulfides[709] is also a useful source of alkalidialkyl or diarylphosphides (for a discussion see Chapter 2), which on

hydrolysis yield secondary phosphines. The choice of solvents is often critical in these reductions. Thus it has been found that tetrahydrofuran is cleaved by the dialkyl-[446] and diarylphosphide ion[425,937,940] to give a 4-hydroxybutyl-substituted tertiary phosphine. Catalytic reduction of biphosphines[100] and biphosphine disulfides with hydrogen and Raney copper produced the secondary phosphines in a yield of up to 90%.[1046,1050] Dimethylphosphine was also obtained by thermal decomposition of the monoborine adduct, $P_2(CH_3)_4 \cdot BH_3$, along with equimolar amounts of polymerized $(CH_3)_2PBH_2$ units.[181,182] The dry reaction of tetraalkylbiphosphine disulfides with MBH_4 (M = Li, Na, K) at 250° gave mainly phosphinoborines and only small amounts of secondary phosphines.[120a]

V. BY CLEAVAGE OF A P-C BOND FROM TERTIARY PHOSPHINES, TERTIARY PHOSPHINE OXIDES, AND SULFIDES

In the presence of at least one aromatic group,[713] tertiary phosphines are cleaved with lithium in tetrahydrofuran,[1438] or in liquid ammonia,[561,1201] and with sodium or potassium in dioxan,[659,713] liquid ammonia,[427] or diglyme[1260] to give alkali phosphides, MPR_2 (M = Li, Na, K). Upon treatment with water or alcohol, these products yield secondary phosphines, for example, diphenylphosphine was obtained in 68.5% yield from triphenylphosphine and lithium.[1438] On the basis of reaction rates of various aromatic tertiary phosphines with potassium, in dioxan under comparable conditions, the following cleavage series was proposed:[713]

α-naphthyl > phenyl > p-tolyl > 2,5-dimethylphenyl > ethyl > cyclohexyl

This order agrees with the cleavage series of tetraarylphosphonium hydroxides[563,575] and tertiary phosphine oxides[618] (p. 117) in which the most electronegative substituent is cleaved first (see Chapters 4 and 6).

The usefulness of Ph_2PLi, prepared by this method, has been extended considerably by selectively eliminating the by-product PhLi through reaction with t-butyl chloride.[11] Bis(phosphino) acetylenes[5] and triphenylmethyl groups containing tertiary phosphines[714] were also readily cleaved by alkali metals.

In a few cases tertiary phosphine oxides have also been successfully cleaved to secondary phosphines.

Triphenylphosphine oxide, when heated with $LiAlH_4$ in dioxan, gave a colored product[543] which was shown to be lithium diphenylphosphide, $LiP(C_6H_5)_2$.[660] Hydrolysis of this product yielded diphenylphosphine (20%). However, tertiary aliphatic phosphine oxides and sulfides, as well as aromatic tertiary phosphine sulfides gave, upon reduction with $LiAlH_4$ or $Ca(AlH_4)_2$, only the tertiary phosphines.[543] Small amounts of diphenylphosphine were also obtained when triphenylphosphine oxide was refluxed with sodium in toluene[608] or tetrahydrofuran,[544] but the main product in this reaction was the sodium derivative of diphenylphosphine oxide.[608] However, the amount of diphenylphosphine might be substantially increased by increasing the ratio of sodium/triphenylphosphine oxide to 4:1 instead of 2:1 used so far, and also by choosing another solvent that does not interfere with the reaction. Yields of about 50% of the secondary phosphines were obtained when tributyl- or trioctylphosphine oxide was heated with sodium in the

$$Ph_3PO + 2Na \longrightarrow Ph_2P(O)Na + PhNa$$

$$Ph_2P(O)Na + 2Na \longrightarrow Ph_2PNa + Na_2O$$

melt to 360° without a solvent. Under these conditions triethylphosphine oxide gave only a small amount of diethylphosphine.[609]

VI. PREPARATION OF $(CF_3)_2PH$ AND $(C_3F_7)_2PH$

Bis(trifluoromethyl)phosphine is best prepared by catalytic reduction of bis(trifluoromethyl)phosphinous iodide with hydrogen in the presence of Raney nickel. A 65% yield was obtained after heating to 100° for 16 hr.[100] Bis(trifluoromethyl)phosphine was also obtained from $MeHP\text{-}P(CF_3)_2$ by pyrolysis at 160°[186] and from tetrakis(trifluoromethyl)-biphosphine and hydrogen (28% yield) or hydrochloric acid (41%).[100] $(CF_3)_2PI$ yielded $(CF_3)_2PH$ in 35% yield by the action of excess protic acid in the presence of mercury,[188,190] and treatment of $(CF_3)_2PR$ with B_2H_6 gave substantial amounts of $(CF_3)_2PH$.[184] The highest yield of $(CF_3)_2PH$ seems to have been obtained in the reaction of $(CF_3)_2PI$ or $[(CF_3)_2P]_2$ with HI in the presence of mercury.[208,265] Bis(heptafluoropropyl)phosphine, $(C_3F_7)_2PH$, was obtained from the basic hydrolysis of $(C_3F_7)_2PI$.[240]

VII. FROM ELEMENTAL PHOSPHORUS OR PHOSPHORUS SULFIDES AND VARIOUS REAGENTS

Lower sulfides of phosphorus react with Grignard reagents and yield a complex mixture of products among which secondary phosphines have been found in yields ranging from 20% (aliphatic secondary phosphines) to 40% (diphenylphosphine).[801,935] The reaction conditions are not well defined yet, and only the lower sulfides such as P_4S_3[628,935] seem to give secondary phosphines, while P_4S_7 and P_2S_5 give mainly tertiary phosphines, the corresponding sulfides, and other products. Interaction of white phosphorus and phenyllithium or phenylmagnesium bromide in ether or dibutyl ether, followed by hydrolysis, gave diphenylphosphine in maximum yield of 15%.[1176] Dibutylphosphine and butylnaphthylphosphine were similarly obtained in 13 and 35% yields, respectively.[1177] Even lower yields of dibutylphosphine (6.3%) were isolated from the electrolysis of white phosphorus in dimethylformamide in the presence of butyl bromide.[745] And the formation of bis(diethylaminoethyl)phosphine in the Mannich-type reaction of white phosphorus with formaldehyde and diethylamine could be demonstrated only by ^{31}P NMR spectroscopy.[927]

VIII. BY MISCELLANEOUS REACTIONS

Normally, all halogen atoms are replaced in the reaction of PCl_3 with Grignard reagents. Only with bulky substituents on the Grignard reagent less substituted products resulted (see Section C.1.I). In the reaction of PCl_3,[581] $PhPCl_2$,[580] and Ph_2PCl[484] with t-BuMgCl, a further complication was observed. In addition to the expected phosphinous chlorides or tertiary phosphines, secondary phosphines were also isolated. The observation that during the addition of Ph_2PCl to the t-butyl Grignard reagent

$$5\text{t-BuMgCl} + 2PCl_3 \longrightarrow \text{t-Bu}_2PCl\ (34\%) + \text{t-Bu}_2PH\ (26\%) + 5MgCl_2 + C_4H_8$$

isobutylene is given off, strongly suggests that the secondary phosphine arises by partial reduction of the chlorophosphine with t-butyl Grignard reagents,[484] for example,

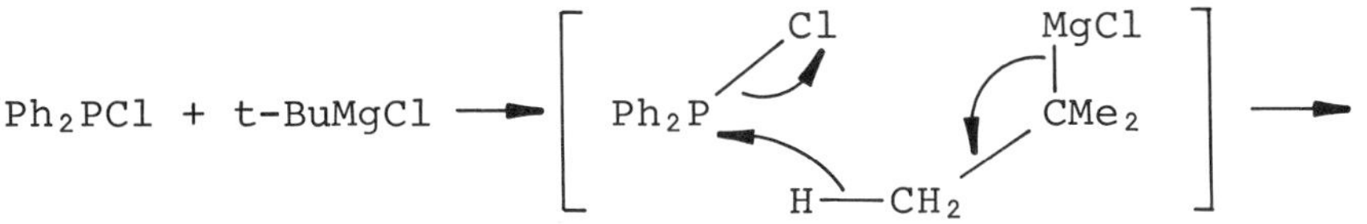

$$Ph_2PH + CH_2{=}CMe_2 + MgCl_2$$

Secondary phosphine oxides are more thermally stable than previously thought. They are isolable entities (see Chapter 11). Upon rapid heating of neat samples, dimethylphosphine oxide decomposed slowly at 100 to 120°, whereas diethylphosphine oxide decomposed at a comparable rate only after a temperature of 180 to 200° was reached. The decomposition seems to be acid catalyzed.[536] The yields of secondary phosphine and phosphinic acid are nearly quantitative.

$$2R_2P(O)H \longrightarrow R_2PH + R_2P(O)OH$$

Secondary phosphines also resulted from the photolysis of aromatic substituted tertiary phosphines[612,768] and phosphonium salts[477] in the presence of protic solvents.

B.2. General Chemistry of Secondary Phosphines

B.2.1. Oxidation

Secondary phosphines are easily oxidized and the lower alkyl members inflame spontaneously when exposed to air. The usual oxidation product is the corresponding phosphinic acid, $R_2P(O)OH$[801] (p. 23). However, careful control of the reaction conditions makes it possible to isolate the first intermediate of oxidation, the secondary phosphine oxide (see Chapter 11). Thus, by treating solutions of the secondary phosphines in isopropyl alcohol with oxygen, the oxides have been obtained as crystalline solids.[1167,1174] Secondary phosphine oxides are also formed when sec-

$$R_2PH + O_2 \xrightarrow{i\text{-PrOH}} R_2\overset{\displaystyle O\text{-}O\cdot}{\overset{|}{P}}H \xrightarrow{+R_2PH} 2R_2P(O)H$$

ondary phosphines, spread in a thin layer, are exposed to dry air at 25° for a few hours without a solvent.[1167,1172] It has been suggested that an equilibrium exists between the two possible structures (10a and 10b) for secondary

$$\underset{\textbf{(10a)}}{R_2\overset{\displaystyle O}{\overset{\|}{P}}H} \rightleftharpoons \underset{\textbf{(10b)}}{R_2POH}$$

phosphine oxides with the equilibrium being far to the left. This is supported by IR and ^{31}P NMR studies[1167] (see Chapter 11). In contrast to this, bis(trifluoromethyl)phosphine oxide apparently exists only as (10b), $(CF_3)_2POH$.[479]

B.2.2. Sulfurization and Selenation

Secondary phosphines readily add sulfur to give dithiophosphinic acids, $R_2P(S)SH$[801] (p. 23). The highest yields are obtained when this reaction is carried out in dilute ammonium hydroxide solution.[1169] A further reaction took place when bis(2-cyanoethyl)phosphine was treated with

$$R_2PH + 2S + NH_4OH \longrightarrow R_2P(S)SNH_4 + H_2O$$

sulfur in dilute ammonium hydroxide solution, and the ammonium salt of bis(2-thiocarbamoylethyl)phosphinic acid was obtained instead of the expected bis(2-cyanoethyl)-dithiophosphinic acid.[1170]

The addition of only one equivalent of sulfur has been achieved by adding the stoichiometric amount of sulfur to a solution of a secondary phosphine in an inert solvent such as benzene or carbon tetrachloride.[1095,1096]

$$R_2PH + S(Se) \xrightarrow[\text{or } C_6H_6]{CCl_4} R_2P(S)H, \quad R_2P(Se)H$$

Secondary phosphine selenides are similarly obtained from the addition of equivalent amounts of elemental selenium to a solution of a secondary phosphine in benzene.[922]

Similar to the secondary phosphine oxides, the secondary phosphine sulfides[1095,1096] and selenides[922] exist in the thiono or seleno (<u>11a</u>), rather than the thiol or selenol (<u>11b</u>) form, as evidenced by IR and ^{31}P NMR spectroscopy.~~[922]~~ [1095,1096] (See Chapter 11.)

$$\underset{(\underline{11a})}{R_2\overset{\overset{\displaystyle S(Se)}{\|}}{P}\text{-}H} \qquad\qquad \underset{(\underline{11b})}{R_2PSH, \quad R_2PSeH}$$

B.2.3. Metalation

Secondary phosphines are readily metalated by alkali metals either by refluxing the secondary phosphines with sodium or potassium in an inert solvent such as ether, hydrocarbons[823,1084,1406,1407] or liquid ammonia,[561,569,1401] or by reaction with methyllithium or phenyllithium[656,669,672,707] (see also Chapter 2, Section A).

B.2.4. Addition to Carbon-Carbon Double Bonds and Carbonyl-Containing Compounds

The addition of secondary phosphines to carbon-carbon double bonds and carbonyl-containing compounds gives tertiary phosphines and is discussed in separate sections (Section C.1.XII and XIII).

B.2.5. Reaction with Acid Halides

Stoichiometric amounts of phosgene converted secondary phosphines into the corresponding phosphinous chlorides in excellent yields. The reaction is best carried out in inert solvents such as methylene dichloride. Dibutylphosphinous chloride was thus obtained in 89% yield.[553,595] When aromatic sulfonyl chlorides were treated with secondary phosphines, reduction occurred and in no case was a product corresponding to the phosphorus analog of a sulfonamide, RSO_2PR_2, detected. The products obtained were dependent on reaction conditions and on the secondary phosphine used. S-Arylphosphinothioates, $R_2P(O)SR$, were formed with bis(2-cyanoethyl)phosphine[171] and diphenylphosphine[1239] when dioxan or benzene with a molar equivalent of pyridine was used as solvent; but this type of product was not detected with di-n-butylphosphine; instead, disulfides and thiophenol were obtained in addition to di-n-butylphosphinic acid.[171]

Treatment of Ph_2PH with BuSCl gave butyldiphenylphosphinothioite.[1112] The same type of products were obtained when secondary phosphines were allowed to react with disul-

$$Ph_2PH + BuSCl \longrightarrow Ph_2PSBu + HCl$$

$$R_2PH + R'SSR' \longrightarrow R_2PSR' + R'SH$$

fides in the presence of radical initiators.[465] In the absence of radical initiators, further abstraction of sulfur occurs and phosphinodithioates are formed.[465] Reduction of 2-hydroxyethyl-substituted disulfides by secondary

$$R_2PH + R'SSR' \longrightarrow R_2P(S)SR' + R'H$$

phosphines occurs by a different route, giving secondary phosphine oxides and ethylene sulfide.[464]

$$R_2PH + HOCH_2CH_2SSCH_2CH_2OH \xrightarrow{-HSCH_2CH_2OH} R_2PSCH_2CH_2OH$$

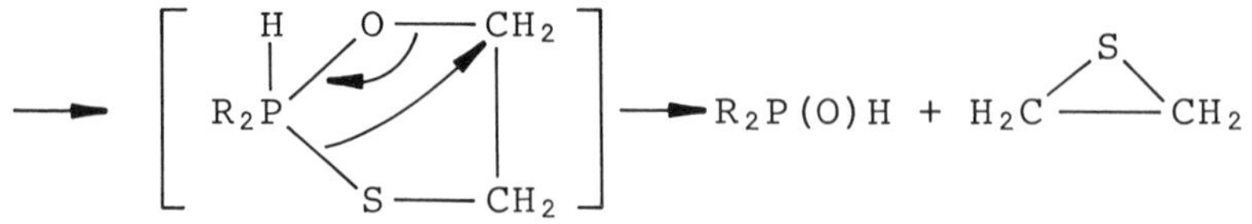

Also, sulfur-free products were isolated in the reaction of secondary phosphines containing a 3-hydroxyalkyl group with disulfides. Spectroscopic studies suggest a 1,2-oxaphospholane structure (<u>12</u>) for the products ob-

tained.[463] Since the intermediate formation of thioite
esters could be demonstrated, the reaction mechanism shown
in the equation below is suggestive.[463]

$$R(HOCH_2CH_2CH_2)PH + RSSR \xrightarrow{-RSH} R-P-SR \longrightarrow R-P-O + RSH$$

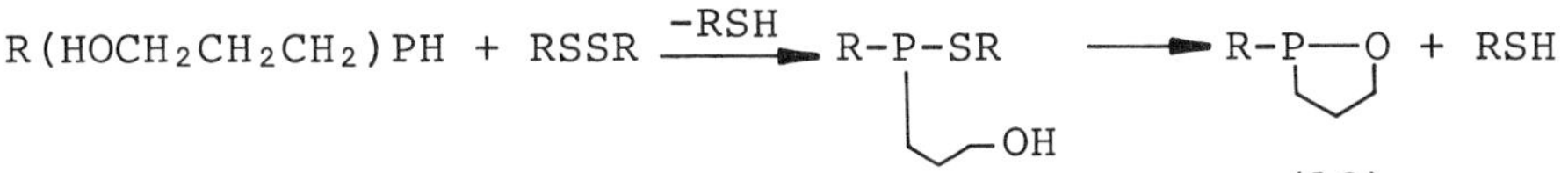

(12)

Chloramine and dimethylchloramine interacted with
Ph_2PH to give tetraphenylbiphosphine and ammonium chloride.[565] Trifluorodiphenylphosphorane, Ph_2PF_3, was obtained when diphenylphosphine was treated with N_2F_4.[394]
In the presence of acid binding agents, secondary phosphines react with acyl halides to give acyl-substituted
tertiary phosphines.[687] Interaction of diphenylphosphine
with $ClSO_3H$ yields diphenylphosphinosulfonic acid, Ph_2-
PSO_3H, which is, however, stable only below $0°$.[1256]
Azides such as Ph_3SiN_3,[1069] $Ph_2P(O)N_3$,[815] PhN_3,[434] and
$PhCON_3$[309] react with secondary phosphines in two stages.
In the first step an aminophosphine is formed which reacts
with a second mole of azide to produce a phosphine imine
derivative.

$$Ph_2PH + Ph_3SiN_3 \longrightarrow Ph_3SiNHPPh_2 \xrightarrow{Ph_3SiN_3}$$

$$Ph_3SiNHP(Ph_2)=NSiPh_3$$

Secondary phosphines also react easily with phosphonous and phosphinous halides, and with metal and transition
metal halides. These reactions are discussed in Chapter 2,
while complexes of secondary phosphines with transition
metals are summarized in Chapter 3A.

B.2.6. Reaction with Halogens

Treatment of secondary phosphines with stoichiometric
amounts of halogen in an inert solvent at 15 to 20° results in the formation of phosphinous halides.[1407] Excess
of halogen must be avoided, otherwise trihalophosphoranes

$$R_2PH + X_2 \longrightarrow R_2PX + HX$$

are formed. Dibutylphosphinous bromide was thus obtained
in 49% yield. Treatment of phosphines with chlorine, however, led only to small and erratic yields of the corresponding chlorophosphines.[553]

B.2.7. Reaction with Lewis Acids

The best known examples of this type of reaction are the addition compounds of secondary phosphines to diborane and trimethylborane. These and other complexes are discussed in Chapter 3A. Diphenylphosphine also gave an addition compound with SO_3, $Ph_2PH \cdot SO_3$, which was, however, stable only at low temperature. In the presence of ether, it rearranged easily to $Ph_2PSO_3H \cdot Et_2O$.[1256]

B.2.8. Phosphonium Salt Formation

Secondary phosphines form with alkyl halides salts that are stable in water but are decomposed by alkaline reagents. The reaction with hydrogen iodide, giving crys-

$$R_2PH + HX \longrightarrow [R_2PH_2]^+X^-$$

$$R_2PH + RX \longrightarrow [R_3PH]^+X^-$$

talline salts, is often used for the characterization of secondary phosphines[706,801] (p. 24). The addition product with alkyl halide has been used as a starting material for the preparation of unsymmetrical tertiary phosphines and unsymmetrical ditertiary phosphines (Section C). A more detailed discussion of the basic and nucleophilic properties of secondary phosphines is given in Section C.2.

B.2.9. Hydrolysis

Similar to the primary phosphines, secondary phosphines containing perfluoroalkyl groups undergo easy carbon-phosphorus bond fission.[99,100,179,385a,385b] When hydrolyzed with aqueous alkali, $(CF_3)_2PH$ yields 50% of its CF_3 content as CHF_3, and 18% is present in the residual solution as fluoride and carbonate.[100] Hydrolysis of $(CHF_2CF_2)_2PH$ gave quantitatively a 1:1 mixture of CHF_2CHF_2 and CHF=CHF.[179]

$$(CF_3)_2PH \xrightarrow{H_2O} CF_3PH_2,\ F^-,\ CO_3^{-2}$$

$$(CF_3)_2PH \xrightarrow{OH^-} CHF_3,\ F^-,\ CO_3^{-2}, \text{ and a } CF_3P \text{ acid}$$

The suggestion that phosphaalkenes are formed as intermediates[179] is supported by the isolation of methyl difluoromethyltrifluoromethylphosphinite from the reaction of methanolic sodium methoxide and $(CF_3)_2PH$.[448,449] At higher temperature the phosphinite reacts further to give a phos-

$$(CF_3)_2PH \xrightarrow{MeO^-/<0°} (CF_3)_2P^- \xrightarrow{-F} [CF_3P{=}CF_2] \xrightarrow{MeOH} CF_3P(OMe)CHF_2$$

phonite, $CHF_2P(OMe)_2$.[448,449] The intermediate phospha-alkene also seems to be formed in the conversion of $(CF_3)_2$-PH into $CF_3P(NH_2)CHF_2$ by reaction with ammonia.[448,449]

$$(CF_3)_2PH + NH_3 \longrightarrow NH_4F + CF_3P{=}CF_2 \xrightarrow{NH_3} CF_3P(NH_2)CHF_2$$

B.2.10. Miscellaneous Reactions

Ethylene carbonate interacts with secondary phosphines to yield ethylenediphosphine dioxides in one step.[468] An attempt to prepare β-(diphenylphosphino)propionic acid by direct union of diphenylphosphine with β-propiolactone failed.[567] None of the acid was isolated, apparently be-

$$R_2PH + 3\overline{OCH_2CH_2OC}{=}O \longrightarrow R_2P(O)CH_2CH_2(O)PR_2 + 2CH_2{=}CH_2 + 3CO_2 + H_2O$$

cause diphenylphosphine catalyzed polymerization of the lactone too effectively. A patent claims[500] that phosphines react with sulfones to give phosphino derivatives of sulfonic acids. The reactions of diphenylphosphine with

$$R_2PH + \overline{CH_2CH_2CH_2SO_2{-}O} \longrightarrow R_2PCH_2CH_2CH_2SO_3H$$

azoisobutyronitrile and tetraphenylhydrazine apparently proceed by a radical mechanism to produce a tertiary phosphine, tetraphenylbiphosphine and diphenylaminodiphenylphosphine, respectively.[870]

$$Ph_2PH \xrightarrow{AIBN} Ph_2PC(CN)Me_2 + [Ph_2P]_2$$

$$Ph_2PH \xrightarrow{[Ph_2N]_2} Ph_2PNPh_2 \text{ and other products}$$

Diphenylphosphine converts acyclic α-halo ketones and α-mesyloxyketones into dehalogenated or demesylated ketones, respectively.[139] A six-centered transition state has been suggested to best explain the results obtained. With cyclic ketones such as cyclohexanone, α-chloro- or α-mesyloxycyclohexanone, α-hydroxy-substituted tertiary phosphines are formed[139] (see also Section C.1.XIII).

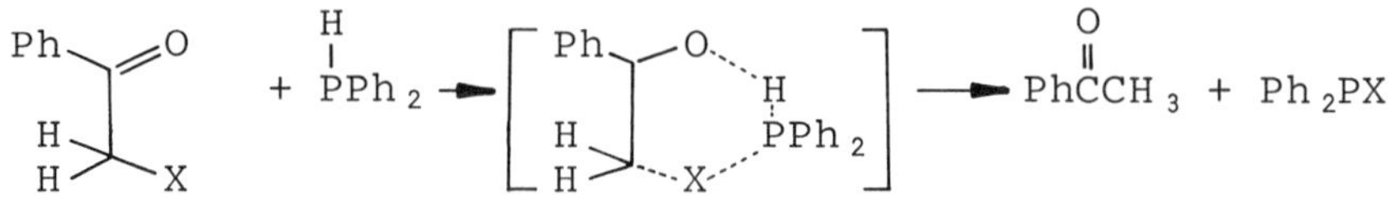

X = Cl, MeO_2SO

C. TERTIARY PHOSPHINES

C.1. Preparation of Tertiary Phosphines

I. BY THE GRIGNARD METHOD

The most convenient laboratory method for the synthesis of tertiary phosphines is the Grignard method.[564,1117] The starting materials may be phosphorus trihalides, phosphonous dihalides, or phosphinous halides. The reaction proceeds by the usual substitution process and, depending on the starting materials and the Grignard reagent used, results in the formation of tertiary phosphines having either identical or different radicals. The reactions are usually conducted in ether and are completed, after the

$$PX_3 + 3RMgX \longrightarrow R_3P + 3MgX_2$$

$$RPCl_2 + 2R'MgX \longrightarrow RR'_2P + 2MgX_2$$

$$RR'PX + R''MgX \longrightarrow RR'R''P + MgX_2$$

addition of the phosphorus halide, by refluxing.[286,291] Refluxing is essential in the preparation of tricyclohexylphosphine, otherwise less substituted products result.[653] Apparently, higher yields of tertiary phosphines are obtained if the reaction is conducted with the Grignard reagent prepared from alkyl chloride and magnesium in diisopropyl ether[312] or in tetrahydrofuran as solvents.[1024,1158,1159] Tetrahydrofuran is also essential for the preparation of vinylphosphines,[115,399,753,934,1154] since the vinyl Grignard reagent cannot be made in ether.[1058] [For a review see refs. 755, 1276.] Tris(trifluorovinyl)phosphine,[255,1336] alkynyl-substituted phosphines,[528,1389,1391] and many others have been obtained similarly using tetrahydrofuran as solvent. It has been stated that the best yields of tertiary phosphines are obtained if a large excess of chloro-Grignard is used and if the phosphorus halide and the Grignard reagent are brought together at as low a temperature as possible, even down to -78°.[753] After this normal Grignard procedures can be followed. The isolation of the tertiary phosphines is usually achieved by hydrolytic treatment of the reaction mixture with ammonium chloride solution followed by distillation of the organic layer.[286] In a few cases tertiary phosphines have also been isolated by omitting the hydrolysis step and subjecting the entire reaction mixture to vacuum distillation.[564] This method does not seem to be a desirable process, however, particularly for the higher boiling products. As a rule, derivatives of the primary halides give the best yields. Alkyl halides with branched chains and, especially, secondary alkyl halides[291] and t-butyl chloride[259,484,580,581,981] give either extremely poor yields or essentially no detectable amounts of the desired products. This takes place even under the best conditions known: a large excess of Grignard reagent. Apparently, steric factors prevent the formation of tertiary phosphines and cause the formation of less substituted products such as phosphonous and phosphinous halides, which have been isolated from these reactions under certain conditions.[1390] t-Butylmagnesium chloride also acted as a reducing agent and produced secondary phosphines also (see Section B.1.VIII).

Unsymmetrical tertiary phosphines containing three different groups have been obtained either from phosphinous chlorides of the type RR'PCl and a Grignard reagent, R"MgX,[761-763] or from phosphonous dihalide and a mixture of different Grignard reagents,[882] for example,

$$PhPCl_2 + EtMgBr + PhCH_2MgCl \longrightarrow PhEt_2P\ (24.5\%) +$$

$$EtPh(PhCH_2)P\ (41\%) + Ph(PhCH_2)_2P\ (18\%).$$

Naturally, the groups of the Grignard reagents must be different enough to allow separation of the products.

The Grignard route has also been used successfully for the synthesis of phenylenediphosphines [e.g., (13)][72,76,77,519,521] bis(phosphino)alkynes [e.g., (14)],[218,220,221,525,890] and cyclic tertiary phosphines [e.g., (15)],[33,42,43,78,492,493,949] by reaction of phosphonous or phosphinous halides with a di-Grignard reagent, respectively.

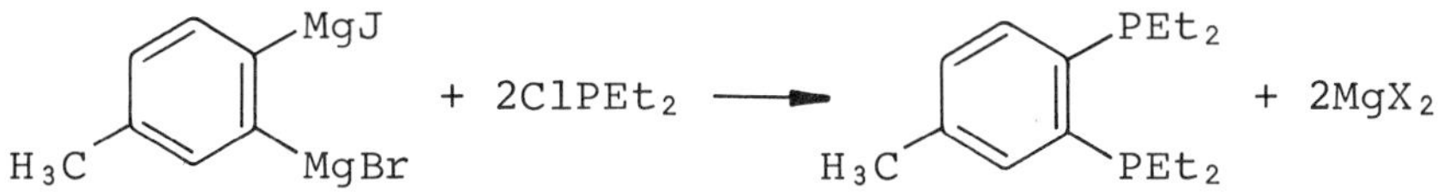

(13)

$$2Ph_2PCl + XMgC{\equiv}CMgX \longrightarrow Ph_2PC{\equiv}CPPH_2 + 2MgX_2$$

(14)

$$PhPCl_2 + XMg(CH_2)_4MgX \longrightarrow PhP\langle\ \rangle + 2MgX_2$$

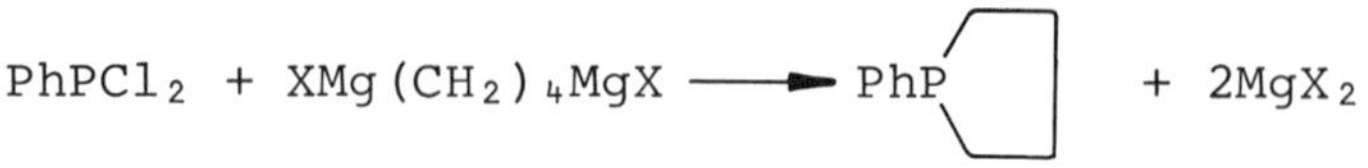

(<u>15</u>)

Reactions of magnesium compounds derived from indole and related substances yield the normally expected tertiary phosphines and, because of the tautomeric nature of the magnesium derivatives, also some phosphorus-nitrogen-bound derivatives. The latter are unstable to warm aqueous alkali and can therefore readily be separated.[1003] A

$$+ 6MgX_2$$

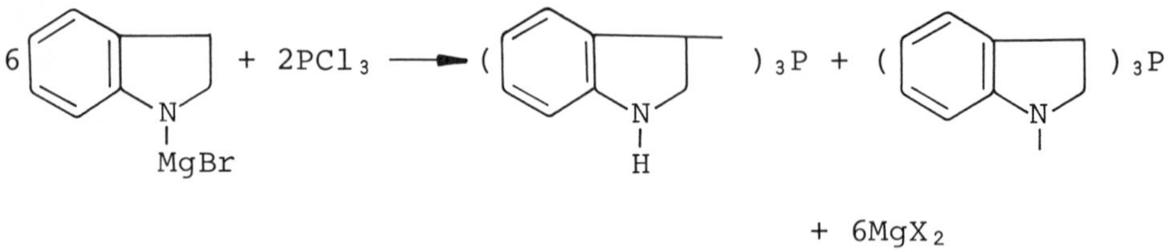

variation of the Grignard procedure has been used to prepare triphenylphosphine in 60% yield from triphenylphosphite and phenylmagnesium chloride.[433] The yields of triphenylphosphine from trimethylphosphite and PhMgBr in ether are nearly quantitative when excess Grignard reagent (4 moles) is used.[114] Tertiary phosphines were also obtained from trimethylphosphite and other Grignard reagents.[1017] The formation of phosphine oxides, reported earlier,[431] was attributed to impurities in the starting materials, such as methyldimethylphosphonate which, with phenylmagnesium bromide, yields methyldiphenylphosphine oxide, and triethylphosphate, which gives triphenylphosphine oxide. [114,1017] Chlorophosphites of the type $(RO)PCl_2$ and $(RO)_2$-PCl were also converted to tertiary phosphines with excess Grignard reagent.[1211] The blocking effect of alkoxy and aryloxy groups at low temperature[749,1211] has been used to prepare alkynyl-substituted tertiary phosphines[23] by the following sequence:

$$(EtO)_2PCl + LiC{\equiv}CR \longrightarrow (EtO)_2PC{\equiv}CR \xrightarrow{2R'MgX} R'_2PC{\equiv}CR$$

Several unsymmetrical tertiary phosphines containing a vinyl group were prepared by the interaction of butyl-dialkylphosphinites with vinylmagnesium bromide (yield 60 to 80%).[747] Only low yields of tertiary phosphines were

$$R_2P(OC_4H_9) + CH_2{=}CHMgBr \longrightarrow R_2PCH{=}CH_2 + C_4H_9OMgBr$$

isolated from the reaction of the lower phosphorus sulfides, such as P_3S_6 or P_4S_7, and Grignard reagents[628,935,801]

(p. 20). This reaction therefore shows no advantage over the usual Grignard-PCl_3 reaction.

II. WITH ALKALI ORGANIC COMPOUNDS

Organolithium compounds, first used in 1941 in place of a Grignard reagent for the preparation of tertiary phosphines,[432] have recently been used more frequently, particularly for the preparation of tertiary phenylenediphosphines (13),[233,518,557,1157] biphenylenediphosphines (16),[72,76] cyclic tertiary phosphines,[91,147,148,293,364,767]

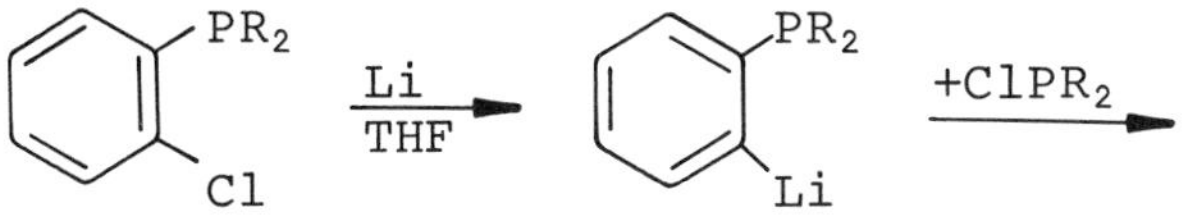

(13)

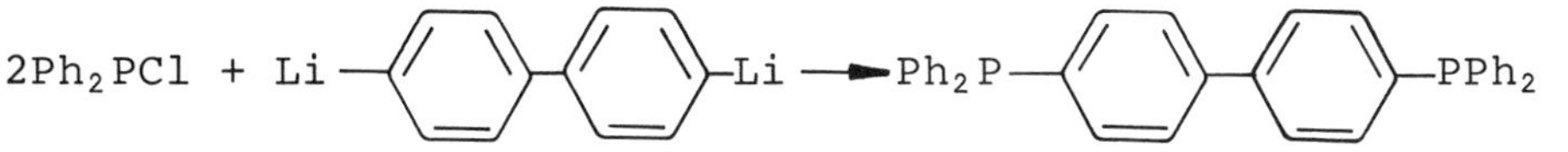

(16) + 2LiCl

[845,951,1349,1444] and the more exotic tertiary phosphines such as tris(9-phenanthryl)phosphine, tris(9-anthryl)phos-

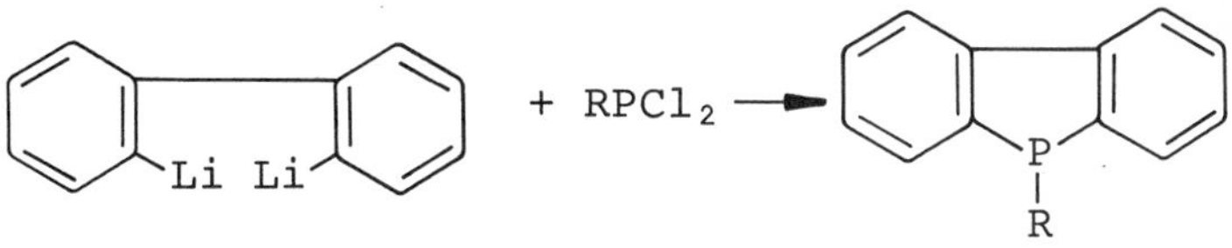

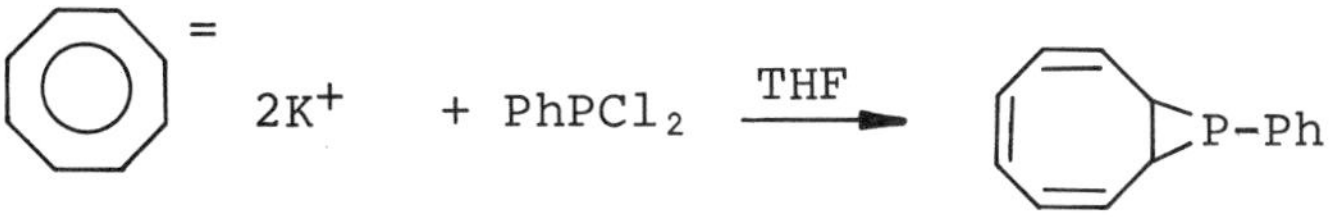

phine,[996,997] tris(2-pyridyl)phosphine,[1122] and tris(o-methoxymethylphenyl)phosphine,[849] tricyclopropylphosphine,[254,306] and many others.[454,1098,1181,1222,1296,1359] The yields of tertiary phosphines obtained by this method are comparable with those from the Grignard route.[1222] The use of the organolithium route was essential in the preparation of tri-t-butylphosphine, since this phosphine could not be made by the Grignard method.[581]

The reaction of propenyllithium with PCl_3 in ether at -20° is stereospecific.[136] Thus trans-propenyllithium gave the trans-phosphine, and cis-propenyllithium the cis-phosphine.[136] Sodium[528,1055] and potassium organic com-

pounds[654] have also been used to convert phosphorus trichloride into tertiary phosphines. The reactions are carried out in a fashion similar to that used with the Grignard reagents. The necessary alkali organic compounds can be made in situ from the organic halide and the alkali metal or by an exchange reaction in ether at low temperature. Subsequent normal procedures give the tertiary phosphines in yields ranging from 40 to 85%. For example, tribenzylphosphine has been obtained in 84.1% yield by the interaction of phosphorus trichloride and benzylsodium.[1055] High yields of tertiary phosphines have also been obtained from triethylphosphite and phenyllithium (80% triphenylphosphine),[1436] and from triphenylphosphite and cyclopropyllithium (80% tricyclopropylphosphine).[306]

III. BY A WURTZ REACTION

The formation of tertiary phosphines from metallic sodium, phosphorus trihalides, or halophosphines and aryl halides, may be considered a forerunner of the organolithium route. This procedure, introduced by Michaelis[986,989,992] was used rather frequently before the introduction of the Grignard reagents. The reaction is conducted at reflux in an organic solvent (frequently ether or benzene) for several hours until all the sodium metal has been used up. Occasionally, this requires 24 hr, or even longer periods. Although the yields are not particularly high (30 to 50%), this method is still being used because of its simplicity.[507,1249] The higher halides, which react rather sluggishly, can be activated by antimony trichloride, added in

$$3RCl + PCl_3 + 6Na \longrightarrow R_3P + 6NaCl$$

small amounts to the mixture.[1447,1448] Addition of malonic diethyl ester has been said to shorten the reaction time and to give higher yields of tertiary phosphines.[1351] Much higher yields were apparently obtained when triphenylphosphite was caused to react with sodium and alkyl or aryl halides in naphtha as solvent,[538,539] for example, Bu_3P

(88%), Ph_3P (85.1%), and $(o\text{-}MeOC_6H_4)_3P$ (78.7%).

IV. FROM PH_3, RPH_2 OR R_2PH, AND ALKYLATING AGENTS

While alkylation of PH_3 or PH_4I with alkylating agents usually yields a mixture of substituted products (Section A.1.VI and B.1.III), high yields of tertiary phosphines were obtained when primary[535] or secondary phosphines[260,580,591] were used as starting materials. The successful alkylation of primary phosphines to tertiary phosphines depends upon close control of several variables such as the alkylating agent, solvent, temperature, time, reactant ratio, the strength of the acid by-product, and the acid concentration.[535] Trialkylphosphines of the type RP'_2P have been obtained in excellent yields (78 to 94%) from primary phosphines and methyl or n-alkyl iodides by using the alcohol corresponding to the alkyl iodide as solvent.

$$RPH_2 + R'X \longrightarrow [RPH_2R']^+X^- \rightleftharpoons HX + RPHR' \xrightarrow{R'X}$$

$$[RPHR'_2]^+X^- \rightleftharpoons HX + RPR'_2 \xrightarrow{R'X} [RPR'_3]^+X^-$$

(17) (18)

When methyl chloride is used as alkylating agent, the addition of excess hydrogen chloride to suppress the dissociation of (17) is essential to the successful preparation of alkyldimethylphosphine. However, even under these conditions some quaternary salt (18) was formed also.[535] Secondary dialkylphosphines are alkylated in the same way. Because of the greater stability of (17) when X = I, alkyl iodides are generally used as alkylating agents.[260,553,580,591] The tertiary phosphines are liberated from the salts (17) with alkaline solutions or, better, with powdered sodium hydroxide.[535] Dialkylphosphinosubstituted carbonic esters, which could not be prepared by the phos-

$$[RPHR'_2]^+X^- + NaOH \longrightarrow RPR'_2 + H_2O + NaX$$

phide route (Section C.1.VII) were obtained in 50 to 60% yield by using this alkylation procedure.[704] Furthermore, using α,ω-dihaloalkanes as alkylating agents and a primary

$$R_2PH + X(CH_2)_nCO_2Et \longrightarrow [R_2PH(CH_2)_nCO_2Et]^+X^- \xrightarrow{NaOR}$$

$$R_2P(CH_2)_nCO_2Et + NaX + HOR$$

or a secondary phosphine, gave cyclic tertiary phosphines[676] or diphosphines,[669,676] respectively. Some limitations may be noted. Cyclic phosphines were obtained only with dibromobutane and dibromopentane, while dibromopropane and dibromohexane gave only polymeric products.[676] And a di-

$$RPH_2 + Br(CH_2)_nBr \longrightarrow [RH\overset{\frown}{P(CH_2)_n}]^+ \xrightarrow{OH^-} R\overset{\frown}{P(CH_2)_n}$$

$$n = 4, 5$$

$$2R_2PH + X(CH_2)_nX \longrightarrow [R_2HP(CH_2)_nPHR]^{2+} \xrightarrow{2OH^-} R_2P(CH_2)_nPR_2$$

phosphine could not be isolated from the reaction of $EtPH_2$ with methylene dibromide or dichloride.[676] Some highly fluorinated chloroolefins such as 1,2-dichlorotetrafluoro-cyclobutene-1,[265,1340] octafluorocyclopentene,[269] 1,2-dichlorohexafluorocyclopentene-1,[1339] and 2,3-dichloro-tetrafluoropropene-1,[124] reacted directly with secondary phosphines. Because of the low basicity of the phosphines produced, the hydrohalides formed as intermediates are unstable and dissociate into the free phosphine and HCl.

Refluxing diphenylphosphine with excess carbon tetrachloride has been claimed to produce trichloromethyldiphenylphosphine.[597]

$$R_2PH + CF_2{=}CClCF_2Cl \longrightarrow [R_2PHCF_2CCl{=}CF_2]^+Cl^- \rightleftharpoons R_2PCF_2CCl{=}CF_2 + HCl$$

V. FROM ELEMENTAL PHOSPHORUS AND VARIOUS REAGENTS

A poor yield of tertiary phosphines is obtained on heating a mixture of alkyl iodide, phosphorus, and zinc in sealed tubes for several hours at 150°. Most of the product consists of quaternary phosphonium compounds.[205,801] (p. 11). Trifluoromethyl iodide, however, undergoes reaction with white or red phosphorus to give substantial amounts of tris(trifluoromethyl)phosphine. Because of the strongly electronegative trifluoromethyl groups which decrease the basicity of the phosphorus atom, no quaternary salts are formed by this phosphine. The reaction is usually carried out in sealed tubes or autoclaves. To obtain the highest yield of tertiary phosphine (84%), heating at 220° for 48 hr is necessary. Lower temperatures give higher yields of the iodophosphines and a decreasing amount of the tertiary phosphine[99] (for reviews see refs. 78a and

835). Tris(trifluoromethyl)phosphine is also obtained by heating a mixture of red phosphorus, silver trifluoroacetate, and iodine for 120 hr at 195°.[191] Refluxing a mixture of alkyl iodide and red phosphorus (ratio 3:1) in the presence of catalytic amounts of iodine until alkyl iodide condensation ceased, followed by reduction of the polyiodides with magnesium at 145 to 150° and then 2 hr at 170°, gave tertiary phosphines in 70 to 80% yield.[383]

Methyl bromide passed over red phosphorus in the presence of copper powder as a catalyst at 350° gives only methylphosphonous dibromide and dimethylphosphinous bromide,[906,913] but methyl chloride gives under these conditions 9% trimethylphosphine, 48% dimethylphosphinous chloride, and 43% methylphosphonous dichloride.[913] Similarly, it has been found that passing trifluoromethyl iodide over red phosphorus heated to 280° and containing copper powder as a catalyst produces tris(trifluoromethyl)phosphine (10%), bis(trifluoromethyl)phosphinous iodide (60%), and trifluoromethylphosphonous diiodide (30%).[913] An elegant one-step synthesis of 2,3,5,6,7,8-hexakis(trifluoromethyl)-1,4-diphosphabicyclo[2.2.2]octa-2,5,7-triene (19) in 43% yield merely involves heating of red phosphorus with hexafluoro-2-butyne and a catalytic amount of iodine at 200° for 8 hr in a closed system, or heating of red phosphorus with 2,3-diiodohexafluoro-2-butene at 210° under pressure.[816,817]

$$CF_3C{\equiv}CCF_3 + P \xrightarrow[\Delta]{I_2}$$

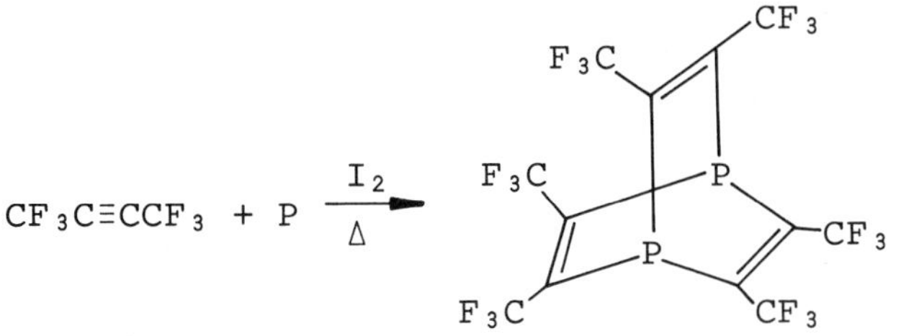

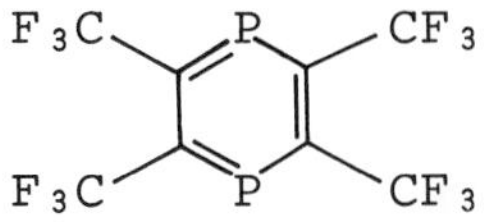

(19)   (20)

Tetrakis(trifluoromethyl)-1,4-diphosphorin (20) might be involved as an intermediate in this synthesis.[894] Heating triphenylarsine,[812,1334] or tetraphenyltin[1269] with phosphorus to 300° results in the formation of triphenylphosphine in rather good yield. Tris(pentafluorophenyl)-phosphine (70%) has similarly been obtained by heating bis-(pentafluorophenyl)Th(III)bromide with phosphorus for 4 days to 190°.[298] Unsymmetrical tertiary phosphines were isolated in a maximum yield of 44% from the reaction of white phosphorus with organolithium compounds or phenylsodium and alkyl halides.[1175 1177] Primary and secondary phosphines were also produced in the last-mentioned reactions (see Sections A.1.III and B.1.VII). Electrolysis of white phosphorus in Me_2NCHO in the presence of butyl bromide produced a very small amount of tributylphosphine (1.46%), isolated as the oxide, in addition to other products.[745] Higher yields of tertiary phosphines are apparently obtained in the electrolysis of a Grignard reagent, preferably RMgCl, using a sacrificial phosphorus anode, preferably from black phosphorus.[560] Finally, small amounts of tertiary phosphines (5%) were formed as by-products in the Mannich-type reaction of white phosphorus with formaldehyde and secondary amines,[927] and in the reaction of white phosphorus with ethyl acrylate.[1164a]

VI. FROM ORGANOMETALLIC REAGENTS OTHER THAN GRIGNARD OR ORGANOLITHIUM COMPOUNDS

Trialkylaluminum compounds react with phosphorus trichloride, yielding all possible substitution products depending on the conditions used.[1064,1426] Thus heating triethylaluminum and phosphorus trichloride in a 1:1 ratio in hexane at 20 to 30° and then, after distilling off the hexane, at a higher temperature, resulted in the formation of 48% triethylphosphine as the aluminum trichloride complex.[1426] The free phosphine was obtained by heating this complex with potassium chloride or sodium chloride,[1426] or by treatment with sodium hydroxide and extraction with ether.[1064] The highest yield (64.6%) of triethylphosphine was obtained when a 1:1 mixture of phosphorus trichloride and triethylaluminum was heated for 19 hr to 200 to 210° in the presence of two equivalents of potassium chloride.[1426]

$$PCl_3 + Al(C_2H_5)_3 + KCl \xrightarrow[19\ hr]{200°} P(C_2H_5)_3 + KAlCl_4$$

A patent claims that tributylphosphine is produced in good yield from tributylaluminum and phosphorus trichloride in the presence of calcium zeolite.[510] Triphenylphosphine has been obtained from the reaction of PCl_3, aluminum, activated by $AlCl_3$, and phenyl halide by heating to 230° for 2 hr (91.5%),[844] or from PCl_3, $AlPr_3$-i and benzene (64.5%).[90] Diphenylalkylphosphines were obtained from the reactions of Ph_2PCl with trialkylboranes.[331a]

Tertiary phosphines were also formed when phosphonous and phosphinous chlorides were heated with tetraethyllead or tetraphenyllead for 40 to 60 hr to temperatures ranging from 125 to 180°.[910] The reactivity of organotin compounds was used in the synthesis of phosphinosubstituted carbonic esters,[1060,1130,1131,1379] phosphinosubstituted ketones[1061] and, in particular, of alkynyl-substituted phosphines.[524]

$$3Bu_3SnCH_2CO_2Me + PCl_3 \longrightarrow P(CH_2CO_2Me)_3 + 3Bu_3SnCl$$

$$R_3SnCH_2COMe + R_2PCl \longrightarrow R_2P(CH_2COMe) + R_3SnCl$$

[1295] Unsaturated organocopper[1305] and mercury compounds,[526] converted PCl_3 and PI_3 into tertiary phosphines, but silver acetylide and p-nitrophenylethynyl copper reacted explosively with PCl_3.[1305]

$$Me_3SnC{\equiv}CSnMe_3 + 2Ph_2PCl \longrightarrow Ph_2PC{\equiv}CPPh_2 + 2Me_3SnCl$$

$$3Me_3SiC{\equiv}CSnMe_3 + PCl_3 \longrightarrow (Me_3SiC{\equiv}C)_3P + 3Me_3SnCl$$

$$3(RC{\equiv}C\text{-}C{\equiv}C)_2Hg + 2PI_3 \longrightarrow 2(RC{\equiv}C\text{-}C{\equiv}C)_3P + 3HgI_2$$

The reaction of organozinc compounds with halogen derivatives of trivalent phosphorus was used extensively before the introduction of the Grignard reagents to prepare tertiary phosphines[280,1432,801] (p. 18). The products are liberated from the zinc halide double salts that form in this reaction by treatment with alkali.[199,280,593]

VII. FROM ALKALI PHOSPHIDES AND ALKYLATING AGENTS

The improvements made in recent years in the preparation of alkali derivatives of phosphines, PM_3, RPM_2, and R_2PM (M = alkali metal), greatly stimulated interest in the chemistry of these compounds (see Chapter 2). The alkali metal derivatives of phosphines, PM_3, RPM_2, and R_2PM, react readily with alkyl halides to give the desired tertiary phosphines in good yields.[11,17,128,566,567,569,664,707,952,953,1084,1401,1407] The typical reactions are illustrated in the equations:

$$PM_3 + 3RX \longrightarrow R_3P + 3MX$$

$$RPM_2 + 2R'X \longrightarrow RR'_2P + 2MX$$

$$RR'PM + R''X \longrightarrow RR'R''P + MX$$

The alkyl halides used in this reaction have been chlorides, bromides, and iodides. Phenyl bromide and phenyl iodide[823,1084] also appear to be suitable but apparently require more drastic reaction conditions. Stoichiometric amounts of alkyl halides are added to a suspension, or a solution, of the metal phosphides in inert solvents. The reaction is usually completed by refluxing the mixture. The metal halide is then removed by filtration or by washing with water, and the phosphine is isolated in the usual manner, by distillation or crystallization from the organic layer. Excess alkyl halide, which would form phosphonium salts, must be avoided. Alkylene dihalides also react readily with the alkali derivatives of primary and secondary phosphines to yield cyclic tertiary phosphines,[148,213,651,

[658,667,702,716,948,949] [e.g., (21) and (22)] or bis(phosphino)alkanes. The latter reaction is either contacted in

$$PhPLi_2 + Cl(CH_2)_4Cl \longrightarrow PhP\!\!\begin{array}{c}\\ \square \end{array} + 2LiCl$$

(21)

$$+ PhPLi_2 \longrightarrow \quad\quad\quad + 2LiX$$

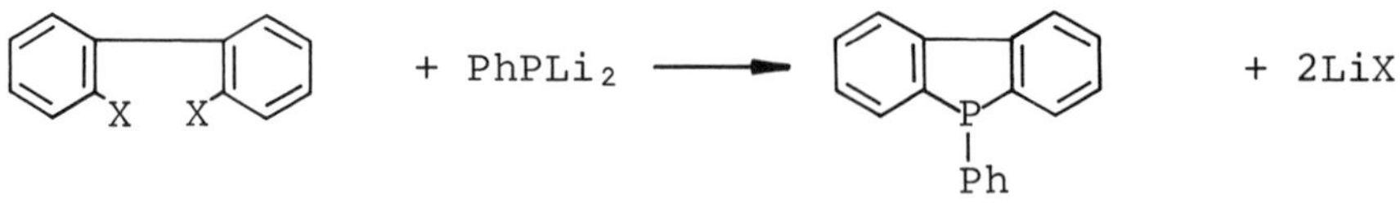

(22)

liquid ammonia,[220,224,569,1410,1451,1452] tetrahydrofuran, dioxan,[20,220,561,672,692] or hexamethylphosphoric amide.[1059] Functionally substituted tertiary phosphines are ob-

$$2R_2PM + X(CH_2)_nX \longrightarrow R_2P(CH_2)_nPR_2 + 2MX$$

tained in the interaction of alkaliphosphides and ethylenimine,[661] epoxides,[691,700,782,1094] episulfides,[428,1271] acyl halides,[687,696,808,859] halogenocarbonic acid esters,[567,703] and sodium salts of halogenosubstituted carbonic[649,717,1237] and sulfonic acids.[1240,1472] The use of a

$$R_2PLi + \begin{array}{c} CH_2\!-\!CH_2 \\ \diagdown\;\diagup \\ NH \end{array} \longrightarrow R_2PCH_2CH_2NHLi \xrightarrow{H_2O} R_2PCH_2CH_2NH_2$$

$$R_2PM + \begin{array}{c} CH_2\!-\!CH_2 \\ \diagdown\;\diagup \\ O \end{array} \longrightarrow R_2PCH_2CH_2OM \xrightarrow{H_2O} R_2PCH_2CH_2OH$$

$$R_2PM + \begin{array}{c} CH_2\!-\!CH_2 \\ \diagdown\;\diagup \\ S \end{array} \xrightarrow{NH_3} R_2PCH_2CH_2SM \xrightarrow{H_2O} R_2PCH_2CH_2SH$$

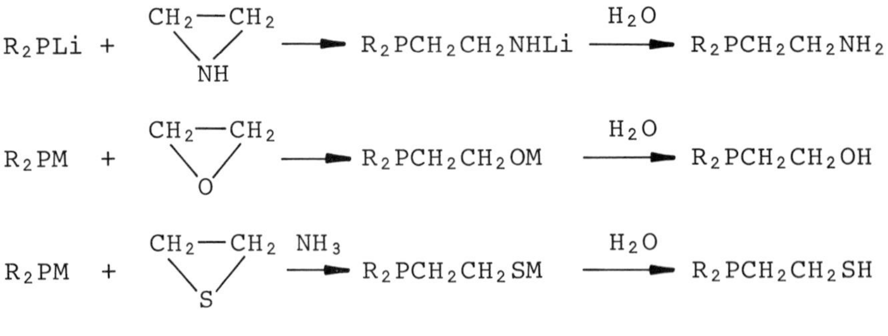

$$R_2PM + ClCOR \longrightarrow R_2PCOR + MCl$$

$$Ph_2PM + Cl(CH_2)_nCO_2R \longrightarrow Ph_2P(CH_2)_nCO_2R + MX$$

$$R_2PM + ClC_6H_4CO_2M \longrightarrow R_2PC_6H_4CO_2M \quad (o, \text{ or } p)$$

$$R_2PM + ClC_6H_4SO_3M\text{-}p \xrightarrow{THF} R_2PC_6H_4SO_3M\text{-}p$$

2:1 ratio in the latter reaction produced a high yield

(92%) of 1,4-phenylenebis(diphenylphosphine).[1240] In contrast to alkali diphenylphosphide, lithium diethyl- and dicyclohexylphosphide interacted differently with halogenocarbonic acid esters. Whereas iodoacetic ethyl ester and bromosuccinic ethyl ester first brought about a metal-halogen interconversion reaction followed by coupling reaction, chloroacetic and γ-bromosuccinic ethyl ester were

$$R_2PLi + ICH_2CO_2Et \longrightarrow R_2PI + LiCH_2CO_2Et$$

$$R_2PI + R_2PLi \longrightarrow R_2P-PR_2 + LiI$$

$$LiCH_2CO_2Et + ICH_2CO_2Et \longrightarrow LiI + EtO_2CCH_2CH_2CO_2Et$$

$$R = Et, c-C_6H_{11}$$

converted to the lithium salts of the parent acid.[703]

$$ClCH_2CO_2Et + R_2PLi \longrightarrow ClCH_2CO_2Li + R_2EtP$$

Thus the halogen-carbon bond did not react with $LiPR_2$, instead, reaction with the ester groups occurred.[703]

    Also, the action of phosgene on MPR_2 did not lead to the formation of the phosphorus analog of substituted urea. Instead, carbon monoxide was evolved and tetraalkylbiphosphines were obtained.[696]

    Replacement of the $-SO_3M$ group by the diphenylphosphino group seems to be a general method for preparing tertiary and ditertiary phosphines.

$$2Ph_2PK + ClC_6H_4SO_3Na-p \longrightarrow Ph_2P-\!\!\left\langle\!\!\bigcirc\!\!\right\rangle\!\!-PPh_2 + KCl + KNaSO_3$$

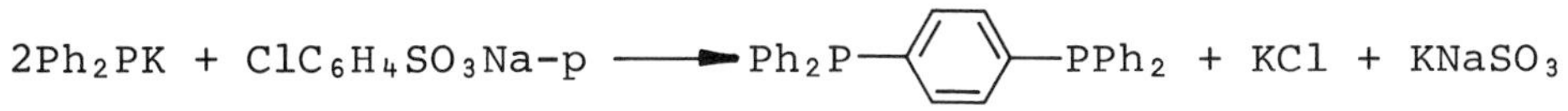

High yields of phosphines have been obtained when 2,2'-diethoxydiethyl ether was used as solvent.[1240,1472] The choice of solvent is also critical in several of the reactions listed above, since some solvents such as tetrahydrofuran,[425,446,937,940] trimethylene oxide,[700] 2,2'dialkoxydiethyl ether,[1247] and dioxan[1244] were found to be cleaved by the phosphide ion under certain conditions.

$$Ph_2PLi + \begin{array}{c} H_2O \\ \overbrace{\hspace{1cm}} \\ \text{(trimethylene oxide)} \end{array} \longrightarrow Ph_2P(CH_2)_4OH$$

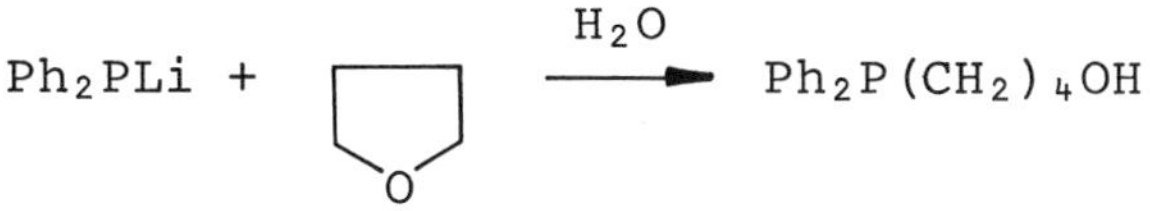

$$2Ph_2PK + ROCH_2CH_2OCH_2CH_2OR \longrightarrow Ph_2PCH_2CH_2PPh_2 + ROK +$$

$$ROCH_2CH_2OK$$

Furthermore, the diphenylphosphide ion rapidly dealkylates certain alkyl aryl ethers, cyclic ethers, alkyl aryl sulfides[941,954,956] (see also Section B.1.I) and alkyl esters of sulfonic acids[1241] to give high yields of alkyldiphenylphosphines.

$$Ph_2PLi + MeOC_6H_5 \longrightarrow Ph_2PMe + LiOC_6H_5$$

$$Ph_2PK + RSO_3R' \longrightarrow Ph_2PR' + RSO_3K$$

The reaction of Ph_2P ion with anisole compared with the reaction with phenetole and other simple alkyl phenyl ethers is so rapid that it is almost selective.

In general, the reaction of lithium dialkyl- or diarylphosphide with organic halogen compounds proceeds stereospecifically[8,12,14] with retention of configuration.[15] In the interaction of alkali diphenylphosphide and aryl halides, no isomers are produced.[18,1471] This suggests that

cis 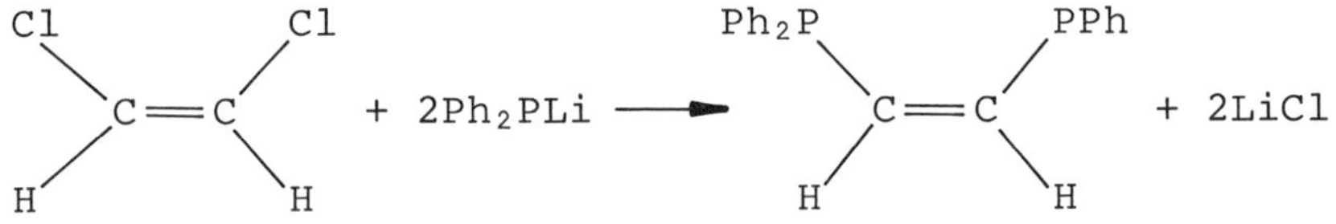

trans 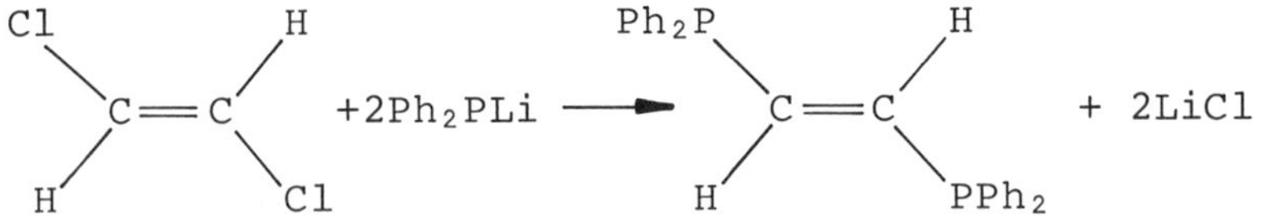

the reaction again proceeds by direct nucleophilic displacement and does not involve an aryne intermediate. The difference in results between normal and inverse addition suggests that the transition state involves two molecules of lithium diphenylphosphide and one of aryl halide. Lithium chloride assists in these reactions. It may take the place of one of the two molecules of phosphide.[18] As an

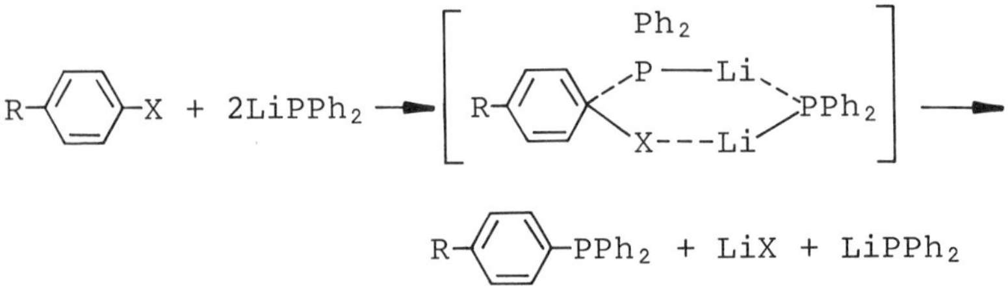

exception, the reaction of the highly nucleophilic lithium di-t-butylphosphide with p-fluorotoluene seems to proceed by an elimination-addition reaction through the intermediate formation of tolyne, since p- and m-tolyldi-t-butylphosphine were isolated.[710] However, the reaction of the less nucleophilic $LiPBu_2$ and $LiPEt_2$ with p-fluorotoluene

$$p\text{-}MeC_6H_4F + LiPBu_2\text{-}t \longrightarrow p\text{-}MeC_6H_4PBu_2\text{-}t + m\text{-}MeC_6H_4PBu_2\text{-}t$$

but $$p\text{-}MeC_6H_4F + LiPEt_2 \longrightarrow p\text{-}MeC_6H_4PEt_2$$

gave only p-tolyldibutyl- or diethylphosphine and thus proceeded by direct nucleophilic displacement.[710]

In the reaction of chloro-2-phenylacetylene with the alkali derivatives of dicyclohexylphosphine and diethylphosphine, some metalhalogen interconversion takes place (28 and 20%, respectively). There is no interconversion when alkali diphenylphosphide is used. Instead, phenylethynyldiphenylphosphine was formed.[664] Also, Ph_2PNa reacts with 3-chloroprop-1-yne and 3-chloro-but-1-yne in liquid ammonia by nucleophilic attack on carbon rather than chlorine, but with the corresponding bromo compounds Ph_2PNa reacts by nucleophilic attack on bromine ($BrCH_2C{\equiv}CH$, $MeCHBrC{\equiv}CH$) in some cases, and in others ($BrCH_2C{\equiv}CMe$) by nucleophilic attack on carbon.[562a]

Similar to the secondary phosphines, alkali phosphides also add easily to olefins, acetylenes, and other unsaturated compounds. These reactions are discussed in Sections C.1.XII and XIII.

VIII. BY REDUCTION OF TERTIARY PHOSPHINE OXIDES AND SULFIDES AND DICHLOROPHOSPHORANES

Tertiary phosphine oxides have been successfully reduced to the corresponding tertiary phosphines with $LiAlH_4$,[543,619,660] $Ca(AlH_4)_2$,[543] CaH_2,[407] silanes,[407,409] boranes [R_3B, $(R_2BH)_2$, $H_3B{\cdot}NR_3$], alanes (R_3Al),[789] and perchloropolysilanes (Si_2Cl_6, Si_3Cl_8).[1033,1034] Phosphine (23), prepared by $LiAlH_4$ reduction of the corresponding optically

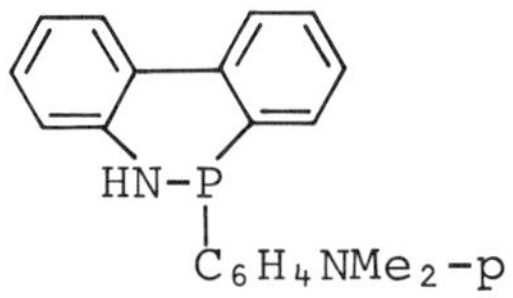

(23)

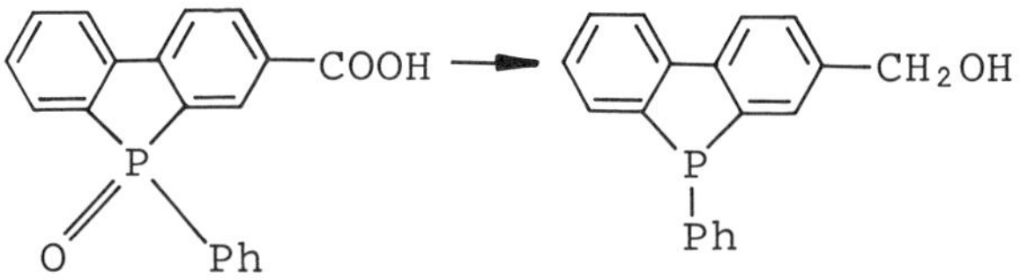

(24) (25)

active phosphine oxide, provided the first example of a trivalent phosphorus compound obtained in an optically active form.[203] Reduction of optically active (24) with LiAlH 4, however, gave racemic (25).[204] It was shown that optically active phosphine oxides undergo rapid stereomutation in the presence of $LiAlH_4$ prior to reduction. Racemization of the phosphine oxide is virtually complete before more than 10% has been reduced to the phosphine.[556] To explain these results pseudorotation of the phosphorane, $Li[Al(OPHR_3)_4]$, formed as an intermediate, followed by dissociation into phosphine oxide and $LiAlH_4$ was suggested.[556] The residual activity of the small amount of phosphine produced in the reduction of (+)-MePrPhPO indicates net retention of configuration.[556] By far the most often used reducing agents for tertiary phosphine oxide to give tertiary phosphines are the silanes ($SiHCl_3$, $R_{4-x}SiH_x$, and methylpolysiloxane).[407,409] Two moles of $SiHCl_3$ are necessary, but in the presence of 1 mole of tertiary amine 1 mole of $SiHCl_3$ suffices to reduce tertiary phosphine oxides to the corresponding tertiary phosphines.[409] Functional groups do not interfere in these reductions. Reduction of optically active tertiary phosphine oxide with $SiHCl_3$, and also in the presence of pyridine or dimethylaniline[605] or other weak bases (pK_b > ca. 7)[1033] proceeds with retention of configuration, but in the presence of triethylamine[605] and other strong bases (pK_b < ca. 5)[1033] with inversion. Similarly, reduction of optically active acyclic phosphine oxides by Si_2Cl_6 or Si_3Cl_8 proceeds with complete or nearly

$$R_3P{=}O + 2SiHCl_3 \longrightarrow R_3P + SiCl_4 + H_2 + (Cl_2SiO)_n$$

$$R_3P{=}O + SiHCl_3 + R_3N \longrightarrow R_3P + (Cl_2SiO)_n + R_3NHCl$$

complete inversion of configuration.[124,1033,1034] Cyclic phosphine oxides are reduced to the corresponding phosphines by silanes[257,534,966,967] and also by Si_2Cl_6[299,1034] with retention of configuration (for mechanism, see Chapter 6). Allylic substituted phosphine oxides may suffer a rearrangement when reduced with $PhSiH_3$.[1218] It could be shown that this rearrangement occurs in the phosphine stage and is catalyzed by $PhSiH_3$ or silanol.[1218]

$$Ph_2P(O)CR_2CH{=}CH_2 \xrightarrow[150°/4\ hr]{PhSiH_3} Ph_2PCR_2CH{=}CH_2 \xrightarrow[210°]{PhSiH_3} Ph_2PCH_2CH{=}CR_2$$

Triphenylphosphine oxide, after alkylation with triethyloxonium tetrafluoroborate to $[Ph_3POEt]^+BF_4^-$, could be reduced with magnesium to Ph_3P (70%).[1189] This method may not be important for preparing tertiary phosphines but may play a role in the synthesis of phosphonites and phosphinites from the corresponding phosphonates and phosphinates, respectively.[1189]

Tertiary phosphine sulfides have been reduced to the corresponding tertiary phosphines with $LiAlH_4$,[543,619] $Ca(AlH_4)_2$,[543] phenyllithium,[1440] NaH in the melt at 300°,[619] Raney nickel in methanol,[619] sodium in naphtalene,[609] iron powder in the melt,[916] tertiary phosphines,[916,1077] and with Si_2Cl_6.[1034,1468] Reduction of optically active acyclic tertiary phosphine sulfides with hexachlorodisilane proceeds with retention of configuration.[1034,1468]

Trialkyl- and triaryldichlorophosphoranes are easily reduced to the corresponding tertiary phosphines, in high yield, by sodium in toluene,[609] $LiAlH_4$ in benzene,[619,900,901,1144] tertiary phosphines,[582] zinc,[1023] magnesium,[1023,1140-1142,1144,1145] aluminum,[719,854,1023] butyllithium,[307,970] hydrazine,[56] and elemental phosphorus.[1450]

Tertiary phosphine oxides could not be reduced to the corresponding phosphines with sodium metal[542,544,608,609] or $NaBH_4$.[556] Since tertiary phosphine oxides can easily be converted to phosphine sulfides by reaction with P_2S_5,[609] and to dichlorophosphoranes by reaction with PCl_5, $SOCl_2$, or $COCl_2$, reduction of tertiary phosphine oxides and sulfides with inexpensive reducing agents, such as the metals (Na, Zn, Mg, Al) or elemental phosphorus seems to be generally applicable.

IX. FROM PHOSPHONIUM SALTS

1. BY THERMAL DECOMPOSITION. Quaternary phosphonium halides decompose on strong heating above 300° with the loss of one radical and the halogen atom to yield tertiary phosphines. In particular, the decomposition of phosphonium chlorides containing an ethyl group has been frequently

$$[R_4P]^{\oplus}Cl^{\ominus} \longrightarrow R_3P + RCl$$

used to prepare asymmetric tertiary phosphines[242,379,757,758,759,978] and cyclic phosphines, for example, (26) and (27).[42,43,93,551,948,953] In these special cases the ethyl group breaks off as ethylene, and tertiary phosphine hydrochlorides are formed.

$$[C_2H_5R_3P]^{\oplus}Cl^{\ominus} \longrightarrow CH_2{=}CH_2 + R_3P\cdot HCl \xrightarrow{NaOH} R_3P$$

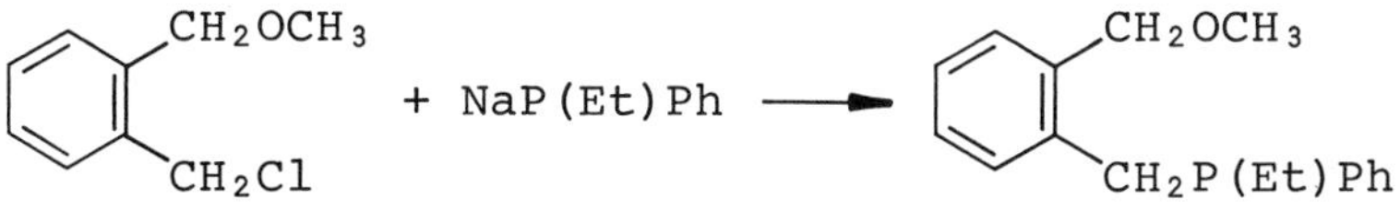

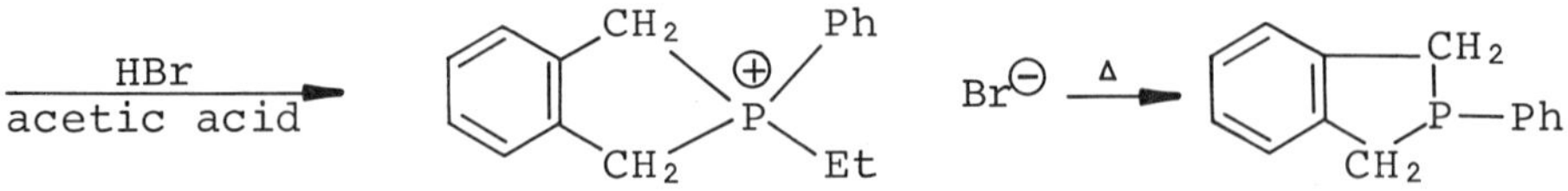

(26)

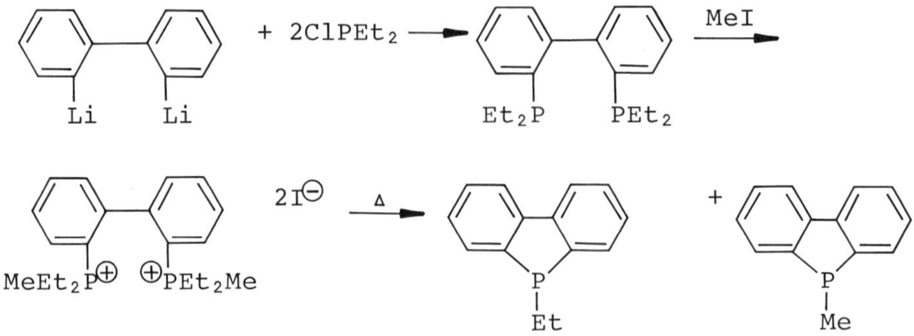

(27)

Addition of alkali liberates the free phosphine which is then extracted by an organic solvent and isolated in the normal way. It should be pointed out that in contrast to the references just given, Fenton et al.[379] reported that they observed no ethylene on decomposition of phosphonium chlorides containing ethyl groups; instead, ethyl chloride was evolved and a tertiary phosphine was formed. Those results still await explanation.

The decomposition of phosphonium chlorides containing ethyl groups proceeds normally with good yields of the expected phosphine, but secondary reactions may occur if benzyl or certain other groups are present, and mixtures of the several possible tertiary phosphines may form.[242,757,758,759,978] A rough order of radical cleavage has been given as: (1) ethyl ∿ benzyl; (2) methyl, propyl, isoamyl; (3) phenyl in descending order.[978]

Thermal decomposition of methylethylphenyl(1,3-diphenyl-3-hydroxypropyl)phosphonium betaine, $PhCH(O^-)CH_2CH(Ph)\overset{+}{P}$-MeEtPh, at a temperature of 190 to 200° produced several products but the major ones were methylethylphenylphosphine and benzylacetophenone.[883] The decomposition of the optically active phosphonium betaine gave racemic methylethylphenylphosphine.[881] This racemization is undoubtedly the result of the thermal racemization of the optically active phosphine at the reaction temperature of 190 to 200°.[635]

Unsymmetrical tertiary phosphines containing perfluoroalkyl groups have been obtained from the reaction of $(CF_3)_3P$ with MeI at 240°,[533] or from that of trialkyl- or triarylphosphines with perfluoroalkyl iodides.[96,532,1138] Thermal decomposition of alkylenebis(trialkylphosphonium) salts of the type $[R_3P(CH_2)_nPR_3]^{2+}2X^-$ produced only tertiary phos-

$$2R_3P + CF_3I \longrightarrow R_2PCF_3 + R_4PI$$

phines, R_3P, and never a ditertiary phosphine.[569] Pyrolysis of n-alkyltriphenylphosphonium alkoxides at 260° afforded an n-alkyldiphenylphosphine and an alkyl phenyl

$$[Ph_3PR]^{\oplus}OR'^{\ominus} \xrightarrow[240-260°]{\Delta} Ph_2PR + PhOR'$$

ether.[363,508] Detailed studies of this reaction suggest a phosphorane mechanism and exclude alternatives involving benzyne.[508] A low yield of bis(2-cyanoethyl)methylphosphine (35%) was isolated when the salt $[(NCCH_2CH_2)_3PMe]^+I^-$ was heated with NaCN to 250°. In addition, acrylonitrile and succinonitrile were obtained in a 2:1 ratio.[470]

2. BY AN ELIMINATION REACTION WITH ALKALI. In general, phosphonium salts are decomposed by aqueous base to give tertiary phosphine oxides and hydrocarbons (see Chapter 6). In some specific cases decomposition of phosphonium salts by strong bases results in the formation of tertiary phosphines. Thus phosphonium salts containing a 2-cyanoethyl group undergo reaction with a strong base such as sodium ethoxide to give a tertiary phosphine and 3-ethoxypropionitrile.[470,576] Starting with tris(2-cyanoethyl)phosphine, which is easily prepared from PH_3 and acrylonitrile,[1173] then quaternization with an alkyl halide, RX, followed by the base-promoted elimination of a 2-cyanoethyl group gives a tertiary phosphine with two different radicals. Since this replacement of the 2-cyanoethyl group can be repeated with a second alkyl halide, R'X, unsymmetrical tertiary phosphines can be prepared in this way, $[RR'PCH_2CH_2CN]$.[470,832] The loss of the 2-cyanoethyl group proceeds probably by a normal E_2 mechanism with subsequent rapid, base-catalyzed formation of a 3-alkoxypropionitrile by addition of the alcohol solvent to acrylonitrile.[470] It is interesting to note that the base-promoted elimination of a 2-cyano-

$$RO^{\ominus} + H_2\underset{\displaystyle CN}{C}\text{-}CH_2\overset{\oplus}{P}R(CH_2CH_2CN)_2 \longrightarrow$$

$$ROH + \underset{\displaystyle CN}{CH}{=}CH_2 + P(R)(CH_2CH_2CN)_2$$

$$ROH + CH_2{=}CHCN \xrightarrow{\text{base}} ROCH_2CH_2CN$$

ethyl group from optically active methyl-n-butylbenzyl-2-

cyanoethylphosphonium salt afforded optically active levorotatory methyl-n-butylbenzylphosphine.[884]

Another interesting application of this procedure resulted in the synthesis of p-nitrophenyl-, and p-acetylphenyl-substituted tertiary phosphines.[1229] Furthermore, ethylenebis[di(2-cyanoethyl)phosphine] can also be prepared by this method.[474]

$$Ph_2PCH_2CH_2CN \xrightarrow{+\ ^{\oplus}N_2C_6H_4NO_2\text{-}4} Ph_2(NCCH_2CH_2)\overset{\oplus}{P}C_6H_4NO_2\text{-}4 \xrightarrow{NaOMe} Ph_2PC_6H_4NO_2\text{-}4$$

$$Ph_2PCH_2CH_2CN \xrightarrow{+\ ^{\oplus}N_2C_6H_4COMe\text{-}4} Ph_2(NCCH_2CH_2)\overset{\oplus}{P}C_6H_4COMe\text{-}4 \xrightarrow{NaOMe} Ph_2PC_6H_4COMe\text{-}4$$

$$2(NCCH_2CH_2)_3P + BrCH_2CH_2Br \longrightarrow [(NCCH_2CH_2)_3\overset{\oplus}{P}CH_2CH_2\overset{\oplus}{P}(CH_2CH_2\text{-}CN)_3]2Br^{\ominus}$$

$$\xrightarrow{+\ 2NaOEt\ +\ EtOH} 2EtOCH_2CH_2CN + 2NaBr + (NCCH_2CH_2)_2PCH_2CH_2P(CH_2CH_2CN)_2$$

Unfortunately, the preparation of ethylenediphosphines according to this procedure is restricted to the presence of three β-cyanoethyl groups on each phosphonium center. The attempted preparation of ethylenebis(dimethylphosphine) from ethylenebis(2-cyanoethyldimethylphosphonium bromide) by base-promoted double elimination of 2-cyanoethyl groups resulted in alkylene-phosphorus bond cleavage and 2-cyanoethyldimethylphosphine was the only product isolated. However, alkylenebis(dialkylphosphines) can be prepared accord-

$$[(NCCH_2CH_2)(CH_3)_2\overset{\oplus}{P}CH_2CH_2\overset{\oplus}{P}(CH_3)_2(CH_2CH_2CN)]2Br^{\ominus} \xrightarrow[EtOH]{2NaOEt}$$

$$(CH_3)_2PCH_2CH_2CN + \underset{\text{(not isolated)}}{[CH_2{=}CH\overset{\oplus}{P}(CH_3)_2CH_2CH_2CN]Br^{\ominus}}$$

ing to this procedure if the two phosphorus atoms are separated by at least three methylene groups. This reaction is quite sluggish and requires 72 hr of refluxing in butanol to force a small yield of the desired trimethylenebis(dimethylphosphine).[470]

$$[NCCH_2CH_2(CH_3)_2\overset{\oplus}{P}CH_2CH_2CH_2\overset{\oplus}{P}(CH_3)_2CH_2CH_2CN]2Br^{\ominus} \xrightarrow[BuOH]{2NaOEt}$$

$$(CH_3)_2PCH_2CH_2CH_2P(CH_3)_2 + 2NaBr + 2EtOCH_2CH_2CN$$

Alkylphosphonium salts,[620,872] vinylphosphonium salts,[1293] and cyclic and acyclic bisphosphonium salts with a two-carbon bridge[153] are cleaved to phosphines in high yield by potassium cyanide in water or dimethyl sulfoxide. Cleavage of S-(+)-Me$(CH_2{=}CHCH_2)(PhCH_2)PhP^+Br^-$ with KCN proceeds with retention of configuration.[620,872] In contrast, ethane- and ethene-1,2-bisphosphonium salts are

$$Me(CH_2{=}CHCH_2)(PhCH_2)PhP^{\oplus} \xrightarrow{CN^{\ominus}} MePhCH_2PhP + CH_3C(CN){=}CH_2$$

$$[Ph_3PCH_2CH_2PPh_3]^{2+} \xrightarrow{2CN^{\ominus}} 2Ph_3P + NCCH_2CH_2CN + 2KBr$$

$$[Ph_3PCH{=}CH_2]^{\oplus} \xrightarrow{CN^{\ominus}} Ph_3P + CH_2{=}CHCN$$

cleaved by alkali into a phosphine and a phosphine oxide with loss of the two-carbon bridge.[151,1442] When the phosphorus atom carries benzyl substituents, loss of benzyl is competitive with loss of the bridge.[3,151] With six-membered 1,4-diphosphonio heterocyclic salts (28), the nature of the products is dependent on whether alkali[3] or phosphonium salt[151] is present in excess.

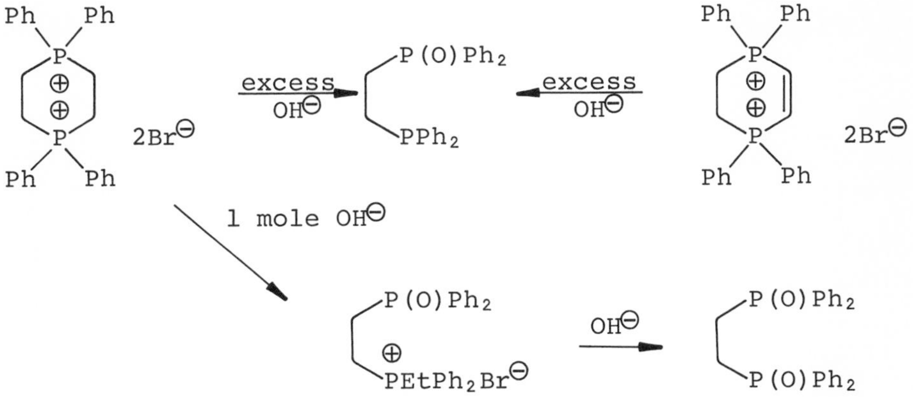

Simple β elimination of triphenylphosphine from phosphonium salts with base was also observed when the α proton was activated by groups other than cyano and phosphonium (see above), such as carbethoxy,[32] carboxylate,[308] and acyl[120,478] groups.

Chloromethylphosphonium chlorides, obtained by the interaction of hydroxymethylphosphonium chloride and phosphorus pentachloride,[547,572] or thionyl chloride,[547,574,1105] yield, when treated with aqueous alkali, tertiary phosphines containing chloromethyl groups.[545-547,572] However, this decomposition is accompanied by side reactions

$$[R_xP(CH_2Cl)_{4-x}]^\oplus X^\ominus + 2NaOH \longrightarrow$$

$$R_xP(CH_2Cl)_{3-x} + CH_2O + 2NaCl + H_2O$$

and the yields of tertiary phosphines are generally low.[545,546] Particularly, aromatic substituted chloromethyl phosphonium salts undergo a rearrangement reaction in which a phenyl group or, more often, a chloromethyl group is shifted to an α-carbon atom to give benzyl- and β-chloroethylphosphine oxides, respectively.[545,546]

$$\begin{bmatrix} Ph & CH_2Cl \\ & P \\ Ph & CH_2Cl \end{bmatrix}^\oplus Cl^\ominus \xrightarrow{NaOH} \begin{array}{c} Ph \quad O \\ \diagdown \parallel \\ PCH_2CH_2Cl \\ \diagup \\ Ph \end{array} + PhP \begin{array}{c} O \quad CH_2Cl \\ \parallel \diagup \\ \diagdown \\ CH_2Ph \end{array}$$

                                    (40 to 48%)   (10%)

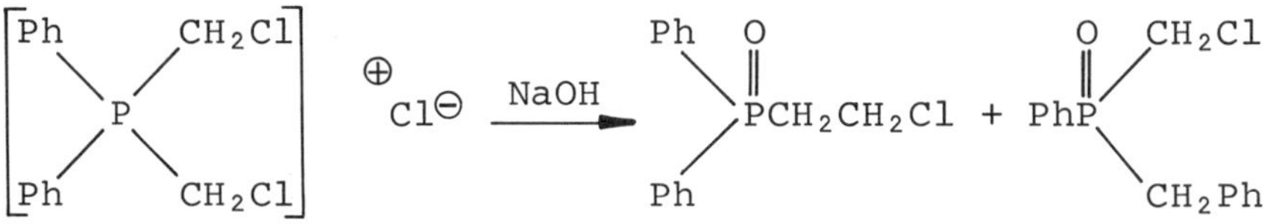

Treatment of phosphonium chlorides containing hydroxymethyl groups with one equivalent of triethylamine[1108,1110,1372,1374] or aqueous alkali results in the formation of tertiary phosphines,[453,459,461,546,547,574,1360] while

$$(HOCH_2)_4PCl + NaOH(Et_3N) \longrightarrow (HOCH_2)_3P + CH_2O +$$

$$H_2O + NaCl(Et_3N \cdot HCl)$$

excess aqueous alkali produces in this same reaction tertiary phosphine oxides.[459,572,1108,1393] Since this replacement of a hydroxymethyl group can be repeated, after

$$(HOCH_2)_4PCl + excess\ NaOH \longrightarrow (HOCH_2)_3PO + CH_2O + H_2 + NaCl$$

alkylation of tris(hydroxymethyl)phosphine with an alkyl halide or a halocarbonic acid ester,[1113] unsymmetrical tertiary phosphines can be prepared by this procedure.[459,546,547,1108,1110,1113,1372,1374] Decomposition of an optically active phosphonium salt containing a hydroxymethyl group with Et_3N gave an optically active tertiary phosphine.[1441]

$$(HOCH_2)_3P + RX \longrightarrow [RP(CH_2OH)_3]X \xrightarrow[\text{or } Et_3N]{NaOH}$$

$$RP(CH_2OH)_2 + CH_2O + H_2O + NaCl(Et_3N \cdot HCl)$$

                                    R = alkyl, RO_2CCH_2

Secondary amines not only decompose tetrakis(hydroxymethyl)phosphonium chloride but interact at the same time with the hydroxy groups to yield tris(dialkylaminomethyl)-phosphines.[235]

$$[(HOCH_2)_4P]^{\oplus}Cl^{\ominus} + 4HNR_2 \longrightarrow (R_2NCH_2)_3P + CH_2O + R_2NH{\cdot}HCl + 3H_2O$$

Interaction of tetrakis(hydroxymethyl)phosphonium chloride with acrylonitrile in the presence of a base resulted in complete replacement of the hydroxymethyl groups and gave tris(β-cyanoethyl)phosphine in good yield.[1392] This reaction certainly goes through the intermediate for-

$$(HOCH_2)_4PCl + 3CH_2{=}CHCN + NaOH \longrightarrow P(CH_2CH_2CN)_3 + 4CH_2O + H_2O + NaCl$$

mation of $(HOCH_2)_3P$, since reaction of this phosphine with acrylonitrile produced tris(β-cyanoethyl)phosphine also.[1187,1373] Hydroxymethylalkylphosphines of type $RP(CH_2OH)_2$ gave, however, with acrylonitrile polymeric materials.[1373]

Acetoxymethyl-substituted tertiary phosphines resulted from the base-promoted decomposition of acetoxymethyl groups containing phosphonium salts obtained from the corresponding hydroxymethyl phosphonium salts and acetic acid,[1004] for example,

$$(HOCH_2)_4PCl + CH_3CO_2H \xrightarrow{H_2SO_4} (CH_3CO_2CH_2)_4PCl \xrightarrow{OH^{\ominus}} (CH_3CO_2CH_2)_3P$$

1-Hydroxyalkyldiphenylphosphines, obtained by the action of aqueous alkali on di(1-hydroxyalkyl)diphenylphosphonium chlorides, proved to be very sensitive to alkali, and the crude products contained about 70% diphenylphosphine. The interaction of aqueous alkali and di(α-hydroxybenzyl)diphenylphosphonium chloride gave, together with the tertiary phosphine, α-hydroxybenzyldiphenylphosphine oxide.[1360] The reactions are summarized below:

$$[Ph_2P(CHROH)_2]^{\oplus}X^{\ominus} \xrightarrow{+OH^{\ominus}} \begin{cases} \xrightarrow{R\,=\,Et,\ n\text{-}Pr} Ph_2PH \\ \xrightarrow{R\,=\,H,\ Et,\ n\text{-}Pr,\ or\ Ph} Ph_2PCHROH \\ \xrightarrow{R\,=\,Ph} Ph_2(PhCHOH)PO \end{cases}$$

All 1-hydroxyalkyldiphenylphosphines rearranged when refluxed with toluene-p-sulfonic acid and gave alkyldiphenylphosphine oxides.[168,1360] Finally, it should be

$$Ph_2PCHROH \xrightarrow{H^{\oplus}} Ph_2(RCH_2)PO$$

mentioned that acetyl-, benzoyl-,[696] and carboxyethyltrialkylphosphonium halides[649] also decompose on treatment with sodium hydroxide to acetic, benzoic, or carbonic acid

$$[R'COPR_3]^{\oplus}X^{\ominus} \xrightarrow{NaOH} NaX + R'COOH + R_3P$$

$$R = Me, EtO, Ph$$

and a tertiary phosphine.[696]

3. BY REDUCTION WITH LITHIUM ALUMINUM HYDRIDE OR SODIUM HYDRIDE. Phosphonium salts containing benzyl groups are reduced with $LiAlH_4$ in tetrahydrofuran to tertiary phosphines[69] in high yield with the removal of a benzyl group as toluene. No appreciable reduction was observed when diethyl ether was used as the solvent.[69] Since only one benzyl group is removed at a time, this reaction can be conveniently adapted to the synthesis of unsymmetrical tertiary phosphines as illustrated below.

$$RPX_2 \xrightarrow[\text{then MeI}]{PhCH_2MgX,} [R(Me)(PhCH_2)_2P]^{\oplus}I^{\ominus} \xrightarrow[\text{then EtI}]{LiAlH_4/THF}$$

$$[R(Me)Et(PhCH_2)P]^{\oplus}I^{\ominus} \xrightarrow[THF]{LiAlH_4} R(Me)EtP$$

$$R = Ph, n\text{-}C_5H_{11}$$

Each step proceeded with greater than 80% yield. The overall yield of ethylmethylphenylphosphine was 59%. Starting with methylphosphonous dichloride, methylethylpentylphosphine was prepared similarly in an overall yield of 50%.[70]

The utility of this method was considerably extended when it was found that reduction of triphenyl-(primary alkyl) phosphonium salts with $LiAlH_4$ afforded diphenylalkylphosphines.[458] Reduction of tetramethylenebis(triphenylphosphonium bromide) similarly gave tetramethylenebis(diphenylphosphine).[458] These findings conform with the idea that, analogous to the attack of hydroxide[378] and alkoxide[467] ions, the hydride ion or AlH_4 ion is first added to the positively charged phosphorus with subsequent or simultaneous expulsion of the group that is most stable as the anion. The preferential removal of benzyl groups, as well as the fact that only racemic methylethylphenylphosphine was obtained on reduction of optically active $[MeEtPh(PhCH_2)P]^+$ salts with $LiAlH_4$,[556,884] and partially racemic (82%) (+)-MePrPhP from optically active $[MePrPh(PhCH_2)P]^+$ and $LiAlH_4$[613] are also explained by this mechanism. The formation of triphenylphosphine from phosphonium salts derived from triphenylphosphine and a secondary halide, $[Ph_3PCHMe_2]^+X^-$,[458] or from cyclopropylmethyl halide,[1272] is, however, not consistent with this mechanism. Steric effects might be responsible for this anomaly, and attack might then occur at carbon.[911] Ylids on reduction with $LiAlH_4$ also gave tertiary phosphines,[1217] for example, $MePPh_2$ from $Ph_3P{=}CH_2$,[1217] i-$PrPPh_2$ from $Ph_3P{=}CMe_2$, and $PhCH_2PPh_2$ from $Ph_3P{=}CHPh$.[458] Thus isopropylidenetriphenylphosphorane is excluded as an intermediate in the reduction of $[Ph_3PCHMe_2]^+X^-$ to triphenylphosphine.[458] Strong support for initial attack at phosphorus comes from the isolation of the phosphorane (30) from the low-temperature reduction of the spirocyclic salt with $LiAlH_4$.[549] Elegant

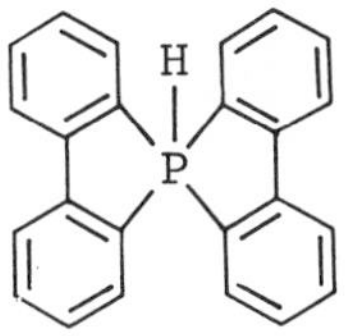

(30)

use of benzyl removal from phosphonium salts by LiAlH was made in the syntheses of unsymmetrical substituted diphosphines (31), diphosphacyclohexanes (32), and triethylenediphosphine (33), starting from tribenzylphosphine and

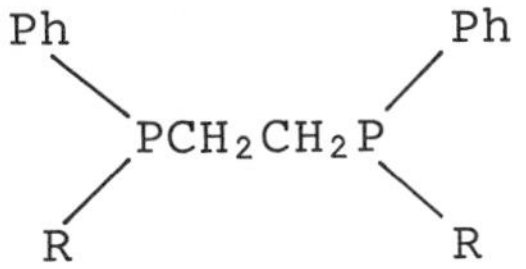

(31)

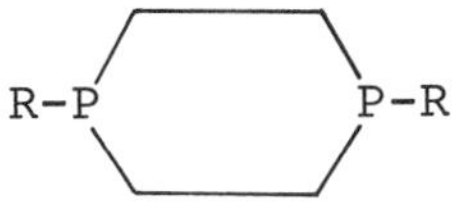

(32)

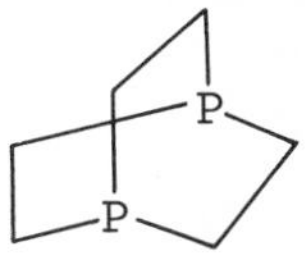

(33)

ethylene dibromide.[566] The products from bisphosphonium salt reductions with $LiAlH_4$ are, however, dependent on the substituents at phosphorus. Thus cleavage of the two-carbon bridge is observed if no benzyl group is present,[152] for example, $[Ph_2MePCH_2CH_2PMePh_2]^{2+}2I^-$ gave Ph_2MeP.[152] Similar results were obtained when the reduction was carried out with NaH,[52,512,513] with the exception that in bisphosphonium salts cleavage of the two-carbon bridge was observed in every case, even when benzyl groups were present.[152] The high-temperature reduction of triphenylhydroxyalkyl phosphonium salts with NaH resulted in the formation of hydroxyalkyldiphenylphosphines and phenoxyalkyldiphenylphosphines,[512,513] for example,

$$[Ph_3P(CH_2)_6OH]^{\oplus}I^{\ominus} \xrightarrow[NaH]{130 \text{ to } 150°} Ph_3P + Ph_2P(CH_2)_6OH + Ph_2P(CH_2)_6OPh$$

β-Hydroxyethyltriphenylphosphonium iodide gave, however, triphenylphosphine oxide and ethylene.[513]

4. BY REDUCTION WITH SODIUM METAL, PHOSPHIDES OR GRIGNARD REAGENTS. Only a limited amount of information is available on the reduction of phosphonium salts with alkali metals. Tribenzylmethylphosphonium bromide was reduced by means of sodium, by using benzene as a solvent, to dibenzylmethylphosphine (isolated as the oxide) in 77% yield. Ethanol and liquid ammonia may also be used as solvents, but the yields of the tertiary phosphine are much lower in these cases.[69] Under the same conditions dibenzyldimethylphosphonium bromide was not reduced to the tertiary phosphine, and no evidence of reduction was observed. Other reducing agents, such as sodium amalgam, which is highly effective in the case of the corresponding ammonium salt,[529] stannous chloride in acetic solution, or hydrogen and palladium, proved to be equally ineffective as reducing agents.[69] In the only other reference available,[609] it was reported that a phenyl group is removed from alkyl- or benzyltriphenyl- and dialkyldiphenylphosphonium salts by sodium, by using toluene or naphtalene as solvents. From trialkylphenylphosphonium salts, however, an alkyl group was split off preferentially (ethyl > methyl)[618] (p. 132). The meager information available indicates the order of cleavage of the various groups in the descending order:

t-butyl > phenyl > benzyl > ethyl > methyl

It is noteworthy that sodium phosphide cleaved a benzyl group from triphenylbenzylphosphonium halide to give triphenylphosphine[618] (p. 132).

Interaction of potassium diphenylphosphide and tetraphenylphosphonium bromide in tetrahydrofuran at reflux produced 71% triphenylphosphine. This means that at least 21% Ph_3P must have been formed by a transfer of phenyl groups.[610] Other salts gave similar results. Certain phosphonium salts produced tertiary phosphines when caused to react with a Grignard reagent.[1025]

$$Ph_4PBr + KP(C_6H_4Me\text{-}4)_2 \longrightarrow [Ph_4P^{\oplus}P^{\ominus}(C_6H_4Me\text{-}4)_2] \longrightarrow Ph_3P + PhP(C_6H_4Me\text{-}4)_2$$

5. BY ELECTROLYTIC REDUCTION OR PHOTOLYSIS. By far the most economical way to reduce phosphonium salts to tertiary phosphines seems to be electrolytic reduction on lead or mercury cathodes. Carbon anodes are usually used, surrounded by a clay diaphragm. The reduction is conveniently carried out in aqueous solution at 70 to 90° with 24 V and 2 to 6 amp. The tertiary phosphines are isolated in high yield (80 to 90%) by extraction with ether, followed by distillation.[623,624] The material of the cathodes has a distinct effect upon the order of cleavage of organic radicals, for example, $[Ph_3PMe]^+Br^-$ gives with a mercury cathode 83% Ph_2MeP and 5.5% Ph_3P, while the same salt yields with a lead cathode 40% Ph_2MeP and 53% Ph_3P.[623,624] A quantitative study suggests the approximate order of radical cleavage on mercury cathodes[615]:

Fluorenyl, Ph_2CH, $PhCH_2$ > t-Bu > 2-$C_{10}H_7$ > 1-$C_{10}H_7$ >

4-PhC_6H_4 > 4-$CF_3C_6H_4$ > Et > Ph > 2-MeC_6H_4 > 3-MeC_6H_4 >

4-$Ph_2CHC_6H_4$ > 4-i-PrC_6H_4 > 4-MeC_6H_4 > Me > 4-$MeOC_6H_4$ >

4-$Me_2NC_6H_4$

The Hammet σ constants of the substituents on the phenyl group decrease parallel to this cleavage series. This means that the P-C bond that exhibits the lowest electron density is cleaved.[615]

The preferential cleavage of benzyl groups has been used successfully for the preparation of a large number of asymmetric tertiary phosphines.[614,629,630,635] The electrolytic reduction proceeds stereospecifically. Thus

$$[(PhCH_2)_3PR]^{\oplus}X^{\ominus} \xrightarrow[\text{then } R'X]{+\,2e^{\ominus}/Hg,} [(PhCH_2)_2RPR']^{\oplus}X^{\ominus}$$

$$\xrightarrow[\text{then } R''X]{+\,2e^{\ominus}/Hg,} [PhCH_2RR'PR'']^{\oplus}X^{\ominus} \xrightarrow{2e^{\ominus}/Hg} RR'R''P$$

reduction of optically active phosphonium salts has produced optically active tertiary phosphines for the first time.[614,629,630,635]

The mechanism of the electrolytic reduction is not yet

clear. The recently isolated quaternary phosphonium amalgams[251] seem to verify our previously suggested mechanism for this type of reduction.[911]

$$[R_4P]^{\oplus} + e^{\ominus} \xrightarrow{Hg} [R_4P] \xrightarrow[+H^{\oplus}]{+e^{\ominus}} R_3P + RH$$

or

$$[R_4P]^{\oplus} + 2e^{\ominus} \xrightarrow{Hg} [R_4P]^{\ominus} \longrightarrow R_3P + R^{\ominus} \xrightarrow{H^{\oplus}} RH$$

Other cathode materials have also been used but with less success. Platinum electrodes produced triphenylphosphine from triphenylbenzylphosphonium chloride, but methyltriphenylphosphonium bromide gave methyldiphenylphosphine (20%), triphenylphosphine (26%), and larger amounts of tertiary phosphine oxide. Tin electrodes were also effective but had the disadvantage of becoming covered with an oily sponge. Copper electrodes produced only phosphine oxides.[624] In one case electrolytic reduction resulted not only in the formation of a tertiary phosphine but also in a phosphorane derivative. Thus phenacyltriphenylphosphonium bromide gave, on reduction with a lead cathode, triphenylphosphine (11%) and benzoylmethylidenetriphenylphosphorane (67%).[624]

$$[Ph_3PCH_2COC_6H_5]^{\oplus}Br^{\ominus} \xrightarrow[\text{electrode}]{Pb} Ph_3P{=}CHCOC_6H_5$$

Electrolysis of alkyldiphosphonium salts using a mercury cathode leads to the formation of alkylenediphosphines. As before, electrolysis is carried out with aqueous solutions of the bisphosphonium salts and the alkylenediphosphines are isolated in about 80% yield by extraction with an organic solvent.[623,624] Reduction again proceeds with

$$[MePh(PhCH_2)PCH_2CH_2P(CH_2Ph)PhMe]^{2+} \xrightarrow[+2H^{\oplus}]{+4e^{\ominus}} MePhPCH_2CH_2PPhMe + 2PhMe$$

retention of configuration, and electrolysis of the optically active bisphosphonium salt gave optically active (+)-ethylenebis(methylphenylphosphine).[107,611]

Electron transfer to the phosphonium cation can also be achieved by photolyzing a solution of the phosphonium salt in alcohol-benzene solution with UV light. As byproducts, secondary phosphines are produced.[477]

X. FROM BIPHOSPHINES OR CYCLOPOLYPHOSPHINES AND VARIOUS REAGENTS

The biphosphines $Me_2P\text{-}PMe_2$, $[(CF_3)_2P]_2$,[182,185] and $Ph_2P\text{-}PPh_2$[264] add across ethylenic or acetylenic multiple bonds in much the same way as a halogen molecule, when heated for extended periods. The addition is catalyzed by

$$Me_2P\text{-}PMe_2 + CH_2{=}CH_2 \xrightarrow{280°} Me_2PCH_2CH_2PMe_2$$

by iodine.[182] o-Phenylenebis(diethylphosphine) has been obtained by adding $Et_2P\text{-}PEt_2$ across the triple bond of benzyne.[222]

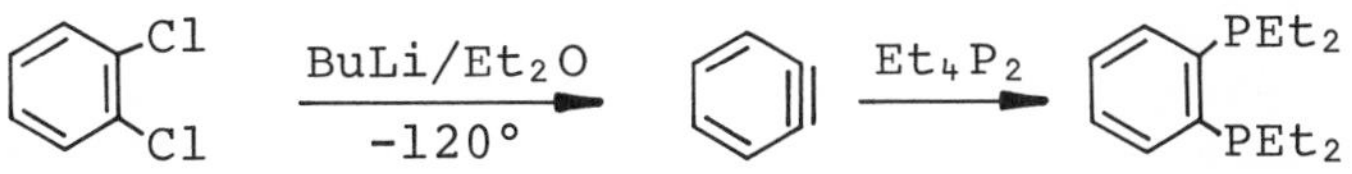

Normally, tetraalkylbiphosphines form monoquaternary phosphonium salts when treated with alkyl iodide (see Chapter 2). However, when they are heated with excess alkyl iodide to 200°,[1056] or with perfluorocyclobutene or 1,2-dichlorotetrafluorocyclobutene-1 to 130°,[265] cleavage of the P-P bonds with formation of tertiary phosphines is observed. Also, cleavage rather than quaternization takes

$$R_2P\text{-}PR_2 + R'I \longrightarrow R_2PI + R_2PR'$$

place when the P-P bond in biphosphines is weakened by electron-withdrawing substituents, such as CF_3[263] or Ph groups,[96,579,708] or when the alkyl groups are replaced by the bulkier t-Bu[668] or $c\text{-}C_6H_{11}$ groups.[701] Several tertiary phosphines were isolated from the reactions of biphosphines with phenyllithium[673] (see Chapter 2) or isothiocyanates.[666] The latter reaction only proceeded when an alcohol was present.

$$R_2P\text{-}PR_2 + PhLi \longrightarrow PhPR_2 + LiPR$$

$$R_2P\text{-}PR_2 + R'NCS \xrightarrow{R''OH} R_2PCSNHR' + R_2POR''$$

An elegant method for the synthesis of phospholene derivatives (34) and tetrahydro-1,2-diphosphorines (35) consists in heating a mixture of a cyclopolyphosphine with a diene at 150 to 180° for 2 hr.[1258] Complete ring cleavage also results in the reaction of $(PhP)_5$ with CF_3I at 185°, giving a tertiary phosphine, $Ph(CF_3)_2P$, a phosphinous iodide, $PhCF_3PI$, and hexafluoroethane[95] (for a discussion, see Chapter 2, Section E).

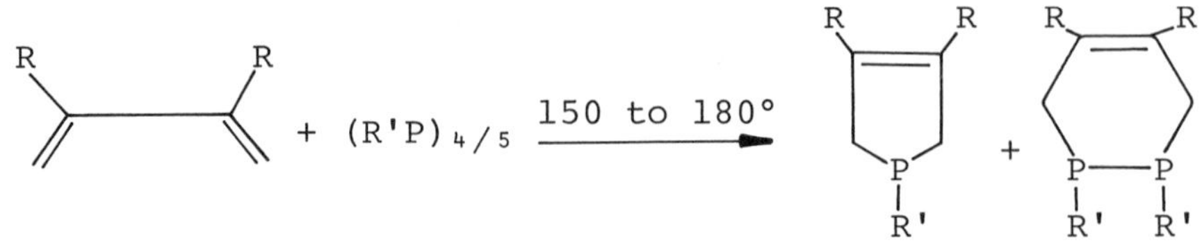

(34) (35)

XI. BY A FRIEDEL-CRAFTS-TYPE REACTION

When PCl_3 is heated with 1,3-dimethoxybenzene[1132,1134] or 1,3,5-trimethoxybenzene[1132,1133] (ratio 1:3) in the presence of equimolar amounts of zinc dichloride to 95 to 110°, good yields of tris(2,4-dimethoxyphenyl)phosphine (14%) or tris(2,4,6-trimethoxyphenyl)phosphine (80%), respectively, are formed. Treatment of the zinc dichloride complexes that are formed, with ammonium hydroxide and extraction with benzene, gives the pure phosphines. Excess phosphorus trichloride produces less substituted products,[1132] and simple phenol ethers, such as anisole

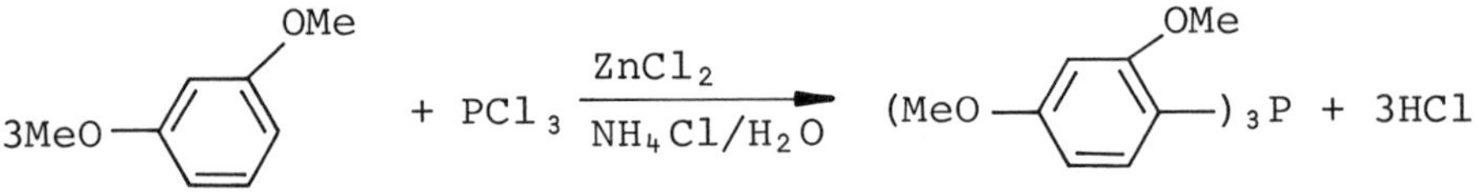

and phenetol, produce under these conditions also only arylphosphonous dichlorides.[1132] The reaction of methoxy-substituted benzenes with phosphorus trichloride is similar to that of dialkylaniline with phosphorus trichloride, which also produces tertiary phosphines, although mono- and disubstituted products are formed simultaneously[801] (p. 20).[1192] It has been shown that in this case substitution takes place in the para position.[1192] The catalyst $AlCl_3$ is unnecessary in the dialkylaniline reaction. Recent reports indicate that the phosphine oxide is actually formed in this reaction.[1225] Tris(4-dimethylaminophenyl)-phosphine can, however, be prepared by the organolithium

$$PCl_3 + 6R_2NC_6H_5 \longrightarrow (4\text{-}R_2NC_6H_4)_3P + 3C_6H_5NR_2 \cdot HCl$$

route from $4\text{-}Me_2NC_6H_4Li$ and PCl_3. The melting points reported for this compound by three different groups, all using the organolithium route, differ, however, widely, m. 254°,[444] m. 282 to 287°,[1225] m. 278 to 282°.[1359] Perhaps this phosphine exists in different crystal modifications. A true Friedel-Crafts reaction was observed in the reaction of PCl_3[1314] phosphonous and phosphinous chlorides[1313,1316] with ferrocene in the presence of $AlCl_3$, giving in every case a ferrocene-substituted tertiary phosphine. Tris(ferrocenyl)phosphine is best prepared from R_2NPCl_2, ferrocene, and $AlCl_3$.[1315]

Cyclization under Friedel-Crafts conditions has been used to prepare derivatives of dibenzophosphole (36),[853] phenoxaphosphine, (37)[471,853,1316] and phenothiaphosphine (38).[472]

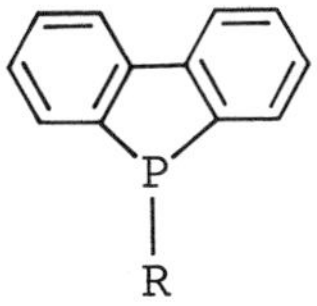

(36)

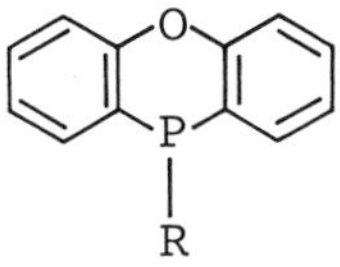

(37)

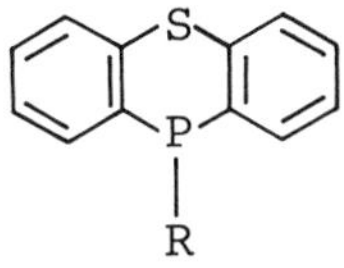

(38)

Usually the phosphinous chloride intermediate is not isolated but is cyclized in situ to the ring systems above,[853] for example,

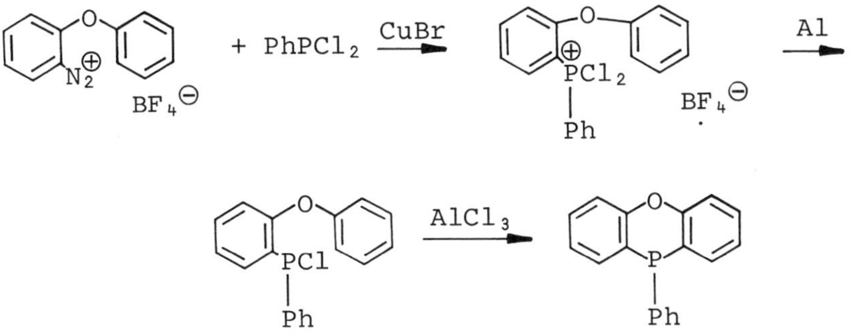

XII. BY ADDITION OF PHOSPHINES OR PHOSPHIDES TO OLEFINS OR ACETYLENES

One of the most economical methods for the preparation of phosphines is the addition of PH_3 to olefins. It is very likely that most tributylphosphine commercially available today is prepared by the interaction of PH_3 and butylene.[1375]

Olefins in which the double bond is activated by electrophilic substituents (e.g., acrylonitrile) add primary and secondary phosphines at elevated temperature without any catalyst to give secondary and/or tertiary phosphines.[58-60,567,568,573,950,1383] In general, however, addition of phosphines to olefins has been effected in the presence of acidic (methylsulfonic acid, phenylsulfonic acid, liquid HF, and boron trifluoride),[157,570] or basic catalysts (KOH, amines)[540,1173] or in the presence of salts such as $NiCl_2$ and $PtCl_4$.[1186] The radical addition of phosphines to olefins has also been reported.

The acid-catalyzed addition of phosphine, PH_3, to olefins probably proceeds by a carbonium ion mechanism. Tertiary olefins, which are known to form the most stable carbonium ion, give in this reaction primary phosphines in high yield.[570] Furthermore, they react selectively in the presence of primary and secondary olefins.[157] A normal Markovnikov addition takes place to yield a product with the phosphorus atom bonded to the tertiary carbon atom. In order to obtain a good conversion, a nearly stoichio-

$$R_2C{=}CH_2 + H^{\oplus} \rightleftharpoons R_2\overset{\oplus}{C}{-}CH_3 \xrightleftharpoons{+PH_3} H_3\overset{\oplus}{P}CR_2CH_3 \rightleftharpoons H_2PCR_2CH_3 + H^{\oplus}$$

metric amount of catalyst is required because the primary phosphine formed is a much stronger base than phosphine, and will tie up the catalyst as a phosphonium ion (see Section C.2). Higher temperatures favor the conversion since the additional protons released by greater dissociation of the alkylphosphonium ion catalyze further reaction.

$$RPH_2 + H^{\oplus} \rightleftharpoons (RPH_3)^{\oplus}$$

The reaction stops after the first step and only minor amounts of dialkylphosphines are formed (∿2%). No tertiary phosphine was detected.[570] These findings reflect the tendency of the monoalkylphosphine to exist as RPH_3^+ in the acidic medium. Therefore, the concentration of free monoalkylphosphine and of the olefin cation (the protons have been captured by the monoalkylphosphine) is low, hence formation of dialkylphosphine is slow.

Base-catalyzed addition of phosphines to olefins was reported in the case of acrylonitrile.[1173] The highest yields of addition products were obtained when acetonitrile, in conjunction with a separate aqueous potassium hydroxide catalyst phase, was employed as the reaction medium. Strong organic bases, such as heptamethylbiquanide and Dowex-2 quaternary ammonium ion exchange resin also catalyze the reaction. The reaction probably proceeds by a Michael reaction mechanism.[646] Alternatively, a nucleophilic addition also seems possible. Although mixtures of

$$PH_3 + OH^{\ominus} \rightleftharpoons PH_2^{\ominus} + H_2O$$

$$^{\ominus}PH_2 + CH_2{=}CHCN \longrightarrow H_2PCH_2\overset{\ominus}{C}HCN$$

$$H_2PCH_2\overset{\ominus}{C}HCN + H_2O \longrightarrow H_2PCH_2CH_2CN + OH^{\ominus}$$

primary, secondary, and tertiary phosphines are always

obtained in the cyanoethylation of phosphine, the degree of substitution can be controlled to a great degree by varying the ratio of phosphine to acrylonitrile present in the reaction mixture. Thus an 80% yield of tris(2-cyanoethyl)phosphine was obtained by adding phosphine to a mixture containing a slight excess of acrylonitrile, while bis(2-cyanoethyl)phosphine was prepared in 63% yield by adding acrylonitrile slowly to a mixture kept saturated with phosphine. By carrying out the reaction in an autocalve under a phosphine pressure of 28 to 32 atm, 2-cyanoethylphosphine was isolated in 52% yield.[1173]

The free-radical addition of phosphine (PH_3, RPH_2, R_2PH $H_2P(CH_2)_xPH_2$) to unsaturated compounds can be initiated by UV light,[1338] x-rays,[1171,1416] organic peroxides,[87,1171,1338] and by α,α'-azobisisobutyronitrile.[58-60,921,1166,1171] The reaction proceeds by a free-radical chain mechanism, as proposed by Kharasch and Mayo (for a review, see 976), and produces primary, secondary, and tertiary

$$PH_3 + R\cdot \text{ (or } h\nu) \longrightarrow \cdot PH_2 + RH$$

$$H_2P\cdot + RCH{=}CH_2 \longrightarrow H_2P\text{-}CH_2\dot{C}HR$$

$$H_2PCH_2\dot{C}HR + H_3P \longrightarrow H_2PCH_2CH_2R + H_2P\cdot$$

phosphines. The product distribution can be controlled to some extent by varying the mole ratio of phosphine to olefin, for example, the reaction of PH_3 with 1-octene and α,α'-azobisisobutyronitrile as a catalyst gave, with a mole ratio of PH_3/octene = 0.33, 0% RPH_2, 0% R_2PH, and 100% R_3P; mole ratio PH_3/octene = 1.1, 32% RPH_2, 38% R_2PH, and 30% R_3P; and mole ratio PH_3/octene = 3.6, 75% RPH_2, 21% R_2PH, and 4% R_3P.[1171]

These data indicate that it is possible to obtain predominantly primary or tertiary phosphines by this procedure, but the yield of secondary phosphine has a relatively low maximum value. Olefins that exhibit steric hindrance shift the product ratio in favor of less substituted phosphines, for example, with cyclohexene, using a mole ratio of PH_3/cyclohexene = 1.5, 63% RPH_2, 37% R_2PH, and 0% R_3P were obtained.[1171]

The effect of steric hindrance is also evident in the reactions of 2-cyanoethylphosphine with cyclohexene and 2,4,4-trimethylpentene-1. Unsymmetrical secondary phosphines are the predominant products.[1171] Unsaturated tertiary phosphines have been obtained by the interaction of dibutylphosphine and 1,3-butadiene and by the action of 2-cyanoethylphosphines[1171] or diphenylphosphine[578] on substituted acetylenes. Acetylene itself also reacted with bis(2-cyanoethyl)phosphine in the presence of α,α'-azobisisobutyronitrile, but no product could be isolated from the reaction mixture.[1171] It has been demonstrated that, in the presence of radical sources,[385a,1087] cis-butene-2 is isomerized to trans-butene-2 during the course of the reaction of cis-butene with various phosphines. Thus the reversibility of a phosphinyl radical addition to a double bond is established. This is in accord with the reversible thiyl radical addition to butene-2.[1408]

The addition of PH_3[178,385,1080] phenylphosphine,[1080] methylphosphine,[1310] diphenylphosphine,[266] dimethyl- and bis(trifluoromethyl)phosphine[385a,385b] to fluorinated olefins or acetylenes was effected thermally[1080] or by initiation with UV light.[178] Primary, secondary, tertiary and, in few cases, diprimary phosphines were obtained.[178,385a,385b,1080] The biphosphine $H_2P\text{-}PH_2$, however, could not be added to $CF_2{=}CF_2$.[178]

$$CF_2{=}CF_2 + PH_3 \longrightarrow F_2CHCF_2PH_2 + H_2PCF_2CF_2PH_2 + (F_2CHCF_2)_2PH$$

$$CF_3C{\equiv}CCF_3 + Ph_2PH \longrightarrow Ph_2PC(CF_3){=}CHCF_3 \quad (\sim 60\%\ \text{trans}, 40\%\ \text{cis})$$ [266]

Another interesting application of the addition of PH_3, primary phosphines, or phosphides to dienes or diacetylenes resulted in the formation of cyclic tertiary phosphines such as phospholanes (15),[282] phospholes (39),[903] phosphorinanes (40),[282,1291] phosphorinanones (41),[63,1423,1424] 1,4-dihydrophosphorin-4-ones (42),[902] 1,4-oxaphosphorinanes (43),[1352] phosphepanes (44),[282,1291] bicyclic phosphines (45),[1291] and tricyclic phosphines (46).[969] Particularly interesting is the intramolecular addition according

Ref.

$$CH_2{=}CH(CH_2)_{n-2}PHR \xrightarrow{h\nu} RP\overbrace{(CH_2)_n}$$

(n = 4) (15) 282,
(n = 5) (40) 1291
(n = 6) (44)

$$R\text{-}C{\equiv}C\text{-}C{\equiv}C\text{-}R \xrightarrow[\text{or } PhPH_2/PhLi]{PhP(CH_2OH)_2/C_5H_5N}$$

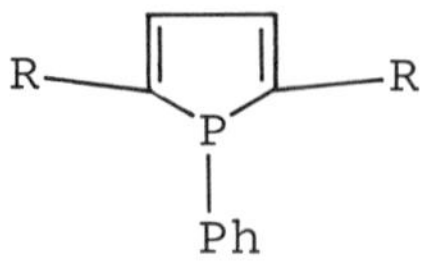

(39)

903

Ref.

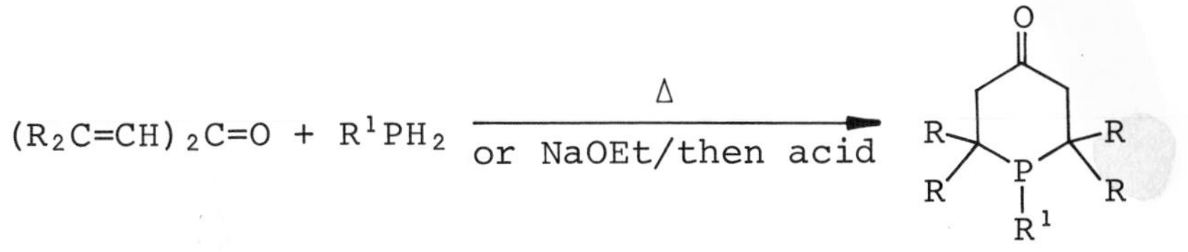

$(R_2C=CH)_2C=O + R^1PH_2 \xrightarrow[\text{or NaOEt/then acid}]{\Delta}$

63,
1424

(41)

$PhC\equiv C-CO-C\equiv C-Ph + PhP(SiMe_3)_2 \xrightarrow{AIBN}$

(42)   902

$CH_2=CHOCH=CH_2 + RPH_2 \xrightarrow[\text{or AIBN}]{h\nu}$

1352

(43)

$+ PhPH_2 \xrightarrow{t-Bu_2O_2}$ $+$ 1291

(45)

$+ PH_3 \xrightarrow[\text{radiation}]{^{60}Co}$ $+$ 969

(46)

ing to the first equation above to give five-, six-, and
seven-membered phosphorus ring compounds.[282] The formation of phosphines (45) and (46) has been mentioned in
patents. These structures have not been established

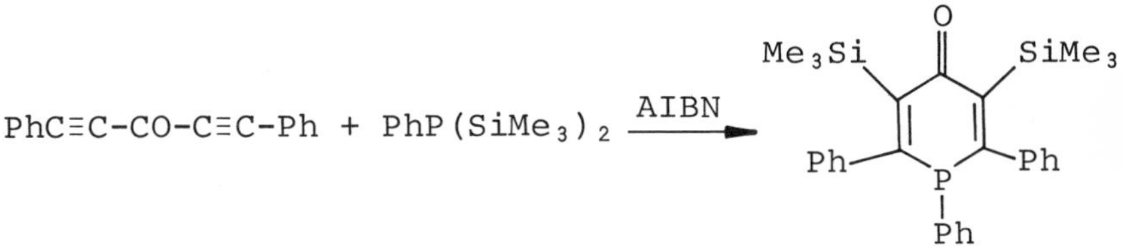

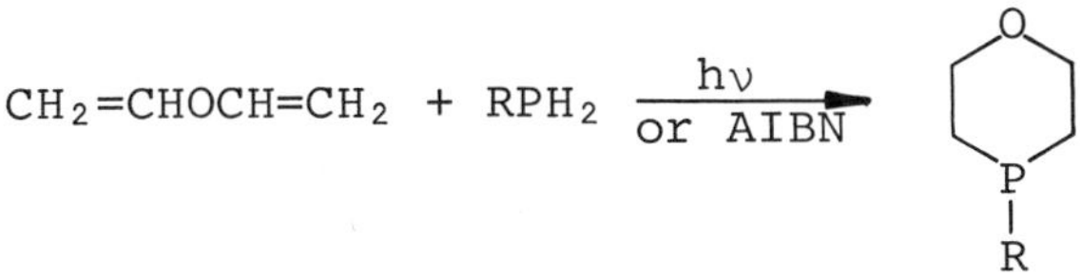

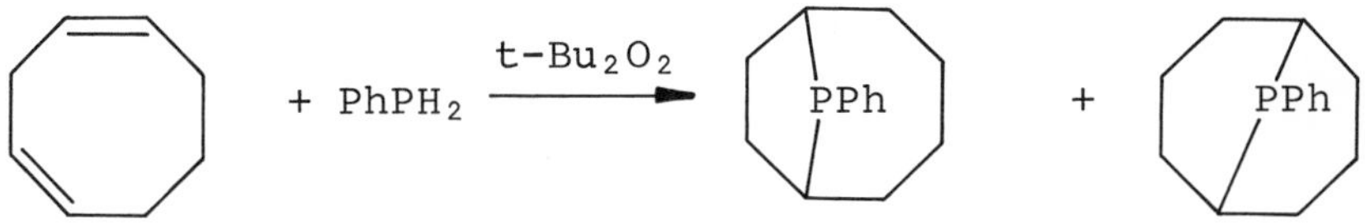

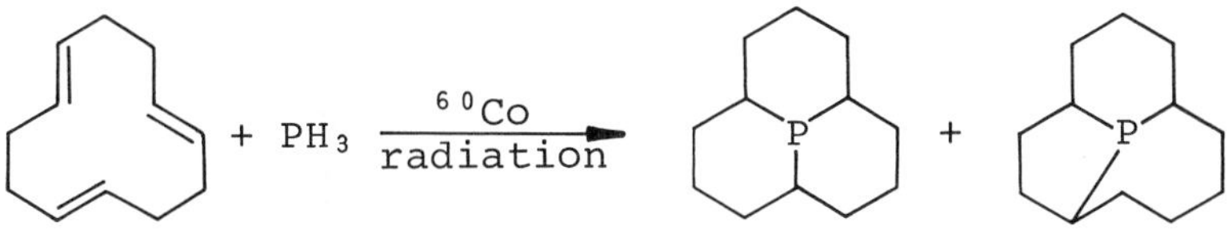

definitively.

Primary and secondary phosphines also add easily to the C=C bond of ketenes to produce acyl-substituted tertiary phosphines.[806,809,810]

$$R_2PH + CH_2{=}C{=}O \xrightarrow{Et_2O/-20°} R_2PCOMe$$

The addition of $LiPR_2$ to unsubstituted vinylacetylenes proceeds in all possible ways and 1,2- as well as 3,4- and 1,4- addition has been observed. With substituted vinylacetylenes of the structure $CH_2{=}CHC{\equiv}CR$, $LiPR_2$ produces the 1,4 compound almost quantitatively.[1104] An attempt to add lithium diphenylphosphide to trans-stilbene in tetrahydrofuran was unsuccessful. No reaction was observed.[11,17] Ready addition of alkali diorganophosphides to activated double bonds[652,655,670,671,777] and triple bonds,[7,8,652,670] however, was achieved. Phenylacetylene adds $LiPPh_2$ in tetrahydrofuran to give mostly trans-β-styryldiphenylphosphine, while cis-β-styryldiphenylphosphine is isolated in 70% yield when a primary or secondary amine is added to the reaction mixture.[7] Tertiary amines have no effect. Complex formation between amine and Ph_2PLi probably causes the change in stereochemistry. Direct addition of Ph_2PH to $PhC{\equiv}CH$[578] gives the cis compound.[7]

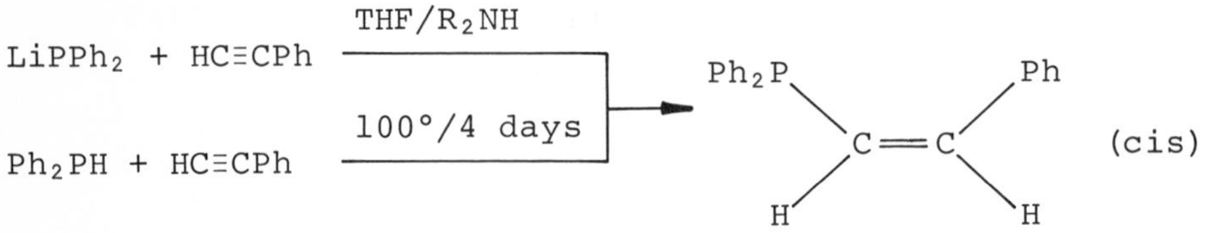

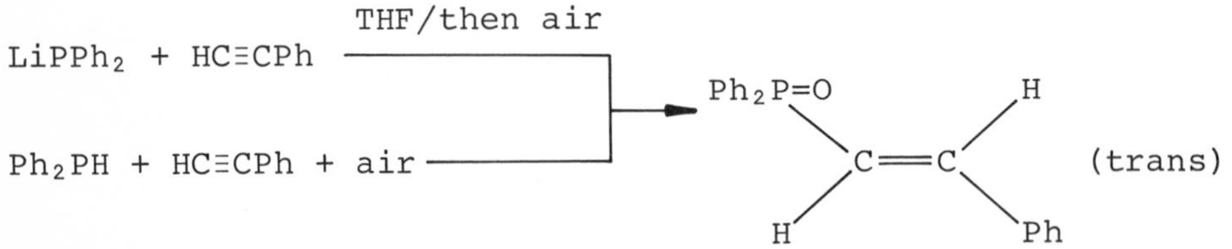

Furthermore, it has been found that Ph_2PLi isomerizes cis-$PhCH{=}CHPPh_2$ to the trans-phosphine, probably by an addition elimination mechanism.[7] With equimolar amounts of Ph_2PLi, the diadduct $Ph_2PCH_2CH{\cdot}PhPPh_2$ is obtained. Acetylene itself may also be added to Ph_2PLi to give the diphosphine $Ph_2PCH_2CH_2PPh_2$.[7] Diphenylacetylene adds Ph_2PLi[8,670] in tetrahydrofuran in the presence of $BuNH_2$ to give trans-1,2-diphenylvinyldiphenylphosphine, but the cis-phosphine is formed when Et_2NH is added to the reaction mixture.[8]

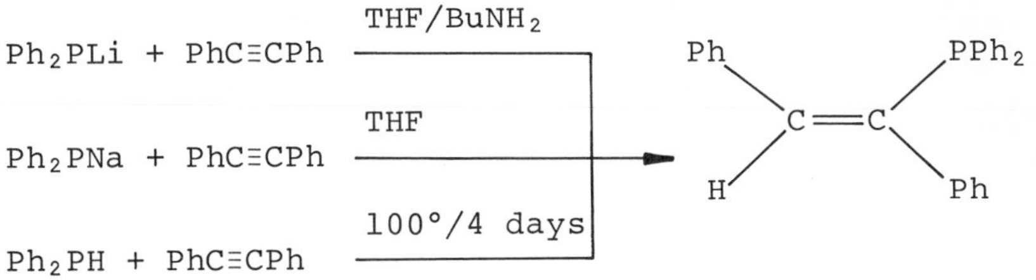

(trans)

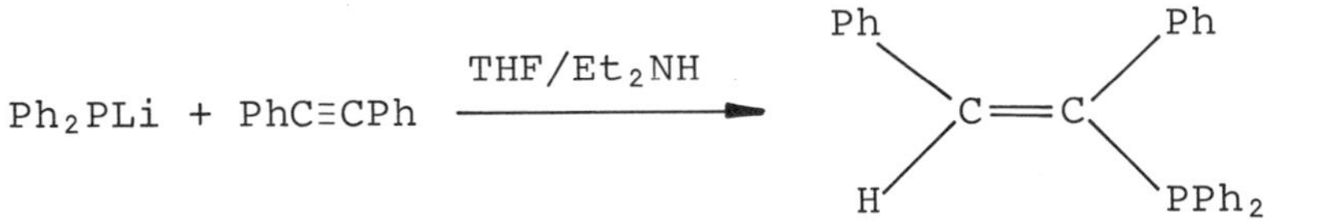

(cis)

From the fact that only the trans-phosphine is obtained with $NaPPh_2$, even in the presence of $BuNH_2$ or Et_2NH, it was concluded that the role of the amine is to complex lithium and make it more ionic and thus more like the sodium reagent.[8] Stannylphosphines add to olefins and acetylenes to give tin-substituted tertiary phosphines,[1268] for example,

$$R_3SnPR_2 + PhCH{=}CH_2 \longrightarrow R_3SnCH\cdot PhCH_2PR_2 + R_3SnCH_2CH\cdot PhPR_2$$

Addition of alkali diorganophosphides to nitriles containing no acidic α-hydrogens,[652] Schiff's bases,[652] and α,β-unsaturated ketones[671] proceeds easily. Nitriles that contain acidic α-hydrogens enter into an exchange reaction.[652]

$$PhCN + LiPR_2 \longrightarrow \underset{\displaystyle \| \atop NLi}{R_2PC\text{-}Ph} \xrightarrow{PhCN} R_2PC\cdot Ph{=}N\text{-}\underset{\displaystyle \| \atop NLi}{C}\text{-}Ph$$

$$PhN{=}CH\cdot Ph + MPR_2 \longrightarrow \underset{\displaystyle | \atop MNPh}{R_2PCHPh} \xrightarrow{H_2O} \underset{\displaystyle | \atop NHPh}{R_2PCHPh}$$

$$RCOCH{=}CH_2 + MPR_2 \xrightarrow{H_2O} RCOCH_2CH_2PR_2$$

$$R_2PLi + CH_3CN \longrightarrow R_2PH + LiCH_2CN$$

With α,β-unsaturated ketones only a 1,2- and never a 1,4-addition was observed.[671] An interesting application of the base-catalyzed addition of P-H bonds to C=C bonds is the addition of PH_3 and primary and secondary phosphines to tertiary phosphines containing vinyl or ethynyl groups. Phenylithium was used as a catalyst in this particular case.[484a,777] Similarly, addition of P-H bonds to vinyl-

or allylsilanes was achieved by radical initiation with

$$4Ph_2PCH{=}CH_2 + H_2PCH_2CH_2PH_2 \xrightarrow[C_6H_6]{PhLi} [Ph_2PCH_2CH_2]_2PCH_2\text{-}CH_2P[CH_2CH_2PPh_2]_2$$

peroxides,[377] by UV light,[794,1045,1048] or by base catalysis with phenyllithium.[489]

XIII. BY ADDITION OF PHOSPHINES TO CARBONYL-CONTAINING COMPOUNDS

The interaction of phosphines with aldehydes and ketones furnishes a variety of products depending upon the reaction conditions and the type of carbonyl compound and phosphine used.

1. PHOSPHINES AND ALIPHATIC ALDEHYDES. Tertiary phosphines containing hydroxymethyl groups are formed when PH_3,[486] primary phosphines, or secondary phosphines[161,486,487,546,547] are heated with formaldehyde to about 100° (when necessary in a bomb tube).[486] The same products are produced when these reactions are carried out in aqueous solution in the presence of metal salts, such as $PtCl_4$, $HgCl_2$, $PdCl_2$, $CdCl_2$, and others, in catalytic amounts.[1184]

$$PH_3 + 3RCHO \longrightarrow (RCHOH)_3P$$

In the early investigation of the phosphine-carbonyl reactions, it was found[980] that tetrakis(1-hydroxyalkyl)-phosphonium halides were produced in variable amounts from phosphine, an n-alkyl aldehyde, and a mineral acid under anhydrous conditions. Subsequent work has resulted in improved reaction conditions (aqueous concentrated HCl solutions) and has shown that the reaction is general with unbranched aldehydes (see Chapter 4).

Chloral and dichloroacetaldehyde gave with PH_3 under similar reaction conditions, however, secondary phosphines,[170,176,358] whereas chloroacetaldehyde produced an amorphous solid of unstated nature.[176] Steric hindrance probably prevents further reaction of these secondary phos-

$$2CXCl_2CHO + PH_3 \xrightarrow{H^{\oplus}} (CXCl_2CHOH)_2PH \qquad (X = Cl, H)$$

phines with more aldehyde. α-Branched aldehydes, such as isobutyraldehyde and 2-ethylhexanal, when treated with PH_3

in aqueous acidic solution, also produced secondary phosphines in which the phosphorus atom was part of a heterocyclic system of the 1,3-dioxa-5-phosphacyclohexane type (47).[175a,176] Obviously, the adduct with two aldehydes reacted immediately with a third molecule of aldehyde to form the acetal (47). An independent experiment verified this assumption.[168,169] However, 2-phenylpropanal gave a

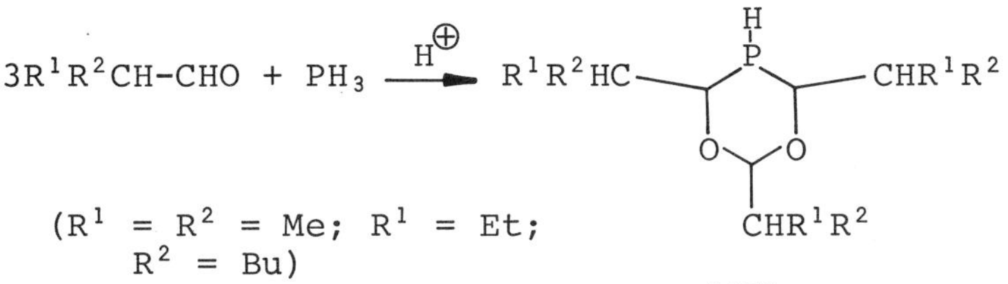
3R¹R²CH-CHO + PH₃
H⊕
R¹R²HC
H
P
CHR¹R²
O
O
CHR¹R²
(R¹ = R² = Me; R¹ = Et;
R² = Bu)

(47)

glasslike material of unknown structure. This abnormal reaction is probably also due to steric repulsion among the substituents, preventing formation of hydroxyalkyl phosphonium salts. Unlike typical secondary phosphines, these phosphines (47) have a mild not unpleasant odor. They are more stable toward atmospheric oxygen than most other dialkylphosphines.[175a,176]

Primary phosphines in which the donor properties of the phosphorus atoms are strongly decreased by substitution with electronegative groups, such as the tetrafluoroethyl group, also produced, when treated with formaldehyde in acidic solution, no phosphonium salts but only tertiary phosphines.[1179]

The condensation of PH_3 and primary and secondary phosphines with formaldehyde in the presence of secondary amines leads in a Mannich-type reaction to tertiary phosphines containing dialkylaminoethyl groups.[659a,660,918,1112] Alternatively, these compounds could also be prepared by treating tris(acetoxymethyl)phosphine[1005] or hydroxymethylphosphonium salts with a secondary amine (see Section B.2.IX.2).[235] Recently, this Mannich-type reaction

$$R_{3-x}PH_x + xCH_2O + xHNR_2 \longrightarrow R_{3-x}P(CH_2NR_2)_x + xH_2O$$

was successfully applied to the synthesis of phosphazolidine derivatives (48).[693,695]

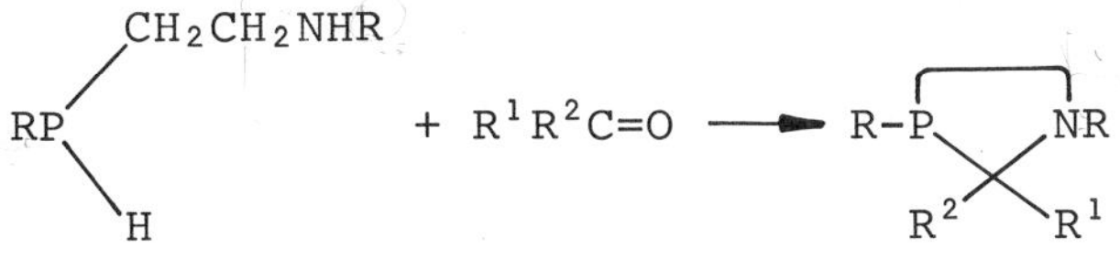
CH₂CH₂NHR
RP
H
+ R¹R²C=O
R-P
NR
R²
R¹

(48)

2. PHOSPHINES AND AROMATIC ALDEHYDES. With aromatic aldehydes a new feature of phosphine-carbonyl reaction in highly acidic solution was uncovered, the transfer of oxygen from carbon to phosphorus to give phosphine oxide products.[168,356,357a] In addition, a secondary phosphine (51) of structure similar to (47) was obtained as a by-

$$3PhCHO + PH_3 \xrightarrow{H^{\oplus}} (PhCHOH)_2P(O)CH_2Ph$$

product in the reaction of PH_3 with benzaldehyde in 10 to 40% yield.[357,357a] When the reaction of PH_3 with benzaldehyde was performed in a medium of alcohol and HCl, the alcohol took part in the reaction to give bis(α-alkoxybenzyl)phosphines (49), and in the presence of methanol with less steric hindrance, tris(α-methoxybenzyl)phosphine (50).[357,357a]

$(PhCH(OR))_2PH$

R = Et, i-Pr

(49)

$(PhCH(OMe)-)_3P$

(50)

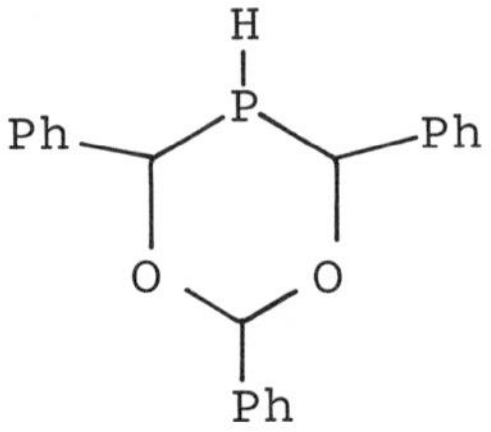

(51)

The reactions of phenylphosphine with benzaldehyde in 1:3 molar ratio provide a good illustration of how changes in reaction conditions can alter the nature of the product obtained. In concentrated hydrochloric acid the phosphine oxide $Ph(PhCH_2)P(O)CHOHPh$ is produced in 54% yield. When the reaction is run in acetonitrile containing a small amount of hydrochloric acid, a 53% yield of the cyclic product (51) is obtained.[356] Furthermore, a tertiary phosphine, phenylbis(α-hydroxybenzyl)phosphine is formed in 87% yield if hydrogen chloride, water, and methanol are used as the reaction medium.[1105,1106]

The formation of phosphine hydrochlorides (52) was noted in the reaction of the more basic isobutylphosphine with benzaldehyde and phenylphosphine with isobutyraldehyde in dilute acid.[356]

$$RPH_2 + 2R^1CHO \xrightarrow{HCl} R(R^1CHOH)_2P \cdot HCl$$

(52)

In the presence of aqueous hydrochloric acid, secondary phosphines produce with benzaldehyde phosphonium salts.[168,1107,1360] Dicyclohexylphosphine and excess benzaldehyde gave, however, the corresponding tertiary phosphine hydrochloride $(c-C_6H_{11})_2P-CHOHPh \cdot HCl$ under the same sort of aqueous conditions, probably because of the steric hindrance toward further addition.[356] A tertiary phosphine, $Ph_2PCHOHR$, was produced from diphenylphosphine and aromatic aldehydes when an organic solvent and dry HCl were used as the reaction medium.[356,610a,1107] These α-hydroxyalkyldiphenylphosphines were found to give phosphine oxides, $Ph_2P(O)CH_2R$, in good yield when heated in concentrated aqueous hydrochloric acid[356] or with p-toluenesulfonic acid in acetic acid,[1360] as had the secondary phosphine and aldehyde when treated under these conditions directly (see Chapter 6).

3. PHOSPHINES AND KETONES. In concentrated hydrochloric acid solution PH_3,[172,173] primary phosphines, and secondary phosphines[356] react with simple ketones in a similar manner as with aromatic aldehydes to give the oxygen transfer products, namely, primary, secondary, and tertiary phosphine oxides, respectively (see Chapters 6, 10, and 11).

$$R_2C=O + H_xPR_{3-x} \longrightarrow R_2HC-\overset{\displaystyle O}{\overset{\|}{P}}H_{x-1}R_{3-x}$$

By using milder conditions propylphosphine and acetophenone gave a secondary phosphine, $Pr(PhMeC \cdot OH)PH$,[1106] whereas phenylphosphine,[1105] or diphenylphosphine[1107] and acetophenone yielded tertiary phosphines containing hydroxy groups.

Ketones that bear electronegative substituents, such as hexafluoroacetone, trifluoroacetone, hexafluorocyclobutanone, pyruvic acid, and 2,4-pentanedione do not undergo oxygen transfer in the course of their phosphine reactions. Instead, the normal carbonyl addition products are obtained. Thus phosphine, CH_3PH_2, and $(CH_3)_2PH$ gave with trifluoroacetone[488] and hexafluoroacetone[163] a 1:1 adduct when heated to 100° for 6 hr.

$$CF_3COCH_3 + PH_3 \longrightarrow H_2P-\overset{\displaystyle CF_3}{\underset{\displaystyle CH_3}{\overset{\displaystyle |}{\underset{\displaystyle |}{C}}}}-OH$$

$$(CF_3)_2CO + Me_2PH \longrightarrow Me_2P-\overset{\displaystyle CF_3}{\underset{\displaystyle CF_3}{\overset{\displaystyle |}{\underset{\displaystyle |}{C}}}}-OH$$

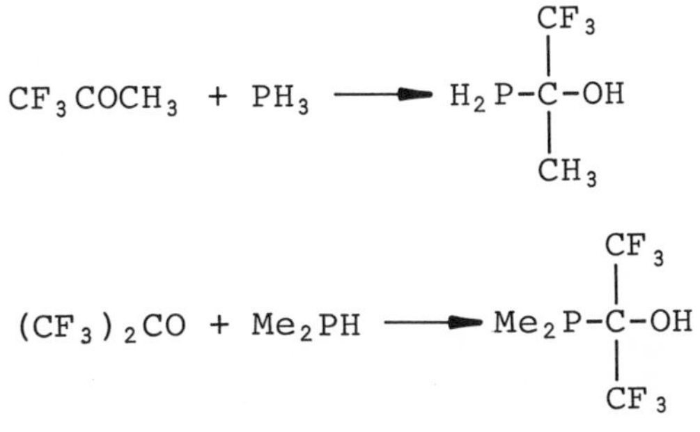

However, diphenylphosphine and dicyclohexylphosphine attacked hexafluoroacetone initially at the carbonyl oxygen to give phosphinites which were isolated after oxidation as phosphinates.[1352]   Depending upon the ratio of reactants, hexafluorocyclobutanone reacts with PH_3 at room temperature to give a primary or secondary phosphine in

$$(CF_3)_2C=O + R_2PH \longrightarrow [(CF_3)_2\overset{\ominus}{C}-O-\overset{\oplus}{P}R_2] \xrightarrow{[O]} (CF_3)_2CHO\overset{\overset{O}{\|}}{P}R_2$$

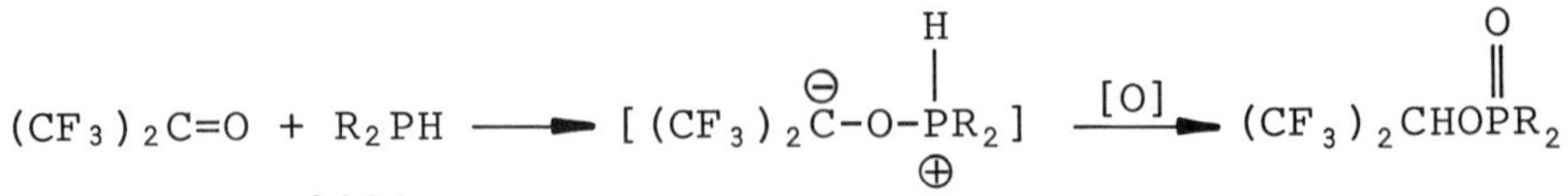

high yield.[1078]

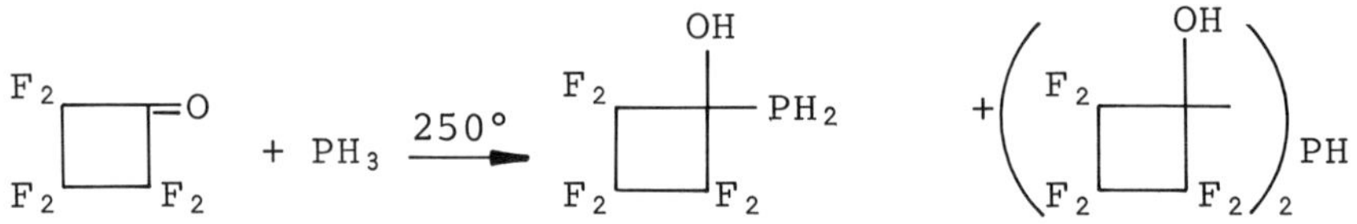

In the interaction of PH_3 and pyruvic acid in ether in the presence of dry HCl, an almost quantitative yield of a compound having the molecular formula $C_9H_9O_6P$ was obtained[980] to which the rigid, polycyclic structure with three lactone rings was assigned (53).[168]

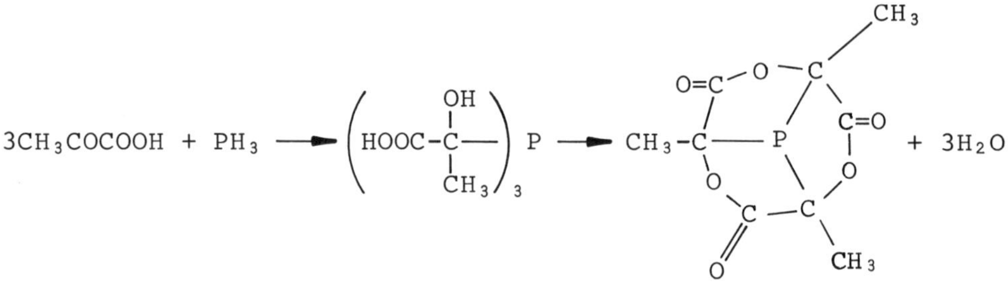

(53)

4.   PHOSPHINES AND DIKETONES.   Solutions of 2,4-pentanedione in aqueous 4 to 6 N hydrochloric acid readily absorb phosphine and primary phosphines to give crystalline solids to which a trioxa-6-phosphoadamantane structure (54) has been assigned.[355]
From the reaction of phenylphosphine with 2,4-pentanedione, materials with two or three phosphorus atoms per molecular formular were isolated, in addition to the expected product (54, R = Ph).   Cyclohexylphosphine and 2-cyanoethylphosphine with the diketone also gave compounds having two phosphorus atoms per molecular formula.   It was suggested[355] that these products might have the same structure as (54) except that one or two of the oxygen atoms

$$2CH_3\overset{\overset{O}{\|}}{C}CH_2\overset{\overset{O}{\|}}{C}CH_3 \; + \; RPH_2 \xrightarrow[-H_2O]{}$$

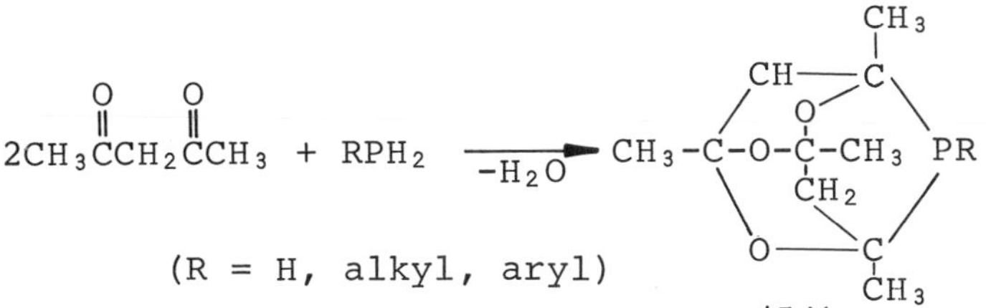

(R = H, alkyl, aryl)

(54)

are replaced by a PR group.

5. PHOSPHINES AND ISOCYANATES OR ISOTHIOCYANATES.
Phosphines react with isocyanic acid[1074,1075,1377] and isocyanates[167,168a,174,175a,355,1290] under mild conditions
in benzene as solvent and in the presence of basic catalysts to give carbamoyl-phosphines. The fact that the
intermediate mono- or dicarbamoyl-phosphine, $RNHCOPH_2$ and
$(RNHCO)_2PH$, were never detected in the reaction of PH_3
with isocyanates indicates that the intermediates are more
reactive toward isocyanates than phosphine itself.[167]

$$R_xPH_{3-x} \; + \; (3-x)RNCO \longrightarrow R_xP(CONHR)_{3-x}$$

Apparently, the nucleophilic reactivity of the phosphorus
atom is increased by substitution of a carbamoyl group for
a hydrogen in the same way the base strength of phosphine
is increased by successive replacement of hydrogen with
methyl groups (see Section C.2). Furthermore, resonance
of the type

$$RNH\overset{\overset{O}{\|}}{C}-PH_2 \rightleftharpoons RNH\overset{\overset{O^{\ominus}}{|}}{C}=\overset{\oplus}{P}H_2$$

is probably not important in these compounds since, if it
were, one would expect to obtain monocarbamoyl derivatives. The yield of the final products increased with the
electronegativity of the substitutent in the para position
of the phenyl group. Tris(arylcarbamoyl)phosphines do not
hydrolyze in boiling water and are only decomposed when
heated above 200° to give carbanilide, elemental phosphorus, carbon monoxide, and carbon dioxide.[167]
     Attempts to react PH_3 with phenylisothiocyanate were
unsuccessful.[167] Primary and secondary phosphines, however, react easily with isothiocyanates without catalyst
to give thiocarbamoylphosphines in high yield.[665,666,1063,1290]

$$RPH_2 \; + \; 2R^1NCS \longrightarrow RP(CSNHR^1)_2$$

$$R_2PH \; + \; R^1NCS \longrightarrow R_2P(CSNHR^1)$$

Phosphinosubstituted group-II and -IV compounds containing Zn-P,[1057] Si-P,[2,718] Ge-P,[1214-1216] Sn-P,[1267] and Pb-P bonds easily undergo an insertion reaction with 1,2-dipolar reagents such as CO_2, CS_2, $CSCl_2$, COS, and many others to give a wide variety of group-II and -IV-substituted phosphines, respectively. These reactions are discussed in detail in Chapter 2, Section D.2, while the end products are listed in the list of compounds at the end of this chapter.

6. MISCELLANEOUS REACTIONS. Phosphine reacts with ethylene oxide under pressure at 100°, preferably in dry ether, to give a mixture of β-hydroxyethyl groups containing primary, secondary, and tertiary phosphines.[782] Propylene oxide reacts similarly. In an attempt to use phenylphosphine in this reaction, only tars were produced.[569] The yields are significantly improved if alkali phosphides are used as starting materials (see Section B.2.VII). The course of the reaction of diphenylphosphine with styrene oxide is largely dependent upon conditions. Without solvent at 120° the products were Ph_2MeP, $PhCH_2CH_2P(O)Ph_2$, and benzaldehyde. However, in refluxing methanol styrene, $Ph_2P(O)H$, benzaldehyde, and Ph_2MeP were obtained.[1361]

Diethylphosphine adds to cyclohexyl isocyanide in the presence of group IB and IIB metal compounds (CuO, $ZnCl_2$, $CdCl_2$) to give N-cyclohexylformimidoyldiethylphosphine.

$$Et_2PH + C{=}NC_6H_{11}\text{-}c \xrightarrow[110°/6hr]{CuO} Et_2PCH{=}NC_6H_{11}\text{-}c$$

This reaction does not proceed without a catalyst.[1209] Since phenylphosphine adds to Schiff's bases such as benzylideneaniline when the mixture is heated at 130 to 135° for 6 hr,[1136] the formimidoyl derivative should add a second mole of phosphine to give a diphosphine compound.

$$PhPH_2 + 2PhN{=}CHPh \xrightarrow[6\ hr]{130°} PhP(CHPhNHPh)_2$$

XIV. BY MISCELLANEOUS METHODS

A new method for the synthesis of tertiary phosphines was discovered in the reaction of triethylphosphite with phenylisocyanate under a CO pressure of 150 lb/sq.in. at 180°. The reaction very likely proceeds through the intermediate formation of tris(ethylphenylamino)phosphine.[61] The yields and the scope of this reaction are unknown.

$$P(OEt)_3 + 3PhNCO \xrightarrow{95°} 3CO_2 + [P(N{\overset{Ph}{\underset{Et}{\diagdown}}})_3 \xrightarrow[180°]{3CO} Et_3P + 3PhNCO]$$

Phosphines containing phosphine ylides or phosphonium cations as substituents have been obtained in the reaction of phosphine ylides with phosphinous halides.[122,683-686,1277]

$$2Ph_3P=CHR + R_2^1PCl \longrightarrow R_2^1PCR=PPh_3 + [RCH_2PPh_3]Cl$$

$$Ph_3P=CR_2 + R_2^1PCl \longrightarrow [R_2^1PCR_2PPh_3]Cl$$

Disagreement exists concerning the reaction of triphenyl-allylenephosphorane and phosphinous halides. While one group proposed attack at the α-carbon atom of the allylene group,[1277] another group suggested that initial attack occurs at the γ-carbon atom of the allylene group.[686] The ylides behave as Wittig reagents and have been used for

$$Ph_3P=CH-CH=CH_2 \rightleftharpoons Ph_3\overset{\oplus}{P}-\overset{\ominus}{CH}-CH-CH_2 \begin{cases} \xrightarrow{R_2PX} [Ph_3PCH(CH=CH_2)-PR_2]^{\oplus}X^{\ominus} \\ \xrightarrow{R_2PX} [Ph_3PCH=CHCH_2-PR_2]^{\oplus}X^{\ominus} \end{cases}$$

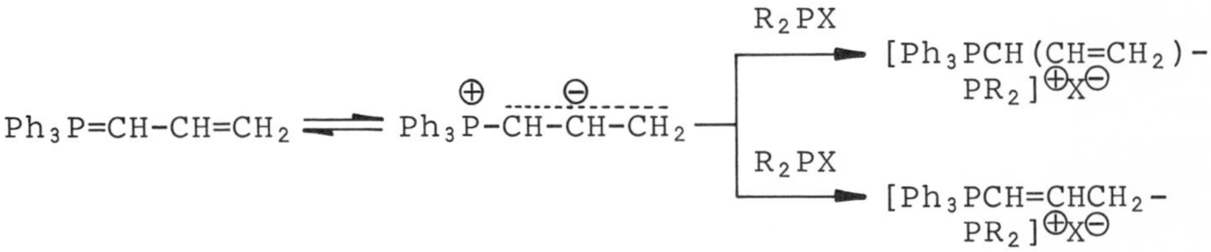

the synthesis of unsaturated tertiary phosphines.[439,683] The interaction of phosphine ylids and arynes produced

$$R_2PCMe=PPh_3 + PhCHO \longrightarrow R_2PCMe=CHPh + Ph_3P=O$$

$$Ph_2PCH_2P(O)Ph_2 + Na \xrightarrow{RCHO} Ph_2PCH=CHR + Ph_2PO_2Na$$

several tertiary phosphines which would otherwise not be easily preparable, for example, (55).[1461] Secondary phosphines react with triphenyl cyanurate at 200° in the ab-

$$Ph_3P=CHR + \text{⟮⟯} \longrightarrow [Ph_2P{\diagdown}\underset{}{\diagup}CHR] \longrightarrow Ph_2P{\diagdown}\underset{}{\diagup}CHRPh$$

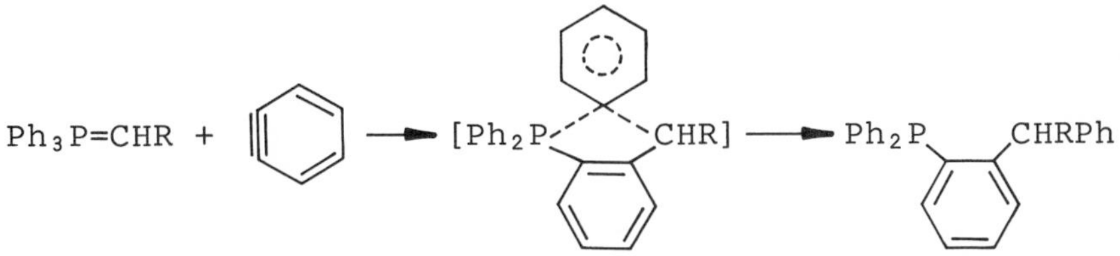

(55)

sence of a solvent to give 2,4,6-trisphosphino-1,3,5-tri-azines.[562] The same products are obtained from secondary

$$N_3C_3(OPh)_3 + 3HPPh_2 \longrightarrow N_3C_3(PPh_2)_3 + PhOH$$

phosphines and cyanuric chloride in the presence of a base.[562] Thermal decomposition of phosphoranes (∿200°) produced tertiary phosphines with unusual structures, for example, decomposition of pentaphenylphosphorane gave 9-phenyl-9-phospha-fluorene (56) in 10% yield,[1443] and decomposition of spirophosphoranes produced the first known nine-membered phosphorus ring compounds (57).[548-550,551,1444]

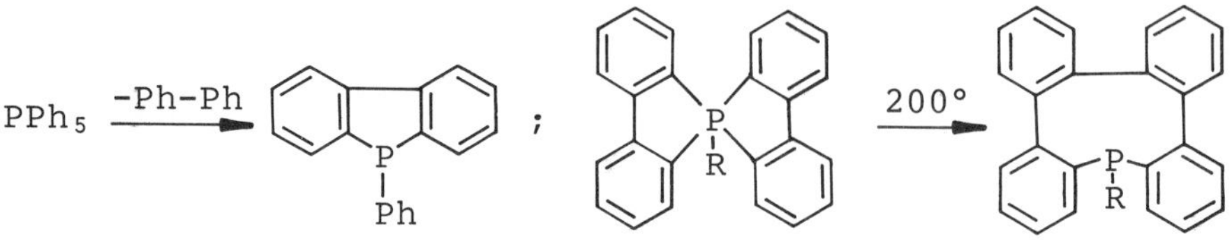

(56) (57)

Dipotassium cyclooctatetraenide reacts with $PhPCl_2$ in tetrahydrofuran at -78° to give (58) which upon warming isomerizes to (59).[767]

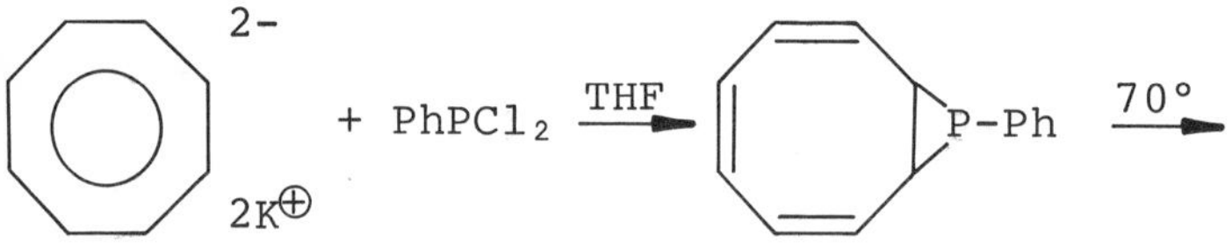

(58)

(59) (60)

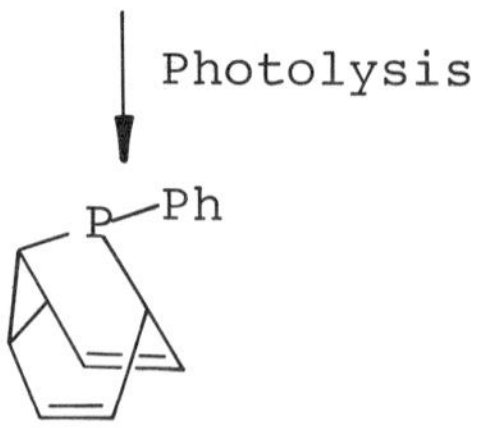

(61)

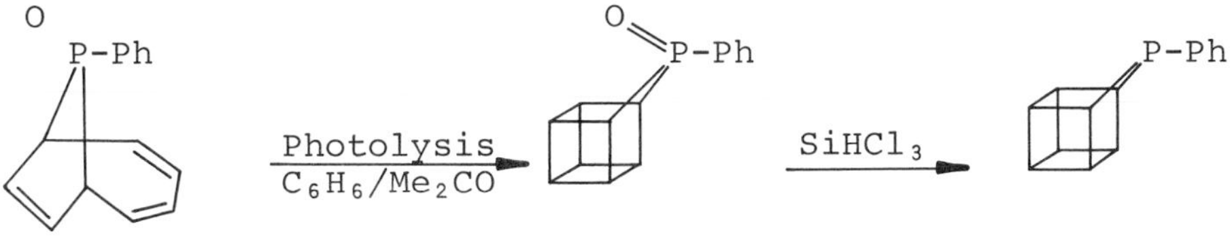

(62) (63) (64)

Upon pyrolysis (59) rearranges to the isomeric phosphine (60).[767] Photolysis of benzene solutions of (59) through pyrex yields as the major product (61) and small amounts of (58) and (60).[766] In contrast, photolysis of the phosphine oxide (62) in benzene-acetone gives the oxide (63) which on reduction with $SiHCl_3/NEt_3$ in benzene yields the phosphine (64).[766] Phosphines obtained by changing the organic rest of a phosphine without changing the phosphine character are discussed in Section C.2.

C.2. General Chemistry of Tertiary Phosphines

C.2.1. General Characteristics of Phosphines

C.2.1.1. The Basicity of Phosphines. The basicity of phosphines increases regularly with the degree of substitution.[156,554a,654a,1342] The following increasing order of base strength toward a proton for the methylphosphines has been given[156]:

$$PH_3 < CH_3PH_2 < (CH_3)_2PH < (CH_3)_3P$$

Unlike methylamines, the phosphines show no evidence of B-strain. This is probably due to the larger dimensions of the phosphorus atom compared to nitrogen, which permit the maintenance of the normal valence angles in tertiary phosphines [C-P-C in $(CH_3)_3P$ is 98.6°[82,1317a] and C-P-C in $(CF_3)_3P$ is 99.6°[141a]]. The increase in base strength also holds for other substituents, such as i-butyl, octyl, and phenyl.[554a,654a] The basicity of phosphines is determined mainly by inductive, resonance, steric, and hybridization effects[554a] and may be expressed in terms of a "Taft equation."[1346a]

$$\log R/R_o = \sigma^*\rho^*$$

A plot of the pK_a values of the phosphines against the sum of the σ* values for the substituent groups shows that the phosphines are grouped in three classes, primary, secondary, and tertiary phosphines (Fig. 1). Phosphine itself, with an estimated pK_a of -14,[1425a] represents a fourth class. The equations for the three lines have been

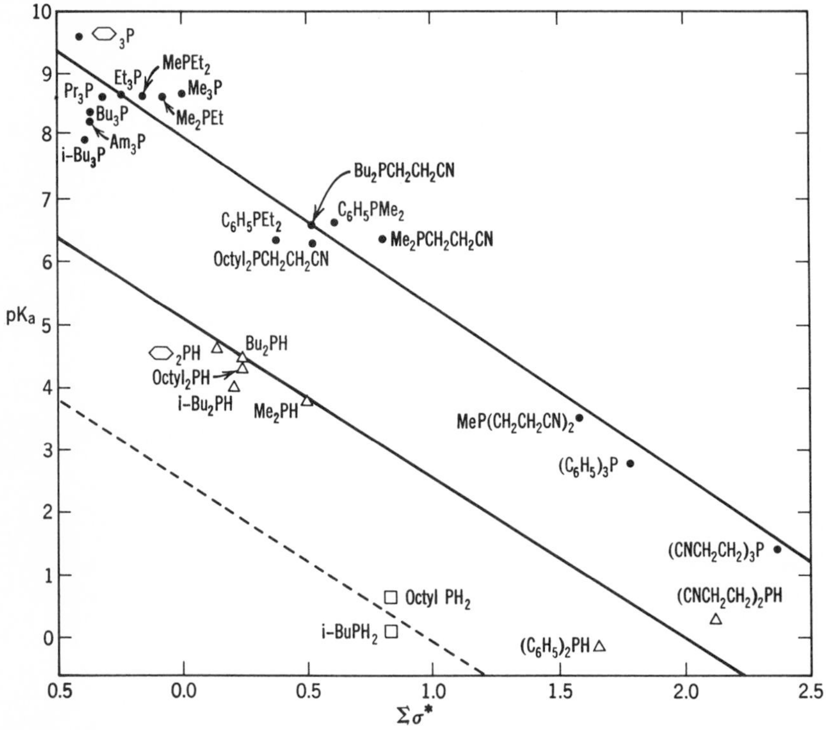

Fig. 1. Plot of pK_a in water versus $\Sigma\sigma^*$ for phosphines.[554a] (Reproduced by permission of the Journal of the American Chemical Society.)

calculated from the pK_a values of the phosphines[1342] by the method of least squares and are given below.[554a] Thus it is possible to predict the pK_a of any phosphine within one unit if the σ^* values for the substituent groups are known. A comparison of the slopes of the equations below with the corresponding ones for the amines[501a] shows that phosphines are slightly less susceptible to inductive

$$\text{Tertiary phosphines} \quad pK_a = 7.85 - 2.67\Sigma\sigma^*$$

$$\text{Secondary phosphines} \quad pK_a = 5.13 - 2.61\Sigma\sigma^*$$

$$\text{Primary phosphines} \quad pK_a = 2.46 - 2.64\Sigma\sigma^*$$

effects than are amines. Furthermore, the pK_a values for

$\sigma^* = 0$ show that a phosphine with a given number of alkyl substituents is a weaker base than an amine of the same $\Sigma\sigma^*$ value. It may be pointed out that the basicity series N > P obtains with other Lewis acids (BMe_3, $AlMe_3$, BF_3, $GaMe_3$) with the exception of BH_3 (P > N).[458a] The basicity of phosphines is largely determined by the inductive effect as measured by $\Sigma\sigma^*$ (Fig. 1). Deviations from the pK_a predicted on the basis of inductive effects are the result of steric effects, for example, triisobutylphosphine. The persistence of steric effects in the secondary and primary phosphines has been ascribed to the varying amount of hindrance to solvation of the phosphonium ion produced. The phenylphosphines fall below the line (Fig. 1) by small but persistent values of the order of 1/2 pK_a unit.[554a] While this may be due to the fact that the steric requirements of a phenyl group are greater than those of an alkyl group, it may also be a result of π-bond formation between the phosphorus atom and the benzene ring.[722]

The resulting electron delocalization would effect a lowering of the basicity of phenylphosphines. The similar lowering of basicity of phosphines toward trimethylborane by vinyl groups has been ascribed to operation of the same effect.[752a] The Hammett equation was also applied to correlate the basicity with the structure of the phosphines.[746a] A plot of $\Sigma\sigma_p$ (σ_p constants used are those for substituents on the phosphorus[746]) of primary, secondary, and tertiary phosphines against pK_a gave a straight line. This result indicates that steric factors do not in general affect the basicity of phosphines. The following parameters for the Hammett equation have been given: $\rho = 3.423$, $pK_a = -3.45$ ($r = 0.950$). The equation that describes the relation of pK_a of protonation of $p\text{-}XC_6H_4PEt_2$ with the substituent constants X (determined from the pK_a of phosphonic acids) is:[134,1321,1322]

$$pK_a = -4.606 - 4.094\Sigma\sigma_p$$

and using normal σ values[440]

$$pK_a = 4.41 - 3.08(\sigma + \Sigma\sigma^*).$$

The basicity of para-substituted phenyldiphenylphosphines, $p\text{-}XC_6H_4PPh_2$, correlates with the equation $pK_a = 2.30 - 1.11\Sigma\sigma$.[445] The basicity constants of ditertiary phosphines, carboxyphosphines,[654a] carboxyalkylphosphines,[705] and some other phosphines[677,1323] toward a proton have also been determined. The basicity of phosphines toward other Lewis acid has been reviewed elsewhere.[29a,1341]

C.2.1.2. The Nucleophilicity of Phosphines. In general, phosphines are stronger nucleophiles than amines,[287] a result mainly of steric effects in the amines.[553] The ease of alkyl halide addition to phosphines is roughly in accord with the basicities of the latter. The rate of S_N2 attack of a phosphine on alkyl halide is mainly determined by inductive effects of the substituents on the phosphorus atom and may be expressed, in the case of tertiary phosphine, by the following "Taft" equation.

$$\log 10^5k = 1.939 - 0.767\sigma^*$$

A plot of log 10^5k against the sum of the σ^* values for the reaction of secondary and tertiary phosphines with ethyl iodide[553] in acetone at 25° gives a straight line (Fig. 2). Any deviation from this straight line may be

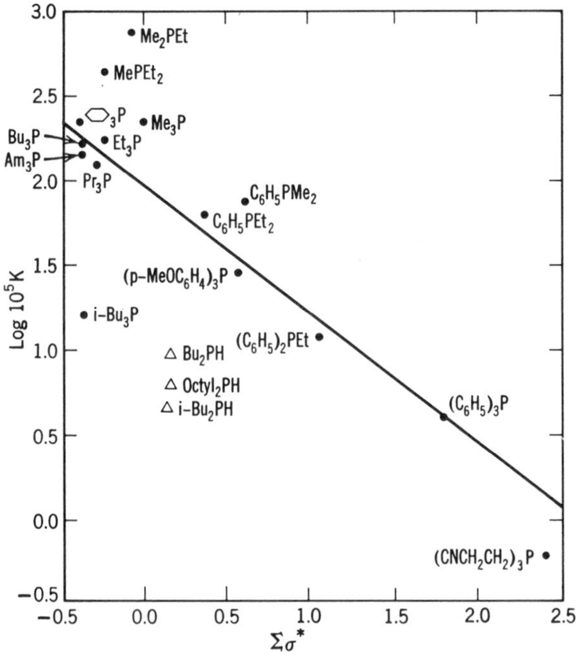

Fig. 2. Plot of log 10^5k versus $\Sigma\sigma^*$ for the reaction of secondary and tertiary phosphines with ethyl iodide in acetone at 35°.[553] (Reproduced by permission of the Journal of the American Chemical Society.)

interpreted in terms of one or more variables other than the inductive effects of the substitutent groups, namely, steric or resonance effects, or changes in hybridization. The reaction is very sensitive to changes in the phosphine structure, the rate of reaction varying from 6 x 10^{-6} to 8 x 10^{-3}, and to the halide of the alkyl group. For example, n-propyl halides gave the following relative rates of reaction with tri-n-butylphosphine: I:Br:Cl = 2660:248:1. The rate depends further on the solvent used. The unusually high reactivity of tertiary ethylmethylphosphines as compared to trimethyl- and triethylphosphine has not yet been satisfactorily explained. The following reactivity series toward alkyl halide has been observed[553]:

$$Me_2EtP > MeEt_2P > Me_3P > Et_3P$$

A similar reactivity series of the methyl-, ethyl-, and vinylphosphines has been observed toward tirmethylborane[752a]:

$$EtMe_2P > Me_3P > (CH_2{=}CH)Me_2P > Et_3P > (CH_2{=}CH)_3P$$

The low reactivity of triisobutylphosphine is most probably the result of a steric effect, and tris(2-cyanoethyl)-phosphine is believed to be less reactive than would be predicted because of interaction between the nitrile group and the phosphorus atom.[553] The inductive effect of the substituent group on the rate of quaternization is also clearly shown in the series of meta- and para-substituted aryldialkylphosphines[287,801] (p. 25).

Quaternization of an optically active tertiary phosphine proceeds with retention of configuration.[635] One can conclude from this result that the attack of a phosphine on an alkyl halide occurs at the carbon-halide bond[553] (see Chapter 4).

The quaternization of tertiary phosphines with alkyl halides goes to completion in polar solvents such as acetone or alcohol. With primary and secondary phosphines, side reactions may occur, since further reaction with an alkyl halide is possible. Under controlled conditions it was possible to isolate the primary adducts $[RR'PH_2]X$[260,535] (see Section A.1.VI) and $[R_2R'PH]X$ (see Section B.1.III),[535,553,591] and kinetic data were obtained in several cases.[553] Other methods for the preparation of phosphonium salts are discussed in Chapter 4.

C.2.2. Addition to Polarized and Polarizable Double Bonds

Tertiary phosphines readily add to polarized and polarizable double bonds.

$$X = Y + PR_3 \longrightarrow \overset{\ominus}{X}\text{-}Y\text{-}\overset{\oplus}{P}R_3$$

X-ray analysis established that there exists a phosphorus-carbon bond in the red adduct of Et_3P and CS_2, and that the structure corresponds to a zwitterion of a quaternary phosphonium derivative of dithioformate, Et_3P^+-CSS^-,[963a] as proposed previously.[292,730a] IR studies indicate that the carbon diselenide adducts, $R_3\overset{+}{P}$-C-$SeSe^-$, have a similar structure.[731] Tertiary phosphines also add to carbonyl and activated C=C bonds with the formation of inner phosphonium salts.[618] (p. 108) These reactions are discussed in Chapter 4.

C.2.3. Phosphines as Reducing Agents

Tertiary phosphines are effective reducing agents, the aliphatic members being much stronger than the aromatic in this respect. Thus they add oxygen (Chapter 6) sulfur, selenium, tellurium (Chapter 7), and halogen (Chapter 5B) directly. The oxygen, sulfur or halogen can also be donated by organic or inorganic molecules containing these elements. Normally, these reactions have not been carried out for the purpose of preparing phosphine oxides, sulfides, or dihalophosphoranes, but rather to deoxygenate, desulfurize, or dehalogenate organic and inorganic molecules. These reactions are discussed in detail, in Chapters 4, 5, 6, and 7, respectively.

C.2.4. Substitution Reactions and Other Reactions

This type of reaction is limited since electrophilic substitution usually forms derivatives of pentavalent phosphorus. Thus the reaction of tertiary phosphines with halogen yields dihalophosphoranes[986] and with nitric acid gives phosphine oxides containing nitro groups.[993a] Aromatic halogen-containing tertiary phosphines have been obtained from the reaction of phosphinous halide with a mono-Grignard or monolithium reagent of a dihalobenzene.

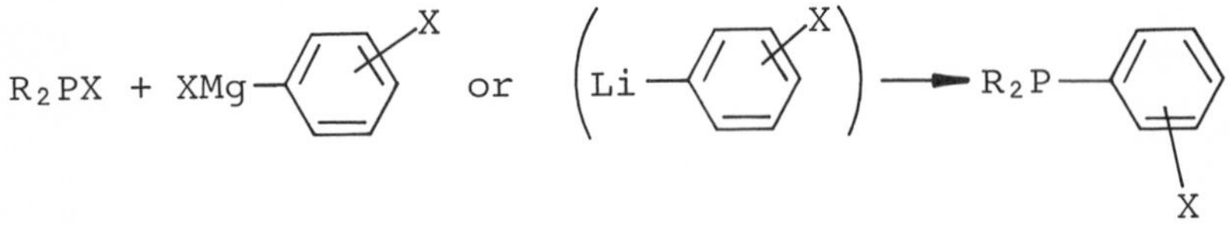

The residual halogen may be attached in the ortho,[518,519,521] meta,[430] or para position.[73,430,445,1228,1234] Interaction of these phosphines with magnesium gives a Grignard reagent, and with butyllithium a metalated phosphine. These products have been successfully used for further synthesis (see Section C.1.I and II). A more direct route to

$$R_2P-\langle C_6H_4\rangle-X \quad + \quad Mg\,(or\ LiBu) \quad \longrightarrow \quad R_2P-\langle C_6H_4\rangle-\underset{(Li)}{MgX}$$

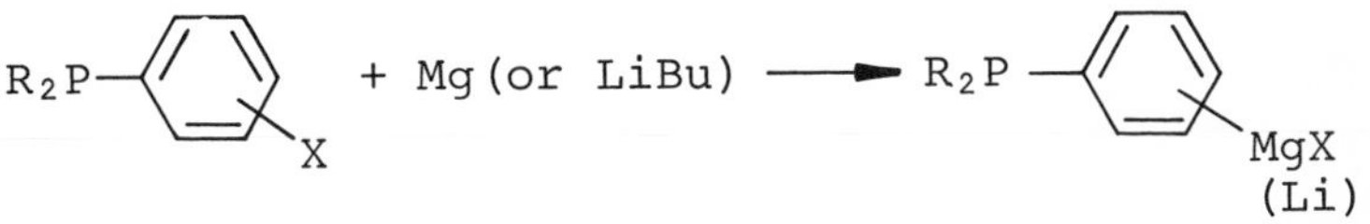

these products consists in the reaction of triphenylphosphine with butyllithium, resulting in meta metalation.[430]

$$Ph_2P-\langle C_6H_5\rangle \quad + \quad C_4H_9Li \quad \longrightarrow \quad Ph_2P-\langle C_6H_4\rangle-Li \quad + \quad C_4H_{10}$$

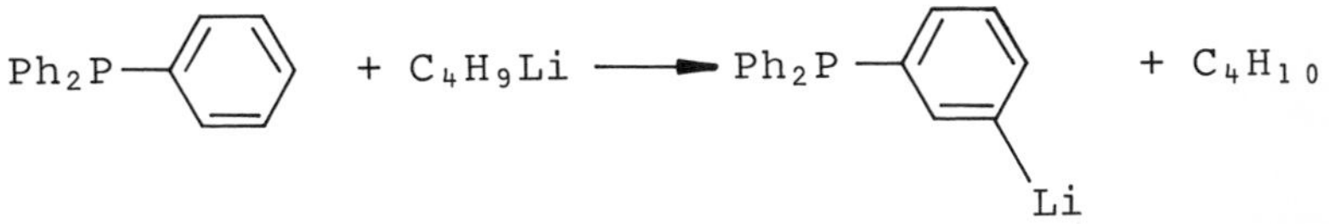

If a methyl or benzyl group is present, the first metalation usually occurs on the methyl[606,1097,1099-1101] or benzyl group,[11,17] but dimetalation is less discriminant.[1100] Dimethylphenylphosphine is monometalated in only the methyl group, whereas the second lithium enters either the methyl or phenyl position.[1100] High yields of the monometalated compound can be obtained by using the 1:1 complex of BuLi

$$PhP\,(CH_3)_2 \quad + \quad t\text{-}BuLi \longrightarrow PhP\,(CH_3)CH_2Li \quad \xrightarrow{\ t\text{-}BuLi\ } \quad \begin{cases} PhP\,(CH_2Li)_2 \\[1em] \langle C_6H_4\rangle \overset{P\,(CH_3)-}{\underset{Li}{}} CH_2Li \end{cases}$$

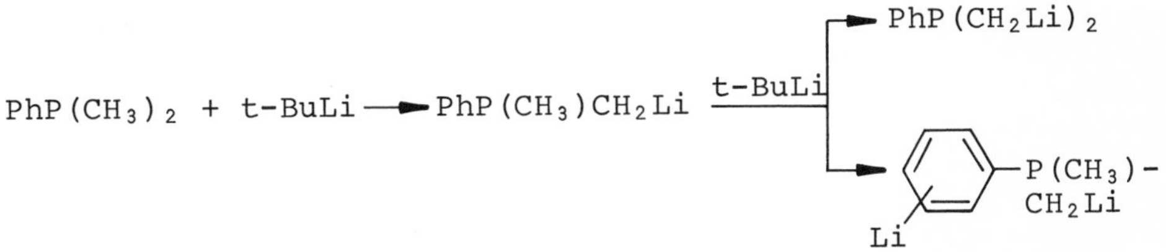

with tetramethylethylenediamine as the metalating reagent.[1099] Metalation of optically active MePrPhP results in retention of configuration.[606] Tri-n-butylphosphine is, however, not metalated.[1099,1101] α-Metalated tertiary phosphines have also been obtained from the Michael-type addition reaction of butyl- or t-butyllithium to diphenylvinylphosphine.[1097] Reaction of BuLi with Bu$_2$PCH=CH$_2$,

$$Ph_2PCH{=}CH_2 \quad + \quad RLi \quad \longrightarrow \quad Ph_2P\underset{Li}{CH}CH_2R$$

however, results in the formation of a telomer.[1097] Other addition reactions are discussed in Section C.1.XII.

Unsaturated tertiary phosphines can be hydrogenated either as such,[12,220,221] or as their nickel chloride complexes,[1148] for example,

$$Et_2PC{\equiv}CPEt_2 \xrightarrow[Ni]{H_2/1.75\ hr} Et_2PCH{=}CHPEt_2 \xrightarrow[Ni]{H_2/15\ hr} Et_2PCH_2CH_2PEt_2$$

Hydroxyalkyl groups attached to phosphorus may be esterified with acetyl chloride,[57,59,163,1004,1115,1165,1329,1364] benzoyl chloride,[1329,1364] ketene and diketene,[1183] orthoformates,[138] and trimethylphosphite[25] without changing the trivalent state of phosphorus.

$$R_2P(CH_2)_xOH + ClC(O)R^1 \longrightarrow R_2P(CH_2)_xOC(O)R^1$$

Hydroxymethylphosphines react with secondary amines to give dialkylaminomethyl-substituted phosphines.[659a,660] These are reactive compounds and yield, for example, with secondary phosphines methylenediphosphines.[916] Phosphines

$$R_2PCH_2OH + HNR_2 \xrightarrow[-H_2O]{} R_2PCH_2NR_2 \xrightarrow[-R_2NH]{R_2PH} R_2PCH_2PR_2$$

containing carbonic acid ester[703,704] and β-cyanoethyl groups[1173] can be hydrolyzed to the corresponding carboxy-substituted phosphines.

$$P(CH_2CH_2CN)_3 \xrightarrow{H_2O} P(CH_2CH_2COOH)_3$$

$$R_2PCH_2CH_2CN \xrightarrow{LiAlH_4} R_2PCH_2CH_2CH_2NH_2$$

Reduction of β-cyanoethylphosphines with $LiAlH_4$ gives γ-aminopropylphosphines.[60,1207] Phosphines, substituted with carboxy,[1228] and keto groups[1061] have been reduced with $LiAlH_4$ to the corresponding phosphinosubstituted alcohols. By reaction with ammonia, ester groups and mixed anhydrides have been converted to the amides[215,705,1228,1234] which can be dehydrated with $POCl_3$ to give the corresponding nitriles.[1228,1234]

$$RCOCH_2CH_2PR_2 + LiAlH_4 \longrightarrow RCHOHCH_2CH_2PR_2$$

p-Cyanophenyldipropylphosphine gives on treatment with MeMgI p-acetophenyldipropylphosphine.[442] Reduction of p-

$$EtOCOC_6H_4PR_2 \xrightarrow[-EtOH]{NH_3} H_2NCOC_6H_4PR_2 \xrightarrow[-H_2O]{POCl_3} NCC_6H_4PR_2 \xrightarrow[H_2O]{MeMgX} MeCOC_6H_4PR_2$$

diphenylphosphinobenzoic acid dimethylamide with lithium triethoxyaluminum hydride produces p-diphenylphosphinobenzaldehyde.[1228] Methoxyphenylphosphines[288,838,1037,1273] and alkoxyalkylphosphines[1,1308] may be converted to the

$$Ph_2PC_6H_4CONMe_2\text{-p} + LiAl(OEt)_3H \longrightarrow Ph_2PC_6H_4CHO\text{-p}$$

corresponding hydroxy compounds by reaction with concentrated HI or HBr solutions. Carboxyphenylphosphines have been transformed to the esters by reaction with methanol

$$MeOC_6H_4PR_2 + HI \longrightarrow HOC_6H_4PR_2 + MeI$$

and an acid.[288,838,1037,1228,1234,1273] Transesterification has also been described.[1381]

$$(HOOCC_6H_4)PR_2 + CH_3OH/HCl \longrightarrow MeO_2CC_6H_4PR_2 + H_2O$$

Perfluoroacylphosphines lose carbon monoxide on heating to 190° and give perfluoroalkylphosphines.[859]

$$C_2F_5COPR_2 \xrightarrow{190°} C_2F_5PR_2 + CO$$

The cleavage of an aromatic group from tertiary phosphines with alkali metals (see Section B.1.V), or with KNH_2 in liquid ammonia,[1261,1262] seems to be a general reaction. Tricyclohexylphosphine and tributylphosphine, however, undergo no ammonolytic cleavage.[1261]

Trifluoromethyl[99,385a,385b,1103] and heptafluoropropyl groups[454] linked to trivalent phosphorus are easily cleaved by sodium hydroxide,[91,385a,454] hydrazine, and substituted hydrazines.[1103] Pentafluorophenyldiphenylphosphine, however, undergoes nucleophilic substitution of the p-fluorine when treated with sodium methoxide or methylamine.[180]

$$Ph_2PC_6F_5 \xrightarrow{NaOMe} Ph_2PC_6F_4OMe\text{-}4 + NaF$$

$$Ph_2PC_6F_5 \xrightarrow{MeNH_2} Ph_2PC_6F_4NHMe\text{-}4 + HF$$

Ethynylphosphines easily condense with ketones to give 3,3-disubstituted 3-hydroxypropynylphosphines (65).[218,219,231,1389]

$$R_2PC{\equiv}CH + R_2C{=}O \longrightarrow R_2PC{\equiv}C\text{-}C(OH)R_2$$

(65)

Ethynylphosphines also add mercaptans;[1389] furthermore,

they can be carboxylated[218,430] using the Grignard route.

$$Bu_2PC{\equiv}CH + HSEt \longrightarrow Bu_2PCH{=}CHSEt$$

$$R_2PC{\equiv}CH + RMgX \longrightarrow R_2PC{\equiv}CMgX \xrightarrow{CO_2} R_2PC{\equiv}CCO_2MgX$$

Triethoxysilane adds to phenylethylallylphosphine when heated to 195° in a pressure vessel in the presence of t-Bu_2O_2 as catalyst.[377a] The beta-oriented addition of piperidine to dibutylvinylphosphine has been accomplished at 120 to 130° in the presence of piperidine hydrochloride as catalyst.[751] Without a catalyst no reaction has been observed. The yield of β-piperidinoethyldibutylphosphine is only 14%.

γ-Aminopropylphosphines yield phosphorus containing Schiff's bases (66) when treated with aldehydes.[1207] Selective catalytic ortho deuteration of Ph_3P was achieved by treatment of a toluene solution of $(Ph_3P)_3RuHCl$ and Ph_3P

$$R_2PCH_2CH_2CH_2NH_2 + RCHO \longrightarrow R_2PCH_2CH_2CH_2N{=}CHR + H_2O$$

(66)

(ratio 1:20) with a stream of deuterium at 100° for 30 hr (yield 83%).[1081] Tri(o-tolyl)phosphine on treatment with

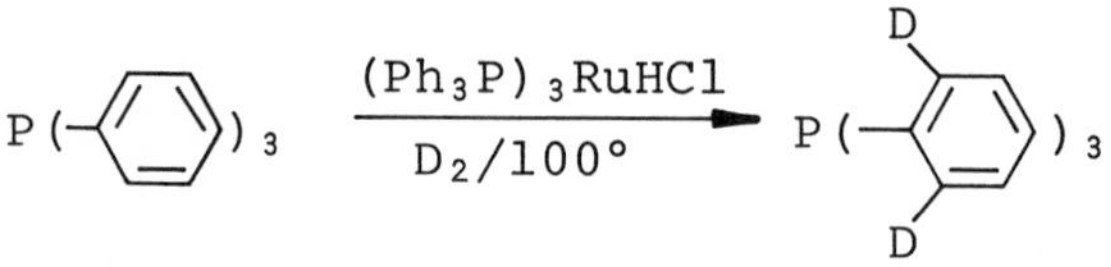

$RhCl_3$ was complexed and oxidatively coupled to trans-2,2'-o-(di-o-tolylphosphino)stilbene. It was liberated from the complex by treatment with KCN.[103]

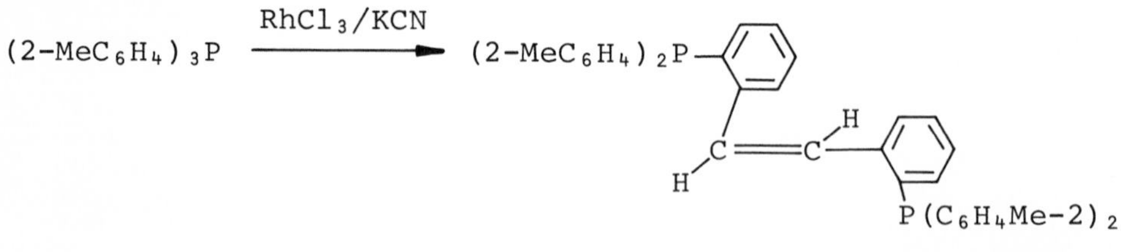

Treatment of triphenylphosphine with 20% SO_3-H_2SO_4 results in sulfonation of one ring without the concomitant formation of a phosphine oxide.[29] The position of sulfonation has not been established exactly. However, the very high heat of solution of triphenylphosphine in sulfuric acid and its resistance to sulfonation (after 2 hr on a

water bath, only a 30% yield was obtained) make it very
likely that it was sulfonated in the form of the phosphonium ion. A meta orientation is thus very probable. In
the presence of bases or basic Al_2O_3, allylphosphines iso-
merize to the corresponding 1-propenylphosphines, for
example,[613a]

$$Ph_2PCH_2CH=CH_2 \xrightarrow{MeONa} Ph_2PCH=CHCH_3$$

The stereochemistry of the reactions discussed above
is in most cases now known (for reviews see McEwen, and
Gallagher and Jenkins). Thus alkylation, oxidation,[635]
sulfurization,[630,635] and phosphine imine formation[630] proceed with retention of configuration. Deoxygenation of
Bz_2O_2 and desulfurization of Bz_2S_2 with (+)-phosphine proceed by a polar mechanism.[633] The extent of racemization
depends upon the polarity of the solvent.[633] Desulfurization proceeds with retention of configuration.[633,1456]
(-)-Phosphine yields with I_2 in water-free solvents upon
hydrolysis inactive oxide, but in water-containing solvents
optically pure oxide.[630]
Formation of polymers from phosphines together with
their properties is described in the list of compounds.
Cyclization of methyl-,[1146] ethyl-, or phenylbis(β-
cyanoethyl)phosphine[1425] with sodium t-butoxide in refluxing toluene results in the formation of 1-substituted 4-
amino-1,2,5,6-tetrahydrophosphorin-3-carbonitriles (67).[1425]

$$RP(CH_2CH_2CN)_2 \longrightarrow R-P\underset{CN}{\overset{}{\bigcirc}}-NH_2 \longrightarrow R-P\bigcirc=O$$

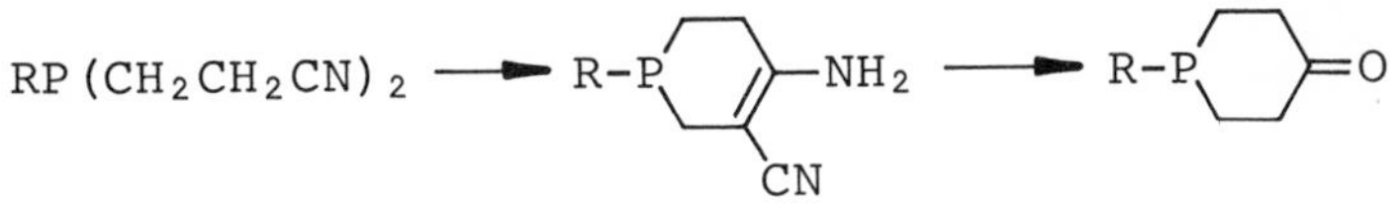

(67)   (68)

Hydrolysis and decarboxylation of these phosphines
yield 4-phosphorinanones (68) which exhibit the properties
of ketones and tertiary phosphines.[420,1146,1425] Thus they
form semicarbazones,[1425] and they can be reduced with
$LiAlH_4$,[971] organozinc compounds,[1292] or Grignard reagents
[1146,1147,1292] to phosphorinanols (69) which can be dehydrated to the corresponding 1,2,5,6-tetrahydrophosphorins
(70).[971,1292] Oxime formation with hydroxylamine causes
oxidation as well to the corresponding phosphine oxide.[968]
4-Phosphorinanones (68) also condense with phenylhydrazine,
o-aminobenzyldehyde, and isatin to yield the corresponding
indolo (71) quinolino (72), and 4'-carboxyquinolino derivatives (73), respectively.[420,421]
9-Phenyl-9-phosphafluorene (74) is obtained from the
reaction of triphenylphosphine with phenylsodium in benzene solution at 20° for 10 hr and then at 70° for 30 hr.[1443]

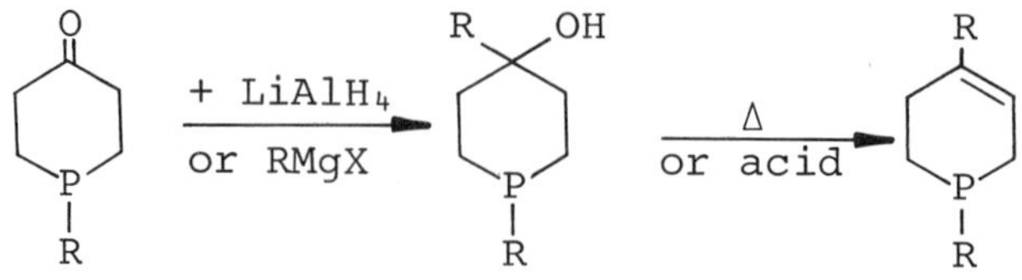

R = alkyl, H

(68) (69) (70)

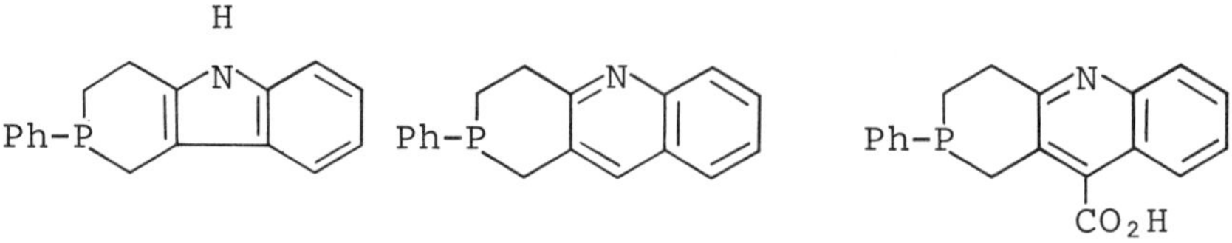

(71) (72) (73)

The same phosphine (74) is produced from o-fluorobromobenzene and magnesium in tetrahydrofuran in the presence of triphenylphosphine.[1439] The proposal that the formation of this phosphine (74) involves addition of triphenylphosphine to benzyne[1439] was verified by the isolation of the betaine, stabilized with triphenylborane,[1439] or by protonation with fluorene,[1445] and by the direct synthesis from aryne and triphenylphosphine.[1279,1462]

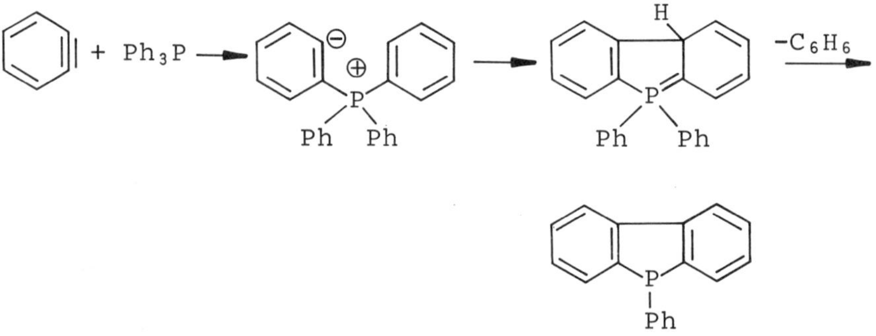

(74)

C.2.5. Applications of Tertiary Phosphines.

Tertiary phosphines are strong nucleophiles and form coordination compounds with many metal halides. They also replace part of the carbon monoxide in metal carbonyls and give mixed carbonyl-metal-phosphine complexes (see Chapter 3A).

Since the discovery by Reppe and Schweckendiek[1181a]

that phosphine complexes of metal halides and metal carbonyls catalyze the cyclization of olefins, acetylene, and derivatives thereof, many of these compounds have been prepared and their catalytic activity investigated.[105a, 848a,979a] The catalytic activity of these compounds has been atributed to dissociation of one ligand then followed by coordination of an acetylene molecule,[848a,979b,1217] for example,

$$Ni(CO)_2(PR_3)_2 \rightleftharpoons [NiCO(PR_3)_2] + CO$$

$$Ni(PX_3)_4 \rightleftharpoons [Ni(PX_3)_n] + (4-n)PX_3$$

Added motivation for research in this field came from a report describing phosphine-stabilized hydrido complexes of platinum(II) and palladium(II).[219a] Since then many other tertiary phosphine stabilized hydrido complexes of transition elements have been reported (see Chapter 3A). Similarly, the isolation of alkyls and aryls of the transition metals became possible through stabilization of these compounds by tertiary phosphines (see Chapter 3A). Tertiary phosphines have been used as starting materials for the preparation of phosphorane derivatives, $R_3P{=}CR_2$, which are obtained either from phosphonium salts and a strong base, or more directly, by addition of a carbene, such as CCl_2, CHCl, CBr_2, or CF_2 to triphenylphosphine (see Chapter 5A). The use of these reagents in the Wittig synthesis, namely, the conversion of an aldehyde or a ketone into an olefin,[1443] has found wide and important applications in organic synthesis (see Chapter 5A).

Reports of other uses of tertiary phosphines are mostly found in the patent literature and are summarized in the following table.

C.2.7. Toxicity of phosphines

Although no direct reports on the toxicity of primary, secondary, and tertiary phosphines are available, it is generally assumed that they resemble phosphine, PH_3, in this respect which is rather toxic. Fatal illness is to be expected for an adult after 1/2 to 1-hr exposure to an atmosphere containing 0.05 mg PH_3/liter (=0.036 vol. %).[1201a] For comparison the fatal doses for a few other known poisons are: 0.6 mg H_2S/liter (=0.431 vol. %); 0.12 mg HCN/liter (=0.109 vol. %), 0.05 mg Br/liter (=0.0076 vol. %), and 1.8 mg HCl/liter (=1.22 vol. %).[1201a] The symptoms of subacute intoxication are pronounced blood and lymph statis accompanied by brain and liver lesions of a special nature. Subchronic poisoning produces degenerative changes in the ganglion cells. This indicates that gaseous PH_3 must be considered a central nervous system

Table 1. Application of Tertiary Phosphines

Phosphine	Suggested Use	Ref.
Ph_3P	Reference substitute for microanalysis	53
$(HOCH_2)_3P$, $(HO_2CCH_2)_3P$	Reducing agents for human gamma globulin and effective agent in the inactivation of rheumatoid factor	852
Ph_3P	Solvent for molecular weight determination; molar depression constant is 12.0 ± 0.5	226
Ph_3P	Inhibits polarographic reduction of $4\text{-}ClC_6H_4NO_2$	765
Ph_3P	Determination of H_2O_2 (see also Chapter 6)	1327
Ph_3P	Structure determination of S linkages in rubber (see also Chapter 7)	1013
R_3P	Catalyst for silane ($\geqslant$Si-H) addition to olefins and acetylenes	1120
Bu_3P	Catalyst for dimerization of acrylates to 2-methylene glutarates	1168, 876
R_3P	Catalyst for dimerization of acrylonitrile	71,210,875,1031,1190, 1289,1437
Ph_3P	Catalyst for hexamerization of acrylonitrile	1348
Ph_3P, Et_3P	Catalyst for polymerization of acrylonitrile	112,248a,621,783,1029, 1377a
Ph_3P	Catalyst for trimerization of isopropenylacetylene in presence of Ni acetylacetonate	227
Ph_3P	Catalyst for ethylene sulfide polymerization	730,1041
Ar_3P	Catalyst for dimerization of vinyl ketones	878

R_3P	Formylation catalyst in the presence of ethylene oxide	149
R_3P	Esterification catalyst	239,331,368,1053,1415, 1454a
R_3P	Catalyst for the addition of carboxylic acids to acetylene	239,1067a
R_3P	Catalyst for the addition of carbon monoxide and alcohol to acetylene [in the presence of Ni(II)]	1055a
Ph_3P	Catalyst for 1-acetoxy-1,1-dicyanoethane formation	366
R_3P	Useful for acid anhydride preparation in the presence of hydrogen acceptors	828
Ph_3P	Catalyst for isomerization of maleate to fumarate	511
Ph_3P	Catalyst for HCl cleavage from 1,2,3-trichlorobutane	1274
R_3P	Catalyst for the preparation of Parathion	1359a
$(C_6H_{11})_3P$	Catalyst for the preparation of vinyl monomers from aldehydes and acryl derivatives	1018
R_3P	Activators for catalysts used in the disproportionation of C_{3-30} alkenes	1118
R_3P	Catalyst for aldehyde polymerization	85,252,344,770,795, 796,798,799,824,880, 1265,1291a,1377a
R_3P	Catalyst for isocyanate polymerization	77b
R_3P	Catalyst for the polymerization of acetylenic compounds	97a
R_3P	Catalyst for the preparation of aromatic isocyanate dimers	1318a
R_3P	Catalyst for the conversion of isocyanates to carbodiimides	204a
R_3P	Catalyst for dialdehyde polymerization	797

(Continued)

Table 1 (Continued)

Phosphine	Suggested Use	Ref.
Ph_3P	Catalyst for CH_2O-PhNCO copolymer preparation	346
Ph_3P	Catalyst for CH_2O-cyclic anhydride copolymer preparation	1346
R_3P	Catalyst for CH_2O-Me_2C=C=O copolymer preparation	1032
Ph_3P	Catalyst for CH_2O-N-fluoroimino-α-cyano composition copolymer preparation	880
Ph_3P	Catalyst for Me_2C=C=O polymerization	1126
R_3P	Catalyst for Me_2C=C=O dimerization	105
R_3P	Catalyst for acrylic and methacrylic ester polymerization	348,641,846a,961,1087a
R_3P	Catalyst for phenylacrylic ester polymerization	244
$(C_9H_{19}C_6H_4)_3P$	Catalyst for cellulose acetate, butyrate polymerization	998
R_2PH	Catalyst for acrylonitrile-styrene block copolymer preparation	642,842,1089a,1090, 1092
Bu_3P	Catalyst for α-cyanostyrene polymerization	643,1093
R_3P	Catalyst for acrylamide polymerization	744
R_3P	Catalyst for polymerization of cyclic anhydrides of hydroxycarboxylic acids	1288
Ph_3P	Catalyst for making fiber-forming polyesters in the presence of ZnO	861
Et_3P, Ph_3P, $(2\text{-}Me_2NCH_2C_6H_4)_3P$ $Me_2PCH_2CH_2PMe_2$ $Ph_2PC_6H_4PPh_2$	Catalyst for β-propiolactone polymerization	964,1397

$(HOCH_2)_3P$	Elimination of induction period in polymerization	728
Ph_2PH, Ph_3P	Site removal and site activation in polymerization of olefins	135,836
Bu_2PH, PH_3	Chain transfer agent for ethylene polymerization	1021
Ph_3P	Regulator for isoprene polymerization	490
R_3P	Change of stereoregularity of polymerization	1027
R_3P	Initiator for photopolymerization of acrylic monomers	347,962,1422
R_3P	Increase sensitivity of photographic emulsions	30,41,559
$(4\text{-}HOC_6H_4)_3P$	Used as additive to photographic developers for the Ag salt diffusion process	126
R_3P	Antifogging agents in light-sensitive systems	604
Ph_3P	Stabilizer for unsaturated polyester resins	345,370
Ph_3P	Stabilizer for polyaldehydes	1091
Ph_3P	Stabilizer for polyethylene in presence of bisphenol-A	1317
R_3P	Stabilizer for polyamides	1357
$PhMe_2P$	Stabilizer for polyvinyl chloride	452
Ph_3P	Flame retardant for styrene-vinyl bromide copolymer	1387
R_3P	Flame retardant for acrylic polymers	68
Ph_3P	Stabilizer for polypropylene in presence of RSSR	541
R_3P	Improve mechanical properties of short-chain polytetrafluoroethylene	647
Me_2PH, Ph_3P	Stabilizers for acetylated polyoxymethylene	1311
Ph_3P	Stabilizer for $(MeSiHO)_4$ and $[Me(CH_2{=}CH)SiO]_4$	211
Ph_3P	Stabilizer for polyurethan	1022

(Continued)

Table 1 (Continued)

Phosphine	Suggested Use	Ref.
Ph_3P	Makes polyethylene in the presence of NH_4NO_3 antistatic	1220
R_3P	As corrosion inhibitors	77a,627,869
R_3P	As poisons of hydrogenation catalysts	626
Ph_3P	Inhibitor of electrode reactions	975
$(C_8H_{17})_3P$	Increases rate of extraction of U(VI)	841
$(RCHOH)_3P$	Gasoline and diesel fuel additive	367
R_3P	Additive to jet fuels	313
Ph_3P, $(C_6F_5)_3P$ $(4\text{-}CF_3C_6F_4)_3P$, $(C_6F_5OC_6F_4)_3P$	Lubricant additive	261,329
R_3P	Reduce leakage of gasoline through cellu-lose packing	86
$(HOCH_2)_3P$	Used in hair waving	729
Ph_3P	Gives with acryl derivatives and aldehydes β,γ-unsaturated compounds	877,1062
Ph_3P	Gives with activated acetylenes and D_2O 1,2-dideuterated olefins	1191
R_3P	Modify nickel catalysts to allow selective hydrogenation of isoprene to monoolefins	46
R_3P	Activator for polyepoxide curing	1061a,1291b
R_3P	Useful as anxiliary froth-flotation agent	1309a
R_3P	Oxidation inhibitor in terpenes	1059a
R_3P	Scavengers for leaded fuel	1310a,1456a
R_3P	Inhibitors for deposit-induced ignition in engines	108a
R_3P	Additives for nonhypergolic hydrocarbons to make them hypergolic	205a,1467a

poison.[865a] The insecticidal activity of triphenylphosphine is, however, not very high compared with that of Parathion.[342a] However, the lower members of the phosphines certainly produce toxic symptoms and great care should be taken in handling these compounds.

D. PHYSICAL PROPERTIES OF PRIMARY, SECONDARY, AND TERTIARY PHOSPHINES

Phosphines are very reactive substances and they are all subject to oxidation. Those containing the lower aliphatic radicals are outstanding in their affinity for atmospheric oxygen and usually ignite in air.

D.1. Thermochemical Data

From the heats of combustion of trimethyl-,[867] triethyl-,[840] tri-n-hexyl,[1124] and triphenylphosphine,[92,121a] the mean dissociation energies were found to be $\bar{D}$ (P-Ph) 70 to 75 kcal/mole and $\bar{D}$ (P-C aliphatic) ∿66 kcal/mole. A higher dissociation energy of P-Ph as compared to P-C aliphatic is to be expected because of the possibility of additional bonding involving 3d orbitals of the phosphorus atom and the π orbitals of the benzene ring.[722] From the heat of hydrogenation, a resonance energy of 19 kcal/mole was found for 9-phenyl-9-phosphafluorene.[91] An estimate of the extra stabilization energy in pentaphenylphosphole (∿28 kcal/mole) was obtained from the difference in the heats of formation of the corresponding phosphine oxide and $Ph_3P{=}O$.[232a] This method, however, leads to a negligible conjugation energy for the phosphafluorene in disagreement with the value obtained from the heat of hydrogenation. The mean dissociation energy of a P-H bond is $\bar{D}$ (P-H) 77 kcal/mole,[522] while the energy for breaking the first P-H bond in PH_3 is somewhat higher D (PH_2-H) 83.9 ± 3 kcal/mole.[874]

Thermal racemization of optically active acyclic phosphines follows cleanly first-order kinetics with $\Delta G^{\ddagger}_{130°}$ falling into the range of 29 to 36 kcal/mole.[66a,632,635] Steric effects appear to be insignificant, but electron-withdrawing substitutents in the para position of aromatic substituted phosphines result in a lowering of the energy barrier to pyramidal inversion.[66a]

D.2. Crystal Structures

X-ray analyses of Ph_3P,[277] 4-$BrC_6H_4PPh_2$,[825] $(PhC{\equiv}C)_3P$,[1016] and Ph_2P-C≡C-PPh_2[80] indicate that all the phosphines have a pyramidal configuration about the phosphorus atoms. The following bond distances and bond angles were found: P-C(Ph) 1.83 Å, P-C(≡C-) 1.76 Å; ∠C-P-C 102° ± 1° (see list of compounds).

D.3. Electron Diffraction

Electron diffraction studies have been made of PH_3,[83] CH_3PH_2,[82] $(CH_3)_2PH$,[82] $(CH_3)_3P$,[82,84] and $(CF_3)_3P$.[141a] The values for bond distances and bond angles agree well with each other[911]: P-C 1.85, P-H 1.44, ∠C-P-C 99°, ∠C-P-H 96.5° (assumed) except for the P-C bond in tris(fluoromethyl)phosphine which is longer (P-CF_3 1.937 Å) than that in the corresponding methylphosphines. The methyl groups in methylphosphines are in a staggered configuration analogous to that in ethane or isobutane and the barrier to rotation of methyl groups is >1 kcal.[84]

D.4. Microwave Spectra

The microwave spectra of CH_3PH_2,[790] CF_3PH_2,[1409] $(CH_3)_2PH$,[1035] phosphirane, $\overline{CH_2CH_2P}H$,[142,142a] and $(CH_3)_3P$[857] have been reported. The values obtained for bond distances and bond angles in methylphosphines agree well with those obtained by electron diffraction: P-C 1.85 ± 0.01 Å, P-H 1.41 Å, ∠C-P-C 99°, ∠C-P-H 97°, H-P-H 93°. The P-C bond distance in CF_3PH_2 is again longer (P-CF_3 1.90 Å) than in the corresponding CH_3PH_2. Phosphirane contains a three-membered ring structure with a small C-P-C angle of 47.4°, P-C 1.867 Å.[142] The dipole moments of these compounds, as determined from Stark effects measurements are given in the list of compounds. The barrier to rotation of methyl groups in Me_3P (2.6 ± 0.5 kcal/mole) and $MePH_2$ (1.957 kcal/mole) and of the trifluoromethyl group in CF_3PH_2 (2.36 kcal/mole) is considerably smaller than that in Me_3N (4.4 ± 0.8 kcal/mole) and is certainly a result of the larger C-P distance as compared to the C-N distance.[857,1235a]

D.5. UV Absorption Spectra

Phosphine and aliphatic phosphines show weak absorption (log ε ∿2) in the frequency range 2300 to 2200 Å, as a result of Rydberg transitions.[97] Aromatic phosphines give strong absorption in the region λ ∿2500 to 2700 Å (α band) in addition to an absorption at λ ∿2280 Å corresponding to the p band of benzene.[443,722,723,1160,1236,1238,1245,1248] The α band has been attributed to n → π* transition by one group[443] and to π→π* transition by another group.[1224,1230] Since the α band is produced by configuration interaction, its interpretation is not simple. Para substituents have little effect on the frequency, although the p-MeO group produces a slight increase while a p-Cl group decreases the frequency slightly. These effects have been ascribed to stabilization of the ground state and excited state, respectively.

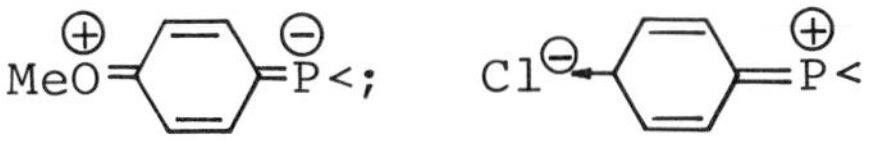

The introduction of two diphenylphosphino groups in the 4-position of oligophenyls gives a strong bathochromic shift.[1248] The PH_2 group acts as a weak auxochrome.[1236] The single peak above 2200 Å in the UV spectra of vinylphosphines[1124,1420,1421] was assigned to an electron-transfer-type of transition in which an electron is removed from the nonbonded orbital on phosphorus and transferred to the empty π orbital of the vinyl group. The significant red shift in trans-1,2-vinylenebis(dibutylphosphine) was attributed to the ability of the second phosphino group to stabilize the electron-transfer excited state of the $R_2PCH{=}CH$ chromophore.[1421]

D.6. IR and Raman Spectra

Several reports dealing with IR and Raman spectroscopic investigations of primary, secondary, and tertiary phosphines have appeared in the literature, and complete vibrational analyses have been made. This subject has recently been reviewed in detail.[249] The P-H stretching vibration in primary and secondary phosphines is usually found in the region of 2280 ± 10 cm^{-1}.[88,89a,194,1252,1326] In only a few compounds, has absorption been observed at higher frequency, for example, CF_3PH_2 at 2370 cm^{-1},[100] CH_3PH_2 at ∿2310 cm^{-1},[839,860,1019,1051] $(CF_3)_2PH$ at 2320 cm^{-1},[100] and $C_6F_5PH_2$ at 2325 cm^{-1}.[389] Absorption resulting from P-H deformation occurs at δP-H = 1077 ± 4 cm^{-1}, and for CH_3PH_2 at somewhat higher frequency δP-H = 1093 cm^{-1}.[860] It has been suggested that a band at 805 ± 10 cm^{-1} may be attributed to γ P-H deformation.[1252] The assignments of P-C vibrations to the 770 to 650 cm^{-1} regions are probably correct but are of very limited value since organophosphorus compounds without P-C linkages usually also absorb in this range.[249] It is frequently possible, however, to identify a P-C linkage from its effect on the vibrations of neighboring linkages, and several characteristic frequencies have been derived by empirical correlation. The presence of a P-Me group in a phosphine is characterized by three absorption bands at 1420, 1295 to 1278, and 930 to 840 cm^{-1}.[249] Tertiary phosphines containing the >PMe group show a sharp band of medium strong intensity in 880 to 875 cm^{-1} region.[942] The P-Et linkage is characterized by two weak bands which usually lie within the limits 1282 to 1227 cm^{-1}. The P-Ph linkage shows two characteristic, sharp absorptions of medium-to-weak intensity in the regions 1450 to 1425 cm^{-1} and 1010 to 990 cm^{-1}.[249]

D.7. Dipole Moments

The pyramidal structures of tervalent phosphorus compounds lead to measurable dipole moments. A comparison between the dipole moments of PH_3, RPH_2, R_2PH, and R_3P shows that the moment increases with alkylsubstitution, as expected from consideration of the inductive effects of substituents.[784,1246,1414,1454b] Furthermore, the dipole moment of aliphatic substituted tertiary phosphines increases with the chain length of the alkyl group, for example, Me_3P 1.19 D,[857] Et_3P 1.35 D,[784] Pr_3P 1.48 D, Bu_3P 1.49 D[270] but soon reaches a constant value. Substitution of an alkyl group for a phenyl group decreases the dipole moment as a result of resonance between the phosphorus atom and the benzene nucleus.[778,1246] The dipole moment of para-substituted phenyldiphenylphosphines is increased by electron-releasing groups, for example, p-R_2N[440,1246] and reduced by electron-attracting groups, for example, p-Cl.[440]

D.8. Phosphorus-31 and Hydrogen-1 Nuclear Magnetic Resonance Spectra of Phosphines

NMR spectroscopy has proved to be a very valuable tool for the identification and characterization of primary, secondary, and tertiary phosphines. Hydrogen directly attached to phosphorus causes splitting of the shift by interaction of the hydrogen nucleus with the phosphorus-31 nucleus. Thus primary phosphines exhibit three peaks with the relative intensities of 1-2-1 and secondary phosphines two peaks with the relative intensities of 1-1. Tertiary phosphines give, of course, only one peak.[262]

The ^{31}P chemical shifts of phosphines vary widely and have been found in the range -62 ppm (t-Bu_3P) to +341 ppm (phosphirane). Normally, the shifts of primary phosphines are found in the range +110 to +160 ppm, secondary phosphines +40 to +100 ppm, and tertiary phosphines +5 ppm to +62 ppm.

Quantum mechanical treatments allow the calculations of phosphorus-31 chemical shifts.[262] However, the predictive utility of this method is rather limited because the parameters, such as bond angles, electronegativities, and average electron excitation energies, that are necessary for the calculations, are generally more tedious to obtain than the chemical shifts. The ^{31}P chemical shifts of phosphines may, however, be predicted rather accurately by using the following empirical equations:

Primary phosphines[924]: $\Delta(\text{ppm}) = 163.5 - 2.5\sigma^P$

Secondary phosphines[483]: $\Delta(\text{ppm}) = 99 - 1.5 \sum_{n=1}^{2} \sigma_n^P$

Tertiary phosphines[483]: $\Delta(\text{ppm}) = 62 - \sum_{n=1}^{3} \sigma_n^P$

The substituent constants σ^P for predicting phosphorus-31 chemical shifts are summarized in Table 2.

Table 2. σ^P Values[413,483,484,924]

Group	σ^P	Group	σ^P
NC	-24.5	$NCCH_2CH_2$	+13
Et_2NCH_2	-1.0	Et	+14
Me	0	$PhCH_2$	+17
Me_3CCH_2	+3	Ph	+18
$BuEtCHCH_2$	+5		
i-Bu	+6	$c\text{-}C_5H_9$	+21
$CH_2{=}CHCH_2$	+9		
n-Pr	+10	$c\text{-}C_6H_{11}$	+23
n-Bu	+10	sec-Bu	+24
n-Am	+10	i-Pr	+27
$n\text{-}C_8H_{17}$	+10	t-Am	+42
$n\text{-}C\ H_{2n+1}\ (n \geq 3)$	+10	t-Bu	+44

The ^1H NMR spectrum of triethylphosphine indicates that the lone electron pair of the phosphorus may be involved in the chemical bonding of the ethyl group to phosphorus.[1030,1318] The hydrogen NMR (and when present ^{19}F NMR) spectra of many primary, secondary, and tertiary phosphines as well as the sign and magnitude of coupling constants have been reported. Reference to this literature is given in the list of compounds (see also 638a).

D.9. Ionization and Appearance Potentials of Phosphines

Although a number of mass spectroscopic investigations of phosphines have been made (see list of compounds), only a few ionization and appearance potentials of phosphines have been reported. In particular, the values listed in Table 3 have been given.

Table 3. Ionization Potentials of Phosphines

Phosphine	Ionization Potential (eV)	Ref.	Phosphine	Ionization Potential (eV)	Ref.
PH_3	9.97[a]	874	Et_2PH	8.87	1394
$\cdot PH_2$	9.83	874	Me_3P[b]	8.6,	396,
$MePH_2$	9.72	1394		9.2	1394
$EtPH_2$	9.61	1394	$(CF_3)_3P$	11.3	268
Me_2PH	9.2	396	Et_3P	8.27	1394
$Me_2P\cdot$	∿8.3	874	Ph_3P	7.36	1378

[a]Appearance potential 13.47 eV.[874]
[b]Appearance potential 11.0 eV.[1347]

E. TRIVALENT PHOSPHORUS COMPOUNDS WITH COORDINATION NUMBER TWO OR ONE

I. PHOSPHACYANINES

Until recently, no phosphorus compound was known in which phosphorus had the coordination number two. In 1961, Burg and Mahler found[189] that the tetramer and pentamer $(CF_3P)_{4/5}$ react easily and reversibly with tertiary phosphines and tertiary amines to form compounds with, apparently, an "ylid" structure. The first isolable compounds in which phosphorus has the coordination number two

$$R_3P{=}PCF_3 \quad \text{and} \quad R_3N{=}PCF_3$$

were obtained in 1964 by Dimroth and Hoffmann.[318] They found that interaction of $(HOCH_2)_3P$ and 2-chlorobenzthiazolium salts or 2-chlorochinolinium salts in the presence of a base produces the orange-red phosphacyanines (75) the analogs of azacyanines. The phosphacyanines are also formed when $(HOCH_2)_3P$ is replaced by $(Me_3Si)_3P$.[893] X-ray

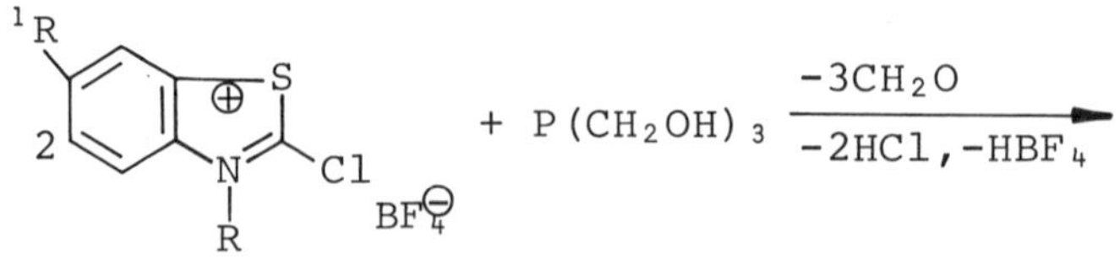

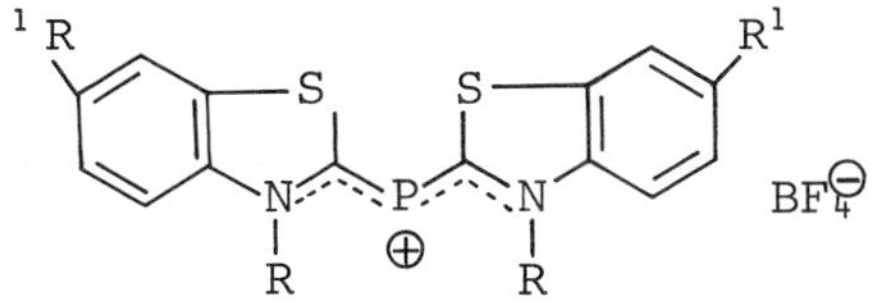

(75)

analysis confirms the structure of these compounds and shows that the cyanine cation in the benzthiazolium derivative (75), (R = Et, R^1 = H, X^- = ClO_4) is nearly planar with equal P-C distances of 1.76 Å and an angle at phosphorus of 104.6°.[45] In bis(1-ethyl-2-chinolyl)phosphamethincyanine perchlorate, however, the cyanine cation departs considerably from a planar structure,[769] probably because of steric interaction. This causes a lengthening of the P-C bonds (1.77 and 1.81 Å) which indicates that there is less resonance here.[769] The electronic spectra of the phosphacyanines are similar to that of the methin- and azamethincyanines, but the absorption maximum is shifted toward longer wavelength.[319] Oxidizing agents destroy the phosphacyanines, and acids seem to protonate the phosphorus atom, thereby interrupting the mesomeric resonance system.[319]

II. PHOSPHORINS

The first derivative of phosphorin, 2,4,6-triphenylphosphorin, was synthesized in 1966 by Märkl.[891] 2,4,6-Triphenylphosphorin (76) was formed in light-yellow needles when 2,4,6-pyrylium tetrafluoroborate was treated with tris(hydroxymethyl)phosphine in refluxing pyridine.[891] More recently, Märkl and coworkers found three more procedures that also produced phosphorin derivatives, that is, (a) interaction of $P(SiMe_3)_3$ and pyrylium iodides in acetonitrile,[892,896] (b) reaction of PH_3 with pyrylium salts in butanol in the presence of acids,[892,897] and (c) pyrolysis

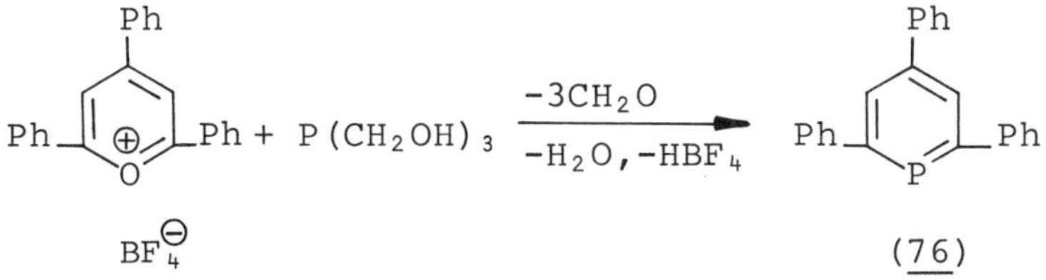

(76)

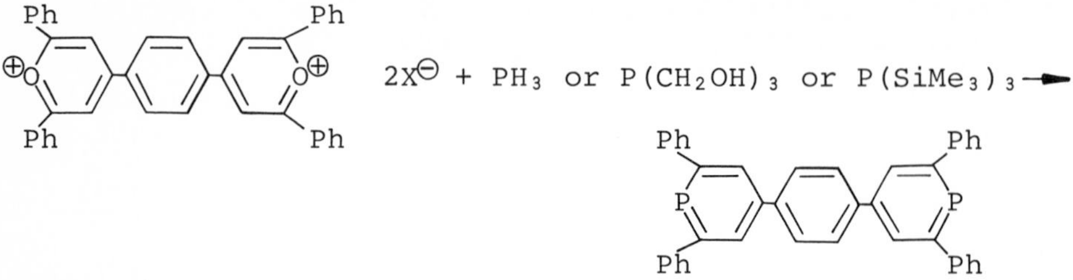

of some benzyl-substituted phosphines at 350°.[892] Numerous phosphorin derivatives have been prepared by using either one of these procedures.[316,320,395,891,892,896,897]

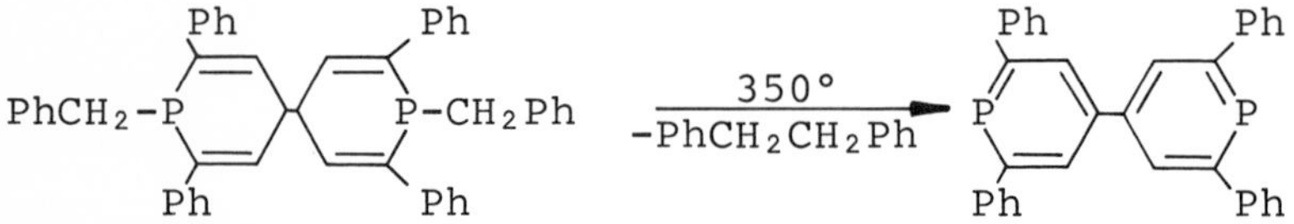

[1358] The ^{31}P chemical shift of (76) at -178.2 ppm indicates strong deshielding of the phosphorus atom. The absorption maximum of (76) in the UV spectrum is shifted toward longer wavelength when compared with 2,4,6-triphenylpyridine.[891] This is in qualitative agreement with the observation made with the phosphacyanines.[319] X-ray analysis of 2,6-dimethyl-4-phenylphosphorin confirms the proposed structure.[81] It shows that the phosphorus ring is planar and that the P-C distances are equal within error limits (1.74 Å),[81] thus indicating an aromatic system. A planar phosphorus ring has also been established in 2-5-butyl-4-(4-methoxyphenyl)-5,6-dihydronaphtho[1.2-b]-phosphorin (77) by x-ray analysis.[395] The P-C distance of 1.751 Å, an angle of 103° for C-P-C and an increase of 3.5° for the rest of the angles indicates an aromatic system.[395]

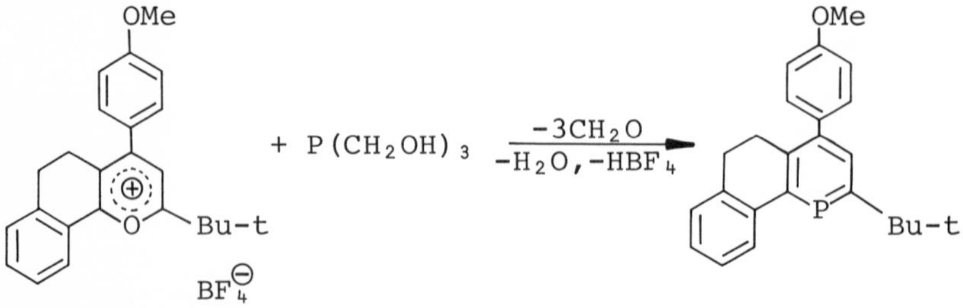

(77)

2,4,6-Triphenylphosphorin was not alkylated by methyl iodide or $[Et_3O]BF_4$. Attack on phosphorus occurs, however, with alkyl or aryllithium reagents and yields a resonance-stabilized anion (78).[895,898] Hydrolysis of the reaction mixture gave a dihydrophosphorin (79) which formed quaternary salts.[895] Dehydrohalogenation of the salt gave a new phosphorin with P(V) (80).[895] Phosphorins with P(V) could be made directly by treating the phosphorus anion with alkyl halides,[895,898] or by treating phosphorin (76) with tetraphenylhydrazine,[317] mercury diaryls,[899] or mercury (II) acetate.[321] As suggested,[1128] the phosphorin

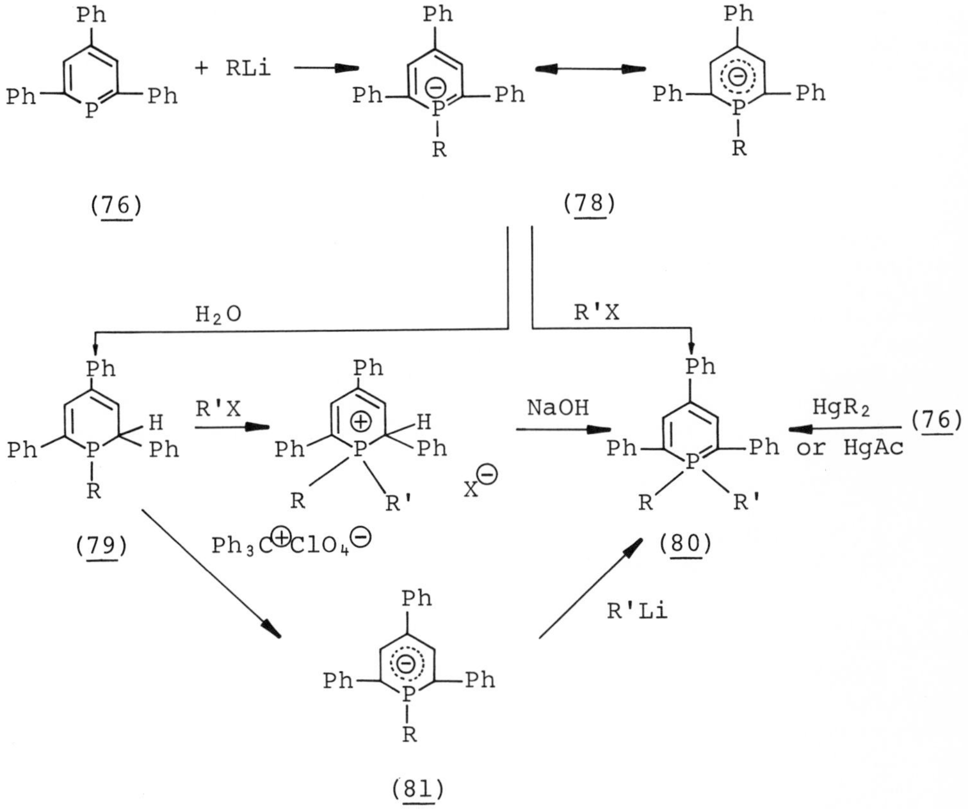

with P(V) (80) was also produced in the reaction of the resonance-stabilized phosphorinium ion (81) with PhLi.[899] Interaction of phosphorin (76) and hexafluorobut-2-yne at 100° gave the new heterocyclic system, 1-phospha-barrelene (82).[894]

2,4,6-Triphenylphosphorin is slowly oxidized by air in benzene solution to give apparently an acid and an anhydride for which structures (83) and (84) have been suggested[323] [but see Ref. 891 which reports that (76) is not autoxidable].

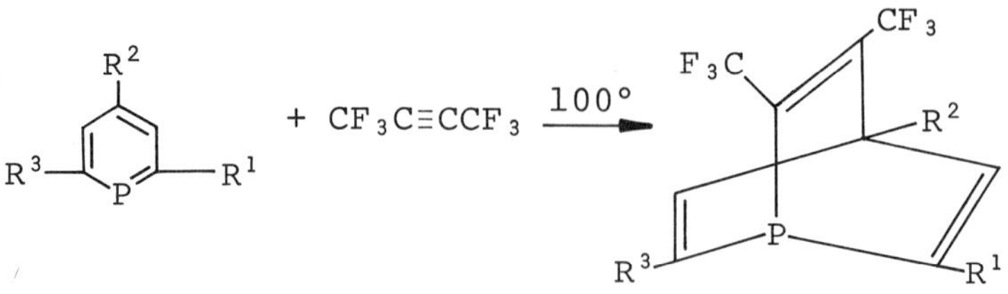

(82)

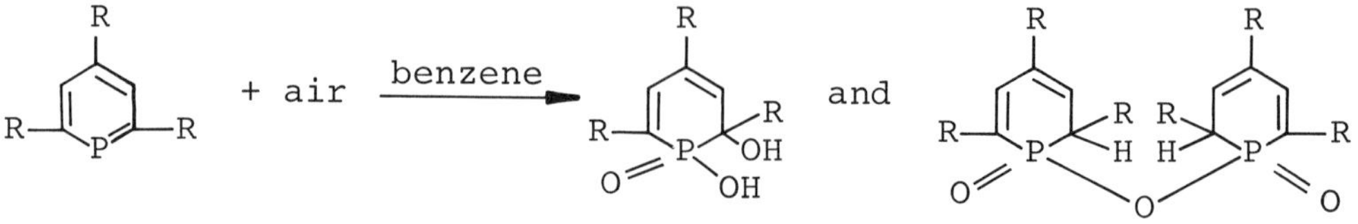

(83) (84)

Substituted phosphorins readily form stable radical cations by oxidation with triphenylphenoxy radical, lead(IV) acetate or benzoate, HgAc, and other oxidizing agents.[315, 316] Reduction of 2,4,6-triphenylphosphorin (76) is effected by metallic potassium or potassium-sodium alloy in THF solution.[322] Analysis of the reaction mixture by ESR spectroscopy is interpreted tentatively to indicate that this phosphorin (76) can take up three electrons in a stepwise fashion to give a monoanion radical, nonradical dianion, and a trianion radical.[322]

III. DIBENZOPHOSPHORINS

9-Phosphaanthracen[303] (85, R = H), the 10-phenyl derivative[305] (85, R = Ph), and 9-phosphaphenanthren (86)[304] were synthesized from the corresponding chlorodihydro derivatives by splitting off hydrogen chloride with a base. Whereas 9-phosphaanthracen (85, R = H) and 9-phosphaphenan-

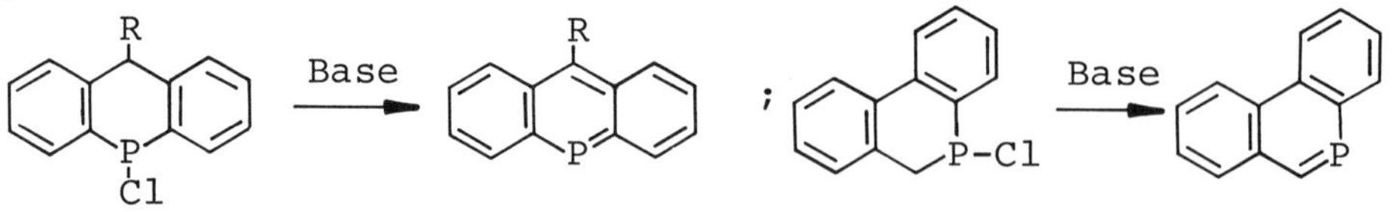

(85) (86)

thren (86) could not be isolated as such and were identified only in solution by their characteristic UV spectra--both show a bathochromic shift in comparison to the corresponding carbon and nitrogen analogs[303,304]--10-phenyl-9-phosphaanthracen (85, R = Ph) was obtained as yellow crystals, m. 173 to 176°. Its UV spectrum is very similar to that of the unsubstituted derivative (85) (R = H) but shows, as expected, a small bathochromic shift. The structure was further confirmed by analysis and mass and ^1H NMR spectroscopy.[305]

IV. METHINOPHOSPHIDE

Methinophosphide is the only phosphorus compound known that contains a P-C triple bond. It has been obtained in low yield by passing phosphine at 40-mm pressure through a low-intensity rotating arc struck between graphite electrodes contained in a water-cooled copper reactor.[429] Methinophosphide is a very reactive colorless gas, stable only below its triple point of -124° ± 2°. The IR spectrum of solid HCP exhibits C-H absorption at 3180 and 1265 cm^{-1} and P-C absorption at 671 cm^{-1}, but no absorption in the 2350 to 2440 cm^{-1} region characteristic of a P-H bond. Thus the structure of methinophosphide is HC≡P and not the isomeric HP≡C.[429] The microwave spectrum is in full agreement with this structure[1367] and gives for P-C 1.5421 Å, C-H 1.0667 Å. The molecule is linear.[1367] Reaction with hydrogen chloride gives methylphosphonous dichloride as the sole reaction product.[429]

$$HC{\equiv}P + 2HCl \longrightarrow CH_3PCl_2$$

F. LIST OF COMPOUNDS

The list of compounds is divided into four major types (primary phosphines, secondary phosphines, tertiary phosphines, and trivalent phosphorus compounds with coordination number two or one) and these primary divisions, very unequal in size, are subdivided into 36 structural classes arranged so as to bring together compounds with similar structures.

Within each structural class the compounds are listed in order of increasing complexity. Compounds containing functionally modified substituents are listed after the unmodified compound, for example, $EtPH_2$, $HCF_2CH_2PH_2$, $HCCl_2CF_2PH_2$, $H_2NCH_2CH_2PH_2$, $Et_2NCH_2CH_2PH_2$, $HOCH_2CH_2PH_2$, $MeOCH_2CH_2PH_2$, $HSCH_2CH_2PH_2$, $HO_2CCH_2PH_2$.

The entry for each compound indicates the methods by which it has been prepared (I to VIII for primary phosphines, and so on. The reference BII listed as a method for making, for example, a primary phosphine, means that this method is discussed in Section B.II), records values of common physical constants, and gives literature references to these and to more complex physical methods. The listing is as comprehensive as possible.

Primary Phosphines

$MePH_2$. I.[124,260,491,609,906,1008] II.[213,361,1401] III.[65,116,1457] VI.[589,592,594,740,935] In method II $(Me_2N)_3P{=}O$[213] and in VI Me_2SO[740] as solvents recommended. Colorless gas, b. -14°,[589] b_{760} -17.1° (calc.),[361] which liquifies at 0° and 1.75 atm,[589] log P_{mm} = 7.402 - (1158/T),[361] IR[860] (also of MePHD and $MePD_2$)[1019] (also of CD_3PH_2 and CD_3PD_2),[839] Raman,[1051] UV,[505] mass spect.,[1394] ionization potential (calc.) 9.72 eV,[1394] ^1H NMR,[380,945,946,1428] sign of geminal coupl. const. in $MePH_2$ and MePHD,[943] magnitude and signs of coupl. const.,[945,946,1428] ^{31}P +163.5 ppm (J_{PH} 188 Hz),[1376] +163.2 ppm (J_{PH} 187.8 Hz),[946] molecular polarizability,[862] μ 1.1 D;[790] electron diffraction gives pyramidal structure[82] with C-P 1.858, P-H 1.445, C-H 1.094 Å,[82] microwave spect.[790] gives C-P 1.863, P-H 1.414, C-H 1.093 Å, ∠C-P-H 97°30', H-P-H 93°23';[790] nonbonded electrons contribute little to the barrier heights of internal rotation;[530] forms crystalline, farily volatile salts with HCl and HI,[589] which dissociate easily;[186] very toxic.

$ClCH_2PH_2$. IV. Liquid, b. 68°,[398] IR,[398] ^1H NMR (J_{PH} 206 Hz, sign of J_{PH} and J_{PCH} is positive);[450] gave with NaOH, $MeP(OH)_2$, and with NaOMe or Et_3N, $MePH_2$,[398] no inversion up to 150°.[450]

CF_3PH_2. I.[208,905] IV.[100,905] VII.[905] A 90% yield is obtained from CF_3PI_2 or $(CF_3P)_{4/5}$ and HI + Hg.[208] Colorless gas, b. -25.5°; vapor pressure at -78.5° is 45 mm;[905] IR,[100] mass spect., UV;[208] flash photolysis suggests CF_2 elimination;[209] ^1H and ^{19}F NMR,[340a] sign of ^{19}F, ^{31}P coupl. const.,[1429] sign of geminal coupl. const. in CF_3PH_2 and CF_3PHD,[943] ^{31}P +129 ppm (J_{PH} 201 Hz);[340a] microwave spect.[1409] gives P-C 1.900, P-H 1.43 Å, ∠C-P-H 91.9°, ∠H-P-H 96.7°, ∠F-C-F 108°,[1409] μ 1.92 D.[1409]

$EtPH_2$. I.[491] II.[35,392,818] [$(Me_2N)_3P{=}O$ as solvent recommended].[213] III.[116] VI.[272,588,594] Liquid, b. 25°,[590] b_{748} 25°,[35,818] μ 1.17 D,[1246] μ 1.15 D,[784] mass spect.,[129,129a,1394] ionization potential (calc.) 9.61 eV,[1394] ^{31}P +128 ppm (J_{PH} 185 Hz);[412] forms crystalline salts with HCl and HI, which are stable only in the concentrated acids.[590]

$HCF_2CH_2PH_2$. C.XII.[178,1080] Liquid, b. 52 to 53.5°,[1080] b_{758} 52 to 53°,[178] ^1H and ^{19}F NMR.[340a]

$HCCl_2CF_2PH_2$. C.XII. Liquid, b. 109.5 to 110.5°.[1080]

$HCFClCF_2PH_2$. C.XII. Liquid, b. 67°,[1080] b_{746} 66 to 67°,[178] ^1H and ^{19}F NMR, ^{31}P +140.3 ppm.[340]

$HCF_2CF_2PH_2$. C.XII. Liquid, b. 20 to 22°,[1080] b_{760} 20.7° (calc.), ΔH_{vap} 6.45 kcal/mole.[178]

$C_2H_2F_4PH_2$. C.XII. Isomer mixture consisting of 85% CHF_2-$CHFPH_2$ and 15% $CH_2FCF_2PH_2$, liquid, b_{759} 56 to 59°,[385] ^1H and ^{19}F NMR.[340a,385]

$H_2NCH_2CH_2PH_2$. II. Liquid, b. 109.5 to 110.5°, ^{31}P +150.4 ppm (J_{PH} 194 Hz), pK_a 8.77.[681]

$EtHNCH_2CH_2PH_2$. II. Liquid, b. 135 to 136°, pK_a 8.85.[681]

$Et_2NCH_2CH_2PH_2$. II. Liquid, b. 155 to 156°, pK_a 8.17.[681]

$HOCH_2CH_2PH_2$. C.XIII.6. Liquid, b. 139 to 140°, b_{45} 70 to 73°, n_D^{20} 1.4950, d_4^{20} 1.004;[782] O-acetyl deriv., from $NaOCH_2CH_2PH_2$ + MeCOCl, b_{69} 73°, $b_{9\text{ to }10}$ 37 to 38°, n_D^{20} 1.4620, d_4^{20} 1.0250;[782] O-benzoyl deriv. similarly obtained, b_{20} 142 to 144°.[782]

$MeOCH_2CH_2PH_2$. C.XII. Liquid, b. 78 to 85.5°.[87]

$BuOCH_2CH_2PH_2$. C.XII. Liquid, b_9 47°, n_D^{25} 1.4478.[1171]

t-$BuOCH_2CH_2PH_2$. II. Liquid, b_{28} 52 to 55°, $L\cdot Me^+BPh_4^-$, m. 133 to 135°.[1308]

$HSCH_2CH_2PH_2$. II. Liquid, b. 131 to 133°, n_D^{20} 1.5442.[428]

$HO_2CCH_2PH_2$. II. (From H_2PNa + $ClCH_2CO_2Na$). Liquid, b_6 70 to 72°,[682] b_{10} 85 to 86,[680] pK_a 4.07, ^1H NMR, ^{31}P +142.6 ppm (J_{PH} 198 Hz);[680] adducts: L·HI, m. 112°,[682] L·HBr, m. 92 to 93°;[680] gives soluble Na, Zn, and Mg salts;[680] lowers hydrogen overvoltage.[690] The following esters, $RO_2CCH_2PH_2$, were prepared:

(R = Me). From acid and CH_2N_2, liquid, b_{10} 27 to 28°,[682] b. 129 to 130°.[680]

(R = Et). From acid and Et_2SO_4, liquid, b_{25} 53 to 55°.[680]

(R = Me_3Si). From acid and $ClSiMe_3$, liquid, b_9 49.5 to 50.5°.[680]

By treating the anhydride $EtO_2COC(O)CH_2PH_2$, obtained from $NaO_2CCH_2PH_2$ and $ClCO_2Et$, with amines, the following amides, $RNHCOCH_2PH_2$ were prepared:[680]

(R = EtO_2CCH_2). Colorless plates, m. 67 to 69° (from AcOEt/hexane), $b_{0.01}$ 130 to 133°.[680]

(R = Ph). Colorless crystals, m. 86 to 87° (from $EtOH/H_2O$).[680]

(R = 2-$HO_2CC_6H_4$). Colorless needles, m. 131 to 133° (from $EtOH/H_2O$).[680]

(R = $HO_2C(PhCH_2)CH$). Colorless crystals, m. 142.3° (from $EtOH/H_2O$).[680]

(R = 1,4-$HO_2CC_6H_4NHCOCH_2$). Colorless crystals, m. 245 to 247° (dec.) (from EtOH).[680]

(R = $HO_2CCH_2NHCOC_6H_4$(-p)-). Colorless crystals, m. 216 to 219° (dec.) (from EtOH).[680]

(R = 1,4-$HO_2CC_6H_4$-). Colorless crystals, m. 212 to 214° (from $EtOH/H_2O$).[680]

$Et_2NCOCH_2PH_2$. II. Liquid, b_2 83 to 86°.[680]

$ClCH_2COPH_2$. C.XIII.6. Unstable powder.[1328]

$Cl_2CHCOPH_2$. C.XIII.6. Yellow powder, dec. 200° (from $Et_2O/EtOH$).[360]

Cl_3CCOPH_2. C.XIII.6. Crystalline powder.[234]

$H_3^{31}SiCH_2CH_2PH_2$. From fast-neutron irradiation of PH_3 with ethylene (tentatively identified).[426]
$PrPH_2$. II.[818] III.[1067] VI.[1086] Liquid, b. 53 to 53.5°,[1086] b_{750} 54°,[818,1067]
i-$PrPH_2$. I.[537] VI.[591] Liquid, b. 41°,[591] ^{31}P +106 ppm (J_{PH} 195 Hz).[537]
$CF_3(CHF_2)CFPH_2$. C.XII. (Thermal addition), liquid, b. 45 to 48°.[1080]
$C_3HF_6PH_2$. C.XII. (UV irradiation), isomer mixture consisting of 66% $CF_3CHFCF_2PH_2$ and 34% $CF_3CF(PH_2)CHF_2$, liquid, b_{768} 47 to 48°.[178]
$ClCF_2CHClCF_2PH_2$. C.XII. Not obtained completely pure; condenses at -35° under vacuum; IR, ^1H NMR (J_{PH} 203 Hz).[124]
$CH_2{=}C{\cdot}MePH_2$. C.XII. From PH_3 and allene, liquid, b. 29° (est.),[447] log P_{mm} = 6.80 - (1185/T), IR, ^1H NMR.[447]
$CH_2{=}CHCH_2PH_2$. II. $(Me_2N)_3P{=}O$ used as solvent; purified by GLC; ^1H NMR.[213]
$HOCH_2CH_2CH_2PH_2$. II.[782] C.XII.[87,1338] Liquid, b. 139 to 140°,[782] b_{32} 79 to 85,[1338] n_D^{20} 1.4950, d_4^{20} 1.004.[782]
$HOCH_2CH{\cdot}MePH_2$. By $LiAlH_4$ reduction of $EtO_2CCH{\cdot}MePH_2$, liquid, b_{15} 54 to 56°.[680]
$Me{\cdot}CHOH{\cdot}CH_2PH_2$. C.XIII.6. Liquid, b_2 37 to 39°, n_D^{20} 1.4863, d_4^{20} 0.9764.[782]
$(CF_3)_2C{\cdot}OHPH_2$. C.XIII.3. Liquid, b. 74 to 76°, n_D^{20} 1.351°, d_4^{20} 1.5636.[163]
$Me(CF_3)C{\cdot}OHPH_2$. C.XIII.3. Liquid, b_{24} 25°, b_{55} 36 to 37°, n_D^{20} 1.4030, d_4^{20} 1.3024;[488] decomposes on storage; O-acetyl deriv. (from phosphine + MeCOCl), liquid, b_2 94 to 95°, n_D^{20} 1.4581, d_4^{20} 1.3593, IR.[488]
$NCCH_2CH_2PH_2$. C.XII. Liquid, b_9 54 to 55°, n_D^{25} 1.4831,[1173] ^{31}P +135 ppm (J_{PH} 195 Hz).[262,924]
$H_2NCH_2CH_2CH_2PH_2$. C.XII. Liquid, b_{52} 58.5 to 62°.[87,1338]
$ClCH_2CH_2CH_2PH_2$. C.XII. Liquid, b. 125°.[87,1338]
$HO_2CCH_2CH_2PH_2$. II. Liquid, b_6 86 to 87°;[680,682] solidifies at room temp.,[680] pK_a 4.23, ^{31}P +136.9 ppm (J_{PH} 194 Hz),[680] L·HI, m. 112°.[680] The following esters were prepared, $RO_2CCH_2CH_2PH_2$:
(R = Me). From acid + CH_2N_2[682] or from PH_3 and acrylate, liquid, b_{53} 74.5°,[87] b_{10} 37 to 38°.[680,682]
(R = Et). From PH_3 and acrylate, liquid, b_9 52 to 53°, n_D^{25} 1.4552.[1171]
(R = Me_3Si). From acid + $ClSiMe_3$, liquid, b_7 47 to 48°.[680]
By treating the anhydride, $EtO_2COC(O)CH_2CH_2PH_2$, obtained from $NaO_2CCH_2CH_2PH_2$ and $ClCO_2Et$, with Et_2NH, the amide was obtained:
$Et_2NCOCH_2CH_2PH_2$. Liquid, $b_{1.5}$ 87 to 88°.[680]
$HO_2CCH{\cdot}MePH_2$. II. Liquid, b_5 73 to 74°,[682] b_6 79 to 80°, pK_a 4.15.[680] Methyl ester, $MeO_2CCH{\cdot}MePH_2$, from acid + CH_2N_2, liquid, b_{13} 33 to 34°,[682] b_{10} 34 to 35°.[680]

By treating the anhydride, $EtO_2COC(O)CH{\cdot}MePH_2$, obtained from $NaO_2CCH{\cdot}MePH_2$ and $ClCO_2Et$, with amines, amides were obtained:

$H_2NCOCH{\cdot}MePH_2$. Colorless plates, m. 138 to 140° (from $EtOH/H_2O$).[680]

$Et_2NCOCH{\cdot}MePH_2$. Liquid, b_2 88 to 90°.[680]

1,4-$HO_2CC_6H_4NHCOCH{\cdot}MePH_2$. Colorless needles, m. 224 to 226° (dec.) (from $EtOH/H_2O$).[680]

$BuPH_2$. I.[410,1287] II.[1252] III.[509,745,1176] V.[1085] C.XII.[87,1338] Liquid, b. 86.7°,[745] b. 86.2 to 87.8°,[87,1338] b_{760} 76°,[1085,1252] (might be an isomeric mixture), n_D^{20} 1.4477,[1338] n_D^{20} 1.4372, d_4^{20} 0.7693,[1085,1252] IR,[1252] ^{31}P +140 ppm,[537] +135 ppm (J_{PH} 195 Hz),[262] μ 1.36 D,[1246] pK_a −0.03 (in H_2O),[1342] mag. rotation.[1385]

i-$BuPH_2$. I.[410] II.[818,1252] VI.[591] C.XII.[87,1171,1338] Liquid, b_{756} 63°,[818] b. 77°,[1171] b. 78 to 79.6°,[1338] b. 62°,[591] b_{760} 60°,[410] n_D^{25} 1.4308,[1171] d_4^{20} 0.7693,[1252] ^{31}P +151 ppm,[537] pK_a (in H_2O) −0.02.[1342]

2-$BuPH_2$. I.[919] C.XII.[87,1338] Liquid, b. 66 to 67°,[87,1338] ^{31}P +120.5 ppm (J_{PH} 182 Hz),[919] +116 ppm.[537]

t-$BuPH_2$. II.[1252] V.[1085] (Probably gives an isomeric mixture[537]). Liquid, b_{760} 54°, n_D^{20} 1.4252, d_4^{20} 0.7360,[1085,1252] IR,[1252] ^{31}P +82 ppm.[537]

$Me_2CHCF_2PH_2$. C.XII. Liquid, b. 75 to 77°.[1080]

$\overline{(CF_2)_3C}{\cdot}OHPH_2$. C.XIII.3. Liquid, b. 96 to 97°, ^1H and ^{19}F NMR.[1078]

$HO_2C(CH_2)_3PH_2$. II. Liquid, b_5 99 to 101°, pK_a 4.53, ^{31}P +138 ppm (J_{PH} 195 Hz);[680] methyl ester (from phosphine and CH_2N_2), liquid, b_8 51 to 52°.[680]

$HO_2CCH{\cdot}EtPH_2$. II. Liquid, b_6 87 to 89°,[680,682] pK_a 4.08, ^{31}P +120.8 ppm (J_{PH} 194 Hz).[680]

The following esters were prepared, $RO_2CCH{\cdot}EtPH_2$:

(R = Me). From phosphine + CH_2N_2, liquid, b_{10} 39 to 41°.[680,682]

(R = Pr). From phosphine + PrOH + H_2SO_4, liquid, b_5 69 to 72°.[680]

$HO_2CCMe_2PH_2$. II. Liquid, b_6 84 to 85°;[680] methyl ester (with CH_2N_2), liquid, b_{10} 35 to 37°.[680]

$AmPH_2$. I.[1083] V.[1085] Liquid, b_{760} 104°,[1083,1085] n_D^{20} 1.4129,[1083] n_D^{20} 1.4431, d_4^{20} 0.7796,[1085] mass spect.,[537] ^{31}P +139 ppm (J_{PH} 192 Hz),[537] mag. rotation;[1385] the phosphine prepared by method V is probably an isomeric mixture.

i-$AmPH_2$. III.[65] IV.[495] VI.[591] Liquid, b. 106 to 107°.[591]

c-$C_5H_9PH_2$. I.[1252] V.[1085] Liquid, b_{760} 121°, n_D^{20} 1.4899, d_4^{20} 0.8818,[1085,1252] IR.[1252]

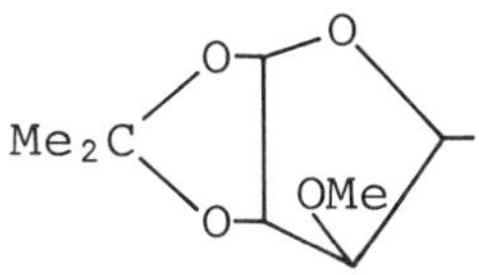

CH_2PH_2. I.[1427] This xylofuranose derivative rearranges in the presence of acids to a xylopyranose derivative,

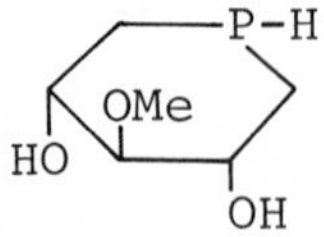

, which was identified as the oxide, m. 208 to 210°, $[\alpha]_D^{25}$ +35°.[1427]

$n-C_6H_{13}PH_2$. I.[1083,1252] Liquid, b_{760} 128°, n_D^{20} 1.4527, d_4^{20} 0.7909,[1083,1252] IR,[1252] μ 1.48 D,[1242] μ 1.34 D,[1246] mag. rotation.[1385]

$c-C_6H_{11}PH_2$. I.[609,619,699,715,1083,1252] II.[715] III.[509,1067] V.[1085] C.XII.[87,1171,1338] Liquid, b. 146.8 to 149.4°,[87,1338] b_{760} 138°,[1085,1252] b_{760} 146°,[609,1083] b_{760} 144 to 145°,[715,1171] b_{160} 97°,[1067] b_9 30 to 31°,[715] b_{15} 43°,[699] n_D^{20} 1.4860,[1085,1252] n_D^{25} 1.4822,[1171] d_4^{20} 0.8750,[1085,1252] IR,[1085] ^{31}P +110 ppm (J_{PH} 162 Hz),[262,924] relative pK_a 32.3 (referred to MeOH).[677]

MeC≡C-C(Me)(OH)CH_2PH_2. II. Liquid, b_1 50 to 51°, n_D^{20} 1.5061, d_4^{20} 0.9880.[1094]

$n-C_7H_{15}PH_2$. II.[1252,1411,1412] Liquid, b_{760} 169.5°,[1411] b_{30} 73 to 74°,[1411] b_{760} 150°,[1252] n_D^{20} 1.4517, d_4^{20} 0.8002,[1252] IR,[1252] mag. rotation.[1385]

$n-C_8H_{17}PH_2$. I.[919] II.[1252] V.[1085] (Probably isomeric mixture). VI.[1010] C.XII.[1087a,1171] Liquid, b_{760} 169°,[1085,1252] b. 184 to 187°,[1010] b_7 66 to 68°,[1171] b_5 58 to 59.5°,[919] n_D^{25} 1.4539,[1171] n_D^{20} 1.4548, d_D^{20} 0.8082,[1085,1252] d_4^{17} 0.8209,[1010] IR,[1252] ^{31}P +128.5 ppm (J_{PH} 183 Hz),[919] pK_a (in H_2O) 0.43;[1342] salt with HI, crystals, soluble in ether;[1010] reduction of $C_8H_{17}P(O)Cl_2$ prepared by the Kinnear-Perren reaction gave four isomeric octylphosphines.[919]

BuEtCH·CH_2PH_2. C.XII. Liquid, b. 170°.[157]

$n-C_9H_{19}PH_2$. II.[1252] V.[1085] Liquid, b_{760} 187°, n_D^{20} 1.4571, d_4^{20} 0.8122,[1085,1252] IR.[1252]

$5-C_9H_{19}PH_2$. I. Liquid, GLC, mass spect. ^{31}P +116 ppm (J_{PH} 190 Hz).[537]

BuC≡C-C(Me)(OH)CH_2PH_2. II. Liquid, $b_{0.5}$ 64 to 65°, n_D^{20} 1.4948, d_4^{20} 0.9360.[1094]

$n-C_{10}H_{21}PH_2$. I.[535] II.[1252] V.[1085] Liquid, b_{760} 203°,[1085] b_{100} 145°,[1252] n_D^{20} 1.4591, d_4^{20} 0.8163,[1085 1252] IR,[1252] GLC,[535] ^{31}P +139 ppm (J_{PH} 190 Hz), ^1H NMR.[535]

(1-Adamantyl)PH_2. I. Liquid, b_1 75°.[1337]

$n-C_{12}H_{25}PH_2$. I.[535,1083,1252] V.[535,1085] C.XII.[1171] Liquid, $b_{0.8}$ 92°,[933] b_{100} 176°,[1252] $b_{0.1}$ 80°,[1171] b_5 141°,[1083] n_D^{25} 1.4585,[1171] n_D^{20} 1.4622, d_4^{20} 0.8227,[1083,1252] ^1H NMR,[535] IR,[535,1252] ^{31}P +135.7 ppm (J_{PH} 180 Hz),[924] +139 ppm (J_{PH} 190 Hz),[535] μ 1.37 D;[1246] method V gave an isomeric mixture of dodecylphosphine, b_2

99 to 102°, n_D^{20} 1.4599, which contained at least seven components and showed the following ^{31}P: +87.8, +95.0, +103.8, and +123.6 ppm.[535]

n-$C_{14}H_{29}PH_2$. I.[1252] V.[1085] (Probably isomeric mixture). Liquid, b_{100} 199°,[1252] b_{50} 182°,[1085] n_D^{20} 1.4632, d_4^{20} 0.8268, IR.[1252]

n-$C_{16}H_{33}PH_2$. I.[1252] V.[1085] (Probably isomeric mixture). Liquid, b_{12} 167°,[1252] b_2 121°,[1085] n_D^{20} 1.4648, d_4^{20} 0.8281, IR.[1252]

n-$C_{18}H_{37}PH_2$. I.[1252] V.[1085] (Probably isomeric mixture). Liquid, b_{10} 188°,[1252] b_{20} 204°, $b_{1.5}$ 138°,[1085] m. -15°,[1085] n_D^{20} 1.4651, d_4^{20} 0.8289, IR.[1252]

$PhCH_2PH_2$. I.[619,924] II.[818] VI.[590,850] Liquid, b_{760} 180°,[590,619,818,850] b_{12} 65°,[933] ^{31}P +120.9 ppm (J_{PH} 184.5 Hz);[924] forms crystalline salts with HCl and HI.[590,850]

$PhCH_2CH_2PH_2$. I.[131] C.XII.[1171] Liquid, b_8 75°,[1171] b_3 73°,[131] n_D^{25} 1.5494,[1171] n_D^{20} 1.5565, d_4^{20} 0.9690.[131]

MeCH•$PhPH_2$. I. Liquid, $b_{2-2.5}$ 44°, ^1H NMR (J_{PH} 186 Hz);[416] MeCH•PhPHD was obtained by using $LiAlH_4/LiAlD_4$ = 1:1 as reducing agent; inversion of the pyramidal structure of the P atom is slow on the NMR scale, and a lower limit of 26 kcal/mole was estimated for $\Delta F^{\ddagger}$.[416]

$PhCH{=}CHPH_2$. I. Liquid, b_2 82°, n_D^{20} 1.6180, d_4^{20} 1.0213, IR, ^1H NMR,[131] storable for brief periods; polymerizes readily.

$PhPH_2$. I.[123,128,400,408,410,609,617,618,636,819,1083,1211,1252] II.[1412] III.[1175,1176,1388] IV.[846,982,983,1418,1419] V.[1085] Liquid, b_{760} 160°,[846,982,983,1252] b_{760} 157 to 158°,[715] b_{14} 60°,[410] b_{10-11} 40 to 41°,[715] n_D^{20} 1.5796,[1252] d_4^{15} 1.001,[846] IR,[48,456,475,1252] Raman,[48,456] UV,[123,1236] ^1H NMR,[415a] sign of geminal coupl. const. in PhPHD and $PhPH_2$,[943] ^2D NMR of $PhPD_2$.[415a] ^{31}P +118.7 ppm (J_{PH} 195 Hz),[412] +122 ppm (J_{PH} 195 Hz)[933] +123.5 ppm (J_{PH} 201 Hz),[300] +123.8 ppm (J_{PH} 198 Hz);[411] deuterated phenylphosphines show an isotope effect; PhPHD: ^{31}P + 125.0 ppm (J_{PH} 203 Hz, J_{PD} 31 Hz)[411] and $PhPD_2$: ^{31}P +127.8 ppm (J_{PD} 31 Hz);[411] in the mass spect. "PhP" is generated;[1464] μ 1.11 D,[1242,1246] ESR upon irradiation of $PhPH_2$,[1345] relative pK_a 24.5 (referred to MeOH).[677]

$C_6F_5PH_2$. I. Liquid, b_{16} 37°,[388] IR,[389] ^1H and ^{19}F NMR,[79] ^{31}P +183.1 ppm (J_{PH} 216 Hz).[388]

3-$FC_6H_4PH_2$. I. Liquid, b_{21} 52 to 53°, n_D^{20} 1.5366,[1251] σ_m (of H_2P group) +0.05[1251] ^{19}F NMR.[1251]

4-$FC_6H_4PH_2$. I. Liquid, b_{760} 154 to 155°,[1243] b_{11} 38 to 40°,[932] n_D^{20} 1.5338,[1243] ^1H NMR,[932] ^{19}F NMR,[1243] ^{31}P +125.4 ppm (J_{PH} 199 Hz),[932] σ_p (of H_2P group) +0.305.[1243]

2-$ClC_6H_4PH_2$. I. Liquid, b_3 54 to 55°, IR, ^1H NMR, ^{31}P +127.5 ppm (J_{PH} 204 Hz).[932]

3-$ClC_6H_4PH_2$. I. Liquid, b_{12} 77 to 78°, n_D^{20} 1.5953, d_4^{20}

1.1798.[864]
4-$ClC_6H_4PH_2$. I.[928,932] IV.[984,1362] Crystals, b_{760} 198 to 200°,[984] b_3 54 to 55°,[932] m. 17°,[984] m. 31 to 32°,[928,1362] IR,[928] ^1H NMR, ^{31}P +124.1 ppm (J_{PH} 201 Hz).[932]
2-$BrC_6H_4PH_2$. I. Liquid, b_1 53 to 56°, ^1H NMR, ^{31}P +130.6 ppm (J_{PH} 204 Hz).[932]
4-$BrC_6H_4PH_2$. I.[932] IV.[984] Crystals, m. 40°,[984] m. 50°,[932] b_{760} 195 to 196°,[984] b_1 53 to 56°, ^1H NMR, ^{31}P +124.0 ppm (J_{PH} 201 Hz).[932]
2-$MeC_6H_4PH_2$. I.[928,932,1252] IV.[991] Liquid, b_{760} 178°,[991,1252] b_{11} 59 to 61°,[932] $b_{0.5}$ 52 to 54°,[928] m. 4°,[991] ^1H NMR, ^{31}P +130.9 ppm (J_{PH} 200 Hz).[932]
4-$MeC_6H_4PH_2$. I. Liquid, b_{11} 59 to 61°,[932] μ 1.41 D,[1242,1246] UV,[1236] ^1H NMR, ^{31}P +124.5 ppm (J_{PH} 200 Hz).[932]
4-$Me_2NC_6H_4PH_2$. I. Crystals, b_1 55 to 60°,[932] ^1H NMR, ^{31}P +126.1 ppm (J_{PH} 202 Hz).[932]
4-$MeOC_6H_4PH_2$. I.[123,932] Liquid, b_{11} 89 to 91°,[932] UV,[123] ^1H NMR, ^{31}P +125.8 ppm (J_{PH} 198 Hz).[932]
4-$EtOC_6H_4PH_2$. I. Liquid, b_{11} 99 to 101°, ^1H NMR, ^{31}P +125.5 ppm (J_{PH} 199 Hz).[932]
4-$EtC_6H_4PH_2$. I.[1252] IV.[984] Liquid, b_{760} 200°,[984] b_{12} 97°,[1252] n_D^{20} 1.5512, UV,[1236] IR;[1252] salt with HI, m. 118°.[984]
4-$PhCH_2C_6H_4PH_2$. IV. Crystals, m. 46°, b_{20} 184°;[986] salt with HI, m. 134°.[986]
4-$PhCH_2CH_2C_6H_4PH_2$. IV. Crystals, m. 75°, b_{45} 190°;[986] salt with HI, crystals.
2-Me-5-i-$PrC_6H_3PH_2$. II. Liquid, b_{10} 68 to 70°.[1343]
3-Me-6-i-$PrC_6H_3PH_2$. II. Liquid, b_{10} 55 to 58°.[1343]
2,4,5-$Me_3C_6H_2PH_2$. IV.[985] Liquid, b. 214 to 218°.[985]
2,4,6-$Me_3C_6H_2PH_2$. IV. Needles, m. 40°, b_{25} 125°.[985]
2-(5,6,7,8-Tetrahydronaphthyl)PH_2. I. Liquid, $b_{0.05-0.1}$ 71°, n_D^{20} 1.590, d_4^{23} 1.038.[638]

Diprimary Phosphines

$H_2PCH_2PH_2$. I. Liquid, b_{720} 76 to 77°,[921] b. 83 to 88°,[537] IR, GLC,[537] ^{31}P +121.8 ppm (J_{PH} 190 Hz),[921] +126 ppm (J_{PH} 199 Hz).[537]
$H_2PCH_2CH_2PH_2$. I. Liquid, b_{760} 113.2° (calc.), log P_{mm} = 8.02813 - (1988.6/T),[921] m. -62.5°, IR, ^1H NMR, ^{31}P +130.8 ppm (J_{PH} 193 Hz);[921] a report that this phosphine is unstable above -78°[847] could not be confirmed.[921]
$H_2PCF_2CF_2PH_2$. C.XII. Liquid, b_{755} 74.5 to 76°.[178]
$H_2PCFClCF_2PH_2$. C.XII. Liquid, b. 107 to 109°.[1080]
$H_2PCCl_2CF_2PH_2$. C.XII. Liquid, b. ∿140°.[1080]
$H_2PCH_2CH_2CH_2PH_2$. I. Liquid, b_{760} 129.1° (calc.), log P_{mm} = 8.11275 - (2104.5/T),[921] IR, ^1H NMR, ^{31}P +138.6 ppm (J_{PH} 190 Hz),[921] flow point -154 to -153°; yields a monosalt with H_3PO_4.[921]

$H_2PCMe_2PH_2$. I. Liquid, ^1H NMR, ^{31}P +74 ppm (J_{PH} 192 Hz).[537]

$H_2P(CH_2)_4PH_2$. I.[921,1212] II.[1399,1400] Liquid, b_{13} 57 to 58°,[1212] $b_{2.2}$ 25°,[1399,1400] b_{13} 64.5°,[921] b_{760} 196.1° (calc.),[921] log P_{mm} = 6.89769 - (1884.7/T), m. -52.5°, IR, ^{31}P +137.0 ppm (J_{PH} 192 Hz).[921]

$H_2PCH_2CH_2OCH_2CH_2PH_2$. C.XII. (From PH_3 and CH_2=CHOCH=CH_2, UV). Liquid, $b_{0.5}$ 83 to 85°,[1352] IR, ^1H NMR (J_{PH} 193 Hz), mass spect.[1352]

$H_2P(CH_2)_5PH_2$. I. Liquid, b_1 32 to 35°, ^{31}P +136.5 ppm (J_{PH} 191 Hz).[933]

$H_2PCH{\cdot}BuPH_2$. I. Liquid, b_9 53 to 55°, ^{31}P +109 ppm (J_{PH} 194 Hz);[537] gave as by-product $AmPH_2$.

$H_2P(CH_2)_6PH_2$. I. Liquid, b_1 78°, n_D^{20} 1.5058.[1212]

$H_2PCH(CO_2H)CH_2CH_2(HO_2C)CHPH_2$. II.[680] Crystals, m. 183 to 184° (from H_2O), pK_a 3.54 and 4.49, and m. 152 to 154° (from H_2O) (diastereomers), pK_a 3.62 and 4.41.[680]

$H_2PCBu_2PH_2$. I. Liquid, b_9 94 to 95°, ^{31}P +91 ppm (J_{PH} 194 Hz).[537]

$H_2P(CH_2)_{10}PH_2$. V. Liquid, $b_{0.5}$ 75°.[1470]

$H_2PCH{\cdot}C_9H_{19}PH_2$. I. Liquid,$_{4-5}$ 103 to 105°, ^{31}P +105 ppm (J_{PH} 191 Hz),[537] GLC.[537]

$H_2PCH{\cdot}C_{10}H_{21}PH_2$. I. Liquid, $b_{0.1}$ 88 to 90°, ^{31}P +108 ppm (J_{PH} 192 Hz), GLC.[537]

$H_2PCH{\cdot}C_{11}H_{23}PH_2$. I. Liquid, $b_{0.1}$ 103 to 105°, ^{31}P +107 ppm (J_{PH} 193 Hz), GLC.[537]

1,4-$H_2PC_6H_4PH_2$. I. Solid, m. 69 to 70.5°, b_{760} 226° (calc.),[362] m. 70°,[77] log P_{mm} (liquid) = 8.168 - (2640/T),[362] IR.[362]

Secondary Phosphines

TYPE: R_2PH

Me_2PH. I.[124,181,182,194,1079] II.[741,1401] III.[260,281,589,935] IV.[709,1050] VII.[913,745] Liquid, b_{720} 20°,[913] vapor pressure at 0° is 338 mm, b_{760} 21.1° (calc.), log P_{mm} = 7.539 - (1370/T),[194,281] IR,[89a,183] Raman,[89a] UV,[385a,505] mass spect.,[396] ionization potential 9.7 eV,[396] ^1H NMR,[380,945,1428] magnitude and sign of coupl. const.[945,946,1428] ^{31}P +99.5 ppm (J_{PH} 188 Hz),[1376] +99.5 ppm (J_{PH} 191 Hz),[946] μ 1.23 D;[784,1035] electron diffraction[82] gives P-C 1.853, C-H 1.097, P-H 1.445 Å, ∠C-P-C 99.2°, ∠C-P-H 96.5° (assumed); configuration of Me groups is staggered;[82] microwave spect.[1035] gives P-C 1.848, C-H 1.093, P-H 1.419 Å, ∠C-P-C 99.4°, ∠C-P-H 96.5°.[1035]

$(CF_3)_2PH$. VI.[100,184,186,188,190] The highest yield was obtained from $(CF_3)_2PI$ or $[(CF_3)_2P]_2$, HI, and mercury (90%).[208,265] Colorless gas, b_{760} 1° (calc.), log P_{mm} = 6.9740 - 0.008310T + 1.75 log T - (1667/T),[184] m.

-137°,[184] IR,[100] UV,[209,385a] ^{19}F NMR,[340a,384,385a,1070] sign of coupl. const.,[944] ^{1}H NMR,[265,340a,384,385a] solvent dependence of coupl. const.[384] ^{31}P +50.7 ppm (J_{PH} 218 Hz),[340a] +49.8 ppm (J_{PH} 217 Hz);[384,385a] flash photolysis suggests "CF_2" elimination.[209]

$(Et_2NCH_2)_2PH$. VII. Liquid, $b_{0.2}$ 70°, ^{31}P +101.9 ppm (J_{PH} 194 Hz).[924,927]

Et_2PH. I.[36,561,619,1386] II.[661] III.[588,594] IV.[706,1050] VII.[132,935] C.XII.[178] Liquid, b. 85°,[588] b_{760} 85 to 86°,[132,661,1386] n_D^{20} 1.447, d_4^{20} 0.7862,[1386] mag. rotation,[1380] μ 1.40 D,[1246] μ 1.36 D,[784] ^{31}P +55.5 ppm (J_{PH} 190 Hz),[1008] Raman,[88] mass spect.,[129,129a] relative pK_a 33.7 (referred to MeOH).[677]

$(HCF_2CF_2)_2PH$. C.XII. Liquid, b_{760} 91 to 93°.[178,1080]

$(HCFClCF_2)_2PH$. C.XII. Liquid, b. 138 to 142°.[1080]

$(HCCl_2CF_2)_2PH$. C.XII. Liquid, b. 180 to 184°.[1080]

$(HCF_2CH_2)_2PH$. C.XII. Liquid, b_{226} 108 to 109°, IR.[178]

$(H_2NCH_2CH_2)_2PH$. Liquid, $b_{1.5}$ 77 to 78°, pK_1 9.33, pK_2 8.22.[681]

$(Et_2NCH_2CH_2)_2PH$. Liquid, b_4 130 to 131°, pK_1 8.68, pK_2 8.15.[681]

$(HOCH_2CH_2)_2PH$. II. Liquid, b. 158 to 160°, n_D^{20} 1.4892, d_4^{20} 1.035.[782]

$(MeOCH_2CH_2)_2PH$. C.XII. Liquid, b_{51} 113 to 125°.[87]

$(BuOCH_2CH_2)_2PH$. C.XII. Liquid, b_9 139°, n_D^{25} 1.4570,[1171] pK_a (H_2O) 4.15.[1342]

$(t\text{-}BuOCH_2CH_2)_2PH$. II. Liquid, b_{22} 138 to 141°, $L\cdot Me^+$ BPh_4^-, m. 129 to 130°.[1308]

$(CCl_3CHOH)_2PH$. C.XIII.1. Liquid, b. 129 to 130°,[176,358] ^{31}P +43 ppm (J_{PH} 226 Hz).[176]

$(HSCH_2CH_2)_2PH$. II. Liquid, b_{10} 126°, n_D^{20} 1.5983.[428]

Pr_2PH. I.[1386] IV.[106] VII.[628] Liquid, b. 131 to 133°,[106] b_{760} 136°,[1386] b_{24} 43 to 46°,[628] n_D^{20} 1.541, d_4^{20} 0.7937,[1386] mag. rotation,[1386] ^{31}P +73 ppm (J_{PH} 178 Hz).[262]

$i\text{-}Pr_2PH$. I.[675] III.[591] Liquid, b_{760} 118 to 119°.[591,640,675]

$(C_3F_7)_2PH$. VI. Liquid, b. 30 to 32°.[240]

$(H_2NCH_2CH_2CH_2)_2PH$. C.XII. Liquid, b_{36} 162 to 168°.[87,1338]

$(NCCH_2CH_2)_2PH$. C.XII. Liquid, $b_{0.3}$ 157 to 159°,[1373] n_D^{25} 1.5070, ^{31}P +75 ppm (J_{PH} 195 Hz),[262] pK_a (H_2O) 0.41.[1342]

$(HOCH_2CH_2CH_2)_2PH$. C.XII. Liquid, b_1 125 to 130°.[87,1338]

$(MeO_2CCH_2CH_2)_2PH$. C.XII. Liquid, b_6 95°.[87]

$(EtO_2CCH_2CH_2)_2PH$. C.XII. Liquid, $b_{0.2}$ 109 to 110°, n_D^{25} 1.4668.[1171]

Bu_2PH. I.[408,410,619,1252,1287,1386] II.[562,1084,1406] IV.[706] VII.[132,745,1177] VIII.[509,1123] C.XII.[87,1338] Liquid, b_{760} 178°,[1386] b. 183 to 186,[562] b_{17} 71°,[619] b_{18} 71 to 78°,[1177] b_{14} 70°,[410] n_D^{20} 1.456, d_4^{20} 0.8083,[1252,1386] mag. rotation[1386] ^{31}P +69.5 ppm (J_{PH} 180 Hz),[553] pK_a (H_2O) 4.51.[1342]

i-Bu_2PH. I.[916] III.[591] C.XII.[87,1171,1338] Liquid, b. 166 to 173°,[916] b. 169 to 171.8°,[1338] b_8 47 to 48°,[1171] n_D^{25} 1.4487,[1171] ^{31}P +82.5 ppm (J_{PH} 194 Hz),[553] +85.1 ppm (J_{PH} 194 Hz),[916] pK_a (H_2O) 4.11.[1342]
2-Bu_2PH. III. Liquid, b_{88} 85.5 to 86°,[1467] m. -70°.[1467]
t-Bu_2PH. I. Liquid, b_{13} 38 to 40°,[580 581] relative pK_a 36.3 (referred to MeOH),[677] ^{31}P -20.1 ppm (J_{PH} 197 Hz).[485]
$(CH_3CHClCCl_2CH{\cdot}OH)_2PH$. C.XIII.1. Crystals, m. 95 to 96°.[170,437]
$((CF_2)_3C{\cdot}OH)_2PH$. C.XIII.3. Crystals, m. 30 to 35°, b_8 74.5 to 75.5°.[1078]
i-Am_2PH. III.[591] Liquid, b. 210 to 215°.[591]
$(c\text{-}C_6H_{11})_2PH$. II.[661] IV.[706,711,1050] VIII.[509] C.XII.[1171] Liquid, b. 281 to 283°,[706] b_8 129°,[1171] b_8 138 to 140°,[674] b_3 105 to 108°,[1050] b_3 118°,[711] n_D^{25} 1.5142,[1171] pK_a (H_2O) 4.55,[1342] relative pK_a 35.7 (referred to MeOH).[677]
$(n\text{-}C_8H_{17})_2PH$. I.[1002,1252] II.[553,1084] VIII.[1002] C.XII.[1170] Liquid, b_{15} 173°,[1252] b_2 120 to 121°,[1084] $b_{1.3}$ 143 to 146°,[1002] $b_{0.25}$ 137°,[1171] n_D^{20} 1.4649, d_4^{20} 0.8302,[1252] $n_D^{26.2}$ 1.4626,[1002] IR,[1252] ^{31}P +71.5 ppm (J_{PH} 196 Hz),[553] pK_a (H_2O) 4.41.[1342]
$(n\text{-}C_{12}H_{25})_2PH$. C.XII. Solid, m. 55 to 60°,[535] $b_{0.35}$ 197 to 199°,[1171] $b_{0.05}$ 170 to 185°,[535] IR, ^1H NMR, ^{31}P +71 ppm (J_{PH} 193 Hz).[535]
$(PhCH_2)_2PH$. IV. Liquid, b_3 115 to 120°;[706] the previously reported m. 205°[590] seems to be that of an oxidation product.
$(PhCH{\cdot}OEt)_2PH$. C.XIII.2. Crystals, m. 107 to 111°.[357]
$(PhCH{\cdot}OPr\text{-}i)_2PH$. C.XIII.2. Crystals, m. 70 to 74°.[357]
$(PhCH_2CH_2)_2PH$. C.XII. Liquid, $b_{0.5}$ 158°, n_D^{25} 1.5815,[1171] pK_a (H_2O) 3.46.[1342]
Ph_2PH. I.[327,408,410,543,609,619,822,823,940,956,1050,1287] II.[1084,1252] IV.[706,1050,1259] V.[427,956,1438] VII.[1174,1175] VIII.[477,612,768,935,989] Liquid, b. 280°,[327,706] b_{25} 155 to 157°,[1050] b_{16} 156 to 157°,[822] b_{15} 130°,[609,619] b_{12} 150 to 151°,[1084,1252] b_1 103°,[427] b_3 119 to 120°,[711] $b_{0.3}$ 110°, $b_{0.1}$ 103 to 106°,[1175] n_D^{20} 1.6269,[940,1252] IR,[1252,1326] Raman,[1326] UV,[1236] mass spect. (generates "PhP");[1464] photolysis yields Ph_2P radical; rose-colored, stable at -140°,[1260] $E_{1/2}$ 0.22 V,[616] ^1H NMR,[956,1202] ^{31}P +41 ppm (J_{PH} 220 Hz),[930] +41.1 ppm (J_{PH} 214 Hz),[1008] +43.8 ppm (J_{PH} 239 Hz),[300] +40.7 ppm (J_{PH} 216 Hz),[1065] +41.0 ppm (J_{PH} 218 Hz);[411] the deuterated phosphine, Ph_2PD, ^{31}P +42.4 ppm (J_{PD} 34 Hz),[411] shows an isotope effect; R_f value,[1250] μ 1.41 D,[1246] pK_a (H_2O) 0.03,[1342] relative pK_a 21.7 (referred to MeOH).[677]
$(C_6F_5)_2PH$. I. Solid, m. 52°,[246,388] $b_{0.8}$ 104°,[388] ^1H and ^{19}F NMR,[79] ^{31}P +143 ppm (J_{PH} 238 Hz).[388]

$(3\text{-}FC_6H_4)_2PH$. I. Liquid, b_{13} 144°, n_D^{20} 1.5355,[1251] ^{19}F NMR σ_m ($3\text{-}FC_6H_4PH$ group) +0.09.[1251]
$(4\text{-}FC_6H_4)_2PH$. I. Liquid, b_{12} 140 to 142°, n_D^{20} 1.5617,[1129] ^{19}F NMR, σ_p ($4\text{-}FC_6H_4PH$ group) +0.25.[1129]
$(2\text{-}ClC_6H_4)_2PH$. VIII. Crystals, m. 67 to 68°, b. 288 to 292°.[1123]
$(4\text{-}ClC_6H_4)_2PH$. VIII. Crystals, m. 39 to 40°, b. 314 to 315°.[1123]
$(2\text{-}MeOC_6H_4)_2PH$. I. Solid, m. 138°.[408]
$(4\text{-}Me_2NC_6H_4)_2PH$. I.[410] V.[1260] Solid, m. 137°, $b_{0.05}$ 220 to 225°,[410] liquid, $b_{0.02}$ 145 to 150°;[1260] photolysis yields green radical, $(Me_2NC_6H_4)_2P$, which is stable at -140°.[1260]
$(4\text{-}NO_2C_6H_4)_2PH$. VIII. Yellow crystals, m. 55 to 56°, b_{15} 210 to 220°.[1123]
$(2\text{-}MeC_6H_4)_2PH$. VIII. Liquid, b. 306 to 307°,[1123] ^{31}P +59.1 ppm (J_{PH} 219 Hz).[485]
$(3\text{-}MeC_6H_4)_2PH$. I. Liquid, $b_{0.01}$ 155°,[408] ^{31}P +40.2 ppm (J_{PH} 214 Hz).[485]
$(4\text{-}MeC_6H_4)_2PH$. I.[610] IV.[1050] VIII.[1123] Liquid, b. 295 to 298°,[1123] b_2 121 to 122°,[610] b_2 122 to 124°,[1050] ^{31}P +42.9 ppm (J_{PH} 212 Hz).[485]
$(4\text{-}t\text{-}BuC_6H_4)_2PH$. ^{31}P +43.9 ppm (J_{PH} 212 Hz).[485]
$(2,6\text{-}Me_2\text{-}4\text{-}Me_2NC_6H_2)_2PH$. VIII. (From PCl_3 + RMgX). Shiny crystals, m. 121 to 123°, IR.[1308]
$(2,4,6\text{-}Me_3C_6H_2)_2PH$. I. Solid, m. 74° (from EtOH),[410] not air-sensitive.
$(1\text{-}C_{10}H_7)_2PH$. VIII. Crude solid.[1123]

TYPE: RR^1PH

$MeCF_3PH$. I. IV. Gas, b_{760} 18° (calc.), $\log P_{mm} = 5.6349 + 1.75 \log T - 0.00556T - (1586/T)$, IR,[187] $L{\cdot}BH_3$, b_{760} 105°.[187]
MeEtPH. IV. Liquid, b_{760} 54.5° (calc.), m. -160°, $\log P_{mm} = 7.6767 - (1671/T)$,[909] ^{31}P +77.0 ppm (J_{PH} 191 Hz).[909]
$Me(HCF_2CF_2)$ PH. C.XII. Liquid, b_{752} 62 to 63°.[164]
$Me(H_2NCH_2CH_2)$ PH. II. Liquid, b. 131 to 132°, pK_a 8.96.[681]
$Me(Me_2NCH_2CH_2)$ PH. II. Liquid, b_{32} 68°, IR.[1398]
$Me(HOCH_2CH_2)$ PH. II. (RPHNa + $\overline{CH_2CH_2O}$). Liquid, b_{20} 74 to 76°, $b_{17.5}$ 72 to 72.5°,[162] n_D^{20} 1.4961, d_4^{20} 0.9815, IR.[162]
MePrPH. IV. Liquid, b_{760} 78.2° (calc.), m. -129.5°, $\log P_{mm} = 8.1138 - (1838/T)$,[909] ^{31}P +87.0 ppm (J_{PH} 196 Hz).[909]
Me(i-Pr)PH. III. Liquid, b. 78 to 80°.[591]
$Me(CH_2{=}CHCH_2)PH$. II. Liquid, b_{760} 81° (calc.), $\log P_{mm} = 7.925 - (1786/T)$,[1398] IR, unstable.[1398]
$Me(CF_2{=}CClCF_2)PH$. II. Liquid, condensing at -24° under vacuum, IR, ^1H and ^{19}F NMR.[124]

Me[Me(CF_3)C·OH]PH. C.XIII.3. Liquid, b_{59} 49 to 51°, n_D^{20} 1.4223, d_4^{20} 1.2797, IR.[488]
Me[(CF_3)$_2$C·OH]PH. C.XIII.3. Liquid, b_{110} 54 to 55°, n_D^{20} 1.3695, d_4^{20} 1.5084.[163]
Me[(CF_3)$_2$C·O_2CMe]PH. From phosphine above and MeCOCl. Liquid, b_1 52 to 54°.[163]
MeBuPH. IV. Liquid, b_{760} 112.7° (calc.), m. -107°, log P_{mm} = 7.5210 - (1790/T),[909] ^{31}P +86 ppm (J_{PH} 202 Hz).[909]
Me(c-C_6H_{11})PH. II. Liquid, b_3 30 to 31°.[656]
Me(n-$C_{10}H_{21}$)PH. III. A.V. Liquid, $b_{0.5}$ 93 to 95°, GLC, IR, ^{31}P +85.3 ppm (J_{PH} 194 Hz).[535]
Me($PhCH_2$)PH. I. Liquid, $b_{0.15}$ 35 to 37°.[1330]
MePhPH. I.[619] II.[1084,1360] IV.[909] Liquid, b_{10} 59 to 60°,[909,1084] b_{11} 62 to 63°,[619] n_D^{20} 1.5695,[1084] ^{31}P +72.3 ppm (J_{PH} 222 Hz).[909]
PhNHCS(c-C_6H_{11})PH. C.XIII.5. Crystals, m. 96° (from EtOH), air-stable.[665]
Et(NC)PH. C.XIII.6. Plates, m. 49 to 50° (from Et_2O) (?), volatile.[279]
Et($H_2NCH_2CH_2$)PH. II. Liquid, 154 to 155°, pK_a 8.81, ^{31}P +72.5 ppm.[681]
Et($EtHNCH_2CH_2$)PH. II. Liquid, b. 173 to 174°, pK_a 9.08.[681]
Et($HOCH_2CH_2$)PH. II. Liquid, b_9 68 to 70°, n_D^{20} 1.4925, d^{20} 0.9580.[162]
Et(HO_2CCH_2)PH. II. Oil; polarography indicates zwitter-ion structure.[682]
Et($HSCH_2CH_2$)PH. II. Liquid, b_{10} 52 to 53°, n_D^{20} 1.5289.[1313]
EtBuPH. II. Liquid, b. 130 to 135°.[1407]
EtPhPH. A.IV.[1123] Liquid, ^1H NMR,[38] ^{31}P +43.7 ppm (J_{PH} 200.6 Hz).[412]
(HCF_2CF_2)PhPH. C.XII. Liquid, $b_{0.5}$ 28°, n_D^{25} 1.4758.[1080]
($H_2NCH_2CH_2$)BuPH. II. Liquid, b_{15} 86.7°, pK_a 8.54, ^{31}P +37.4 ppm.[681]
($H_2NCH_2CH_2$)(c-C_6H_{11})PH. II. Liquid, b_{10} 115 to 116°, pK_a 8.54, ^{31}P +37.4 ppm.[681]
($H_2NCH_2CH_2$)($PhCH_2$)PH. II. Liquid, $b_{3.5}$ 129 to 131°, pK_a 8.58.[681]
($H_2NCH_2CH_2$)PhPH. II. Liquid, b_7 115°, pK_a 8.66 (in EtOH/H_2O).[694]
($EtHNCH_2CH_2$)PhPH. II. Liquid, b_3 111 to 112°, pK_a 8.75.[681]
($Et_2NCH_2CH_2$)BuPH. II. Liquid, b_{18} 116 to 117°, pK_a 8.23.[681]
($Et_2NCH_2CH_2$)PhPH. II. Liquid, b_2 109 to 110°, pK_a 8.18.[681]
($HOCH_2CH_2$)(c-C_6H_{11})PH. II. Pale-yellow oil, b_3 119 to 125°.[700]
($HOCH_2CH_2$)PhPH. II. Colorless liquid, b_4 132 to 136°.[700]
(t-$BuOCH_2CH_2$)PhPH. II. Liquid, $b_{0.15}$ 65 to 66°.[1]
(HO_2CCH_2)(c-C_6H_{11})PH. II. Oil.[682]
(HO_2CCH_2)PhPH. II. Crystals, m. 35 to 37°; polarography indicates zwitterion structure.[682]
(MeO_2CCH_2)PhCOPH. C.XIII.6. Yellow liquid, $b_{0.05}$ 113 to

114°.[678]
MeCO(MeO_2CCH_2)PH. C.XIII.6. Liquid, $b_{0.1}$ 58 to 60°, cleaved by NaOH.[678]
(MeCO)BuPH. C.XIII.6. Liquid, b_{14} 64 to 65°.[678]
MeCO(n-C_6H_{13})PH. C.XIII.6. Liquid, $b_{0.5}$ 55 to 56°.[678]
MeCO(Ph)PH. C.XIII.6. Liquid, b_1 71 to 72°.[678]
($HSCH_2CH_2$)PhPH. II. Liquid, b_{10} 135 to 137°, pK_a 10.7.[712]
($EtSCH_2CH_2$)PhPH. (From phosphine above and EtBr). Liquid, b_8 130 to 132°,[712] L•$NiBr_2$, m. 318 to 324°.
[$(EtO)_3SiCH_2CH_2$]PhPH. C.XII. Liquid, $b_{2.5}$ 157 to 161°, n_D^{25} 1.4844.[377,1121]
Pr(n-C_6H_{13})PH. II. Liquid, b_{760} 194°,[1084,1252] n_D^{20} 1.4579, IR.[1252]
i-Pr(i-Bu)PH. III. Liquid, b. 139 to 140°.[591]
(CH_2=$CHCH_2$)PhPH. II. Liquid, b_3 60 to 65°,[422] $b_{0.1}$ 57°, n_D^{20} 1.5734;[1332] polymerizes on standing even at 0°.[1332]
(CF=$CClCF_2$)PhPH. II. Liquid, $b_{0.1}$ 57°, IR, ^1H and ^{19}F NMR.[124]
$NCCH_2CH_2$($H_2NCH_2CH_2CH_2$)PH. C.XII. Liquid, $b_{0.4}$ 86 to 96°.[1171]
$NCCH_2CH_2$($HOCH_2CH_2CH_2$)PH. C.XII. Liquid, $b_{0.2}$ 131 to 135°, n_D^{25} 1.5090.[1171]
$NCCH_2CH_2$($MeCOCH_2Me_2C$)PH. C.XII. Liquid, $b_{0.8}$ 127°.[1171]
$NCCH_2CH_2$(c-C_6H_{11})PH. C.XII. Liquid, $b_{0.3}$ 97°, n_D^{25} 1.5088.[1171,1453]
$NCCH_2CH_2$(n-C_8H_{17})PH. C.XII. Liquid, $b_{0.45}$ 118°, n_D^{25} 1.4745.[1171,1453]
$NCCH_2CH_2$(t-$BuCH_2$CH•$MeCH_2$)PH. C.XII. Liquid, $b_{0.3}$ 102°, n_D^{25} 1.4735.[1171]
$NCCH_2CH_2$(Ph)PH. C.XII. Liquid, $b_{0.5}$ 104°, n_D^{20} 1.5649, d_4^{20} 1.0710,[59,60] b_1 107°, n_D^{20} 1.5735, d_4^{20} 1.0710.[1383]
$HOCH_2CH_2CH_2$(c-C_6H_{11})PH. II. Pale-yellow oil, b_3 128 to 132°.[700]
$HOCH_2CH_2CH_2$(Ph)PH. II.[700] C.XII.[59] Liquid, b_3 146 to 149°,[700] b_3 107 to 108°,[800] b_2 107 to 110°, n_D^{20} 1.5745, d_4^{20} 1.0696.[59]
CH_2=$CHCH_2O(CH_2)_3$(Ph)PH. C.XII. Liquid, b_2 99 to 101°, IR.[800]
HOCH•$MeCH_2$(c-C_6H_{11})PH. II. Liquid, b_3 130 to 134°.[700]
HOCH•$MeCH_2$(Ph)PH. II. Liquid, b_3 135°.[700]
HO_2CCH•Me(Ph)PH. II. Solid, m. 29 to 31°.[682]
$MeO_2CCH_2CH_2$(Ph)PH. C.XII. Liquid, b_2 109°, n_D^{20} 1.5485, d_4^{20} 1.0982.[59,1383]
$EtO_2CCH_2CH_2$(n-C_8H_{17})PH. C.XIII. Liquid, $b_{0.25}$ 106°, n_D^{25} 1.4620.[1171]
EtCO(Ph)PH. C.XIII.6. Liquid, $b_{1.5}$ 89 to 91°, ^{31}P +29.1 ppm (J_{PH} 231 Hz).[678]
HSCH•$MeCH_2$(Ph)PH. II. Liquid, b_8 145°, pK_a 11.[712]
EtSCH•$MeCH_2$(Ph)PH. From phosphine above and EtBr. Liquid, b_8 133 to 135°.[712]
Bu(n-$C_{12}H_{25}$)PH. II. Liquid, b_2 122°, n_D^{20} 1.4593, d_4^{20} 0.8221,[1084,1252] IR.[1252]

Bu(PhCO)PH. C.XIII.6. Yellow liquid, $b_{0.2}$ 99 to 101°; cleaved by acids and $PhNHNH_2$.[678]
BuPhPH. I.[408] II.[1084] Liquid, $b_{0.25-0.3}$ 85 to 90°,[408] b_9 102°, n_D^{20} 1.5400.[1084]
Bu(1-$C_{10}H_7$)PH. VII. Liquid, $b_{0.1}$ 111 to 113°.[1177]
t-BuPhPH. I. A.IV. Liquid, $b_{0.05}$ 40 to 41°,[580] ^{31}P +5.7 ppm (J_{PH} 200 Hz).[485]
CH_2=$CHCH_2CH_2$(Ph)PH. II. Liquid, b_{12} 109 to 110°, n_D^{20} 1.5614, IR;[282] cyclizes on heating with UV light.[282]
NC$(CH_2)_3$(Ph)PH. C.XII. Liquid, b_1 112 to 113°, n_D^{20} 1.5595, d_4^{20} 1.0520.[59]
MeO_2CCH•$MeCH_2$(Ph)PH. C.XII. Liquid, $b_{0.8}$ 91 to 92°, n_D^{20} 1.5372, d_4^{20} 1.0744.[1383]
CH_2=CH$(CH_2)_3$(Ph)PH. II. Liquid, $b_{1.5}$ 90 to 92°, n_D^{20} 1.5545, IR;[282] cyclizes on heating with UV light.[282]
Me_2CHCH_2CO(Ph)PH. C.XIII.6. Liquid, $b_{0.4}$ 95 to 96°, ^{31}P +39.2 ppm (J_{PH} 218 Hz).[678]
CH_2=CH$(CH_2)_4$(Ph)PH. II. Liquid, $b_{1.5}$ 106°, n_D^{20} 1.5474, IR;[282] cyclizes on heating with UV light.[282]
(c-2-HOC_6H_{10})(c-C_6H_{11})PH. II. Liquid, b_3 160 to 164°.[700]
(c-2-HOC_6H_{10})(Ph)PH. II. Liquid, b_5 170°.[700]
CH_2=CH$(CH_2)_9$(Ph)PH. II. Liquid, $b_{0.4}$ 142 to 143°, n_D^{20} 1.5245, IR;[282] polymerizes on heating with UV light.[282]
$PhCH_2$(Ph)PH. II. Solid, m. 75 to 80°, $b_{0.3}$ 107 to 110°,[1330] ^1H NMR.[37]
PhC•Me•OH(Ph)PH. C.XIII.2. Liquid, b_4 162 to 163°, n_D^{20} 1.5863.[1105]
4-$BrC_6H_4(CH_2)_3$(Ph)PH. C.XII. Liquid, $b_{0.009}$ 135 to 136°, n_D^{20} 1.6126, d_4^{20} 1.2772.[59]
CH_2=C•Ph•CH_2(Ph)PH. II. Liquid, $b_{0.1}$ 53 to 55°.[422]
EtO_2CCH_2CH•Ph(Ph)PH. C.XII. Liquid, b_1 147°, n_D^{20} 1.5680, d_4^{20} 1.0921.[59]
Ph(C_6F_5)PH. I. Liquid, $b_{0.5}$ 87°, ^1H NMR, ^{31}P +92.2 ppm,[388] (J_{PH} 225 Hz).[390]
Ph(2-ClC_6H_4)PH. I. Liquid, $b_{0.4}$ 113 to 117°, $b_{0.25}$ 105 to 115°,[419] $b_{0.25}$ 110 to 115°,[940] n_D^{20} 1.641.[419]
Ph(4-ClC_6H_4)PH. I. Liquid, $b_{0.25}$ 115 to 120°, n_D^{20} 1.6307.[940]
Ph(2-BrC_6H_4)PH. I. Liquid, $b_{0.1}$ 128 to 130°, n_D^{23} 1.656.[419]
Ph(2-$MeOC_6H_4$)PH. I. Liquid, $b_{0.3}$ 116 to 118°, ^1H NMR.[956]
Ph(3-$MeOC_6H_4$)PH. I. Liquid, $b_{0.5}$ 135 to 140°,[567] $b_{0.3}$ 115 to 117°,[956] $b_{0.4}$ 130°, n_D^{20} 1.6228,[941] ^1H NMR, IR.[956]
Ph(4-$MeOC_6H_4$)PH. I. Crystals, m. 35.5 to 38°, $b_{0.4}$ 116 to 119°, $b_{0.5}$ 122 to 130°,[567] ^1H NMR.[956]
Ph(3-$EtOC_6H_4$)PH. I. Liquid, $b_{0.5}$ 135 to 136°,[567] $b_{0.2}$ 111 to 113°, ^1H NMR, IR.[956]
Ph(4-$EtOC_6H_4$)PH. I. Liquid, $b_{0.8}$ 140 to 150°,[567] $b_{0.2}$ 119 to 124°, IR, ^1H NMR.[956]
Ph(3-MeC_6H_4)PH. I. Liquid, $b_{0.16}$ 102 to 103°, IR, ^1H NMR.[419]
Ph(2-$HO_2CC_6H_4$)PH. II. Colorless needles, m. 133 to 134°,[717]

L•MeI, m. 120 to 122°, L•HCl, dec. 120°.[717]
$Ph(3,5-Me_2C_6H_3)PH$. I. Liquid, $b_{0.25}$ 112 to 113° (not pure).[419]

Disecondary Diphosphines

TYPE: $RHP(CH_2)_xPHR$

$MeHPCH_2PHMe$. II. Liquid, m. -20 to -13°,[1400] (has been said to be unstable at ambient temperature).[1400]
$EtHPCH_2PHEt$. II. Liquid, b_5 54°,[657] disulfide, m. 150 to 153°.[657]
$EtHPCH_2CH_2PHEt$. I.[716] II.[658] Liquid, b_{18} 90°,[658] b_{10} 81°, ^{31}P +57.3 ppm (J_{PH} 191 Hz).[716]
$EtBuPCH_2CH_2PHEt$. II. Liquid, $b_{0.5}$ 85 to 85°,[716] L•2MeI, m. 209 to 213°.[716]
$(c\text{-}C_6H_{11})HPCH_2CH_2PH(C_6H_{11}\text{-}c)$. I.[716] II.[658] Liquid, b_4 162 to 168°,[658] $b_{0.8}$ 152 to 154°,[716] L•2MeI, m. 274°.
$PhHPCH_2CH_2PHPh$. I.[716] II.[702] Liquid, $b_{0.8}$ 161°, ^{31}P +46.3 ppm (J_{PH} 205 Hz);[716] oxidation gives diphosphinic acid, m. 230°;[702] pyrolysis at 150 to 160° yields $PhPH_2$, EtPhPH, and $(PhP)_x$.[702]
$EtHP(CH_2)_3PHEt$. II. Liquid, b_{17} 103°,[657] L•2HI, m. 116 to 118°, L•2MeI, m. 108 to 110°.[657]
$(H_2NCH_2CH_2)HP(CH_2)_3PH(CH_2CH_2NH_2)$. II. Liquid, $b_{1.5}$ 159 to 161°.[681]
$(Et_2NCH_2CH_2)HP(CH_2)_3PH(CH_2CH_2NEt_2)$. II. Liquid, b_6 205 to 212°, pK_1 8.58, pK_2 7.90.[681]
$(c\text{-}C_6H_{11})HP(CH_2)_3PH(C_6H_{11}\text{-}c)$. II. Liquid, b_2 151°.[656]
$PhHP(CH_2)_3PHPh$. II. Liquid, b_5 191 to 200°,[669] $b_{9.5}$ 222 to 224°.[596]
$EtHP(CH_2)_4PHEt$. II. Liquid, b_{16} 119°,[657] L•2HI, m. 102 to 103°, L•2MeI, m. 115 to 117°.[657]
$(H_2NCH_2CH_2)HP(CH_2)_4PH(CH_2CH_2NH_2)$. II. Liquid, $b_{0.1}$ 145 to 148°, pK_1 9.14, pK_2 8.17.[681]
$(Et_2NCH_2CH_2)HP(CH_2)_4PH(CH_2CH_2NEt_2)$. II. Liquid, b_6 215 to 221°, pK_1 8.43, pK_2 8.01.[681]
$(c\text{-}C_6H_{11})HP(CH_2)_4PH(C_6H_{11}\text{-}c)$. II. Liquid, b_2 165 to 166°, m. 4°.[656]
$PhHP(CH_2)_4PHPh$. II. Liquid, b_4 208 to 212°,[669] b_8 223°.[596]
$EtHP(CH_2)_5PHEt$. II. Liquid, b_2 79°,[657] L•2HI, m. 86 to 88°, L•2MeI, m. 98 to 100°.[657]
$(c\text{-}C_6H_{11})HP(CH_2)_5PH(C_6H_{11}\text{-}c)$. II. Liquid, b_2 177°.[656]
$PhHP(CH_2)_5PHPh$. II. Liquid, b_{4-5} 211 to 214°.[672]
$EtHP(CH_2)_6PHEt$. II. Liquid, b_2 92°,[657] L•2HI, m. 99 to 110°, L•2MeI, m. 110 to 112°.[657]
$(c\text{-}C_6H_{11})HP(CH_2)_6PH(C_6H_{11}\text{-}c)$. II. Liquid, b_2 200 to 201°, m. 13 to 14°.[656]
$PhHP(CH_2)_6PHPh$. II. Liquid, b_4 207 to 211°,[672] $b_{0.2}$ 178 to 181°, n_D^{20} 1.5835.[1330]
$(n\text{-}C_6H_{13})HP(CH_2)_{10}PH(C_6H_{13}\text{-}n)$. II. Liquid, $b_{0.05}$ 140 to

150°, n_D^{20} 1.4631, d_4^{20} 0.8283.[1474]

$PhHP(CH_2)_{10}PHPh$. II. C.XII. Crystals, m. 40 to 40.5°,[800] m. 37 to 38°, b_3 244 to 245°, b_1 220 to 221°,[800] n_D^{46} 1.5530, d_D^{46} 0.9940, IR, ^1H NMR.[800]

$(2\text{-}MeC_6H_4)HP(CH_2)_{19}PH(C_6H_4Me\text{-}2)$. II. Liquid, n_D^{20} 1.5368, d_4^{20} 0.9491.[1474]

$PhHPCH_2CH_2SiMe_2OSiMe_2CH_2CH_2PHPh$. C.XII. Liquid, b_3 223 to 228°, IR.[800]

$PhHP(CH_2)_3O(CH_2)_3PHPh$. C.XII. Liquid, b_2 205 to 207°, IR.[800]

$PhHPCH_2CH_2SCH{\cdot}Ph{\cdot}SCH_2CH_2PHPh$. From $HSCH_2CH_2PHPh$ and PhCHO. Colorless oil; decomposes on dist.[712]

$PhHPCH_2CH{\cdot}Me{\cdot}SCH{\cdot}PhSCH{\cdot}Me{\cdot}CH_2PHPh$. From $HSCH_2CH{\cdot}MePHPh$ and PhCHO. Colorless oil.[712]

$1,4\text{-}MeHPCH_2C_6H_4CH_2PHMe$. I. Viscous, colorless oil, $b_{0.001}$ 100°.[1330]

$1,4\text{-}PhHPCH_2C_6H_4CH_2PHPh$. II. Colorless, waxy product.[1330]

$1,4\text{-}PhHPCH_2CH_2C_6H_4CH_2CH_2PHPh$. C.XII. Liquid, b_2 210 to 220°.[800]

$1,4\text{-}MeHPC_6H_4PHMe$. II. Liquid, b. 50° (high vacuum), IR.[362]

$1,4\text{-}PhHPC_6H_4PHPh$. II. Liquid, $b_{0.05}$ 200 to 207°.[596]

Cyclic Secondary Phosphines

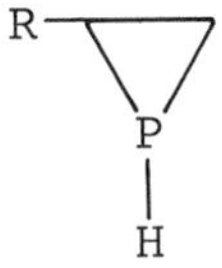

(R = H). II.[1399,1400,1402] (From $NaPH_2$ + $ClCH_2CH_2Cl$ in NH_3). I.[1402] (From $BrCH_2CH_2PBr_2$ + $LiAlH_4$). Liquid, b_{760} 36.5° (calc.), $\log P_{mm} = 7.753 - (1509/T)$,[1399,1400,1402] m. -121.4 to -120.9°,[1399,1402] IR,[214,1006,1402] Raman,[1006] ^1H NMR, ^{31}P +341 ppm (J_{PH} 155 Hz),[1402] mass spect.[1402] unstable in the liquid phase; decomposes completely within 24 hr at 25° to give $EtPH_2$ (15%), ethylene (6%), and a viscous nonvolatile liquid;[1402] it is probably a weaker base than $MePH_2$ since it undergoes no rapid exchange with MeOH hydrogen atoms;[1402] microwave spect.,[142] P-C 1.863, C-C 1.501, C-H_{cis} 1.097, C-H_{trans} 1.088, P-H 1.387 Å, ∠C-P-C 47.4°, ∠H-P-C 98.6°,[142] μ 1.121 D.[142]

(R = D). II. From $NaPH_2$ and $ClCDHCH_2Cl$, GLC,[142] deuterated >PD also prepared,[142] IR.[214]

(R = Me). II. Condenses at -63° under vacuum, IR, ^1H NMR;[213] microwave spect.[142a] rotational barriers of cis and trans isomer are 3.01 kcal/mole, and 3.05 kcal/mole[142a] μ 1.28 D (cis), μ 1.29 D (trans); the trans isomer is favored by ΔF = 130 ± 30 cal/mole.[142a]

(R = Et). II. Condenses at -63° under vacuum, ^1H NMR.[213]

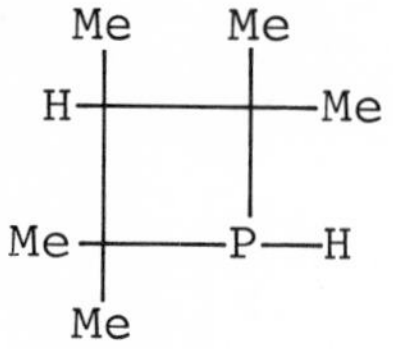

I. Liquid, b_6 32°.[1309]

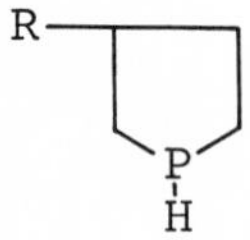

(R = H). I.[193] II.[1400] Liquid, b_9 0°,[1400] b_{760} 105.4° (calc.), m. -88°,[193] log $P_{mm} = 4.7472 - 0.003059T + 1.75 \log T - (1976/T)$.[193] Mass spect.[388] IR.[193]
(R = Me). I. Liquid, b_{760} 119°, n_D^{20} 1.4964, d_4^{20} 0.90018, IR.[133]

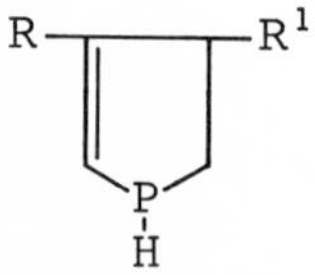

I.[410] (R = R^1 = H). Liquid, b_{760} 75°.[410]

(R = Me; R^1 = H). Liquid, b_{760} 80°.[410]
(R = Me; R^1 = Me). Liquid, b_{760} 146 to 148°.[410]

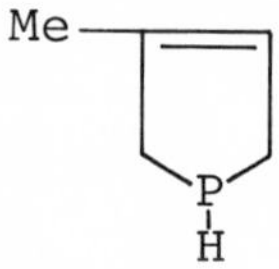

I.[133] Liquid, b_{45} 49.5°, n_D^{20} 1.5182, d_4^{20} 0.9448, IR.[133]

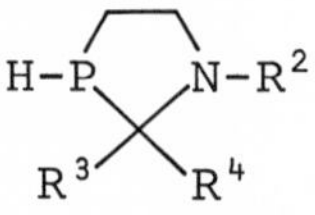

VIII. Were obtained from $H_2PCH_2CH_2NHR^2$ and R^3-COR^4.
(R^2 = H; R^3 = Ph; R^4 = H). Liquid, b_2 108 to 111°, pK_a 6.93, IR.[695]
(R^2 = H; R^3R^4 = $-(CH_2)_5-$). Liquid, b_2 79 to 82°, pK_a 7.82, IR.[695]
(R^2 = Et; $R^3 = R^4$ = H). Liquid, b_{15} 61 to 64°, pK_a 7.58.[695]
(R^2 = Et; R^3 = Ph; R^4 = H). Liquid, $b_{0.5}$ 88 to 90°, pK_a 6.34.[695]

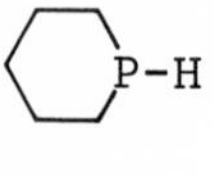

I.[637,837] Liquid, b. 110°, m. 19°,[637] ^1H NMR

(J_{PH} 200 Hz);[837] exists almost entirely in the conformation with the phosphorus proton axial (chair conformation).[837]

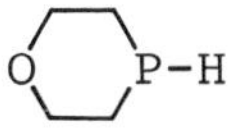

C.XII. (From PH_3 and $CH_2=CHOCH=CH_2$, UV). Liquid, b_{16} 45 to 48°, IR, ^1H NMR,[1352] ^{31}P +79 ppm (J_{PH} 185 Hz), mass spect.[1352]

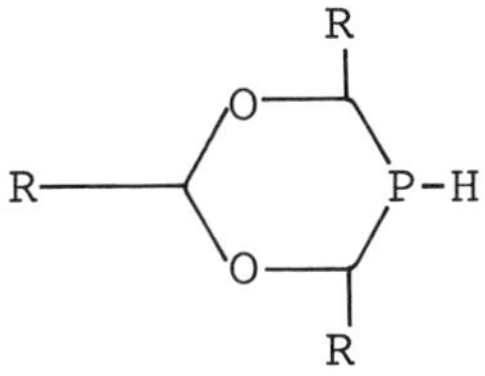

C.XIII.1. (R = i-Pr). Liquid, b_{14} 110°,[176] b_8 100 to 101°,[469] n_D^{20} 1.4602.[176,469] (R = EtBuCH-). Liquid, $b_{0.02}$ 149 to 150°, n_D^{20} 1.4709.[176]

C.XIII.2. (R = Ph). Solid, m. 110°.[357]

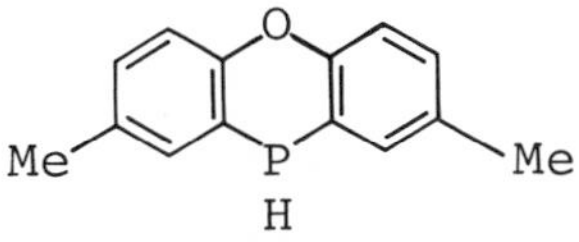

I. Liquid, $b_{0.25}$ 128°.[410]

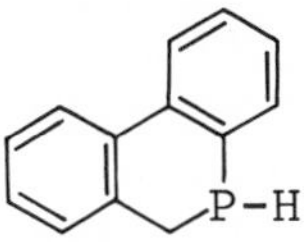

II. Liquid, $b_{0.01}$ 132°.[304]

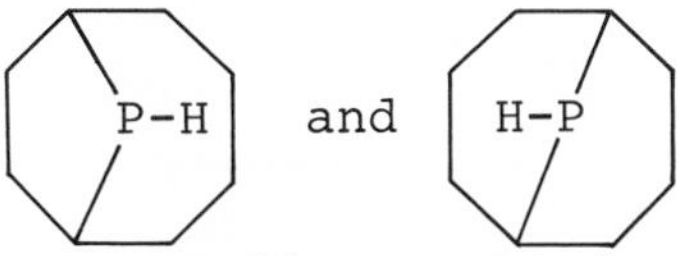

C.XII. From PH_3 and cyclooctadiene, obtained as a mixture, solid, m. 117°.[1291]

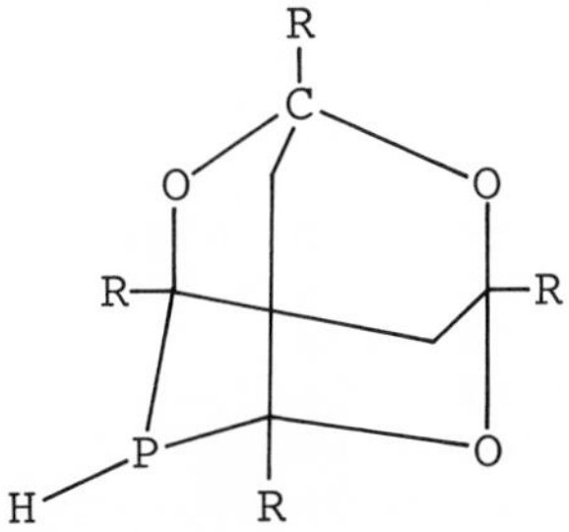

C.XIII.4. (R = Me). Crystals,

m. 88 to 90°, ^{31}P +51.5 ppm (J_{PH} 178 Hz).[355]
(R = CF_3). Crystals, m. 72 to 75°, $b_{1.5}$ 83 to 85°.[355]

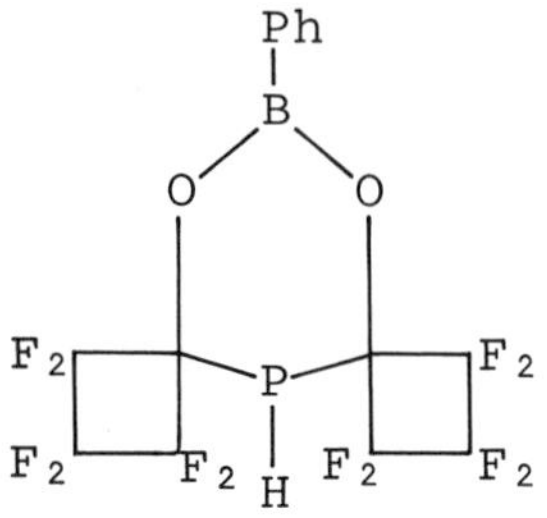

From $[(CF_2)_3C{\cdot}OH)_2PH$ and $PhB(OEt)_2$, crystals, m. 109 to 110°, IR.[1078]

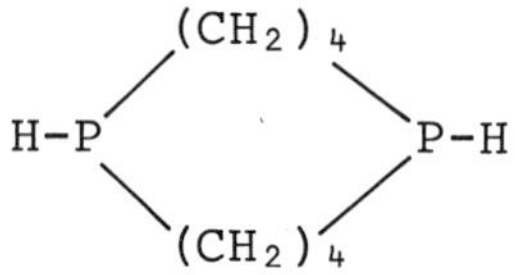

II. Liquid, $b_{0.3}$ 120 to 220° (crude?).[598]

Tertiary Phosphines

TYPE: R_3P

Me_3P. I.[94,1197,1466] II.[1222] IV.[260,332,393,586,587] V.[913] VI.[199,280,593,915] VII.[199,1102,1319] VIII.[719] IX.1.[925] May be purified through formation of AgI complex, $[AgI{\cdot}PMe_3]_4$, m. 137 to 140°,[359,808,960,1195] from which Me_3P is liberated by heating to 140 to 260°,[1356] or more conveniently through the $AgNO_3$ complex, $[Me_3PAgNO_3]_x$, which on boiling with thiourea liberates Me_3P;[359] liquid, b_{740} 37 to 39°,[1466] b. 40 to 41°,[199,719,913] b. 37.8°,[1195] b_{760} 38.4° (calc.),[1194] log P_{mm} = 7.7627 - (1518/T),[1195] m. -85.3 to -84.3°,[280] m. -85°,[1280] m. -85.9°,[1195] d_4^{20} 0.748,[94] UV,[505] IR,[455,504,1076,1403] Raman,[140,455,1076,1194] force constant (2.99 dynes/cm x 10^{-5}),[140,866,1286,1294] mass spect.,[268,396,429a,803,1347 1394] ionization potential 9.2 eV,[396] 11.0 eV,[1347] dissoc. energy $\overline{D}$ (P-C) 65.3 kcal/mole,[867] 66.2 kcal/mole,[911] ^1H NMR,[555,802,879,945,1371,1428] sign of coupl. const.,[352,380,945,946] ^{13}C NMR,[166] ^{13}C-H coupl. const.,[499] ^{31}P +62 ppm,[391,553] +63.0 ppm,[915] +62.5 ppm,[413] +63.3 ppm,[1358a] pK_a (H_2O) 8.65,[1342] μ 1.192 D;[857,1414] electron diffraction gives pyramidal structure:[82,84] P-C 1.846, C-H 1.091 Å, ∠C-P-C 98.6°; configuration of Me groups is staggared;[82,84] microwave spect. gives P-C 1.841, C-H 1.090 Å, ∠C-P-C 99.1°;[857] inversion barrier (calc.) ∿20 kcal/mole;[788] radioactive Me_3P has been made by neutron capture;[503,506] vapor pressure of the salt with HCl is 0.4 mm at 75° and 14 mm at 120°;[156] ignites in air; gives red adducts with polynitroaryls.[94a]

$(CF_3)_3P$. V. Gas, b_{760} 17.3° (calc.), log P_{mm} = 7.323 - (1289.6/T),[99] IR,[97] ^{19}F NMR, ^{31}P +2.6 ppm,[1070] mass spect., ionization potential 11.3 eV;[268] electron diffraction gives pyramidal structure:[141a] P-C 1.937, C-F 1.342 Å, ∠C-P-C 99.6°;[141a] it is cleaved by OH^-[99] and N_2H_4, Me_2NNH_2.[1103]

$(ClCH_2)_3P$. IX.2. Liquid, b_7 100°, d_4^{20} 1.414,[546,571] b_2 56 to 57°, n_D^{20} 1.5530, d_4^{20} 1.4204,[1361a] IR;[402] acid[750,931] or base[261,750] treatment gives $MeP(O)(CH_2Cl)_2$.

$(HOCH_2)_3P$. IX.2.[453,461,1110] XIII.1.[486,487,1184] Crystals, m. 52.8°,[453,486,1184] m. 55°,[1372] $b_{2.5}$ 111 to 113°,[487,1110,1184] n_D^{20} 1.5320, d_4^{20} 1.1610[1110] (undercooled melt), n_D^{20} 1.5593, d_4^{20} 1.3160,[487] K_b 4.5 x 10^{-6},[487] pK_a 5.5,[871] ESR of radical,[871] ^1H NMR,[1393] ^{31}P +24.8 ppm,[1393] +29.1 ppm,[1393] +31 ppm;[262] from a ^{31}P NMR study it was concluded that base decomposition of $(HOCH_2)_4PCl$ gives $(HOCH_2)_3P$ and $HOCH_2OCH_2P(CH_2OH)_2$;[1393] gives with diketene at 45° $(MeCOCH_2CO_2CH_2)_3P$;[1183] gives no CS_2 adduct.[487]

$(MeCO_2CH_2)_3P$. XIII.2. Liquid, b_6 152 to 153°, n_D^{20} 1.4765, d_4^{20} 1.1988,[1004] IR,[354,1004] ^1H NMR;[1115] ^1H NMR was also given of $(EtCO_2CH_2)_3P$.[1115]

$(Et_2NCH_2)_3P$. V.[917] XIII.1.[236,917,927] Liquid, b_2 121°,[236] $b_{1.2}$ 110 to 115°,[917,927] $b_{0.01}$ 61 to 63°,[1112] n_D^{25} 1.4781,[236] d_4^{20} 1.4679, d_4^{20} 0.9757,[1192] ^1H NMR,[974a] ^{31}P +65.7 ppm,[911] +65.3 ppm.[927]

$(Bu_2NCH_2)_3P$. XIII.1. Liquid, $b_{0.01}$ 170°, n_D^{25} 1.4725, d_4^{20} 0.9837.[236]

$(C_5H_{10}NCH_2)_3P$. V. ^{31}P +62.7 ppm.[927]

$(\overline{CH_2CH_2OCH_2CH_2N}CH_2)_3P$. XIII.1. Crystals, m. 55°.[236]

$(PhNHCO)_3P$. XIII.5. Crystals, m. 212 to 213°.[167,176]

$(4\text{-}ClC_6H_4NHCO)_3P$. XIII.5. Solid, m. 245° (dec.).[167]

$(4\text{-}O_2NC_6H_4NHCO)_3P$. XIII.5. Solid, m. 277 to 278° (dec.).[167]

$(MeSCH_2)_3P$. II. Liquid, $b_{0.3}$ 120 to 126°, ^1H NMR;[1098] characterized as sulfide (Chapter 7).

$(EtSCH_2)_3P$. XIV. From $(ClCH_2)_3P$ and NaSEt. Liquid, b_2 137 to 138°, n_D^{20} 1.5665, d_4^{20} 1.0749.[1361a]

Et_3P. I.[94,564,628,753,955,1149] II.[1222] IV.[393,586] V.[584] VI.[199,332,583,584] VII.[111,198,915] VIII.[719] IX.1.[241,546,851] XII.[178,753] XIV.[61] Liquid, b_{760} 127.5°,[753,955,1463] b. 125 to 126°,[833,973,1149] b. 129 to 130°,[61,270,546,719] log P_{mm} = 8.035 - (2065/T),[753] n_D^{20} 1.456, d_4^{20} 0.7999,[973] d_4^{25} 0.8,[94] $d_4^{18.6}$ 0.80006,[1463] d_4^{15} 0.812, n_D^{18} 1.45799,[1463] mag. susceptibility,[67,834] μ 1.35 D,[1414] 1.48 D,[270] 1.84 D,[778] 2.9 D,[973] oxidation potential -0.3 V,[625,969a] IR,[88,603,732,754] Raman,[88] mass spect.,[129,129a,1394] ionization potential (calc.) 8.27 eV,[1394] dissoc. energy $\overline{D}$ (P-C) 66.7

kcal/mole,[840,911] ^1H NMR,[555,1030,1318] magnitude and sign of coupl. const.[945] ^{31}P +20.4 ppm,[391,553,1026] +19.7 ppm,[413] +21.0ppm,[1358a] pK_a (H_2O) 8.69,[1342] 8.86;[134] CS_2 adduct, red solid, m. 121 to 122° (from EtOH),[584,1430] m. 118 to 120°;[1222] this adduct adds HCl, yielding an acid stable product, $Et_3P(Cl)CS_2H$,[584] which decomposes in H_2O yielding sulfur and H_2S; photopolymerization of white P in the presence of triethylphosphine or triphenylphosphine gave solid, insoluble polymers which contained organic radicals as terminal groups of the red P network; oxidation of these polymers with nitric acid and treatment with lead ion resulted in the formation of lead ethyl or p-nitrophenylphosphonate; from these results it was concluded that commercial red P contains O and HO terminal groups and is thus a compound and not an element in the true sense.[812a]

$(CH_2{=}CH)_3P$. I. Liquid, b_{760} 116.6°,[399,753,934] b_{50} 44°,[1420] log Pmm = 7.868 - (1944/T),[753] UV,[1420] IR,[754] ^1H NMR,[50,945] ^{31}P +20.7 ppm.[1140]

$(CF_2{=}CF)_3P$. I.[255,1221] From PBr_3 and $CF_2{=}CFMgI$; PCl_3 and method II failed to give the product.[255] Liquid, b_{760} 99 to 101°,[1336] b_{760} 102.7° (calc.),[255] log Pmm = 8.689 - (2184/T),[255] $n_D^{23.5}$ 1.3909, $d_4^{23.5}$ 1.6150,[1336] IR, Raman,[1335] mass spect.[336] ^{19}F NMR;[255,337] is devoid of Lewis basicity.[255]

$(HC{\equiv}C)_3P$. I. Solid, m. 36 to 37°, b_{30} 52°,[1391] IR,[999,1391] Raman,[999] ^1H NMR,[1391] P-C bond cleavage by dilute alkali.[1391] Caution: detonates violently by strong friction.[1391]

$(Et_2NCH_2CH_2)_3P$. VII. Liquid, $b_{1.5}$ 147 to 148°, pK_1 8.0, pK_2 8.34, pK_3 7.29.[681]

$(HOCH_2CH_2)_3P$. XIII.6. Liquid, b. 183 to 185°, n_D^{20} 1.4780, d_4^{20} 1.053.[782]

$(MeOCH_2CH_2)_3P$. XII. Liquid, b_{51} 162 to 166°.[87]

$(BuOCH_2CH_2)_3P$. XII. Liquid, $b_{0.25}$ 148 to 152°, n_D^{25} 1.4617,[1171] pK_a (H_2O) 8.03.[1342]

$(t\text{-}BuOCH_2CH_2)_3P$. VII. Liquid, $b_{0.08}$ 98 to 101°, $L{\cdot}Me^+BPh_4^-$, m. 168 to 169°.[1308]

$(ClCH_2CHOH)_3P$. XIII.1. Solid, m. 97°.[1185]

$((HO)_2CHCHOH)_3P$. XIII.1. Solid, dec. 112°.[1185]

$(MeO_2CCH_2)_3P$. VI. Liquid, b_2 141°, n_D^{20} 1.4865, d_4^{20} 1.2025,[1131] oxide, m. 38 to 40°.

$(EtO_2CCH_2)_3P$. VI. Liquid, b_1 160 to 161°, n_D^{20} 1.4713, d_4^{20} 1.1346,[1131] oxide, m. 74 to 75°.

$(HSCH_2CH_2)_3P$. VII. Liquid, $b_{0.005}$ 143°, n_D^{20} 1.6169,[428] pK 4.2.[1271]

Pr_3P. I.[94,291,378,628,955,1148,1466] II.[1222] IX.2.[628] XII.[1067] Liquid, b. 187.5,[94,291] b. 185 to 186°,[1067] b_{740} 180°,[1466] b_{50} 103.5°,[291] b_{31} 88 to 93°,[1222] b_{24} 85.5 to 87°,[955] b_{20} 81.8 to 81.9°,[270] b_{16} 76°,[973]

b_{13} 77°,[1198] n_D^{20} 1.458,[973] n_D^{27} 1.5071,[1466] d_4^{25} 0.807,[94,291] d_4^{20} 0.8065,[973] $\Delta H_{vap.}$ 11.17 kcal/mole,[94] pK_a 8.64,[306,1342] mag. susceptibility,[833,834] μ 1.48 D,[270] 2.6 D,[973] ^{31}P +33 ppm,[553] enthalpy of formation (calc.) -90.52 kcal/mole;[1052] CS_2 adduct, m. 108°.[291]

$(c\text{-}C_3H_5)_3P$. I.[254] II.[254,306] Liquid, $b_{0.6}$ 38 to 42°,[306] $b_{7.6}$ 53 to 70°,[254] log P_{mm} = 7.072 - (2021.2/T),[254] IR, ^1H NMR,[254,306] sulfide, m. 53°; is a stronger Lewis base than $i\text{-}Pr_3P$,[254] pK_a 7.6;[306] attempts to obtain an anion radical were unsuccessful.[254]

$i\text{-}Pr_3P$. I.[254,283a,868] IV.[591] Liquid, $b_{0.4}$ 48°,[254] b_{12} 64°,[868] b_{22} 81°,[283a] IR, ^1H NMR,[254] sign of coupl. const.,[1073] ^{31}P -19.3 ppm,[1358a] -19.1 ppm,[413] -19.4 ppm;[254,483,1065] CS_2 adduct, m. 111°,[283a] sulfide, m. 35°.

$(CH_2{=}CHCH_2)_3P$. I. Liquid, b_{13} 69°,[743] ^{31}P +34.3 ppm,[413] +34.5 ppm;[933] CS_2 adduct, m. 32.5°; adduct with benzoquinone, yellow, dec. 100°.[743]

$(cis\text{-}MeCH{=}CH)_3P$. II. Liquid, b_{20} 76 to 77°, n_D^{20} 1.5172, d_4^{20} 0.8698, IR;[136] cis structure unchanged by heating to 100° for 12 hr, but irradiation with UV light causes 40% conversion to trans.[136]

$(trans\text{-}MeCH{=}CH)_3P$. II. Liquid, b_{10} 69 to 70°, n_D^{20} 1.5160, d_4^{20} 0.8577, IR;[136] trans structure unchanged in 12 hr with UV light or heating at 100 to 105°, reaction rate with MeI and $HgCl_2$ different from cis.[136]

$(CH_2{=}C{\cdot}Me)_3P$. II. Liquid, b_{30} 74 to 75°, n_D^{20} 1.5085, d_4^{20} 0.8703.[136]

$(NCCH_2CH_2)_3P$. XI.[1187,1373,1392] XII.[1173,1186] Also from $(HOCH_2)_4Cl$ and $CH_2{=}CHCN$ in the presence of a base,[1392,1393a] or from $(HOCH_2)_3P$ and acrylonitrile.[1187,1373] Crystals, m. 98 to 99°,[1173,1186] m. 97°,[1373] m. 95.8 to 97°,[1392] ^{31}P +23 ppm,[553] $pK_a(H_2O)$ 1.37.[1342]

$(HOCH_2CH_2CH_2)_3P$. XII. Liquid, b_1 196 to 198°,[87,334a,1338] d_4^{25} 1.053.[94]

$(MeCHOHCH_2)_3P$. XIII.6. Liquid, b_5 65°, n_D^{20} 1.4620, d_4^{20} 1.035.[782]

$(EtO_2CCH_2CH_2)_3P$. XII. Liquid, b_1 193 to 194°, n_D^{25} 1.4748.[1171,1173]

$(H_2NCOCH_2CH_2)_3P$. XIV. From ester and NH_3. Crystals, m. 191.5 to 192.5°,[215] textile auxiliary; used in combination with CH_2O to make textile fibers flame-, crease-, and rot-resistant.[215]

Bu_3P. I.[286,378,441,955,971,1024,1459,1466] II.[1222] III.[539] V.[1177] VI.[510] VII.[1084] VIII.[407,409,916] IX.1.[925] XII.[87,1338] Liquid, b_{760} 240 to 242.4°,[87,94,1338] b_{50} 149.5°,[286] b_{43} 135 to 144°, b_{32} 136 to 137°, b_{22} 129 to 130°,[955] b_{17} 118 to 122°,[1177] b_{16} 121 to 122°,[378] b_{12} 122 to 123°,[441] b_{10} 109 to 110°,[925,1459] b_{14} 111°,[409] b_{14} 116°,[973] b_8 109°,[916] b_1 83 to 84°,[1004] $b_{0.8}$ 63°,[1466] $b_{0.85}$ 77°,[270] n_D^{20} 1.4635,[539,

[925,1004,1459] d_4^{20} 0.8165,[973] d_4^{20} 0.8201,[1004] d_4^{25} 0.8118,[286] mag. susceptibility,[834] μ 1.49 D,[270] μ 2.4 D,[973] μ 2.22 D,[778] IR,[354] ^{1}H NMR,[974] ^{31}P +32.3 ppm,[391,553] +33.3 ppm,[1358a] enthalpy of formation (calc.) -122.9 kcal/mole,[1052] $\Delta H_{vap.}$ 12.82 kcal/mole,[94] pK_a 8.43,[306,1342] TLC,[451] CS_2 adduct, red, m. 65.5°;[286,1210] adduct with p-benzoquinone, yellowish, m. 180 to 190°,[292] with B_2S_3, m. 140° (dec.);[936] association equilibrium const. with PhOH;[143] red adducts with polynitroaryls.[94a]

i-Bu_3P. I.[291] IV.[591] XII.[1171] Liquid, b. 215°,[591] b_{50} 126°,[291] b_7 85°, n_D^{25} 1.4530,[1171] $\Delta H_{vap.}$ 11.83,[218] pK_a 7.97,[1342] ^{31}P +40 ppm,[484,553] +45.3 ppm.[413]

2-Bu_3P. I. Liquid, b_{11} 108°,[283a] ^{31}P -7.8 ppm,[413] -8.0 ppm;[485] CS_2 adduct, m. 66°.[283a]

t-Bu_3P. II. Crystals, m. 30°, b_{13} 102 to 103°,[581] L•MeI, m. >360°; forms no CS_2 adduct;[581] ^{31}P -61.9 ppm,[1065] -63.0 ppm,[1270] -61.1 ppm[1358a] (lowest value for a tertiary phosphine).

$(CH_2{=}C{\cdot}MeCH_2)_3P$. I. Liquid, b_{15} 112°;[743] adduct with benzoquinone, dec. 200°.[743]

$(MeCHOHCH_2CH_2)_3P$. XII. Liquid, $b_{0.1}$ 197°;[462] has been claimed to be useful in hair-waving, smoothing, or depilating compositions.[462]

$(MeCHClCCl_2CH{\cdot}OH)_3P$. XIII.1. Solid, m. 88°.[1185]

$(MeCH{=}CHCH{\cdot}OH)_3P$. XIII.1. Solid, dec. 300°.[1185]

Am_3P. I.[270,291,441,955,1466] XIII.2.[1004] Solid, when pure, m. 29°,[270,955] b_{50} 185.5°,[291] b_{19} 165 to 166°,[955] $b_{0.75}$ 107°,[1466] $b_{0.6}$ 100°,[270] b_3 121 to 122°,[1004] b_{20} 165 to 166°,[441] n_D^{20} 1.4642, d_4^{20} 0.8325,[1004] mag. susceptibility,[67,834] μ 1.48 D,[270] $pK_a(H_2O)$ 8.33,[1342] ^{31}P +34 ppm,[553] enthalpy of formation (calc.) -149.9 kcal/mole;[1052] CS_2 adduct, red solid, m. 55°.[291]

i-Am_3P. I.[291,1319] Liquid, b_{11} 131°,[291] b_{11} 131 to 132°;[1319] CS_2 adduct, red solid, m. 79.5°.

$(MeEtCH{\cdot}CH_2)_3P$. I. Liquid, b_{10} 113 to 117°.[291]

$(c\text{-}C_5H_9)_3P$. I. ^{31}P -4.7 ppm,[413] -0.8 ppm.[484]

$(MeC{\equiv}C{-}C{\equiv}C)_3P$. VI. Solid, m. 117° (dec.), IR;[526] not sensitive to humidity but turns dark on exposure to light.

$(i\text{-}BuCH{\cdot}OH)_3P$. XIII.1. Solid, m. 54°.[1185]

$(n\text{-}C_6H_{13})_3P$. I.[720,829,1459] VIII.[382] Liquid, b_{50} 227°,[720] b_{15} 192 to 193°,[829] b_5 163.5 to 165°,[382,1459] m. 20°,[720] m. 19.7°,[1459] mag. susceptibility,[834] μ 1.48 D,[1242] pK_a 9.70, enthalpy of formation -177.8 kcal/mole.[1052]

$(c\text{-}C_6H_{11})_3P$. I.[653] II.[1222] VIII.[382,407] XII.[1067] Crystals, m. 76 to 78°,[382,653] ^{31}P -7 ppm,[553] -9.4 ppm (in C_6H_6),[413] -11.3 ppm (in $CHCl_3$),[1065] $pK_a(H_2O)$ 9.70;[1342] CS_2 adduct, m. 116 to 118°.[407,653,1222]

$(t\text{-}BuC{\equiv}C)_3P$. II. Crystals, m. 83 to 84°, IR, mass spect.,[1181] dec. 283°.[1181]

$(EtC{\equiv}C\text{-}C{\equiv}C)_3P$. VI. Solid, m. 71° (dec.), IR.[526]
$(n\text{-}C_7H_{15})_3P$. I.[720 829] Liquid, b_2 181 to 183°,[829] b_{50} 260°, m. 20°, d_4^{25} 0.833.[720]
$(n\text{-}C_8H_{17})_3P$. I.[720,829] III.[539] VII.[1084] VIII.[609] XII.[1171] Crystals, m. 48°,[609] m. 30°,[720,1084] b_{50} 291,[720] b_1 194 to 195°,[1084] b_1 234°,[609] b_1 190 to 210,[539] b_2 197 to 200°,[829] $b_{0.3}$ 178 to 180°,[1171] n_D^{25} 1.4666,[1171] n_D^{20} 1.4683,[1084] ^{31}P +31.8 ppm,[979] +32.8 ppm,[413] TLC.[451]
$(BuEtCH{\cdot}CH_2)_3P$. I. ^{31}P +48.5 ppm.[413]
$(n\text{-}C_9H_{19})_3P$. I.[829] VIII.[382] Crystals, m. 45 to 46°, $b_{0.01}$ 160 to 162°,[382] b_2 225 to 227°.[829]
$(n\text{-}C_{10}H_{21})_3P$. I.[829] VIII.[382] Crystals, m. 48 to 50°, $b_{0.09}$ 210 to 212°,[382] b_2 243 to 244°.[829]
$(3,5,5\text{-}Me_3C_7H_{12})_3P$. VIII. Liquid, $b_{0.01}$ 153 to 155°, n_D^{20} 1.4730, d_4^{20} 0.8714.[382]
$(n\text{-}C_{12}H_{25})_3P$. XII. Liquid, $b_{0.05}$ 255 to 275°, GLC,[535] ^{31}P +36 ppm.[535]
$(n\text{-}C_{16}H_{33})_3P$. VIII. Crystals, m. 60 to 62°, $b_{0.02}$ 170 to 172°.[382]
$(PhCH_2)_3P$. I.[566,1055,1211] IV.[850] Crystals, m. 92 to 95°,[566] m. 185 to 188°, (oxide?),[1211] $b_{2.5}$ 208 to 212°,[1211] $b_{0.5}$ 203 to 210°,[566] UV,[1236] ^{31}P +12.9 ppm (in C_6H_6),[413] +10.4 ppm,[484] -23.0 ppm (oxide?).[843]
$(PhCHOH)_3P$. XIV. From $(PhCO)_3P$ and $LiAlH_4$. Crystals, m. 156°.[1366]
$(PhCH{\cdot}OMe)_3P$. XIII.2. Crystals, m. 80 to 83°.[357]
$(PhCO)_3P$. VII.[1121a] XIII.6.[1365] Crystals, m. 147°,[1121a] m. 49°,[1365] IR UV;[1365] is an acylating agent for RNH_2 and ROH.[1365]
$(3\text{-}MeC_6H_4CO)_3P$. XIII.6. (From PH_3, RCOCl and pyridine). Crystals, m. 136°, IR, UV.[1365]
$(4\text{-}MeC_6H_4CO)_3P$. XIII.6. Crystals, m. 137°, IR, UV.[1365]
$(1\text{-}C_{10}H_7CO)_3P$. XIII.6. Crystals, m. 163°, IR, UV.[1365]
$(2\text{-}C_{10}H_7CO)_3P$. XIII.6. Crystals, m. 190°, IR, UV.[1365]
$(PhCH_2CH_2)_3P$. XII. Liquid, $b_{0.35}$ 218 to 224°, n_D^{25} 1.5950,[1171] pK_a (H_2O) 6.60.[1342]
$(PhCH{=}CH)_3P$. VIII. Solid, m. 116.5 to 118.5°.[374]
$(PhCCl{=}CH)_3P$. VIII. Solid, m. 167 to 168°.[374]
$(PhC{\equiv}C)_3P$. I.[231,528] VI.[524,1305] (From $(PhC{\equiv}C)Cu$ or $R_3SnC{\equiv}CR$ and PCl_3). Crystals, m. 91 to 92°,[231,528,1305] m. 92°,[524] μ 2.02 D,[130] x-ray analysis[1016] shows space groups R3, pyramidal configuration about P, but possesses no symmetry because of asymmetric orientation of benzene rings, P-C 1.765 A, ∠C-P-C 100.7°,[1016] $d\pi\text{-}p\pi$ bonds postulated.[1016]
$(4\text{-}ClC_6H_4C{\equiv}C)_3P$. I. Crystals, m. 192° (dec.), IR;[527] decomposed by strong alkaline sol. and by $EtOH/AgNO_3$.[527]
$(4\text{-}MeC_6H_4C{\equiv}C)_3P$. I. Crystals, m. 125°, IR,[527] μ 2.30 D;[130] decomposed by strong alkaline sol. and by $EtOH/AgNO_3$.[527]
$(9\text{-}Anthryl\text{-}CH_2C{\equiv}C)_3P$. I. Solid, m. 179°, resolidified,

m. 193°.[231]

Ph_3P. I.[114,326,791,947,1117,1222,1459] III.[539,989,992,994,1175,1249] V.[812,1176,1269,1334] VI.[844] VII.[200,270,409,789,916,1175,1240] VIII.[56,119,407,409,789,916,1023,1189,1450] IX.[610,623,624] Monoclinic crystals,[277,639] m. 79 to 81° (from EtOH or MeOH or hexane),[1404a] and all other references, b_{760} 384° (calc.),[94] b. 360° (dec.),[1404a] b_{25} 250 to 260°,[844] b_1 188°,[1404a] $b_{0.05}$ 170 to 177°,[1175] $b_{0.0001}$ 123 to 125°,[789] H_{vap}. 17 kcal/mole,[398a] UV,[443,722,723,793,1135,1160,1230,1236,1284] IR,[51,155,296,297,343,443,457,481,516,888,1325,1326] Raman,[296,343,457,1326] mass spect.,[165,1378,1435,1464] ionization potential 7.36 eV,[1378] mass spect. also of $(2,4,6\text{-}D_3C_6H_2)_3P$, $(C_6D_5)_3P$, and $(C_6D_5)_2PhP$,[1435] ^1H NMR,[517,771,974,1284] ^{13}C NMR,[1182] ^{31}P +5.4 ppm (in CCl_4),[1065] +5.9 ppm,[391,553,1026,1358a] +8 ppm (in Et_2O),[413] ^{31}P signal enhancement by radicals;[339,1125] x-ray analysis[277] shows pyramidal configuration about the phosphorus atom but, owing to the unequal rotation of the benzene rings about the P-C bonds, possesses no symmetry; space group $P2_1$[277] P-C 1.828, C-C 1.3996 Å; ∠C-P-C 103°,[277] molecular polarizability,[62] μ 1.42 D,[443] 1.44 D,[110,778,1119] μ 1.49 D,[62,270] μ 1.54 D[1242,1246] pK_a 2.30,[445] pK_a (H_2O) 2.73,[971,1342] 2.61,[134] pK_b (acetone/dioxan) 9.89,[792] σ_p (of Ph_2P group) +0.68 (from ^{13}C NMR)[1182] (does not agree with values obtained by other methods), σ_p -0.01[73] (from titration), σ_p +0.31,[1243] σ_p 0.16,[738] σ_p +0.18,[1156] (from ^{19}F NMR), σ^* +0.03,[73] dissoc. energy $\bar{D}$ (P-C) 71.3 kcal/mole,[92] 75.6 kcal/mole,[911] $E_{1/2}$ 0.08 V,[625] 0.12 V,[616] relaxation time,[49,531,780,1127] π-inversion barrier 1.36 kcal/mole (calc.),[727] TLC and R_f value,[108,330,375,451,1250,1384] paper chromatography,[776,1304] GLC,[72,74,113,376,494] dec. 371°,[1007] dec. products at 400°,[1127] isotope effect on hydrogen bonding $K_H/K_D = 1.1$,[1303] ESR of radical anion, $Ph_2PM^{\overline{\cdot}}$[150] (not attributable to $Ph_3P^{\overline{\cdot}}$ as reported previously[515]); ESR of radical anion generated electrochemically[735,1213] (undergoes cleavage of a phenyl group to give $Ph_2P^{\overline{\cdot}}$)[1213,1413] radioactive Ph_3P;[1028] because of the low basicity Ph_3P is easily precipitated from solution in fuming HCl by dilution; HI salt, m. 215°; does not form an adduct with CS_2 but forms complexes with many metal salts (Chapter 3A); adducts: p-benzoquinone, crystals, m. 253°,[1263] with B_2S_3, dec. 180°,[936] with C_6Me_6, m. 145°;[1285] yields with PCl_5[1200] and $SbCl_5$[1201b] salts of the type $[Ph_3PCl]^+PCl_6^-$; red adducts with polynitroaryls;[94a] reduces PI_3 to P_2I_4.[382a]

$(2\text{-}DC_6H_4)_3P$. I. Crystals, m. 82 to 83° (from EtOH),[865] m. 76 to 77° (from hexane).[104]

$(3\text{-}DC_6H_4)_3P$. I. Crystals, m. 82 to 83° (from EtOH).[865]

$(4\text{-}DC_6H_4)_3P$. I. Crystals, m. 78°,[752] m. 80 to 81°,[508] m. 82 to 83° (from EtOH),[865] UV,[508] rate of isotopic exchange.[865]

$(C_6D_5)_3P$. I. Crystals, m. 76° (from hexane), isomorphous with Ph_3P,[104] m. 80° (from EtOH), IR.[857a]

$(C_6F_5)_3P$. I.[353,386,645,1405] (also from $PSCl_3$ and C_6F_5MgBr, no sulfide formed).[386] II. (High yield).[774] V.[298] Crystals, m. 116 to 117°,[1405] m. 115 to 116°,[774] m. 105 to 107°,[645] ^{19}F NMR,[79,600] mass spect.,[1001] ^{31}P +77.9 ppm (in $CHCl_3$),[353] +75.2 ppm (in C_6H_6),[388] +75.1 ppm (J_{PF} 56 Hz).[1358a]

$(C_6Cl_5)_3P$. I. Pale-yellow solid, ^{31}P +9.2 ppm, mass spect.[1001]

$(3\text{-}FC_6H_4)_3$. I. Crystals, m. 62°,[302] m. 64° (from MeOH),[738,1251] $b_{0.1}$ 150°,[59] IR, ^1H NMR,[59] ^{19}F NMR,[302,738,1251] ^{31}P +6.5 ppm,[302] σ_{meta} (of $(3\text{-}FC_6H_4)_2P$ group) +0.31.[1251]

$(4\text{-}FC_6H_4)_3P$. I. Crystals, m. 77 to 80° (from MeOH),[1243] m. 80 to 81° (from MeOH),[738] m. 79 to 80° (from EtOH),[302] $b_{0.1}$ 160°,[302] IR,[302] magnitude and sign of coupl. const.,[886a] ^1H NMR,[302,886a] ^{19}F NMR,[302,738,1243] ^{31}P +9.0 ppm (in CH_2Cl_2),[485] +9.4 ppm,[886a] +8.8 ppm,[302] σ_p (of $(4\text{-}FC_6H_4)_2P$ group) +0.338,[1243] +0.25.[738]

$(2\text{-}ClC_6H_4)_3P$. I.[947] VIII.[916] Crystals, m. 185° (from EtOH),[947] $b_{0.03-0.05}$ 192 to 200°,[916] ^{31}P +9.2 ppm.[916]

$(3\text{-}ClC_6H_4)_3P$. I. Crystals, m. 67°.[947] ^{31} P +4.4 ppm (in CH_2Cl).[485]

$(4\text{-}ClC_6H_4)_3P$. I.[445,947,1228] VIII.[916] Crystals, m. 103° (from EtOH),[947] m. 103 to 104°,[445] m. 102 to 105°,[1228] $b_{0.03-0.05}$ 192 to 200°,[916] pK_a 2.86,[445] μ 0.65 D,[443,778] ^{31}P +9.2 ppm,[916] +8.5 ppm (in CH_2Cl_2)[485] -22.0 ppm (in THF, oxide?),[843] $E_{1/2}$ 0.24 V.[616]

$(4\text{-}BrC_6H_4)_3P$. ^{31}P +8.2 ppm (in CH_2Cl_2).[485]

$(2\text{-}MeC_6H_4)_3P$. I.[947] VII.[407] Crystals, m. 125° (from EtOH),[407,947] UV,[1238] ^1H NMR,[1232] ^{31}P +30.0 ppm (in CH_2Cl_2),[485] +30.2 ppm,[1358a] σ_o (of $(2\text{-}MeC_6H_4)_2P$ group) +0.25,[1232] μ 0.53 D,[1238] R_f values.[1250]

$(3\text{-}MeC_6H_4)_3P$. I. Crystals, m. 100° (from EtOH),[947] UV,[1238] ^1H NMR,[517,1232] ^{31}P +5.3 ppm (in CH_2Cl_2),[485] σ_m (of $(3\text{-}MeC_6H_4)_2P$ group) > -0.21,[1232] μ 1.65 D,[1238,1242] R_f values.[1250]

$(4\text{-}MeC_6H_4)_3P$. I.[779,939,947] III.[986] VIII.[916] Prisms, m. 146° (from EtOH),[939,947,986] m. 144 to 146°,[779] UV,[1238] ^1H NMR,[517,771,1233] ^{31}P +8.0 ppm (in CH_2Cl_2),[485] σ_p [of $(4\text{-}MeC_6H_4)_2P$ group] > -0.16,[1233] $E_{1/2}$ 0.10 V,[616] μ 1.92 D,[778] μ 2.14 D,[1238] R_f values;[451,1250] red adducts with polynitroaryls.[94a]

$(2\text{-}CF_3C_6H_4)_3P$. II. Crystals, m. 164 to 165°,[1000] ^{19}F NMR,[1000] ^{31}P +18.5 ppm (in CH_2Cl).[485,1000]

$(3\text{-}CF_3C_6H_4)_3P$. I. Liquid, $b_{0.2}$ 130 to 132°, ^{19}F NMR,[1000] ^{31}P +5.0 ppm (in CH_2Cl_2).[485,1000]

$(4\text{-}CF_3C_6H_4)_3P$. I. Crystals, m. 71 to 73°,[1465] m. 68 to

70° (subl.),[1000] ^{19}F NMR,[1000] ^{31}P +6.0 ppm (in CH_2Cl_2).[485,1000]

$(2\text{-}ClCH_2C_6H_4)_3P$. II. Crystals, m. 140 to 141°.[849]

$(2\text{-}MeOCH_2C_6H_4)_3P$. I. Crystals, m. 105 to 106°.[849]

$(2\text{-}CH_2{=}CHC_6H_4)_3P$. I. ^{1}H NMR.[501b]

$(4\text{-}i\text{-}PrC_6H_4)_3P$. ^{31}P +7.8 ppm (in CH_2Cl_2).[485]

$(4\text{-}t\text{-}BuC_6H_4)_3P$. ^{31}P +9.1 ppm (in CH_2Cl_2).[485]

$(3\text{-}HO_2CC_6H_4)_3P$. VIII. Crystals, m. 276 to 279° (from MeOH/H_2O).[1234] Methyl ester. VIII. Crystals, m. 115 to 116° (from MeOH).[1234]

$(4\text{-}HO_2CC_6H_4)_3P$. VIII. Crystals, m. 270 to 274° (from MeOH/H_2O).[1234] Methyl ester. I. Crystals, m. 155 to 156°.[1037]

$(2\text{-}HOC_6H_4)_3P\cdot H_2O$. I. Crystals, m. 182 to 183°.[1037] (From MeO compound and HI).

$(3\text{-}HOC_6H_4)_3P$. From methoxy compound and HI. Crystals, m. 186 to 188°.[838]

$(4\text{-}HOC_6H_4)_3P\cdot H_2O$. From methoxy compound and HI. Crystals, m. 130 to 138°,[1273] m. 134 to 137° and 188 to 189° (two forms),[1037] ESR of ion radical produced by oxidation.[1039]

$(2\text{-}MeOC_6H_4)_3P$. I.[947] III.[539] VIII.[407] Crystals, m. 204° (from EtOH or benzene),[407,947] m. 203 to 205°,[539] ^{31}P +38.5 ppm (in CH_2Cl_2).[485]

$(3\text{-}MeOC_6H_4)_3P$. I. Crystals, m. 115° (from EtOH),[947] m. 112 to 114°,[838] ^{31}P +2.1 ppm (in CH_2Cl_2).[485]

$(4\text{-}MeOC_6H_4)_3P$. I.[445,939,947] VIII.[409] Crystals, m. 131° (from EtOH),[445,939,947] m. 135°,[409] IR,[1227] ^{1}H NMR,[476,886a] ^{31}P +10.2 ppm (in CH_2Cl_2),[485] +9.8 ppm,[886a] $E_{1/2}$ 0.29 V,[616] pK_a (H_2O) 4.46,[1342] pK_a 3.15,[445] σ_p (of $(4\text{-}MeOC_6H_4)_2P$ group) +0.14, σ^* 0.19 (from IR measurements),[1227] ESR of ion radical produced by oxidation,[1039] magnitude and sign of coupl. const.[886a]

$(4\text{-}PhOC_6H_4)_3P$. I. Crystals, m. 111° (from benzene/EtOH).[289]

$(3\text{-}NCC_6H_4)_3P$. VIII. Crystals, m. 154 to 155° (from MeOH).[1234]

$(4\text{-}NCC_6H_4)_3P$. VIII. Crystals, m. 186 to 189° (from EtOH).[1234]

$(3\text{-}H_2NC_6H_4)_3P$. VIII. Crystals, m. 200° (from Ph_2O).[407]

$(2\text{-}Me_2NC_6H_4)_3P$. II. Crystals, m. 109° (from EtOH),[405] ^{1}H NMR.[406]

$(4\text{-}Me_2NC_6H_4)_3P$. II.[444,1225,1359] XI.[141,514,787,993,1161] Needles (from dilute EtOH), m. 275°,[1161] m. 273°,[993] m. 308(?),[787] m. 254° (from C_6H_6),[444] m. 282 to 287° (from C_6H_6),[1225] m. 278 to 282°,[1359] UV,[1224] ^{31}P +11.5 ppm (in CH_2Cl_2);[485] phosphorus acts as an electron acceptor toward the $Me_2NC_6H_4$ group,[1224] σ' +0.68;[1226] is oxidized by $AgClO_4$ to a green radical;[1359] develops blue color on air exposure, caused by traces of crystal violet;[787] soluble in dilute HCl.[993]

$(4\text{-}Et_2NC_6H_4)_3P$. XI. Needles, m. 274° (from EtOH).[787]
$(2\text{-}MeSC_6H_4)_3P$. II. White crystals, used as a complexing ligand.[341]
$(4\text{-}MeSC_6H_4)_3P$. ^{31}P +8.3 ppm (in CH_2Cl_2).[485]
$(2\text{-}MeSeC_6H_4)_3P$. II. Crystals, m. 159 to 161° (low yield).[341,342]
$(2,6\text{-}D_2C_6H_3)_3P$. XIV. From Ph_3P by selective ortho deuteration with D_2, using $(Ph_3P)_3RuHCl$ as catalyst.[1081] Crystals, m. 79°, IR, ^1H NMR, mass spect.[1081]
$[2,4\text{-}(MeO)_2C_6H_3]_3P$. XI. Crystals, m. 185 to 186°,[1132] m. 187 to 188°.[1134]
$[2,5\text{-}(MeO)_2C_6H_3]_3P$. II. Solid, m. 131.5 to 132.5°.[838]
$(2,4,6\text{-}D_3C_6H_2)_3P$. I. Crystals, purified by sublimation, m. 78.5 to 79.5°,[1279] mass spect.[1464]
$(2,4\text{-}Me_3C_6H_2)_3P$. III. Needles, m. 216 to 217° (from $CHCl_3$/ligroin).[986]
$(2,4,6\text{-}Me_3C_6H_2)_3P$. I.[1324] III.[507] White needles, m. 191 to 192° (from EtOH),[507] m. 205 to 206°,[986] m. 192 to 193°,[1324] UV, μ 1.37 D,[507] IR,[1324] ^1H NMR.[1324]
$[2,4,6\text{-}(MeO)_3C_6H_2]_3P$. XI. Crystals, m. 147 to 148°.[1132,1133]
$(3,5\text{-}Me_2\text{-}4\text{-}Me_2NC_6H_2)_3P$. II. Crystals, m. 169.5 to 171° (from EtOH).[444]
$(4\text{-}CF_3C_6F_4)_3P$. I. Crystals, m. 103 to 105° (from i-PrOH).[1349a]
$(4\text{-}C_6F_5OC_6F_4)_3P$. I. Crystals, m. 135 to 137° (from i-PrOH).[1349a]
$(2\text{-}PhC_6H_4)_3P$. I.[1007] III. (Using a little Sb metal as catalyst).[585] VII.[1007] Plates, m. 151 to 152° (from EtOH),[1448] m. 245°,[1007] dec. temp. 364°,[1007] ^{31}P +27 ppm.[444]
$(3\text{-}PhC_6H_4)_3P$. I. Glass. Dec. 372°.[1007]
$(4\text{-}PhC_6H_4)_3P$. III. Needles, m. 173°,[1007] m. 172° (from benzene),[1447] displays no trace of disproportionation with PCl_3 at 250°,[1447] dec. 405°.[1007]
$4(4'\text{-}Me_2NC_6H_4C_6H_4)_3P$. I. Needles, m. 270 to 272° (from Me_2NCHO).[1359]
$(1\text{-}C_{10}H_7)_3P$. I.[1123] II.[996,997,1355] Crystals, m. 278 to 280°,[996,997] m. 263 to 265° (from EtOH);[1355] m. rose to 270° when the sample was heated 1 hr at 280°;[1355] m. 189 to 190°(?),[1123] m. 282° (from dioxan);[55] adduct with $CHCl_3$, m. 262°.[55]
$(4\text{-}MeC_{10}H_6)_3P$. II. Crystals, m. 285.5° (from BuOH),[1355] m. 282 to 284° (from C_6H_6).[1225]
$(4\text{-}Me_2NC_{10}H_6)_3P$. I. Crystals, m. 305 to 308° (from diglyme).[1359]
$(9\text{-}Phenanthryl)_3P$. II. Crystals, m. 274 to 276°.[996,997]
$(9\text{-}Anthryl)_3P$. II. Crystals, m. 270 to 273°.[996,997]
$(1,2\text{-}Benz\text{-}10\text{-}anthryl)_3P$. II. Crystals, m. 192 to 194°.[996,997]
$(9\text{-}Bromo\text{-}10\text{-}anthryl)_3P$. II. Crystals, m. 206 to 208°.[996,997]

(2-Furyl)$_3$P. II. Crystals, m. 63° (from C_6H_6/hexane), b_4 136°, UV,[1054] ^{1}H NMR.[726a]
(2-pyrryl)$_3$P. II. Liquid, b_{10} 150°.[654]
(2-pyridyl)$_3$P. I. Crystals, m. 115°,[957] m. 113 to 114° (from MeOH),[288] $b_{0.15}$ 210°,[288] ^{1}H NMR;[478a] trihydrochloride, m. 207.5 to 209.5°; dipicrate, m. 142 to 143°.[957]
(2-pyridyl-CH_2)$_3$P. II. Crystals, m. 115 to 120°.[228]
(3-indolyl)$_3$P. I. Crystals, m. 195 to 196° (from acetone/EtOH),[1003] stable to hot aqueous alkali; forms a N-Ag derivative with NH_3/$AgNO_3$.[1003]
(3-Me-2-indolyl)$_3$P. I. Crystals, m. 156 to 158° (from ligroin),[1003] similar to previous compound.
(2-thienyl)$_3$P. II. Liquid, b_{1-2} 205°,[654] ^{1}H NMR.[725,726a]
(3-thienyl)$_3$P. II. Crystals, m. 69 to 70° (from C_6H_6), $b_{0.2}$ 150 to 156°,[726] ^{1}H NMR.[726,726a]

TYPE: R_2R^1P.

Me_2CF_3P. V. Liquid, b_{760} 46.8° (calc.), log P_{mm} = 7.630 - (1519/T),[532] ^{31}P +26.9 ppm (J_{PF} 65 Hz).[1358a]
$Me_2(HOCH_2)P$. XIII.1. Liquid, b_{10} 50 to 51°, b_{30} 70°, n_D^{20} 1.5011, d_4^{20} 0.9931,[486] K_b 9 x 10^{-7}.[487]
Acetyl deriv. XIV. (From Me_2HOCH_2P and MeCOCl). Liquid, $b_{10.5}$ 40 to 42°, n_D^{20} 1.4592, d_4^{20} 0.9940.[1099]
$Me_2(MeSCH_2)P$. VII. Liquid, b_5 28°;[873] inflames in air.
$Me_2(CH_2{=}CH)P$. I. Liquid, b_{760} 67.9° (calc.), log P_{mm} = 7.846 - (1693/T),[753] ^{1}H NMR.[555]
Me_2EtP. I.[753] VIII.[719] IX.1.[242] XII.[385a] Liquid, b. 83.4 to 84.5°,[242,719] b_{754} 63°,[385a] b_{760} 71.2° (calc.), log P_{mm} = 7.850 - (1712/T),[753] ^{1}H NMR,[385a] ^{31}P +48.5 ppm,[553,1008] pK_a (H_2O) 8.61.[1342]
$Me_2(HCFClCF_2)P$. I. XII. Liquid, b_{400} 108°,[178,385a] ^{1}H and ^{19}F NMR, ^{31}P +26.8 ppm.[340]
$Me_2(CH_2FCF_2)P$. XII. (Addition of Me_2PH to $CF_2{=}CHF$ gives also 48% $Me_2PCHF-CF_2H$). Liquid, b_{760} 96.7°, GLC, mass spect., ^{1}H and ^{19}F NMR, ^{31}P +32 ppm.[385b]
$Me_2(CHF_2CHF)P$. XII. (Addition of Me_2PH to $CF_2{=}CHF$ gives also 52% $Me_2PCF_2CH_2F$). Liquid, b_{760} 88°, GLC, mass spect., ^{1}H and ^{19}F NMR, ^{31}P +61 ppm.[385b]
$Me_2(CHF_2CF_2)P$. XII. Liquid, b_{758} 89°, ^{1}H and ^{19}F NMR,[385a] resistant to hydrolysis.
$Me_2(C_2F_5)P$. IX.1. Liquid, b_{760} 57° (calc.), log P_{mm} = 6.551 - (1218/T), IR.[1138]
$Me_2(HOCH_2CH_2)P$. IV. VII. Liquid, b_{20} 80 to 84°, n_D^{20} 1.4900, d_4^{20} 0.9421,[162] IR, oxide, m. 85 to 86.3°, K_b 2.40 x 10^{-6}.[162]
Acetyl deriv. XIV. (From phosphine above plus MeCOCl). Liquid, b_6 55 to 57°, n_D^{20} 1.4635, d_4^{20} 0.9910.[162]
$Me_2(MeCO)P$. XIII.6. Colorless liquid, b. 130.5°,[806] b_{58} 58°, n_D^{20} 1.4750, d_4^{20} 0.9841,[810] IR, UV, ^{1}H NMR, mass

spect.,[810] is oxidized in air.
$Me_2(CF_2{=}CClCF_2)P$. IV. Liquid, condensing at -35° under vacuum, IR, ^1H and ^{19}F NMR.[124]
$Me_2(NCCH_2CH_2)P$. IX.2. Liquid, $b_{0.1}$ 34 to 35°,[470] n_D^{25} 1.4800,[470] pK_a (H_2O) 6.37.[1342]
$Me_2(MeCF_3C{\cdot}OH)P$. XIII.3. Liquid, b_{44} 44 to 46°, n_D^{20} 1.4311, d_4^{20} 1.3255, IR.[488]
$Me_2[(CF_3)_2C{\cdot}OH]P$. XIII.3. Liquid, b_{100} 39 to 40°, n_D^{20} 1.3712, d_4^{20} 1.4191;[163] Ph_2PH and $(c\text{-}C_6H_{11})_2PH$ attack $(CF_3)_2C{=}O$ initially at the carbonyl oxygen to give $(CF_3)_2CHOPR_2$.[1352]
Acetyl deriv. XIV. From phosphine above and AcCl. Liquid, b_3 39 to 41°, n_D^{20} 1.3520, d_4^{20} 1.3369.[163]
Me_2BuP. I.[967] IX.1.[759] Liquid, b_{100} 69 to 70°,[759] b_{72} 56 to 60°,[967] n_D^{20} 1.4458, d_4^{20} 0.8455.[759]
$Me_2(EtC{\equiv}C)P$. I. II. Liquid, b_{90} 76 to 80°, IR, ^1H NMR.[23]
$Me_2((CF_3)_2CHCO)P$. XIII.3. [$Me_2PH + (CF_3)_2C{=}C{=}O$]. Liquid, b_{62} 61°, n_D^{20} 1.3955, d_4^{20} 0.9408, IR, ^1H NMR.[809]
$Me_2(n\text{-}C_{10}H_{21})P$. IV. (From MeCl + $C_{10}H_{21}PH_2$ than OH^-). Liquid, $b_{0.4-0.5}$ 85°, IR, ^1H NMR, ^{31}P +52.5 ppm.[535]
$Me_2(n\text{-}C_{12}H_{25})P$. IV. Liquid, $b_{0.03}$80 to 83°, IR, ^1H NMR, ^{31}P +52.5 ppm.[535]
$Me_2(PhCH_2)P$. I. Liquid, b_{12} 93 to 96°.[733]
$Me_2(PhC{\equiv}C)P$. I. II. Liquid, $b_{0.6}$ 62 to 65°, IR, ^1H NMR.[23]
Me_2PhP. I.[34,94,441,752,947,977,1101,1454] IV.[438] VI.[983] Liquid, b_{760} 192°,[94,983] b_{760} 190°,[977] b_{42} 116°,[1149] b_{15} 85°,[1101] b_{14} 84°,[441] $b_{13.5}$ 83 to 84°,[977] b_{20} 82°,[947] b_8 65°,[34] b_{30} 91 to 92°,[752] $n_D^{19.5}$ 1.5673,[34] n_D^{20} 1.5620, d_4^{20} 0.9670,[752,1454] d_4^{11} 0.9678,[983] IR,[51] UV,[1160,1236] ^1H NMR,[300,764,802] sign of coupl. const.,[885] ^{13}C-P coupl. const. -14 Hz,[885] ^{31}P +47 ppm,[553,1008] +46.9 ppm,[482] μ 1.31 D,[1242,1246] μ 1.22 D,[778] pK_a (H_2O) 6.50,[1342] ΔH_{vap}. 10.38 kcal/mole.[94] ESR of radical anion,[349] D/H exchange in the Me_2P group takes place twice as fast as in the Me group of toluene, 120 times faster as in C_6H_6, while 10^4 faster than in the Me_2N group;[1454] Me_2PhP carbanion stabilized as a result of 2p-3d conjugation;[1454] the deuterated compound $(CD_3)_2$-PhP shows a small isotope effect in salt formation when compared with Me_2PhP;[764] red adducts with poly-nitroaryls.[94a]
$Me_2C_6F_5P$. I. Liquid, $b_{0.4}$ 47 to 48°,[387] IR,[387,389] ^1H and ^{19}F NMR,[79,599,600] ^{31}P +47.8 ppm.[390]
$Me_2(3\text{-}DC_6H_4)P$. I. Liquid, b_8 64 to 65°, n_D^{20} 1.5631, d_4^{20} 0.9809.[865]
$Me_2(4\text{-}DC_6H_4)P$. I. Liquid, b_8 65 to 66°, n_D^{20} 1.5627, d_4^{20} 0.9757.[865]
$Me_2(3\text{-}FC_6H_4)P$. I. Liquid, b_9 67 to 69°,[1251] $b_{1.5}$ 48 to 50°,[1156] n_D^{20} 1.5394,[1251] ^{19}F NMR, σ_m (of Me_2P group) +0.03.[1156,1251]

$Me_2(4\text{-}FC_6H_4)P$. I. Liquid, $b_{1.0}$ 49 to 50°,[1156] b_{12} 72 to 72.3°, n_D^{20} 1.5369,[1243] ^{19}F NMR, σ_p (of Me_2P group) +0.23.[1243]
$Me_2(2\text{-}ClC_6H_4)P$. I. Liquid, $b_{0.25}$ 43°.[518]
$Me_2(4\text{-}ClC_6H_4)P$. I. Liquid, b_4 78 to 80°, n_D^{20} 1.5795, d_4^{20} 1.1240.[1362]
$Me_2(4\text{-}BrC_6H_4)P$. I. Liquid, $b_{17.5}$ 99°, b_4 55°; CS_2 adduct, m. 96°;[292] mixture of isomers separated by GLC.[1199]
$Me_2(4\text{-}MeC_6H_4)P$. I.[292,947] VI.[274] Liquid, b. 210°,[274] b_{12} 93 to 95°,[947] ^1H NMR, σ_p (of Me_2P group) > -0.02;[1233] CS_2 adduct, red plates, m. 110° (open tube),[274] m. 118° (sealed tube);[292] adduct with p-benzoquinone, cream-colored solid, m. above 250°.[292]
$Me_2(3\text{-}EtC_6H_4)P$. I. Liquid, b_{18} 100 to 102°.[74]
$Me_2(4\text{-}EtC_6H_4)P$. I. Liquid, b_{18} 101 to 105°.[74]
$Me_2(4\text{-}Me_2NC_6H_4)P$. I. Liquid, b. 265°,[993] UV,[1224] (the phosphorus atom acts here as an electron acceptor); CS_2 adduct, m. 162°.[993]
$Me_2(4\text{-}MeOC_6H_4)P$. I. CS_2 adduct, m. 119°.[292]
$Me_2(4\text{-}PhOC_6H_4)P$. I. Liquid, b_{13} 183°; d_4^{20} 1.037;[289] CS_2 adduct, m. 88°,[292] m. 87.5°.[290]
$Me_2(4\text{-}PhCH_2C_6H_4)P$. VI. Liquid, b_{20} 197°.[986]
$Me_2(4\text{-}PhCH_2CH_2C_6H_4)P$. VI. High-boiling liquid.[986]
$Me_2(2,5\text{-}Me_2C_6H_3)P$. I.[292,721] VI.[274] Liquid, b_{12} 106°, d_4^{20} 0.9541;[721] CS_2 adduct, red plates, m. 76°,[721] m. 72° (sealed tube).[292]
$Me_2(2,4\text{-}Me_2C_6H_3)P$(?). VI. Liquid, b. 230°,[274] b. 233;[245] the precise structure of this phosphine is in doubt; probably also contains $(3,5\text{-}Me_2C_6H_3)PMe_2$; CS_2 adduct, red plates, m. 115°.[274]
$Me_2(2,4,6\text{-}Me_3C_6H_2)P$. I. Liquid, b_{16} 133°,[284] b_6 100°,[292] n_D^{25} 1.5554, d_4^{25} 0.9570;[284] CS_2 adduct, red solid, m. 58 to 59°,[284] m. 46° (sealed tube).[292]
$(CF_3)_2MeP$. II.[454] VI.[480] XIV.[533] [From MeI and $(CF_3)_3P$]. Liquid, b_{760} 35.2° (calc.), log P_{mm} = 7.356 - (1379/T),[533] IR,[454] ^1H and ^{19}F NMR,[192,340a] ^{31}P +5.76 ppm.[192]
$(CF_3)_2EtP$. XII. Liquid, b_{763} 62 to 63°,[385a] ^1H and ^{19}F NMR;[340a,385a] CF_3 groups split off by OH^-.
$(CF_3)_2(CHF_2CHF)P$. XII. [Contained 1% isomer $(CF_3)_2P(CF_2\text{-}CH_2F)$]. Liquid, b_{760} 66.5°, GLC, ^1H and ^{19}F NMR, $\Delta H_{vap.}$ 7.16 kcal $mole^{-1}$, ^{31}P +7.8 ppm;[385b] hydrolysis with OH^- gives CHF_3 and cis-CHF=CHF.[385b]
$(CF_3)_2(CHF_2CF_2)P$. Liquid, b_{755} 61°, ^1H and ^{19}F NMR;[385a] hydrolysis with OH gives CHF_3 and CHF_2CHF_2.
$(CF_3)_2PrP$. XII. Liquid, b_{748} 85°, GLC,[385a] ^{19}F NMR;[340a,385a] CF_3 groups split off by OH^-.
$(CF_3)_2C_3F_7P$. II. Liqui^, b_{760} 74° (calc.), log P_{mm} = 7.33 - (1543/T);[454] CF_3 groups split off by OH^-; IR, ^{19}F NMR;[454] inflames in air.
$(CF_3)_2BuP$. II.[454] XII.[385a] Liquid, b_{750} 107°,[385a] b_{760} 110° (calc.), log P_{mm} = 7.89 - (1920/T);[454] freezes to

a glass at -90 to -100°;[454] GLC,[385a] IR,[454] ^{1}H and ^{19}F NMR;[385a] CF_3 groups split off by OH^-.[385a,454]
$(CF_3)_2$(2-Bu)P. XII. Liquid, b_{764} 108°, GLC, ^{1}H and ^{19}F NMR;[385a] CF_3 groups split off by OH^-.
$(CF_3)_2$i-BuP. II. Liquid, b_{760} 98° (calc.), log P_{mm} = 7.73 - (1800/T),[454] IR, CF_3 groups split off by OH^-.[454]
$(CF_3)_2(CF_3CH{=}C{\cdot}CF_3)$P. XII. Condensing at -46° in the vacuum, IR.[264]
$(CF_3)_2$(cyclopent-3-enyl)P. XII. (Ni catalysis). Liquid, vapor pressure 520 mm at 12°, IR, ^{1}H and ^{19}F NMR, mass spect.[324]
$(CF_3)_2$PhP. II.[454] X.[95] Liquid, b_{20} 62 to 65°,[95] b. 146 to 148°,[454] $\Delta H_{vap.}$ 9.05 kcal/mole.[94]
$(CF_3)_2(3\text{-}FC_6H_4)$P. I. Liquid, b_{60} 69°,[1156] GLC, ^{19}F NMR, σ_m [of $(CF_3)_2$P group] +0.69.[1156]
$(CF_3)_2(4\text{-}FC_6H_4)$P. I. Liquid, b_{60} 70°, ^{19}F NMR, GLC.[1156]
$(ClCH_2)_2$MeP. I. Liquid, b_5 44°; ^{1}H NMR, identified as sulfide, yellow oil.[933]
$(ClCH_2)_2$EtP. IX.2. Liquid, b_4 56 to 58°,[546] b_5 55 to 57°,[545] not stable at room temp.; gives off HCl.[546,933]
$(ClCH_2)_2$PhP. IX.2. Decomposes on heating under vacuum.[546,933]
$(Et_2NCH_2)_2(Et_2NCH_2CH_2)$P. XIII.1. Liquid, b_3 160 to 162°, pK_1 9.09, pK_2 8.46, pK_3 7.85.[681]
$(Et_2NCH_2)_2(MeO_2CCH_2)$P. XIII.1. Liquid, $b_{0.01}$ 91 to 93°.[679]
$(Et_2NCH_2)_2$PrP. XIII.1. Liquid, $b_{0.01}$ 74 to 76°, n_D^{20} 1.4662, d_4^{20} 0.9046.[1112]
$(Et_2NCH_2)_2(MeO_2CCH_2CH_2)$P. XIII.1. Liquid, $b_{0.3}$ 140 to 142°.[679]
$(Et_2NCH_2)_2(MeO_2CCH{\cdot}Me)$P. XIII.1. Liquid, $b_{0.01}$ 99 to 101°.[679]
$(Et_2NCH_2)_2(MeO_2CCH{\cdot}Et)$P. XIII.1. Liquid, $b_{0.01}$ 106 to 108°.[679]
$(Et_2NCH_2)_2$PhP. XIII.1. Liquid, $b_{0.05}$ 104 to 106°,[236] $b_{0.1}$ 136°,[918] $b_{0.001}$ 170 to 175° (bath temp.), n_D^{20} 1.5300,[918] n_D^{20} 1.5192, d_4^{20} 0.9369,[1112] IR,[918] ^{31}P +51.3 ppm.[911,918]
$(HOCH_2)_2$MeP. IX.2.[460,461] XIII.1.[486,487] Liquid, b_3 89 to 91°,[486] $b_{0.5}$ 67 to 69°,[460,461] b_2 87 to 89°,[1372] n_D^{20} 1.5325,[486] n_D^{20} 1.5270,[1372] d_4^{20} 1.157,[486] d_4^{20} 1.1545,[1373] K_b 1.5 x 10^{-7}.[487]
$(HOCH_2)_2$EtP. IX.2.[546 1372] XIII.1.[487] Liquid, $b_{0.01}$ 73°,[546] b_1 82 to 83°,[487] b_2 102 to 105°,[1372] n_D^{20} 1.5255, d_4^{20} 1.0855,[487] n_D^{20} 1.5182, d_4^{20} 1.093.[1372]
Diacetyl deriv. VI.[1379] IX.2.[1004] Liquid, $b_{0.5}$ 95 to 97°,[1379] b_3 93.5 to 94.5°, n_D^{20} 1.4702, d_4^{20} 1.0950,[1004] n_D^{20} 1.4832, d_4^{20} 1.1213,[1379] ^{1}H NMR.[1115]
$(HOCH_2)_2(HCF_2CF_2)$P. XIII.1. Liquid, $b_{0.018}$ 100 to 110°, n_D^{20} 1.4342, d_4^{20} 1.4428.[1179]
$(HOCH_2)_2$PrP. IX.2. Liquid, b_8 113 to 115°,[1372] b_1 84°, n_D^{20} 1.5061, d_4^{20} 1.0690,[1108,1110] n_D^{20} 1.5108, d_4^{20} 1.0670.[1372]
Diacetyl deriv. IX.2. Liquid, b_2 108 to 109.5°, n_D^{20}

1.4694, d_4^{20} 1.0627.[1004]
$(HOCH_2)_2(CH_2{=}CHCH_2)P$. IX.2. Liquid, b_7 120 to 122°, n_D^{20} 1.5260, d_4^{20} 0.860;[1374] underwent some polymerization during dist.[1374]
Diacetyl deriv. IX.2. Liquid, b_3 118 to 119°, n_D^{20} 1.4855, d_4^{20} 1.0935.[1004]
$(HOCH_2)_2(MeCF_3C{\cdot}OH)P$. XIII.1. Oil, decomposes on dist., n_D^{20} 1.5040, d_4^{20} 1.5193, IR.[488]
$(HOCH_2)_2[(CF_3)_2C{\cdot}OH]P$. XIII.1. Liquid, decomposes on dist., n_D^{20} 1.4501, d_4^{20} 1.5193, IR.[488]
$(HOCH_2)_2BuP$. IX.2. Liquid, b_5 120 to 122°, n_D^{20} 1.5010, d_4^{20} 1.0252.[1372]
Diacetyl deriv. VI.[1379] IX.2.[1004] Liquid, b_1 102°,[1379] b_3 121 to 122°, n_D^{20} 1.4680, d_4^{20} 1.0501,[1004] n_D^{20} 1.4760, d_4^{20} 1.0687,[1379] ^1H NMR.[1115]
$(HOCH_2)_2(n\text{-}C_7H_{15})P$. IX.2. Undistillable liquid, n_D^{20} 1.4850.[1111]
Diacetyl deriv. IX.2. Liquid, b_2 153 to 154°, n_D^{20} 1.4666, d_4^{20} 1.10081.[1004]
$(HOCH_2)_2PhP$. IX.2.[1110] XIII.1.[546] Liquid, b_1 84 to 85°,[1110] $b_{0.1-0.15}$ 93 to 96° (partial dec.); characterized as sulfide, m. 84 to 86°.[546]
Diacetyl deriv. IX.2. Liquid, b_1 140°, n_D^{20} 1.5475, d_4^{20} 1.1830,[1379] ^1H NMR.[1115]
$(MeCO_2CH_2)_2AmP$. IX.2. Liquid, b_3 133 to 134°, n_D^{20} 1.4681, d_4^{20} 1.0354.[1004]
$(MeCO_2CH_2)_2(n\text{-}C_6H_{13})P$. IX.2. Liquid, b_3 142 to 143°, n_D^{20} 1.4670, d_4^{20} 1.0181.[1004]
$(KO_2C)_2(c\text{-}C_6H_{11})P$. VII. Solid, dec. 257 to 260°.[715]
$(NaO_2C)_2PhP$. VII. Solid, dec. 235 to 240°, free acid not stable.[715]
$(KO_2C)_2PhP$. VII. Solid, dec. 215 to 220°, free acid not stable.[715]
$(EtO_2C)_2(c\text{-}C_6H_{11})P$. VII. Liquid, b_{4-5} 138 to 139°,[715] stable in air, fruitlike odor.[715]
$(EtO_2C)_2PhP$. VII. Liquid, b_{4-5} 150 to 153°,[737] b_{5-6} 152 to 155°.[715]
$(H_2NCO)_2PhP$. XIII.5. Crystals, m. 159 to 161°, IR;[1075] has been claimed to be useful for flameproofing cloth.[1075]
$(PhNHCO)_2BuP$. XIII.5. Crystals, m. 84°.[1287]
$(PhNHCO)_2i\text{-}BuP$. XIII.5. Crystals, m. 142 to 143°,[174] useful as gasoline additive.
$(PhNHCO)_2(n\text{-}C_8H_{17})P$. XIII.5. Viscous liquid, useful as gasoline additive.[174]
$(PhNHCO)_2PhP$. XIII.5. Viscous liquid, useful as gasoline additive.[174]
$(MeNHCS)_2(c\text{-}C_6H_{11})P$. VII. Yellow crystals, m. 153.5° (from EtOH), IR.[666]
$(MeNLiCS)_2(c\text{-}C_6H_{11})P$. VII. Solid, contains $2MeNCS{\cdot}3Et_2O$, m. 150° (dec.).[666]

$(PhNHCS)_2(c\text{-}C_6H_{11})P$. XIII.5. Yellow needles, m. 118.5°.[665]
$(PhNHCS)_2PhP$. XIII.5. Orange oil, decomposes on dist.[666]
$(4\text{-}BrC_6H_4NHCS)_2PhP$. XIII.5. Crystals, m. 153° (from EtOH).[666]
$(2\text{-}C_{10}H_7NHCS)_2(c\text{-}C_6H_{11})P$. XIII.5. Yellow crystals, m. 136° (from EtOH), IR.[666]
Et_2MeP. VIII.[719] IX.1.[242] Liquid, b. 110 to 112°,[242] b. 111 to 112°,[719] ^{31}P +34 ppm,[553,1008] pK_a (H_2O) 8.61.[1342]
Et_2CF_3P. IX.1. Liquid, b_{760} 97°, log P_{mm} = 6.813 - (1457/T), IR.[1138]
$Et_2(HOCH_2)P$. IX.2. XIII.1. Liquid, b_{11} 72°,[487] b_{10} 70°, n_D^{20} 1.4938,[546] n_D^{20} 1.4971, d_4^{20} 0.9342;[487] isomerizes on heating to 250 to 260° to $Et_2MeP{=}O$.[546]
Acetyl deriv. IX.2. Liquid, b_4 50 to 51°, n_D^{20} 1.4635, d_4^{20} 0.9626,[1004] ^1H NMR.[1115]
$Et_2(EtO_2C)P$. VII. Liquid, b_3 53°,[649] b_{13} 68.5 to 70°, n_D^{20} 1.4633, d_4^{20} 0.9561, IR.[814]
$Et_2[M^+(\bar{S}_2C)_2]P^+$. From Et_2PH and CS_2.
(M = Et_3NH). Red crystals, m. 53 to 57° (dec.).
(M = K). Carmine-red crystals, m. ∿95° (dec.).
(M = Ph_4P). Red crystals, m. ∿110° (dec.), ^1H NMR.[275a]
$Et_2(PhNHCS)P$. XIII.5. Yellow-orange oil; adduct with $HgBr_2$, m. 144°.[665]
$Et_2(CH_2{=}CH)P$. I. Liquid, b_{744} 125°,[399] b_{54} 48°, UV.[1420]
$Et_2(HC{\equiv}C)P$. I. Liquid, b_{760} 128°,[218,219] b_{77} 66 to 67°, n_D^{20} 1.4808,[1389] IR,[218,219,1389] ^1H NMR.[218,219,334,1298]
$Et_2(H_2NCH_2CH_2)P$. VII. Liquid, b_9 124 to 128°;[661] MeI quaternizes N atom, dec. 80°.[661]
$Et_2(Et_2NCH_2CH_2)P$. VII. Colorless liquid, b_8 86°, pK 7.93, IR, ^{31}P +20.4 ppm;[697] EtI quaternizes phosphorus atom, dec. 86°.[697]
$Et_2(HOCH_2CH_2)P$. VII. Liquid, b_{24} 106 to 109°,[691] b_{15} 85 to 87°, n_D^{20} 1.4911, d_4^{20} 0.9152.[162]
$Et_2(NaO_2CCH_2)P$. IX.2. Solid, dec. 260 to 290°.[704]
Methyl ester: VI. Liquid, b_1 48 to 49°, n_D^{20} 1.4735, d_4^{20} 0.9771.[1130]
Ethyl ester: IX.2. Liquid, b_{12} 90 to 91°.[704]
$Et_2(H_2NCOCH_2)P$. XIV. (From ester and alcoholic NH_3). Colorless plates, m. 98 to 99.5° (from MeOH), pK 4.05 ($EtOH/H_2O$), ^1H NMR.[705]
$Et_2(MeCO)P$. VII.[696] XIV. (From R_3GePEt_2 and Ac_2O).[1216] Liquid, b. 165 to 170°,[696] b_{10} 63°, n_D^{20} 1.4746, IR, ^1H NMR.[1216]
$Et_2(HSCH_2CH_2)P$. VII. Liquid, $b_{0.001}$ 43 to 44°, n_D^{20} 1.5202,[428] pK_a 7.4,[1271] useful for extraction of metals from aqueous solution and also as antioxidants; oxygen catalyzes isomerization to Et_3PS.[1271]
$Et_2(EtSCH_2CH_2)P$. VII. Liquid, $b_{0.4}$ 59°.[1297]
Et_2PrP. IX.1. Liquid, b. 146 to 149°.[242]
$Et_2C_3F_7P$. II. Liquid, b. 124 to 125°, b_{760} 126° (calc.), log P_{mm} = 7.98 - (2034/T),[454] ^{19}F NMR;[454] C_3F_7 group

split off by OH^-.

$Et_2(CH_2=C=CH)P$. I. Et_2PCl and XMgC≡CMe give a mixture of $Et_2PC≡CMe$ and $Et_2PCH=C=CH_2$, b. of mixture: b_{12} 48°,[1302] ^1H NMR and sign of coupl. const.[1302]

$Et_2(MeC≡C)P$. I.[23,218,219] II.[23] Liquid, b_{760} 165°,[218,219] b_{50} 84 to 85°,[23] IR,[23,218] ^1H NMR.[23,218,219,1298]

Et_2(cis-NCCH=CH)P. IV. Liquid, $b_{0.8}$ 75 to 77°, n_D^{20} 1.5128; L•MeI, m. 113°.[496]

Et_2(trans-NCCH=CH)P. IV. Liquid, $b_{0.8}$ 68 to 72°, b_3 80°, n_D^{20} 1.5103, IR;[496] L•MeI, m. 249°.[496]

$Et_2(NCCH_2CH_2)P$. XII. Liquid, b_3 110°; L•MeI, m. 213°.[496]

$Et_2[NCCH_2(PhCH_2S)CH]P$. XII. Liquid, $b_{0.6}$ 150 to 170°, n_D^{20} 1.5647; L•MeI, m. 121°.[496]

$Et_2(H_2NCH_2CH_2CH_2)P$. XII. Liquid, b_4 74 to 83°.[462]

$Et_2(HOCH_2CH_2CH_2)P$. VII. Liquid, b_3 81 to 82°, IR; L•MeI, m. 151 to 152°.[700]

$Et_2(HOCH•MeCH_2)P$. VII. Pale-yellow oil, b_3 64 to 66°, IR; L•MeI, m. 173 to 174°.[700]

$Et_2(HO_2CCH_2CH_2)P$. XII. Liquid, b_4 120 to 126°.[462] Na salt, m. 220 to 221°.[704]
Ethyl ester: IX.2. Liquid, b_1 70 to 71°.[704]

$Et_2(H_2NCOCH_2CH_2)P$. XIV. From ester and NH_3. Viscous oil, pK 5.89 ($EtOH/H_2O$).[705]

$Et_2(NaO_2CCH•Me)P$. IX.2. Solid, m. 310 to 312° (dec.).[704]
Ethyl ester: IX.2. Liquid, b_1 53 to 54°.[704]

Et_2BuP. IX.1. Liquid, b_{100} 110 to 111°, n_D^{20} 1.4596, d_4^{20} 0.8094.[759]

$Et_2(CH_2=CHCH_2CH_2)P$. I. II. Liquid, b. 170 to 172°; L•MeI, m. 189 to 191°.[662]

Et_2(trans-$CF_3CH=C•CF_3$)P. XII. Liquid, b. 132°, IR, ^1H and ^{19}F NMR.[264]

Et_2-1-(2-chlorotetrafluorocyclobut-1-enyl)P. IV. Liquid, decomposes above 20°, IR, ^1H NMR.[265]

Et_2(1-pentafluorocyclobutenyl)P. IV. Solid, m. 101 to 103°, IR, ^1H NMR.[265]

$Et_2(NaO_2CCH_2CH_2CH_2)P$. IX.2. Solid, m. 240 to 243°.[704]
Ethyl ester: IX.2. Liquid, b_1 76.5 to 77°.[704]

$Et_2(H_2NCOCH_2CH_2)P$. XIV. From ester and NH_3. Crystals, m. 34 to 36° (from MeOH), pK 6.2 ($EtOH/H_2O$).[705]

$Et_2(EtO_2CCH_2Me•CH)P$. IX.2. Liquid, b_1 59 to 60°.[704]

$Et_2(EtO_2CEt•CH)P$. IX.2. Liquid, b_1 54.5 to 57°.[704]

Et_2(PrCO)P. VII. XIII.6. Liquid, b_{15} 84 to 87°.[687]

Et_2(i-PrCO)P. VII. Liquid, b_{760} 183 to 186°.[687]

$Et_2((CF_3)_2CHCO)P$. XII. Liquid, b_{10} 53 to 54°, IR, ^1H NMR.[809]

Et_2(i-Am)P. IX.1. Liquid, b. 185 to 187°; hydrochloride is volatile at 270°.[242]

$Et_2(MeCH=C=CHCH_2)P$. XII. Liquid, b_5 60 to 61°, n_D^{20} 1.5050, d_4^{20} 0.867.[1104]

Et_2(MeC≡C-C≡C)P. I. Liquid, b 85°,[163,219] UV,[218] IR.[218,219]

$Et_2(MeCOCH_2Me \cdot CH)P$. VII. Colorless oil, b. 155 to 160°.[671]
$Et_2[\overline{(CH_2)_4C} \cdot OH]P$. XIII. Liquid, ^{31}P +53 ppm.[1216a]
$Et_2(BuCO)P$. VII. Liquid, b_{15} 90°.[687]
$Et_2(i\text{-}BuCO)P$. VII. Liquid, b_{15} 85°.[687]
$Et_2(t\text{-}BuCO)P$. VII. Liquid, b_{38} 94 to 96°.[687]
$Et_2(n\text{-}C_6H_{13})P$. IX.2. Liquid, b_1 73 to 74°, n_D^{20} 1.4630, d_4^{20} 0.8144.[1004]
$Et_2(EtCH{=}C{=}CHCH_2)P$. XII. Liquid, b_5 78 to 79°, n_D^{20} 1.5025, d_4^{20} 0.8569.[1104]
$Et_2(c\text{-}2\text{-}HOC_6H_{10})P$. VII. Liquid, b_3 100°, IR; L•MeI, m. 110°.[700]
$Et_2(1\text{-cyclohexenyl-}C{\equiv}C)P$. I. Liquid, b_2 120°,[163,219] UV,[219] IR.[218]
$Et_2(n\text{-}C_{12}H_{25})P$. IV. Liquid, $b_{0.6}$ 117 to 135°,[535] IR, ^1H NMR, ^{31}P +25 ppm.[535]
$Et_2(PhCH_2)P$. IX.1. Liquid, b. 250 to 255°; hydrochloride is volatile at 325°;[242] crude product b. 240 to 260°.[1406]
$Et_2(3\text{-}MeC_6H_4CH_2)P$. I. Liquid, $b_{1.8}$ 97 to 114° (impure); L•EtI, m. 133 to 135°.[294]
$Et_2(2\text{-}Cl\text{-}5\text{-}MeC_6H_3CH_2)P$. I. Liquid, b_4 123 to 127°; L•EtI, m. 123 to 123.5°.[294]
$Et_2(PhCO)P$. VII. Liquid, b_2 98 to 100°.[696]
$Et_2(3\text{-}ClC_6H_4CO)P$. VII. Liquid, b_{12} 175°.[687]
$Et_2(4\text{-}ClC_6H_4CO)P$. VII. Liquid, b_{12} 165 to 168°.[687]
$Et_2(4\text{-}BrC_6H_4CO)P$. VII. Liquid, b_{12} 185°, IR.[687]
$Et_2(3\text{-}MeC_6H_4CO)P$. VII. Liquid, b_{10} 138 to 141°.[687]
$Et_2(4\text{-}MeC_6H_4CO)P$. VII. Liquid, b_{13} 144 to 146°.[687]
$Et_2(4\text{-}MeOC_6H_4CO)P$. VII. Liquid, b_{14} 200 to 204°, IR.[687]
$Et_2(1\text{-}C_{10}H_7CO)P$. VII. Liquid, $b_{2.5}$ 165°, UV, IR;[687] L•MeI, dec. 118°.[687]
$Et_2(2\text{-furyl-}CO)P$. VII. Liquid, b_2 85°, IR.[687]
$Et_2(2\text{-thienyl-}CO)P$. VII. Liquid, b_2 102 to 103°, UV, IR;[687] L•MeI, dec. 73°.[687]
$Et_2(PhN{=}C \cdot Ph)P$. VII. Liquid, b_3 180°.[687]
$Et_2(HN{=}C \cdot Ph\text{-}N{=}C \cdot Ph)P$. VII. Crystals, m. 181 to 183°, IR.[652]
$Et_2(Ph_3C)P$. I. Viscous, yellow oil, b_3 194 to 198°;[714] L•MeI, m. 205 to 207° (dec.).[714]
$Et_2(PhCH_2CH \cdot OH)P$. VII. Pale-yellow oil, b_3 138 to 140°.[700]
$Et_2(PhCH_2CO)P$. VII. Liquid, b_{12} 168°.[687]
$Et_2(Ph_2CHCH_2)P$. VII. (From $Ph_2C{=}CH_2$ + $LiPR_2$). Liquid, b_2 165 to 170°.[670]
$Et_2(Ph_2CHCO)P$. XIV. (From R_3GePEt_2 + $(Ph_2CHCO)_2O$). Liquid, IR, ^1H NMR.[1216]
$Et_2(PhCH{=}C \cdot Me)P$. XIV. (From $Et_2PC \cdot Me{=}PPh_3$ and PhCHO). Liquid, b_{12} 142 to 145°,[683] stereochemistry not known.
$Et_2(PhCBr{=}CH)P$. XIV. (From $Et_2PC{\equiv}CPh$ and HBr). Light-yellow powder, m. ∿27°, IR, ^1H NMR,[24] stereochemistry not known.
$Et_2(trans\text{-}PhCH{=}C \cdot Ph)P$. VII. Colorless oil, b_3 165 to 167°.[670]

Et_2[trans-Ph(HO_2C)C=C·Ph]P. VII. Crystals, m. 180 to 183° (dec.) (from acetone/ligroin).[670]
Ethyl ester: VII. Liquid, b_3 200 to 205°.[670]

Et_2(PhC≡C)P. I.[23,219] II.[23] Liquid, $b_{0.7}$ 105 to 107°,[23] $b_{0.8}$ 105°,[219] b_1 105°,[163] UV,[218,219] IR,[23,218,219] ^1H NMR.[23]

Et_2($MeCOCH_2$Ph·CH)P. VII. Pale-yellow oil, b_4 135 to 140° (crude); L·MeI, m. 184 to 185°.[671]

Et_2(t-Bu$COCH_2$Ph·CH)P. VII. Colorless oil, b_4 150 to 155°; L·MeI, m. 159°.[671]

Et_2(Ph$COCH_2$Ph·CH)P. VII. Colorless oil, b_2 190 to 195° (crude); L·MeI, m. 177 to 177.5°.[671]

Et_2PhP. I.[290,440,882,947,977,1466] IV.[1241] VI.[926,983,987,1012] X.[674] Liquid, b. 221.9°,[94,983] b_{740} 223°,[1466] b_{29} 120 to 121°,[287] b_{20} 108 to 109°,[947] b_{10} 96 to 98°,[757,977] b_{11} 97 to 99°,[440] b_8 95°,[674] b_5 85 to 90°,[882] b_2 60 to 65°,[1012] b_2 57 to 60°,[1323] n_D^{20} 1.5458,[287] n_D^{20} 1.5460,[1323] d_4^{20} 0.9545, d_4^{13} 0.9571,[287] d_4^{25} 0.9545,[94] d_4^{20} 0.9445,[1323] $\Delta H_{vap.}$ 12.46 kcal/mole,[94] pK_a (in 80% EtOH) 3.21,[440] pK_a (H_2O) 6.25,[1342] 6.41,[1323] μ 1.35 D,[778] μ 1.40 D,[440] μ 1.42 D,[1246] UV,[1284] IR,[275] ^{31}P +16 ppm, +15.1 ppm,[553,1008] +17.1 ppm;[482] mono- and dihydrochlorides have been isolated;[983] CS_2 adduct, red solid, m. 45°.[292]

$Et_2C_6F_5P$. I. Liquid, b_{14} 92 to 94°, IR,[387] ^1H and ^{19}F NMR,[79,600] ^{31}P +23.4 ppm.[390]

Et_2(4-FC_6H_4)P. I. Liquid, b_{13} 95°,[440] μ 1.82 D, pK_a (80% EtOH) 3.02.[440]

Et_2(2-ClC_6H_4)P. I. Liquid, $b_{0.4}$ 83 to 85°.[518]

Et_2(4-ClC_6H_4)P. I.[287] VI.[984] Liquid, b. 255 to 257°,[984] b_2 105°,[440] b_4 89 to 90°,[1323] b_{15} 129 to 130°,[287] n_D^{20} 1.5603, d_4^{20} 1.0708,[94,287] n_D^{20} 1.5610, d_4^{20} 1.067,[1323] pK_a (80% EtOH) 2.79,[440] pK_a (H_2O) 5.68,[134,1323] μ 1.90 D,[440] $\Delta H_{vap.}$ 13.13 kcal/mole.[94]

Et_2(2-BrC_6H_4)P. II. Liquid, $b_{0.15}$ 90 to 91°.[518]

Et_2(4-BrC_6H_4)P. I.[287] VI.[984] Liquid, b. 265°,[984] $b_{3.5}$ 118°,[440] b_{15} 141 to 143°, n_D^{20} 1.5821, d_4^{20} 1.2886;[287] CS_2 adduct, very unstable solid;[292] pK_a 2.70,[440] μ 1.97 D,[440] $\Delta H_{vap.}$ 14.18.[94]

Et_2(2-MeC_6H_4)P. VI. Liquid, b. 263°.[984]

Et_2(3-MeC_6H_4)P. I. II. Liquid, $b_{0.6}$ 70 to 90°, L·EtI, m. 151°.[294]

Et_2(4-MeC_6H_4)P. I.[94,287,947] VI.[274] VII.[710] Liquid, b. 240°,[274] b_{13} 113.5 to 114.5°,[287] b_{12} 114 to 115°,[947] b_{1-2} 90 to 95°,[710] $\Delta H_{vap.}$ 12.74,[94] n_D^{20} 1.5428, d_4^{20} 0.9373,[287] pK_a (H_2O) 6.80;[1321] CS_2 adduct, m. 55° (sealed tube).[292]

Et_2(4-EtC_6H_4)P. VI. Liquid, b. 268 to 270°, d_0^{25} 0.929.[984]

Et_2(2-$MeOCH_2CH_2C_6H_4$)P. I. Liquid, b_{26} 119 to 133°.[948]

Et_2(4-$PhCH_2C_6H_4$)P. VI. Liquid, b_{20} 235°.[986]

Et_2(4-$PhCH_2CH_2C_6H_4$)P. VI. High-boiling liquid.[986]

$Et_2(4\text{-}HOC_6H_4)P$. XIV. From 4-methoxy derivative and HI at 135°. Liquid, b_{19} 168 to 176°; L•MeI, m. 168 to 169°.[288]

$Et_2(4\text{-}MeOC_6H_4)P$. I.[287,288,292,440] VI.[984] Liquid, b. 266 to 267°,[984] b_{40} 166 to 171°,[288] b_{10} 130 to 131°,[287] $b_{1.5}$ 108°,[440] n_D^{20} 1.5498,[287] d_4^{20} 1.0015,[287] d_0^{18} 0.9978,[984] pKa (80% EtOH) 3.88,[440] pK_a (H_2O) 7.11,[1321] μ 1.99 D,[440] σ_p (of Et_2P group) +0.03.[1321]

$Et_2(4\text{-}EtOC_6H_4)P$. VI. Liquid, b. 275°.[984]

$Et_2(4\text{-}PhOC_6H_4)P$. I. Liquid, b_{13} 208°,[289] b_{10} 199 to 200°,[287] n_D^{20} 1.5968,[287] d_4^{20} 1.0743,[287] d_4^{20} 1.0711;[289] CS_2 adduct, m. 69°,[292] m. 67°.[290]

$Et_4(4\text{-}MeO_2CC_6H_4)P$. XIV. (From acid and MeOH/HCl). Liquid, b_2 117 to 122°,[1323] n_D^{20} 1.5570, d_4^{20} 1.053, pK_a (H_2O) 5.27.[1321,1323]

($Et_2(4\text{-}EtO_2CC_6H_4)P$. XIV. (See above). Liquid, b_2 122 to 125°, n_D^{20} 1.5455, d_4^{20} 1.030;[1320] L•picrate, m. 131 to 132°.[1320]

$Et_2(2\text{-}Me_2NC_6H_4)P$. I. Liquid, b_{16} 132 to 135°.[959]

$Et_2(4\text{-}Me_2NC_6H_4)P$. I.[292,440] VI.[993] Liquid, b. 298°,[993] $b_{0.8}$ 138°,[440] m. 12.5°,[993] pK_a (80% EtOH) 5.08,[440] pK_a (H_2O) 8.45;[1321] CS_2 adduct, m. 107°,[993] m. 103°.[292]

$Et_2(4\text{-}Et_2NC_6H_4)P$. μ 2.83 D.[778]

$Et_2(2,5\text{-}Me_2C_6H_3)P$. I.[721] VI.[274] Liquid, b. 260°, b_{52} 157°, d_4^{25} 0.9392.[721]

$Et_2[2,5\text{- or }5,2\text{-}Me(i\text{-}Pr)C_6H_3]P$. VI. Liquid, b. 260 to 270°.[985]

$Et_2(2\text{-}Cl\text{-}5\text{-}MeC_6H_3)P$. I. Liquid, $b_{0.4}$ 99 to 101°, L•EtBr, m. 181°.[294]

$Et_2(2,4,6\text{-}Me_3C_6H_2)P$. VI. Liquid, b. 270°.[985]

$Et_2(2,4,5\text{-}Me_3C_6H_2)P$. VI. Liquid, b. 274 to 275°.[985]

$Et_2(4\text{-}Ph\text{-}2\text{-}ClC_6H_3)P$. I. Liquid, $b_{0.3}$ 147 to 163° (crude); $L{\cdot}MeO_3SC_6H_4Me$, m. 151 to 151.5°.[294]

$Et_2(1\text{-}C_{10}H_7)P$. VI. Yellow liquid, b. 360° (dec.).[773]

Et_2-9-(9,10-dihydroacridyl)P. VII. Crystals, m. 98 to 100° (from Et_2O).[655]

Et_2(2-thienyl)P. VI. Liquid, b. 225°.[1208]

$(CHF_2CF_2)_2MeP$. XII. Liquid, b_{734} 123°.[164]

$(CHF_2CF_2)_2PhP$. XII. Liquid, $b_{0.5}$ 42°.[1080]

$(ClCH_2CH_2)_2MeP$. XIV. (From hydroxy compound and $SOCl_2$).[1307,1308] Apparently not the hydrochloride[1308] but the phosphorane, $(ClCH_2CH_2)_2MePCl_2$[1307] is formed in this reaction; identified as oxide;[1307] shows no mustardlike activity as tumor growth inhibitor.[1308]

$(ClCH_2CH_2)_2PhP$. XIV. From hydroxy compound and $SOCl_2$. Isolated only as the hydrochloride, yellow, semicrystalline paste; identified as oxide, m. 92°.[1]

$(CH_2{=}CH)_2EtP$. I. Liquid, b_{751} 121°,[399] b_{54} 48°,[1420] UV,[1420] ^{31}P +20.8 ppm.[1140]

$(CH_2{=}CH)_2BuP$. I. Liquid, b_{11} 48 to 50°, UV.[1420]

$(CH_2{=}CH)_2PhP$. I. Liquid, $b_{0.5}$ 55°.[747,934]

$(H_2NCH_2CH_2)_2PhP$. VII. Stable only as dihydrochloride, dec. 255°,[694] pK_1 8.24, pK_2 8.98, ^{31}P +35.8 ppm.[694]
$(EtHNCH_2CH_2)_2PhP$. VII. Liquid, b_3 160 to 163°, pK_1 9.26, pK_2 8.48.[681]
$(Et_2NCH_2CH_2)_2(Et_2NCH_2)P$. XIII.1. Liquid, b_2 152 to 153°.[681]
$(Et_2NCH_2CH_2)_2PhP$. VII. Liquid, $b_{0.03}$ 135 to 137°,[325,681] pK_1 8.53, pK_2 7.66.[681]
(N-phthalimidyl-$CH_2CH_2)_2PhP$. XII. Solid, m. 173 to 174°.[59]
$(HOCH_2CH_2)_2MeP$. XIV. From $(t\text{-}BuOCH_2CH_2)_2MeP$ and HCl. Isolated as hydrochloride, colorless syrup;[1308] $L\cdot HBPh_4$, m. 158 to 160°.[1308]
$(HOCH_2CH_2)_2(c\text{-}C_6H_{11})P$. VII. Pale-yellow oil,[700] $E_{1/2}$ -0.2 V.[689]
$(HOCH_2CH_2)_2PhP$. VII.[700] XIV. From $(t\text{-}BuOCH_2CH_2)_2PhP$ and HCl.[1] Pale-yellow oil, b_3 242°,[700] viscous oil, dec. on dist.;[1] hydrochloride, viscous gum;[1] $E_{1/2}$ -0.11 V.[689] Dibenzoyl deriv. XIV. From phosphine and PhCOCl. Cloudy yellow-green syrup, ^1H NMR;[1307] shows no "mustardlike" effect.[1307]
$(EtOCH_2CH_2)_2PhP$. VII. Liquid, $b_{0.5}$ 118 to 122°.[569]
$(t\text{-}BuOCH_2CH_2)_2MeP$. VII. Colorless oil, b_{22} 146 to 147°; $L\cdot MeI$, m. 84 to 85°.[1308]
$(t\text{-}BuOCH_2CH_2)_2PhP$. VII. Colorless, viscous liquid, $b_{0.2}$ 113 to 115°, n_D^{25} 1.508; $L\cdot MeI$, m. 146 to 147°.[1]
$(MeCH\cdot OH)_2PhP$. XIII.1. Liquid, $b_{0.3}$ 86 to 87°, n_D^{20} 1.5381, d_4^{20} 1.0186.[1105]
$(CH_2{=}CHCH_2O_2CCH_2)_2MeP$. VI. Liquid, $b_{0.025}$ 86°, n_D^{20} 1.4933, d_4^{20} 1.0858.[1382]
$(MeO_2CCH_2)_2EtP$. VI. Liquid, $b_{0.5}$ 91 to 92°, n_D^{20} 1.4735, d_4^{20} 1.0620.[1130]
$(CH_2{=}CHCH_2O_2CCH_2)_2EtP$. VI. Liquid, $b_{0.06}$ 112°, n_D^{20} 1.4930, d_4^{20} 1.0663.[1382]
$(MeO_2CCH_2)_2BuP$. VI. Liquid, b_1 103 to 104°, n_D^{20} 1.4773, d_4^{20} 1.0782.[1130]
$(MeO_2CCH_2)_2PhP$. VI. Liquid, b_1 149 to 150°, n_D^{20} 1.5480, d_4^{20} 1.1742.[1130]
$(CH_2{=}CHCH_2O_2CCH_2)_2PhP$. VI. Liquid, $b_{0.03}$ 137°, n_D^{20} 1.5418, d_4^{20} 1.1281.[1382]
$(MeCO)_2MeP$. XII. $(MePH_2 + CH_2{=}C{=}O)$. Pale-yellow liquid, b_{21} 92.5 to 94°, n_D^{20} 1.5095, d_4^{20} 1.088,[810] UV, IR, ^1H NMR, mass spect.;[810] oxidizes in air.
$(MeCO)_2(MeO_2CCH_2)P$. XIII.6. Colorless oil, $b_{0.05}$ 76 to 78°.[678]
$(HSCH_2CH_2)_2EtP$. VII. Liquid, $b_{0.1}$ 90 to 91°, n_D^{20} 1.5717,[428] pK_a 6.1.[1271]
$(HSCH_2CH_2)_2PhP$. VII. Liquid, $b_{0.35}$ 164 to 166°, n_D^{20} 1.6211.[428]
$Pr_2(HOCH_2)P$. IX.2. Liquid, b_3 73 to 76°, n_D^{20} 1.4835, d_4^{20} 0.9956.[1110]
Acetyl deriv. IX.2. Liquid, b_4 70 to 71°, n_D^{20} 1.4610, d_4^{20} 0.9322.[1004]

$Pr_2(Et_2NCH_2)P$. XIII.1. Liquid, $b_{0.01}$ 110 to 120°, n_D^{20} 1.4592, d_4^{20} 0.9700.[1112]
$Pr_2(CH_2{=}CH)P$. I. Liquid, b_{23} 68 to 69°, n_D^{20} 1.4680, d_4^{20} 0.8144.[747]
$Pr_2(HC{\equiv}C)P$. I. Liquid, $b_{0.2}$ 36°, b_{11} 55°, n_D^{20} 1.4770, IR,[1389] pKa 3.27.[333]
$Pr_2(MeCO)P$. VII. Liquid, b. 196 to 199°.[687]
$Pr_2(EtCO)P$. VII. XIII.6. Liquid, b_{12} 88 to 90°.[687]
Pr_2BuP. IX.1. Liquid, b_{33} 107 to 108°, n_D^{20} 1.4579, d_4^{20} 0.8130.[759]
$Pr_2(PrCO)P$. VII. Liquid, b_{15} 100°.[687]
$Pr_2(i\text{-}PrCO)P$. VII. Liquid, b_{12} 103 to 105°.[687]
$Pr_2(PhCO)P$. VII. Liquid, b_3 128 to 130°, UV.[687]
$Pr_2(1\text{-}C_{10}H_7CO)P$. VII. Liquid, $b_{2.5}$ 193°, UV.[687]
$Pr_2(2\text{-furyl-CO})P$. VII. Liquid, b_2 100 to 101°, UV.[687]
$Pr_2(2\text{-thienyl-CO})P$. VII. Liquid, b_2 122 to 123°, UV.[687]
$Pr_2(PhCH{=}C{\cdot}Me)P$. XIV. From $Ph_2PCMe{=}PPh_3$ and PhCHO. Liquid, b_{14} 162 to 164°,[683] stereochemistry not known.
Pr_2PhP. I. Liquid, b_{50} 159°,[291] b_{11} 126°,[444] d_5^{25} 0.925, ^{31}P +27.7 ppm.[1349]
$Pr_2(4\text{-}ClC_6H_4)P$. I. Liquid, b_{11} 155°.[444]
$Pr_2(4\text{-}BrC_6H_4)P$. I. Liquid, b_{11} 172°.[444]
$Pr_2(4\text{-}MeC_6H_4)P$. I. Liquid, b_{50} 174°,[291] b_{13} 136 to 137°, n_D^{20} 1.5315, d_4^{20} 0.9228,[760] d_4^{25} 0.921.[291]
$Pr_2(4\text{-}EtC_6H_4)P$. I. Liquid, b_{21} 157°, n_C^{25} 1.5208, n_D^{25} 1.5255, n_F^{25} 1.5370, d_4^{25} 0.9147.[720a]
$Pr_2(4\text{-}NCC_6H_4)P$. I. Liquid, $b_{1.5}$ 139°.[442]
$Pr_2(4\text{-}Me_2NC_6H_4)P$. I. Liquid, b_{11} 185°.[444]
$Pr_2(4\text{-}MeOC_6H_4)P$. I. Liquid, b_{17} 165°, n_C^{25} 1.5301, n_D^{25} 1.5352, n_F^{25} 1.5477, d_4^{25} 0.9738.[720a]
$Pr_2(4\text{-}PhOC_6H_4)P$. I. Liquid, b_{13} 218°;[289] CS_2 adduct, m. 57°.[290]
$Pr_2(4\text{-}MeCOC_6H_4)P$. XIV. From $Pr_2(4\text{-}NCC_6H_4)P$ and MeMgI. Liquid, b_3 156°.[442]
$Pr_2(2,5\text{-}Me_2C_6H_3)P$. I. Liquid, b_{25} 161°, d_4^{25} 0.9281.[721]
$i\text{-}Pr_2(HC{\equiv}C)P$. II. Liquid, b_{11} 43 to 45°, IR, ^1H NMR.[23]
$i\text{-}Pr_2(MeC{\equiv}C)P$. II. pK_a 5.07.[333]
$i\text{-}Pr_2(PrC{\equiv}C)P$. II. Liquid, $b_{1.25}$ 68 to 69°, IR, ^1H NMR.[23]
$i\text{-}Pr_2(BuC{\equiv}C)P$. II. pKa 5.12.[333]
$i\text{-}Pr_2(C_6H_{13}C{\equiv}C)P$. II. Liquid, $b_{0.4}$ 89 to 93°, IR, ^1H NMR.[23]
$i\text{-}Pr_2(PhC{\equiv}C)P$. II. Liquid, $b_{0.75}$ 116 to 120°, IR, ^1H NMR.[23]
$i\text{-}Pr_2PhP$. I. ^1H NMR,[886] (conformers unequally populated), ^{31}P -9.3 ppm.[413]
$i\text{-}Pr_2(4\text{-}PhOC_6H_4)P$. I. Liquid, b_{13} 209°, n_D^{25} 1.5826, d_4^{25} 1.0423.[283a]
$(C_3F_7)_2CF_3P$. II. Liquid, b. 108 to 115°,[454] inflammable in air; P-C bonds cleaved with OH^-.[454]
$(C_3F_7)_2BuP$. II. Liquid, b_{760} 173° (calc.), log P_{mm} = 7.83 - (2210/T),[454] IR, ^{19}F NMR;[454] C_3F_3-P bond cleaved

by OH^-.
$(C_3F_7)_2PhP$. II. Liquid, b_{760} 204° (calc.), log P_{mm} = 7.96 - (2425/T),[454] C_3F_7-P bond cleaved by OH^-.
$(CH_3CH{=}CH)_2PhP$. From $(CH_2{=}CHCH_2)_2PhP$ and MeONa. Liquid, $b_{0.2}$ 84 to 86°, ^1H NMR.[613a]
$(CH_2{=}CHCH_2)_2(MeCO_2CH_2)P$. IX.2. Liquid, b_3 66 to 67°, n_D^{20} 1.4948, d_4^{20} 0.9845.[1004]
$(CH_2{=}CHCH_2)_2(t\text{-}Bu)P$. ^{31}P +6.0 ppm.[485]
$(CH_2{=}CHCH_2)_2PhP$. I. Liquid, b_{14} 127°, n_D^{25} 1.5670, d_4^{25} 0.9693,[743] ^{31}P +27.5 ppm.[485]
$(CH_2{=}CHCH_2)_2(4\text{-}BrC_6H_4)P$. I. Liquid, b_{37} 186°, d_4^{25} 1.2783.[743]
$(CH_2{=}CHCH_2)_2(4\text{-}MeOC_6H_4)P$. I. Liquid, b_{15} 162°, n_D^{25} 1.5705, d_4^{25} 1.0189.[743]
$(CH_2{=}CHCH_2)_2(4\text{-}PhOC_6H_4)P$. I. Liquid, b_{15} 238°, n_D^{25} 1.6040, d_4^{25} 1.0847.[743]
$(CH_2{=}CHCH_2)_2(4\text{-}MeC_6H_4)P$. I. Liquid, b_{14} 138°, n_D^{25} 1.5545, d_4^{25} 0.9651.[743]
$(CH_2{=}CHCH_2)_2(4\text{-}EtC_6H_4)P$. I. Liquid, b_{10} 145°, n_D^{25} 1.5545, d_4^{25} 0.9484.[743]
$(CH_2{=}CHCH_2)_2(4\text{-}i\text{-}PrC_6H_4)P$. I. Liquid, b_{11} 153°, n_D^{25} 1.5435, d_4^{25} 0.9361.[743]
$(CH_2{=}CHCH_2)_2(2,5\text{-}Me_2C_6H_3)P$. I. Liquid, b_{13} 144°, n_D^{25} 1.5540, d_4^{25} 0.9584.[743]
$(NCCH_2CH_2)_2MeP$. IX.2. Liquid, b_3 174 to 176°,[832] $b_{0.35}$ 159 to 160°, n_D^{25} 1.5030,[470] n_D^{20} 1.5056, d_4^{20} 1.0779,[832] pK_a (H_2O) 3.61.[1342]
$(NCCH_2CH_2)_2(4\text{-}ClC_6H_4NHCO)P$. XIII.5. Solid, m. 125°,[1162, 1163] useful as flame retardant.[1162]
$(NCCH_2CH_2)_2EtP$. IX.2 and .3. Liquid, b_2 173 to 174°,[832] $b_{0.55}$ 153 to 155°,[470] b_{3-4} 178 to 180°,[624] n_D^{25} 1.5040,[470] n_D^{20} 1.5067, d_4^{20} 1.0425,[832] pK_a (H_2O) 3.80.[1342]
$(NCCH_2CH_2)_2(MeCO_2CH_2CH_2)P$. XII. Could not be distilled since it self-quaternized below its dist. temp.[1165]
$(NCCH_2CH_2)_2(BuEtCHCH_2OCH_2CH_2)P$. XII. Liquid, $b_{0.2}$ 204 to 207°, n_D^{25} 1.4860.[1171]
$(NCCH_2CH_2)_2(CH_2{=}CHCH_2)P$. IX.2. Liquid, $b_{1.65}$ 178 to 179°, n_D^{25} 1.4980.[470]
$(NCCH_2CH_2)_2PrP$. IX.2. Crystals, m. 34 to 36°, b_1 162 to 163°.[832]
$(NCCH_2CH_2)_2(HOCH_2CH_2CH_2)P$. XII. Liquid, $b_{0.5}$ 227 to 229°, n_D^{20} 1.5202.[1171]
$(NCCH_2CH_2)_2BuP$. IX.2.[832] Liquid, b_3 184 to 185°,[832] n_D^{20} 1.4989, d_4^{20} 0.9775.[832]
$(NCCH_2CH_2)_2[EtO_2CCH_2(EtO_2C)\cdot CH]P$. XII. Liquid, $b_{0.9}$ 225 to 228°, n_D^{25} 1.4922.[1171]
$(NCCH_2CH_2)_2(c\text{-}C_5H_9)P$. IX.2. Liquid, b. 165 to 169° (pressure not given in C.A.).[466]
$(NCCH_2CH_2)_2(c\text{-}C_6H_{11})P$. XII. Liquid, $b_{0.4}$ 185 to 186°, n_D^{25} 1.5241.[1171]
$(NCCH_2CH_2)_2(AmCH{=}CH)P$. XII. Liquid, $b_{0.4}$ 170 to 171°, n_D^{25} 1.5008.[1171]

$(NCCH_2CH_2)_2(n\text{-}C_8H_{17})P$. XII. Liquid, $b_{0.2}$ 183 to 186°, n_D^{25} 1.4885,[1171] ^{31}P +25.7 ppm.[924]
$(NCCH_2CH_2)_2(t\text{-}BuCH_2CH{\cdot}MeCH_2)P$. XII. Liquid, $b_{0.15}$ 157°, n_D^{25} 1.4920.[1171]
$(NCCH_2CH_2)_2(n\text{-}C_6H_{13}CH{=}CH)P$. XII. Liquid, $b_{0.6}$ 187 to 190°, n_D^{25} 1.5000.[1171]
$(NCCH_2CH_2)_2(PhCH_2CH_2)P$. XII. Liquid, $b_{0.6}$ 212°, n_D^{25} 1.5580.[1171]
$(NCCH_2CH_2)_2PhP$. XII. Crystals, m. 72 to 73°,[60,950] $b_{0.5}$ 176.5 to 178°,[60] $b_{0.2}$ 195 to 205°,[950] n_D^{20} 1.5672, d_4^{20} 1.1043,[60,950] UV,[1230,1236] pKa (H_2O) 3.20.[1342]
$(H_2NCH_2CH_2CH_2)_2(NCCH_2CH_2)P$. XII. Liquid, $b_{0.5}$ 163 to 167°, n_D^{25} 1.5250.[1171]
$(H_2NCH_2CH_2CH_2)_2(HOCH_2CH_2CH_2)P$. XII. Liquid, $b_{0.25}$ 170 to 183°.[462]
$(H_2NCH_2CH_2CH_2)_2BuP$. XII. Liquid, $b_{0.3}$ 95 to 97°, n_D^{20} 1.5048, d_4^{20} 0.9213.[57,59]
$(H_2NCH_2CH_2CH_2)_2i\text{-}BuP$. XII. Liquid, $b_{0.2}$ 94°, n_D^{25} 1.5014;[1089] gives a polymer with adipic acid.
$(H_2NCH_2CH_2CH_2)_2(n\text{-}C_8H_{17})P$. XII. Liquid, $b_{0.5}$ 155 to 163°, n_D^{25} 1.4948.[1089]
$(H_2NCH_2CH_2CH_2)_2PhP$. XII. XIV. From $PhP(CH_2CH_2CN)_2$ + $LiAlH_4$.[60] Liquid, b_1 144°,[60] b_1 148.5 to 149°,[58] n_D^{20} 1.5728,[60] d_4^{20} 1.0292,[58,60] b_1 140 to 150°, n_D^{25} 1.5722.[1089]
$(HOCH_2CH_2CH_2)_2MeP$. XII. Liquid, $b_{0.07}$ 125 to 129°.[462]
$(HOCH_2CH_2CH_2)_2EtP$. XII. Liquid, $b_{0.15}$ 141 to 147°;[462] has been claimed to be useful in hair-waving, smoothing, or depilating compositions.[462]
$(HOCH_2CH_2CH_2)_2(NCCH_2CH_2)P$. XII. Liquid, $b_{0.7}$ 190 to 196°, n_D^{25} 1.5230.[1171]
$(HOCH_2CH_2CH_2)_2(H_2NCH_2CH_2CH_2)P$. XII. Liquid, $b_{0.2}$ 175 to 176°.[462]
$(HOCH_2CH_2CH_2)_2BuP$. XII. Liquid, $b_{2.5}$ 140 to 142°, n_D^{20} 1.5029, d_4^{20} 0.9873.[57,59]
Diacetyl deriv. XII. Liquid, b_2 133 to 135°, n_D^{20} 1.4718, d_4^{20} 1.0086.[57,59]
$(HOCH_2CH_2CH_2)_2(c\text{-}C_6H_{11})P$. VII.[700] XII.[462] Liquid, $b_{0.3}$ 185 to 192°,[462] pale-yellow oil.[700]
$(HOCH_2CH_2CH_2)_2PhP$. VII.[700] XII.[60] Pale-yellow oil, b_4 246 to 250°,[700] b_1 175°, n_D^{20} 1.5740, d_4^{20} 1.018.[60]
Diacetyl deriv. XII. Liquid, $b_{1.5}$ 178 to 179°, n_D^{20} 1.5262, d_4^{20} 1.005.[58]
$(EtO_2CCH_2CH_2)_2[EtO_2CCH_2CH_2(EtO_2C){\cdot}CHCH_2]P$. XII. Liquid, $b_{0.2}$ 195 to 198°, n_D^{25} 1.4730.[1171]
$(MeO_2CCH_2CH_2)_2PhP$. XII. Liquid, b_1 149 to 150°, n_D^{20} 1.5361, d_4^{20} 1.1388.[60]
$(CH_2{=}CHCH_2O_2CCH_2CH_2)_2PhP$. XII. Liquid, $b_{0.014}$ 171 to 172°, n_D^{20} 1.5354, d_4^{20} 1.0970;[1381] also obtained by transesterification of methyl ester with allyl alcohol.

$(HOCH{\cdot}MeCH_2)_2(c\text{-}C_6H_{11})P$. VII. Slightly yellow oil,[700] $E_{1/2}$ -0.27 V.[689]
$(HOCH{\cdot}MeCH_2)_2PhP$. VII. Slightly yellow oil, b_3 166°, $E_{1/2}$ -0.14 V.[689]
Bu_2CF_3P. II. Liquid, b_{760} 167° (calc.), log P_{mm} = 8.10 - (2300/T),[454] b_{62} 106 to 108°; P-CF_3 bond partially cleaved by OH^-.
$Bu_2(HOCH_2)P$. IX.2. Liquid, b_{14} 105 to 107°,[1111,1329] $b_{1.5}$ 80 to 81°,[1372] n_D^{20} 1.4848,[1111] n_D^{20} 1.4851,[1329] n_D^{20} 1.4835, d_4^{20} 0.9150.[1372]
Acetyl deriv. IX.2. Liquid, b_4 99 to 100°, n_D^{20} 1.4632, d_4^{20} 0.9320.[1004]
Benzoyl deriv. Liquid, $b_{0.2}$ 121 to 122°, n_D^{20} 1.4993.[1329]
$Bu_2(PhNHCO)P$. XIII.5. Crystals, m. 98 to 99°,[1287] m. 102°.[1290]
$Bu_2(1\text{-}C_{10}H_7NHCO)P$. XIII.5. Solid, m. 80 to 81°;[1162,1163] suggested as a flame retardant.[1162]
$Bu_2(Et_2NCH_2)P$. XIII.1. Liquid, b_3 107°, n_D^{25} 1.4675.[236]
$Bu_2(C_5H_{10}NCH_2CH_2)P$. XII. Liquid, b_3 128 to 128.5°, n_D^{20} 1.4891, d_4^{20} 0.8854.[751]
$Bu_2(EtO_2CCH_2)P$. IX.2. Liquid, $b_{0.02-0.03}$ 150 to 160°, $n_D^{21.5}$ 1.4625;[1113] amide (from ester and NH_3) is an oil.
$Bu_2(MeCO)P$. VII. Liquid, b. 188 to 193°.[687]
$Bu_2(MeCO_2CH_2CH_2)P$. XII. Liquid, $b_{0.005}$ 62 to 66°;[1164,1165] self-quaternizes easily to give a cyclic phosphonium salt at 105°.[1164,1165]
$Bu_2(CH_2{=}CH)P$. I. Liquid, b_{10} 120 to 125°,[1097] $b_{0.7}$ 38°,[1420] b_2 48 to 49°, n_D^{20} 1.4710, d_4^{20} 0.8210,[747] UV.[1124,1420]
$Bu_2(EtSCH{=}CH)P$. XIV. By addition of EtSH to $Bu_2PC{\equiv}CH$. Liquid, $b_{0.001}$ 75 to 77°, n_D^{20} 1.5170, IR.[1389]
$Bu_2(HC{\equiv}C)P$. I. Liquid, $b_{0.3}$ 37°,[1389] b_3 91°(?),[273] b_{10} 85°,[1389] n_D^{20} 1.4765, IR,[1389] ^1H NMR.[334,1298]
$Bu_2(EtSC{\equiv}C)P$. II. Liquid, $b_{0.001}$ 78°, n_D^{20} 1.5265, IR.[1389]
$Bu_2(MeC{\equiv}C)P$. II. XIV. Liquid, b_1 66 to 67°, n_D^{20} 1.4867, IR.[1389]
$Bu_2(MeCOCH_2)P$. ^1H NMR.[1115]
$Bu_2(n\text{-}C_{12}H_{25}NHCOCH_2CH_2)P$. XII. Liquid, $b_{0.025}$ 198 to 208°, n_D^{25} 1.4770;[1395] solid,[1396] useful as a corrosion inhibitor in steel.[1396]
$Bu_2(HOCH_2CH_2CH_2)P$. XII. Liquid, b_{14} 154 to 155°, n_D^{20} 1.4850.[1329]
Acetyl deriv. XII. Liquid, $b_{0.1}$ 98 to 102°,[1165] no self-quaternization up to 300°.[1165]
Benzoyl deriv. Liquid, $b_{0.2}$ 150 to 152°, n_D^{20} 1.5179.[1329]
$Bu_2(EtCO)P$. VII. Liquid, b_{14} 137 to 141°.[687]
$Bu_2(NCCH_2CH_2)P$. XII. pK_a (H_2O) 6.48.[1342]
$Bu_2(t\text{-}Bu)P$. ^{31}P +4.6 ppm.[485]
$Bu_2(MeCH{=}CHCH_2)P$. XII. Liquid, b_{13} 118°, b_5 100°, n_D^{25} 1.4725.[1171]
$Bu_2(HC{\equiv}CCH_2CH_2)P$. XII. Liquid, b_5 95 to 96°, n_D^{20} 1.4830, d_4^{20} 0.8448.[1104]

$Bu_2(CH_2=CH-C\equiv C)P$. I. Liquid, b_3 92°, n_D^{20} 1.5110, d_4^{20} 0.8602, IR.[963]
$Bu_2(EtC\equiv C)P$. XIV. Liquid, $b_{0.25}$ 58°, n_D^{20} 1.4848, IR,[1389] pK_a 5.34.[333]
$Bu_2(PrCO)P$. VII. Liquid, b_{12} 117 to 123°.[687]
$Bu_2(Me_2C{\cdot}OHC\equiv C)P$. II. XIV. From $Bu_2PC\equiv CH$ + Me_2CO. Liquid, $b_{0.03}$ 91°, n_D^{20} 1.4871, IR.[1389]
$Bu_2(EtCH=C=CHCH_2)P$. XII. Liquid, b_5 125 to 126°, n_D^{20} 1.4952, d_4^{20} 0.8534.[1104]
$Bu_2(c\text{-}1\text{-}HOC_6H_{10}C\equiv C)P$. XIV. From $Bu_2PC\equiv CH$ and cyclohexanone. Liquid, $b_{0.001}$ 105°, m. ∿25°, n_D^{20} 1.5090, IR.[1389]
$Bu_2(n\text{-}C_{12}H_{25})P$. IV. $C_{12}H_{25}PH_2$ + BuCl, then OH^-. Liquid, $b_{0.5}$ 135 to 145°, IR, GLC,[535] ^1H NMR, ^{31}P +32.4 ppm.[535]
$Bu_2(PhCO)P$. VII. Liquid, b_2 140 to 144°.[687]
$Bu(1\text{-}C_{10}H_7CO)P$. VII. Liquid, $b_{2.5}$ 218°, UV;[687] L·MeI, dec. 141°.[687]
$Bu_2(2\text{-furyl-CO})P$. VII. Liquid, b_6 131°.[687]
$Bu_2(2\text{-thienyl-CO})P$. VII. Liquid, b_2 143 to 144°.[687]
$Bu_2(PhC\equiv C)P$. pK_a 4.68.[333]
Bu_2PhP. I.[286] V.[1177] XIV.[1389] (From $Bu_2PC\equiv CPBu_2$ and LiPh). Liquid, b_{50} 184.5 to 185.5°,[286] $b_{0.2}$ 83°,[1177] b_{17} 163°,[1177] n_D^{20} 1.5249,[1389] d_4^{25} 0.9115,[286] UV,[1236] GLC,[1177] ^{31}P +25.8 ppm,[413] +26.2 ppm,[483] μ 1.33 D,[778] μ 1.45 D.[1246]
$Bu_2(3\text{-}ClC_6H_4)P$. I. Liquid, b_3 123 to 123.5°, n_D^{20} 1.5363, d_4^{20} 1.0064.[864]
$Bu_2(4\text{-}ClC_6H_4)P$. I. Liquid, b_5 144 to 146°, n_D^{20} 1.5373, d_4^{20} 1.0045.[1362]
$Bu_2(4\text{-}MeC_6H_4)P$. I.[286,760] VII.[710] Liquid, b_{50} 197°,[286] b_{17} 165 to 166°,[760] b_{1-2} 125 to 130°,[710] n_D^{20} 1.5245, d_4^{20} 0.9097,[760] d_4^{25} 0.9076,[286] μ 1.55 D.[1246]
$Bu_2(3\text{-}CF_3C_6H_4)P$. V. Liquid, b_{18} 147°.[1177]
$Bu_2(4\text{-}EtC_6H_4)P$. I. Liquid, b_{15} 176°, n_C^{25} 1.5162, n_D^{25} 1.5208, n_F^{25} 1.5319, d_4^{25} 0.9042.[720a]
$Bu_2(4\text{-}Me_2NC_6H_4)P$. μ 2.58 D.[778]
$Bu_2(4\text{-}MeOC_6H_4)P$. I.[720a] V.[1177] Liquid, b_{16} 190°, n_C^{25} 1.5226, n_D^{25} 1.5274, n_F^{25} 1.5389, d_4^{25} 0.9600,[720a] n_D^{25} 1.5277.[1177]
$Bu_2(4\text{-}PhOC_6H_4)P$. I. Liquid, b_{13} 235°, d_4^{20} 1.0310.[289]
$Bu_2(2,5\text{-}Me_2C_6H_3)P$. I. Liquid, b_{16} 171°, d_4^{25} 0.9124.[721]
$i\text{-}Bu_2(HC\equiv C)P$. I. pK_a 3.46.[333]
$i\text{-}Bu_2(MeC\equiv C)P$. I. pK_a 4.84.[333]
$i\text{-}Bu_2(MeCO_2CH_2CH_2)P$. XII. Liquid, $b_{0.3}$ 80 to 84°,[1165] $b_{0.35}$ 75 to 84°;[1164] undergoes self-quaternization when heated to 120°.[1164,1165]
$i\text{-}Bu_2PhP$. I. Liquid, b_{50} 168°,[291] b_3 108 to 109°,[484] d_4^{25} 0.910,[291] UV,[1236] ^{31}P +34.2 ppm,[484] μ 1.29 D.[1246]
$i\text{-}Bu_2(4\text{-}MeC_6H_4)P$. I. Liquid, b_{50} 182.5 to 184.5°, d_4^{25} 0.915,[291] μ 1.45 D, UV.[1236]
$i\text{-}Bu_2(2,5\text{-}Me_2C_6H_3)P$. I. Liquid, b_{20} 184°.[721]
$2\text{-}Bu_2(HC\equiv C)P$. pK_a 3.30.[333]

2-Bu_2PhP. I. Liquid, b_2 107 to 110°, ^{31}P -1.8 ppm.[484]
t-Bu_2MeP. IV. IX.2. Liquid, b_{760} 170 to 172°;[580] L•MeI, m. > 365°.
t-$Bu_2(MeCO)P$. XII. Liquid, $b_{0.5}$ 45 to 47°, IR, ^1H NMR;[1013] adduct L•MeI, m. 155 to 156°.[805]
t-$Bu_2(HC{\equiv}C)P$. pK_a 3.68,[333] ^1H NMR.[334]
t-$Bu_2(i\text{-}Pr)P$. I. Liquid, b_{13} 85 to 90° (crude);[581] L•MeI, m. > 360°.[581]
t-$Bu_2(MeC{\equiv}C)P$. pK_a 4.98.[333]
t-$Bu_2(Bu)P$. ^{31}P -26.6 ppm.[485]
t-$Bu_2[(CF_3)_2CHCO]P$. XII. Liquid, b_1 54 to 55°, IR, ^1H NMR,[805] oxide m. 36 to 40°.
t-$Bu_2(BuC{\equiv}C)P$. pK_a 5.56.[333]
t-$Bu_2(PhCH_2)P$. II. Liquid, $b_{0.25}$ 126 to 130°, IR, ^1H NMR; L•MeI, m. 205.[1337a]
t-$Bu_2(3\text{-}MeC_6H_4)P$. II. Liquid, b_{1-2} 105°, IR, GLC.[710]
t-$Bu_2(4\text{-}MeC_6H_4)P$. II. Solid, m. 24°, b_{1-2} 108°, IR;[710] the reaction of t-Bu_2PLi with 4-MeC_6H_4F gave a mixture of 4- and 3- tolyl di-t-butylphosphines.[710]
$(CH_2{=}C{\cdot}MeCH_2)_2PhP$. I. Liquid, b_{13} 148°, n_D^{25} 1.5485, d_4^{25} 0.9484.[743]
$(CH_2{=}C{\cdot}MeCH_2)_2(4\text{-}BrC_6H_4)P$. I. Liquid, b_{18} 189°, n_D^{25} 1.5752, d_4^{25} 1.2094.[743]
$(CH_2{=}C{\cdot}MeCH_2)_2(4\text{-}MeC_6H_4)P$. I. Liquid, b_{23} 168°, n_D^{25} 1.5465, d_4^{25} 0.9426.[743]
$(CH_2{=}C{\cdot}MeCH_2)_2(4\text{-}EtC_6H_4)P$. I. Liquid, b_{20} 178°, n_D^{25} 1.5435, d_4^{25} 0.9360.[743]
$(CH_2{=}C{\cdot}MeCH_2)_2(4\text{-}i\text{-}PrC_6H_4)P$. Liquid, b_{19} 182.5°, n_D^{25} 1.5350, d_4^{25} 0.9279.[743]
$(CH_2{=}C{\cdot}MeCH_2)_2)_2(4\text{-}MeOC_6H_4)P$. I. Liquid, b_{20} 192°, n_D^{25} 1.5450, d_4^{25} 0.9402.[743]
$(CH_2{=}C{\cdot}MeCH_2)_2(2,5\text{-}Me_2C_6H_3)P$. I. Liquid, b_{16} 166°, n_D^{25} 1.5450, d_4^{25} 0.9402.[743]
$(Me_2CHCHOH)_2PhP$. XIII.1. Isolated as hydrochloride, crystals, m. 123 to 125° (from CH_3CN).[356]
$[\overline{(CF_2)_3C}{\cdot}OH]_2(C_6F_5)P$. XIII.3. Solid, m. 68 to 70°.[385c]
$(NCCH_2CH_2CH_2)_2BuP$. XII. Liquid, $b_{2.5}$ 150 to 152°, n_D^{20} 1.4932, d_4^{20} 0.9649.[57,59]
$(NCCH_2CH_2CH_2)_2PhP$. XII. Liquid, b_1 190 to 191°, n_D^{20} 1.5560, d_4^{20} 1.0692.[59]
$(HO_2CCH{\cdot}MeCH_2)PhP$. XII. Acid not prepared, but the following esters were isolated:[60]
Me ester. XII. Liquid, $b_{1.5}$ 139 to 140°, n_D^{25} 1.5242, d_4^{20} 1.1076.[60]
Et ester. Liquid, b_1 150 to 151°, n_D^{20} 1.5172, d_4^{20} 1.0764.[60]
Pr ester. Liquid, b_1 170 to 171°, n_D^{20} 1.5061, d_4^{20} 1.0394.[60]
i-Pr ester. Liquid, b_1 152 to 153°, n_D^{20} 1.5038, d_4^{20} 1.0394.[60]
Allyl ester. Liquid, $b_{0.5}$ 158 to 160°,[1381] $b_{0.025}$

154 to 155°,[59] n_D^{20} 1.5232, d_4^{20} 1.0643,[1381] n_D^{20} 1.5239, d_4^{20} 1.0636;[59] also obtained by transesterification.[1381]
Bu ester. Liquid, b_1 185 to 186°, n_D^{20} 1.5038, d_4^{20} 1.0267.[60]
i-Bu ester. Liquid, b_1 178 to 179°, n_D^{20} 1.5001, d_4^{20} 1.0181.[60]

$[EtO_2CCH_2(EtO_2C)\cdot CH]_2PhP$. XII. Liquid, $b_{0.01}$ 194 to 196°, n_D^{20} 1.5076, d_4^{20} 1.1659.[59]

$Am_2(MeCO_2CH_2)P$. IX.2. Liquid, b_4 126 to 127°, n_D^{20} 1.4648, d_4^{20} 0.9205.[1004]

$Am_2(CH_2{=}CH)P$. I. Liquid, $b_{1.5}$ 73 to 74°, n_D^{20} 1.4708, d_4^{20} 0.8232.[747]

$Am_2(HC{\equiv}C)P$. I. Liquid, $b_{0.35}$ 65°, n_D^{20} 1.4759, IR.[1389]

Am_2PhP I. Liquid, b_{50} 210°, d_4^{25} 0.902.[291]

$Am_2(4\text{-}MeC_6H_4)P$. I. Liquid, b_{50} 220°, d_4^{25} 0.898.[291]

$Am_2(4\text{-}EtC_6H_4)P$. I. Liquid, b_{18} 201°, d_4^{25} 0.9022.[720a]

$Am_2(4\text{-}MeOC_6H_4)P$. I. Liquid, b_{18} 202°, n_C^{25} 1.5132, n_D^{25} 1.5178, n_F^{25} 1.5289, d_4^{25} 0.9382;[720a] $L\cdot HgCl_2$, m. 114°.[720a]

$Am_2(2,5\text{-}Me_2C_6H_3)P$. I. Liquid, b_{23} 214°;[721] $L\cdot HgCl_2$, m. 117°.[721]

$i\text{-}Am_2PhP$. I. Liquid, b_{50} 198.5°, d_4^{25} 0.900;[291] $L\cdot HgCl_2$, m. 152°.

$i\text{-}Am_2(4\text{-}MeC_6H_4)P$. I. Liquid, b_{50} 210°, d_4^{25} 0.824;[291] $L\cdot HgCl_2$, m. 107°.

$(c\text{-}C_5H_9)_2PhP$. I. Liquid, $b_{0.7}$ 112 to 115°, ^{31}P -1.6 ppm.[484]

$(MeEtCH\cdot CH_2)_2PhP$ I. Liquid, b_{50} 198°, d_4^{25} 0.906;[291] $L\cdot HgCl_2$, m. 120°.

$(MeEtCH\cdot CH_2)_2(4\text{-}MeC_6H_4)P$. I. Liquid, b_{50} 210 to 211°, d_4^{25} 0.902;[291] $L\cdot HgCl_2$, m. 99°.

$(CH_2{=}CHCH_2CH_2CH_2)_2PhP$. I. Liquid, $b_{0.2}$ 98 to 101°; $L\cdot HgCl_2$, m. 86 to 87°.[102]

$(n\text{-}C_6H_{13})_2MeP$. I. Liquid, $b_{0.01\text{-}0.02}$ 110 to 120°, n_D^{26} 1.4602.[1101]

$(n\text{-}C_6H_{13})_2(MeCO_2CH_2)P$. IX.2. Liquid, b_3 149 to 150°, n_D^{20} 1.4645, d_4^{20} 0.9124.[1004]

$(n\text{-}C_6H_{13})_2(c\text{-}C_{12}H_{23})P$. XII. Liquid, $b_{0.1}$ 165 to 185°.[969]

$(n\text{-}C_6H_{13})_2(PhCH{=}CH)P$. I. Liquid, b_3 190 to 193°, n_D^{20} 1.4348.[372]

$(n\text{-}C_6H_{13})_2PhP$. I. Liquid, b_{50} 236°,[720] b_1 162 to 166°,[882] d_4^{20} 0.901.[720]

$(Me_2CH\cdot CH_2CH_2CH_2)_2PhP$. I. Liquid, b_{50} 219°.[291]

$(Me_2CH\cdot CH_2CH_2CH_2)_2(4\text{-}MeC_6H_4)P$. I. Liquid, b_{50} 234 to 235°, d_4^{25} 0.888,[291] $L\cdot HgCl_2$, m. 110.5°.

$(c\text{-}C_6H_{11})_2(HOCH_2)P$. XIII.1. Liquid, characterized as sulfide, m. 111 to 113°;[546] rearranged on heating to 240 to 250° to $(C_6H_{11})_2MeP{=}O$, $(C_6H_{11})_2PH$, and CH_2O.[546]

$(c\text{-}C_6H_{11})_2(EtO_2C)P$. VII. Liquid, b_3 165 to 169°.[649]

$(c\text{-}C_6H_{11})_2(H_2NCO)P$. XIII.5. Solid, m. 121 to 125°,[1074, 1075] IR,[1075] useful for flame-proofing cloth.[1075]

$(c\text{-}C_6H_{11})_2(RNHCS)P$. XIII.5.

(R = Me). Crystals, m. 119° (from EtOH), IR.[666]
(R = Et). Crystals, m. 123 to 125° (from EtOH), IR.[666]
(R = CH_2=$CHCH_2$). Pale-yellow crystals, m. 104° (from ligroin), IR.[666]
(R = Ph). Crystals, m. 118° (from EtOH), IR;[665,666] shows in polarogram an anodic wave at -0.08 ± 0.03 V, which has been interpreted to indicate the presence of tautomeric form $(C_6H_{11})_2$(PhN=C•SH)P;[665] L•$HgBr_2$, m. 164°.[665]
(R = 4-BrC_6H_4). Yellow, viscous oil, IR.[666]
(R = 2-$C_{10}H_7$). Crystals, m. 119° (from EtOH), IR.[666]

(c-C_6H_{11})$_2$(PhN=C•SEt)P. XIV. From R_2(PhN-C•SLi)P and EtI. Crystals, m. 81° (from EtOH);[665] the lithium salt reacts with another mole of PhNCS to give $(C_6H_{11})_2$-(PhN=C(SLi)-N•PhCS) P•1.5Et_2O, m. 59° and with 1.5 THF, m. 110°.[665]

(c-C_6H_{11})$_2$EtP. VII. Liquid, b_{10} 155 to 157°, b_4 130°.[691, 703]

(c-C_6H_{11})$_2$($H_2NCH_2CH_2$)P. VII. Liquid, b_9 181 to 183°;[661] MeI quarternizes N, dec. 94 to 96°.[661]

(c-C_6H_{11})$_2$($Et_2NCH_2CH_2$)P. VII. Liquid, b_3 159°;[697] EtI quaternizes P atom, dec. 83°.[697]

(c-C_6H_{11})$_2$(MeCO)P. VII. Liquid, b_2 122 to 128°.[696]

(c-C_6H_{11})$_2$($HOCH_2CH_2$)P. VII. Solid, m. 110 to 112°,[691] $E_{1/2}$ -0.43 V.[689]
Acetyl deriv. XII. Could not be distilled since it self-quaternized below its dist. temp.[1165]

(c-C_6H_{11})$_2$(NaO_2CCH_2)P. IX.2. Solid, m. 348 to 352° (dec.).[704]
IX.2. Ethyl ester. Liquid, b_1 147 to 148°.[704]

(c-C_6H_{11})$_2$(H_2NCOCH_2)P. XIV. From ester and NH_3. Crystals, m. 159 to 160° (from MeOH),[705] pK_a ∿3.[705]

(c-C_6H_{11})$_2$($NCCH_2CH_2$)P. XII. Liquid, $b_{0.4}$ 149°, n_D^{25} 1.5235,[1171] pK_a (H_2O) 7.13.[1342]

(c-C_6H_{11})$_2$($HOCH_2CH_2CH_2$)P. VII. Liquid,[700] $E_{1/2}$ -0.42 V.[689]

(c-C_6H_{11})$_2$($EtO_2CCH_2CH_2$)P. IX.2. Liquid, b_1 158 to 160°.[704]

(c-C_6H_{11})$_2$(CH_2=$CHCH_2CH_2$)P. I. II. Liquid, b_7 190 to 192°;[662] L•MeI, m. 128°.

(c-C_6H_{11})$_2$(c-2-Cl-hexafluoropent-1-enyl)P. VII. Liquid, $b_{0.3}$ 114 to 116°,[1340] IR, ^1H and ^{19}F NMR, ^{31}P +19.1 ppm.[1339]

(c-C_6H_{11})$_2$(c-2-HOC_6H_{10})P. VII. Crystals, m. 134 to 136° (from EtOH);[700] L•MeI, m. 173 to 174°.

(c-C_6H_{11})$_2$(PhCH•OH). XIII.2. Isolated as hydrochloride, crystals, m. 146 to 149° (from THF).[356]

(c-C_6H_{11})$_2$(PhCO)P. VII. Crystals, m. 75°, b_2 185 to 189°.[696]

(c-C_6H_{11})$_2$(HN=C•Ph-N=C•Ph)P. VII. Crystals, m. 217 to 218° (from H_2O/EtOH), IR.[652]

(c-C_6H_{11})$_2$[NH=C(C_6H_4Me-4)-N=C•(C_6H_4Me-4)]P. VII. Crystals, m. 236 to 238.5°, IR.[652]

$(c-C_6H_{11})_2(Ph_3C)P$. I. Crystals, m. 165 to 167° (from Me_2CO);[714] L•MeI, m. 203 to 205° (dec.).

$(c-C_6H_{11})_2(Ph_2CHCH_2)P$. VII. Colorless needles, m. 178 to 180° (from benzene/ligroin).[670]

$(c-C_6H_{11})_2$(trans-PhCH=C•Ph)P. VII. Crystals, m. 128 to 129° (from EtOH).[670]

$(c-C_6H_{11})_2$[trans-Ph(HO_2C)C=C•Ph]P. VII. Crystals, m. 206 to 216° (dec.) (from Et_2O),[670] L•MeI, m. 220 to 222° (dec.).

$(c-C_6H_{11})_2$(PhC≡C)P. VII. Crystals, m. 74°.[664]

$(c-C_6H_{11})_2$PhP. II.[1222] VII.[710,713] Crystals, m. 57 to 58° (from acetone),[713,1222] ^{31}P -2.5 ppm.[413,483,484]

$(c-C_6H_{11})_2(4-MeC_6H_4)P$. VII. Colorless oil, b_{1-2} 175 to 180°.[710]

$(c-C_6H_{11})_2$-9-(9,10-dihydroacridyl)P. VII. Crystals, m. 219 to 221° (from dioxan).[655]

$(c-C_6H_{11})_2$(9-acridyl)P. VII. Crystals, m. 184 to 185° (from toluene).[655]

$(c-2-HOC_6H_{10})_2(c-C_6H_{11})P$. VII. Crystals, m. 178 to 180° (from EtOH),[700] $E_{1/2}$ -0.44 V.[689]

$(c-2-HOC_6H_{10})_2$PhP. VII. Crystals, m. 128 to 130° (from EtOH),[700] $E_{1/2}$ -0.17 V.[689]

$(n-C_7H_{15})_2(HOCH_2)P$. IX.2. Undistillable oil, n_D^{20} 1.4780.[1111]

$(n-C_7H_{15})_2(EtO_2CCH_2)P$. IX.2. Liquid, $n_D^{18.5}$ 1.4650.[1113]

$(n-C_7H_{15})_2$(PhCH=CH)P. I. Liquid, b_3 205 to 207°.[372]

$(n-C_7H_{15})_2$PhP. I. Liquid, b_{50} 260°, d_4^{20} 0.895.[720]

$(AmCH=CH)_2(NCCH_2CH_2)P$. XII. Liquid, $b_{0.25}$ 140°, n_D^{25} 1.4930.[1171]

$(n-C_8H_{17})_2(NCCH_2CH_2)P$. XII. Liquid, $b_{0.15}$ 165 to 167°, n_D^{25} 1.4735,[1171] pK_a (H_2O) 6.29.[1342]

$(n-C_8H_{17})_2(MeCO_2CH_2CH_2)P$. XII. Could not be distilled since it self-quaternized below its dist. temp.[1165]

$(n-C_8H_{17})_2(EtO_2CCH_2CH_2)P$. XII. Liquid, $b_{0.2}$ 165°, n_D^{25} 1.4655.[1171]

$(n-C_8H_{17})_2$BuP. I. ^{31}P +33.3 ppm.[413]

$(n-C_8H_{17})_2$(PhCH=CH)P. I. Liquid, b_3 207 to 210°.[372]

$(n-C_8H_{17})_2$PhP. I. Liquid, b_{50} 277°, d_4^{20} 0.890.[720]

$(t-BuCH_2CH•MeCH_2)_2(NCCH_2CH_2)P$. XII. Liquid, $b_{0.8}$ 156°, n_D^{25} 1.4712.[1171]

$(n-C_9H_{19})_2$(PhCH=CH)P. I. Liquid, b_3 230 to 235°, n_D^{20} 1.5208.[372]

$(n-C_9H_{19})_2$PhP. I. Liquid, b_3 221 to 223°, n_D^{20} 1.5005.[373]

$(n-C_{10}H_{21})_2$(PhCH=CH)P. I. Liquid, b_3 225 to 227°, n_D^{20} 1.5132.[372]

$(n-C_{10}H_{21})_2$PhP. I. Liquid, b_4 241 to 243°, n_D^{20} 1.4990.[373]

$(n-C_{12}H_{25})_2$MeP. XII. Liquid, $b_{0.022}$ 188 to 195°, IR, ^1H NMR, ^{31}P +34 ppm.[535]

$(n-C_{12}H_{25})_2$PhP. I. Crystals, m. 36°,[1246] μ 1.55 D (in C_6H_6).[1246]

$(n-C_{16}H_{33})_2(4-MeC_6H_4)P$. I. Crystals, m. 28 to 31°, μ 1.55 D (in C_6H_6).[1246]

$(PhCH_2)_2MeP$. I. Liquid, $b_{0.2}$ 95 to 124° (crude).[914]
$(PhCH_2)_2EtP$. IX.1 and .2. Liquid, b. 320 to 330°,[242] b_1 143 to 145°.[623,624]
$(PhCH_2)_2(MeC{\equiv}C)P$. I. II. Liquid, $b_{0.6}$ 168 to 171°, IR, ^1H NMR.[23]
$(PhCH_2)_2BuP$. IX.1. Liquid, b_5 163 to 178°, n_D^{20} 1.5678.[759]
$(PhCH_2)_2PhP$. I.[160,566,952] IX.1.[757] Crystals, m. 71 to 72°,[566,952] m. 73° (dec.),[160] $b_{0.3}$ 176 to 177°,[952] b_{10} 170°(?),[757] ^{31}P +12.1 ppm;[484] an attempt to prepare this phosphine by modified VI resulted in a substance, $C_{13}H_{13}P$, m. 169 to 170°, which did not possess the expected phosphine like properties.[989]
$(PhCH{\cdot}OH)_2PrP$. XIII.2. Crystals, m. 174 to 175°.[1106]
$(PhCH{\cdot}OH)_2$i-BuP. XIII.2. Isolated as hydrochloride, crystals, m. 109 to 111° (from CH_3CN).[356]
$(PhCH{\cdot}OH)_2PhP$. XIII.2. Liquid, b_3 140 to 143°.[1105]
$(PhCMe{\cdot}OH)_2PhP$. XIII.3. Crystals, m. 105 to 107°.[1105]
$(PhCH_2CH_2)_2(NCCH_2CH_2)P$. XII. pK_a (H_2O) 3.43.[1342]
(3-Pyridyl-$CH_2CH_2)_2PhP$. XII. Liquid, b_1 227 to 228°, n_D^{20} 1.6022, d_4^{20} 1.0854.[58,59]
$(PhCCl{=}CH)_2(PhCH{=}CH)P$. VIII. Solid, m. 96.5 to 97.5°.[374]
$(PhCH{=}CH)_2PhP$. VIII. Solid, m. 70.5 to 71.5°.[374]
$(PhCCl{=}CH)_2PhP$. VIII. Solid, m. 97.5 to 98.7°.[374]
$(PhC{\equiv}C)_2PhP$. Crystallizes in the space group $Pca2_1$ (Pcam).[1014]
$(PhC{\equiv}C)_2C_6F_5P$. Solid, m. 102°.[385c]
$(4\text{-}BrC_6H_4CH_2CH_2CH_2)_2PhP$. XII. Solid, m. 65 to 66.5°.[59]
Ph_2MeP. I.[34,752,914,939,1101,1454,1466] II.[1278] VI.[990] VII.[562,956,1120,1241,1247] IX.3-5.[458,609,623,624] Also obtained from Ph_2PLi and C_6H_5OMe by demethylation.[956] Liquid, b. 284°,[609,990] b_{50} 184 to 186°,[752] b_{12} 160°,[624] b_{11} 154 to 160°,[1247] b_8 148°,[34] b_7 146 to 147°,[1454] $b_{0.5}$ 125 to 128°,[1101] $b_{0.5}$ 108 to 110°,[939] $b_{0.35}$ 98 to 100°,[956] $b_{0.2}$ 87 to 90°,[954,1120] $b_{0.15}$ 111 to 112°,[562] $b_{0.15}$ 108 to 110°,[1278] $b_{0.15}$ 120 to 122°,[458] $b_{0.05}$ 80 to 83°,[914] n_D^{25} 1.6230,[1466] n_D^{20} 1.6257,[1454] n_D^{20} 1.6252,[1241] n_D^{20} 1.6232,[752] n_D^{20} 1.6245,[956] n_D^{20} 1.6241,[956] $n_D^{19.5}$ 1.6261,[34] d_4^{20} 1.0646,[1454] d_4^{20} 1.0779,[752] d_4^{15} 1.0784,[990] IR,[942,956] ^1H NMR,[418,484] ^{31}P +28 ppm,[1008] +28.1 ppm,[482] μ 1.39 D;[1242] L·MeI, m. 242 to 243°;[458] reacts with CCl_4 to give $CHCl_3$.[248]
Ph_2CD_3P. VIII. Liquid, b. (under vacuum, pressure not given) 145 to 146°, IR;[1253] oxide, m. 112.5 to 113.5°.
Ph_2CF_3P. X.[96] XIV. (By heating of CF_3COPPh_2).[858] Liquid, b_{760} 255 to 257°,[96] $b_{0.001}$ 105°,[858] log P_{mm} = 7.781 - (2598/T),[96] IR,[97,858] stable in air up to 300°; hydrolyzed by KOH.[96]
$Ph_2(ClCH_2)P$. I.[125] IX.2. Viscous, colorless oil,[546] yellow oil;[125] decomposes on heating under vacuum;[546] dimerized on attempted dist.; oxide, m. 132°;[125] yields with excess PhMgBr $CH_2{=}PPh_3$.[125]

$Ph_2(Me_2NCH_2)P$. VII. Liquid, $b_{0.2}$ 130°, ^1H NMR,[20] oxide, m. 189 to 190°.

$Ph_2(Et_2NCH_2)P$. X.[930] XIII.1.[918,933] Liquid, $b_{0.3}$ 133 to 138°,[933] n_D^{20} 1.5912,[918] IR,[918] ^{31}P +27.8 ppm,[918] +28.8 ppm,[930] +27.8 ppm,[911] +27.3 ppm.[924]

$Ph_2(HOCH_2)P$. IX.2.[546,1107,1360] XIII.1.[546] Liquid, b_1 113°, n_D^{20} 1.6137, d_4^{20} 0.9316;[1107,1360] isomerizes on heating to 250 to 260° to $Ph_2MeP{=}O$,[546] and decomposes to Ph_2PH and CH_2O.

$Ph_2(MeOCH_2)P$. VII.[20] VIII.[1360] Liquid, $b_{0.1}$ 138 to 139°,[1360] $b_{0.65}$ 126 to 127°;[20] L•MeI, m. 153 to 155°.

$Ph_2(2\text{-}HOC_6H_4OCH_2)P$. VII. Liquid, $b_{0.25}$ 167 to 168°.[954]

$Ph_2(NaO_2C)P$. VII. Solid, m. 240° dec.,[823] free acid not stable.

Ethyl ester. VII. Liquid, b_4 163 to 165°,[649] b_{5-6} 185 to 186°.[737]

$Ph_2(RHNCO)P$. XIII.5.

(R = H). Crystals, m. 115 to 116°,[1074,1075] m. 118 to 120°,[1377] IR,[1075] ^1H NMR,[1377] useful for flameproofing cloth.

(R = $ClCH_2CO$). Crystals, m. 130 to 131°.[1290]

(R = Cl_3CCO). Crystals, m. 78 to 79°.[1290]

(R = Ph). Crystals, m. 136 to 137°,[1287] m. 139°,[1290] m. 142.5 to 143°, mass spect.[807]

(R = PhCO). Crystals, m. 133°.[1290]

$Ph_2(Ph{\cdot}N{=}C{\cdot}NHPh)P$. XIII.5. Plates, m. 130°,[718] (from Me_3SiPPh_2 and PhNCO).

$Ph_2(RHNCS)P$. XIII.5.

(R = H). Colorless needles, m. 114 to 116° (from i-PrOH), IR, UV, mass spect.[1063]

(R = Me). Crystals, with $2MeNCS{\cdot}3Et_2O$, m. 132.5° (from Et_2O), IR.[666]

(R = Et). XIII.5. Crystals, m. 125°, 167.5° (from EtOH),[666] IR;[2,666] K salt, m. 232° (dec.).[666]

(R = Ph). X.[666] XIII.5. Crystals, m. 118°,[1290] m. 119° (from EtOH).[665]

(R = $4\text{-}BrC_6H_4$). XIII.5. Crystals, m. 138° (from EtOH), IR.[666]

(R = $2\text{-}C_{10}H_7$). Crystals, m. 188.5° (from EtOH), IR.[666]

$Ph_2(PhN{=}C{\cdot}SEt)$. XIV. From K salt and EtI. Crystals, m. 96° (from EtOH).[665]

$Ph_2(MS_2C)P$. VII.

(M = Li, Na, K). Light orange-yellow solids with 1 mole of dioxan, free acid not stable.[813]

(M = Et_3NH). Orange-red crystals, m. 62 to 66°.[275a]

(M = K). Brick-red crystals, m. 180° (dec.).[275a]

(M = Ph_4P). Orange-red crystals, m. 127 to 130°, IR.[275a]

$Ph_2(MeSCH_2)P$. II.[1098] VII.[20] Liquid, $b_{1.0}$ 152 to 160°,[20] $b_{0.5}$ 160 to 165°,[1098] IR,[20] ^1H NMR,[20,1098] ^{31}P +21 ppm;[1098] L•MeI, m. 161.3 to 163.5°.[1098]

Ph_2EtP. I.[440,977,1466] IV.[990,1241] VI.[1012] VII.[940,1059,1237] [$OP(NMe_2)_3$ as solvent recommended]. IX.3-5.[458,609,623,624] Liquid, b. 293°,[990] b_{22} 184°,[977] b_{14} 182°,[609] b_{10} 167 to 168°,[1241] b_{10} 159 to 161°,[1237] $b_{1.7}$ 142°,[440] $b_{1.4}$ 112°,[1466] $b_{0.3}$ 108 to 111°,[939] $b_{0.3}$ 114 to 116°,[940] $b_{0.1}$ 112 to 116°,[1012] n_D^{20} 1.6096,[940] μ 1.35 D,[778] μ 1.46 D,[440] pK_a 2.62,[440] UV,[1284] ^{31}P +13.5 ppm,[553,1008] +12.5 ppm,[482] +13.1 ppm;[1358a] L•MeI, m. 186.7°.[939]

$Ph_2C_2F_5P$. XIV. From $C_2F_5COPPh_2$ by heating to 190°. Liquid, $b_{0.001}$ 114 to 116°, IR.[859]

$Ph_2(ClCH_2CH_2)P$. VII. Crystals, m. 40 to 42° (from ligroin), $b_{0.05}$ 110 to 117°, IR;[1344] displayed cytotoxic activity at a concentration of 100 μg/ml against Eagle's KB cells; cell growth was less than 50% of the growth of controls.[1344]

$Ph_2(CH_2{=}CH)P$. I.[115,1097,1112,1154,1449] XIV. From Ph_2-($HOCH_2CH_2$)P by heating to 330°.[1364] Liquid, $b_{0.6}$ 117.5 to 119°,[1154] $b_{0.5}$ 116 to 118°,[1097] $b_{0.3}$ 123 to 124°,[1449] $b_{0.25}$ 104°,[115] $b_{0.24}$ 103.5 to 104.2°,[1072] n_D^{25} 1.6215,[1097] n_D^{27} 1.6229,[1072] n_D^{30} 1.6247,[1449] ^1H NMR,[1449] ^{31}P +13.8 ppm,[1449] +11.7 ppm,[484] could not be homopolymerized by radical initiation[115,1152] but yields on polymerization with $BF_3 \cdot Et_2O$ a low-molecular-weight, oily compound;[1072] the corresponding oxide and sulfide, however, gave a polymer;[1152] L•MeI, m. 124 to 125°.[1364]

$Ph_2(HC{\equiv}C)P$. I. Crystals, m. 35° (from EtOH),[218,219,231] IR,[218,219,972] ^1H NMR,[218,219] pK_a 1.57;[333] yields on treatment with HBr in AcOH a cyclic diphosphonium salt.[19]

$Ph_2(H_2NCH_2CH_2)P$. VII. Yellow liquid, b_3 176 to 177°,[694] b_9 220°;[661] MeI quaternizes the N aton, m. 146 to 148° (dec.).[661]

$Ph_2(Et_2NCH_2CH_2)P$. VII. Colorless oil, b_3 185°,[697] $b_{0.05}$ 140 to 143°,[325] IR;[697] EtI quaternizes P atom, dec. 139°.[697]

$Ph_2(HOCH_2CH_2)P$. VII.[691,700] XIV.[1364] (From t-$BuOCH_2CH_2$-PPh_2 and HCl). Liquid, b_{18} 178 to 184°,[691] b_3 220 to 222°,[691] b_3 178 to 180°,[1364] b_{15} 204 to 206°,[1364] $E_{1/2}$ -0.13 V,[689] IR;[1364] L•MeI, m. 186 to 188°;[1364] on heating to 330°, $Ph_2PCH{=}CH_2$ and $Ph_2P(O)CH_2CH_2(O)PPh_2$ are formed.[1364]

Acetyl deriv. From phosphine and MeCOCl. Colorless oil; L•MeI, m. 107 to 108°;[1364] heating to 147° gave $Ph_2PCH_2CH_2PPh_2$ and $Ph_2PCH{=}CH_2$.[1364]

Benzoyl deriv. From phosphine and PhCOCl. Pale-yellow oil, IR;[1364] L•MeI, m. 130 to 131°;[1364] heating to 156° gave $Ph_2PCH_2CH_2PPh_2$ and $Ph_2PCH{=}CH_2$; underwent self-quaternization at room temp.[1364]

$Ph_2(EtOCH_2CH_2)P$. VII. Liquid, $b_{0.09}$ 131 to 135°, IR.[1344]

$Ph_2(t\text{-}BuOCH_2CH_2)P$. VII. Liquid, $b_{1.0}$ 146 to 148°; L•MeI,

m. 212 to 214°.[1364]

$Ph_2(2\text{-}HOC_6H_4OCH_2CH_2)P$. VII. Oil.[954]

$Ph_2(MeCO)P$. VII.[691] XII.[808] (From Ph_2PH and $CH_2{=}C{=}O$). Liquid, $b_{2.5}$ 143 to 146°,[691] b_1 133°, $b_{1.5}$ 139 to 141°, n_D^{20} 1.6220, d_4^{20} 1.1294,[808] UV,[806] IR, ^1H NMR;[808] L•MeI, m. 121 to 124°.[808]

$Ph_2(MeC^{18}O)P$. VII. Liquid, $b_{1.5}$ 140 to 142°, n_D^{20} 1.6625, IR, ^1H NMR;[804] two conformers are present.

$Ph_2(CF_3CO)P$. VII. Light-yellow liquid, $b_{0.001}$ 102°,[858,859] b_2 122 to 123°, n_D^{20} 1.5315, d_4^{20} 1.0599;[808] is not oxidized in air and does not react with sulfur or MeI;[808] ref. 858, however, reports oxidation to $Ph_2CF_2P{=}O$ and CO_2; at 190° yields Ph_2CF_3P and CO, IR.[858]

$Ph_2(HO_2CCH_2)P$. VII. Solid, m. 120 to 121°,[703,704] pK_a 5.49 (in 50% EtOH) and 5.90 (in 66.7% EtOH),[684] ^1H NMR;[705] catalyzes hydrogen evolution in electrolysis.[688]
Na salt. VII. Solid, m. 264 to 267°.[703]
Ethyl ester. VII. Liquid, b_5 183 to 185°.[703]

$Ph_2(H_2NCOCH_2)P$. XIV. From ester and NH_3. Crystals, m. 173 to 174° (from MeOH).[705]

$Ph_2(HO_2CC{\equiv}C)P$. XIV. From $Ph_2PC{\equiv}CMgBr$ and CO_2. Crystals, m. 145° (from Et_2O).[218]

$Ph_2(MeSCH_2CH_2)P$. VII. Crystals, m. 49 to 51°.[1196]

Ph_2PrP. I. Liquid, b_1 138°, μ 1.43 D, pK_a 2.62,[440] ^{31}P +17.6 ppm.[484]

$Ph_2i\text{-}PrP$. VII.[823] IX.3.[458] IX.5.[623,624] Liquid, b_{13} 165°,[624] b_{17} 165°,[823] $b_{0.5}$ 145 to 147°,[458,623,624] ^{31}P -0.2 ppm.[483]

$Ph_2(c\text{-}C_3H_5)P$. II.[254] IX.5.[117] Solid, m. 47°, $b_{0.15}$ 100°,[254] IR, ^1H NMR.[254]

$Ph_2(C_3F_7)P$. II.[454] VII.[1283] Liquid, b_1 94 to 95°,[1283] $b_{0.001}$ 40°,[454] IR, ^{19}F NMR;[454] C_3F_7 group slowly attacked by H_2O and OH^-;[454] has been claimed useful for end-capping azide-terminated polymers and thus increasing their thermal stability.[1283]

$Ph_2(CH_2{=}CHCH_2)P$. I.[160] VII.[698,954] IX.1.[609] Liquid, b_{15} 194 to 200°,[609] b_4 150°,[698] $b_{0.9}$ 144 to 145°,[954] $b_{0.45}$ 102 to 103°,[160] IR,[698] ^{31}P +17.1 ppm;[448] L•MeI, m. 142.5 to 143.5°.[954]

$Ph_2(CH_3CH{=}CH)P$. From above and MeONa. Liquid, $b_{0.1}$ 110 to 112°, ^1H NMR; L•MeI, m. 162°.[613a]

$Ph_2(CH_2{=}C{=}CH)P$. I. From Ph_2PCl and $XMgC{\equiv}CMe$. Gives a mixture of $Ph_2PC{\equiv}CMe$ and $Ph_2PCH{=}C{=}CH_2$, liquid, $b_{0.05}$ 105° (b. of mixture),[1302] ^1H NMR,[1301] sign of coupl. const.[1302]

$Ph_2(CF_2{=}CClCF_2)P$. IV. Liquid, b_2 142°, $b_{0.01}$ 91 to 95°,[124] IR, ^1H and ^{19}F NMR;[124] oxide, m. 80 to 81°.[124]

$Ph_2(MeC{\equiv}C)P$. I.[163,219] II.[22,28] Crystals, m. 33° (from Et_2O),[163,219] $b_{0.1}$ 143°,[219] $b_{0.1}$ 120 to 123°,[22] $b_{0.3}$ 130 to 131°,[28] UV,[218] IR,[28,218,972] ^1H NMR,[28,218,219,1298] $^{13}C\text{-}H$ coupl. const.,[1299] pK_a 3.30;[333] contrary to the sulfide and oxide, the tertiary phosphine is not reduced by $LiAlH_4$ to the vinylphosphine,[22] and does not add Grignard reagents in the presence of $CuCl_2$;[21] the phosphine gives a cyclic diphosphonium salt on treatment with HCl or HBr.[28]

$Ph_2(H_2NCH_2CH_2CH_2)P$. XIV. From $Ph_2PCH_2CH_2CN$ and $LiAlH_4$. Dense, yellow liquid, $b_{0.4}$ 150°.[1207]

$Ph_2(Me_2NCH_2CH_2CH_2)P$. VII. Liquid, $b_{1.2}$ 175 to 178°;[1434] oxide, m. 87.5°;[1434] the phosphine, the oxide, and the sulfide caused a marked depression in spontaneous activity in mice at dose levels well below the LD_{50} range.[1434]

$Ph_2[(Et_2NCH_2)_2CH]P$. VII. Straw-colored oil, $b_{0.04}$ 146 to 149°.[1197,1353]

$Ph_2(2\text{-}MeNHC_6H_4CH{=}NCH_2CH_2CH_2)P$. XIV. From $Ph_2P(CH_2)_3NH_2$ and aldehyde. Crystals, m. 93 to 95°.[1207]

$Ph_2(2\text{-}MeOC_6H_4CH{=}NCH_2CH_2CH_2)P$. XIV. (See above). Crystals, m. 58 to 58.5°.[1207]

$Ph_2(2\text{-}MeSC_6H_4CH{=}NCH_2CH_2CH_2)P$. XIV. (See above). Crystals, m. 78 to 79°.[1207]

$Ph_2(NCCH_2CH_2)P$. VII.[1229] VIII.[1229] XII.[567,950,1229] Crystals, m. 64 to 65.4°,[567,950,1229] $b_{0.66}$ 175 to 178°,[1229] UV.[1230,1236]

$Ph_2(HOCH_2CH_2CH_2)P$. VII. Crystals, m. 60 to 61° (from ligroin).[700]

$Ph_2(PhOCH_2CH_2CH_2)P$. IX.3. Isolated as the oxide.[512]

$Ph_2(MeCHOHCH_2)P$. XIV. From Ph_2PCH_2COMe and $LiAlH_4$. Liquid, $b_{0.01}$ 127 to 128°,[1061] n_D^{20} 1.6145, d_4^{20} 1.1000, ^1H NMR,[1061] $E_{1/2}$ −0.15 V.[689]

$Ph_2(HOCH{\cdot}MeCH_2)P$. VII. Liquid, b_3 192 to 195°;[700] L·MeI, m. 150 to 160°.

$Ph_2(\overset{O}{\overbrace{CH_2CH}}CH_2)P$. VII. Pale-yellow oil; polymerizes easily and on heating splits off H_2O to give $Ph_2(CH_2{=}CHCH_2)P$ and other products; IR.[698]

$Ph_2(MeCOCH_2)P$. VI. Liquid, $b_{0.01}$ 125 to 126°,[1061] $b_{0.03}$ 138°,[1060] n_D^{20} 1.6149, d_4^{20} 1.1233.[1061]

$Ph_2(C_2F_5CO)P$. VII. Liquid, $b_{0.001}$ 109 to 111°, IR;[972] decomposes at 190° to give $Ph_2PC_2F_5$ plus CO.[859]

$Ph_2(HO_2CCH_2CH_2)P$. VII. Crystals, m. 127 to 128°,[703,950] UV,[1236] ^1H NMR,[705] pK_a 5.03,[1116] stability constant of Ag complex, $-M_{\beta LAg}$ 3.80.[1116]
Methyl ester. VII. Liquid, $b_{0.4}$ 151°.[567]

Ph_2BuP. V.[1177] II.[518] VII.[823] IX.3.[458] IX.5.[624] Liquid, b_4 180 to 182°,[823] b_{13} 172 to 174°,[624] $b_{0.3}$ 117 to 120°,[518] $b_{0.45}$ 140°,[458] $b_{0.15}$ 114 to 115°,[1177] GLC,[1177] ^{31}P +17.1 ppm.[483]

$Ph_2i\text{-}BuP$. VII. From Ph_2PK and $PhSO_3Bu\text{-}i$. Liquid, b_{50} 202°,[1241] $b_{0.5}$ 113 to 115°,[484] n_D^{20} 1.5906,[1241] ^{31}P +21 ppm.[484]

$Ph_2(2\text{-}Bu)P$. VII. Liquid, $b_{1.5}$ 141 to 145°, ^{31}P +3.2 ppm.[484]

Ph_2(t-Bu)P. I. Liquid, b_2 144 to 146°,[484] $b_{0.2}$ 118 to 122°,[580] ^1H NMR and sign of coupl. const.,[352,484] ^{31}P -17.1 ppm;[484] L•MeI, m. 185 to 187°.[580]

Ph_2(CH_2=CHCH•Me)P. VIII. Liquid, IR;[1218] identified as oxide, m. 90 to 91°; rearranges at 200° in the presence of $PhSiH_3$ to Ph_2($MeCH_2$CH=CH)P; oxide, m. 115 to 116°.[1218]

Ph_2(EtCBr=CH)P. XIV. From Ph_2(EtC≡C)P and HBr. Red-yellow liquid, IR, ^1H NMR; oxide, m. 118 to 120°.[24]

Ph_2(CH_2=C=C•Me)P. I. From Ph_2PCl and XMgC≡CEt. Obtained as a mixture of Ph_2(EtC≡C)P and Ph_2(CH_2=C=C•Me)P, $b_{0.01}$ 110°,[1302] ^1H NMR.[1300,1302]

Ph_2[CH_2=C•(CH=CH_2)]P. I. Pale-yellow solid, m. 35 to 36.5°, IR.[64]

Ph_2(HC≡C•C≡C)P. I. Crystals, m. 52° (from pentane),[218,219] IR,[218,219] UV,[218] ^1H NMR.[218,219,1298]

Ph_2(EtC≡C)P. II. Liquid, $b_{0.45}$ 133 to 136°,[24] $b_{0.001}$ 105 to 108°,[1389] $b_{0.45}$ 133 to 134°,[28] n_D^{20} 1.5570,[1389] IR,[24,28,1389] ^1H NMR;[24,28] the phosphine gives a cyclic diphosphonium salt with HCl or HBr.[28]

Ph_2(CF_3CH=C•CF_3)P. XII. Liquid, $b_{0.001}$ 127°, IR, ^1H NMR[266] (a mixture of 60% trans and 40% cis).

Ph_2(2-Cl-tetrafluorocyclobuten-1-yl)P. IV. Liquid, $b_{0.1}$ 114 to 116°, IR, GLC, ^{31}P +25.1 ppm.[1340]

Ph_2(2-F-cyclobuten-1-one-4)P. X. Solid, m. 120 to 124°, IR.[265]

Ph_2(Me_2C•CN)P. XIV. From Ph_2PH and AIBN. Characterized as sulfide, m. 123 to 125°.[870]

Ph_2[HO(CH_2)$_4$]P. VII.[940] XIV.[425] Solid, m. 40°,[425] $b_{0.003-0.004}$ 120 to 135°, $b_{0.16}$ 170°,[425] $b_{0.2}$ 160 to 163°,[937,940] n_D^{20} 1.6101,[937,940] n_D^{25} 1.6103,[425] IR;[425] L•MeI, m. 164.5 to 166°.[940]

Ph_2[(CF_3)$_2$CHCO]P. XII. From Ph_2PH and (CF_3)$_2$C=C=O. Liquid, b_1 194 to 195°, n_D^{20} 1.3955, d_4^{20} 0.9408, IR, ^1H NMR.[809]

Ph_2($HO_2CCH_2CH_2CH_2$)P. VII. Crystals, m. 97 to 98°.[703]

Ph_2($Me_2CHCH_2CH_2$)P. VII. Liquid, $b_{11.5}$ 181 to 181.5°, n_D^{20} 1.5832.[1241]

Ph_2(t-BuCH_2)P. VII. Liquid, $b_{0.1}$ 116 to 122°, ^{31}P +23.9 ppm.[484]

Ph_2(t-Am)P. I. ^{31}P -15.4 ppm.[484]

Ph_2(c-C_5H_9)P. I. Liquid, b_2 160 to 163°, ^{31}P +3.9 ppm.[484]

Ph_2(CH_2=CH-C•Me_2)P. VIII. Liquid, b_1 160 to 165°;[1218] this phosphine rearranged at 200° in the presence of $PhSiH_3$ to Ph_2(Me_2CCH=CH_2)P, identified as oxide.[1218]

Ph_2(CH_2=$CHCH_2CH_2CH_2$)P. I. Liquid, $b_{0.15}$ 118 to 121°; L•MeI, m. 105°.[102]

Ph_2(PrC≡C)P. II. Liquid, $b_{0.4}$ 132 to 134°, IR, ^1H NMR;[28] gives cyclic salt with HCl or HBr.[28]

Ph_2(MeC≡C-C≡C)P. I. Crystals, m. 61° (from EtOH),[218,219] UV,[218] IR,[218,219,972] ^1H NMR.[218,219,1298]

Ph_2(2-Cl-hexafluorocyclopenten-1-yl)P. IV. Solid, m. 49 to 50°, b_2 148 to 149°, IR, ^1H and ^{19}F NMR, ^{31}P +19.6 ppm.[1339]
$Ph_2(Me_2C{\cdot}OHC{\equiv}C)P$. XIV. From $Ph_2PC{\equiv}CH$ and $Me_2C{=}O$. Crystals, m. 76° (from hexane)[218,219,231] IR.[218,972]
$Ph_2[HO(CH_2)_5]P$. VII. From Ph_2PLi and tetrahydropyran. Oil, $b_{0.3}$ 160 to 188°, IR.[954]
$Ph_2(MeCOCH_2CH{\cdot}Me)P$. VII. Colorless oil, b_2 185 to 190°.[671]
$Ph_2(n\text{-}C_6H_{13})P$. I. Liquid, $b_{0.09}$ 138°, n_D^{20} 1.5760.[1097]
$Ph_2(t\text{-}BuCH_2CH_2)P$. VII. Liquid, $b_{0.3}$ 110°, ^1H NMR;[1429] exists as one trans and two gauche conformers; energy difference between gauche and trans conformers is 1.99 kcal/mole (from ^1H NMR).[1429]
$Ph_2(Et_2MeC)P$. ^{31}P -11.4 ppm.[485]
$Ph_2(c\text{-}C_6H_{11})P$. VI.[331a] VII. Crystals, m. 60 to 61°,[713] ^{31}P +4.4 ppm.[483]
$Ph_2[HO(CH_2)_6]P$. IX.3. Liquid, $b_{0.1}$ 152°;[513] from $[Ph_3P(CH_2)_6OH]I$ and NaH, gave also $Ph_2[PhO(CH_2)_6]P$, isolated as oxide.
$Ph_2(2\text{-}HOC_6H_{10}\text{-}c)P$. VII. Crystals, m. 137° (from EtOH),[700] L·MeI, m. 218 to 223°.
$Ph_2(PrC{\equiv}C{-}C{\equiv}C)P$. I. Crystals, m. 44°, IR,[218,219] UV.[218]
$Ph_2[Me(C{\equiv}C)_3]P$. I. Crystals, m. 131°,[219] UV, IR,[218] ^1H NMR.[218,219,1298]
$Ph_2(c\text{-}C_5H_9C{\equiv}C)P$. II. Liquid, $b_{0.25}$ 175 to 178°, IR, ^1H NMR.[28]
Ph_2(1-cyclopentenyl-C≡C)P. I. Crystals, m. 42° (from EtOH).[218,219]
$Ph_2(c\text{-}C_6H_{11}C{\equiv}C)P$. II. Liquid, $b_{0.75}$ 212 to 214°, IR, ^1H NMR.[28]
Ph_2(1-cyclohexenyl-C≡C)P. I. Crystals, m. 60°, IR, UV.[218,219]
$Ph_2(c\text{-}1\text{-}HOC_6H_{10}C{\equiv}C)P$. XIV. From $Ph_2PC{\equiv}CH$ and cyclohexanone. Solid, m. 101°.[218]
$Ph_2(n\text{-}C_6H_{13}C{\equiv}C)P$. II. Liquid, $b_{0.16}$ 152 to 152.5°, IR, ^1H NMR.[28]
$Ph_2(n\text{-}C_6H_{13}CH{=}CH)P$. XII. Liquid, $b_{1.25}$ 190 to 195°, ^1H NMR.[7]
$Ph_2[Me(C_6H_{13})CH]P$. VII. Liquid, $b_{1.5}$ 180 to 200°,[256] optically active because of the asymmetric carbon atom.[256]
Ph_2(1-cycloheptenyl-C≡C)P. I. Characterized as oxide, m. 114°.[218,219]
$Ph_2[Me_2C{=}CHCH_2CH_2CH_2(CH_2{=}CH){\cdot}C{\cdot}Me]P$. VIII. Liquid, $b_{0.4}$ 180 to 200°;[1218] this phosphine rearranged at 200° in the presence of $PhSiH_3$ to Ph_2(geranyl)P, identified as oxide, m. 113 to 114°.[1218]
$Ph_2[HO(CH_2)_{11}]P$. IX.3. Solid, m. 52°.[513] From $[Ph_3P(CH_2)_{11}OH]I$ and NaH; gave also $Ph_2(PhO(CH_2)_{11})P$.[513]
$Ph_2(n\text{-}C_{16}H_{33})P$. VII. Solid, m. 36 to 42°.[1241]
$Ph_2(PhCH_2)P$. I.[160] VII.[954] ($LiPPh_2$ + $PhOCH_2Ph$,[954] or

$PhCH_2Cl$).[364] IX.2. and .4.[609,1360] Crystals, m. 142 to 143°,[160] m. 74°,[1360] $b_{1.5}$ 205 to 208°,[609] $b_{0.15}$ 146 to 148°,[160] $b_{0.75}$ 185 to 190°,[364] ^1H NMR,[364] ^{31}P +10.4 ppm;[484] L·MeI, m. 239 to 240°.[954]

$Ph_2(2\text{-}MeOCH_2C_6H_4CH_2)P$. VII. Liquid, $b_{0.15}$ 180 to 196°.[952]

Ph_2(PhNH·CH·Ph)P. XII. Crystals, m. 34 to 35° (from EtOH).[1136]

Ph_2(PhN=C·Ph)P. VII. Crystals, m. 96 to 99° (from EtOH).[687]

Ph_2(PhCH·OH)P. XIII.2. (From Ph_2PH and PhCHO in MeOH with little HCl). Crystals, m. 186°;[1107] L·MeI, m. 153 to 155°.[356]

$Ph_2(4\text{-}ClC_6H_4CH\cdot OH)P$. XIII.2. (As above). Solid, isolated as oxide, m. 168 to 170°.[356]

$Ph_2(1\text{-}C_{10}H_7CH\cdot OH)P$. XIII.2. (As above). Solid, isolated as oxide, m. 158 to 159°.[356]

Ph_2(PhC·MeOH)P. XIII.3. Crystals, m. 116 to 116.5°.[1107]

Ph_2(PhCO)P. VII.[696] XIV.[687] Yellow crystals, m. 68 to 81°,[696] m. 89 to 92° (from EtOH),[687] $b_{2.5}$ 199 to 201°,[696] (from $PhCOPEt_2$ and $HPPh_2$).[687]

$Ph_2(Ph_3C)P$. I. Crystals, m. 156 to 158° (from acetone);[714] L·MeI, dec. 242 to 244°.

$Ph_2(PhCH_2CH_2)P$. VII. Colorless oil, b_{11} 223 to 224°, n_D^{20} 1.6312,[1241] m. ∿-10°.[670]

$Ph_2(PhCHOHCH_2)P$. VII. Colorless liquid, b_3 218 to 222°.[700]

$Ph_2(Ph_2CHCH_2)P$. VII. From $Ph_2C=CH_2$ + $LiPPh_2$. Colorless needles, m. 92 to 93° (from C_6H_6/ligroin).[670]

$Ph_2(Ph_2C\cdot OHCH_2)P$. II. From Ph_2PCH_2Li and $Ph_2C=O$. Characterized as sulfide, m. 141 to 145°.[1099]

$Ph_2(2\text{-}pyridyl\text{-}CH_2CH_2)P$. VII. XII. Crystals, m. 58°.[1370]

Ph_2(cis-PhCH=CH)P. VII.[15] XII. From Ph_2PH and PhC≡CH[7,578] and from Ph_2PLi and PhC≡CH in the presence of amine.[7] In the absence of amine, trans is formed.[7] XIV. From $Ph_2PCHNaP(O)Ph_2$ and PhCHO.[439] (Probably gives a mixture of geometrical isomers). Crystals, m. 90 to 92°,[7] m. 89 to 90°,[578] m. 57 to 58°,[439] (isomer), ^1H NMR;[7] oxide, m. 103 to 104°.[15]

Ph_2P(trans-PhCH=CH)P. VII.[14,154] XII.[7] Liquid, $b_{1.6}$ 198 to 200°,[7] $b_{0.25}$ 175 to 180°,[154] ^1H NMR;[7] oxide, m. 168 to 169°.[14]

$Ph_2(2\text{-}MeOC_6H_4CH=CH)P$. XIV. From $Ph_2PCHNaP(O)Ph_2$ and 2-$MeOC_6H_4CHO$. Crystals, m. 58 to 59°.[439]

$Ph_2(CH_2=C\cdot Ph)P$. VII. Crystals, m. 91.5 to 92.5° (from MeOH).[1219]

Ph_2(2-pyridyl-CH=CH)P. XIV. From $Ph_2PCHNaP(O)Ph_2$ and RCHO. Crystals, m. 82 to 83°.[439]

Ph_2(2-thienyl-CH=CH)P. XIV. As above. Crystals, m. 58 to 59°.[439]

Ph_2(cis-PhCH=C·Ph)P. VII. XII. From $LiPPh_2$ and PhC≡CPh in the presence of Et_2NH. Crystals, m. 114 to 115° (from hexane),[8] ^1H NMR,[8] μ 1.96 D.[9]

Ph_2(trans-PhCH=C·Ph)P. VII.[670] XII.[8] From $LiPPh_2$ and PhC≡CPh in the presence of $BuNH_2$ or from Ph_2PH or $NaPPh_2$ and PhC≡CPh.[8] Crystals, m. 120 to 121° (from MeOH; EtOH causes isomerization to cis);[8] m. 116 to 117°,[670] μ 2.59 D,[670] μ 0.99 D,[9] ^1H NMR.[8]

Ph_2(PhC≡C)P. I.[218,219,231] VII.[664] Crystals, m. 44°,[664] m. 43° (from EtOH);[218,219,231] crystallizes in the space group $P2_12_12_1$,[1014] IR.[218,219]

Ph_2(9-HO-fluorenyl·C≡C)P. XIV. From $Ph_2PC≡CH$ and fluorenone. Crystals, m. 133°.[218,231]

Ph_2(CH_2=CHCH·Ph)P. VIII. Crystals, m. 74 to 76° (from MeOH), $b_{0.6}$ 190 to 195°, IR;[1218] the pure phosphine is stable, but in the presence of catalytic amounts of $PhSiH_3$ it rearranges at 210° to Ph_2(PhCH=CHCH_2)P;[1218] the phosphine also rearranges on salt formation to [Ph_2MePC·Ph=CHMe]I.[1218]

Ph_2(EtO_2CCH=C·Ph)P. XII. Isolated as methiodide, m. 105° (dec.).[578]

Ph_2(PhCOCH_2CH·Ph)P. VII. Colorless needles, m. 172 to 174° (sealed tube).[671]

Ph_2(Ph_2C·OHC≡C)P. XIV. From $Ph_2PC≡CH$ and Ph_2C=O. Crystals, m. 143° (from THF).[218,231]

Ph_2(9-anthranyl-CH_2·C≡C)P. I. Solid, m. 178° (from AcOH/toluene).[218,219]

Ph_2(anthrafuchsonyl·OHC≡C)P. From $Ph_2PC≡CH$ and anthrafuchsonone. Solid, m. 202°.[218,231]

Ph_2(MeCOCH_2CH·Ph). VII. Colorless crystals, m. 120° (from Et_2O) (sealed tube).[671]

Ph_2(MeCHOHCH_2CH·Ph). XIV. Viscous oil, b_5 220 to 230°;[671] L·MeI, m. 238 to 239°.

Ph_2(PhCH_2CH_2C≡C)P. II. Liquid, $b_{0.60-0.75}$ 180 to 184° (molecular still), IR, ^1H NMR.[28]

Ph_2(PhCH=CH-CH=CH)P. XIV. (From Ph_2PCHNaP(O)Ph_2 and PhCH=CHCHO). Crystals, m. 90 to 91°.[439]

Ph_2(Ph_2C=CH-C≡C)P. I. Characterized as oxide, m. 136°,[218,219] IR.[219]

Ph_2(PhC≡C-C≡C)P. I. Solid, m. 75°, IR,[218 219] UV.[218]

Ph_2(PhC≡CCOCH=C·Ph)P. XII. Isolated as oxide, m. 207 to 209°.[578]

Ph_2(t-BuCOCH_2CH·Ph)P. VIII. Colorless crystals, m. 210 to 213° (from EtOH) (sealed tube).[671]

Ph_2(2- or 3- or 4-DC_6H_4)P. I. All three compounds had same m. 82 to 83° (from EtOH);[865] rate const. of protophilic isotopic exchange of deuterium.[865]

Ph_2(3-FC_6H_4)P. I. Crystals, m. 56 to 58°,[1156] m. 57°,[1251] ^{19}F NMR,[1156,1251] σ_m (of Ph_2P group) +0.23.[1251]

Ph_2(4-FC_6H_4)P. I. Crystals, m. 36 to 37°,[1156] m. 37 to 39°, b_9 207 to 208°,[1243] ^{19}F NMR,[1156,1243] σ_p (of Ph_2P group) +0.18,[1156] +0.25,[738] +0.31.[1243]

Ph_2(2-ClC_6H_4)P. I. Crystals, m. 107 to 108°.[518]

Ph_2(3-ClC_6H_4)P. I. Crystals, m. 30 to 31° (from i-PrOH),

b_4 195 to 198°.[864]
$Ph_2(4\text{-}ClC_6H_4)P$. I.[443,1228,1362] III.[986] Crystals, m. 44 to 45° (from EtOH),[443] m. 41.5 to 42.5°,[1362] m. 39 to 42°,[1228] b_4 196 to 197°,[1362] $b_{0.05}$ 187 to 194°,[1228] UV, IR, μ 2.08 D,[443] pK_a 2.18.[445]
$Ph_2(2\text{-}BrC_6H_4)P$. II. Crystals, m. 115°.[518]
$Ph_2(3\text{-}BrC_6H_4)P$. I.[72,1234] Viscous liquid, $b_{0.1}$ 160 to 182°,[1234] $b_{0.01}$ 157 to 162°, n_D^{22} 1.6740 to 1.6754;[72] L·MeI, m. 150 to 153°.[1234]
$Ph_2(4\text{-}BrC_6H_4)P$. I.[430,443,1228,1233a] II.[72] Crystals, m. 79 to 80° (from MeOH),[72,443,1228] $b_{0.02-0.01}$ 181 to 185°,[1228] IR, UV, μ 2.05 D,[443] pK_a 2.09;[445] x-ray analysis[825] gives pyramidal configuration about P atom, orthorhombic crystals, P-C 1.83, C-Br 1.86 Å, ∠C-P-C 103°.[825]
$Ph_2(2\text{-}Me_2NC_6H_4)P$. II.[405] VII.[1240] Crystals, m. 121 to 122° (from Et_2O/EtOH),[1240] m. 113 to 114°,[405] IR,[1240] ^1H NMR.[406]
$Ph_2(3\text{-}Me_2NC_6H_4)P$. VII. Crystals, m. 77 to 78° (from EtOH/Et_2O), IR.[1240]
$Ph_2(4\text{-}Me_2NC_6H_4)P$. I.[443,1465] II.[1225] III.[993] Crystals, m. 152° (from EtOH/Et_2O),[993] m. 153°,[443] m. 149 to 150°,[1465] m. 154 to 156°,[1225] soluble in 1:1 HCl, precipitated on dilution,[993] UV, IR, μ 3.35 D,[443] pK_a 2.95;[445] P acts as an electron acceptor toward the $4\text{-}Me_2NC_6H_4$ group.[1224]
$Ph_2(4\text{-}NO_2C_6H_4)P$. IX.2. Yellow plates, m. 96 to 97° (from MeOH).[1229]
$Ph_2(3\text{-}HOC_6H_4)P$. XIV. From $Ph_2(3\text{-}MeOC_6H_4)P$ and HI. Solid, m. 55 to 56°.[838]
Acetyl deriv. m. 85 to 86°.[838]
$Ph_2(4\text{-}HOC_6H_4)P$. XIV. Crystals, m. 113 to 114° and 102 to 103° (two forms),[1038,1273] m. 114 to 115° (from C_6H_6/petrol ether),[1363] pK_a 10.46, σ^- 0.26, σ_p (of Ph_2P group) +0.19.[1363]
$Ph_2(3\text{-}MeOC_6H_4)P$. I. Solid, m. 60 to 61°;[838] oil did not crystallize.[1011]
$Ph_2(4\text{-}MeOC_6H_4)P$. I. Crystals, m. 78° (from MeOH),[443] m. 62 to 64°,[1011] m. 78 to 79° and 68 to 69° (two forms),[1038,1273] UV, IR, μ 2.55 D,[443] pK 2.58.[445]
$Ph_2(2\text{-}PhOC_6H_4)P$. XIV. From $[Ph_3PC_6H_4OH\text{-}2]I$ and OH^-, then heating to 350 to 400°. Yellow liquid, b_{15} 350 to 400°; oxide, m. 164 to 165°.[118]
$Ph_2(2\text{-}MeSC_6H_4)P$. II. White crystals, m. 101 to 102° (from EtOH),[863] m. 115 to 116°.[1446]
$Ph_2(3\text{-}HO_3SC_6H_4)P$. XIV. From Ph_3P and SO_3/H_2SO_4.[29] Na salt·$2H_2O$ stable up to 335°; K salt·$2H_2O$ loses H_2O at 50°;[1210] benzylisothiuronium salt, m. 146 to 148°;[29] stability const. of Ag complex log K = 8.15,[29] of Hg(II) complex log K = 14.3,[1210] of Cd(II) complex log K = ∿0.9,[29b] of Cu(I) complex log K = 5.76,[428a] of

Au(I) complex log K = 7.5,[534a] of Bi(III) complex log K = 3.7.[1448a]

$Ph_2(4\text{-}NaO_3SC_6H_4)P$. VII. White solid, IR.[1240]

$Ph_2(2\text{-}MeC_6H_4)P$. I.[103] VII.[1240] Crystals, m. 73°,[103] m. 67 to 68°, IR,[1240] ^1H NMR,[1232] σ_o (of Ph_2P group) +0.20.[1232]

$Ph_2(3\text{-}MeC_6H_4)P$. I.[1011] II.[1370,1454] Crystals, m. 51 to 52°,[1370,1454] m. 48.8 to 50.2°,[1011] ^1H NMR,[1232] σ_m (of Ph_2P group) > -0.72.[1232]

$Ph_2(4\text{-}MeC_6H_4)P$. I.[443,1011,1454] III.[327,734] VII.[1240] Crystals, m. 66.5° (from EtOH),[443,1011] m. 68 to 69°,[1240] m. 68°,[327,1454] b_{14} 250°,[734] IR, UV,[443] ^1H NMR,[1233] σ_p (of Ph_2P group) > -0.41.[1233]

$Ph_2(3\text{-}CD_3C_6H_4)P$. I. Crystals, m. 51 to 52° (from EtOH),[865] rate const. of isotopic exchange of deuterium.[865]

$Ph_2(4\text{-}CD_3C_6H_4)P$. I. Crystals, m. 67 to 68° (from EtOH),[865] rate const. of isotopic exchange of deuterium.[865]

$Ph_2(2\text{-}CF_3C_6H_4)P$. II. Crystals, m. 85 to 87° (from EtOH), ^{19}F NMR,[1000] ^{31}P +10.9 ppm.[1000]

$Ph_2(4\text{-}CF_3C_6H_4)P$. I. Crystals, m. 55 to 57°.[1465]

$Ph_2(2\text{-}BrCH_2C_6H_4)P$. II. Crystals, m. 85 to 87°;[889] quaternizes at higher temp. intermolecularly.[889]

$Ph_2(2\text{-}MeOCH_2C_6H_4)P$. I. Crystals, m. 94 to 96°,[889] m. 92.5 to 93.5° (from EtOH).[889]

$Ph_2(4\text{-}HOCH_2C_6H_4)P$. XIV. From acid and $LiAlH_4$. Oil, isolated as oxide, m. 189 to 191°.[1228]

$Ph_2(4\text{-}\overline{OCH_2CH_2OC}\cdot MeC_6H_4)P$. I. Crystals, m. 73 to 74°.[1273]

$Ph_2(4\text{-}CH_2{=}CHC_6H_4)P$. I. Crystals, m. 77 to 78° (from i-PrOH),[144,1150,1151,1273] UV;[1153] could be polymerized and homopolymerized by AIBN initiation; polymer more thermally stable than polystyrene;[1153] polymerizes 10 times faster than styrene;[144,1153] viscosity of polymer 0.28/30°.[1151]

$Ph_2(2\text{-}CH_2{=}CHC_6H_4)P$. ^1H NMR.[1020]

$Ph_2(2\text{-}MeOCH_2CH_2CH_2C_6H_4)P$. I. Crystals, m. 43 to 44°;[890] quaternizes on treatment with HBr to a cyclic salt.[890]

$Ph_2(2\text{-}CH_2{=}CHCH_2C_6H_4)P$. I. Crystals, m. 63 to 67°,[101] m. 65 to 68° (from pentane), IR, ^1H NMR.[648]

$Ph_2(cis\text{-}2\text{-}MeCH{=}CHC_6H_4)P$. I. Crystals, m. 72 to 77° (purified by subl.), IR, ^1H NMR.[648]

$Ph_2(trans\text{-}2\text{-}MeCH{=}CHC_6H_4)P$. Viscous oil, IR, ^1H NMR.[648]

$Ph_2(2\text{-}PhCH_2C_6H_4)P$. XIV. From $Ph_3P{=}CH_2$ and benzyne. Liquid, $b_{0.001}$ 180 to 200°,[1461] L•MeI, m. 297 to 302°.[1461]

$Ph_2(2\text{-}PhCH\cdot MeC_6H_4)P$. XIV. As above. Viscous, refractive oil, $b_{0.001}$ 180 to 220°;[1461] L•MeI, m. 312 to 315°.[1461]

$Ph_2(2\text{-}PhC\cdot Me_2C_6H_4)P$. XIV. As above. Glassy, refractive oil, $b_{0.001}$ 190 to 230°.[1461]

$Ph_2(2\text{-}PhCH\cdot PrC_6H_4)P$. XIV. As above. Viscous, yellow oil, $b_{0.001}$ 200 to 215°;[1461] L•MeI, m. 258 to 260°.[1461]

$Ph_2(2\text{-}Ph_2CHC_6H_4)P$. XIV. As above. Crystals, m. 158 to 160°, $b_{0.001}$ 210 to 240°, ^1H NMR, mass spect.;[1461]

L•MeI, m. 296°.

$Ph_2(4\text{-}Ph_2CHC_6H_4)P$. VII. Crystals, m. 140 to 142° (from acetone/MeOH), ^1H NMR.[915]

$Ph_2(3\text{-}NCC_6H_4)P$. II.[1370] VIII.[1234] XIV.[1234] From amide and $POCl_3$. Crystals, m. 97 to 98° (from MeOH),[1234] m. 98 to 99°.[1370]

$Ph_2(4\text{-}NCC_6H_4)P$. XIV. From amide and $POCl_3$. Crystals, m. 86 to 87° (from MeOH).[1228]

$Ph_2(2\text{-}HO_2CC_6H_4)P$. II. Pale-yellow needles, m. 187 to 188° (from $EtOH/H_2O$);[717] L•MeI, m. 169 to 173° (dec.); gives a Ni(II) salt.[717]

$Ph_2(3\text{-}HO_2CC_6H_4)P$. VII.[1237] VIII.[1234] XIV.[430,1234] From $Ph_2PC_6H_4Li + CO_2$,[430] $Ph_2PC_6H_4MgBr + CO_2$;[430] and from $Ph_2PC_6H_4CN$ + HCl.[1363,1370] Crystals, m. 157° (from EtOH),[430] m. 157 to 160° (from AcOH),[1237] m. 159 to 160.5°,[1363] m. 160 to 164°,[1234] pK_a 6.52 (in $EtOH/H_2O$ = 4:1),[1363] pK_a 5.51, σ_m (of Ph_2P group) +0.17,[1370] +0.11;[1363] ratio of K_{meta}/K_{para} for Ph_2P group is <1, indicating that it is a meta-orienting substituent.[1370]

$Ph_2(4\text{-}HO_2CC_6H_4)P$. VII.[1237] VIII.[1235] XIV.[73,430,1228,1363] From $Ph_2PC_6H_4MgBr + CO_2$,[430] or from PhMgBr + $Cl_2PC_6H_4$-CN.[1363] Crystals, m. 156° (from AcOH),[430] m. 156 to 158°,[73,1228,1237] m. 156 to 157°,[1363] m. 153 to 154°,[1235] pK_a 6.34,[1363] pK_a 6.38, σ_p (of Ph_2P group) -0.01, σ +0.03,[73] σ_p +0.19.[1363]

Methyl ester (with MeOH/HCl). Crystals, m. 99 to 100° (from MeOH).[1228]

$Ph_2(3\text{-}NH_2COC_6H_4)P$. XIV. (From $Ph_2PC_6H_4CO_2Et + NH_3$). Crystals, m. 122 to 123° (from C_6H_6) and m. 97 to 100° (two forms).[1234]

$Ph_2(4\text{-}NH_2COC_6H_4)P$. XIV. From $Ph_2C_6H_4CO_2Et + NH_3$). Crystals, m. 150° (from benzene/ligroin).[1228]

$Ph_2(4\text{-}Me_2NCOC_6H_4)P$. XIV. (As above and Me_2NH). Crystals, m. 109 to 111° (from benzene/ligroin).[1228]

$Ph_2(4\text{-}MeCOC_6H_4)P$. I.[1273] IX.2.[1229] Blunt crystals, m. 121 to 122°,[1273] m. 121 to 123° (from MeOH).[1229]

$Ph_2(2,5\text{-}Me_2C_6H_3)P$. VII. Crystals, m. 81 to 83° (from MeOH).[1240]

$Ph_2(2\text{-}PhCH{\cdot}Pr\text{-}6\text{-}MeOC_6H_3)P$. XIV. From ylid and aryne. Viscous, brown oil, $b_{0.001}$ 210 to 240°;[1461] L•MeI, m. 197 to 204°.

$Ph_2(2,4,6\text{-}Me_3C_6H_2)P$. I. VII. Liquid, $b_{0.3}$ 201 to 203°;[1240] sulfide, m. 166 to 167°.

$Ph_2(C_6F_5)P$. I.[387] II.[774] (Gives high yield). Solid, m. 70°,[387] m. 68 to 70°,[774] ^{19}F NMR,[599,600] ^{31}P +26.3 ppm.[390]

$Ph_2C_6Cl_5P$. II. Crystals, m. 131.8 to 132.4°, IR, ^1H NMR.[1178]

$Ph_2(2\text{-}BrC_6F_4)P$. II. Crystals, m. 65 to 66° (from MeOH).[349a]

$Ph_2(4\text{-}MeOC_6F_4)P$. XIV. From $Ph_2PC_6F_5$ and NaOMe. Liquid, $b_{0.1}$ 200° (bath temp.), ^1H and ^{19}F NMR.[180]

$Ph_2(4\text{-}MeNHC_6F_4)P$. XIV. From $Ph_2PC_6F_5$ and $MeNH_2$. Crystals, m. 79 to 80° (from MeOH), ^1H and ^{19}F NMR.[180]

$Ph_2(2\text{-}MeSC_6F_4)P$. II. Light-tan crystals, m. 70 to 71° (from EtOH), ^1H NMR.[349a]

$Ph_2(2\text{-}PhC_6H_4)P$. I. Crystals, m. 61°, dec. 337°.[1007]

$Ph_2(2,2'\text{-}MeOCH_2C_6H_4C_6H_4)P$. I. Crystals, m. 104.5 to 105.5° (from EtOH).[247]

$Ph_2(3\text{-}PhC_6H_4)P$. I. Glass, IR,[1250] dec. 360°.[1007]

$Ph_2(4\text{-}PhC_6H_4)P$. VII. Colorless needles, m. 84°, $b_{0.05-0.06}$ 202 to 206°,[1007] IR,[1250] dec. 384°.[1007]

$Ph_2(1\text{-}C_{10}H_7)P$. VII. Crystals, m. 122 to 124°,[713] m. 124°,[1469,1471,1472] m. 118 to 119°,[1355] IR,[1472] UV.[1245]

$Ph_2(2\text{-}C_{10}H_7)P$. VII. Crystals, m. 117°,[1469] m. 118 to 119°,[1471,1472] IR,[1472] UV,[1245] dec. 387°.[1007]

$Ph_2(4\text{-}MeC_{10}H_7)P$. II.[1355] VII.[1240] Crystals, m. 111 to 112° (from MeOH or EtOH).[1240,1355]

Ph_2(o-terphenylyl-(2))P. XIV. From thermal dec. of o-diphenylenetriphenyl phosphorane at 200°. Colorless crystals, m. 133 to 133.5° (from EtOH).[1444]

Ph_2(1-anthracenyl)P. VII. Crystals, m. 195 to 196°,[1469,1472] IR,[1472] UV.[1245]

Ph_2(2-anthracenyl)P. VII. Crystals, m. 164°,[1472] m. 157°,[1469] UV,[1245] IR.[1472]

$Ph_2(2\text{-}C_5H_4N)P$. I. Crystals, m. 84 to 85° (from $MeOH/H_2O$), $b_{0.05}$ 132 to 138° (crude); picrate, m. 137 to 138° (from Me_2CO/EtOH).[957]

Ph_2-9'(9,10-dihydroacridyl)P. VII. Crystals, m. 233 to 235° (from Et_2O).[655]

Ph_2(9-acridyl)P. VII. Yellow crystals, m. 192° (from toluene).[655]

Ph_2(2,4-diphenyl-triazinyl)P. XIV. From triazine ester and Ph_2PH. Crystals, m. 130 to 132°.[562]

$(4\text{-}DC_6H_4)_2PrP$. IX.2. Liquid, $b_{0.2}$ 120 to 125°.[508]

$(C_6D_5)_2(CF_3CO)P$. VII. Pale-yellow liquid, $b_{0.001}$ 103°, IR; yields on oxidation a phosphinate.[857a]

$(C_6F_5)_2MeP$. I. Liquid, $b_{0.4}$ 92 to 94°, IR,[387] ^{31}P +52.2 ppm.[390]

$(C_6F_5)_2EtP$. I. Liquid, $b_{0.4}$ 105 to 107°, IR,[387] ^{31}P +44.0 ppm.[390]

$(C_6F_5)_2[(CF_2)_3C{\cdot}OH]P$. XIII.3. Solid, m. 74°.[385c]

$(C_6F_5)_2(C_6H_5C{\equiv}C)P$. Solid, m. 105°.[385c]

$(C_6F_5)_2PhP$. I.[387] II.[774] (Gives high yield). Crystals, m. 69 to 70°,[387] m. 68 to 70°,[774] IR,[387] ^{19}F NMR,[599,600] ^{31}P +48.7 ppm.[390]

$(3\text{-}FC_6H_4)_2MeP$. I. Liquid, b_{10} 152 to 154°, n_D^{20} 1.5856, ^{19}F NMR,[1251] σ_m (of $3\text{-}FC_6H_4MeP$ group) +0.20.[1251]

$(3\text{-}FC_6H_4)_2PhP$. I. Crystals, m. 51° (from MeOH), ^{19}F NMR,[1251] σ_m (of $Ph(3\text{-}FC_6H_4)P$ group) +0.25.[1251]

$(4\text{-}FC_6H_4)_2MeP$. I. Liquid, b_{11} 149 to 151°, n_D^{20} 1.5812, ^{19}F NMR,[1129] σ_p (of $4\text{-}FC_6H_4MeP$ group) +0.28.[1129]

$(4\text{-}FC_6H_4)_2CF_3P$. I. Liquid, b_{20} 98°, GLC, ^{19}F NMR.[1156]

$(4\text{-}FC_6H_4)_2PhP$. I. Crystals, m. 53°, ^{19}F NMR.[1129]
$(2\text{-}ClC_6H_4)_2EtP$. I. Crystals, m. 85° (from EtOH);[293] methopicrate, m. 126 to 127°.[293]
$(4\text{-}ClC_6H_4)_2PrP$. I. Liquid, b_1 182°, μ 1.81, pK_a 2.59.[440]
$(2\text{-}ClC_6H_4)_2PhP$. I. Crystals, m. 133 to 134° (from PrOH).[293]
$(4\text{-}ClC_6H_4)_2PhP$. I. Crystals, m. 43 to 44° (from MeOH),[440] b_2 215°.[826]
$(2\text{-}BrC_6H_4)_2PhP$. XIV. Crystals, m. 122 to 123° (from EtOH).[419]
$(4\text{-}BrC_6H_4)_2PhP$. I. Liquid, b_2 190 to 230° (not pure).[826]
$(4\text{-}H_2NC_6H_4)_2PhP$. II. Undistillable oil. (From $PhPCl_2$ and $H_2NC_6H_4Li$).[432]
Diacetyl deriv. m. 169° (from 50% EtOH).[432]
Di(p-acetyl aminobenzenesulfonyl) deriv. m. 186 to 187° (EtOH).[432]
Di(p-aminobenzenesulfonyl) deriv. By hydrolysis of the above compound with hot, dilute NaOH, m. 202 to 204° (from EtOH).[432]
$(4\text{-}Me_2NC_6H_4)_2(PhNHCO)P$. XIII.5. Crystals, m. 155°.[1290]
$(2\text{-}Me_2NC_6H_4)_2PhP$. II. Crystals, m. 84.5° (from EtOH),[405] ^1H NMR.[406]
$(4\text{-}Me_2NC_6H_4)_2PhP$. II. Crystals, m. 150 to 153° (from EtOH),[1225] UV,[1224] σ' +0.87.[1226]
$(4\text{-}Me_2NC_6H_4)_2(4\text{-}MeOC_6H_4)P$. II. Crystals, m. 173 to 175° (from EtOH).[1359]
$(4\text{-}HOC_6H_4)_2PhP$. XIV. From methoxy compound and HI. Crystals, m. 159 to 161°,[1038,1273] ESR of radical ion produced by oxidation.[1039]
$(2\text{-}MeOC_6H_4)_2MeP$. I. Crystals, m. 128 to 129°,[775] m. 121 to 123° (from EtOH),[939,956] IR.[942]
$(4\text{-}MeOC_6H_4)BuP$. V. Liquid, $b_{0.3}$ 175 to 178°.[1177]
$(4\text{-}MeOC_6H_4)PhP$. I. Crystals, m. 89 to 90°,[1038,1273] m. 90°,[615] $E_{1/2}$ 0.08 V,[616] IR,[1227] σ_p (of $Ph(4\text{-}MeOC_6H_4)P$ group) +0.18, σ' +0.24,[1227] ESR of ion radical produced by oxidation.[1039]
$(4\text{-}MeOC_6H_4)_2(4\text{-}Me_2NC_6H_4)P$. I. Crystals, m. 139 to 140° (from EtOH).[1359]
$(2\text{-}MeC_6H_4)_2MeP$. I. Crystals, m. 54 to 54.5° (from EtOH), IR;[939,942,956] L•MeI, m. 226 to 228°.[939]
$(2\text{-}MeC_6H_4)_2PhP$. I. White crystals, m. 83°,[103] μ 1.53 D,[1242] ^1H NMR,[1232] σ (of PhP< group) +0.23.[1232]
$(3\text{-}MeC_6H_4)_2PhP$. I. Crystals, m. 53 to 53.5°, $b_{0.5}$ 210 to 220°,[520] μ 1.70 D,[1242] ^1H NMR,[1232] σ (of PhP< group) > -0.43.[1232]
$(4\text{-}MeC_6H_4)_2MeP$. I.[942,956] VI.[986] Liquid, b. 345°,[986] $b_{0.45}$ 125 to 127°,[942,956] ^1H NMR,[1233] IR,[942] σ (of MeP< group) > -0.09;[1233] L•MeI, m. 159.5 to 161°.[939]
$(4\text{-}MeC_6H_4)_2PhCH_2P$. III. Needles, m. 187°(?).[986]
$(4\text{-}MeC_6H_4)_2PhP$. I.[415,610,882] III.[327] Crystals, m. 57°,[327] m. 53 to 54°,[415,610] m. 56 to 58°,[882] b_2 140 to 145°,[415] ^1H NMR,[1233] σ (of PhP< group) > -0.16.[1233]

$(4\text{-}MeC_6H_4)_2(4\text{-}ClC_6H_4)P$. III. Crystals, m. 115° (from ligroin).[986]
$(4\text{-}MeC_6H_4)_2(4\text{-}MeOC_6H_4)P$. III. Liquid.[986]
$(4\text{-}MeC_6H_4)_2[2\text{-},(4\text{-}MeC_6H_4CH{\cdot}Me)\text{-}6\text{-}MeOC_6H_3]P$. XIV. From ylid and aryne. Liquid, $b_{0.001}$ 190 to 210°.[1461]
$(3\text{-}CF_3C_6H_4)_2BuP$. V. Liquid, b_{18} 174 to 175°.[1177]
$(2\text{-}CF_3C_6H_4)_2PhP$. II. Crystals, m. 126 to 127° (from EtOH), ^{19}F NMR, ^{31}P +14.5 ppm.[1000]
$(3\text{-}CF_3C_6H_4)_2PhP$. I. Liquid, $b_{0.2}$ 123 to 126°;[202] claimed to be useful as force transmission fluid.
$(2\text{-}MeOCH_2CH_2CH_2C_6H_4)_2PhP$. Liquid, b_2 215 to 218°.[502]
$(3\text{-}NCC_6H_4)_2PhP$. VIII. Crystals, m. 107 to 109° (from MeOH).[1224]
$(4\text{-}NCC_6H_4)_2PhP$. VIII. Crystals, m. 110 to 111° (from MeOH).[1234]
$(3\text{-}HO_2CC_6H_4)_2PhP$. VIII. Crystals, m. 228 to 231° (from MeOH)/H_2O).[1234]
$(4\text{-}HO_2CC_6H_4)_2PhP$. VIII. From oxide and $SiHCl_3$. XIV. From ester and HCl. Crystals, m. 184 to 186°, m. 175 to 178° (from MeOH/AcOH).[1234]
$(2\text{-}Cl\text{-}5\text{-}MeC_6H_3)_2EtP$. I. Needles, m. 137 to 138° (from acetone);[294] L•EtI, m. 218°.
$(3,5\text{-}Me_2C_6H_3)_2PhP$. I. Crystals, m. 98°, b_1 150 to 195°.[520]
$(2\text{-}MeSC_6F_4)_2PhP$. II. White crystals, m. 97, ^1H NMR.[349a]
$(2\text{-}PhC_6H_4)_2PhP$. I. Crystals, m. 147° (from EtOH), dec. 347°.[1007]
$(3\text{-}PhC_6H_4)_2PhP$. I. Glass, dec. 369°.[1007]
$(4\text{-}PhC_6H_4)_2PhP$. I. Crystals, m. 112° (from EtOH), dec. 395°.[1007]
$(1\text{-}C_{10}H_7)_2BuP$. V. Liquid, $b_{0.2}$ 210°.[1177]
$(1\text{-}C_{10}H_7)_2PhP$. I.[520,882] II.[1355] Crystals, m. 207.5 to 208.5°,[520] m. 208°,[882] m. 210 to 212° (from BuOH).[1355]
$(4\text{-}MeC_{10}H_6)_2PhP$. II. Crystals, m. 176 to 177° (from EtOH).[1355]
$(2\text{-}C_5H_4N)_2PhP$. I. Crystals, m. 96° (from EtOH/ligroin), $b_{0.4}$ 196 to 210°; dihydrochloride, m. 185 to 187°; dipicrate, m. 131°.[957]

Tertiary Phosphines

TYPE: RR^1R^2P

$Me(HOCH_2)EtP$. XIII.1. Liquid, b_{10} 64 to 65°, n_D^{20} 1.4998, d_4^{20} 0.9726.[487]
$Me(HOCH_2)HCF_2CF_2P$. XIII.1. Liquid, b_{20} 71°, n_D^{20} 1.4240, d_4^{20} 1.4540.[164]
$Me(HOCH_2)[(CF_3)_2C{\cdot}OH]P$. XIII.1. Oil, undistillable.[163]
$Me(HOCH_2)(MeCF_3C{\cdot}OH)P$. XIII.1. and .3. Oil, decomposes on dist., n_D^{20} 1.4787, d_4^{20} 1.3517, IR.[488]
$Me(LiCH_2)PhP$. XIV. From Me_2PhP and BuLi. ^{31}P +22.9 ppm;[1100] metalation proceeds with retention of configuration.[606]

MeEtPrP. VIII. Isolated as oxide, $b_{0.01}$ 75°, $[\alpha]_{300}$ +1.2°.[1033]
MeEt($NCCH_2CH_2$)P. IX.2. Liquid, $b_{1.75}$ 68 to 70°, n_D^{25} 1.4805.[470]
MeEtAmP. IX.3. Liquid, b_{22} 74 to 74.5°, n_D^{20} 1.4574.[70]
MeEtPhP. I. VI.[761,910,926,1012] IX.3.[70] IX.4.[609,623,625] IX.5.[635] Liquid, b_{15} 96 to 97°,[624,635,926,1012] b_9 82 to 84°, n_D^{20} 1.5555, d_4^{20} 0.9603,[761] n_D^{25} 1.5524, d_4^{25} 0.9540; obtained in optically active form, $[\alpha]_D$ +1.5°.[635]
MeEt(4-MeC_6H_4)P. I. Liquid, b_{11-12} 95 to 98°, n_D^{20} 1.5533, d_4^{20} 0.9480.[763]
Me(CH_2=CH)PhP. I. Liquid, b_{13} 90 to 90.5°, n_D^{20} 1.5732, d_4^{20} 0.9714.[747]
Me($HOCH_2CH_2$)($PhCO_2CH_2CH_2$)P. XIV. From Me($PhCO_2CH_2CH_2$)$_2$P and Na_2CO_3. Oil, identified as salt; L•Me BPh_4, m. 163 to 165°.[1307]
MePr(c-C_6H_{11})P. VIII. Liquid, $b_{0.01}$ 40°, (R)-$[\alpha]_{325}$ +124°, $\Delta G^{\ddagger}_{130°}$ 35.6 kcal/mole.[66a]
MePr$PhCH_2$P. IX.5. Liquid, obtained in optically active form $[\alpha]_D$ +7.3° (29.5% optically pure);[622] arylation of this phosphine according to the "complex salt" method leads to a phosphonium salt with retention of configuration in spite of the high temp. used (∿200°).[622]
MePrPhP. VIII.[66a,207,605,1033,1468] IX.5.[614,635] Liquid, b_3 86 to 88°;[635] $b_{0.02}$ 35°;[66a] obtained in optically active form by cathodic cleavage of optically active MePrPh($PhCH_2$)PBr with retention of configuration,[614,635] $[\alpha]_D^{20}$ -13.45° and $[\alpha]_D^{20}$ +16.8°; absolute rotation should be 19.5°[207] (deduced from the ^{1}H NMR of the salt with PhCH•$OMeCD_2$Br), UV, rotation dispersion and circular dichroism of (+)-phosphine;[607] reduction of the optically active oxide with $SiHCl_3$ (also in the presence of pyridine or dimethylaniline) proceeds with retention of configuration, but in the presence of Et_3N with inversion;[605] oxide can also be reduced with Si_2Cl_6 with inversion;[207,1034] reduction of (+)-MePrPhP=O with $LiAlH_4$ resulted in complete racemization of the oxide before more than 10% had been reduced;[556] reduction of the optically active benzyl salt with $LiAlH_4$ gave also inactive MePrPhP, but the recovered salt had suffered no change in optical activity;[556] another group reports 82% racemization in this reduction;[613] (+)MePrPhP has (S)-configuration and is stable in boiling toluene but racemizes at 230°,[635] with first-order kinetics; activation energy of racemization E_a = 29.6 ± 1 kcal/mole;[66a,632] alkylation, oxidation,[635] sulfurization,[630,635] and phosphine imine formation[630] proceed with retention of configuration; deoxygenation of Bz_2O_2 and desulfurization of Bz_2S_2 with (+)-phosphine proceed by a polar mechanism;[633] extent of racemization depends upon the polarity

of the solvent;[633] desulfurization proceeds with retention of configuration;[633,1456] optically active (-)-phosphine yields with I_2 in water-free solvents upon hydrolysis inactive oxide, but in water-containing solvents optically pure oxide;[631] hydrogenation of hydratropic acid with $RhCl_3$[(-)-MePrPhP]$_3$ as catalyst gives optically active compounds.[781]

Me(CH_2=$CHCH_2$)(t-Bu)P. ^{31}P +18.2 ppm.[485]

Me(CH_2=$CHCH_2$)PhP. VIII.[66,66a,1468] IX.5.[635] Liquid, $b_{0.1}$ ∿45°,[1468] $b_{0.05}$ 40°;[66a] obtained in optically active form;[66,66a,635] [α] ∿-10°;[635] a report that it racemizes more readily than other optically active phosphines (allylic rearrangement)[635] could not be confirmed;[66] racemizes with first-order kinetics, E_a 32.3 kcal/mole;[66,66a] in Me(CH_2=$CHCD_2$)PhP no change observed when heated up to 150°.[66]

Me($NCCH_2CH_2$)($NCCH_2CH_2CH_2$)P. IX.2. Liquid, $b_{1.3}$ 174°, n_D^{25} 1.4995.[470]

Me($NCCH_2CH_2$)PhP. IX.2. Liquid, $b_{0.3}$ 100 to 101°, n_D^{20} 1.5690,[419] IR.[942]

Me($HO_2CCH_2CH_2$)PhP. XIV. By hydrolysis of nitrile. Clear, yellow oil, IR; benzylthiouronium salt, m. 130 to 131° (from H_2O); L•MeI, m. 176 to 177°.[419]

Me-t-BuPhP. I.[212] IV.[580] VIII.[66a,856] IX.2.[580] Liquid, b_{13} 108 to 112°,[580] $b_{0.05}$ 60°,[66a] optically resolved through complex formation with $PtCl_2$ and resolution with deoxyephedrine; the phosphine was liberated from the complex with KCN, identified as oxide;[212] reduction of the oxide with Si_2Cl_6 proceeds with inversion; phosphine obtained as (+)-(R), $[\alpha]_D$ +29.5°[66a,856] (62% optically pure);[856] racemizes with first-order kinetics; $\Delta G^{\ddagger}_{130°}$ 32.7 kcal/mole.[66a]

Me($PhCH_2$)PhP. VIII.[1033,1034] IX.5.[623,624,635] Liquid, b_{12} 156 to 158°,[623,624,635] $b_{0.01}$ 90 to 95°,[1033,1034] ^{1}H NMR;[1033] reduction of (+)-(R)-MePhCH_2PhP=O, $[\alpha]_D$ +51.4°, with Si_2Cl_6 proceeds with inversion $[\alpha]_D$ +81°;[1033] oxidation of this phosphine with H_2O_2 proceeds with retention, $[\alpha]_D$ -46.9°.[1034]

MePh(PhCH=$CHCOCH_2$CH•Ph)P. IX.2. From 4-oxophosphorinanium iodide and base. Cream-colored solid, m. 119 to 121°, IR, UV, ^{31}P +17 ppm.[1424]

MePh(3-HOC_6H_4)P. VII. From Ph(3-$MeOC_6H_4$)PLi in THF by demethylation. Liquid, $b_{0.3}$ 135 to 142°, n_D^{20} 1.6446,[941] ^{1}H NMR; L•MeI, m. 173.5 to 175°.[941]

MePh(2-$MeOC_6H_4$)P. XIV. As above. Solid, m. 45 to 46.5°, $b_{0.3}$ 122 to 123°, ^{1}H NMR,[956] IR.[942,956]

MePh(3-$MeOC_6H_4$)P. VII.[567] XIV.[956] Liquid, $b_{0.5}$ 133°,[567] $b_{0.3}$ 114 to 117°, n_D^{22} 1.6173, ^{1}H NMR,[956] IR.[942,956]

MePh(4-$MeOC_6H_4$)P. VII.[567] VIII.[66a,619] Liquid, $b_{0.5}$ 135 to 138°,[567,619] $b_{0.05}$ 140°,[66a] IR;[942] obtained in optically active form by reduction of oxide, (S)-$[\alpha]_{330}$

+112°; racemizes with first-order kinetics, $\Delta G^{\ddagger}_{130°}$ 30.8 kcal/mole.[66a]

MePh(4-MeC_6H_4)P. I.[1155] VII.[956] VIII.[66a,635] Liquid, $b_{0.8}$ 153 to 156°,[635] $b_{0.1}$ 120°,[66a] $b_{0.35}$ 116 to 119°, m. 16.2 to 17.5°, ^1H NMR;[956] obtained in optically active form by reduction of oxide, (S)-$[\alpha]_{330}$ +25°; racemizes with first-order kinetics, $\Delta G^{\ddagger}_{130°}$ 30.3 kcal/mole.[66a]

MePh(4-$CF_3C_6H_4$)P. VIII. Liquid, $b_{0.01}$ 90°; obtained in optically active form, (S)-$[\alpha]_{340}$ -85°, $\Delta G^{\ddagger}_{130°}$ 29.1 kcal/mole.[66a]

MePh(C_6F_5)P. Liquid, $b_{2.7}$ 121°.[385c]

MePh(2-$C_{10}H_7$)P. VIII. Liquid; obtained in optically active form, (S)-$[\alpha]_{350}$ +144°; racemizes with first-order kinetics, $\Delta G^{\ddagger}_{130°}$ 29.7 kcal/mole.[66a]

$HOCH_2$EtPhP. IX.2. Liquid, b_{14} 116 to 117°, n_D^{20} 1.5815.[1329]

PhNHCO(CH_2=$CHCH_2CH_2CH_2$)PhP. XIII.5. Crystals, m. 64 to 65° (from ligroin), IR, TLC.[282]

Et(MeCO)BuP. VII. Liquid, b. 188 to 193°, IR.[687]

Et(MeCO)PhP. VII. Liquid, b_{12} 153 to 158°, IR.[687]

Et($EtOCH_2CH_2$)PhP. VII. Liquid, b_{18} 146.5°.[569]

EtPrPhP. I. Liquid, b_9 109 to 111°, n_D^{20} 1.5360, d_4^{20} 0.9320.[761]

EtPr(4-MeC_6H_4)P. I. Liquid, b_3 106 to 109°, n_D^{20} 1.5369, d_4^{20} 0.9294.[763]

Et(i-Pr)i-BuP. IV. Liquid, b. 190°.[591]

Et(CH_2=$CHCH_2$)PhP. IX.5. Liquid, b_{15} 113 to 115°.[623,624,635]

Et(EtCO)PhP. VII. Liquid, b_{12} 157 to 160°, IR.[687]

Et($HOCH_2CH_2CH_2$)PhP. XII. Liquid, b_{14} 172 to 173°, n_D^{20} 1.5649.[1329]

Benzoyl deriv. Prepared with $PhCO_2H$ or PhCOCl. Liquid, $b_{0.2}$ 155 to 160°, n_D^{20} 1.5560.[1329]

EtBu(n-$C_{12}H_{25}$)P. IX.5. Liquid, b_{13} 188 to 191°.[623,624,635]

EtBuPhCH_2P. IX.1.[758,759] IX.5.[623,624,635] Liquid, b_{16} 137 to 139°,[624] b_6 125 to 126°, n_D^{20} 1.5310, d_4^{20} 0.9338.[758,759]

EtBuPhP. I. Liquid, b_{11} 121 to 123°, n_D^{20} 1.5325, d_4^{20} 0.9255.[761]

EtBu(4-MeC_6H_4)P. I. Liquid, b_5 115 to 118°, n_D^{20} 1.5319, d_4^{20} 0.9246.[763]

Et(t-Bu)PhP. IV. IX.2. Liquid, b_{13} 108 to 110°; L•MeI, m. 206 to 208°.[580]

Et(PrCO)PhP. VII. Liquid, b_{12} 171 to 173°, IR.[687]

EtAmPhP. I. Liquid, b_{10} 129 to 131°, n_D^{20} 1.5299, d_4^{20} 0.9237.[761]

EtAm(4-MeC_6H_4)P. I. Liquid, b_2 107 to 109°, n_D^{20} 1.5273, d_4^{20} 0.9164.[763]

Et(n-C_6H_{13})PhP. I. Liquid, b_1 120 to 124°,[703] b_5 123 to 125°, n_D^{20} 1.5255, d_4^{20} 0.9212.[761]

$Et(n-C_6H_{13})(4-MeC_6H_4)P$. I. Liquid, b_5 136.5 to 137.5°, n_D^{20} 1.5230, d_4^{20} 0.9110.[763]
$Et(n-C_7H_{15})PhP$. I. Liquid, b_3 137 to 138°, n_D^{20} 1.5195, d_4^{20} 0.9088.[762]
$Et(n-C_7H_{15})(4-MeC_6H_4)P$. I. Liquid, b_4 144 to 147°, n_D^{20} 1.5188, d_4^{20} 0.9079.[763]
$Et(n-C_8H_{17})PhP$. I. Liquid, b_1 132 to 134°, n_D^{20} 1.5160, d_4^{20} 0.9062.[762]
$Et(PhCH_2)PhP$. I.[882] VII.[515a] VIII.[605] IX.1.[757,758,977] IX.5.[623,624,635] Liquid, b_{33} 204 to 206°,[977] b_{13} 165 to 168°,[758] b_{10} 156 to 160°,[757] b_5 85 to 90°,[882] b_1 115 to 120°,[515a] b_{12} 168°,[624] n_D^{20} 1.5960, d_4^{20} 1.0393;[758] reduction of the oxide with $SiHCl_3$ (also in the presence of pyridine or dimethylaniline) proceeds with retention, and in the presence of Et_3N with inversion.[605]
Et(PhCO)PhP. VII. Liquid, b_5 183 to 185°, IR.[687]
$Et(2-MeOCH_2C_6H_4CH_2)PhP$. VII. Liquid, b_{13} 202 to 206°.[953]
$EtPh(4-BrC_6H_4)P$. I. Liquid, $b_{0.05}$ 136 to 138°.[288]
$EtPh(4-Me_2NC_6H_4)P$. VIII. Crystals, m. 45 to 48°.[619]
$EtPh(3-MeOC_6H_4)P$. VII. Liquid, $b_{0.5}$ 140°.[567]
$EtPh(4-MeOC_6H_4)P$. I.[288] VII.[567] Liquid, $b_{0.3}$ 125 to 135°,[567] $b_{0.1}$ 137°, b_{20} 210 to 211°.[288]
$EtPh(4-HOC_6H_4)P$. XIV. From methoxy analog with HI. Oil, $b_{0.1}$ 160 to 175°.[288]
Benzoyl deriv. m. 79 to 80° (from EtOH).[288]
$EtPh(4-MeC_6H_4)P$. I.[882,1417] VI.[986] Liquid, b. 340°,[986] b_5 138 to 145°.[882]
$EtPh(2,4,5-Me_3C_6H_2)P$. VI. Liquid, b. 352°, b_{10} 225 to 230°.[986]
$EtPh(C_6F_5)P$. Liquid, $b_{1.2}$ 115°.[385c]
$H_2NCH_2CH_2(MeCHOHCH_2)PhP$. VII. Liquid, b_4 212 to 214°.[694]
$H_2NCH_2CH_2(Bu)PhP$. VII. Liquid, b_3 130 to 132°, pK_a 8.80.[694]
$H_2NCH_2CH_2(c-C_6H_{11})PhP$. VII. Liquid, b_3 166 to 168°, pK_a 8.76.[694]
$HO_2CCH_2(Ph)(3-MeOC_6H_4)P$. XIV. By alkaline hydrolysis of ester. Crystals, m. 99 to 100.5°;[956] acidic hydrolysis causes decarboxylation and gives $MePh(3-MeOC_6H_4)P$; benzylthiouronium salt, m. 128 to 129°.[567,956]
Methyl ester. VII. Liquid, $b_{0.6}$ 160 to 164°, n_D^{22} 1.6000,[956] IR.[942,956]
Ethyl ester. VII. Liquid, $b_{0.1}$ 170°,[567] $b_{0.8}$ 162 to 164°, n_D^{20} 1.5902, IR.[956]
$HO_2CCH_2(Ph)(3-EtOC_6H_4)P$. VII. Isolated as benzylthiouronium salt, m. 134°.[567]
Methyl ester. VII. Liquid, $b_{0.5}$ 175°.[567]
$PrBuPhCH_2P$. IX.1. Liquid, b_1 113 to 115°, n_D^{20} 1.5090, d_4^{20} 0.9467.[758,759]
PrBuPhP. IX. Liquid, b_{10} 127 to 130°.[546]
$Pr(n-C_6H_{13})(n-C_9H_{19})P$. VII. Liquid, $b_{1.5}$ 130 to 133°, n_D^{20} 1.4660.[1084]

$Pr(PhCH_2)PhP$. IX.1.[758] IX.5.[623,624,635] Liquid, b_{12} 172 to 175°, n_D^{20} 1.5680,[758] b_{12} 173 to 174°.[624]
$PrPh(4\text{-}MeOC_6H_4)P$. I. Liquid, $b_{0.3}$ 163.5°.[288]
$i\text{-}Pr(CH_2{=}CHCH_2)(t\text{-}Bu)P$. ^{31}P -14.4 ppm.[485]
$(CH_2{=}CHCH_2)(PhCH_2)PhP$. ^{31}P +20.7 ppm.[485]
$NCCH_2CH_2(H_2NCH_2CH_2CH_2)PhP$. XII. Liquid, b_3 189 to 193°, n_D^{25} 1.5695, d_4^{25} 1.0606.[59,1383]
$NCCH_2CH_2(NCCH_2CH_2CH_2)PhP$. XII. Liquid, b_1 187 to 188°, n_D^{20} 1.5620, d_4^{20} 1.0839.[59,1383]
$NCCH_2CH_2(MeO_2CCH{\cdot}MeCH_2)PhP$. XII. Liquid, b_1 153.5 to 154°, n_D^{20} 1.5425, d_4^{20} 1.0997.[1383]
$NCCH_2CH_2Ph(4\text{-}ClC_6H_4)P$. XII. Liquid, $b_{0.005}$ 183 to 185°, n_D^{19} 1.629.[419]
$NCCH_2CH_2Ph(4\text{-}BrC_6H_4)P$. XII. Crystals, m. 61 to 62° (from cyclohexane), $b_{0.03}$ 188 to 191°;[419] could not be cyclized with Li but with CuCN.[419]
$NCCH_2CH_2Ph(3\text{-}MeOC_6H_4)P$. XII. Liquid, $b_{0.35}$ 192 to 193°,[567] $b_{0.002}$ 159 to 162°, n_D^{19} 1.6130, IR.[956]
$NCCH_2CH_2Ph(2\text{-}NCC_6H_4)P$. XIV. From 2-$BrC_6H_4$-deriv. and CuCN. Liquid, $b_{0.35}$ 183 to 185°;[419] methopicrate, m. 158 to 159°; could be cyclized with t-BuONa.[419]
$MeO_2CCH_2CH_2(NCCH_2CH_2CH_2)PhP$. XII. Liquid, b_1 170 to 171°, n_D^{20} 1.5468, d_4^{20} 1.1011.[1383]
$MeO_2CCH_2CH_2(MeO_2CCH{\cdot}MeCH_2)PhP$. XII. Liquid, b_{15} 147 to 150°, n_D^{20} 1.5305, d_4^{20} 1.1174.[1383]
Diallyl ester. By transesterification of dimethyl ester above. Liquid, $b_{0.01}$ 153 to 155°, n_D^{20} 1.5265, d_4^{20} 1.0755.[1381]
$HO_2CCH_2CH_2Ph(2\text{-}ClC_6H_4)P$. XIV. By hydrolysis of ester. Benzylthiouronium salt, m. 148 to 148.5°.[419]
Ethyl ester. XII. Liquid, $b_{0.45}$ 189 to 190°, n_D^{23} 1.598;[419] ester could not be cyclized.
$HO_2CCH_2CH_2Ph(2\text{-}MeOC_6H_4)P$. XIV. By hydrolysis of ester. Crystals, m. 114.5 to 116.5° (from C_6H_{12}).[956]
Ethyl ester. XII. Liquid, $b_{0.25}$ 166 to 168°, n_D^{20} 1.5820.[956]
$HO_2CCH_2CH_2Ph(3\text{-}MeOC_6H_4)P$. XIV. By hydrolysis of ester or nitrile (alkaline). Colorless syrup; benzylthiouronium salt, m. 134 to 135°.[956]
Methyl ester. VII. XII. Liquid, $b_{0.2}$ 168 to 170°,[567] $b_{0.5}$ 174°, n_D^{20} 1.5920, IR.[956]
Ethyl ester. VII. XII. Liquid, $b_{0.1}$ 155 to 158°,[567] $b_{0.3}$ 173 to 175°, n_D^{20} 1.5820, IR.[956]
$HO_2CCH_2CH_2Ph(4\text{-}MeOC_6H_4)P$. XIV. By hydrolysis of ester. Benzylthiouronium salt, m. 143°.[567]
Methyl ester. XII. Liquid, $b_{0.5}$ 185°.[567]
Ethyl ester. XII. Liquid, $b_{0.5}$ 190°.[567]
$HO_2CCH_2CH_2Ph(3\text{-}EtOC_6H_4)P$. XIV. Hydrolysis of ester. Crystals, m. 95 to 95.5°;[567] benzylthiouronium salt, m. 130 to 131°.[567]
Methyl ester. XII. Liquid, $b_{0.2}$ 178 to 185°.[567]

Ethyl ester. XII. Liquid, $b_{0.5}$ 250°.[567]
$HO_2CCH_2CH_2Ph(4\text{-}EtOC_6H_4)P$. XIV. By hydrolysis of ester. Benzylthiouronium salt, m. 151 to 152°.[567]
Ethyl ester. VII. Liquid, $b_{0.5}$ 170 to 175°.[567]
$HO_2CCH_2CH_2Ph(3\text{-}MeC_6H_4)P$. XIV. By hydrolysis of ester. Crystals, m. 89 to 90°;[419] could not be cyclized with P_2O_5.[419]
Methyl ester. XII. Liquid, $b_{0.15}$ 152 to 154°.[419]
Ethyl ester. XII. Liquid, $b_{0.2}$ 160 to 163°.[419]
$HO_2CCH_2CH_2Ph(3,5\text{-}Me_2C_6H_3)P$. XIV. By hydrolysis of ester. Crystals, m. 119 to 119.5° (from H_2O/EtOH);[419] cyclization with polyphosphoric acid gave only a small yield of 4-oxophosphinoline.[419]
Ethyl ester. XII. Liquid, $b_{0.05}$ 149 to 150°; L•MeI, m. 103.5 to 105°.[419]
$Bu(n\text{-}C_{16}H_{33})(1\text{-}C_{10}H_7)P$. VII. Pale-yellow oil,[1469,1472] IR.[1472]
$Bu(PhCH_2)PhP$. IX.1.[758] IX.5.[623,624] Liquid, b_3 151 to 152°,[623,624] b_{11} 184 to 190°, n_D^{20} 1.5729, d_4^{20} 1.0112.[758]
$Bu(PhCH_2)(4\text{-}MeC_6H_4)P$. I. Liquid, b_4 196 to 200°, n_D^{20} 1.5653, d_4^{20} 0.9920.[760]
$BuPh(4\text{-}MeOC_6H_4)P$. I. Liquid, $b_{0.025}$ 139 to 141°, $b_{0.5}$ 176 to 179°.[288]
$BuPh(4\text{-}HOC_6H_4)P$. XIV. Could not be isolated after refluxing the above substance with HI, followed by neutralization; the benzoate (with PhCOCl), however, was obtained pure, m. 91° (from EtOH).[288]
$i\text{-}Bu(4\text{-}MeC_6H_4)(2\text{-}anthracenyl)P$. VII. Viscous, yellow liquid, IR.[1472]
$t\text{-}BuPh(PhCOCH_2CH{\cdot}Ph)P$. XII. Crystals, m. 149 to 150° (from MeOH).[580]
$NCCH_2CH_2CH_2(MeO_2CCH_2CH_2)PhP$. XII. Liquid, b_1 170 to 171°, n_D^{20} 1.5468, d_4^{20} 1.1011.[59]
$HO(CH_2)_4Ph(3,5\text{-}Me_2C_6H_3)P$. VII. Liquid, $b_{0.2}$ 170 to 172°, IR; L•MeI, m. 146 to 147.5°.[419,937]
$HO_2CCH_2CH_2CH_2Ph(3\text{-}MeOC_6H_4)P$. XIV. Hydrolysis of ester. Benzylthiouronium salt, m. 119 to 120°.[567]
Methyl ester. VII. Liquid, $b_{0.15}$ 172 to 180°.[567]
$PhCH_2(2\text{-}MeOC_6H_4CH_2)PhP$. IX.1. Liquid, $b_{0.5}$ 275 to 295°.[953]
$Ph(2\text{-}MeOCH_2C_6H_4CH_2CH_2)(4\text{-}BrC_6H_4)P$. I. Green oil, $b_{0.1}$ 214 to 216°.[720]
$Ph(2\text{-}MeOCH_2C_6H_4CH_2CH_2)(4\text{-}MeOC_6H_4)P$. I. Liquid, $b_{0.05}$ 208°.[720]
$Ph(4\text{-}BrC_6H_4)(4\text{-}MeOC_6H_4)P$. I. Crystals, m. 71° (from MeOH), $b_{0.01}$ 204°.[288]
$Ph(4\text{-}BrC_6H_4)(4\text{-}Me_2NCC_6H_4)P$. I. II. (Gives higher yield). Crystals, m. 107 to 108° (from EtOH), $b_{0.05}$ 218 to 220°; $L{\cdot}PdCl_2$, dec. 247 to 249°.[288]
$Ph(4\text{-}BrC_6H_4)(3\text{-}C_5H_4N)P$. I. Oil, $b_{0.15}$ 202 to 210°; picrate, m. 143 to 144°.[288]
$Ph(4\text{-}BrC_6H_4)(2\text{-}C_5H_4N)P$. I. Crystals, m. 90 to 91° (from

MeOH), $b_{0.01}$ 180 to 230° (crude); picrate, m. 132°.[288]
$Ph(4\text{-}MeOC_6H_4)(4\text{-}MeC_6H_4)P$. I. Crystals, m. 116 to 118° (from MeOH), $b_{0.1}$ 197 to 200°, $b_{0.08}$ 176 to 183°.[288]
$Ph(4\text{-}MeOC_6H_4)(1\text{-}C_{10}H_7)P$. I. Crystals, m. 124°;[629] $L \cdot PhCH_2Br$, m. 273° (dec.); this salt resolved into the optical antipodes with dibenzoyl hydrogen tartrate and electrolyzed gave optically active phosphine $[\alpha]_{578}$ +5° (concentration of 0.09 g/100 ml toluene), m. 120°.[629]
$Ph(3\text{-}MeC_6H_4)(3\text{-}CF_3C_6H_4)P$. I. Liquid, $b_{0.5}$ 150 to 180°;[202] claimed to be useful as force transmission fluid.
$Ph(4\text{-}PhC_6H_4)(1\text{-}C_{10}H_7)P$. I. Crystals, m. 192 to 193°;[1433] m. 190.1°;[515a] this phosphine was resolved into the optically active antipodes by forming a salt with CH_2O and (+)-camphor-10-sulfonic acid and using the method of half-quantities;[1433] (−)-(ℓ), m. 194 to 195°, $[\alpha]_{578}^{24}$ −5.2° and (+)-(ℓ), m. 195 to 196°, $[\alpha]_{578}^{25.4}$ +8.7°.[1433]
$Ph(2\text{-}MeOC_6H_4)(2\text{-}C_{10}H_7)P$. VIII. Obtained in optically active form, (S)-$[\alpha]_D$ −10.0°.[1033]

Cyclic Tertiary Phosphines

Three-Membered Rings

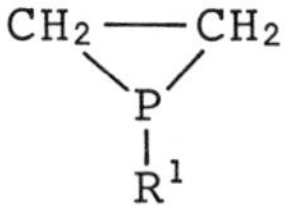

VII. (R^1 = Ph). Liquid, $b_{1.5}$ 44 to 48°, IR, ^1H NMR, ^{31}P +234 ppm.[213]
(R^1 = Me). Condenses at −82° in the high vacuum,[213] ^1H NMR,[213] IR,[213,214] ^{31}P +251 ppm;[213] inversion at phosphorus is slow on the NMR scale.

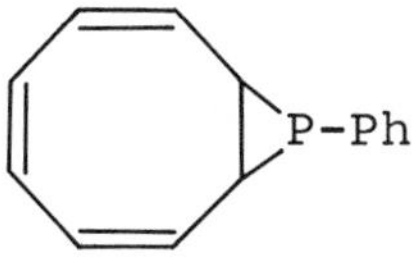

II. White crystals at −78° but somewhat oily at room temp; purified by subl. at 55°/10^{-6} torr;[767] UV, ^1H NMR, ^{31}P +181 ppm;[767] compound darkens rapidly upon exposure to air; in $CHCl_3$ at 70° or at room temp. in the course of a few months, this phosphine is transformed into 9-phenyl-9-phosphabicyclo[4.2.1]nonatriene.[767]

Four-Membered Rings

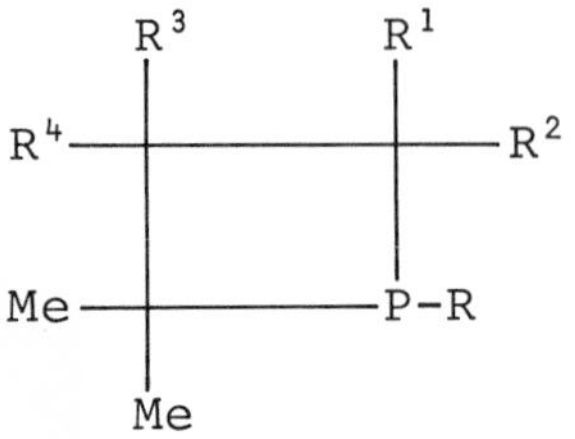

VIII. Reduction of oxides with Si_2Cl_6[299] and $SiHCl_3/Et_3N$ proceeds here with retention of configuration;[257,534] in every case a mixture of geometrical isomers, also present in the oxides, was obtained, which could not be separated by GLC;[257] the chemical shift of the 3-H in the trans isomer is about 0.4 to 0.7 ppm downfield from that of the cis isomer; ^{31}P shifts for cis are about 20 to 30 ppm to higher field of the trans isomers.[257a]

($R = Me$; $R^1 = R^2 = R^3 = Me$; $R^4 = H$). Showed no inversion (cis → trans) after 4 days at 162°;[258] characterized as L·MeBr, m. > 305°.[257]

($R = t$-Bu; $R^1 = R^2 = R^3 = Me$; $R^4 = H$). Activation energy for inversion is 28.2 ± 0.9 kcal/mole.[258]

($R = Ph$; $R^1 = R^2 = R^3 = H$; $R^4 = Me$).[257,397] Liquid, $b_{0.3}$ 71°, ^1H NMR;[257] L·MeBr, m. 184 to 185°,[257] L·MeI.[397]

($R = Ph$; $R^1 = R^2 = H$; $R^3 = R^4 = Me$). Liquid, $b_{0.2}$ 74 to 76°, ^1H NMR.[257]

($R = Ph$; $R^1 = R^2 = R^3 = Me$; $R^4 = H$). Solid, m. ∿50°, $b_{1.5}$ 111°,[534] $b_{0.05}$ 75 to 77°,[257] $b_{0.1}$ ∿80°,[299] ^1H NMR;[257 534] activation energy for inversion is 29.8 ± 0.1 kcal/mole;[258] reaction with $HC{\equiv}CCO_2Et$ proceeds with ring expansion.[534]

Five-Membered Rings

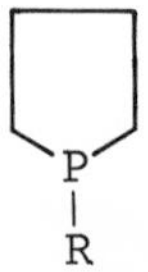

($R = Me$). VIII.[967] XIV.[1148] (By hydrogenation of 3-phospholene). Liquid, b_{760} 123°;[967] $L{\cdot}PhCH_2Br$, m. 184 to 184.5°.[1148]

($R = Et$). I.[78] VII.[667] Liquid, b. 145 to 147°;[78,667] L·MeI, m. 289 to 291°.

($R = Bu$). VIII. Liquid, b_{10} 67 to 68°, n_D^{20} 1.4920, d_{20}^{20} 0.8835.[310]

($R = c$-C_6H_{11}). VII.[667] IV.[676] Liquid, $b_{3.5}$ 94°,[667] b_3 90°;[676] L·MeI, m. 249°,[667] m. 243°.[676]

(R = Ph). I.[33,493] VII.[667] XII.[282] From CH_2=$CHCH_2$-CH_2PHPh and UV. Liquid, b_{16-18} 132 to 133°,[493] b_{10} 123°,[33] b_3 97°,[667] b_{14} 125°,[282] n_D^{20} 1.5909,[33] n_D^{20} 1.5918,[282] $n_D^{22.5}$ 1.5894,[493] d_4^{20} 1.033,[33] d_4^{20} 1.0354,[493] d_4^0 1.0502,[493] IR;[282] L•MeI, m. 130°;[667] sulfide, m. 77°,[667] m. 71 to 73°.[282]

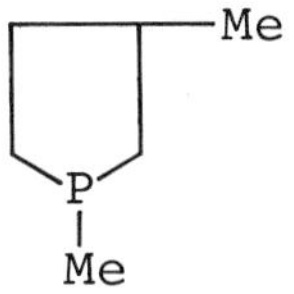

VIII. Reduction of cis and trans oxides with $SiHCl_3$[966] or Si_2Cl_6[967] (retention of configuration) gave mixture of cis- and trans-phosphines. Liquid, b. 138.5 to 141°,[966,967] ^{1}H NMR, ^{31}P +33.8 ppm;[967] careful fractionation gave one pure isomer, b. 141°,[966,967] ^{1}H NMR; isomers are configurationally stable at 150°;[966] adducts: one isomer gave L•$PhCH_2Br$, m. 168.5 to 169.5°, and the other L•$PhCH_2Br$, m. 180 to 181°.[966]

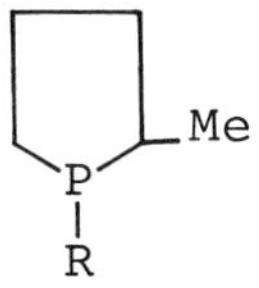

(R = Me). XIV. By hydrogenation of Ni complexes of 3-phospholene deriv.; isolated as L•$PhCH_2Br$ (mixture of cis/trans), m. 195.5 to 197.5°.[945]
(R = Ph). XIV. As above. Isolated as L•$PhCH_2Br$ (mixture of cis/trans) m. 202.1 to 202.5°.[945]

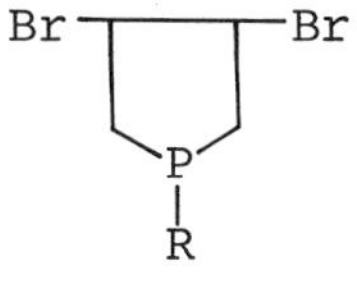

(R = Me). VIII. Crystals, m. 47.5 to 49°,[139] ^{1}H NMR, ^{31}P +41.5 ppm.[1140]
(R = Ph). VIII. Characterized as L•$PhCH_2Br$, m. 156 to 157°.[904]

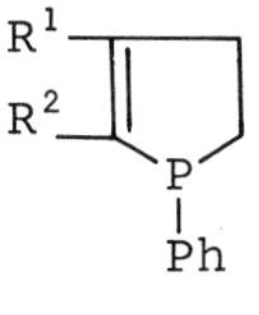

(R^1 = R^2 = H). VIII. Liquid, b_{14} 120°; L•MeI, m. 122°.[407]
(R^1 = Me; R^2 = H). VIII. Liquid, $b_{0.05}$ 79 to 80°,[407]

$b_{0.02}$ 64 to 67°,[409] ^{1}H NMR;[1141] L•PhCH Br, m. 184 to 185°.[1141]
(R^1 = H; R^2 = Ph). VIII. Crystals, m. 63°;[407] L•MeI, m. 198°.

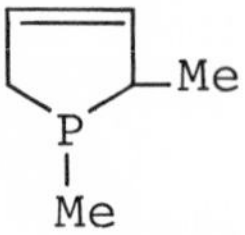
VIII. Liquid, b_{760} 135 to 136°,[1144] b. 130 to 134°,[1141,1142] IR, ^{1}H NMR;[1142] the geometrical isomers could not be separated by GLC;[1142] L•$PhCH_2Br$, m. 158 to 159°;[1142] J_{PCH} 10 Hz (cis) and 18 Hz (trans).[78b]

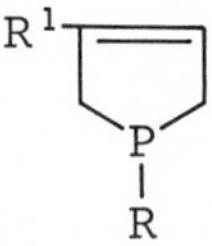
VIII. (R = Me; R^1 = H). Liquid, b. 114 to 115°, ^{1}H NMR,[1141,1145] ^{31}P +41.3 ppm;[1139] L•$PhCH_2Br$, m. 185 to 186.5°.[1145]
(R = R^1 = Me). Liquid, b_{760} 135 to 138°; L•$PhCH_2Br$, m. 142 to 143°,[1144] ^{1}H NMR.[1141]
(R = Me; R^1 = Cl). Liquid, b_{760} 159 to 160°; L•$PhCH_2$-Br, m. 153 to 154°.[1144]
(R = Ph; R^1 = Me). Liquid, b_{13} 125°,[1141] b_{16} 133 to 134°,[1144] ^{1}H NMR;[1141] L•EtI, m. 84 to 85.5°.[1144]
(R = Ph; R^1 = Cl). Liquid, b_{22} 179 to 180°; L•$PhCH_2Br$, m. 168 to 170°.[1144]
(R = Ph; R^1 = H). Liquid, ^{1}H NMR;[417,904] L•$PhCH_2Br$, m. 233 to 234°.[904]
(R = 4-MeC_6H_4; R^1 = Me). Liquid, b_{20} 165.6°, L•$PhCH_2Br$, m. 181 to 182.5°.[1144]

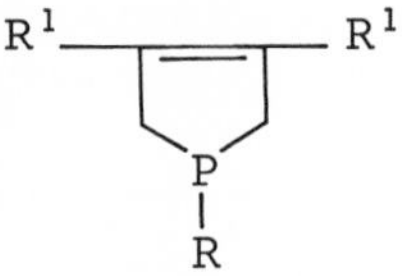
(R = Me; R^1 = H). X. From $(RP)_5$ and diene. Liquid, b_{760} 115°; L•$PhCH_2Br$, m. 189°.[1257]
(R = R^1 = Me). VIII.[1141] X.[1257 1258] Liquid, b. 160 to 161°,[1141] b_{12} 55 to 57°,[1257,1258] ^{1}H NMR;[37,1257,1258] L•$PhCH_2Br$, m. 183.5°,[1257] m. 185.5 to 187.5°.[1141]
(R = Ph; R^1 = Me). VIII. From oxide and silane,[407] or from dibromophosphorane and BuLi.[970] X.[1257,1258] Liquid, $b_{0.7}$ 97°,[407] $b_{0.3}$ 89 to 91°, $b_{0.03}$ 70 to 71°,[970] $b_{0.05}$ 65 to 67°,[1257,1258] b_{20} 158 to 160°,[1144] b_{14} 138°,[407] GLC, ^{31}P +34 ppm;[970] L•$PhCH_2Br$, m. 219 to 221°;[1144,1257,1258] L•MeI, m. 210°.[407]

(R = PhNHCO; R^1 = Me). XIII.5. Solid, m. 110°.[1287,1290]

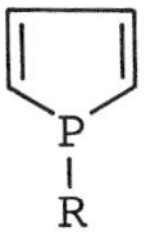

(R = Me). XIV. From dibromide and KOBu-t.

Liquid, b_{317} 82 to 85° (not completely pure),[1139,1140] UV, mass spect., ^1H NMR, ^{31}P +8.7 ppm,[1140] +8.0 ppm;[1139] forms no CS_2 adduct, but gives a salt, L•MeI, m. 190 to 194°, whose structure is not clear;[1140] has very low basicity, pK_a 0.5;[1140] extensive electron delocalization suggested (heteroaromatic system).[1139,1140]
(R = Ph). XIV. From dibromide and 1,5-diazabicyclo-[5.4.0]undec-5-ene. Liquid, $b_{0.4}$ 64 to 65°, ^1H NMR;[904] monomeric oxide and sulfide could not be made; only dimeric Diels-Adler product isolated;[904] attempts to obtain this phosphole from $PhPH_2$ and HC≡C-C≡CH were unsuccessful.[904]

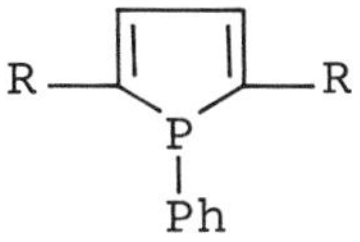

XII. From $PhPH_2$ or $PhP(CH_2OH)_2$ and RC≡C-C≡C-R with PhLi as catalyst.
(R = Me). XII. Liquid, $b_{0.2}$ 66 to 69°, ^1H NMR.[903]
(R = Ph). XII.[903] XIV. From $PhPCl_2$ and PhCH=CH-CH=CHPh at 214 to 217°.[200,201] Yellow needles, m. 187 to 189° (from $CHCl_3$),[200,201] m. 184 to 186°,[903] ^1H NMR,[903] ^{31}P -3 ppm;[1404] in solution shows a strong blue fluorescence; L•MeI, m. 218.5 to 222°;[201] shows little or no aromatic character;[201,644] reaction with $MeCO_2$-$CC{\equiv}CCO_2Me$ gives a spirobiphosphole in addition to other products;[644] more recently, a tricycloallylene-phosphorane structure was assigned to this adduct;[1404] photodimerizes (2 + 2) in ethereal solution.[1034a]
(R = 4-MeC_6H_4). XII. Crystals, m. 194 to 196°, ^1H NMR.[903]
(R = 4-BrC_6H_4). XII. Crystals, m. 200 to 202°, ^1H NMR.[903]
(R = 2-$C_{10}H_7$). XII. Crystals, m. 230 to 231°, ^1H NMR.[903]

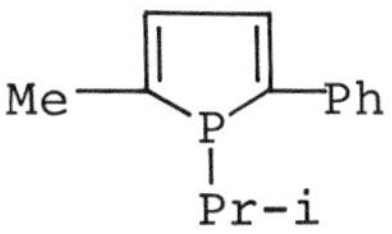

XII. From $PhPH_2$ and $CH_3C{\equiv}C$-$C{\equiv}CPh$, then cleavage of P-Ph bond with Li and treatment with i-PrBr. Liquid, $b_{0.04}$ 85°, ^1H NMR;[343a] the barrier to pyramidal inversion at P has the low value of ($\Delta G_{25}^{\ddagger}$) 16 kcal/mole, which indicates (3p-2p) π delocalization and aromaticity in phosphole systems.[343a]

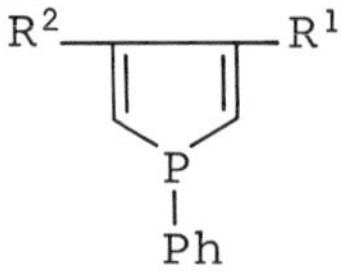

(R^1 = R^2 = Me). XIV. From 3-phospholene-P-dibromide and 1,5-diazabicyclo[5.4.0]undec-5-ene. Liquid, $b_{0.05}$ 59 to 60°, ^{31}P +2 ppm.[970]

(R^1 = Me; R^2 = H). XIV. As above. Liquid, $b_{0.03}$ 56.5 to 57°, ^{31}P -10 ppm.[970]

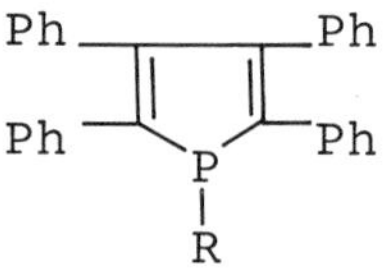

(R = Ph). II.[147,148,845] VII.[148] XIV. From $PhPCl_2$ and PhCH=C•Ph-C•Ph=CHPh[201] or from $Fe_2(CO)_6$-$(PhC{\equiv}CPh)_2$ and $PhPCl_2$.[147] Greenish-yellow fluorescent needles, m. 256 to 257°,[845] m. 255 to 256° (from CH_2Cl_2 or toluene),[147,148] m. 254.5 to 255.5°,[201] UV,[147,148] fluorescence spect.;[147,148] slowly oxidized in solution but stable in the crystalline state;[147,148] adduct with maleic anhydride, m. 260 to 264°;[174] behaves as a conjugated diene and possesses little or no aromatic character;[148] yields with K radicals of Ph• and not of phosphole.[1356a]

(R = $PhCH_2$). II. Greenish-yellow crystals, m. 203 to 213°.[148]

(R = Bu). XIV. From Bu_3P, PhC≡CPh, and I_2 at 250°/4 hr). Crystals, m. 189 to 190° (from CH_2Cl_2), IR.[739]

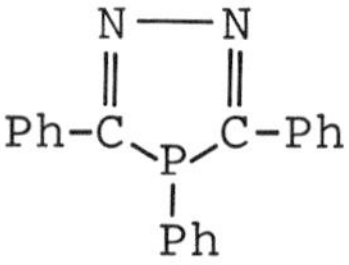

VII. From Li_2PPh and PhCCl=N-N=C•ClPh. Yellow crystals, m. 195 to 196° (from acetone).[651]

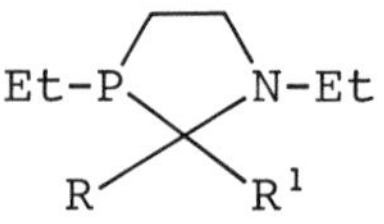

XIII.1. and .2. From $EtHPCH_2CH_2NHEt$ and

aldehyde.
(R = R^1 = H). Liquid, b_{21} 75 to 80°, pK_a 7.57.[695]
(R = Ph; R^1 = H). Liquid, b_{15} 159 to 161°, pK_a 6.14.[695]

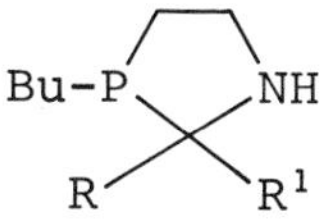

XIII.1-3. From $BuHPCH_2CH_2NH_2$ and aldehyde or ketone. HCl protonates nitrogen.
(R = R^1 = H). Liquid, b_3 80 to 82°, pK_a 8.05.[695]
(R = Ph; R^1 = H). Liquid, $b_{0.5}$ 138 to 139°, pK_a 6.70.[695]
(R = R^1 = $-(CH_2)_5-$). Liquid, b_3 129 to 130°, pK_a 8.01.[695]

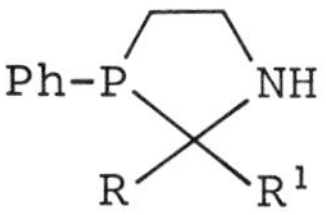

XIII.1-3. From $PhHPCH_2CH_2NH_2$ and aldehyde or ketone. HCl protonates nitrogen.
(R = R^1 = H). Liquid, b_2 109 to 112°.[693,694]
(R = Me; R^1 = H). Liquid, b_2 120°.[693,694]
(R = Et; R^1 = H). Liquid, b_3 127 to 130°.[693,694]
(R = Ph; R^1 = H). Crystals, m. 78.5 to 79.5°,[693,694] ^{31}P +2.9 ppm; L•HCl, m. 211 to 213°.[694]
(R = 2-C_4H_3O; R^1 = H). Liquid, b_2 155 to 159°.[693,694]
(R = R^1 = Me). Liquid, b_2 106 to 108°.[693,694]
(R = R^1 = Et). Liquid, b_2 130 to 133°,[693,694] ^{31}P ±0.0 ppm, L•HCl (on N), m. 218 to 221°.[694]
(R = Ph; R^1 = Me). Liquid, b_2 173 to 175°;[693,694] L•HCl (on N), m. 195 to 198°.[694]
(R = Ph; R^1 = Et). Liquid, $b_{2.5}$ 182 to 184°.[693,694]
(R = R^1 = $-(CH_2)_4-$). Liquid, b_3 146 to 147°.[693,694]
(R = R^1 = $-(CH_2)_5-$). Liquid, b_3 159 to 161°.[693,694]
(R = R^1 = $-(CH_2)_6-$). Liquid, b_3 165 to 167°,[693,694] ^{31}P -3.5 ppm; L•HCl (on N), m. 214 to 217°.[694]

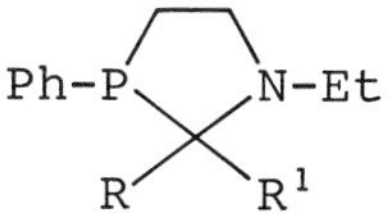

XIII.1-3. From $PhHPCH_2CH_2NHEt$ and aldehyde or ketone.
(R = R^1 = H). Liquid, $b_{3.5}$ 117 to 118°, pK_a 6.91.[695]
(R = Ph; R^1 = H). Liquid, b_2 172 to 174°, pK_a 5.90.[695]
(R = R^1 = $-(CH_2)_5-$). Liquid, $b_{0.5}$ 139 to 141°, pK_a 7.00.[695]

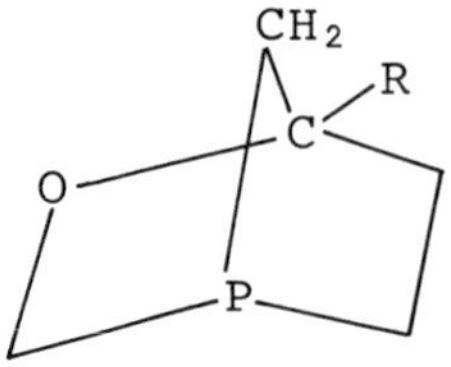

XIV. From $(HOCH_2)_3P$ and $RCOCH_2Br$ →

$$[(HOCH_2)_3PCH_2COR]^+ + \xrightarrow[Et_3N]{(HOCH_2)_3P} \text{product.}$$

(R = Ph). Crystals, m. 72 to 74° (from Et_2O), IR, ^1H NMR.[811]
(R = 4-MeC_6H_4). Crystals, m. 84 to 86° (from Et_2O).[811]
(R = 4-BrC_6H_4). Crystals, m. 128 to 129° (from Et_2O).[811]
(R = 4-$MeOC_6H_4$). Crystals, m. 71 to 73° (from i-PrOH).[811]
(R = 4-$NO_2C_6H_4$). Crystals, m. 154 to 155° (from i-PrOH).[811]

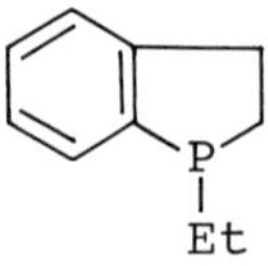

IX.1. Colorless oil, b_{13} 104 to 106°; L·EtI, m. 123°; $L_2 \cdot PdBr_2$.[948]

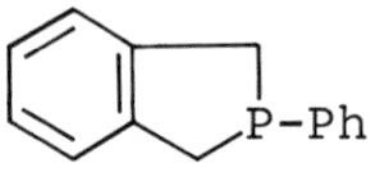

VII. IX.1. Liquid, b_{15} 182 to 186°, $b_{0.2}$ 110 to 113°;[953] gives crystalline methiodide and complexes of the type L_2PdCl_2 and L_3PdCl_2.[958]

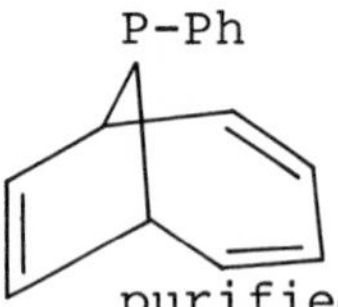

II. Needlelike crystals, m. 85.5 to 86.5°; purified by dist.; $b_{0.4}$ 160 to 170°,[767] UV, ^1H NMR, ^{31}P +79 ppm; L·MeI, m. 239 to 240°;[767] also obtained from 9-Ph-9-phosphabicyclo[6.1.0]nonatriene.

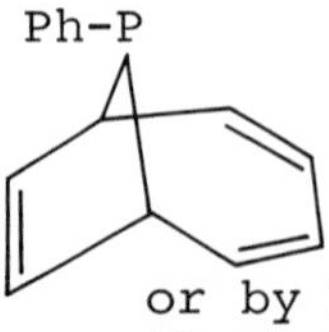

XIV. From phosphine above by heating to 480° or by heating in acidified $CHCl_3$ solution; purified through the $PdCl_2$ complex and liberated with KCN and sublimed. White crystals, m. 84.5 to 85.5°, UV, ^1H NMR, ^{31}P +14 ppm;[767] L·MeI, m. 184 to 190°.[767]

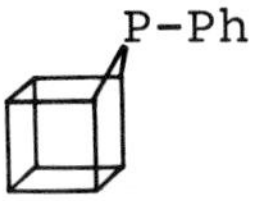

VIII. Solid, m. 58 to 59°;[766] the oxide of this phosphine was obtained by photolysis of the oxide of 9-Ph-9-phosphabicyclo[4.2.1]nonatriene (above) through Corex.[766]

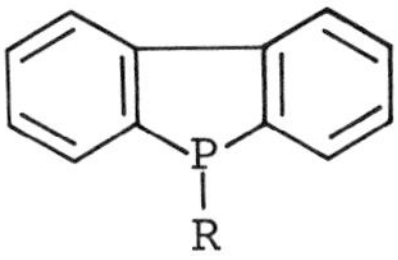

(R = Me). I.[42,43] VII.[364] IX.1.[43] Liquid, $b_{0.01}$ 110°,[364] $b_{0.2}$ 103°,[43] GLC;[43] L·MeI, m. 280 to 280.5°.[42,43]
(R = Et). I.[42,43] VII.[364] IX.1.[42] Liquid, $b_{0.15}$ 105 to 106°.[42,43] Crystals, m. 42 to 44°,[43] GLC,[43] $b_{0.2}$ 103°, $b_{0.01}$ 110°,[364] ^1H NMR.[364]
(R = $PhCH_2$). VII. Crystals, m. 76 to 78° (from MeOH), $b_{0.01}$ 165 to 170°,[364] ^1H NMR.[364]
(R = Ph). I.[293] II.[91,1444] VIII.[1439] XI.[853] XIV. From PPh_5 by dec.;[1443] dec. in pyridine gives higher yield,[1180] or from Ph_3P and PhNa[1443] or benzyne;[1439] also obtained from Ph_4PBr and Et_2NLi[577] or MeLi.[1279] Crystals, m. 93 to 94°,[1180,1443] m. 91° (from MeOH),[293] m. 92 to 94° (from $EtOH/H_2O$),[853,1279,1444] $b_{0.01}$ 165°,[1439] $b_{0.0005}$ 180°,[853] GLC,[43] dec. 380°,[1007] ^{31}P +10.2 ppm,[548] heat of formation ΔH_f +44.9 kcal/mole;[91] has a stabilization or resonance energy of 19 kcal/mole;[91] the previously reported formation of the oxide by method II[1439] was apparently due to admittance of O_2;[91] reaction of $(2,4,6\text{-}D_3C_6H_2)_4PBr$ and MeLi gave hepta-deuterated 9-Ph-9-phosphafluorene, $L{\cdot}Me^+BPh_4^-$, m. 173 to 176°.[1279]
(R = $4\text{-}Me_2NC_6H_4$). II. Crystals, m. 177 to 177.5° (from EtOH);[1444] $L{\cdot}HgCl_2$, m. 234 to 235°.[1444]
(R = $4\text{-}MeC_6H_4$). II. Crystals, m. 110 to 113° (from $EtOH/H_2O$), $b_{0.0005}$ 180°.[853]

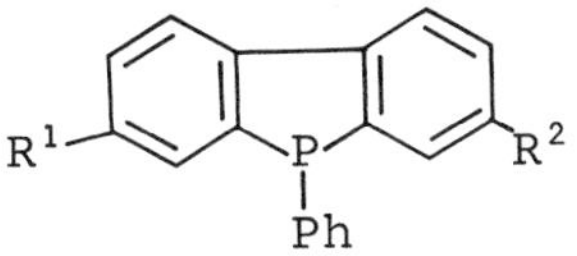

($R^1 = R^2 = Me_2N$). II. Yellow crystals, m. 226 to 227° (from acetone);[1444] $L{\cdot}HgCl_2$, m. 264 to 265°.[1444]
($R^1 = HOCH_2$; R^2 = H). VIII. Reduction of carboxylic acid with $LiAlH_4$. Crystals, m. 52 to 54° (with 1EtOH).[204]

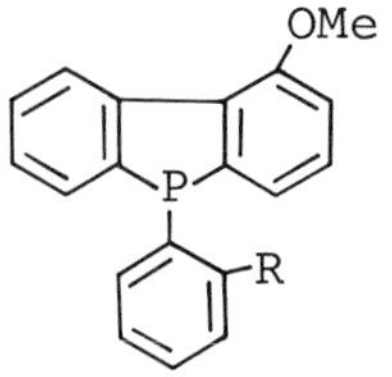

(R = H). XIV. From Ph_3P and methoxybenzyne. Oil.[1462]
(R = Ph). XIV. From 9-Ph-9-phosphafluorene and methoxybenzyne. Solid, m. 210°.[1462]

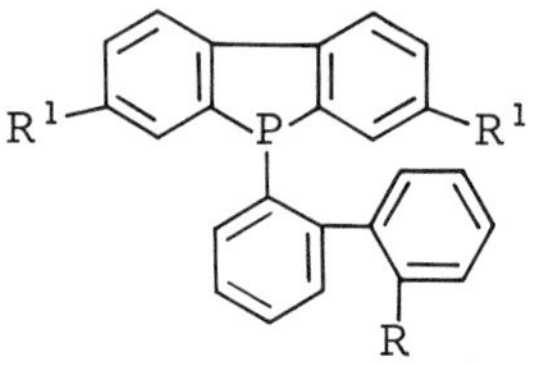

(R^1 = H; R = Cl). IX.1. and .3. Crystals, m. 221 to 223° (from $Me_2CO/EtOH$).[551]
(R^1 = H; R = Br). IX.1. and .3. Crystals, m. 202.5 to 203° (from $CH_2Cl_2/MeOH$).[551]
(R^1 = H; R = I). IX.1. and .3. Crystals, m. 154 to 158° (from EtOH, not pure).[551]
(R^1 = R = H). XIV. From dec. of bis-2,2'-biphenylylenephosphorane. Crystals, m. 162 to 164°, IR, ^{31}P +19 ppm.[550]
(R^1 = Me_2N; R = Ph). XIV. Thermal dec. of phosphorane. Yellow crystals, m. 200 to 207°; $L{\cdot}HgCl_2$, dark-red needles, m. 255 to 256°.[1444]
(R^1 = H; R = 2,2'-biphenylylenephosphorane). XIV. Thermal dec., crystals, m. 220 to 222°,[550] ^{31}P +19 ppm [P(III)] +85 ppm [P(V)].[550]

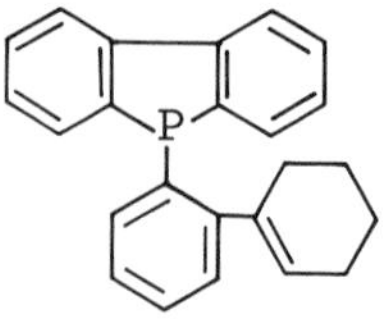 and 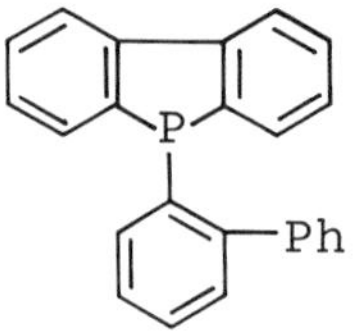

XIV. From dec. of bis-2,2'-biphenylylenephosphorane. Obtained as a mixture, m. 152 to 156°.[550]

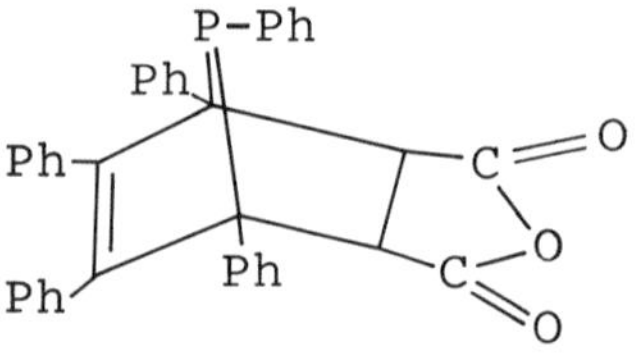

XIV. From pentaphenylphosphole and

and maleic anhydride. Solid, m. 260 to 264°.[148]

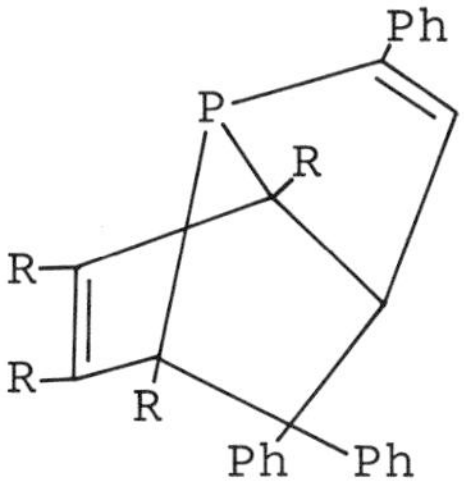

XIV. From Diels-Alder adduct of triphenylphosphole and $MeO_2CC{\equiv}CCO_2Me$ by thermal dec.[644] (R = MeO_2C). Crystals, m. 220° (from $CHCl_3$), IR, UV, ^1H NMR;[644] structure not definitely established; recently, a nine-membered ring structure was assigned to this phosphine (see 9-membered rings).[1404]

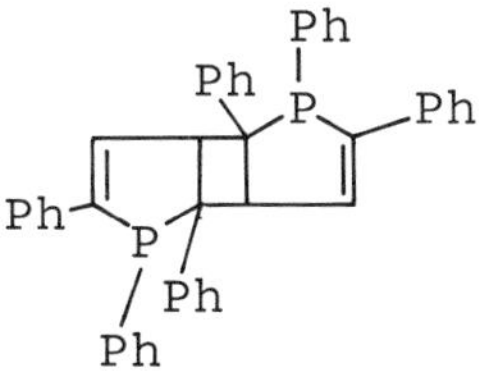

XIV. By irradiation of 1,2,5-triphenylphosphole in ether. Crystals, m. 229 to 230°.[84a]

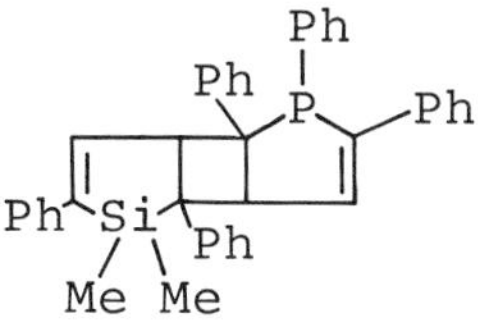

XIV. By irradiation of a mixture of 1,2,5-triphenylphosphole and 1,1-dimethyl-2,5-diphenylsilacyclopenta-2,4-diene in ether. Crystals, m. 248 to 249°.[84a]

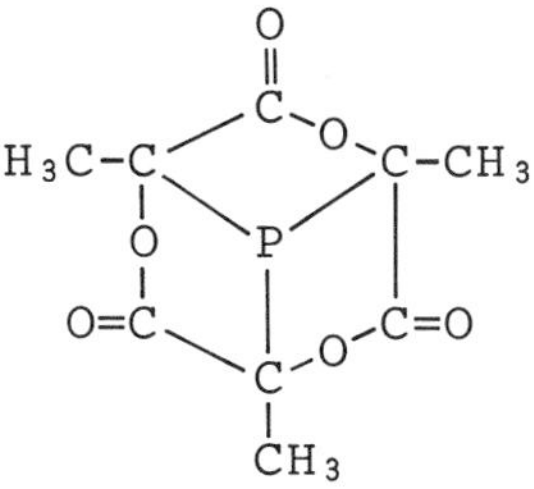

XIV. From pyruvic acid, $MeCOCO_2H$, and PH_3 plus HCl.[168,980] Crystals, m. 273 to 274°,[168] IR, ^{31}P +21 ppm;[168] oxidative degradation with HNO_3 produced $(HO_2CCMe{\cdot}OH)_2PO_2H$.[168]

Six-Membered Rings

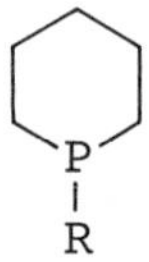

(R = Et). VII. Liquid, b. 170°; L•MeI, m. 293 to 296°.[667]

(R = c-C_6H_{11}). VII.[667] IV.[676] Liquid, b_3 115°,[667] b_2 112°;[676] L•MeI, m. 230 to 232°.[667]

(R = $PhCH_2$). Liquid, b_{30} 153 to 156°, n_D^{20} 1.5612.[1306]

(R = Ph). I.[33,493] VII.[667] XII. From CH_2=CH$(CH_2)_3$-PHPh and UV.[282] Liquid, b_{22-24} 154 to 155°,[493] b_{16-18} 143 to 144°,[493] b_9 128°,[33] b_3 119°,[667] b_{14} 125°,[282] $b_{0.5-1.0}$ 75 to 85°,[1291] n_D^{20} 1.5886,[493] n_D^{20} 1.5882,[33] n_D^{20} 1.5890,[282] d_4^{21} 1.0306,[493] d_4^{20} 1.029,[33] IR;[282] L•MeI, m. 176°.[667]

(R = 4-MeC_6H_4). I. Liquid, b_{26} 167 to 168°, n_D^{22} 1.5729, d_4^{20} 1.0007;[493] L•EtI, m. 163 to 164°.[493]

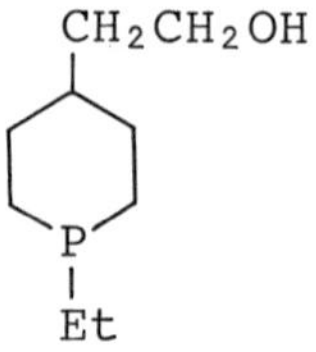

XIV. By hydrogenation of Ni complex of 1,2,-5,6-tetrahydrophosphorin deriv. Liquid, $b_{0.15}$ 83 to 85°.[1148]

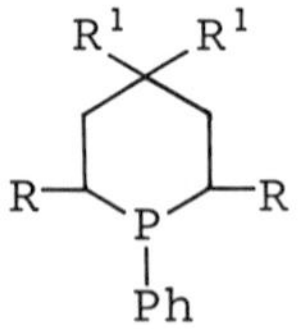

(R^1 = Me; R = H). XII. From $PhPH_2$ and CH_2=CHCMe_2CH=CH_2 with Bz_2O_2.[1291] Liquid, b_1 105 to 107°.[1291]

(R^1 = H; R = Ph). XIV. From phosphorinanone by reduction with N_2H_4. White crystals, m. 167 to 168° (from EtOH); L•MeI, m. 240°.[1424]

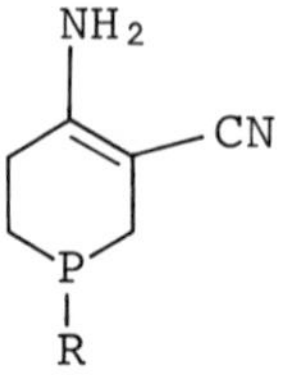

XIV. By cyclizing alkyl- or arylbis(2-

cyanoethyl)phosphine with t-BuONa in refluxing toluene.[1425]
(R = Et). Crystals, m. 74.5 to 75°.[1425]
(R = Ph). Crystals, m. 139.5 to 140°.[1425]

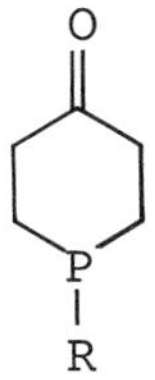

XIV. Hydrolysis and decarboxylation of the above phosphines yield (4-phosphorinanones);[1425] also obtained from RPH_2 and CH_2=CHCOCH=CH_2.[1423]
(R = Me). Liquid, $b_{0.7}$ 57 to 58°,[1146,1292] IR, ^1H NMR;[1292] attempts to prepare the oxime gave the oxime of the phosphine oxide.[968]
(R = Et). Liquid, b_7 92°,[1425] semicarbazone derivative, m. 167 to 169°;[1425] thiosemicarbazone derivative, m. 153 to 154.5° (dec.);[968] attempts to prepare the oxime gave the oxime of the oxide (NH_2OH acted as oxidizing agent);[968] phosphine has been claimed to be useful as a sequestering agent for metal salts.[1423]
(R = Ph). Crystals, m. 43 to 44°,[971,1425] b_1 185 to 190°,[1425] $b_{0.4}$ 110 to 112°,[971] IR,[971] ^1H NMR;[37] adduct: L•HCl•$2H_2O$, m. 231 to 233°;[420] semicarbazone deriv., m. 155.5 to 156.5°.[1425]

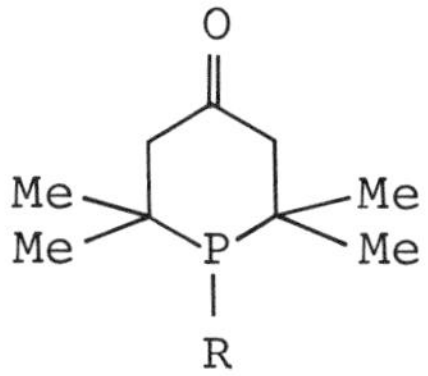

XII. From phorone, $(Me_2C=CH)_2CO$, and RPH_2.[1423,1424]
(R = $NCCH_2CH_2$). Solid, m. 48 to 50°, b_1 140 to 150°; L•MeI, m. 290 to 295° (dec.); pK_a 0.79.[1424]
(R = i-Bu). Liquid, $b_{0.3}$ 90 to 97°, n_D^{20} 1.4962; L•MeI, m. 275 to 278°; pK_a 3.58.[1424]
(R = c-C_6H_{11}). Solid, m. 63 to 65°, b_1 135 to 140°; L•MeI, m. 275 to 278°; pK_a 3.86.[1424]
(R = n-C_8H_{17}). Liquid, $b_{0.8}$ 146 to 148°; L•MeI, m. 150 to 152°.[1424]
(R = Ph). Crystals, m. 91 to 92°,[63,1424] $b_{0.5}$ 130 to 140°,[1424] $b_{0.5}$ 137 to 139°;[63] showed polymorphism;[1424] pK_a 1.85, IR, ^1H NMR;[1424] L•MeI, m. 229 to 230°; semicarbazone deriv., m. 198.5 to 199.5°.[1424]

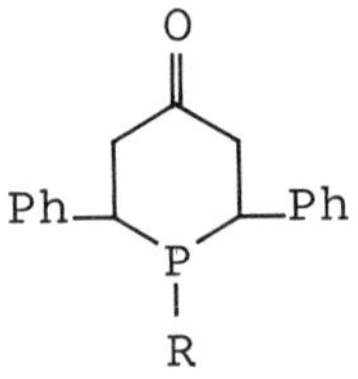

XII. From $(PhCH{=}CH)_2C{=}O$ and RPH_2 or PH_3 at 130°.[1424] Catalyzing the reaction first with NaOEt and then with acids (cyclization) gave higher yields.[63]
(R = Me). Crystals, m. 144 to 145°, IR, ^1H NMR.[892]
(R = $NCCH_2CH_2$). Crystals, m. 126 to 128° (from EtOH).[1424]
(R = i-Bu). Crystals, m. 118 to 119°.[1424]
(R = $c\text{-}C_6H_{11}$). Crystals, m. 120 to 121° (from $MeOH/H_2O$), pKa 7.00.[1424]
(R = $n\text{-}C_8H_{17}$). Crystals, m. 80 to 81° (from $MeOH/H_2O$).[1424]
(R = $PhCH_2$). Crystals, m. 167 to 169°, IR, ^1H NMR;[892] can be dehydrogenated with SeO_2 to the dienone.[892]
(R = $PhCH{=}CHCOCH_2{\cdot}CHPh$). Amorphous resin, m. 130 to 150°.[1424]
(R = Ph). Crystals, m. 178°,[901] m. 176.5 to 177.5°,[892,1424] pKa 4.61,[1424] IR, ^1H NMR;[892] L·MeI, m. 137° (dec.);[1424] semicarbazone deriv., m. >270°.[1424]

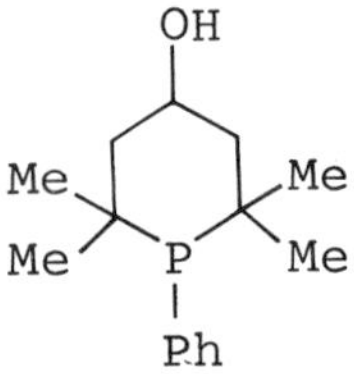

XIV. From phosphorinanone and $LiAlH_4$, α and β isomers are obtained. Crystals, m. 123 to 124.5° (from hexane).[1424]

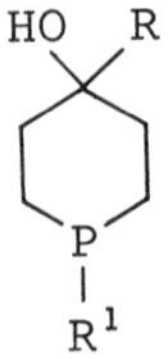

XIV. Starting from phosphorinanone the following compounds were prepared:
(R^1 = Me; R = H). (With $LiAlH_4$). Liquid, $b_{0.5}$ 65° (cis and trans isomer), IR, ^1H NMR;[1292] $L{\cdot}PhCH_2Br$, m. 185° up to 225° (depending upon cis/trans ratio).[1292]

(R^1 = Me; R = Et). (With EtMgBr). Liquid, $b_{0.2}$ 45 to 62°, cis and trans (1:1) separated by GLC and dist., cis $b_{0.55}$ 62°, trans $b_{0.6}$ 68 to 69°, IR, ^{1}H NMR;[1146, 1292] isomers are stable when refluxed in benzene or toluene.[1146]
(R^1 = Me; R = Ph). (With PhMgBr). Liquid, $b_{0.1}$ 110 to 118°, cis and trans separated by GLC and dist., cis $b_{0.2}$ 102 to 110°, m. 92 to 100°, trans $b_{0.2}$ 114 to 116°, IR, ^{1}H NMR;[1292] cis L·$PhCH_2ClO_4$, m. 228 to 229°; trans L·$PhCH_2ClO_4$, m. 215 to 217°.
(R^1 = Me; R = EtO_2CCH_2). (With EtO_2CCH_2ZnBr). Liquid, $b_{0.15}$ 75 to 90°, mixture of cis-trans, IR, ^{1}H NMR;[1292] L·$PhCH_2BPh_4$, m. 145.5 to 146.5°.[1292]
(R^1 = Me; R = $HOCH_2CH_2$). (Reduction of above ester with $LiAlH_4$). Liquid, $b_{0.15}$ 107 to 120° (cis-trans), IR, ^{1}H NMR; L·$PhCH_2BPh_4$, m. 169.5 to 171°.[1292]
(R^1 = Et; R = Et). (With EtMgBr). Liquid, $b_{0.1}$ 47 to 63° (cis-trans), IR, ^{1}H NMR; L·$PhCH_2ClO_4$, m. 139°.[1292]
(R^1 = Et; R = EtO_2CCH_2). (With EtO_2CCH_2ZnBr). Liquid, $b_{0.05}$ 97 to 101°, IR.[1143]
(R^1 = Et; R = $HOCH_2CH_2$). (Reduction of above ester with $LiAlH_4$). Liquid, $b_{0.15-0.2}$ 128 to 130°, IR;[1143, 1147] conversion of this phosphine into a bicyclic salt[1143] could not be confirmed.[1147]
(R^1 = Ph; R = H). (With $LiAlH_4$). Crystals, m. 68 to 69°, $b_{0.6}$ 133.4°.[971]

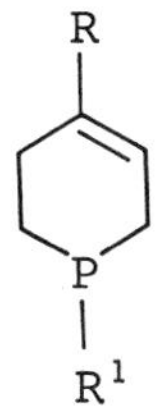

XIV. From phosphorinanol (above) and acid.

(R^1 = Me, R = Ph). Liquid, $b_{0.1}$ 84 to 85°, UV, IR, ^{1}H NMR; L·$PhCH_2Br$, m. 222 to 223°.[1292]
(R^1 = Me; R = Et). Liquid, $b_{2.3}$ 59 to 74°; L·$PhCH_2ClO_4$, m. 108 to 110°.[1147]
(R^1 = R = Et). Isolated as salt, L·$PhCH_2ClO_4$, m. 134°.[1147]
(R^1 = Ph; R = H). (With Al_2O_3 at 380°). Liquid, $b_{0.7}$ 86°, n_D^{20} 1.6084, IR.[971]

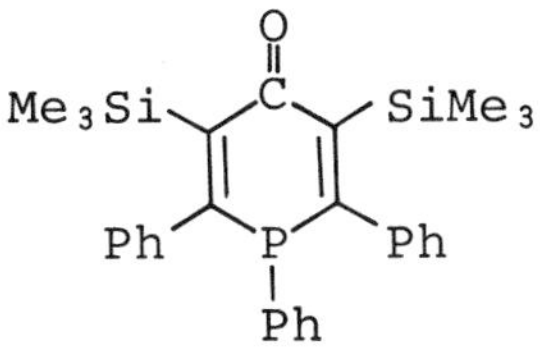

XII. From $PhP(SiMe_3)_2$ and PhC≡CCOC≡CPh

with AIBN initiation. Lemon-yellow crystals, m. 170 to 171°, IR, UV, ^{1}H NMR;[902] L•MeI;[902] a dipolar structure was not excluded, but alkylation occurs exclusively at P.[902]

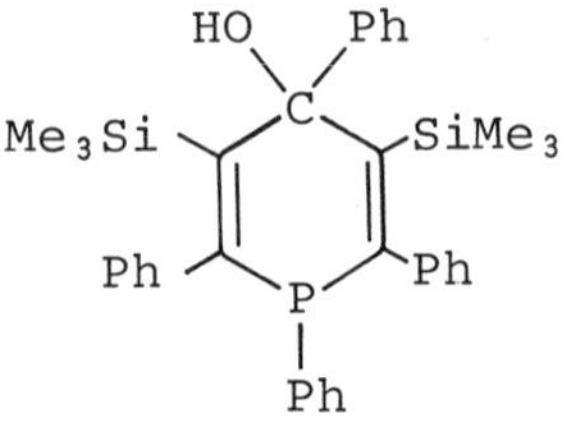

XIV. From ketone above and PhLi.

Crystals, m. 156 to 157°, IR, UV, ^{1}H NMR;[902] $HClO_4$ protonates at P and does not give a phosphapyrylium salt.[902]

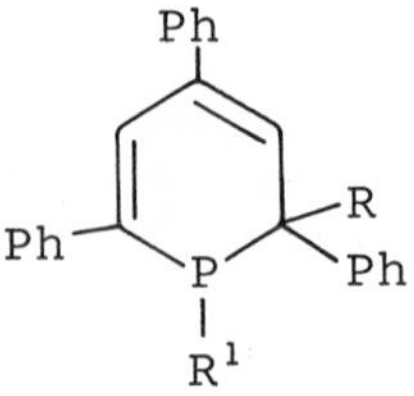

XIV. From phosphorin and R^1Li. Work-up

up with H_2O (R = H) or alkylation with RX in benzene.[895,898]
(R^1 = Bu; R = H). Light-yellow oil, ^{1}H NMR.[895]
(R^1 = Ph; R = H). Crystals, m. 144 to 145° (from EtOH), UV, ^{1}H NMR.[895]
(R^1 = Ph; R = Et). Crystals, m. 184 to 185°, UV, ^{1}H NMR.[898]
(R^1 = Ph; R = CH_2=$CHCH_2$). Crystals, m. 155 to 156°, UV, ^{1}H NMR.[898]
(R^1 = Ph; R = $PhCH_2$). Crystals, m. 168 to 169.5°, UV, ^{1}H NMR.[898]

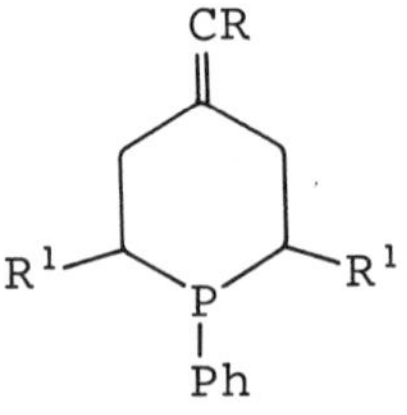

XIV. From dichlorophosphorane and $LiAlH_4$.

(R = Ph; R^1 = Ph). Yellow crystals, m. 238 to 240°, UV; L•MeI, m. 257 to 258°;[900] P alkylation and not C alkylation.
(R-fluorenyl; R^1 = Ph). Crystals, m. 223 to 225°, UV.[900]

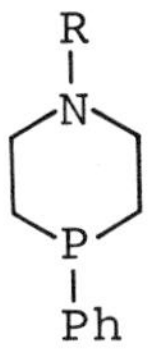

(R = Ph). VII. Colorless solid, m. 89 to 90°;[949] phosphine is air-stable and only a weak base; forms a monohydroiodide and monosalts with MeI and EtBr; L•2MeI is unstable, but L•2HCl has been isolated.[949]
(R = H). From $Ph(H_2NCH_2CH_2)PLi$ and $ClCH_2CH_2NH_2$. Liquid, b_3 145 to 147°, pK_a 8.45.[694]

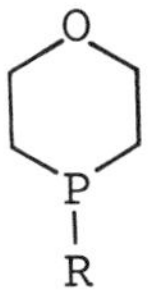

XIV. From RPH_2 and $CH_2=CHOCH=CH_2$ with UV irradiation.[1352]
(R = n-C_8H_{17}). Liquid, $b_{0.5}$ 109 to 110°; ^{31}P +52 ppm.[1352]
(R = n-$C_{12}H_{25}$). Liquid, $b_{0.5}$ 155 to 157°; ^{31}P +52 ppm.[1352]
(R = n-$C_{14}H_{29}$). Liquid, $b_{0.1}$ 150 to 152°.[1352]
(R = n-$C_{18}H_{37}$). Liquid, $b_{0.5}$ 211 to 212°.[1352]
(R = Ph). [From $PhP(MgX)_2$ and $(XCH_2CH_2)_2O$]. Crystals, m. 135 to 137° (from EtOH).[846]

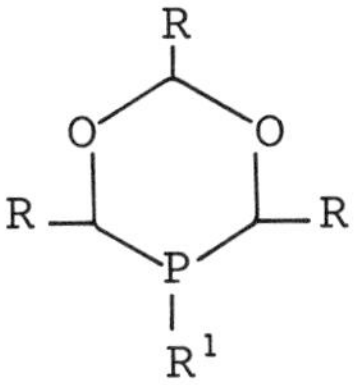

XII. From phosphine and aldehyde in CH_3CN with a little HCl.[356]
(R^1 = $MeCO_2CH_2CH_2$; R = i-Pr). Liquid, $b_{0.3}$ 123°;[1165] did not self-quaternize when heated at 300° for 3 hr (steric reason).[1165]
(R^1 = $NCCH_2CH_2$; R = 4-ClC_6H_4). Crystals, m. 197 to 200° (from EtOH).[469]
(R^1 = R = Ph). Crystals, m. 195 to 198° (from MeOH),[469] (or CH_3CN).[356]

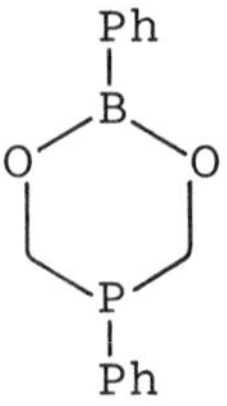

XIV. From $[(HOCH_2)_3PhP]Cl$ and $PhBCl_2$. Solid, m. 106 to 108°, sublimes under vacuum.[1071]

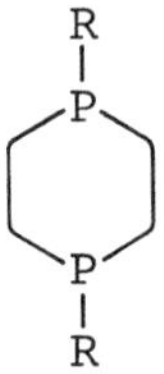

(R = Et). VII. Pale-yellow oil, b_4 135 to 145°;[658] disulfide, m. 225 to 235°.
(R = c-C_6H_{11}). VII. Liquid, b_2 225 to 230°,[658] disulfide, m. 250 to 255° and 325 to 326° (cis and trans).
(R = $PhCH_2$). IX.3. Crystals, m. 128 to 130°;[566] $BrCH_2CH_2Br$ affords bicyclic salt.
(R = Ph). VII.[702,716] IX.3.[566] Crystals, m. 92 to 95°,[566] m. 104 to 105° (from EtOH);[716] liquid, b_7 300 to 320°,[702] (probably not correct, see ref. 716); disulfide, m. 253° and 154°;[702] in ref. 716 only one disulfide reported, m. 253 to 255°; $BrCH_2CH_2Br$ affords bicyclic salt.[566]

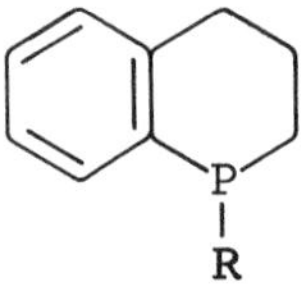

IX.1. (R = Et). Colorless liquid, b_{18} 141 to 143°; L•MeI, m. 184 to 185°.[93]
(R = 2-$MeOCH_2CH_2C_6H_4$). Oily liquid, b. ∿250 to 320° (crude); this phosphine on treatment with HBr in boiling $CHCl_3$ underwent cyclization by quaternization, giving the first known spirocyclic phosphonium salt,[520] which was optically resolved.[520]

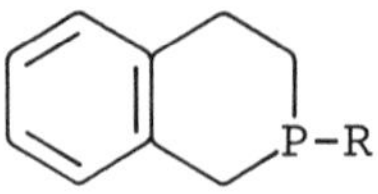

IX.1. (R = Et). Colorless liquid, b_{15} 129 to 130°; L•MeI, m. 93 to 94°;[93] this phosphine,

P-ethyl-tetrahydroisophosphinoline, is more basic than the corresponding P-ethyltetrahydrophosphinoline (see above).
(R = Ph). III. Oily liquid, $b_{0.2}$ 130 to 160°; characterized as L•MeI, m. 116 to 118°.[601]
(R = 4-MeC_6H_4). III. Oil, $b_{0.1}$ 150 to 180°;[602] L•4-$ClC_6H_4COCH_2Br$, m. 227 to 230°.

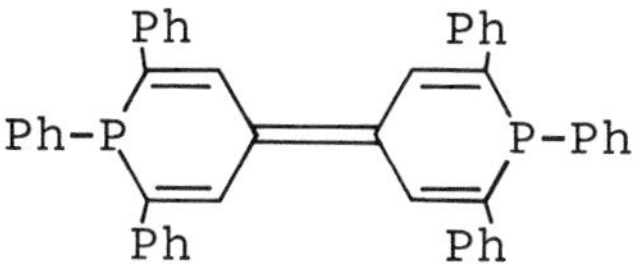

XIV. From bisdichlorophosphorane by reduction with $LiAlH_4$ or from Cl_2CPCl_2(Ph) and $LiAlH_4$. Brownish-red crystals, m. 335 to 337°, UV, mass spect.;[901] not autoxidizable; gives no salt but gives a disulfide, m. 403 to 405°.[901]

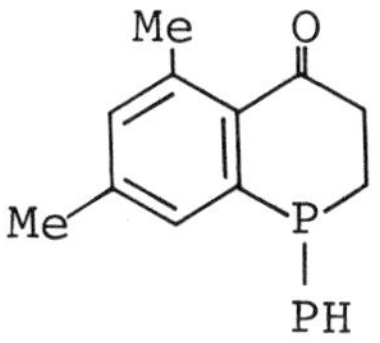

XIV. By cyclization of $(HO_2CCH_2CH_2)$Ph(3,4-$Me_2C_6H_3$)P with polyphosphoric acid.[419] Liquid, $b_{0.2}$ 150 to 152°, IR;[419] characterized as methopicrate, m. 170 to 171° (from EtOH).[419]

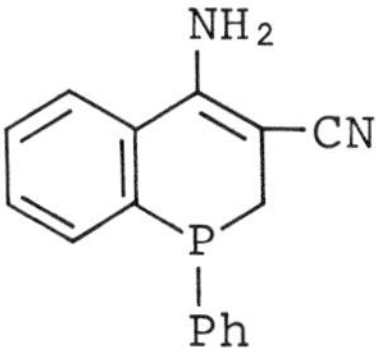

XIV. By cyclization of $(NCCH_2CH_2)$Ph(2-NCC_6H_4)P with t-BuONa. Crystals, m. 180 to 181° (from EtOH), $b_{0.2}$ 205 to 206°.[419]

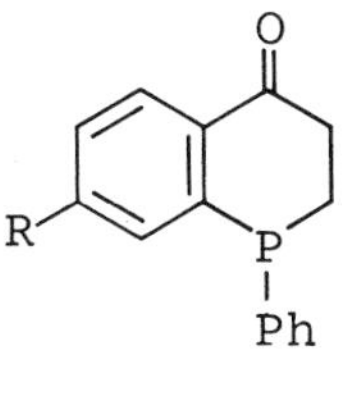

XIV. (R = H). By hydrolysis of the

aminonitrile above with concentrated HCl. Crystals, m. 46 to 47° (from H_2O/EtOH), $b_{0.05}$ 143 to 145°, UV;[419] derivatives: 4-phenylsemicarbazone, m. 225 to 226° (from C_6H_6); perchlorate, m. 179.5 to 181°; 2,4-dinitrophenylhydrazone, two forms, m. 156 to 157° and red crystals, m. 160 to 160.5°;[419] phenylhydrazone, yellow solid, m. 55 to 60°.[421]
XIV. (R = EtO). By cyclization of $(HO_2CCH_2CH_2)Ph(3\text{-}EtOC_6H_4)P$ with P_2O_5 in toluene. Liquid, $b_{0.1}$ 220 to 230°; L•AgI, m. 120°;[567] structure of product not definitely established.[567]

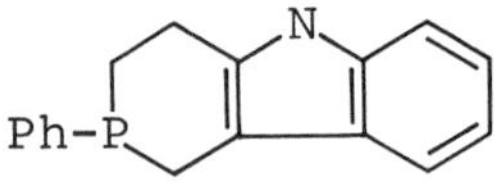

XIV. From phosphorinanone and $PhNHNH_2$ by Fischer's indolization. Crystals, m. 113 to 114°.[420]

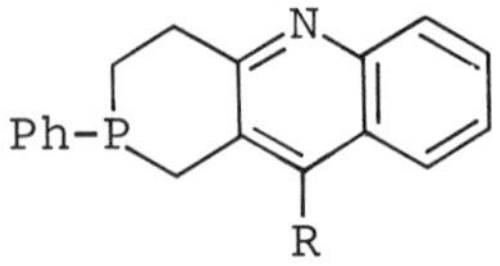

(R = H). XIV. From 1-phenylphosphorinanone and o-aminobenzaldehyde. Crystals, m. 66.5 to 67° (from ligroin);[420] L•2HCl•$2H_2O$, m. 171 to 173°.[420]
(R = COOH). XIV. From phosphorinanone and isatin. Solid, m. 248 to 249°;[420] gives two forms of benzylthiouronium salts, m. 138.5 to 139.5° and m. 212.5 to 213.5°.[420]

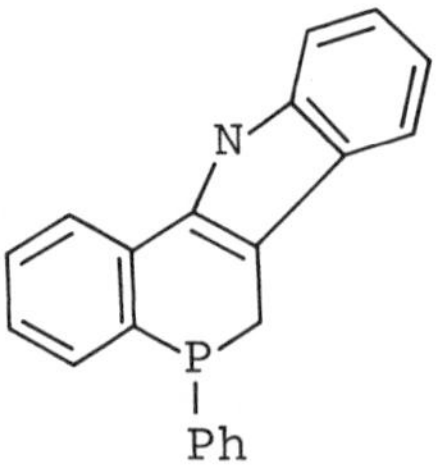

XIV. From the phenylhydrazone of the ketophosphine by Fischer's indolization with ethanolic hydrogen chloride. Yellow crystals, m. 191 to 192° (from EtOH), IR;[421] L•MeI•H_2O, m. 278 to 280°.

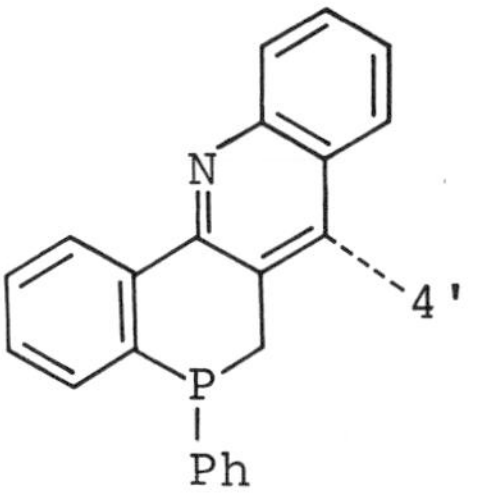

XIV. From ketophosphine and o-aminobenzaldehyde in alkaline ethanolic solution. Crystals, m. 127.5 to 128.5° (from EtOH);[421] L•picrate•H_2O, m. 168 to 169°; L•HCl•H_2O, m. 218 to 219°;[421] the ketophosphine gave with isatin apparently the 4'-carboxylic acid of the quinolinophosphinoline, but this was not isolated pure.[421]

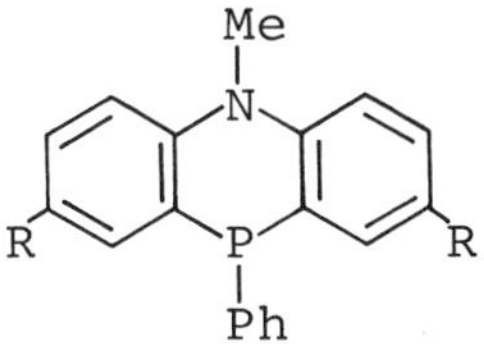

II.[1349] IV.[765] (R = H). Crystals, m. 159 to 160° (from EtOH),[765,1349] IR, UV, ^1H NMR:[827] claimed to be useful as antioxidants and antiwear agent.[1349]

II. IV. (R = Me). Crystals, m. 144 to 146° (from EtOH), IR, UV, ^1H NMR.[827]

II. IV. (R = Br). Crystals, m. 227 to 229° (from EtOH), IR, UV, ^1H NMR.[827]

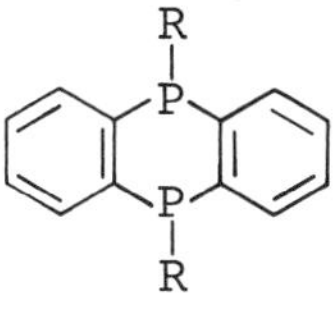

By combination of I and II. 2-ClC_6H_4MgBr + $RPCl_2$ → $(2\text{-}ClC_6H_4)_2PEt$, then BuLi + $RPCl_2$.

(R = Et). Two isomeric forms, m. 52 to 53° (from MeOH, probably cis) and m. 96 to 97° (trans form); molecule is folded about the P-P axis, thus giving two alternative relative positions for the ethyl groups;[283,293] monoxide, m. 140° (of cis form); dioxide•H_2O_2•H_2O, m. 128°.[293]

VII. (R = Ph). Only one form isolated, m. 184 to 187° (from MeEtC=O);[283,293] L•2$PhCH_2Br$, m. 387°; monoxide, m. 231 to 232.5°; dioxide, m. >400°.[293]

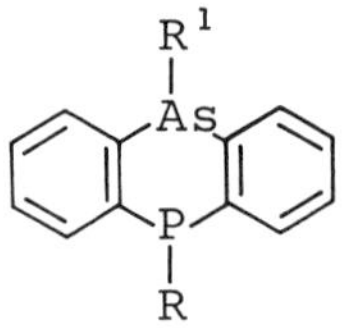

I. and II. (R = Ph; R^1 = Et). Isolated as dibenzyl dipicrate, m. 206° (dec.).[293]
(R = Ph; R^1 = Ph). Crystals, m. 189 to 190° (from MeEtC=O);[293] L·MeI (on P), m. 300 to 302°; P oxide, m. 214 to 216°; dioxide, m. 360°.[293]

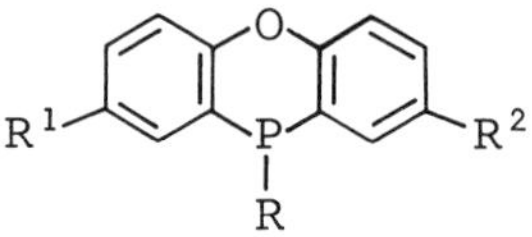

(R = Me; R^1 = R^2 = Me). XI. Crystals, m. 61 to 62° (from EtOH); L·MeI, m. 260°.[471]
(R = $Me_2NCH_2CH_2CH_2$; R^1 = R^2 = Me). VIII. Isolated as dihydrochloride $2H_2O$, m. 71°, IR.[471]
(R = Ph; R^1 = R^2 = F). VIII. Crystals, m. 57°, UV, IR;[472a] L·MeI, m. 275°.[472a]
(R = Ph; R^1 = Cl; R^2 = H). XI. ($RN_2^+BF_4^-$ + $PhPCl_2$ and Al). Crystals, m. 41.5 to 42° (from $EtOH/H_2O$), $b_{0.005}$ 160°, IR.[854]
(R = Ph; R^1 = R^2 = H). XI. Crystals, m. 97.5 to 98° (from EtOH), $b_{0.0005}$ 150°.[853]
(R = Ph; R^1 = R^2 = Me). VIII. Crystals, m. 80°, IR.[471]
(R = $4\text{-}MeC_6H_4$; R^1 = R^2 = H). XI. Crystals, m. 53 to 57° (from $EtOH/H_2O$), $b_{0.0005}$ 180°.[853]

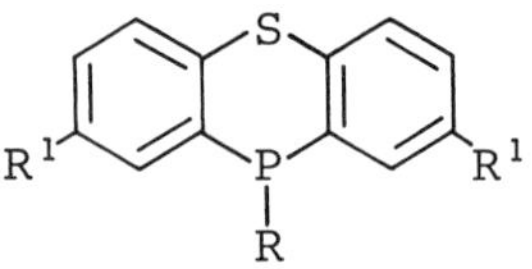

(R = NCH_2CH_2; R^1 = H). II. Solid, m. 152 to 153°.[146]
(R = $H_2NCH_2CH_2CH_2$; R^1 = H). II. Liquid, $b_{0.001}$ 160°.[146]
(R = $Me_2NCH_2CH_2CH_2$; R^1 = H). II.[146] VIII.[1433] Liquid, $b_{0.01}$ 140°;[146] semisolid product, UV;[1433] shows depression of spontaneous activity in mice (the oxide is less active) but has only about 1/10 the activity of the corresponding nitrogen compound "promazine";[1433] useful as a fungicide.[146]
(R = $Me_2NCH_2CH_2CH_2$; R^1 = Me). VIII. Isolated as L·2HCl·H_2O, hygroscopic needles, m. 95°, IR.[472]
(R = $PhCH_2$; R^1 = H). II. Solid, m. 82 to 89°.[146]
(R = $PhCH_2$; R^1 = F). II. Solid, m. 92.5 to 95.0°,[146] useful as fungicide.

(R = Ph; R^1 = H). II. VIII. From 5,5'-dioxide + $LiAlH_4$. Crystals, m. 92 to 93°, oxide, m. 112 to 113°; L•MeI, m. 250 to 260°.[146]
(R = Ph; R^1 = Me). XI. Crystals, m. 147° (from AcOH), UV, IR.[472]
(R = 4-MeC_6H_4; R^1 = Me). XI. Crystals, m. 167° (from EtOH), UV, IR; L•MeI, m. 348°.[472]

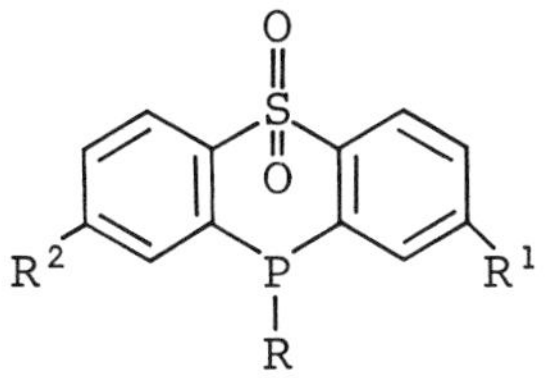

(R = Me; R^1 = R^2 = Me). XI. Isolated as oxide, m. 256°.[472]
(R = $BrCH_2CH_2$; R^1 = R^2 = H). II. Solid, m. 142 to 143°;[146] dimerized on heating to give a cyclic diphosphonium salt.
(R = Ph; R^1 = R^2 = H). II. Solid, m. 155.5 to 156.5°;[146] oxide, m. 246 to 247°.
(R = Ph; R^1 = Cl; R^2 = H). II. Solid, m. 193 to 194°.[146]

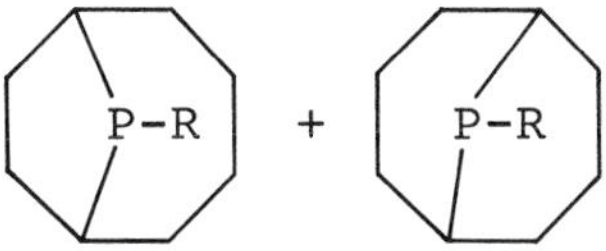

XII. Obtained as a mixture from RPH_2 and 1,4-cyclooctadiene.
(R = n-$C_{20}H_{41}$). Liquid, $b_{0.8}$ 164 to 210°.[1291]
(R = Ph). Liquid, $b_{0.3}$ 134 to 135°.[1291]

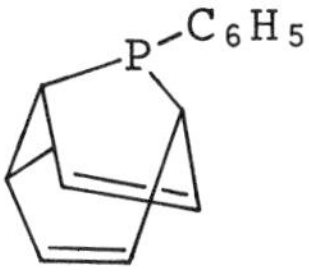

XIV. VIII. By photolysis of 9-phenyl-9-phosphabicyclo[4.2.1]nonatriene in benzene solution through Pyrex.[766] Solid, m. 43 to 45°, ^1H NMR;[766] oxide, m. 101 to 102.5°.

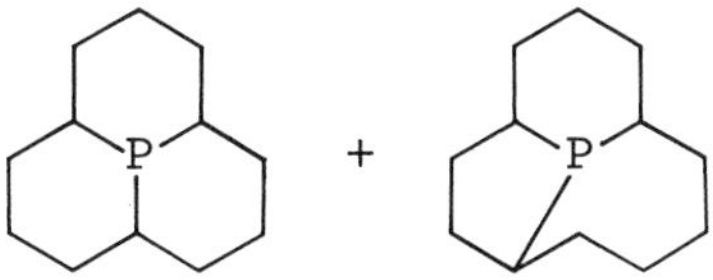

XII. From PH_3 and 1,5,9-cyclododecatriene. Liquid, b_2 98 to 112°.[969]

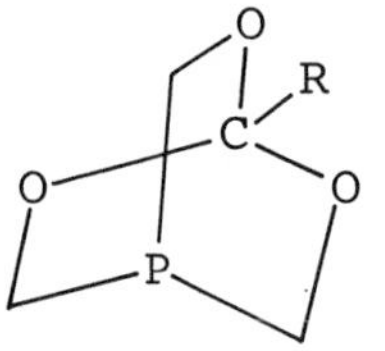

XIV. From R-C(OR')$_3$ and $(HOCH_2)_3P$.

(R = H). Colorless crystals, m. 88 to 89°,[138] ^{1}H NMR.[138,166]

(R = Me). Colorless solid, μ 1.53 D, mass spect., ^{1}H NMR.[166]

(R = Bu). Sweet-smelling liquid, $b_{0.75}$ 63°, ^{1}H NMR, ^{31}P +81 ppm.[772]

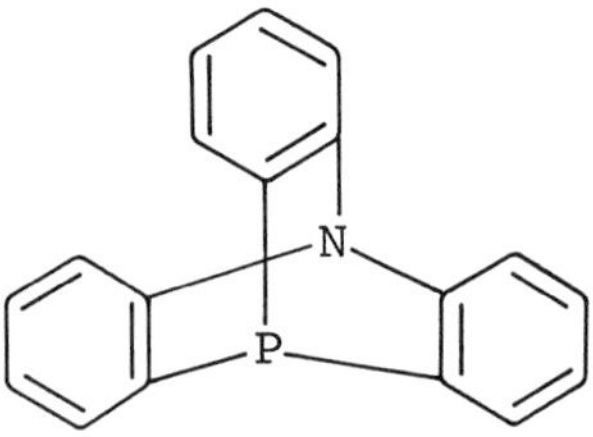

II. Colorless crystals, m. 255 to 255.5°, mass spect.,[552] ^{31}P +80 ppm (in $CHCl_3$), +79 ppm (in THF), UV, C-P-C angle calc. ∿90°;[552] yields a phosphonium salt and gives an oxide, m. ∿300° (dec.).[552]

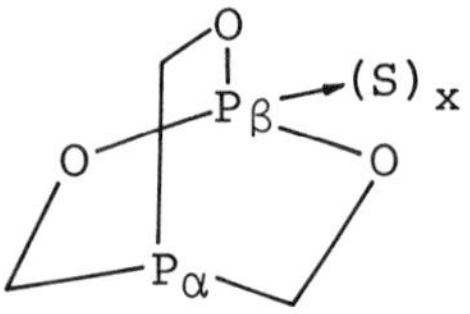

XIV. From $(HOCH_2)_3P$ and $(MeO)_3P$ in THF.

(x = 0). Crystals, m. 75 to 76°, IR,[250] ^{31}P +67 ppm (P_α) and -90.0 ppm (P_β), ^{1}H NMR;[138] magnitude and sign of coupl. const.[887]

(x = 1). From above and sulfur. Crystals, m. 235 to 237°, ^{31}P +70 ppm (P_α) and -51.8 ppm (P_β); disulfide could not be made.[250]

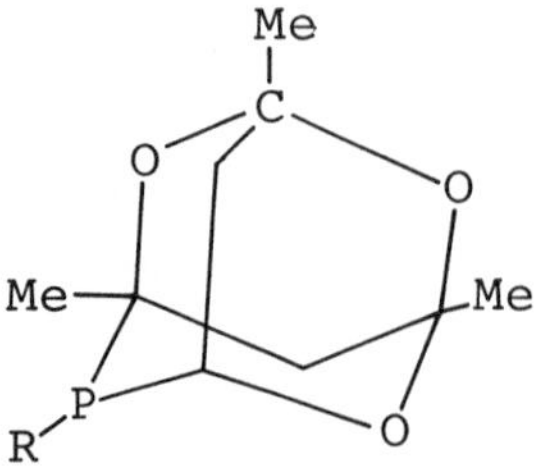

XIII.4. From 2MeCOCH$_2$COMe and RPH_2 in

4 to 6N HCl.[355]
(R = $NCCH_2CH_2$). Only that product isolated in which one O atom of formula is replaced by the $NCCH_2CH_2P<$ group. Crystals, m. 188 to 189° (from i-PrOH).[355]
(R = c-C_6H_{11}). Only that product isolated in which one O atom of formula is replaced by the c-C_6H_{11} group. Crystals, m. 123 to 125° (from CH_3CN).[355]
(R = i-Bu). Crystals, m. 75 to 77°.[355]
(R = n-C_8H_{17}). Crystals, m. 42 to 43°.[355]
(R = Ph). Crystals, m. 105 to 107°.[355] In addition, two other crystalline compounds, m. 204 to 205° and m. 238 to 239°, were isolated in which one or two O atoms of formula are replaced by the PhP< group, respectively.[355]
(R = PhNHCO). Crystals, m. 108 to 109°.[355]

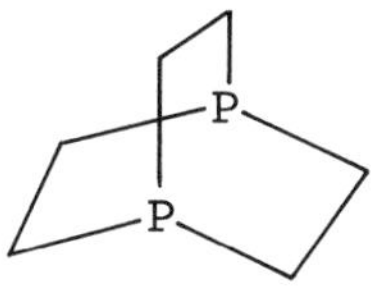

IX.3. Crystals, m. 252° (under N_2 in a sealed tube).[566] Forms crystalline dimethiodide, dimethopicrate, and bis(aurous)chloride deriv.; easily retains traces of solvent; readily sublimable at low pressures; forms a dioxide and disulfide.[566]

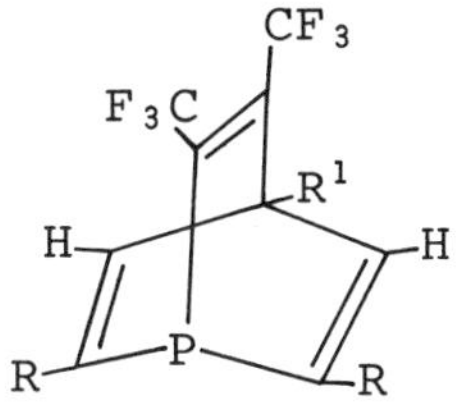

XIV. From substituted phosphorines and $CF_3C{\equiv}CCF_3$.[894]
(R = Me; R^1 = Ph). Crystals, m. 105 to 106°, UV, IR, ^1H NMR.[894]
(R = t-Bu; R^1 = Me). Crystals, m. 92 to 93° (from EtOH/H_2O), UV, IR, ^1H NMR.[894]
(R = R^1 = Ph). Crystals, m. 189°; UV, IR, ^1H NMR, ^{31}P +65 ppm.[894]

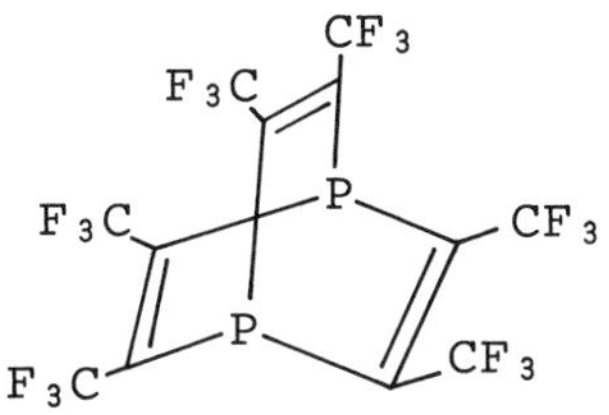

XIV. From red P and $CF_3C{\equiv}CCF_3$ with

catalytic amounts of iodine at 200°/8 hr.[816,817] Crystals, m. 118 to 119° (sublimes), purified by subl.,[816,817] UV, IR, ^{19}F NMR;[816] $L \cdot 2BH_3$ (unstable at room temp.); does not react with O_2, Br_2, MeI, or $PhCH_2Cl$.[816]

Seven-Membered Rings

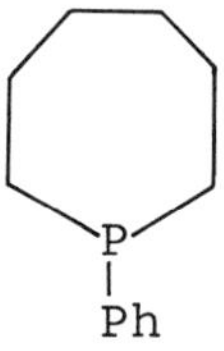

XII. From $Ph(CH_2{=}CH(CH_2)_4)PH$ and UV light.

Liquid, $b_{1.5}$ 106 to 120°, n_D^{20} 1.5730, IR; sulfide, m. 90.5 to 91.5°.[282]

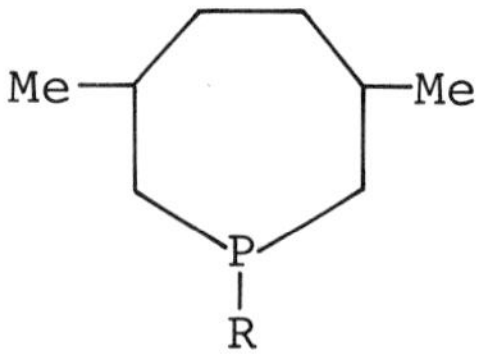

XII. From RPH_2, bimethyllyl, and Bz_2O_2.

(R = n-$C_{10}H_{21}$). Crystals, $b_{0.3-0.5}$ 226 to 236°, m. 47 to 48°.[1291]
(R = n-$C_{18}H_{37}$). Liquid, $b_{0.9}$ 223 to 225°.[1291]
(R = Ph). Liquid, $b_{0.2}$ 88 to 92°.[1291]

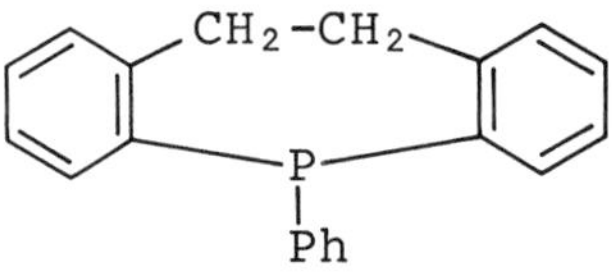

II. Crystals, m. 75 to 75.5° and 94.5 to 95°, $b_{0.01}$ 170 to 190°,[951] UV; this phosphine shows isodimorphism and exists in a low- and a high-melting form; the high-melting form is obtained on heating the low-melting phosphine to 80°; it also precipitates from a boiling aqueous ethanolic solution of the low-melting phosphine hydroiodide on cooling;[951] both forms give the same crystalline methiodide, m. 251 to 252°, and $L_2PdCl_2 \cdot$dioxan, m. 278 to 280°.[951]

Eight-Membered Rings

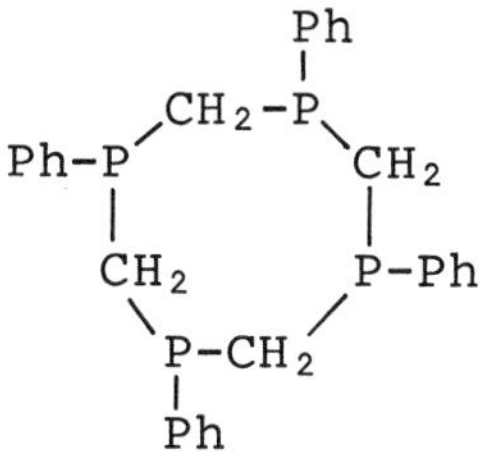

XIV. From $PhPH_2$ and $PhP(CH_2NEt_2)_2$ by heating. Crystals, m. 125 to 127° (from EtOH), IR,[918] stable at reflux temp. ∿300° under vacuum.

Nine-Membered Rings

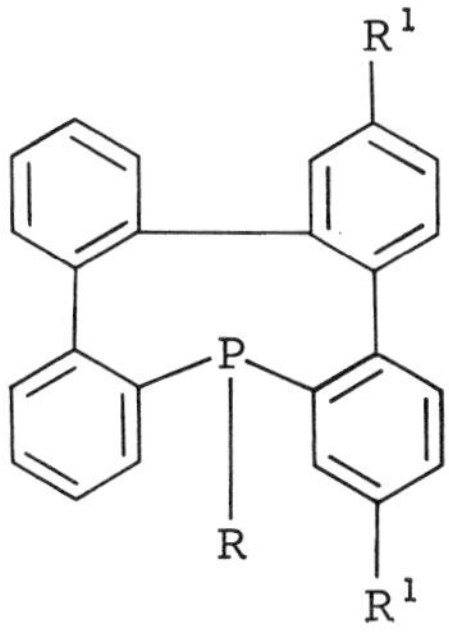

XIV. From thermal dec. of phosphoranes.

($R = Me$; $R^1 = H$). Crystals, m. 130 to 131° (from EtOH), ^1H NMR; L•MeI, m. 370° (dec.).[548]
($R = 2\text{-}PhC_6H_4$; $R^1 = H$). Crystals, m. 193 to 195° (from EtOH).[548]
($R = Ph$; $R^1 = H$). Colorless crystals, m. 131 to 132° (from EtOH);[1444] L•MeI, m. 322 to 324°.[1444]
($R = Ph$; $R^1 = Me_2N$). Nearly colorless crystals, m. 207 to 208° (from EtOH).[1444]
($R = 4\text{-}Me_2NC_6H_4$; $R^1 = H$). Crystals, m. 235 to 236° (from MeOH); L•MeI, m. 296 to 297°.[1444]

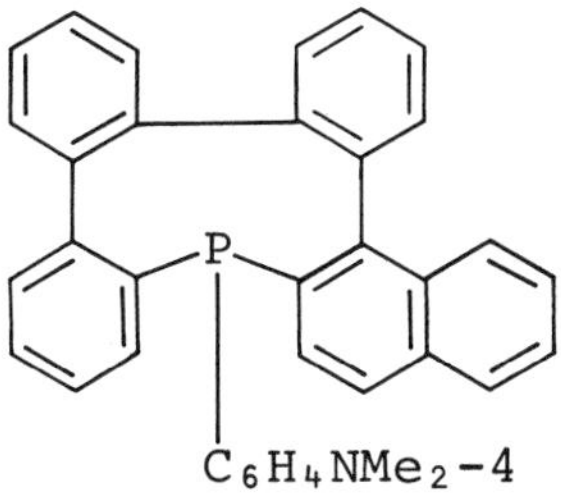

XIV. From thermal dec. of phosphorane. Crystals, m. 168 to 170° (from EtOH).[1444]

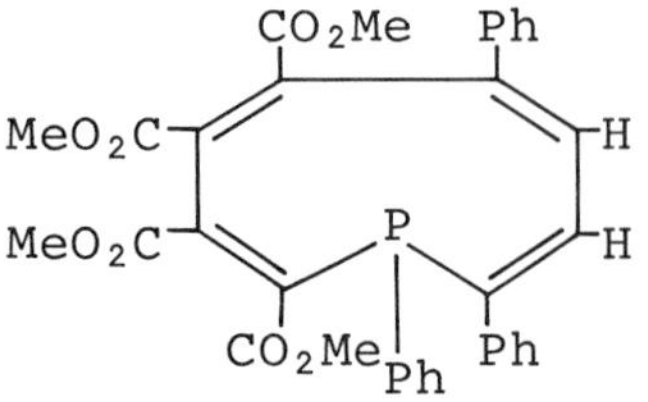

XIV. From Diels-Alder adduct of triphenylphosphole and dimethyl acetylenedicarboxylate by thermal dec.[1404] Crystals, m. 218 to 220°,[1404] (m. 220°,[644] proposed different structure); sublimes at 150°/1 torr; decomposes at 245° in diglyme;[1404] mass spect., ^{1}H NMR, ^{31}P +35 ppm;[1404] cis,cis,trans,cis-phosph(III)onin structure proposed;[1404] oxide, m. 252 to 254°,[1404] m. 256 to 259°,[644] (different structure proposed, see 5-membered rings).

Di- to Hexatertiary Phosphines

TYPE: $RR^1P(CH_2)PR^1R$

$(Et_2NCH_2)_2PCH_2P(CH_2NEt_2)_2$. XIII.1. Liquid, $b_{0.001}$ 139 to 141°, n_D^{20} 1.5006,[921] IR, ^{1}H NMR, ^{31}P +56.5 ppm.[921]

$(Et_2NCH_2)_2PCH_2PPh_2$. XIV. From $(Et_2NCH_2)_3P$ and $HPPh_2$. Liquid, $b_{0.01}$ 142 to 143°, IR;[918] dioxide, m. 98 to 100°;[918] useful as flameproofing agent.[920]

$Et_2NCH_2PhPCH_2PPh_2$. XIV. From $PhP(CH_2NEt_2)_2$ and $HPPh_2$. Liquid, $b_{0.05}$ 180 to 200° (partial dec.), IR; dioxide, m. 183 to 184.5°;[918] useful as polymerization catalyst and flameproofing agent.[920]

$Ph_2PCH_2PPh_2$. II.[1099] From Ph_2PCH_2Li + $ClPPh_2$. VII.[10,561,692] VIII.[650] XIV.[918,933] From $Ph_2PCH_2NEt_2$ + $HPPh_2$.[918] Crystals, m. 122° (from EtOH),[692] m. 120.5 to 121.5° (from PrOH),[561] m. 120 to 121°,[10] m. 117.5 to 119°,[1099] IR,[918] ^{31}P +23.6 ppm (in acetone),[918] +23.6 ppm,[1008] mass spect.;[243] is cleaved by PhN_3 to give Ph_2MePO and $Ph_2P(O)NHPh$,[10] but only if traces of acid are present,[4] otherwise no cleavage occurs.[4,435]

$Me_2PCH_2CH_2PMe_2$. VII.[224] VIII.[1077] X.[182,185] Liquid, b. 188.1°, m. -1 to 0°,[182,185] b_{60} 104°,[1077] b_{26} 81 to 82°,[244] ^{31}P +49.4 ppm.[1358a]

$Me_2PCF_2CHFPMe_2$. X. Also from Me_2PH + $CF_2{=}CHF$, by-product. Liquid, b_{760} 190 to 191°, ^{1}H and ^{19}F NMR, mass spect.[385b]

$MePhPCH_2CH_2PPhMe$. IX.2.[702] IX.5.[611] Liquid, b. 250 to 260°; L·2MeI, m. 234 to 235°;[702] by electrolytic cleavage of the salts obtained as meso form, m. 90°, $b_{0.2}$ ∿130°, racemate, an oil[611] and optically active (+) $[\alpha]_D^{20}$ +27.87°;[107,611] on dist., $b_{0.3}$ 134 to 136°, the optically active (+) becomes inactive meso.[611]

$MePhP_{\alpha}CH_2CH_2P_{\beta}Ph_2$. XII. ^{31}P +31.7 ppm (P_{α}) and 13.2 ppm (P_{β}).[484a]

$(CF_3)_2PCH_2CH_2P(CF_3)_2$. X. From $[(CF_3)_2P]_2$ and $CH_2{=}CH_2$. Liquid, m. -49.7 to -49.6°,[185] m. -51.4 to -51.3°,[185] IR.[196]

$(Et_2NCH_2)_2PCH_2CH_2P(CH_2NEt_2)_2$. XIII.1. Liquid, $b_{0.005-0.001}$ 152 to 155°, n_D^{20} 1.5012, IR, ^1H NMR, ^{31}P +48.7 ppm.[921]

$Et_2PCH_2CH_2PEt_2$. VII.[220,221,566,658,1451] IV. IX.2.[676] XIV. From $Et_2PC{\equiv}CPEt_2 + H_2$.[220,221] Liquid, b_{760} 220 to 230°,[658] b_{760} 250 to 255°,[676] b_{10} 124 to 126°,[561] $b_{0.001}$ 56°,[220] n_D^{20} 1.5100,[220] ^{31}P +19.3 ppm;[979] L·CS_2, m. 102 to 105°;[676] L·2HBr, m. 85° (dec.);[676] L·2HI, m. 181 to 183°; L·EtI, m. 100°.[658]

$EtPhPCH_2CH_2PPhEt$. VII. Crystals, m. 69 to 70° (from EtOH), $b_{0.4}$ 148 to 165°.[569]

$EtPhP_{\alpha}CH_2CH_2P_{\beta}Ph_2$. XII. ^{31}P +17.0 ppm (P_{α}) and +13.7 ppm (P_{β}).[484a]

$(H_2NCH_2CH_2)PhPCH_2CH_2PPh(CH_2CH_2NH_2)$. VII. Viscous oil, decomposes on dist.[694]

$(NCCH_2CH_2)_2PCH_2CH_2P(CH_2CH_2CN)_2$. IX.2. Crystals, m. 101 to 102°,[470] ^{30}P +21.4 ppm.[979]

$PrPhP_{\alpha}CH_2CH_2P_{\beta}Ph_2$. XII. ^{31}P +21.7 ppm (P_{α}) and +13.2 ppm (P_{β}).[484a]

i-$PrPhP_{\alpha}CH_2CH_2P_{\beta}Ph_2$. XII. ^{31}P +4.1 ppm (P_{α}) and +13.3 ppm (P_{β}).[484a]

i-$BuPhP_{\alpha}CH_2CH_2P_{\beta}Ph_2$. XII. ^{31}P +25.4 ppm (P_{α}) and +13.5 ppm (P_{β}).[484a]

2-$BuPhP_{\alpha}CH_2CH_2P_{\beta}Ph_2$. XII. ^{31}P +7.1 ppm (P_{α}) and +12.9 ppm (P_{β}).[484a]

$AmPhP_{\alpha}CH_2CH_2P_{\beta}Ph_2$. XII. Viscous oil. ^{31}P +21.2 ppm (P_{α}) and +13.7 ppm (P_{β}).[484a]

$(c\text{-}C_6H_{11})_2PCH_2CH_2P(C_6H_{11}\text{-}c)_2$. VII.[658] IV. IX.2.[676] Crystals, m. 96 to 97° (from THF/EtOH);[658,676] CS_2 adduct, m. 99 to 100°.[676]

$(PhCH_2)_2PCH_2CH_2P(CH_2Ph)_2$. IX.3. Liquid, $b_{0.01}$ 180°.[566]

$(PhCH_2)PhPCH_2CH_2PPh(CH_2Ph)$. IX.3. Crystals, m. 85 to 98°, $b_{0.0005}$ 200°.[566]

$Ph_2PCH_2CH_2PPh_2$. VII.[10,76,542,561,658,692,954,1244,1247,1344] IX.2.[676,1244] XII. From $Ph_2PCH{=}CH_2$ + $HPPh_2$[484a,777] or from Ph_2PLi + $HC{\equiv}CH$.[7] XIV. By hydrogenation of $Ph_2PCH{=}CHPPh_2$.[12] Pyrolysis of $Ph_2PCH_2CH_2O_2CPh$.[1364] Crystals, m. 143 to 144° (from benzene),[220,561,1244,1344] m. 143 to 145° (from EtOH),[954] m. between 139 to 143°,[10,777,1247,1364] m. 137 to 139°,[76] m. 161 to 163° (seems to be incorrect),[676] IR,[1244] mass spect.,[243] ^1H NMR;[10,206] displayed cytotoxic activity;[1344] TLC,[451] ^{31}P +13.2 ppm.[484a]

$MePhPCH_2CH_2CH_2PPhMe$. IV. IX.2. Liquid, b_{4-5} 199 to 201°.[676]

$(Et_2NCH_2)_2PCH_2CH_2CH_2P(CH_2NEt_2)_2$. XIII.1. Liquid, $b_{0.0005}$ 165 to 168°, n_D^{20} 1.5018, IR, ^1H NMR, ^{31}P +52.9 ppm.[921]

$Et_2PCH{\cdot}CNCH_2PEt_2$. XII. Liquid, b_3 110°; L·2MeI, m. 213°.[496]

$EtPhPCH_2CH_2CH_2PPhEt$. VII. Liquid, b_3 196 to 200°.[672]

$(H_2NCH_2CH_2)PhPCH_2CH_2CH_2PPh(CH_2CH_2NH_2)$. VII. Viscous, colorless oil, decomposes on dist., pK_1 8.06, pK_2 8.93.[694]

$(NCCH_2CH_2)_2PCH_2CH_2CH_2P(CH_2CH_2CN)_2$. IV. IX.2. Crystals, m. 53 to 54°.[470]

$(c\text{-}C_6H_{11})_2PCH_2CH_2CH_2P(C_6H_{11}\text{-}c)_2$. VII. Oil.[692]

$(n\text{-}C_8H_{17})_2PCH_2CH_2CH_2P(C_8H_{17}\text{-}n)_2$. XII. Liquid, $b_{0.001}$ 175 to 180°, n_D^{20} 1.4837,[921] IR, ^{31}P +33.7 ppm; dioxide, m. 145 to 150°.[921]

$Ph_2PCH_2CH_2CH_2PPh_2$. VII. Crystals, m. 61 to 62°,[1455] useful as catalyst in the manufacture of 1,4-hexadiene from ethylene and butadiene.[1455]

$Ph_2PCH_2CH{\cdot}OHCH_2PPh_2$. VII. Colorless oil, IR; L•2MeI, dec. 135 to 137°.[698]

$Ph_2PCMe_2PPh_2$. VII. Crystals, m. 123 to 123.5° (from PrOH).[561]

$MePhP(CH_2)_4PPhMe$. IV. IX.2.[676] IX.5.[623,624,635] Solid, m. 47 to 48°,[623,624] m. 40°,[676] b_{4-5} 218 to 220°,[676] b_1 178 to 180°.[623,624,635]

$(Et_2NCH_2)_2P(CH_2)_4P(CH_2NEt_2)_2$. XIII.1. Liquid, $b_{0.001}$ 190 to 193°, n_D^{20} 1.4990, IR,[921] ^1H NMR, ^{31}P +53.2 ppm.[921]

$(PhNHCS)PhP(CH_2)_4PPh(SCNHPh)$. XIII.5. Crystals, m. 154° (from benzene).[665]

$(CF_3)_2PCH{\cdot}CF_3CH{\cdot}CF_3P(CF_3)_2$. XII. Liquid; condensing at -36° under vacuum; IR, ^{19}F NMR.[264]

$EtPhP(CH_2)_4PPhEt$. VII.[672] IV. IX.2.[676] Liquid, b_4 225 to 230°.[672,676]

$(H_2NCH_2CH_2)PhP(CH_2)_4PPh(CH_2CH_2NH_2)$. VII. Viscous, colorless oil, decomposes on dist.; pK_1 8.06, pK_2 8.93;[694] L•2HCl (quaternizes nitrogen), dec. 285°, ^{31}P +29 ppm (of salt).[694]

$Bu_2P(CH_2)_4PBu_2$. XII. Liquid, $b_{0.07}$ 144°, n_D^{25} 1.4895.[1171]

$(c\text{-}C_6H_{11})_2P(CH_2)_4P(C_6H_{11}\text{-}c)_2$. VII. Crystals, m. 98 to 100° (from dioxan).[692]

$(c\text{-}C_6H_{11})PhP(CH_2)_4PPh(C_6H_{11}\text{-}c)$. VII. Crystals, m. 104 to 105°.[672]

$(n\text{-}C_{12}H_{25})_2P(CH_2)_4P(C_{12}H_{25}\text{-}n)_2$. XII. Liquid, $b_{0.01}$ 190 to 210°; dioxide, m. 91 to 91.8°.[921]

$Ph_2P(CH_2)_4PPh_2$. VII.[121,1204,1455] IX.3.[458] Crystals, m. 135 to 136° (from EtOH),[458] m. 137 to 139°,[1204] m. 134.5 to 135°,[121] m. 132 to 133.5°,[1455] useful in the manufacture of 1,4-hexadiene from ethylene and butadiene.[1455]

$MePhP(CH_2)_5PhPMe$. IV. IX.2. Liquid, b_{6-7} 257 to 258°.[676]

$Et_2P(CH_2)_5PEt_2$. VII.[692] IV. IX.2.[676] Liquid, b. 295 to 315°, b_3 136 to 139°,[692] b_4 143°;[676] L•2HBr, dec. 222 to 224°.[676]

$EtPhP(CH_2)_5PPhEt$. VII. Liquid, b_3 211 to 213°.[672]

$(c\text{-}C_6H_{11})_2P(CH_2)_5P(C_6H_{11}\text{-}c)_2$. VII.[692] IV. IX.2.[676] Crystals, m. 62 to 63° (from EtOH),[692] m. 63°.[676]

$(c\text{-}C_6H_{11})PhP(CH_2)_5PPh(C_6H_{11}\text{-}c)$. VII. Colorless oil.[672]

$Ph_2P(CH_2)_5PPh_2$. VII. Clear, yellow oil.[1204]

$MePhP(CH_2)_6PPhMe$. IV. IX.2. Solid, m. 45°, b_{2-3} 205°.[676]
$EtPhP(CH_2)_6PPhEt$. VII. Solid, m. 40 to 41°, $b_{0.1}$ 193 to 196°.[672]
$(c\text{-}C_6H_{11})PhP(CH_2)_6PPh(C_6H_{11}\text{-}c)$. VII. Colorless oil.[672]
$Bu(n\text{-}C_{12}H_{25})P(CH_2)_8P(C_{12}H_{25}\text{-}n)Bu$. VII. Liquid, $b_{0.001}$ 230°, n_D^{20} 1.4843, d_4^{20} 0.8670.[1474]
$Bu(n\text{-}C_6H_{13})P(CH_2)_{19}P(C_6H_{13}\text{-}n)Bu$. VII. Liquid, n_D^{20} 1.4852, d_4^{20} 0.8678.[1474]
$(n\text{-}C_6H_{13})PhP(CH_2)_{19}PPh(C_6H_{13}\text{-}n)$. VII. Liquid, $b_{0.001}$ 250°,[1354] n_D^{20} 1.5225, d_4^{20} 0.9270.[1474]
$(n\text{-}C_6H_{13})(4\text{-}MeC_6H_4)P(CH_2)_{19}P(C_6H_4Me\text{-}4)(C_6H_{13}\text{-}n)$. VII. Liquid, $b_{0.001}$ 230°,[1354] n_D^{20} 1.5181, d_4^{20} 0.9130,[1474] UV,[1236] useful as lubricant and additive.[1474]
$Ph_2P(CH_2)_{19}PPh_2$. VII. Liquid, n_D^{20} 1.5794, d_4^{20} 0.9748.[1474]
$(4\text{-}MeC_6H_4)_2P(CH_2)_{19}P(C_6H_4Me\text{-}4)_2$. VII. Liquid, n_D^{20} 1.5181, d_4^{20} 0.9130.[1474]

Diphosphines Containing Nitrogen, Oxygen, or Sulfur in the Bridge

$Ph_2PCH_2CH_2N{\cdot}EtCH_2CH_2PPh_2$. VII. White solid, m. 43 to 45°.[325]
$(c\text{-}C_6H_{11})_2PC{\cdot}Ph{=}N{-}N{=}C{\cdot}PhP(C_6H_{11}\text{-}c)_2$. VII. Crystals, m. 132 to 137°.[651]
$Ph_2PC{\cdot}Ph{=}N{-}N{=}C{\cdot}PhPPh_2$. VII. Red needles, m. 162 to 164° (from dioxan).[651]
$(c\text{-}C_6H_{11})_2PCSNH(CH_2)_6NHCSP(C_6H_{11}\text{-}c)_2$. XIII.5. Yellow crystals, m. 144° (from EtOH), IR.[666]
$(c\text{-}C_6H_{11})_2PCSNH(4,4'\text{-}C_6H_4C_6H_4)NHCSP(C_6H_{11}\text{-}c)_2$. XIII.5. Yellow crystals, m. 160.5° (from benzene), IR.[666]
$Ph_2PCH_2OCH_2PPh_2$. VII. Crystals, m. 86 to 88° (from MeOH), IR, ^1H NMR;[20] $L{\cdot}2PhCH_2Br$, m. 286 to 288°.[20]
$Ph_2PCH_2CH_2OCH_2CH_2PPh_2$. VII. Clear, yellow oil.[1204]

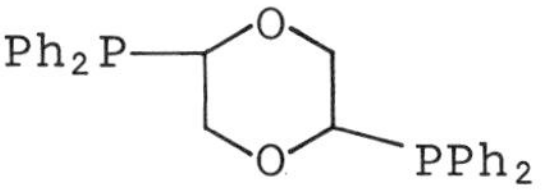

VII. From $KPPh_2$ and $ClCH_2\overset{O}{CH{-}CH_2}$.

Viscous oil, b_4 210 to 220°.[698]
$Ph_2PCH_2CH_2SCH_2CH_2PPh_2$. VII. Solid, m. 63°.[301]
$Ph_2PCH_2CH_2SCH_2CH_2SCH_2CH_2PPh_2$. VII. Solid, m. 121°.[301]
$Ph_2P(CH_2)_3S(CH_2)_3S(CH_2)_3PPh_2$. VII. From Ph_2PLi and $\overline{CH_2CH_2CH_2S}$, then $Br(CH_2)_3Br$. Oil;[335] sulfide also an oil.

Bis(phosphino)alkenes

$(CF_3)_2PCH{=}CHP(CF_3)_2$. X. Liquid, m. −54°.[185]
$Et_2PCH{=}CHPEt_2$. XIV. From $Et_2PC{\equiv}CPEt_2$ and H_2. Liquid, $b_{0.03}$ 67°.[221]

trans-$Bu_2PCH=CHPBu_2$. VII. Liquid, $b_{0.05}$ 123 to 125°, UV, IR, ^1H NMR;[1421] L·2MeI, m. 130°.

cis-$Ph_2PCH=CHPPh_2$. VII. Crystals, m. 123 to 125°,[1006a] m. 125 to 126°, IR, ^1H NMR,[12] μ 1.96 D;[25] could not be hydrogenated (complex formation with catalyst); dioxide, m. 244 to 245°;[12] yields with HBr in AcOH $Ph_2PCH_2CH_2$-$P(O)Ph_2$.[13]

trans-$Ph_2PCH=CHPPh_2$. VII.[12] XII.[777] From $Ph_2PC{\equiv}CH$ + $HPPh_2$, base-catalyzed, gives trans addition. Crystals, m. 125 to 126°, IR, ^1H NMR,[12] μ 0.99 D;[25] could not be hydrogenated; dioxide, m. 310 to 311°.[12]

$Ph_2PC(CF_3)=C(CF_3)PPh_2$. X. Liquid, $b_{0.001}$ 138 to 141°, IR, ^{19}F NMR.[264]

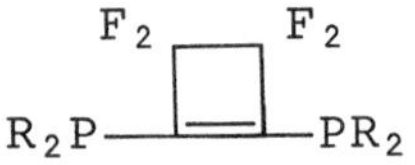

X. (R = Me). Solid, m. 137 to 139°, IR, ^1H and ^{19}F NMR.[265]
IV. (R = Ph). Solid, m. 129.5 to 130.5°,[265] m. 127 to 127.5°,[1340] IR,[265,1340] ^1H and ^{19}F NMR,[265] GLC, ^{31}P +22.7 ppm.[1340]

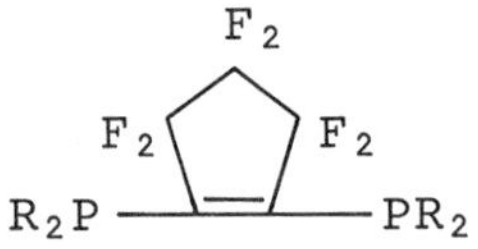

IV. (R = c-C_6H_{11}). Yellow solid, m. 194 to 194.5° (from $Me_2C=O$), IR, ^1H and ^{19}F NMR, ^{31}P +17.0 ppm.[1339]
IV. (R = Ph). White crystals, m. 97 to 99°, ^{19}F NMR.[269]

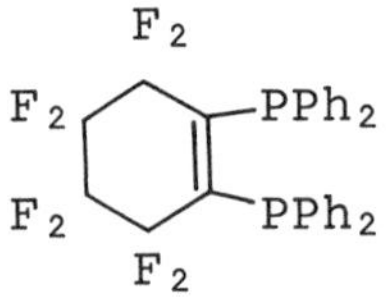

IV. Orange crystals, m. 135° (from hexane), IR, ^{19}F NMR,[267] stable in air and light.

$Ph_2PCH=CH$-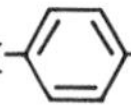-$CH=CHPPh_2$. XIV. $Ph_2PCHNaP(O)Ph_2$ and $OCHC_6$-H_4CHO. Crystals, m. 221 to 223° (sealed tube).[439]

trans-2,2'-$(2\text{-}MeC_6H_4)_2PC_6H_4CH=CHC_6H_4P(C_6H_4Me\text{-}2)_2$. From $(2\text{-}MeC_6H_4)_3P$ by oxidative coupling with $RhCl_3$ and liberation of the diphosphine from the complex with KCN. White crystalline solid, m. 263°.[103]

Bis(phosphino)alkynes

$Et_2PC{\equiv}CPEt_2$. I. Liquid, $b_{0.07}$ 54°, n_D^{20} 1.5332.[220,221]

$Bu_2PC{\equiv}CPBu$. I. II. XIV. Crystals, m. 33° (from EtOH),[1389] m. 36°,[273] $b_{0.02}$ 121 to 125°,[1389] b_3 162°.[273]

$Ph_2PC{\equiv}CPPh_2$. I.[218,525] VI.[137,524] Crystals, m. 86° (from EtOH),[218,524,525] m. 84 to 84.5° (from MeOH);[137] x-ray analysis[80] gives pyramidal configuration about both P atoms; monoclinic crystals, space group $P2_1/n$; benzene rings have asymmetric orientation; P-C(Ph) 1.832, P-C≡ 1.765 Å; <(Ph)C-P-C≡ 101.1°;[80] π interaction of phosphorus atoms with C≡C bond postulated.[80]

$Ph_2PCH_2C{\equiv}CCH_2PPh_2$. VII. Crystals, m. 79 to 80° (from benzene/EtOH), IR, mass spect.[777a]

$Ph_2PC{\equiv}C{-}C{\equiv}CPPh_2$. I. Crystals, m. 105° ($Et_2O$).[218]

$Ph_2PC{\equiv}C$-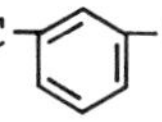-$C{\equiv}CPPh_2$. I. Crystals, m. 199° (from cyclohexane), IR.[218]

$Ph_2PC{\equiv}C$--$C{\equiv}CPPh_2$. I. Crystals, m. 204° (from benzene), IR.[218]

Phenylenediphosphines

TYPE: o-Phenylenediphosphines

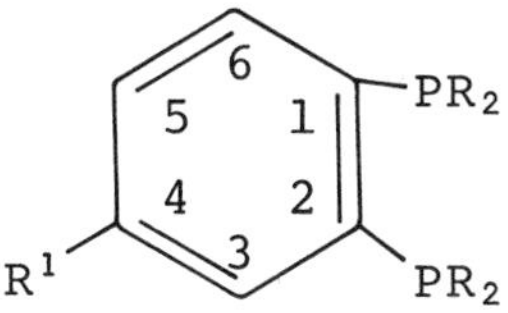

I. II. All these compounds were prepared by $XC_6H_4MgX + R_2PCl \rightarrow XC_6H_4PR_2 \xrightarrow[+\ R_2PX]{BuLi} C_6H_4(PR_2)_2$.

(R = Me, R^1 = H). Liquid, $b_{0.3}$ 98 to 101°,[518] $b_{0.4-0.5}$ 92°.[232]

(R = Et; R^1 = H). Liquid, $b_{0.2}$ 100 to 103°,[221,518] $b_{0.09}$ 110 to 111°.[223]

(R = Et; R^1 = Me). Liquid, $b_{0.25}$ 107 to 115°,[519,521] $b_{0.6}$ 115 to 120°,[294] $L{\cdot}CH_2CH_2{\cdot}$dipicrate, m. 178.5 to 179°.[294]

(R = Et; R^1 = Ph), Liquid, $b_{0.1}$ 133 to 153° (impure); $L_2{\cdot}AuI$, m. 313 to 315°.[294]

(R = Ph; R^1 = H). Crystals, m. 186.5 to 187.5°.[218,518]

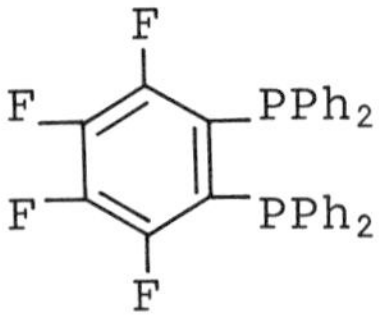

II. Light-yellow crystals, m. 108 to 109° (from EtOH).[349a]

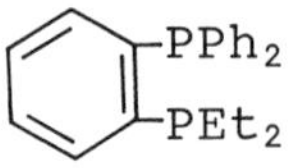

I + II. Crystals, m. 103.5 to 105.5°.[518]

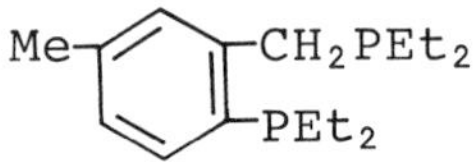

I + II. Liquid, $b_{1.2}$ 100 to 104°; $L \cdot CH_2Br_2$, m. 308 to 309°.[294]

TYPE: m-Phenylenediphosphines

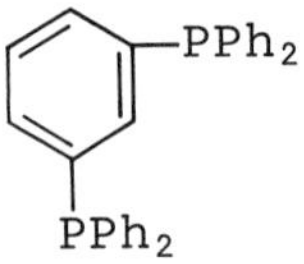

II.[72] VII.[1472] Oil,[72] IR,[1472] UV;[1238,1245] $L \cdot 2MeI$, m. 199 to 200°,[1472] m. 282 to 284°.[72]

TYPE: p-Phenylenediphosphines

(R = R^1 = Me). II. Liquid, $b_{0.01}$ 66 to 70°, n_D^{24} 1.5890, IR;[77] $L \cdot 2BBr_3$, m. 305 to 315°.[77]
(R = R^1 = Et). II. Liquid, b_9 172 to 174°, n_D^{25} 1.5666,[233,1157] pK_a 3.35 and 6.57,[134] IR.[1157]
(R = n-C_6H_{13}; R^1 = 4-MeC_6H_4). II. Pale-yellow, viscous liquid, flow point -29°, n_D^{20} 1.5726, d_4^{23} 1.0098,[1471] μ 4.2 D.[1472]
(R = R^1 = Ph). I.[72,76] II.[72,557] (Gives higher yield than I).[72] VII.[1240,1248,1471] White crystals, m. 166 to 167°,[557] m. 162 to 168°,[72] m. 166 to 168.5°,[76] m. 170 to 171° (from PrOH),[1248,1471,1473] m. 171 to 172°,[1240] UV,[1236,1238,1245,1248,1473] IR,[1472,1473] μ 2.4 D,[1472] 2.54 D,[1248,1473] dec. 367°;[1007] rearranges with excess PCl_3 at 300° to give $Cl_2PC_6H_4PCl_2$.[77]

Other Diphosphines

VII. Isolated as the dioxide, m.

283°.[26]

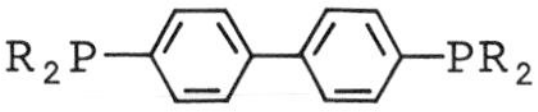 (R = Me). VII. Radical anion.[253]

(R = Ph). I.[72] II.[72,76,121] VII.[1248] Crystals, m. 192.5 to 194° (from THF/MeOH),[76] m. 194 to 196°,[1248] m. 191 to 193°,[72,121] UV,[1248] μ 2.06 D;[1248] method II with THF gives highest yield.[72]

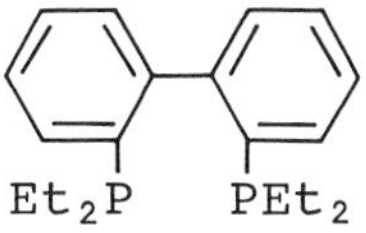 II. Crystals, m. 28 to 30°, $b_{0.25}$ 152°,[42,43] ^1H NMR;[44] L·2MeI, m. 255 to 256°; forms with $Br(CH_2)_nBr$ cyclic diphosphonium salts.[42,43]

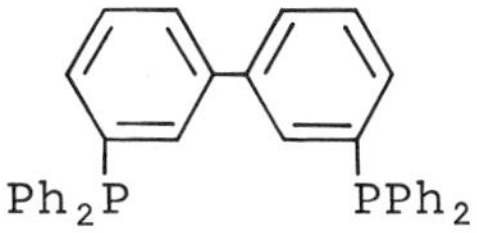 II. Crystals, m. 125 to 127°.[230]

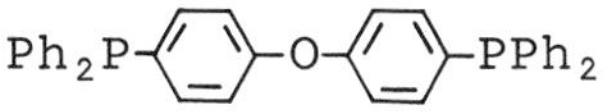 II. Crystals, m. 115 to 116° (from ethyl acetate); dioxide, m. 191 to 193°.[72]

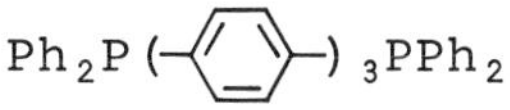 VII. Crystals, m. 136 to 137°, UV, μ 2.25 D.[1248]

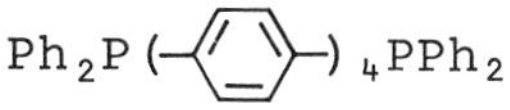 VII. Crystals, m. 184 to 186°, UV, μ 1.77 D.[1248]

Ph_2P(4,4'-o-terphenylene)PPh_2. VII. Crystals, m. 178 to 179°, UV, μ 2.14 D.[1248]

Ph_2P(4,4'-m-terphenylene)PPh_2. VII. Crystals, m. 73 to 74°, UV, μ 2.82 D.[1248]

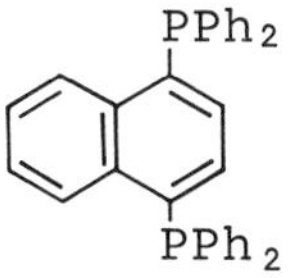 VII. Solid, m. 214 to 215°,[699] IR,[1472] UV.[1245]

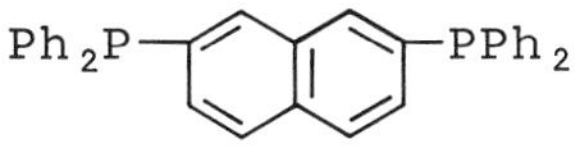 VII. Crystals, m. 141°,[1469,1472] IR,[1472] UV.[1245]

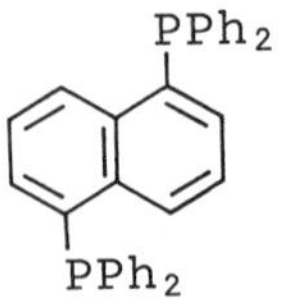

VII. Crystals, m. 261° (from benzene/ligroin),[1469,1472] IR,[1472] UV.[1245]

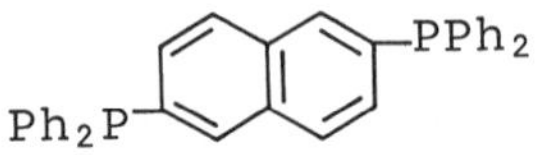

VII. Crystals, m. 221° (from benzene/ligroin),[1472] m. 219 to 220°,[1469] IR,[1472] UV,[1245] dec. 371°.[1007]

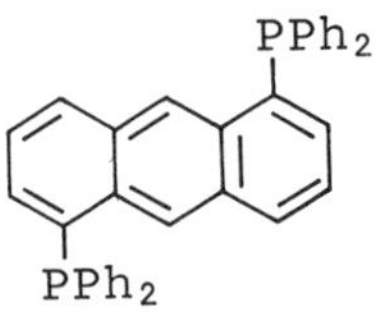

VII. Crystals, m. 326°, IR,[1472] UV.[1245]

Tritertiary Phosphines

$MeC(CH_2PEt_2)_3$. VII. Liquid, $b_{0.1}$ 120°.[561]
$MeC(CH_2PPh_2)_3$. VII. Crystals, m. 100 to 101° (from EtOH or ligroin).[561]
$(Ph_2PCH_2)_2PPh$. XIV. From Ph_2PH and $PhP(CH_2NEt_2)_2$. Crystals, m. 107 to 110° (from EtOH), b_{720} 412°;[918] lost 1.6% weight when refluxed for 2 hr at 412°;[918] polymerization catalyst.[920]
$(Ph_2PCH_2CH_2)_2PPh$. VII.[561] XII.[777] Crystals, m. 131 to 132° (from MeOH),[561] m. 129 to 130°.[777]
$(2\text{-}Et_2PC_6H_4)_2PPh$. II. Crystals, m. 106 to 107.5°.[518]
$(2\text{-}Ph_2PC_6H_4)_2PPh$. I and II. Crystals, m. 226 to 228° (monosolvate with Me_2NCHO),[523] ^{31}P +14.5 ppm (all P are equivalent).[295]
$(4\text{-}Ph_2PC_6H_4)_2PPh$. II. White solid, m. 154 to 158°,[72] trisulfide, m. 295 to 298°.[72]
$(2\text{-}Ph_2PC_6F_4)_2PPh$. II. Yellow needles, m. 173.5 to 174.5° (from CH_2Cl_2/EtOH).[349a]
$(Ph_2PCH_2CH_2)_3N$. VII. Colorless crystals;[1203] compounds of the type $(Et_2NCH_2CH_2)_2PPh$ and $RN(CH_2CH_2PPh_2)_2$[1205] were used as complexing ligands, but properties of the phosphines have not been given.
$N_3C_3(PBu_2)_3$. XIV. (From $N_3C_3(OPh)_3$ + $HPBu_2$). Liquid, $b_{0.01}$ 115°, n_D^{25} 1.5160.[562]
$N_3C_3(PPh_2)_3$. XIV. From $N_3C_3X_3$ + $HPPh_2$ (X = Cl, OPh). Crystals, m. 143 to 144°.[562]

Tetratertiary Phosphines

$C(CH_2PPh_2)_4$. VII. Colorless needles, m. 176 to 178° (from EtOH),[350] IR.

$C(CH_2PPhCO_2Na)_4$. VII. From >PNa + CO_2. Free acid not stable;[351] not able to fix N; the corresponding reaction product of $C(CH_2PPhNa)_4$ and CS_2, for which the following structure was proposed: $C(CH_2P(Ph)=C=S)_4$ (with SNa on P), is able to fix 2 moles of N.[351]

$Ph_2PCH_2CH_2P{\cdot}Ph{\cdot}CH_2CH_2P{\cdot}Ph{\cdot}CH_2CH_2PPh_2$. XII. From $Ph_2PCH{=}CH_2$ and $HP{\cdot}Ph{\cdot}CH_2CH_2PPhH$. Solid, m. 155 to 158°.[777]

Hexatertiary Phosphines

$(Ph_2PCH_2CH_2)_2PCH_2CH_2P(CH_2CH_2PPh_2)_2$. XII. Solid, m. 138 to 140°.[777]

Polymeric Phosphines

PH_3 gave with CH_2O and urea a polymeric material which was self-extinguishing.[238]

PH_3 gave with epoxides in the presence of $BF_3{\cdot}Et_2O$ a polymer of type $P[(CH_2CHXO)_nH]_3$ ($n > 1$; X = H, Me, $ClCH_2$, Ph), which was said to be useful as a flame retardant in polyurethanes.[237]

i-$BuPH_2$ gave with 1,3-$(OCN)_2C_6H_4$ a polymer of m. 265 to 270°.[174,175]

n-$C_8H_{17}PH_2$ gave with 1,3-$(OCN)_2C_6H_4$ a polymer of m. 160 to 165°, η 0.10 (30°).[174,175]

$PhPH_2$ gave with $PhP(CH_2NEt_2)_2$ a polymer, when heated, of composition $[-PhPCH_2-]_x$, m. 137 to 139°, $b_{0.00001}$ 300°;[918] it lost 0.47% weight when refluxed at 300°/10^{-5} torr for 2 hr.[918]

$PhPH_2$ gave with $PhCH{=}CHCOCH{=}CHPh$ a polymer of m. 156 to 159°,[1136] degree of polymerization 2 to 3; when the polymer is sublimed, a phosphorinanone is obtained[1424] (monomeric ring compound).

$PhPH_2$ gave with $PhCH{=}NC_6H_4C_6H_4N{=}CHPh$ a polymer of m. 245 to 250°,[1136] degree of polymerization 2 to 3.

$PhPH_2$ gave with 1,4-diacryloylpiperazine a polyphosphine amide, $[-CH_2CH_2CON\langle\rangle NCOCH_2CH_2PPh-]_x$ of η 0.25 dl/g (30°/$CHCl_3$).[381]

$PhPH_2$ gave with diisocyanates polymers of the type $H(PPhCONHRNHCO)_{n-1}PPhCONHRNCO$;[1137] [R = $-(CH_2)_6-$], viscous mass; (R = 4-Me-3-phenylene), m. 240 to 243°; [R = $-(CH_2)_2-$], m. 88 to 91°; [R = $-(CH_2CH_2O)_6-$], m. 62 to 65°; which were said to be fire-resistant.[1137]

$PhPH_2$ and other primary phosphines add to dienes,[498] dialkylsilanes, and divinylsilane,[497,794,1048] in the presence of radical initiators to give polymeric products.

$PhPH_2$ gave with $CH_2{=}CHSiMe_2OSiMe_2CH{=}CH_2$ an oligomeric product of composition $PhHP[CH_2CH_2SiMe_2OSiMe_2CH_2CH_2$-$PPh]_3H$, a viscous mass, in addition to the monomeric adduct.[800]

$PhPH_2$ and other primary phosphines add to diallyl,[423] p-diallylbenzene,[1333] and diallyl sulfide,[1044] in the presence of radical initiators, to give self-extinguishing polymeric materials.

$PhPH_2$ adds to propargyl alcohol with radical initiation to give a slightly yellow, brittle, hard polymer.[217]

$Ph(CH_2{=}CHCH_2)PH$ gave on irradiation with UV light a polymeric phosphine.[424]

$Bu(CH_2{=}C{\cdot}PhCH_2)PH$ gave on irradiation with UV light a glasslike polymer.[424]

$Ph[CH_2{=}CH(CH_2)_9]PH$ gave on irradiation with UV light a clear, viscous polymeric phosphine,[282] the oxide of which was a glassy solid.[282]

$(HOCH_2)_3P$ yields with epichlorohydrin and hydroxyalkylmelamines[1188] or ammonia[371] unburnable resins which may be used for making cotton flame-resistant.

$(HOCH_2)_3P$ and epoxides give heat-resistant polymers.[831]

$(HOCH_2)_3P$ and 1,2-$(ClO_2S)_2C_6H_4$ give a polymer which has good adhesion to metal, glass, and paper.[830]

$(HOCH_2)_2RP$ and phenols give polymers which possess ion-exchange properties in alkaline solution.[1109]

$(HOCH_2)_2PrP$ gave polymers with diamines of the type H_2NXNH_2.[1114]

(X = CO). White solid, softening point to 85 to 95°, mol. wt. 54,000.

[X = $-(CH_2)_6-$]. White solid resin, softening point 115 to 125°, mol. wt. 45,500.

(X = 1,4-C_6H_4). Light-brown polymer, softening point 80 to 90°, mol. wt. 49,500.[1114]

$(HOCH_2)_2C_6H_{13}P$ gave with adipic acid dichloride a brown polyester.[1331]

$(HOCH_2)_2PhP$ gave with $C_6H_4(COCl)_2$ a light-brown polyester, softening point 45°, which was said to be useful as a hydraulic fluid, lubricant, additive, impregnating agent, and thermoplastic material.[225,1331]

$(H_2NCH_2CH_2CH_2)_2$i-BuP gave with adipic acid a hard, tough, transparent colorless polymer, softening point 100 to 110°, flow temp. 120°,[1089] glass transition temp. 71°.[1088]

$(H_2NCH_2CH_2CH_2)_2$n-$C_8H_{17}P$ gave with MeO_2CCO_2Me a polymer, softening point 40 to 42°, flow temp. 65 to 70;[1089] and with adipic acid a soft, tough, white, opaque polymer, softening point 135°, flow temp. 160 to 170°.[1089]

$(H_2NCH_2CH_2CH_2)_2PhP$ gave polymers with adipic acid, hard, tough, transparent, colorless solid, softening point 72°, flow temp. 200°,[1089] less flammable than nylon,[1089] glass transition temp., 49°;[1088] and also with $HO_2C(CH_2)_8CO_2H$, softening point 37°, flow temp. 50 to 60°;[1089] with MeO_2CCO_2Me, softening point 74 to 75°, flow temp. 80 to 100°;[1089] with H_2NCONH_2, softening point 84°, flow temp. 105°;[1089] with 1,4-$HO_2CC_6H_4CO_2H$, softening point 135°, flow temp. 165 to 200°;[1089] with $HO_2C(CH_2)_xCO_2H$.[1380]

$(NCCH_2CH_2)_2PhP$ gave with $H_2N(CH_2)_6NH_2$ a transparent, greenish polymer, softening point 42 to 48°, flow temp. 52 to 95°.[1089]

$(HOCH_2CH_2CH_2)_2C_6H_{13}P$ gave with $C_6H_4(COCl)_2$ a yellow, solid polyester, softening point 115°.[1331]

$(HOCH_2CH_2CH_2)_2PhP$ gave with $MeO_2CCH{=}CHCO_2Me$ a viscous yellow mass;[1331] and with $H_2N(CH_2)_6NH_2$ a solid polymer.[1380]

$Ph_2(4\text{-}CH_2{=}CHC_6H_4)P$ could be polymerized and copolymerized by AIBN initiation;[1153] the polymer was more thermally stable than polyester.[1153]

$Ph_2(CH_2{=}CH)P$ and other vinylphosphines add in the presence of radical initiators stannanes of the type R_2SnH_2 to give tin-containing polymeric phosphines.[1043]

$Ph_2(CH_2{=}CH)P$ could be polymerized and copolymerized with Ziegler catalysts to low-molecular-weight, oily products.[1072]

$Ph_2(CH_2{=}CH)P$ grafted onto polypropylene improved adhesion toward metals.[855]

$Ph_2(CH_2{=}CHCH_2CH_2)P$ and α-olefins gave copolymers which have high notch-impact strength, low stress-cracking, and high resistance to UV light and thermal oxidative degradation.[369]

$(NCCH_2CH_2)_3P$ when heated with $Et_3N{\cdot}BH_3$ to 80 to 140° gave a polymer of m. >500°.[365]

p-$Ph_2PC_6H_4PPh_2$ and 1,4-$(N_3)_2C_6H_4$ when heated gave a polymeric product which is useful as a dielectric and semiconductor material.[558]

$Ph_2PCH_2CH_2PPh_2$ and p-$Ph_2PC_6H_4PPh_2$ when heated with $Ph_2(N_3)$-$PB_{10}H_{12}P(N_3)Ph_2$ gave polymers of softening point 225° and m. >300°, respectively, which were said to be resistant to hydrolysis and solvents.[1264]

[p-C_6H_4PPh-]$_{10}$ when heated with $ZnCl_2$ to 300° gave a hard, yellow cross-linked polymer.[127]

Tertiary Phosphines Containing Other Phosphorus Functions or a Metal

TYPE P-C-B

$Et_2PCH_2CH_2\overline{BOCH_2CH_2O}$. XII. Liquid, $b_{0.6}$ 72 to 73.5°.[145]

$(Ph_2PCH_2CH_2)_3B_3N_3Ph_3$. XII. Crystals, m. 160 to 162.5°.[1163]

$Ph_2PC{\equiv}CBPh_2$. I. ($Ph_2PC{\equiv}CMgBr$ + $ClBPh_2$). Crystals, m. 280°, IR, UV.[218]

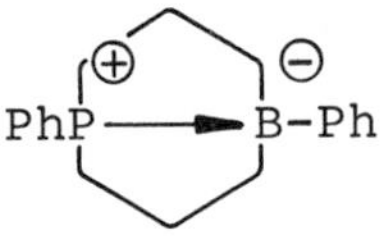
From $PhP(CH_2CH{=}CH_2)_2$ and $PhBH_2$. Solid, m. 75.5 to 76.5°, ^1H NMR, ^{11}B NMR;[197] retained 95% weight at 350°, and 30% weight at 500°.[197]

o-$B_{10}H_{10}C_2(PEt_2)_2$. II. Crystals, m. 56 to 57.5°, $b_{0.01}$ 150°.[1193]

o-$B_{10}H_{10}C{\cdot}PhCP(C_6H_{13}$-n$)_2$. II. Liquid, m. 9 to 10° (from hexane).[1458]

o-$(B_{10}H_{10}C_2PPh)_2$. II. Crystals, m. 356 to 358° (from benzene).[39]

o-$B_{10}H_{10}C_2(PPh_2)_2$. II. Crystals, m. 219° (from ligroin),[39] m. 208 to 210° (from C_6H_6/EtOH).[1460]

o-$B_{10}H_{10-n}Cl_nC_2(PPh_2)_2$. II.
(n = 1). Crystals, m. 201 to 203° (from C_6H_6/EtOH).[1460]
(n = 2). Crystals, m. 228 to 230°.[1460]
(n = 3). Crystals, m. 215 to 217°.[1460]

TYPE: P-C-M (M = Group IV metal)

$P(CH_2SiMe_3)_3$. I. Crystals, m. 66 to 69°.[1275]

$P(C_6H_4SiMe_3$-4$)_3$. I. Crystals, m. 95 to 96°, b_{31} 112 to 117°,[403,404] ^{31}P +5.8 ppm (in CH_2Cl_2).[485]

$P(C_6H_4SiMe_3$-3$)_3$. ^{31}P +5.6 ppm (in CH_2Cl_2).[485]

$P(C{\equiv}CSiMe_3)_3$. VI. (From $R_3SnC{\equiv}CSiMe_3$ + PCl_3). Crystals, m. 49 to 50° (from hexane), $b_{0.02}$ 90 to 92°, ^1H NMR.[1295]

$MeP(CH_2SiMe_3)_2$. II. Liquid, b_1 52 to 54°, m. -13 to -15°, ^1H NMR.[1255]

$PhP(CH_2CH_2SiMe_2Cl)_2$. XII. Liquid, $b_{0.3-0.9}$ 134 to 156°,[995] gives on hydrolysis a polymeric silicone.[995]

$PhP(C{\equiv}CSiMe_3)_2$. II. Liquid, $b_{0.02}$ 103 to 105°, ^1H NMR.[1295]

$Me_2PCH_2SiMe_3$. II. Liquid, b_{40} 59 to 60°, m. -23°, ^1H NMR.[1255]

$Me_2PCH_2CH_2SiMe_3$. XII. Liquid, b_{760} 174° (calc.), log P_{mm} = 7.02 - (1860/T),[489] IR, ^1H NMR, mass spect., stable at its b.[489]

$Me_2P(CH_2)_4OSiMe_3$. From $Me_2P(CH_2)_2OLi$ + $ClSiMe_3$. Liquid, b_{20} 108°,[446] IR, ^1H NMR, ^{31}P +53.6 ppm.[446]

$Et_2PCH_2CH_2SiEt_3$. XII. Liquid, b_{10} 123 to 124°.[1048]

$Et_2PCH_2CH_2Si(OEt)_3$. XII. Liquid, b_{10} 123 to 124°;[1045] gives on hydrolysis a polymer; useful as biocide, anti-foam agent, and others.[1045]

$Et_2PCH_2CH_2SiPhCl_2$. XII. Liquid, b_2 126 to 127°.[1045]

trans-$Et_2PC{\cdot}Ph{=}C{\cdot}PhSiMe_3$. VII. Liquid, b_3 190 to 192°.[670]

$EtPhP(CH_2)_3Si(OEt)_3$. VII. Liquid, $b_{0.55}$ 129 to 130°, n_D^{20} 1.4840;[377] hydrolysis gives a resin which is useful for coating metals.[377]

$BuPhPCH_2CH_2Si(OEt)_3$. XII. Liquid, $b_{0.25}$ 126°, n_D^{25} 1.4911.[1121]

$Ph_2PCH_2SiMe_3$. I.[177] VII.[248] Liquid, $b_{0.3}$ 110°,[248] $b_{1.5}$ 124 to 125°, n_D^{20} 1.581, d_4^{20} 1.0046,[177] ^1H NMR;[248] reacts with CCl_4 to give Me_3SiCl, $CHCl_3$, and an unknown P compound.[248]

$Ph_2PCO_2SiMe_3$. XIII.5. Liquid, decomposes on dist., n_D^{23} 1.5665, IR.[2]

$Ph_2PCS_2SiMe_3$. XIII.5. Yellow, viscous, foul-smelling liquid, decomposes on dist., IR.[2]

$Ph_2PCS \cdot NPhSiMe_3$. XIII.5. Yellow crystals, IR;[2] in contrast, PhNCO gave only dimeric $(RNCO)_2$.[2,718]

$Ph_2PCS \cdot NEtSiMe_3$. XIII.5. Yellow crystals, IR.[2]

$Ph_2PCH_2CH_2SiMe(OEt)_2$. XII. Liquid, $b_{0.05}$ 137 to 151°, n_D^{25} 1.5496.[1121]

$Ph_2PCH_2CH_2Si(OEt)_3$. XII. Liquid, b_2 178 to 179°.[1048]

$Ph_2PCH_2CH_2SiMeCl_2$. XII. Liquid, $b_{0.08-0.35}$ 140 to 153°; hydrolysis gives a silicone.[1068]

$Ph_2PCF{=}CFSiEt_3$. VII. Liquid, $b_{0.35}$ 153 to 155°, n_D^{25} 1.5619.[1282]

$Ph_2PC{\equiv}CSiMe_3$. II. Liquid, $b_{0.05}$ 114 to 115°, m. 17 to 18°,[1296] IR, ^1H NMR, mass spect., dec. 320 to 330°.[1296]

$Ph_2PC{\equiv}CSiPh_3$. II. Crystals, m. 100 to 101° (from MeOH/acetone), IR.[1296]

$Ph_2PCOCH_2SiMe_3$. XIII.3. (From $\geq$P-Si$\leqq$ and $CH_2{=}C{=}O$). Yellow liquid, decomposes on dist., IR.[2]

$Ph_2PCH_2CH_2SiCl_3$. XII. Liquid, $b_{0.45}$ 159 to 160°; hydrolysis yields a silicone.[1068]

$Ph_2PC(CF_3)_2OSiMe_3$ and $Ph_2P(O)C(CF_3)_2SiMe_3$ (main product). [From $\geq$P-Si$\leqq$ and $(CF_3)_2CO$]. Mixture could not be separated; n_D^{20} 1.5048, IR, ^1H NMR.[2]

$(Me_2PCH_2CH_2)_2SiMe_2$. XII. Liquid, b_{760} 256° (calc.), log P_{mm} = 8.76 - (3100/T),[489] IR, ^1H NMR, mass spect., stable at its b.[489]

$(Et_2PCH_2CH_2)_2SiMe_2$. XII. Liquid, b_3 155 to 160°,[1045] b_9 164 to 169°.[1048]

$(Et_2PCH_2CH_2)_2SiCl_2$. XII. Liquid, b_2 139 to 140.5°,[1045,1048] b_3 167 to 170°.[1047]

$(Et_2PCH_2CH_2)_2Si(OMe)_2$. XII. Liquid, b_3 182 to 184°.[1045]

$[Et_2P(CH_2)_3]_2SiMe_2$. XII. Liquid, b_4 170 to 171°.[1045,1048]

$[Ph_2P(CH_2)_3]_2SiMe_2$. XII. Liquid, b_2 280 to 285°.[1048]

$(4\text{-}Ph_2PC_6H_4)_2SiPh_2$. II. Crystals, m. 250 to 252° (from toluene);[72] disulfide, m. 305 to 308°.

$(Et_2PCH_2CH_2)_4Si$. XII. Liquid, b_2 224.5 to 228°.[1045,1048]

$(Ph_2PCH_2CH_2)_4Si$. XII. Crystals, m. 208 to 211° (from benzene).[1045,1048]

$P(C{\equiv}CGeMe_3)_3$. VI. (From $Me_3GeC{\equiv}CSnR_3$ + PCl_3). Crystals, m. 81 to 82° (from hexane), $b_{0.02}$ 112 to 114°, ^1H NMR.[1295]

$PhP(C{\equiv}CGeMe_3)_2$. II. Liquid, $b_{0.02}$ 120 to 122°, ^1H NMR.[1295]

$Et_2PCH_2CH \cdot CNGeEt_3$. XII. Liquid, $b_{0.1}$ 100°, n_D^{20} 1.5006, d_4^{20} 1.0730, IR.[1214]

$Et_2PC \cdot Ph{=}CHGeEt_3$ and $Et_2PCH{=}C \cdot PhGeEt_3$. XII. (Mixture, in each case cis and trans). Liquid, $b_{0.1}$ 108 to 111°, ^1H NMR.[1214]

$Ph_2PC{\equiv}CGeMe_3$. II. Solid, m. 35°, $b_{0.05}$ 126 to 127°, IR, ^1H NMR, mass spect.,[1296] dec. 280 to 300°.
$Ph_2PC{\equiv}CGePh_3$. II. Crystals, m. 102 to 103° (from MeOH/acetone), IR.[1296]
$Et_2PCSNPhGeEt_3$. XIII.5. Liquid, $b_{0.2}$ 92°, n_D^{20} 1.5745, IR, ^1H NMR.[1214]
$Et_2PCONPhGeEt_3$. XIII.5. Liquid, $b_{0.1}$ 83°, n_D^{20} 1.5390, IR, ^1H NMR.[1214]
$Et_2PC(=CH_2)OGeEt_3$. XIII.3. Liquid, $b_{0.35}$ 89°, IR, ^1H NMR.[1216]
$Et_2PCH{\cdot}PrOGeEt_3$. XIII.1. (From Ge-P + RCHO). Liquid, $b_{0.15}$ 85°, IR, ^1H NMR.[1215]
$Et_2PCH{\cdot}MeCH{=}CHOGeEt_3$. XIII.1. Liquid, $b_{0.15}$ 93 to 95°, IR, ^1H NMR,[1215] (cis and trans present).
$Et_2PCH{\cdot}PhCH{=}CHOGeEt_3$. XIII.1. (Cis and trans). Liquid, $b_{0.15}$ 113°, IR, ^1H NMR.[1215]
$Et_2PCH{\cdot}PhOGeEt_3$. XIII.1. Liquid, $b_{0.09}$ 117°, IR, ^1H NMR.[1215]
$Et_2PC(=CPh_2)OGeEt_3$. XIII.3. (From Ge-P + $R_2C{=}C{=}O$). Liquid, $b_{0.004}$ 154°, n_D^{20} 1.5792, IR, ^1H NMR.[1216]
$Et_2PC(CF_3)_2OGeEt$. XIII. ^{19}F NMR. $Et_2P(O)C(CF_3)_2GeEt_3$ also present.[1216a]
Et_2(3-Et_3GeO-cyclopent-2-enyl)P. XIII. Liquid, $b_{0.2}$ 125°, IR,[1216a]
$Ph_2PC(CF_3)_2OGeEt_3$. XIII. ^{19}F NMR. $Ph_2P(O)C(CF_3)_2GeEt_3$ also present.[1216a]
$Et_2PCS_2GeEt_3$. XIII.5. Liquid, $b_{0.1}$ 98°, n_D^{20} 1.5493, d_4^{20} 1.1694, IR.[1214]
$PhP(C{\equiv}CSnMe_3)_2$. II. Crystals, m. 72 to 74° (from hexane), $b_{0.02}$ 146 to 148°, ^1H NMR.[1295]
$Ph_2PC{\equiv}CSnMe_3$. II. XIV. (R_3SnNR_2 + $HC{\equiv}CPR_3$). Solid, m. 35 to 36°, $b_{0.05}$ 146 to 147°, IR, ^1H NMR, mass spect.,[1296] dec. 190 to 210°.
$Ph_2PC{\equiv}CSnPh_3$. II. XIV. Crystals, m. 89 to 90.5° (from Et_2O), IR.[1296]
$Ph_2PC{\cdot}Ph{=}CHSnPh_3$ and $Ph_2PCH{=}C{\cdot}PhSnPh_3$. XII. Mixture could not be separated. Crystals, m. 45° (from pentane).[1268]
$Ph_2PCH{\cdot}PhCH_2SnPh_3$ and $Ph_2PCH_2CH{\cdot}PhSnPh_3$. Mixture could not be separated. Crystals, m. 59° (from pentane).[1268]
$Ph_2PCH{\cdot}CH_2ClCH_2SnPh_3$ and $Ph_2PCH_2CH{\cdot}CH_2ClSnPh_3$. XII. Mixture could not be separated. Crystals, m. 39° (from pentane).[1268]
$Ph_2PCH(CH{=}CH_2)OSnMe_3$. XIII.1. From Ph_2PSnMe_3 + $CH_2{=}CHCHO$. Crystals, m. 90 to 95°.[1266]
$Ph_2PCONPhSnPh_3$. XIII.5. Crystals, m. 58° (dec.) (from benzene/pentane).[1267]
$Ph_2PCSNPhSnPh_3$. XIII.5. Crystals, m. 87° (from benzene/pentane).[1267]
$Ph_2PCSOSnPh_3$. XIII.5. Crystals, m. 97° (from benzene/pentane), IR.[1267]

$Ph_2PCCl_2SSnPh_3$. XIII.5. Crystals, m. 93° (from benzene/pentane), IR.[1267]
$Ph_2PC(NH_2)_2SSnPh_3$. XIII.5. Crystals, m. 115° (from benzene/pentane), IR.[1267]
$Ph_2PCS_2SnPh_3$. XIII.5. Crystals, m. 68° (from benzene/pentane), IR.[1267]
$PhP(C{\equiv}CPbMe_3)_2$. II. Crystals, m. 94 to 95° (from hexane), ^1H NMR; explodes at 130°.[1295]
$Ph_2PC{\equiv}CPbMe_3$. II. Crystals, m. 58 to 60° (from Et_2O), IR, ^1H NMR,[1296] dec. 130 to 140°.
$Ph_2PC{\equiv}CPbPh_3$. II. Crystals, m. 83 to 84.5° (from Et_2O), IR.[1296]

TYPE: P-C-M (M = Group V element)

trans-$Bu_2PCH{=}CHP(O)Bu_2$. XII. (From $Bu_2PC{\equiv}CH$ + $HP(O)Bu_2$). Crystals, m. 107 to 109°, UV, ^1H NMR;[1421] L•MeI, m. 82 to 83°.[1421]
$Ph_2PCH_2CH_2P(O)Ph_2$. IX.2.[6] XIV.[13] Crystals, m. 190.5 to 193°,[13] IR, ^1H NMR, GLC.[13]
$Ph_2PCH{\cdot}PhCH_2P(O)Ph_2$. VII. From $PhCHBrCH_2P(O)Ph_2$ + $LiPPh_2$, isolated as dioxide, m. 277 to 278°.[15]
4-$Ph_2PC_6H_4P(O)Ph(OH)$. II. (From $Ph_2PC_6H_4Li$ + $PhP(O)NR_2Cl$). Crystals, m. 192.5 to 193.5°,[73] pK_a 3.89, σ_p (of Ph_2P group) -0.01, σ^* +0.03.[73]
4-$Ph_2PC_6H_4P(O)(OH)C_6H_4PPh_2$-4. II. (From $Ph_2PC_6H_4Li$ + $R_2NP(O)Cl_2$). Solid, m. 258 to 261°.[73]
$Et_2PC{\cdot}Me{=}PPh_3$. XIV. (From ylid, $MeCH{=}PPh_3$ + $ClPEt_2$). Yellow solid, m. 173 to 176° (dec.).[683]
$Et_2P\overset{\ominus}{C}H\text{-}CH\text{-}\overset{\oplus}{C}HPPh_3$. XIV. From ylid + $ClPEt_2$. Yellow solid, m. 65 to 67°.[686]
$(c\text{-}C_6H_{11})_2PC{\cdot}Me{=}P(C_6H_{11}\text{-}c)_3$. XIV. (From ylid and R_2PCl). Solid, dec. at 130°.[683]
$(c\text{-}C_6H_{11})_2PC{\cdot}Me{=}PPh_3$. XIV. Yellow solid, m. 226 to 229°.[683]
$Ph_2PCR{=}P(C_6H_{11}\text{-}c)_3$. XIV.[683] (R = H). Solid, dec. 160°.[683]
(R = Me). Solid, dec. 155 to 156°.[683]
(R = Ph). Solid, dec. 170°.[683]
$Ph_2PCR{=}PPh_3$. XIV.[683] (R = H). Slightly yellow solid, m. 113 to 115°.[683]
(R = Me). Yellow solid, m. 148 to 150°.[683]
(R = CO_2Me). Colorless solid, m. 192 to 194°.[683]
(R = CO_2Et). Colorless solid, m. 151 to 153°.[683]
(R = COMe). Pale-yellow crystals, m. 188 to 191°.[684]
(R = Ph). Yellow solid, m. 214 to 216°.[683]
$Ph_2P\overset{\ominus}{C}H\text{-}CH\text{-}\overset{\oplus}{C}HPPh_3$. XIV. From ylid and $ClPPh_2$. Orange solid, m. 223 to 225°.[686]
$(Ph_2P)_2C{=}PPh_3$. XIV. Colorless solid, m. 239 to 243° (dec.).[683]

$(Ph_2P)_2\overset{\ominus}{C}\text{-}CH\text{-}\overset{\oplus}{CHPPh_3}$. XIV. Slightly yellow solid, m. 226 to 228°.[686]

$(Ph_2P)_2\overset{\ominus}{C}\text{-}CH\text{-}\overset{\oplus}{CMePPh_3}$. XIV. Yellow solid, m. 229 to 233°.[686]

$PhP[\overset{\ominus}{CH}\text{-}CH\text{-}\overset{\oplus}{CHPPh_3}]_2$. XIV. From ylid and Cl_2PPh. Orange-red solid, m. 108 to 110°.[686]

$P[\overset{\ominus}{CH}\text{-}\overset{\oplus}{PPh_3}]_3$. XIV. From ylid and PCl_3. Dark-red solid, m. 104 to 106°.[686]

$P[\overset{\ominus}{CH}\text{-}CH\text{-}\overset{\oplus}{CHPPh_3}]_3$. XIV. Brown-red solid, m. 118 to 120°.[686]

$[Ph_2P\text{-}\overset{\bullet}{C}Me\overset{\begin{array}{c}\diagup CH_2\diagdown\end{array}}{\underset{\begin{array}{c}\diagdown CH_2\diagup\end{array}}{}}\overset{+}{PPh_2}]Cl^-$. II. White crystals, m. 268

to 272° (from EtOH).[109]

$[4\text{-}Ph_2PC_6H_4PPh_2\cdot NH_2]^+Cl^-$. From diphosphine and NH_2Cl. Solid, m. 135° (dec.).[724]

$[trans\text{-}Ph_2PCH\text{=}CHPPh_2\cdot NH_2]^+Cl^-$. XIV. From diphosphine + NH_2Cl. Solid, m. 215 to 218°, IR.[724]

$[Ph_2P_\alpha CH\text{=}CHP_\beta Ph_2\cdot CH_2Ph]^+Br^-$. XIV. (Cis). From diphosphines + $PhCH_2Br$. Crystals, m. 218 to 219° (from MeOH/acetone), ^{31}P +3.2 ppm (P_α) and -4.6 ppm (P_β).[3]

$[Et_2PCMe_2PPh_3]^+BPh_4^-$. XIV. From ylid and Et_2PCl. Solid, m. 152 to 154°.[685]

$[(c\text{-}C_6H_{11})_2PCMe_2PPh_3]^+X^-$. XIV. (X = BPh_4); m. 134 to 139°; (X = ClO_4) m. 117 to 119°;[685] an isomeric compound $[Me(c\text{-}C_6H_{11})_2PCMePPh_3]$ + BPh_4^-, obtained from ylid and RX, has m. 211 to 213°.[683,685]

$[Ph_2PCRR^1PPh_3]^+X^-$. XIV. Obtained from $Ph_2PCR\text{=}PPh_3$ and HCl,[683,685] or from $Ph_3P\text{=}CRR^1$ and $XPPh_2$.[683,685,1277] The reaction of $Ph_2PCR\text{=}PPh_3$ with R'X gives $[R'Ph_2\overset{\oplus}{P}CRPPh_3]X^-$ [685] and not $[Ph_2PCRR'PPh_3]^+X^-$ as reported previously.[683] (R = R^1 = H; X = Cl), m. 228 to 230° (dec.);[683] (X = BPh_4), m. 203 to 205°,[683] m. 185.5 to 186°;[1277] (X = Br), oil;[1277]
(R = H; R^1 = Me; X = BPh_4), m. 184 to 185°;[683]
(R = H; R^1 = $CH_2\text{=}CH$; X = BPh_4), m. 132 to 133°; (X = Br), oil;[1277] structure recently questioned; it was said attack occurs at the γ-carbon;[686]
(R = R^1 = Me; X = Cl), m. 173 to 178°,[683] ^{31}P -5.5 and -2.8 (P, III) and -32.5 ppm [broad, (P, IV)];[685] (X = BPh_4), m. 182.5 to 184.5°;[683,685]
[R = R^1 = (MeO)MeC =; X = BPh_4], m. 152 to 154°;[684]
(R = H; R^1 = Ph, X = BPh_4), m. 179 to 181°,[683] reineckeate, m. 183°.[683]

$[Ph_2PCH_2CH\text{=}CHPPh_3]^+Cl^-$. XIV. From ylid and HCl. Solid, m. 205 to 208°.[686]

$[Ph_2PCH_2CH\text{=}CMePPh_3]^+Cl^-$. XIV. As above. Solid, m. 244

to 246°.[686]
$[PhP(CH_2PPh_3)_2]^{2+}2X^-$. XIV. From $Ph_3P=CH_2$ + Br_2PPh. (X = Br). Oil; (X = BPh_4), m. 132 to 133°.[1277]
$[P(CH_2PPh_3)_3]^{3+}3Cl^-$. XIV. From ylid and HCl. Solid, m. 110 to 112°.[686]

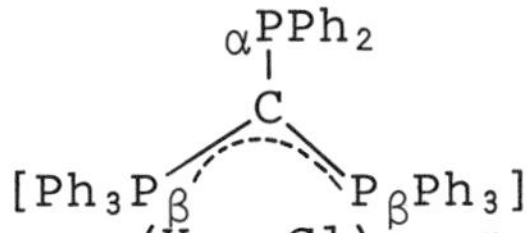

$[Ph_3P_\beta C(P_\alpha Ph_2) P_\beta Ph_3]^+X^-$. XIV. (From $Ph_3P=C=PPh_3$ + Ph_2PCl).[122] (X = Cl). m. 251 to 254°, ^1H NMR, ^{31}P +1.5 ppm (P_α) and -26 ppm (P_β) ($J_{P_\alpha P_\beta}$ 75 Hz).[122]
(X = I), m. 264 to 267°; (X = PF_6), m. 264 to 265°; (X = BF_4), m. 261 to 263°; (X = BPh_4), m. 252 to 254°.[122]
$P[(CH_2)_3AsMe_2]_3$. I. White crystals; tetrasulfide, m. 271 to 271.5°.[98]
$P(C_6H_4AsPh_2\text{-}2)_3$. II. Crystals, m. 238 to 238.5° (from anisole),[640] ^{31}P +5.3 ppm.[295]
$PhP(C_6H_4AsPh_2\text{-}2)_2$. II. Crystals, m. 238 to 240° (from anisole).[640]
$PhP(CH_2CH_2AsPh_2)_2$. XII. Solid, m. 160 to 162°.[777]
$Et_2PC_6H_4AsMe_2$-2. II. Liquid, $b_{0.6}$ 100 to 102°.[742]
$Et_2PC_6H_4AsEt_2$-2. II. Liquid, $b_{1.0}$ 136°.[742]
$(c\text{-}C_6H_{11})_2P(CH_2)_4As(C_6H_{11}\text{-}c)_2$. VII. Crystals, m. 78°.[1368]
$(c\text{-}C_6H_{11})_2P(CH_2)_4AsPh_2$. VII. Crystals, m. 59 to 61°.[1369]
$Ph_2PCH_2CH_2AsPh_2$. XII. Solid, m. 116 to 118°.[777]
trans-$Ph_2PCH=CHAsPh_2$. XII. Solid, m. 95 to 96°, ^1H NMR;[777] addition is stereospecific.
$Ph_2PC_6H_4AsPh_2$-2. II. Crystals, m. 190.5 to 192.5°.[1042]
$(2\text{-}Ph_2PC_6H_4)_3As$. II. ^{31}P +11.4 ppm.[295]
$(c\text{-}C_6H_{11})_2P(CH_2)_4SbEt_2$. (From $R_2P(CH_2)_4Cl$ + $LiSbEt_2$). Oil, decomposes on dist.; disulfide, m. 125 to 126°.[663]
$(2\text{-}Ph_2PC_6H_4)Sb$. II. ^{31}P +6.0 ppm.[295]

TYPE: P-C-M (M = transition metal)

$Ph_2PCONRZnR^1$. XIII.5.[1057] (From Ph_2PZnR^1 + RNCO). (R = Ph; R^1 = Et). Crystals, m. 98 to 105° (from C_6H_6/pentane).[1057]
(R = Me; R^1 = Bu). Crystals, m. 125 to 130° (from C_6H_6/pentane).[1057]
(R = Ph; R^1 = Ph). Crystals, m. 136 to 140° (from C_6H_6/pentane).[1057]
$Ph_2PCSNRZnR^1$. XIII.5.[1057] (R = Ph; R^1 = Et). Crystals, m. 139 to 142° (from C_6H_6/pentane).[1057]
(R = Me; R^1 = Ph). Crystals, m. 125° (dec.).[1057]
(R = Ph; R^1 = Ph). Crystals, m. 166 to 175° (from C_6H_6/pentane).[1057]
Ph_2PCS_2ZnPh. XIII.5. Crystals, m. 193 to 195° (from C_6H_6/pentane).[1057]
Triferrocenylphosphine. XI. (Best obtained from R_2NPCl_2

+ $AlCl_3$ + ferrocene). Fine, yellow needles, m. 271 to 273° (closed capillary), IR.[1314,1315]

Diferrocenyl-PhP. XI. Orange crystals, m. 191 to 193° (dec.),[1313] m. 194 to 195.5° (from hexane).[1040]

Diferrocenyl-(4-MeC_6H_4)P. XI. Solid, m. 165 to 167° (dec.) (from EtOH).[1316]

Diferrocenyl-(4-ClC_6H_4)P. XI. Solid, m. 166 to 168° (dec.) (from EtOH).[1316]

Diferrocenyl-(4-$MeOC_6H_4$)P. XI. Solid, m. 158 to 160° (dec.) (from EtOH).[1316]

Ferrocenyl-Ph_2P. XI. Orange crystals, m. 122 to 124°,[1313,1316] m. 112 to 115°.[1036]

Ferrocenyl$(PPh_2)_2$. XI. Solid, m. 182 to 184°.[1316]

Trivalent Phosphorus Compounds with Coordination Number Two or One

TYPE: Coordination Number Two

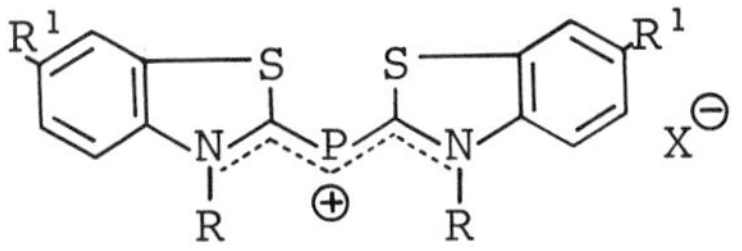

From chlorobenzthiazolium salts and $P(CH_2OH)_3$[318,319] or $(Me_3Si)_3P$.[893] All salts are colored; oxidizing agents destroy the salts, and acids protonate at P.[319]

(R^1 = H; R = Me; X = BF_4). Crystals, m. 222 to 230°,[318] m. 224 to 227°,[893] UV,[318,893] ^1H NMR.[318]

(R^1 = H; R = Et; X = BF_4). Orange-red crystals, m. 214 to 220°,[318] m. 214 to 219°,[893] UV,[318,893] ^{31}P -26 ppm.

(R^1 = H; R = Et; X = ClO_4). m. 224 to 226°,[319] UV, ^{31}P -24.9 ppm;[319] x-ray analysis[45] shows cyanine cations nearly planar, P-C bonds equal 1.76 Å, and <C-P-C 104.6°.[45]

(R^1 = MeO; R = Me; X = BF_4). Crystals, m. 208 to 220°, UV.[318]

(R^1 = MeO; R = Et; X = BF_4). Crystals, m. 215 to 223°, UV.[318]

(R^1 = MeO; R = Et; X = ClO_4). Crystals, m. 213 to 215°, UV.[319]

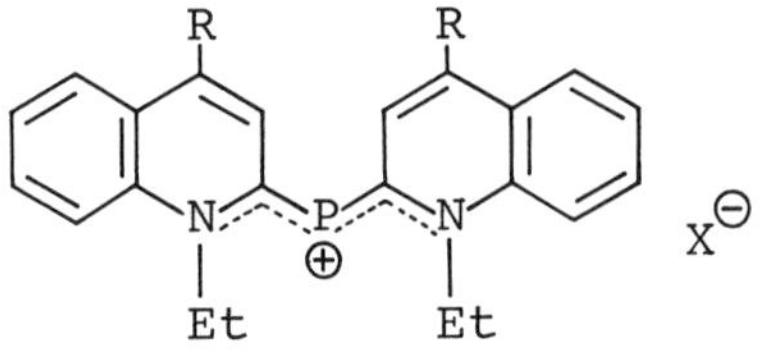

From chlorochinolinium salts

and $P(CH_2OH)$.[319]
($R = H$; $X = BF_4$). Crystals, m. 126° (dec.), UV, electronic spect.[318]
($R = H$; $X = ClO_4$). Crystals, m. 197 to 200°, UV, ^{31}P -48.8 ppm.[319]
($R = Me$; $X = BF_4$). Crystals, m. 178 to 185°, UV.[318]
($R = Me$; $X = ClO_4$). Crystals, m. 187 to 189°, UV;[319] x-ray analysis[769] shows that the cyanine cation is not planar, P-C 1.81 and 1.77 Å, <C-P-C 100.4°.[769]

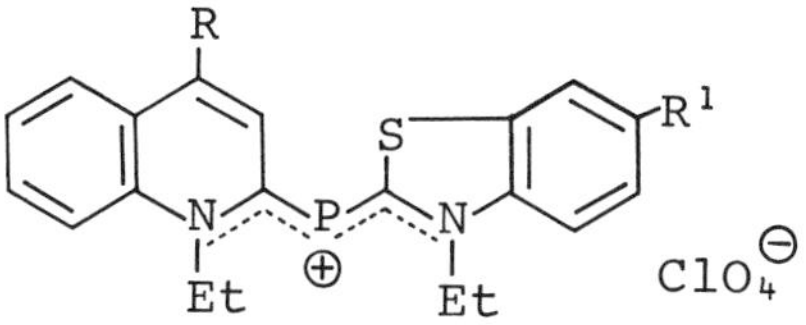

From chlorobenzthiazolium salts, chlorochinolinium salts, and $P(CH_2OH)_3$.[319]
($R = R^1 = H$). Crystals, m. 163 to 167°, UV.[319]
($R = Me$; $R^1 = H$). Crystals, m. 204 to 210°, UV.[319]
($R = H$; $R^1 = MeO$). Crystals, m. 162 to 168°, UV.[319]

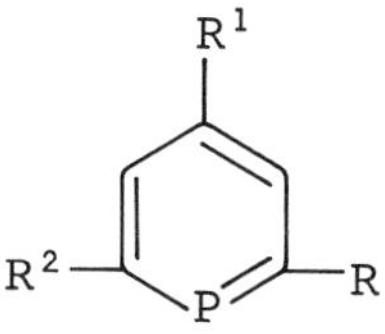

Four methods are available for synthesizing phosphorin derivatives.
(a) From $P(CH_2OH)_3$ and pyrylium fluoroborate in pyridine.[891]
(b) From $P(SiMe_3)_3$ and pyrylium iodide in acetonitrile.[896]
(c) From $PH_4^+X^-$ and pyrylium salts in BuOH.[897]
(d) Decomposition of particular benzyl-substituted phosphines.[892]
($R = R^1 = R^2 = Ph$). Light-yellow needles, m. 172 to 173° (from EtOH or $CHCl_3$/EtOH),[891] UV, mass spect., ^1H NMR, ^{31}P -178.2 ppm (in pyridine);[891] x-ray analysis,[1066] only preliminary report, shows crystals are orthorhombic;[1066] ESR spect. of cation radical which is stable;[315] is also able to accept up to three electrons;[322] is oxidized in air to an acid.[323]
($R = R^2 = Ph$; $R^1 = 4\text{-}MeOC_6H_4$). Crystals, m. 110 to 111°,[316,1358] m. 106°,[896,897] UV,[316,896,897,1358] ESR spect. of radical cation.[316]
($R = R^2 = 4\text{-}MeOC_6H_4$; $R^1 = Ph$). Crystals, m. 131 to 132°,[316,1358] m. 136 to 137°,[896] UV,[316,896,1358] ESR spect. of radical cation.[315,316]
($R = R^1 = R^2 = 4\text{-}MeOC_6H_4$). Crystals, m. 106 to 107°, UV.[896,1358]

($R = R^2 = Ph$; $R^1 = 4\text{-}Me_2NC_6H_4$). Crystals, m. 116 to 117°, UV.[1358]
($R = R^2 = Ph$; $R^1 = Me$). Crystals, m. 118 to 120°, UV, ^1H NMR.[897]
($R = R^2 = 4\text{-}MeC_6H_4$; $R^1 = Ph$). Crystals, m. 133 to 134°, UV.[896,897]
($R = R^1 = R^2 = C_6D_5$). Crystals, m. 168 to 171°, UV, ESR spect. of radical cation.[316]
($R = R^2 = C_6D_5$; $R^1 = Ph$). Crystals, m. 167°, UV, ESR spect. of radical cation.[316]
($R = R^2 = Me$; $R^1 = Ph$). Crystals, m. 62 to 63°, x-ray analysis[81] shows crystals are orthorhombic and the phosphorus ring is planar, P-C distances are equal 1.74 Å, <C-P-C 102.9° (aromatic system).[81]
($R^1 = R^2 = Ph$; $R = Me$). Crystals, m. 79 to 81°, UV, ^1H NMR.[897]
($R = 4\text{-}ClC_6H_4$; $R^1 = R^2 = Ph$). Crystals, m. 166 to 167°, UV, ESR spect. of radical cation.[316]
($R = R^1 = R^2 = 4\text{-}ClC_6H_4$). Crystals, m. 181 to 182°, UV, ESR spect. of radical cation.[316]
($R = 4\text{-}MeOC_6H_4$; $R^1 = R^2 = Ph$). Crystals, m. 161.5 to 163°, UV, ESR spect. of radical cation.[316]
($R = R^1 = 4\text{-}MeOC_6H_4$; $R^2 = Ph$). Crystals, m. 134 to 136°, UV, ESR spect. of radical cation.[316]
($R = 4\text{-}MeC_6H_4$; $R^1 = R^2 = Ph$). Crystals, m. 155 to 156.5°, UV.[316]
($R = 4\text{-}MeOC_6H_4$; $R^1 = 4\text{-}PhC_6H_4$; $R^2 = Ph$). Crystals, m. 148 to 150.5°, UV.[316]
($R = 1\text{-}C_{10}H_7$; $R^1 = R^2 = Ph$). Crystals, m. 163 to 164°, UV.[316]
($R = R^2 = Ph$; $R^1 = t\text{-}Bu$). Crystals, m. 87.5 to 88.5°, UV, ESR spect. of radical cation.[316]
($R = R^2 = t\text{-}Bu$; $R^1 = Ph$). Crystals, m. 104 to 105°, UV, ESR spect. of radical cation.[316]
($R = R^2 = t\text{-}Bu$; $R^1 = 4\text{-}MeOC_6H_4$). Crystals, m. 116 to 116.5°, UV, ESR spect. of radical cation.[316]
($R = R^1 = R^2 = t\text{-}Bu$). Crystals, m. 88°, UV, ESR spect. of radical cation,[316,320] ^1H NMR.[323]
($R = R^1 = R^2 = Ph$). Yields with $HgAc_2$ in the presence of alcohols 1,1-dialkoxy- and 1,1-diaryloxyphosphorins.[321]

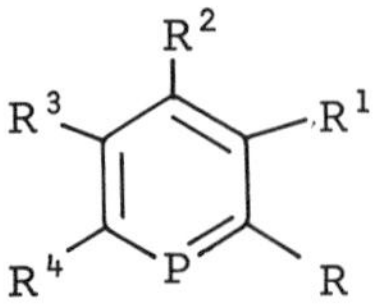

From pyrylium fluoroborate and $P(CH_2OH)_3$.

($R = R^1 = R^2 = R^3 = R^4 = Ph$). Crystals, m. 253 to 255°,[896,897] m. 216 to 217° (different crystal modification?)[316] UV, ESR spect. of radical cation.[316]
($R^3 = H$; $R = R^1 = R^2 = R^4 = Ph$). Crystals, m. 209 to

210°, UV,[896] m. 188.5 to 189.5°, UV,[316] ESR spect. of radical cation.[316]

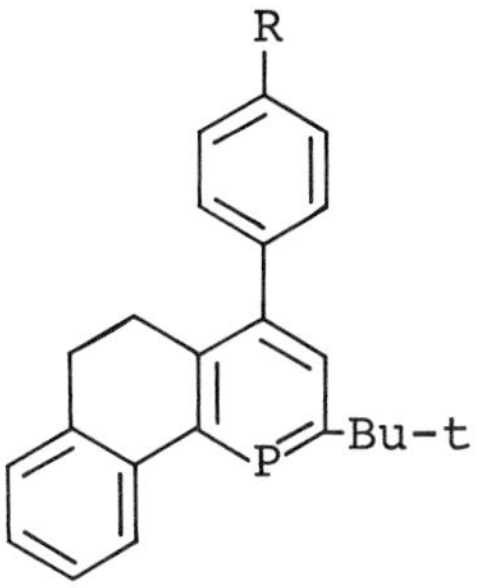

From pyrylium fluoroborate and $P(CH_2OH)_3$.[316]

(R = H). Crystals, m. 99 to 101°.[395]
(R = Cl). Crystals, m. 150 to 152°.[395]
(R = Me). Crystals, m. 115 to 117°.[395]
(R = OMe). Crystals, m. 160 to 163°;[395] x-ray analysis[395] shows phosphorin ring is planar, P-C 1.75 Å, <C-P-C 103° (aromatic system).[395]

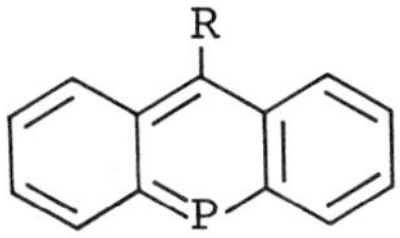

(R = H). From 5-chloro-5,10-dihydrodibenzo[b,e]phosphorin by splitting off HCl with 1,5-diazabicyclo[4.3.0]non-5-ene. Not isolated pure; obtained in solution, UV.[303]
(R = Ph). From phenyl derivative as above.[305] Yellow crystals, m. 173 to 176°, UV, ^1H NMR, mass spect.[305]

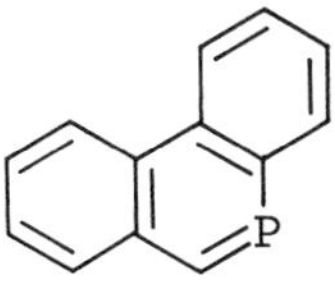

From 5-chloro-5,6-dihydrodibenzo[b,d]phosphorin by splitting off HCl as above.[304] Not isolated pure; obtained only in solution, UV.[304]

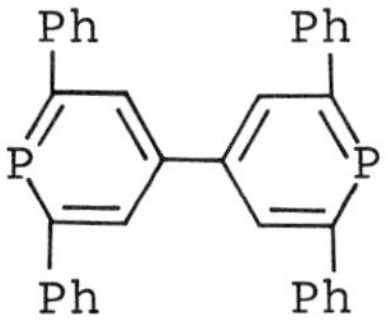

By thermal dec. at 350° of the benzyl-substituted phosphine.[892] Solid, m. 236 to 238° (sealed tube), UV, ^1H NMR;[892] gives in EtOH in air a phosphinate ester.[892]

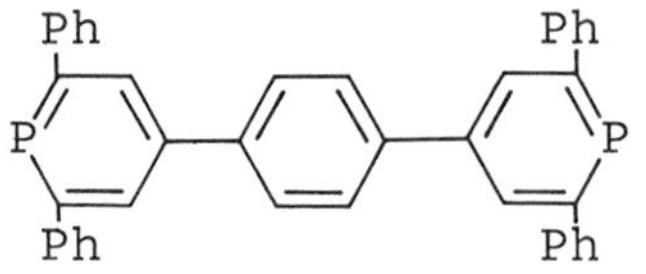

From pyrylium salt and PR_3 (R = H, CH_2OH, $SiMe_3$).[892] Solid, m. 218°.[892]

TYPE: Coordination Number One

HC≡P. From PH_3 and a rotating arc struck between graphite electrodes.[429] Reactive colorless gas, stable only below its triple point of -124 ± 2°; vapor pressure ∿30 mm at -124°;[429] the monomer polymerizes slowly to a black solid at -130° but more rapidly at -78°;[429] with HCl it gives CH_3PCl_2;[429] IR, mass spect.,[429] microwave spect.[1367] gives P-C 1.5421, H-C 1.0667 Å; the molecule is linear; μ 0.397 D,[1367] ionization potential 14.1 eV,[396] 13.0 ± 0.6 eV;[1394] polymeric HCP has been claimed to be useful as a fertilizer (it slowly releases phosphorus).[429]

(received July 9, 1970)

BIBLIOGRAPHY

Series

Trippett, S., R. S. Davidson, D. W. Hutchinson, R. Keat, R. A. Shaw, and J. C. Tebby, Organophosphorus Chemistry, Vol. 1, The Chemical Society, 1970.

Grayson, M., and E. J. Griffith, Eds., Topics in Phosphorus Chemistry, Vols. 1 to 6, Interscience.

Monographs

Hudson, R. F., "Structure and Mechanism in Organo-Phosphorus Chemistry," in Organic Chemistry, A. T. Blomquist, Ed., Vol. 6, Acadmic, 1965.

Kirby, A. J., and S. G. Warren, "The Organic Chemistry of Phosphorus," in Reaction Mechanisms in Organic Chemistry, C. Eaborn and N. B. Chapman, Eds., Vol. 5, Elsevier, 1967.

Kosolapoff, G. M., Organophosphorus Compounds, Wiley, 1950.

Mann, F. G., "The Heterocyclic Derivatives of Phosphorus, Arsenic, Antimony and Bismuth" (sec. ed.) in the Chemistry of Heterocyclic Compounds, A. Weissberger and E. C. Taylor, Eds., Wiley-Interscience, 1970.

Purdela, D., and R. Vilceanu, Chimia Compusilor Organici al Fosforului si al Acizilor Lui, Editura Academici Republicii Socialiste Romania, 1965.

Sasse, K., Organische Phosphorverbindungen in Methoden der Organischen Chemie (Houben-Weyl), E. Müller, Ed., Vol. XII/1, Thieme, 1963.

Review Articles

Bellaart, A. C., "Reduction of Aromatic Nitro-Compounds with Phosphine," Chem. Weekbl., 64, 8 (1968); C. A., 69, 76755v (1968).

Berlin, K. D., T. H. Austin, M. Peterson, and M. Nagabhushanam, "Nucleophilic Displacement Reactions on Phosphorus Halides and Esters by Grignard and Lithium Reagents," in Topics in Phosphorus Chemistry, M. Grayson and E. J. Griffith, Eds., Vol. 1, Interscience, 1964, p. 17.

Berlin, K. D., and D. M. Hellwege, "Carbon-Phosphorus Heterocycles," in Topics in Phosphorus Chemistry, M. Grayson and E. J. Griffith, Eds., Vol. 6, Interscience, 1969, p. 1.

Bregadze, V. I., and O. Yu. Okhlobystin, "Organoelement Derivatives of Barenes," Organometal. Chem. Rev., 4A, 345 (1969).

Burg, A. B., "Chemical Consequences of Fluorocarbon Phosphines," Accounts Chem. Res., 2, 353 (1969).

Cadiot, P., W. Chodkiewicz, B. Borecka, C. Charrier, and M. P. Simonnin, "Dérivés Acétyléniques et Alléniques du Phosphore," in Composés Organiques du Phosphore, Tolouse, Collog. Nat. Centre Nat. Rech. Sci., 1965, p. 99.

Cadiot, P., and W. Chodkiewicz, "Acetylenic Derivatives of Groups IIIb, IVb, and Vb" in Chemistry of Acetylenes, H. G. Viehe, Ed., Dekker, 1969, p. 913.

Daly, J. J., "Stereochemical Aspects of Organophosphorus Compounds," in Perspectives in Structural Chemistry, J. D. Dunitz and J. A. Ibers, Eds., Vol. 3, Interscience, 1970, p. 165.

Davidsohn, W. E., and M. C. Henry, "Organometallic Acetylenes of the Main Groups III to V," Chem. Rev., 67, 73 (1967).

Fild, M., and O. Glemser, "Phosphorus, Arsenic, and Antimony Pentafluorophenyl Compounds," Fluorine Chem. Rev., 3, 129 (1969).

Fluck, E., and V. Novobilsky, "Die Chemie des Phosphins," Fortschr. Chem. Forsch., 13, 125 (1969).

Gallagher, M. J., and I. D. Jenkins, "Stereochemical Aspects of Phosphorus Chemistry," in Topics in Stereochemistry, N. L. Allinger and E. L. Eliel, Eds., Vol. 3, Interscience, 1968, p. 1.

Horak, J., "Reactions of Phosphine with Organic Compounds," Chem. Listy, 55, 1278 (1961).

Horner, L., and H. Hoffmann, "Präperative und Analytische Bedeutung von Phosphinen und Verwandten Verbindungen," in Neuere Methoden der Präparativen Organischen Chemie, Vol. II, Verlag Chemie, 1960, p. 108.

Horner, L., "Darstellung und Eigenschaften von Optisch Aktiven Tertiären Phosphinen," Pure Appl. Chem., 9, 225 (1964).

Horner, L., "Stereochemische Untersuchungen Organischer Verbindungen von Elementen der 5. Reihe des Periodensystems," in Helv. Chim. Acta, Werner Celebration--IX.-ICCC, 1966, 93.

Hughes, A. N., and C. Sirvanavit, "Chemistry of Phosphol Derivatives," J. Heterocycl. Chem., 7, 1 (1970).

Ionin, B. I., G. M. Bogolyubov, and A. A. Petrov, "Organophosphorus Compounds Containing Acetylenic and Diene Substituents," Usp. Khim., 36, 587 (1967); Engl. Transl., p. 248.

Issleib, K., "Zur Synthese von Organo-Phosphor-Verbindungen unter Verwendung von P-Substituierten Metallphosphiden," Pure Appl. Chem., 9, 205 (1964).

Märkl, G., "Phosphor-Heterocyclen," Angew. Chem., 77, 1109 (1965); Angew. Chem. Int. Ed., 4, 1023 (1965).

Maier, L., "Preparation and Properties of Primary, Secondary, and Tertiary Phosphines," in Progress in Inorganic Chemistry, F. A. Cotton, Ed., Vol. 5, Interscience, 1963, p. 27.

Mann, F. G., "The Heterocyclic Derivatives of Phosphorus, Arsenic, and Antimony," Progr. Org. Chem., 4, 217 (1958).

Mavel, G., "Studies of Phosphorus Compounds Using the Magnetic Resonance Spectra of Nuclei Other Than Phosphorus-31," in Prog. Nucl. Magnetic Resonance Spectrosc., 1, 251 (1966).

McEwen, W. E., "Stereochemistry of Reactions of Organophosphorus Compounds," in Topics in Phosphorus Chemistry, M. Grayson and E. J. Griffith, Eds., Vol. 2, Interscience, 1965, p. 1.

Paddock, N. L., "Structure and Reactions in Phosphorus Chemistry," Royl. Inst. Chem. (London), Lecture Ser., 2, 1 (1962).

Petrov, K. A., and V. A. Parshina, "Hydroxyalkylphosphines and Hydroxyalkylphosphine Oxides," Usp. Khim., 37, 1218 (1968); Engl. trans., p. 532.

Quin, L. D., "Trivalent Phosphorus Compounds as Dienophiles," in 1,4-Cycloaddition Reactions, J. Hamer, Ed., Organic Chemistry, Vol. 8, Academic, 1967, p. 47.

REFERENCES

1. Abbiss, T. P., A. H. Soloway, and V. H. Mark, J. Med. Chem., 7, 763 (1964).
2. Abel, E. W., and I. H. Sabherwal, J. Chem. Soc., A, 1968, 1105.
3. Aguiar, A. M., and H. Aguiar, J. Am. Chem. Soc., 88, 4090 (1966).

4. Aguiar, A. M., H. J. Aguiar, and T. G. Archibald, Tetrahedron Lett., 1966, 3187.
5. Aguiar, A. M., H. J. Aguiar, T. G. Archibald, and D. Daigle, J. Organomet. Chem., 5, 205 (1966).
6. Aguiar, A. M., H. Aguiar, and D. Daigle, J. Am. Chem. Soc., 87, 671 (1965).
7. Aguiar, A. M., and T. G. Archibald, Tetrahedron Lett., 1966, 5471.
8. Aguiar, A. M., and T. G. Archibald, Tetrahedron Lett., 1966, 5541.
9. Aguiar, A. M., and T. G. Archibald, and L. A. Kapicak, Tetrahedron Lett., 1967, 4447.
10. Aguiar, A. M., and J. Beisler, J. Org. Chem., 29, 1660 (1964).
11. Aguiar, A. M., J. Beisler, and A. Miller, J. Org. Chem., 27, 1001 (1962).
12. Aguiar, A. M., and D. Daigle, J. Am. Chem. Soc., 86, 2299 (1964).
13. Aguiar, A. M., and D. Daigle, J. Am. Chem. Soc., 86, 5354 (1964).
14. Aguiar, A. M., and D. Daigle, J. Org. Chem., 30, 2826 (1965).
15. Aguiar, A. M., and D. Daigle, J. Org. Chem., 30, 3527 (1967).
16. Aguiar, A. M., J. Giacin, and H. J. Greenberg, J. Org. Chem., 28, 3545 (1963).
17. Aguiar, A. M., J. Giacin, and A. Mills, J. Org. Chem., 27, 674 (1962).
18. Aguiar, A. M., H. J. Greenberg, and K. E. Rubinstein, J. Org. Chem., 28, 2091 (1963).
19. Aguiar, A. M., and K. C. Hansen, J. Am. Chem. Soc., 89, 4235 (1967).
20. Aguiar, A. M., K. C. Hansen, and J. T. Magne, J. Org. Chem., 32, 2383 (1967).
21. Aguiar, A. M., and J. R. S. Irelan, 21, 4031 (1969).
22. Aguiar, A. M., J. R. S. Irelan, and N. S. Bhacca, J. Org. Chem., 34, 3349 (1969).
23. Aguiar, A. M., J. R. S. Irelan, C. J. Morrow, J. J. John, and G. W. Prejean, J. Org. Chem., 34, 2684 (1969).
24. Aguiar, A. M., J. R. S. Irelan, G. W. Prejean, J. P. John, and C. J. Morrow, J. Org. Chem., 34, 2681 (1969).
25. Aguiar, A. M., J. T. Magne, H. J. Aguiar, T. G. Archibald, and G. Prejean, J. Org. Chem., 33, 1681 (1968).
26. Aguiar, A. M., and M. G. R. Nair, J. Org. Chem., 33, 579 (1968).
28. Aguiar, A. M., G. W. Prejean, J. R. Irelan, and C. J. Morrow, J. Org. Chem., 34, 4024 (1969).
29. Ahrland, S., J. Chatt, N. R. Davies, and A. A. Williams, J. Chem. Soc., 1958, 276.

29a. Ahrland, S., J. Chatt, and N. R. Davies, Quart. Rev. (London), 12, 265 (1958).
29b. Ahrland, S., J. Chatt, N. R. Davies, and A. A. Williams, J. Chem. Soc., 1958, 1403.
30. Akhmedzyanov, M. A., V. A. Kukhtin, V. I. Slesareva, A. V. Borin, M. S. Khaikin, and K. M. Kirillova, U.S.S.R., 142,524 (1961); C. A., 56, 13702c (1962); C. A., 71, 17513h (1969).
31. Akitt, J. W., R. H. Cragg, and N. N. Greenwood, Chem. Commun., 1966, 134.
32. Aksnes, G., Acta Chem. Scand., 15, 438 (1961).
33. Aksnes, G., and K. Bergesen, Acta Chem. Scand., 19, 931 (1965).
34. Aksens, G., and L. J. Brudvik, Acta Chem. Scand., 17, 1616 (1963).
35. Albers, H., and W. Schuler, Chem. Ber., 76, 23 (1943).
36. Albers, H., W. Künzel, and W. Schuler, Chem. Ber., 85, 239 (1952).
37. Albrand, J. P., D. Gagnaire, and J. B. Robert, Chem. Commun., 1968, 1469.
38. Albrand, J. P., D. Gagnaire, and J. B. Robert, J. Mol. Spectroscopy, 27, 428 (1968); C. A., 69, 63375w (1968).
39. Alexander, R. P., and H. Schroeder, Inorg. Chem., 2, 1107 (1963).
40. Alexander, R. P., and H. Schroeder, Inorg. Chem., 5, 493 (1966).
41. All Union Sci. Res. Cinema Phot. Inst., USSR, 193,299 (1967); C. A., 60, 188497z (1968).
42. Allen, D. W., F. G. Mann, and I. . Millar, Chem. Ind. (London), 1966, 196.
43. Allen, D. W., I. T. Millar, and F. G. Mann, J. Chem. Soc., C, 1967, 1869.
44. Allen, D. W., I. T. Millar, F. G. Mann, R. M. Canadine, and J. Walker, J. Chem. Soc., A, 1969, 1097.
45. Allman, R., Angew. Chem., 77, 134 (1965); Chem. Ber., 99, 1332 (1966).
46. Allum, K. G., C. J. L. Metcalfe, and D. J. Thomasson, Ger. 1,815,631 (1969); C. A., 71, 123495p (1969).
47. American Cyanamid, Brit., 923,532 (1963); 912,269 (1962); C. A., 59, 7560h (1963).
48. Amster, R. L., and N. B. Colthup, Spectrochim. Acta, 19, 1849 (1963).
49. Anderson, J. E., and C. P. Smyth, J. Chem. Phys., 42, 473 (1965).
50. Anderson, W. A., R. Freeman, and C. A. Reilly, J. Chem. Phys., 39, 1518 (1963).
51. Angelelli, J. M., R. I. C. Brownlee, A. R. Katritzky, R. D. Topsom, and L. Yakhontov, J. Am. Chem. Soc., 91, 4500 (1969).
52. Anker, M. W., R. Colton, and I. B. Tomkins, Aust. J. Chem., 21, 1143 (1968).

53. Anonymous, Analyst, 87, 304 (1962).
54. Anschütz, L., H. Kraft, and K. Schmidt, Ann. Chem., 542, 14 (1939).
55. Anschütz, L., and H. Wirth, Naturwissenschaften, 43, 16 (1956).
56. Appel, R., and R. Schoellhorn, Ger. 1,221,220 (1966); C. A., 65, 15427c (1966).
57. Arbuzov, B. A., and G. M. Vinokurova, Izv. Akad. Nauk SSSR., 1963, 502; C. A., 59, 3947d (1963).
58. Arbuzov, B. A., G. M. Vinokurova, and I. A. Alexandrova, Izv. Akad. Nauk SSSR, 1962, 290.
59. Arbuzov, B. A., G. M. Vinokurova, I. A. Aleksandrova, and S. G. Fattakhov, Nekotorye Vopr. Organ. Khim., Sb., 1964, 244; C. A., 65, 3899g (1966).
60. Arbuzov, B. A., G. M. Vinokurova, and I. A. Perfil'eva, Dokl. Akad. Nauk SSSR, 127, 1217 (1959); C. A., 54, 1377h (1960).
61. Aries, R. S., U.S. 3,053,871 (1962); C. A., 58, 12602e (1963).
62. Aroney, M. J., R. J. W. LeFevre, and J. Saxby, J. Chem. Soc., 1963, 1739.
63. Asinger, F., A. Saus, and E. Michel, Monatsh. Chem., 99, 1695 (1968).
64. Aufdermarsh, C. A., Jr., J. Org. Chem., 29, 1994 (1964).
65. Auger, V., C. R. Acad. Sci., Paris, 139, 639, 671 (1904).
66. Baechler, R. D., W. B. Farnham, K. Mislow, J. Am. Chem. Soc., 91, 5686 (1969).
66a. Baechler, R. D., and K. Mislow, J. Am. Chem. Soc., 92, 3090 (1970).
67. Baeteman, N., and J. Baudet, C. R. Acad. Sci., Paris, Ser. C, 265, 288 (1967).
68. Bailey, W. J., U.S. 3,334,064 (1967); C. A., 67, 74108w (1967).
69. Bailey, W. J., and S. A. Buckler, J. Am. Chem. Soc., 79, 3567 (1957).
70. Bailey, W. J., S. A. Buckler, and F. Marktscheffel, J. Org. Chem., 25, 1996 (1960).
71. Baizer, M. M., and J. D. Anderson, J. Org. Chem., 30, 1357 (1965).
72. Baldwin, R. A., and M. T. Cheng, J. Org. Chem., 32, 1572 (1967).
73. Baldwin, R. A., M. T. Cheng, and G. D. Homer, J. Org. Chem., 32, 2176 (1967).
74. Baldwin, R. A., K. A. Smitheman, and R. M. Washburn, J. Org. Chem., 26, 3547 (1961).
75. Baldwin, R. A., and R. M. Washburn, J. Am. Chem. Soc., 83, 4466 (1961).
76. Baldwin, R. A., and R. M. Washburn, J. Org. Chem., 30, 3860 (1965).

77. Baldwin, R. A., C. O. Wilson, Jr., and R. I. Wagner, J. Org. Chem., 32, 2172 (1967).
77a. Balezin, S. A., and M. A. Ignat'eva, Dokl. Akad. Nauk SSSR, 109, 771 (1956); C. A., 51, 3433q (1957).
77b. Balon, W. J., and O. Stallmann, U.S. 2,683,144 (1954); C. A., 48, 12465c (1954).
78. Balon, W. J., U.S. 2,853,518 (1958); C. A., 53, 5202b (1959).
78a. Banks, R. E., and R. N. Haszeldine, Advan. Inorganic Chem. Radiochem., 3, 367 (1961).
78b. Barket, T. P., and L. D. Quin, A.C.S. 158th Meeting, Sept. 1969, New York, Abstr. ORGN.71.
79. Barlow, M. G., M. Green, R. N. Haszeldine, and H. G. Higson, J. Chem. Soc., B, 1966, 1025.
80. Bart, J. C. J., Acta Crystallogr., B25, 489 (1969).
81. Bart, J. C. J., and J. J. Daly, Angew. Chem., 80, 843 (1968); J. Chem. Soc., A, 1970, 567.
82. Bartell, L. S., J. Chem. Phys., 32, 832 (1960).
83. Bartell, L. S., J. Appl. Phys., 32, 252 (1960).
84. Bartell, L. S., and L. O. Brockway, J. Chem. Phys., 32, 512 (1960).
84a. Barton, T. J., and A. J. Nelson, Tetrahedron Lett., 1969, 5037.
85. Bastian, B. N., Belg. 624,174 (1963); C. A., 60, 13347h (1963).
86. Bataafsche Petroleum Maatschappij, N.V., Neth. 105,547 (1963), C. A., 62, 10276c (1965).
87. Bataafsche Petroleum Maatschappij, N.V., Brit. 673,451 (1952); C. A., 47, 5426b (1953).
88. Baudler, M., and H. Grundlach, Naturwissenschaften, 42, 152 (1955).
89. Baudler, M., K. Kipker, and H. W. Valpertz, Naturwissenschaften, 53, 612 (1960).
89a. Beachel, H. C., and B. Katlafsky, J. Chem. Phys., 27, 182 (1957).
90. Becker, S. B., U.S. 3,036,132 (1962); C. A., 57, 11237e (1962).
91. Bedford, A. F., D. M. Heinekey, I. T. Millar, and C. T. Mortimer, J. Chem. Soc., 1962, 2932.
92. Bedford, A. F., and C. T. Mortimer, J. Chem. Soc., 1960, 1622.
93. Beeby, M. H., and F. G. Mann, J. Chem. Soc., 1951, 411.
94. Beg, M. A. A., Bull. Chem. Soc. (Japan), 40, 15 (1967).
94a. Beg, M. A. A., and M. S. Siddiqui, Pakistan Sci. Ind. Res., 12, 19 (1969).
95. Beg, M. A. A., and H. C. Clark, Can. J. Chem., 39, 564 (1961).
96. Beg, M. A. A., and H. C. Clark, Can. J. Chem., 40, 283 (1962); C. A., 57, 9877f (1962).

97. Beg, M. A. A., and H. C. Clark, Can. J. Chem., 40, 393 (1962); C. A., 57, 2995h (1962).
97a. Beneš, M., J. Peška, and O. Wichterle, Chem. Ind. (London), 1962, 562.
98. Benner, G. S., W. E. Hatfield, and D. W. Meek, Inorg. Chem., 3, 1544 (1964).
99. Bennett, F. W., H. J. Emeléus, and R. N. Haszeldine, J. Chem. Soc., 1953, 1565.
100. Bennett, F. W., H. J. Emeléus, and R. N. Haszeldine, J. Chem. Soc., 1954, 3896.
101. Bennett, M., L. V. Interrante, and R. S. Nyholm, Z. Naturforsch., B, 20, 633 (1965).
102. Bennett, M. A., H. W. Konwenhoven, J. Lewis, and R. S. Nyholm, J. Chem. Soc., 1964, 4570.
103. Bennett, M. A., and P. A. Longstaff, J. Am. Chem. Soc., 91, 6266 (1969).
104. Bennett, M. A., and D. L. Milner, J. Am. Chem. Soc., 91, 6983 (1969).
105. Bentrude, W. G., and W. D. Johnson, J. Am. Chem. Soc., 90, 5924 (1968).
105a. Benson, R. E., and R. V. Lindsey, Jr., J. Am. Chem. Soc., 81, 4247 (1959).
106. Benzing, E., private communication.
107. Bercz, J. P., L. Horner, and C. V. Bercz, Ann. Chem., 703, 17 (1967).
108. Berei, K., J. Chromatogr., 20, 406 (1965); Kozlemen, 13, 49 (1965); Marg. Kem. Foly., 73, 313 (1967); C. A., 67, 96623n (1967).
109. Berglund, D., and D. W. Meek, J. Am. Chem. Soc., 90, 518 (1968).
110. Bergmann, E., and W. Schütz, Z. Phys. Chem. (Leipzig), B19, 401 (1932).
111. Berlé, F., Jahresber., 1855, 590; Ann. Chem., 97, 334 (1856).
112. Berlin, A. A., E. V. Kochetov, and N. S. Enikolopyan, Vysokomolekul. Soedin, 7, 187 (1965); C. A., 62, 16391 (1965).
113. Berlin, K. D., T. H. Austin, M. Nagebhushanam, M. Peterson, J. Calvert, L. A. Wilson, and D. Hopper, J. Gas Chromatogr., 3, 256 (1965); C. A., 63, 17155b (1965).
114. Berlin, K. D., T. H. Austin, and K. L. Stone, J. Am. Chem. Soc., 86, 1787 (1964).
115. Berlin, K. D., and G. B. Butler, J. Org. Chem., 26, 2537 (1961).
116. Berthaud, J., C. R. Acad. Sci. (Paris), 143, 1166 (1906).
117. Bestmann, H. J., H. Hartung, and I. Pils, Angew. Chem., 77, 1011 (1965).
118. Bestmann, H. J., and G. Hofmann, Ann. Chem., 716, 98 (1968).

119. Bestmann, H. J., and O. Klein, Tetrahedron Lett., 1966, 6181.
120. Bestmann, H. J., F. Seng, and H. Schulz, Chem. Ber., 96, 465 (1963).
120a. Bidduph, R. H., M. P. Brown, R. C. Cass, R. Long, and H. B. Silver, J. Chem. Soc., 1961, 1822.
121. Bilbo, A. J., and C. M. Sharts, J. Polym. Sci., Part A, 5, 2891 (1967).
121a. Birr, K. H., Z. Anorg. Allg. Chem., 306, 21 (1960).
122. Birum, G. H., and C. N. Matthews, J. Am. Chem. Soc., 88, 4198 (1966).
123. Bissey, J. E., and H. Goldwhite, Tetrahedron Lett., 1966, 3247.
124. Bissey, J. E., H. Goldwhite, and D. G. Rowsell, J. Org. Chem., 32, 1542 (1967).
125. Bloch, B., and D. Charrier, C. R. Acad. Sci. Paris, Ser. C, 263, 1160 (1966).
126. Bloom, S. M., Ger. 1,903,103 (1968); C. A., 72, 37736g (1970).
127. Bloomfield, P. R., Brit. 983,698 (1965); C. A., 62, 13270f (1965).
128. Bloomfield, P. R., and K. Parvin, Chem. Ind. (London), 1959, 541.
129. Bogolyubov, G. M., N. N. Grishin, and A. A. Petrov, Zh. Obshch. Khim., 39, 1808 (1969).
129a. Bogolyubov, G. M., N. N. Grishin, and A. A. Petrov, Zh. Obshch. Khim., 39, 1808 (1969).
130. Bogolyubov, G. M., K. S. Mingaleva, and A. A. Petrov, Zh. Obshch. Khim., 35, 1566 (1965).
131. Bogolyubov, G. M., and A. A. Petrov, Zh. Obshch. Khim., 33, 3774 (1963); C. A., 60, 8059b (1964).
132. Bogolyubov, G. M., and A. A. Petrov, Dokl. Akad. Nauk SSSR, 173, 1076 (1967); C. A., 67, 90887e (1967).
133. Bogolyubov, G. M., N. A. Razumova, and A. A. Petrov, Zh. Obshch. Khim., 33, 2419 (1963); C. A., 59, 14018h (1963).
134. Bokanov, A. I., B. A. Korelov, and B. I. Stepanov, Zh. Obshch. Khim., 35, 1879 (1965); C. A., 64, 1938c (1966).
135. Boor, J., Jr., J. Polym. Sci., Part A, 3, 995 (1965).
136. Borisov, A. E., A. N. Abramov, and A. N. Nesmeyanov, Izv. Akad. Nauk SSSR, Otd. Khim. Nauk, 1962, 1258; C. A., 58, 9121b (1963).
137. Borisov, A. E., A. I. Borisova, and L. V. Kudryavtseva, Izv. Akad. Nauk SSSR, 1968, 2287.
138. Boros, E. J., K. J. Coskran, R. W. King, and J. G. Verkade, J. Am. Chem. Soc., 88, 1140 (1966).
139. Borowitz, I. J., K. C. Kirby, Jr., P. E. Rusek, and E. Lord, J. Org. Chem., 34, 2687 (1969).
140. Bouquet, G., and M. Bigorgne, Spectrochim. Acta, 23A, 1231 (1967).

141. Bourneuf, M., Bull. Soc. Chim. Fr., (4), 33, 1808 (1923).
141a. Bowen, H. J. M., Trans. Faraday Soc., 50, 463 (1954).
142. Bowers, M. I., R. A. Beaudet, H. Goldwhite, and R. Tang, J. Am. Chem. Soc., 91, 17 (1969).
142a. Bowers, M. T., R. A. Beaudet, H. Goldwhite, and S. Chan, J. Chem. Phys., 52, 2831 (1970).
143. Brandt, W. W., and J. Chojnowski, Spectrochim. Acta, 25A, 1639 (1969).
144. Braun, D., H. Daimon, and G. Becker, Makromol. Chem., 62, 183 (1963).
145. Braun, J., C. R. Acad. Sci., Paris, Ser. C, 260, 218 (1965).
146. Braye, E. H., U.S., 3,449,426 (1969); C. A., 71, 91649y (1969).
147. Braye, E. H., and W. Hübel, Chem. Ind. (London), 1959, 1250.
148. Braye, E. H., W. Hübel, and I. Caplier, J. Am. Chem. Soc., 83, 4406 (1961); U.S. 3,151,140 (1964).
149. Brewis, S., W. T. Dent, and R. D. Smith, J. Chem. Soc., 1965, 1539; Brit. 970,072 (1964).
150. Britt, A. D., and E. T. Kaiser, J. Phys. Chem., 69, 2775 (1965).
151. Brophy, J. J., and M. J. Gallagher, Aust. J. Chem., 22, 1385 (1969).
152. Brophy, J. J., and M. J. Gallagher, Aust. J. Chem., 22, 1399 (1969).
153. Brophy, J. J., and M. J. Gallagher, Aust. J. Chem., 22, 1405 (1969).
154. Brophy, J. J., K. L. Freeman, and M. J. Gallagher, J. Chem. Soc., C, 1968, 2760.
155. Brown, D. H., A. Mohamed, and D. W. A. Sharp, Spectrochim. Acta, 21, 659 (1965).
156. Brown, H. C., J. Am. Chem. Soc., 67, 503 (1945).
157. Brown, H. C., U.S. 2,584,112; C.A., 46, 9580h (1952).
158. Brown, H. C., J. Chem. Soc., 1956, 1248.
159. Brown, H. C., E. A. Fletcher, E. Lawton, and S. Sujushi, A.C.S., 121st Meeting, Buffalo, 1952, Abstr., p. 9N.
160. Browning, M. C., J. R. Mellor, D. J. Morgan, S. A. J. Pratt, L. E. Sutton, and L. M. Venanzi, J. Chem. Soc., 1962, 693.
161. Bruker, A. B., M. K. Baranaev, E. I. Grinshtein, R. I. Novoselova, V. V. Prokhorova, and L. Z. Soborovskii, Zh. Obshch. Khim., 33, 1919 (1963); C. A., 59, 11207h (1963).
162. Bruker, A. B., E. I. Grinshtein, and L. Z. Soborovskii, Zh. Obshch. Khim., 36, 484 (1966); C. A., 65, 740d (1966).

163. Bruker, A. B., E. I. Grinshtein, and L. Z. Soborovskii, Zh. Obshch. Khim., 36, 1133 (1966); C. A., 65, 12230f (1966).
164. Bruker, A. B., Kh. R. Raver, and L. Z. Soborovskii, Probl. Organ. Sinteza, Akad. Nauk SSSR, Otd. Obshch. Tekhn. Khim., 1965, 285; C. A., 64, 6681c (1966).
165. Bublitz, D. E., and A. Baker, J. Organometal. Chem., 9, 383 (1967).
166. Bucci, P., J. Am. Chem. Soc., 90, 252 (1968).
167. Buckler, S. A., J. Org. Chem., 24, 1460 (1959).
168. Buckler, S. A., J. Am. Chem. Soc., 82, 4215 (1960).
168a. Buckler, S. A., U.S. 2,969,390 (1961); C. A., 55, 14381e (1961).
169. Buckler, S. A., U.S. 3,005,020 (1959); C. A., 56, 6002f (1962).
170. Buckler, S. A., and L. Doll, U.S. 2,999,882; C. A., 56, 2473f (1962).
171. Buckler, S. A., L. Doll, F. K. Lind, and M. Epstein, J. Org. Chem., 27, 794 (1962).
172. Buckler, S. A., and M. Epstein, J. Am. Chem. Soc., 82, 2076 (1960).
173. Buckler, S. A., and M. Epstein, Tetrahedron, 18, 1211 (1962).
174. Buckler, S. A., and M. Epstein, Tetrahedron, 18, 1221 (1962); U.S. 3,052,719 (1962); C. A., 58, 551d (1963).
175. Buckler, S. A., and M. Epstein, U.S. 3,213,042 (1965); C. A., 64, 2190f (1966).
175a. Buckler, S. A., and V. P. Wystrach, J. Am. Chem. Soc., 80, 6454 (1958).
176. Buckler, S. A., and V. P. Wystrach, J. Am. Chem. Soc., 83, 168 (1961); Buckler, S. A., U.S. 2,894,683 (1961); C. A., 55, 22347a (1961).
177. Bugerenko, E. F., E. A. Chernyshev, and A. D. Petrov, Izv. Akad. Nauk SSSR, Ser. Khim., 1965, 286; C. A., 62, 14721h (1965).
178. Burch, G. M., H. Goldwhite, and R. N. Haszeldine, J. Chem. Soc., 1963, 1083.
179. Burch, G. M., H. Goldwhite, and R. N. Haszeldine, J. Chem. Soc., 1964, 572.
180. Burdon, J., I. N. Rozhkov, and G. M. Perry, J. Chem. Soc., C, 1969, 2615.
181. Burg, A. B., J. Inorg. Nucl. Chem., 11, 258 (1959).
182. Burg, A. B., J. Am. Chem. Soc., 83, 2226 (1961).
183. Burg, A. B., Inorg. Chem., 3, 1325 (1964).
184. Burg, A. B., and G. Brendel, J. Am. Chem. Soc., 80, 3198 (1958).
185. Burg, A. B., and L. R. Grant, U.S. 3,118,951 (1964); C. A., 60, 10718b (1964).
186. Burg, A. B., and K. K. Joshi, J. Am. Chem. Soc., 86, 353 (1964).

187. Burg, A. B., K. K. Joshi, and J. F. Nixon, J. Am. Chem. Soc., 88, 31 (1966).
188. Burg, A. B., and W. Mahler, J. Am. Chem. Soc., 79, 4242 (1957).
189. Burg, A. B., and W. Mahler, J. Am. Chem. Soc., 83, 2388 (1961).
190. Burg, A. B., and W. Mahler, U.S. 2,866,824 (1958); C. A., 53, 10037e (1959).
191. Burg. A. B., W. Mahler, A. J. Bilbo, C. P. Haber, and D. L. Herring, J. Am. Chem. Soc., 79, 247 (1957).
192. Burg, A. B., and I. B. Mishra, Inorg. Chem., 8, 1199 (1969).
193. Burg, A. B., and P. J. Slota, Jr., J. Am. Chem. Soc., 82, 2148 (1960).
194. Burg, A. B., and P. J. Slota, Jr., J. Am. Chem. Soc., 82, 2145 (1960).
195. Burg, A. B., and P. J. Slota, Jr., U.S. 2,877,272 (1959); C. A., 53, 16062g (1959).
196. Burg, A. B., and G. B. Street, J. Am. Chem. Soc., 85, 3522 (1963).
197. Butler, G. B., and G. L. Statton, J. Am. Chem. Soc., 86, 5045 (1964).
198. Cahours, A., Ann. Chem., 122, 331 (1862).
199. Cahours, A., and A. W. Hofmann, Ann. Chem., 104, 1, 29 (1857).
200. Campbell, I. G. M., R. C. Cookson, and M. B. Hocking, Chem. Ind. (London), 1962, 359.
201. Campbell, I. G. M., R. C. Cookson, M. B. Hocking, and A. N. Hughes, J. Chem. Soc., 1965, 2184.
202. Campbell, J. R., and R. E. Hatton, U.S. 3,020,315 (1962); C. A., 56, 12948c (1962).
203. Campbell, I. G. M., and J. K. Way, J. Chem. Soc., 1960, 5034.
204. Campbell, I. G. M., and J. K. Way, J. Chem. Soc., 1961, 2133.
204a. Campbell, T. W., and J. Verbanc, U.S. 2,853,473 (1958); C. A., 53, 10126e (1959); J. Am. Chem. Soc., 84, 3673 (1962).
205. Carius, L., Ann. Chem., 137, 117 (1866).
205a. Carmody, D. R., and A. Zletz, U.S. 2,857,736 (1958); C. A. 53, 4703c (1959).
206. Carty, A. J., R. K. Harris, Chem. Commun., 1967, 234.
207. Casey, J. P., R. A. Lewis, and K. Mislow, J. Am. Chem. Soc., 91, 2789 (1969).
208. Cavell, R. G., and R. C. Dobbie, J. Chem. Soc., A, 1967, 1308.
209. Cavell, R. G., R. C. Dobbie, and W. J. R. Tyerman, Can. J. Chem., 45, 2849 (1967).
210. Chabardes, P., C. Grard, P. Lafont, and M. Thiers, Fr. 1,366,081 (1964); C. A., 61, 14538h (1964).
211. Chalk, A. J., U.S. 3,188,300 (1965); C. A., 63, 7043d (1965).

212. Chan, T. H., Chem. Commun., 1968, 895.
213. Chan, S., H. Goldwhite, H. Keyzer, D. G. Rowsell, and R. Tang, Tetrahedron, 25, 1097 (1969).
214. Chan, S., H. Goldwhite, H. Keyzer, and R. Tang, Spectrochim. Acta., 26A, 249 (1970).
215. Chance, L. H., W. A. Reeves, and G. L. Drake, Jr., U.S. 3,270,052 (1966); C. A., 65, 13867e (1966).
216. Chance, L. H., D. J. Daigle, and G. L. Drake, J. Chem. Eng. Data, 12, 282 (1967).
217. Chaplin, E. C., and A. Y. Garner, U.S. 3,158,642 (1964); C. A., 62, 6592e (1965).
218. Charrier, C., W. Chodkiewicz, and P. Cadiot, Bull. Soc. Chim. Fr., 1966, 1002.
219. Charrier, C., M. P. Simonnin, W. Chodkiewicz, and P. Cadiot, C. R. Acad. Sci., Paris, Ser. C, 258, 1537 (1964).
219a. Chatt, J., L. A. Duncanson, and B. L. Shaw, Proc. Chem. Soc., 1957, 343.
220. Chatt, J., and F. A. Hart, J. Chem. Soc., 1960, 1378.
221. Chatt, J., F. A. Hart, and H. C. Fielding, Brit. 859,391 (1961); C. A., 55, 23345a (1961).
222. Chatt, J., F. A. Hart, and H. C. Fielding, U.S. 2,922,819 (1960); C. A., 54, 9847b (1960).
223. Chatt, J., F. A. Hart, and H. C. Fielding, Brit. 877,592; C. A., 56, 6002a (1962).
224. Chatt, J., and R. G. Hayter, J. Chem. Soc., 1961, 896.
225. Chemische Fabrik Kalk GmbH., Fr. 1,400,892 (1965); C. A., 64, 2238f (1966).
226. Chen, Y. T., Sci. Sinica (Peking), 14, 936 (1965); C. A., 63, 6331d (1965).
227. Chini, P., G. De Venuts, T. Salvatori, and M. De Malde, Chim. Ind. (Milan), 46, 1049 (1964).
228. Chiswell, B., Aust. J. Chem., 20, 2533 (1967).
229. Chiswell, B., and L. M. Venanzi, J. Chem. Soc., A, 1966, 417.
230. Chiswell, B., and L. M. Venanzi, J. Chem. Soc., A, 1966, 901.
231. Chodkiewicz, W., P. Cadiot, and A. Willemart, C. R. Acad. Sci., Paris, Ser. C., 250, 866 (1960).
232. Clark, R. J. H., R. H. U. Negrotti, and R. S. Nyholm, Chem. Commun., 1966, 486.
232a. Clayton, A. P., P. A. Fowell, and C. T. Mortimer, J. Chem. Soc., 1960, 3284.
233. Clifford, A. F., and R. R. Olsen, A.C.S., 135th Meeting, Boston, 1959, Abstr., p. 16M.
234. Cloez, S., Ann. Chim. (Paris), (3), 17, 311 (1846).
235. Coates, H., and P. A. T. Hoye, Brit. 842,593 (1960); C. A., 55, 4363c (1961); Coates, H., and P. A. T. Hoye, Ger. 1,077,214 (1960).

236. Coates, H., and P. A. T. Hoye, Ger. 1,096,905 (1961).
237. Coates, H., P. A. T. Hoye, and J. W. Wallis, Brit. 970,815 (1964); C. A., 61, 14712c (1964).
238. Coates, H., and J. J. Lawless, U.S. 3,050,522 (1962); C. A., 58, 6975f (1963).
239. Cockerille, F. O., U.S. 2,462,163 (1949); C. A., 43, 3838e (1949).
240. Codell, M., J. Chem. Eng. Data, 8, 460 (1963).
241. Coffmann, D. D., and C. S. Marvel, J. Am. Chem. Soc., 51, 3496 (1929).
242. Collie, N., Trans. Chem. Soc., 53, 636, 714 (1888).
243. Colton, R., and Q. N. Porter, Aust. J. Chem., 21, 2215 (1968).
244. Compagnie de Saint-Gobain, Fr. 1,357,679 (1964); C. A., 61, 12167d (1964).
245. Conen, F., Chem. Ber., 31, 2919 (1898).
246. Cooke, M., M. Green, and D, Kirkpatrick, J. Chem. Soc., A, 1968, 1507.
247. Cookson, E. A., and P. C. Crofts, J. Chem. Soc., C, 1966, 2003.
248. Cooper, B. E., and W. J. Owen, J. Organometal. Chem., 21, 329 (1970).
248a. Coover, H. W., and J. B. Dickey, U.S. 2,675,372 (1954); C. A., 48, 8587d (1954).
249. Corbridge, D. E. C., in Topics in Phosphorus Chemistry, M. Grayson and E. J. Griffith, Eds., Vol. 6, Interscience, 1969, p. 235.
250. Coskran, K. J., and J. G. Verkade, Inorg. Chem., 4, 1655 (1965).
251. Cottrell, R. T., and R. A. N. Morris, Chem. Commun., 1968, 409.
252. Courtaulds Ltd., Brit. 902,602 (1962); C. A., 57, 12739b (1962).
253. Cowley, A. H., and M. H. Hnoosh, J. Am. Chem. Soc., 88, 2595 (1966).
254. Cowley, A. H., and J. L. Mills, J. Am. Chem. Soc., 91, 2915 (1969).
255. Cowley, A. H., and M. W. Taylor, J. Am. Chem. Soc., 91, 1929 (1969).
256. Cram, D. J., and R. D. Partos, J. Am. Chem. Soc., 85, 1093 (1963).
257. Cremer, S. E., and R. J. Chorvat, J. Org. Chem., 32, 4066 (1967).
257a. Cremer, S. E., Chem. Commun., 1970, 616.
258. Cremer, S. E., R. J. Chorvat, C. H. Chang, and D. W. Davis, Tetrahedron Lett., 1968, 5799.
259. Crofts, P. C., and G. M. Kosolapoff, J. Am. Chem. Soc., 75, 3379 (1953).
260. Crosbie, K. D., and G. M. Sheldrick, J. Inorg. Nucl. Chem., 31, 3684 (1969).

261. Crosly, G. W., and A. F. Millikan, U.S. 2,983,681 (1957); C. A., 56, 2659e (1962).
262. Crutchfield, M. M., C. H. Dungan, J. H. Letcher, V. Mark, and J. R. Van Wazer, in Topics in Phosphorus Chemistry, M. Grayson and E. J. Griffith, Eds., Vol. 5, Interscience, 1967.
263. Cullen, W. R., Can. J. Chem., 38, 439 (1960).
264. Cullen, W. R., and D. S. Dawson, Can. J. Chem., 45, 2887 (1967).
265. Cullen, W. R., D. S. Dawson, and P. S. Dhaliwal, Can. J. Chem., 45, 683 (1967).
266. Cullen, W. R., D. S. Dawson, and G. E. Styan, Can. J. Chem., 43, 3392 (1965).
267. Cullen, W. R., D. F. Dong, and J. A. J. Thompson, Can. J. Chem., 47, 4671 (1969).
268. Cullen, W. R., and D. C. Frost, Can. J. Chem., 40, 390 (1962).
269. Cullen, W. R., D. A. Harbourne, B. V. Liengme, and J. R. Sams, Inorg. Chem., 8, 95 (1969).
270. Cumper, C. W. N., A. A. Foxton, J. Read, and A. I. Vogel, J. Chem. Soc., 1964, 430.
271. Cuneo, G., Atti Accad. Naz. Lincei, Rend., 32, II, 230 (1923).
272. Cuneo, G., Atti Reale Accad. Lincei (5), 32, II, 357 (1923).
273. Czempik, H., Doctoral thesis, University of Braunschweig, 1957.
274. Czimatis, L., Chem. Ber., 15, 2014 (1882).
275. Daasch, L. W., and D. C. Smith, Anal. Chem., 23, 853 (1951).
275a. Dahl, O., N. C. Gelting, and O. Larsen, Acta Chem. Scand., 23, 3369 (1969).
276. Dahl, O., and O. Larson, Acta Chem. Scand., 22, 2037 (1968).
277. Daly, J. J., J. Chem. Soc., 1964, 3799; Z. Kristallogr., 118, 332 (1963).
278. Daly, J. J., and L. Maier, Nature, 208, 383 (1965).
279. Darmstädter, L., and A. Henninger, Chem. Ber., 3, 179 (1870).
280. Davidson, N., and H. C. Brown, J. Am. Chem. Soc., 64, 316 (1942).
281. Davidson, N., and H. C. Brown, J. Am. Chem. Soc., 64, 718 (1942).
282. Davies, J. H., J. D. Downer, and P. Kirby, J. Chem. Soc., C, 1966, 245; Fr. 1,488,936 (1967).
283. Davies, M., and F. G. Mann, Chem. Ind. (London), 1962, 1539.
283a. Davies, W. C., J. Chem. Soc., 1933, 1043.
284. Davies, W. C., J. Chem. Soc., 1935, 462.
285. Davies, W. C., and H. W. Addis, J. Chem. Soc., 1937, 1622.

286. Davies, W. C., and W. J. Jones, J. Chem. Soc., 1929, 33.
287. Davies, W. C., and W. P. G. Lewis, J. Chem. Soc., 1934, 1599.
288. Davies, W. C., and F. G. Mann, J. Chem. Soc., 1944, 276.
289. Davies, W. C., and C. J. O. R. Morris, J. Chem. Soc., 1932, 2880.
290. Davies, W. C., and C. J. O. R. Morris, Bull. Soc. Chim. Belges, 53, 980 (1933).
291. Davies, W. C., P. L. Pearse, And W. J. Jones, J. Chem. Soc., 1929, 1262.
292. Davies, W. C., and W. P. Walters, J. Chem. Soc., 1935, 1786.
293. Davis, M., and F. G. Mann, J. Chem. Soc., 1964, 3770.
294. Davis, M., and F. G. Mann, J. Chem. Soc., 1964, 3786.
295. Dawson, J. W., and L. M. Venanzi, J. Am. Chem. Soc., 90, 7229 (1968).
296. Deacon, G. B., and J. H. S. Green, Spectrochim. Acta, 24A, 845 (1968).
297. Deacon, G. B., and R. A. Jones, Aust. J. Chem., 16, 499 (1963).
298. Deacon, G. B., and J. C. Parrott, J. Organometal. Chem., 17, P17 (1969); 22, 287 (1970).
299. DeBruin, K. E., G. Zon, K. Naumann, and K. Mislow, J. Am. Chem. Soc., 91, 7027 (1969).
300. Deeming, A. J., and B. L. Shaw, J. Chem. Soc., A, 1969, 597.
301. Degischer, G., and G. Schwarzenbach, Helv. Chim. Acta, 49, 1927 (1966).
302. DeKetelaere, R. M. E., W. Vanermen, E. Claeys, and G. P. Van der Kelen, Bull. Soc. Chim. Belges, 78, 219 (1969); C. A., 71, 50109d (1969).
303. DeKoe, P., and F. Bickelhaupt, Angew. Chem., 79, 533 (1967); Angew. Chem. Int. Ed., 6, 567 (1967).
304. DeKoe, P., R. Van Veen, and F. Bickelhaupt, Angew. Chem., 80, 486 (1968); Angew. Chem. Int. Ed., 7, 465 (1968).
305. DeKoe, P., and F. Bickelhaupt, Angew. Chem., 80, 912 (1968); Angew. Chem. Int. Ed., 7, 889 (1968).
306. Denney, D. B., and F. C. Gross, J. Org. Chem., 32, 2445 (1967).
307. Denney, D. B., and F. J. Gross, J. Org. Chem., 32, 3710 (1967).
308. Denney, D. B., and L. C. Smith, J. Org. Chem., 27, 3404 (1962).
309. Derkach, G. I., and E. S. Gunitskaya, Zh. Obshch. Khim., 34, 604 (1964).

310. Derkach, N. Ya., and A. V. Kirsanov, Zh. Obshch. Khim., 38, 331 (1968); C. A., 69, 96832w (1968).
311. Dessy, R. E., T. Chivers, and W. Kitching. J. Am. Chem. Soc., 88, 467 (1966).
312. Deutsche Advance Produktion GmbH, Neth. 6,614,945 (1967); C. A., 67, 108757z (1967).
313. DeVault, A. N., U.S. 3,074,230 (1963); C. A., 59, 9715f (1963).
314. Dickson, R. S., and B. O. West, Aust. J. Chem., 15, 710 (1962).
315. Dimroth, K., N. Greif, H. Perst, F. W. Steuber, W. Sauer, and L. Duttka, Angew. Chem., 79, 58 (1967).
316. Dimroth, K., N. Greif, W. Städe, and F. W. Steuber, Angew. Chem., 79, 725 (1967); Angew. Chem. Int. Ed., 6, 711 (1967).
317. Dimroth, K., A. Hettche, W. Städe, and F. W. Steuber, Angew. Chem., 81, 784 (1969).
318. Dimroth, K., and P. Hoffmann, Angew. Chem., 76, 433 (1964).
319. Dimroth, K., and P. Hoffmann, Chem. Ber., 99, 1325 (1966).
320. Dimroth, K., and W. Mach, Angew. Chem., 80, 489 (1968); Angew. Chem. Int. Ed., 7, 460 (1968).
321. Dimroth, K., and W. Städe, Angew. Chem., 80, 966 (1968); Angew. Chem. Int. Ed., 7, 881 (1968).
322. Dimroth, K., F. W. Steuber, Angew. Chem., 79, 410 (1967); Angew. Chem., Int. Ed., 6, 445 (1967).
323. Dimroth, K., K. Vogel, W. Mach, and U. Schoeler, Angew. Chem., 80, 359 (1968); Angew. Chem. Int. Ed., 7, 371 (1968).
324. Dobbie, R. C., M. Green, and F. G. A. Stone, J. Chem. Soc., A, 1969, 1881.
325. Dobson, G. R., R. C. Taylor, and T. D. Walsh, Chem. Commun., 1966, 281.
326. Dodonow, J., and H. Medox, Chem. Ber., 61, 907 (1928).
327. Dörken, C., Chem. Ber., 21, 1505 (1888).
329. Dolle, R. E., Jr., F. J. Harsacky, and C. Tamborski, U.S. 3,481,872, 3,483,129 (1969); C. A., 72, 34097p, 34086j (1970).
330. Donner, R., and Kh. Lohs, J. Chromatogr., 17, 349 (1965).
331. Doyle, T. E., F. Fekete, P. J. Keenan, and W. J. Plant, U.S. 3,317,465 (1967); C. A., 67, 12096g (1967).
331a. Draper, P. M., T. H. Chan, and D. N. Harpp, Tetrahedron Lett., 1970, 1687.
332. Drechsei, E., and E. Finkelstein, Chem. Ber., 4, 352 (1871).
333. Drenth, W., and A. Hogervorst, Rec. Trav. Chim., 87, 41 (1968).

334. Drenth, W., and D. Rosenberg, Rec. Trav. Chim., 86, 26 (1967); 81, 635 (1962).
334a. Drucker, A., and M. Grayson, U.S. 3,489,811 (1970); C. A., 72, 67098t (1970).
335. Du Bois, T. D., and D. W. Meek, Inorg. Chem., 8, 146 (1969).
336. Dubov, S. S., F. N. Chelobov, and R. N. Sterlin, Zh. Vses. Khim. Obshchestva, 7, 585 (1962); C. A., 58, 2970e (1963).
337. Dubov, S. S., B. I. Tetel'baum, and R. N. Sterlin, Zh. Vses. Khim. Obshchestva, 7, 691 (1962); C. A., 58, 8538c (1963).
338. Duffield, A. M., H. Budzikiewicz, and C. Djerassi, J. Am. Chem. Soc., 87, 2920 (1965).
339. Dwek, R. A., and R. E. Richards, Chem. Commun., 1966, 581.
340. Dyer, J., and J. Lee, Trans. Faraday Soc., 62, 257 (1966); Proc. Chem. Soc., 1963, 275.
340a. Dyer, J., and J. Lee, J. Chem. Soc., B, 1970, 409.
341. Dyer, G., and D. W. Meek, J. Am. Chem. Soc., 89, 3983 (1967).
342. Dyer, G., and D. W. Meek, Inorg. Chem., 6, 149 (1967).
342a. Eaton, J. K., and R. G. Davies, Ann. Appl. Biol., 37, 92 (1950); C. A., 44, 9613f (1950).
343. Edgell, W. F., and M. P. Dunkle, Inorg. Chem., 4, 1629 (1965).
343a. Egan, W., R. Tang, G. Zon, and K. Mislow, J. Am. Chem. Soc., 92, 1442 (1970).
344. Eifert, R. L., and B. M. Marks, Brit. 872,331 (1961); C. A., 58, 3522a (1963).
345. Eimers, E., H. Schirmer, and K. Prater, Belg. 659,101 (1965); C. A., 64, 6844c (1966).
346. Eisenbraun, A. A., and B. E. Lloyd, Fr. 1,353,446 (1964); C. A.; 61, 8443a (1964).
347. Eldred, R. J., U.S. 3,450,616 (1969); C. A., 71, 61884d (1969).
348. Eldred, R. J., J. Polym. Sci., Part A, 7, 265 (1969).
349. Eliseeva, N. V., V. A. Sharpatyi, and A. N. Pravednikov, Zh. Strukt. Khim., 7, 511 (1966); C. A., 65, 17946e (1966).
349a. Eller, P. G., and D. W. Meek, J. Organometal. Chem., 22, 631 (1970).
350. Ellermann, J., and K. Dorn, Chem. Ber., 99, 653 (1966).
351. Ellermann, J., F. Poersch, R. Kunstmann, and R. Kramolowsky, Angew. Chem., 81, 183 (1969).
352. Elser, H., and H. Dreeskamp, Ber. Bunsenges. Phys. Chem., 73, 619 (1969).
353. Emeléus, H. J., and J. M. Miller, J. Inorg. Nucl. Chem., 28, 662 (1966).
354. Epshtein, L. M., Z. S. Novikova, L. D. Ashkinadze, and V. N. Kostylev, Dokl. Akad. Nauk SSSR, Ser. Khim., 184, 1346 (1969).

355. Epstein, M., and S. A. Buckler, J. Am. Chem. Soc., 83, 3279 (1961); U.S. 3,050,531 (1962); C. A., 57, 16659c (1962).
356. Epstein, M., and S. A. Buckler, Tetrahedron, 18, 1231 (1962).
357. Ettel, V., and J. Horák, Collect. Czech. Chem. Commun., 25, 2191 (1960); C. A., 55, 5402g (1961).
357a. Ettel, V., and J. Horák, Collect. Czech. Chem. Commun., 26, 1949 (1961).
358. Ettel, V., and J. Horák, Collect. Czech. Chem. Commun., 26, 2087 (1961).
359. Evans, J. G., P. L. Goggin, R. J. Goodfellow, and J. G. Smith, J. Chem. Soc., A, 1968, 464.
360. Evans, P. N., and C. E. Vanderkleed, Am. Chem. J., 27, 142 (1902).
361. Evers, E. C., E. H. Street, Jr., and S. L. Jung, J. Am. Chem. Soc., 73, 5088, 2038 (1951).
362. Evleth, E. M., L. V. D. Freeman, and R. I. Wagner, J. Org. Chem., 27, 2192 (1962).
363. Eyles, C. T., and S. Trippett, J. Chem. Soc., C, 1966, 67; Proc. Chem. Soc., 1963, 19.
364. Ezzell, B. R., and L. D. Freedman, J. Org. Chem., 34, 1777 (1969).
365. Farbenfabriken Bayer, A.G., Belg. 654,413 (1965); C. A., 64, 16017d (1966).
366. Farbwerke Hoechst A.G., Belg. 633,671 (1963); C. A., 61, 580b (1964).
367. Farbwerke Hoechst A.G., Brit. 951,858 (1964); C. A., 61, 6842a (1964).
368. Farbwerke Hoechst A.G., Neth. 6,508,185 (1965); C. A., 64, 15747h (1966).
369. Farbwerke Hoechst A.G., Belg. 660,900 (1965); Neth. 6,507,477 (1965); C. A., 64, 14310g (1966).
370. Farbenfabriken Bayer A.G., Neth. 6,506,043 (1965); C. A., 65, 17808f (1966).
371. Farbwerke Hoechst A.G., Neth. 6,512,332 (1966); C.A., 65, 10728d (1966).
372. Fedorova, G. K., and O. V. Kirsanov, Zh. Obshch. Khim., 33, 1011 (1963); C. A., 59, 8785a (1963).
373. Fedorova, G. K., and G. A. Lanchuk, Zh. Obshch. Khim., 34, 511 (1964); C. A., 60, 12048f (1964).
374. Fedorova, G. K., L. S. Moskalevskaya, and A. V. Kirsanov, Zh. Obshch. Khim., 39, 1227 (1969); C. A., 71, 124585e (1969).
375. Feigl, F., and D. Goldstein, Michrochem. Acta, 1966, 1.
376. Feinland, R., J. Sass, and S. A. Buckler, Anal. Chem., 35, 920 (1963); C. A., 59, 2167c (1963).
377. Fekete, F., U.S. 3,019,248 (1962); C. A., 57, 11238c (1962); U.S. 3,067,229 (1962); C. A., 58, 10239b (1963).

377a. Fekete, F., U.S. 2,995,594 (1961); C. A., 56, 3516d (1962).
378. Fenton, G. F., and C. K. Ingold, J. Chem. Soc., 1929, 2342.
379. Fenton, G. F., L. Hey, and C. K. Ingold, J. Chem. Soc., 1933, 989.
380. Ferretti, J. A., and L. Paolillo, Ric. Sci., 36, 1008 (1966); C. A., 66, 104561n (1967).
381. Ferruti, P., and R. Alimardanov, Chim. Ind. (Milan), 49, 831 (1967); C. A., 67, 91154a (1967).
382. Feshchenko, N. G., T. I. Alekseeva, and A. V. Kirsanov, Zh. Obshch. Khim., 38, 122 (1968); C. A., 69, 52210n (1968).
382a. Feshchenko, N. G., Zh. K. Gorbatenko, and A. V. Kirsanov, Zh. Obshch. Khim., 39, 2596 (1969).
383. Feshchenko, N. G., I. K. Mazepa, Yu. P. Makovetskii, and A. V. Kirsanov, Zh. Obshch. Khim., 39, 1886 (1969).
384. Fields, R., M. Green, and A. Jones, J. Chem. Soc., A, 1969, 2740.
385. Fields, R., H. Goldwhite, R. N. Haszeldine, and J. Kirman, J. Chem. Soc., C, 1966, 2075.
385a. Fields, R., R. N. Haszeldine, and J. Kirman, J. Chem. Soc., C, 1970, 197.
385b. Fields, R., R. N. Haszeldine, and N. F. Wood, J. Chem. Soc., C, 1970, 744.
385c. Fild, M., and O. Glemser, Fluorine Chem. Rev., 3, 129 (1969).
386. Fild, M., O. Glemser, and G. Christoph, Angew. Chem., 76, 953 (1964).
387. Fild, M., O. Glemser, and J. Hollenberg, Naturwissenschaften, 52, 590 (1965).
388. Fild, M., O. Glemser, and I. Hollenberg, Z. Naturforsch., B, 21, 920 (1966).
389. Fild, M., I. Hollenberg, and O. Glemser, Z. Naturforsch., B, 22, 248 (1967).
390. Fild, M., I. Hollenberg, and O. Glemser, Z. Naturforsch., B, 22, 253 (1967).
391. Finegold, H., Ann. N. Y. Acad. Sci., 70, 875 (1958).
392. Finholt, A. E., C. Helling, V. Imhof, L. Nielsen, and E. Jacobson, Inorg. Chem., 2, 504 (1963).
393. Fireman, P., Chem. Ber., 30, 1088 (1897).
394. Firth, W. C., S. Frank, M. Gerber, and V. P. Wystrach, Inorg. Chem., 4, 765 (1965).
395. Fischer, W., E. Hellner, A. Chatzidakis, and K. Dimroth, Tetrahedron Lett., 1968, 6227.
396. Fischler, J., and M. Halmann, J. Chem. Soc., 1964, 31.
397. Fishwick, S. E., and J. A. Flint, Chem. Commun., 1968, 182.

398. Fontal, B., H. Goldwhite, and D. G. Rowsell, J. Org. Chem., 31, 2424 (1966).
398a. Forward, M. V., S. T. Bowden, and W. J. Jones, J. Chem. Soc., 1949, S121.
399. Foster, D. J., Brit. 870,425 (1961).
400. Freedmann, L. D., and G. O. Doak, J. Am. Chem. Soc., 74, 3414 (1952).
401. Frank, A. W., Chem. Rev., 61, 389 (1961).
402. Frank, A. W., and I. Gordon, Can. J. Chem., 44, 2593 (1966).
403. Frisch, K. C., and H. Lyons, J. Am. Chem. Soc., 75, 4078 (1953).
404. Frisch, K. C., and H. Lyons, U.S. 2,673,210 (1954); C. A., 49, 4018h (1955).
405. Fritz, H. P., I. R. Gordon, K. E. Schwarzhans, and L. M. Venanzi, J. Chem. Soc., 1965, 5210.
406. Fritz, H. P., and K. E. Schwarzhans, J. Organometal. Chem., 5, 103 (1966).
407. Fritzsche, H., U. Hasserodt, and F. Korte, Chem. Ber., 97, 1988 (1964).
408. Fritzsche, H., U. Hasserodt, and F. Korte, Ger. 1,223,838 (1966); C. A., 65, 18619d (1966).
409. Fritzsche, H., U. Hasserodt, and F. Korte, Chem. Ber., 98, 171 (1965); Fr. Add. 87882 (1966).
410. Fritzsche, H., U. Hasserodt, and F. Korte, Chem. Ber., 98, 1681 (1965); Ger. 1,203,773 (1965).
411. Fluck, E., and H. Binder, Z. Naturforsch., B, 22, 1001 (1967).
412. Fluck, E., and K. Issleib, Chem. Ber., 98, 2674 (1965).
413. Fluck, E., and J. Lorenz, Z. Naturforsch., B, 22, 1095 (1967).
414. Fluck, E., and R. M. Reinisch, Chem. Ber., 95, 1388 (1962).
415. Frunze, T. M., V. V. Korshak, V. V. Kurashev, G. S. Kolesnikov, and B. A. Zhubanov, Izv. Akad. Nauk SSSR, Otd. Khim. Nauk, 1958, 783; C. A., 52, 20001f (1958).
415a. Fung, B. M., and I. Y. Wei, J. Am. Chem. Soc., 92, 1497 (1970).
416. Gagnaire, D., and M. St. Jacques, J. Phys. Chem., 73, 1678 (1969).
417. Gagnaire, D., J. B. Robert, and J. Verrier, Chem. Commun., 1967, 819.
418. Gallagher, M. J., Aust. J. Chem., 21, 1197 (1968).
419. Gallagher, M. J., E. C. Kirby, and F. C. Mann, J. Chem. Soc., 1963, 4846.
420. Gallagher, M. J., and F. G. Mann, J. Chem. Soc., 1962, 5110.
421. Gallagher, M. J., and F. G. Mann, J. Chem. Soc., 1963, 4855.

422. Garner, A. Y., U.S. 3,010,999 (1961); C. A., 56, 5002c (1962).
423. Garner, A. Y., U.S. 3,010,946 (1959); 3,127,357 (1964); C. A., 60, 16005f (1964).
424. Garner, A. Y., U.S. 3,235,536 (1966); C. A., 64, 17748d (1966).
425. Garner, A. Y., and A. A. Tedeschi, J. Am. Chem. Soc., 84, 4734 (1962); U.S. 3,267,149 (1966).
426. Gaspar, P. P., S. A. Bock, and W. C. Eckelman, J. Am. Chem. Soc., 90, 6914 (1968).
427. Gee, W., R. A. Shaw, and B. C. Smith, Inorg. Syn., 9, 19 (1967).
428. Geigy, A. G., Fr. 1,401,930 (1965), C. A., 63, 11615g (1963).
428a. George, R., and J. Bjerrum, Acta Chem. Scand., 22, 497 (1968).
429. Gier, T. E., J. Am. Chem. Soc., 83, 1769 (1961); U.S. 3,051,956 (1962); C. A., 58, 9104b (1963).
429a. Gillis, R. G., G. J. Long, Org. Mass Spectrometry, 2, 1315 (1969); C. A., 72, 60496n (1970).
430. Gilman, H., and G. E. Brown, J. Am. Chem. Soc., 67, 824 (1945).
431. Gilman, H., and J. D. Robinson, Rec. Trav. Chim., 48, 328 (1929).
432. Gilman, H., and C. G. Stuckwisch, J. Am. Chem. Soc., 63, 2844 (1941).
433. Gilman, H., and C. C. Vernon, J. Am. Chem. Soc., 48, 1063 (1926).
434. Gilyarov, V. A., The Chemistry and Applications of Organophosphorus Compounds (in Russian), Vol. II, Izd. HN SSSR, 1962, p. 75.
435. Gilyarov, V. A., V. Yu. Kovtun, and M. I. Kabachnik, Izv. Akad. Nauk SSSR, Ser. Khim., 1967, 1159.
436. Ginsburg, V. A., and N. F. Privezentseva, Zh. Obshch. Khim., 28, 736 (1958); C. A., 52, 17092g (1958).
437. Girard, A., Ann. Chim. (Paris), (6), 2, 1 (1884).
438. Gladshtein, B. M., V. G. Noskov, and L. Z. Soborovskii, U.S.S.R. 166,681 (1964); C. A., 62, 10461g (1965).
439. Gloyne, D., and H. G. Henning, Angew. Chem., 78, 907 (1966).
440. Goetz, H., and S. Domin, Ann. Chem., 704, 1 (1967).
441. Goetz, H., and H. Juds, Ann. Chem., 698, 1 (1966).
442. Goetz, H., and B. Klabuhn, Ann. Chem., 724, 1 (1969).
443. Goetz, H., F. Nerdel, and K. H. Wiechel, Ann. Chem., 665, 1 (1963).
444. Goetz, H., and D. Probst, Ann. Chem., 715, 1 (1968).
445. Goetz, H., and A. Sidhu, Ann. Chem., 682, 71 (1965).
446. Goldberry, R. E., D. E. Lewis, and K. Cohn, J. Organometal. Chem., 15, 491 (1968).
447. Goldwhite, H., J. Chem. Soc., 1965, 3901.

448. Goldwhite, H., R. N. Haszeldine, and D. G. Rowsell, J. Chem. Soc., 1965, 6875.
449. Goldwhite, H., R. N. Haszeldine, and D. G. Rowsell, Chem. Commun., 1965, 83.
450. Goldwhite, H., and D. G. Rowsell, J. Phys. Chem., 72, 2666 (1968).
451. Gonnet, C., and A. Lamotte, Bull. Soc. Chim. Fr., 1969, 2932.
452. Gordon, E. G., and O. Koenig, U.S. 3,050,499 (1962); C. A., 57, 16885e (1962).
453. Gordon, I., and G. M. Wagner, U.S. 3,257,460 (1966); C. A., 65, 8962b (1966).
454. Gosling, K., D. J. Holman, J. D. Smith, and B. N. Ghose, J. Chem. Soc., A, 1968, 1909.
455. Goubeau, J., R. Baumgärtner, W. Koch, and U. Müller, Z. Anorg. Allg. Chem., 337, 174 (1965).
456. Goubeau, J., and D. Langhardt, Z. Anorg. Allg. Chem., 338, 163 (1965).
457. Goubeau, J., and G. Wenzel, Z. Phys. Chem., 45, 31 (1965).
458. Gough, S. T. D., and S. Trippett, J. Chem. Soc., 1961, 4263.
458a. Graham, W. A. G., and F. G. A. Stone, J. Inorg. Nucl. Chem., 3, 164 (1956).
459. Grayson, M., J. Am. Chem. Soc., 85, 79 (1963).
460. Grayson, M., Ger. 1,151,255 (1963);C. A., 60, 554h (1964).
461. Grayson, M., U.S. 3,243,450 (1966); C. A., 64, 19678e (1966).
462. Grayson, M., Fr. 1,522,660 (1968); C. A., 71, 42171c (1969).
463. Grayson, M., and C. E. Farley, Chem. Commun., 1967, 830.
464. Grayson, M., and C. E. Farley, Chem. Commun., 1967, 831.
465. Grayson, M., and C. E. Farley, J. Org. Chem., 32, 236 (1967).
466. Grayson, M., and P. T. Keough, U.S. 3,005,013 (1958); C. A., 56, 4798b (1962).
467. Grayson, M., and P. T. Keough, J. Am. Chem. Soc., 82, 3919 (1960).
468. Grayson, M., and P. T. Keough, A.C.S., 140th Meeting, Chicago, 1961, Abstr., p. 95Q; J. Org. Chem., 27, 1817 (1962).
469. Grayson, M., and P. T. Keough, Fr. 1,333,818 (1963); C. A., 60, 3013b (1964).
470. Grayson, M., P. T. Keough, and G. A. Johnson, J. Am. Chem. Soc., 81, 4803 (1959); U.S. 3,005,013 (1958); C. A., 56, 4798b (1962); U.S. 3,148,206 (1962).
471. Granoth, I., A. Kalir, and Z. Pelah, Israel J. Chem., 6, 651 (1968).

472. Granoth, I., A. Kalir, Z. Pelah, and E. D. Bergmann, Tetrahedron, 25, 3919 (1969).
472a. Granoth, I., A. Kalir, Z. Pelah, and E. D. Bergmann, Tetrahedron, 26, 813 (1970).
473. Green, M., R. N. Haszeldine, B. R. Iles, and D. G. Rowsell, J. Chem. Soc., 1965, 6879.
474. Green, M., and R. F. Hudson, Proc. Chem. Soc., 1961, 145.
475. Green, J. H. S., and W. Kynaston, Spectrochim. Acta, 25A, 1677 (1969).
476. Griffin, C. E., Tetrahedron, 20, 2399 (1964).
477. Griffin, C. E., and M. L. Kaufman, Tetrahedron Lett., 1965, 773.
478. Griffin, C. E., and G. Witschard, J. Org. Chem., 29, 1001 (1964).
478a. Griffin, G. E., and W. A. Thomas, J. Chem. Soc., B, 1970, 477.
479. Griffiths, J. E., and A. B. Burg, J. Am. Chem. Soc., 82, 1507 (1960).
480. Griffiths, J. E., and A. B. Burg, J. Am. Chem. Soc., 84, 3442 (1962).
481. Griffiths, V. S., and G. A. Derwish, J. Mol. Spectrosc., 13, 393 (1964); C. A., 61, 7843f (1964).
482. Grim, S. O., R. L. Keiter, and W. McFarlane, Inorg. Chem., 6, 1133 (1967).
483. Grim, S. O., and W. McFarlane, Nature, 208, 995 (1965).
484. Grim, S. O., W. McFarlane, and E. F. Davidoff, J. Org. Chem., 32, 781 (1967).
484a. Grim, S. O., R. P. Molenda, and R. L. Keiter, Chem. Ind. (London), 1970, 1378.
485. Grim, S. O., A. W. Yankowsky, S. A. Bruno, W. J. Bailey, E. F. Davidoff, and T. J. Marks, J. Chem. Eng. Data, 15, 497 (1970).
486. Grinshtein, E. I., A. B. Bruker, and L. Z. Soborovskii, Dokl. Akad. Nauk SSSR, 139, 1359 (1961); C. A., 56, 1475g (1962); U.S.S.R. 138,617.
487. Grinshtein, E. I., A. B. Bruker, and L. Z. Soborovskii, Zh. Obshch. Khim., 36, 302 (1966).
488. Grinshtein, E. I., A. B. Bruker, and L. Z. Soborovskii, Zh. Obshch. Khim., 36, 1138 (1966); C. A., 65, 12230h (1966).
489. Grobe, J., and U. Möller, J. Organometal. Chem., 17, 263 (1969).
490. Groene, H., and G. Pampus, Ger. 1,120,695 (1961); C. A., 56, 11771 (1962).
491. Groenweghe, L. C. D., U.S. 3,223,737 (1965); C. A., 64, 5137e (1966).
492. Grüttner, G., and E. Krause, Chem. Ber., 49, 437 (1916).
493. Grüttner, G., and M. Wiernik, Chem. Ber., 48, 1473 (1915).

494. Gudzinowicz, B. J., and R. H. Campbell, Anal. Chem., 33, 1510 (1961); C. A., 56, 4076h (1962).
495. Guichard, F., Chem. Ber., 32, 1572 (1899).
496. Gunderman, K. D., and A. Garming, Chem. Ber., 102, 3023 (1969).
497. Gutweiler, K., and H. Niebergall, U.S. 3,070,582 (1962); C. A., 58, 12600h (1963).
498. Gutweiler, K., M. Sander, and C. Schneider, Ger. 1,131,412 (1962); C. A., 57, 12722e (1962).
499. Haake, P., W. B. Miller, and D. A. Tyssee, J. Am. Chem. Soc., 86, 3577 (1964).
500. Haas, H., Ger. 937,949 (1956); C. A., 50, 12108i (1956).
501. Haber, C. P., and C. O. Wilson, Jr., U.S. 2,892,873 (1959); C. A., 54, 3203h (1960).
501a. Hall, H. K., Jr., J. Am. Chem. Soc., 73, 5441 (1957).
501b. Hall, D. I., and R. S. Nyholm, Chem. Commun., 1970, 488.
502. Hall, R. E., A. Kessler, and A. R. McLain, U.S. 3,459,808 (1969); C. A., 71, 81524q (1969).
503. Halmann, M., Proc. Chem. Soc., 1960, 289.
504. Halmann, M., Spectrochim. Acta, 16, 407 (1960).
505. Halmann, M., J. Chem. Soc., 1963, 2853.
506. Halmann, M., and L. Kugel, J. Inorg. Nucl. Chem., 25, 1343 (1963).
507. Halpern, E. J., and K. Mislow, J. Am. Chem. Soc., 89, 5224 (1967).
508. Hamid, A. M., and S. Trippett, J. Chem. Soc., C, 1967, 2625.
509. Hamilton, L. A., U.S. 3,352,925 (1967); C. A., 68, 49762t (1968).
510. Hamilton, L. A., U.S. 3,392,180 (1968); C. A., 69, 77475r (1968).
511. Hands, A. R., J. Chem. Soc., 1964, 1181.
512. Hands, A. R., and A. J. H. Mercer, J. Chem. Soc., C, 1967, 1099.
513. Hands, A. R., and A. J. H. Mercer, J. Chem. Soc., C, 1968, 1331.
514. Haniman, J., Chem. Ber., 9, 845 (1876).
515. Hanna, M. W., J. Chem. Phys., 37, 685 (1962).
515a. Harmon, R. E., J. L. Parsons, and S. K. Gupta, Org. Preparative Procedures, 2, 19 (1970); C. A., 72, 90605q (1970).
516. Harrah, L. A., M. T. Ryan, and C. Tamborski, Spectrochim. Acta, 18, 21 (1962).
517. Harrocks, W. D., Jr., R. C. Taylor, and G. N. LaMar, J. Am. Chem. Soc., 86, 3031 (1964).
518. Hart, F. A., J. Chem. Soc., 1960, 3324.
519. Hart, F. A., and F. G. Mann, Chem. Ind. (London), 1956, 574.
520. Hart, F. A., and F. G. Mann, J. Chem. Soc., 1955, 4107.

521. Hart, F. A., and F. G. Mann, J. Chem. Soc., 1957, 3939.
522. Hartley, S. B., W. S. Holmes, J. K. Jacques, M. F. Mole, and J. C. McCoubrey, Quart. Rev., 17, 204 (1963).
523. Hartley, J. G., L. M. Venanzi, and D. C. Goodall, J. Chem. Soc., 1963, 3930.
524. Hartmann, H., Ann. Chem., 714, 1 (1968).
525. Hartmann, H., C. Beermann, and H. Czempik, Z. Anorg. Allg. Chem., 287, 261 (1956).
526. Hartmann, H., and H. Fratzscher, Naturwissenschaften, 51, 213 (1964).
527. Hartmann, H., and A. Meixner, Naturwissenschaften, 50, 403 (1963).
528. Hartmann, H., H. Niemöller, W. Reiss, and B. Karbstein, Naturwissenschaften, 46, 321 (1959).
529. Hartung, W. H., and R. Simonoff, in Organic Reactions, Vol. VII, Wiley, 1953, p. 263.
530. Harvey, A. B., J. Phys. Chem., 70, 3370 (1966).
531. Hassell, W. F., A. Jarvie, and S. Walker, J. Chem. Phys., 46, 3159 (1967).
532. Haszeldine, R. N., and B. O. West, J. Chem. Soc., 1956, 3631.
533. Haszeldine, R. N., and B. O. West, J. Chem. Soc., 1957, 3880.
534. Hawes, W., and S. Trippett, J. Chem. Soc., C, 1969, 1465.
534a. Hawkins, C. J., O. Mønsted, and J. Bjerrum, Acta Chem. Scand., 24, 1059 (1970).
535. Hays, H. R., J. Org. Chem., 31, 3817 (1966); Ger. 1,257,143 (1967); Fr. 1,482,434; Neth. 6,607,892 (1966).
536. Hays, H. R., J. Org. Chem., 33, 3690 (1968).
537. Hays, H. R., and T. J. Logan, J. Org. Chem., 31, 3391 (1966); U.S. 3,445,522 (1969).
538. Hechenbleikner, I., and K. R. Molt, Belg. 635,189 (1964); C. A., 61, 10708b (1964).
539. Hechenbleikner, I., and K. R. Molt, U.S. 3,223,736 (1965); C. A., 64, 5137d (1966); U.S. 3,470,254; C. A., 71, 12464 3d (1969).
540. Hechenbleikner, I., and M. M. Rauhut, U.S. 2,822,376 (1958); C. A., 52, 10147d (1958).
541. Hecker, A. C., N. L. Perry, and O. S. Kauder, Ger. 1,236,186 (1967); C. A., 67, 12149b (1967).
542. Hein, Fr., and H. Hecker, Chem. Ber., 93, 1339 (1960).
543. Hein, Fr., K. Issleib, and H. Rabold, Z. Anorg. Chem., 287, 208 (1956).
544. Hein, Fr., H. Plust, and H. Pohlemann, Z. Anorg. Allg. Chem., 272, 25 (1953).
545. Hellmann, H., and J. Bader, Tetrahedron Lett., 1961, 724.

546. Hellmann, H., J. Bader, H. Birkner, and O. Schumacher, Ann. Chem., 659, 49 (1962).
547. Hellmann, H., and O. Schumacher, Angew. Chem., 72, 211 (1960).
548. Hellwinkel, D., Chem. Ber., 98, 576 (1965).
549. Hellwinkel, D., Angew Chem., 78, 985 (1966).
550. Hellwinkel, D., Chem. Ber., 102, 528 (1969).
551. Hellwinkel, D., Chem. Ber., 102, 548 (1969).
552. Hellwinkel, D., and W. Schenk, Angew. Chem., 81, 1049 (1969).
553. Henderson, W. A., Jr., and S. A. Buckler, J. Am. Chem. Soc., 82, 5794 (1960).
554. Henderson, W. A., Jr., S. A. Buckler, N. E. Day, and M. Grayson, J. Org. Chem., 26, 4770 (1961).
554a. Henderson, W. A., Jr., and C. A. Streuli, J. Am. Chem. Soc., 82, 5791 (1960).
555. Hendrickson, J. B., M. L. Maddox, J. J. Sims, and H. D. Kaesz, Tetrahedron, 20, 449 (1964).
556. Henson, P. D., K. Nauman, and K. Mislow, J. Am. Chem. Soc., 91, 5645 (1969).
557. Herring, D. L., J. Org. Chem., 26, 3998 (1961).
558. Herring, D. L., U.S. 3,182,028 (1965); C. A., 63, 1897d (1965).
559. Herz, A., Deferred Publ., U.S. Pat. Off. 808,669 (1969); C. A., 71, 66 039s (1969).
560. Hettinger, W. P., U.S. 3,079,311 (1963); C. A., 59, 2859f (1963).
561. Hewertson, W., and H. R. Watson, J. Chem. Soc., 1962, 1490.
562. Hewertson, W., R. A. Shaw, and B. C. Smith, J. Chem. Soc., 1964, 1020.
562a. Hewerton, W., and I. C. Taylor, Chem. Commun., 1970, 119.
563. Hey, L., and C. K. Ingold, J. Chem. Soc., 1933, 531.
564. Hibbert, H., Chem. Ber., 39, 160 (1906); Dissertation, Leipzig, 1906.
565. Highsmith, R. E., and H. H. Sisler, Inorg. Chem., 7, 1740 (1968).
566. Hinton, R. C., and F. G. Mann, J. Chem. Soc., 1959, 2835.
567. Hinton, R. C., F. G. Mann, and D. Todd, J. Chem. Soc., 1961, 5454.
568. Hinton, R. C., F. G. Mann, and D. Todd, Proc. Chem. Soc., 1959, 365.
569. Hitchcock, C. H. S., and F. G. Mann, J. Chem. Soc., 1958, 2081.
570. Hoff, M. C., and P. Hill, J. Org. Chem., 24, 356 (1959).
571. Hoffman, A., J. Am. Chem. Soc., 43, 1684 (1921); C. A., 52, 2995 (1930).
572. Hoffman, A., J. Am. Chem. Soc., 52, 2995 (1930).

573. Hoffmann, H., Doctoral thesis, University of Mainz, 1956.
574. Hoffmann, H., Angew, Chem., 72, 77 (1960).
575. Hoffmann, H., Ann. Chem., 634, 1 (1960).
576. Hoffmann, H., Chem. Ber., 94, 1331 (1961).
577. Hoffmann, H., Chem. Ber., 95, 2563 (1962).
578. Hoffmann, H., and H. J. Diehr, Chem. Ber., 98, 363 (1965).
579. Hoffmann, H., R. Grünewald, and L. Horner, Chem. Ber., 93, 861 (1960).
580. Hoffmann, H., and P. Schellenbeck, Chem. Ber., 99, 1134 (1966).
581. Hoffmann, H., and P. Schellenbeck, Chem. Ber., 100, 692 (1967).
582. Hoffmann, H., and P. Schellenbeck, Chem. Ber., 101, 2203 (1968).
583. Hofmann, A. W., Phil. Trans., 150, 409 (1860).
584. Hofmann, A. W., Ann. Chem. Suppl., 1, 1, 59 (1861).
585. Hofmann, A. W., Jahresber., 1861, 490.
586. Hofmann, A. W., Chem. Ber., 4, 205 (1871).
587. Hofmann, A. W., Chem. Ber., 4, 372 (1871).
588. Hofmann, A. W., Chem. Ber., 4, 430 (1871).
589. Hofmann, A. W., Chem. Ber., 4, 605 (1871).
590. Hofmann, A. W., Chem. Ber., 5, 100 (1872).
591. Hofmann, A. W., Chem. Ber., 6, 292 (1873).
592. Hofmann, A. W., Chem. Ber., 6, 301 (1873).
593. Hofmann, A. W., and A. Cahours, Quart. J. Chem. Soc., 11, 56 (1859); 13, 289 (1861); 14, 73, 316 (1861).
594. Hofmann, A. W., and F. Magla, Chem. Ber., 25, 2436 (1892).
595. Hofmann, E., Brit. 904,086 (1962); C. A., 57, 16661a (1962).
596. Hofmann, E., Brit. 921,463 (1963); C. A., 59, 1682f (1963).
597. Hofmann, E., Brit. 928,207 (1963); C. A., 59, 12844b (1963).
598. Hofmann, E., Brit. 941,557 (1963); C. A., 60, 4184g (1964).
599. Hogben, M. G., R. S. Gay, and W. A. G. Graham, J. Am. Chem. Soc., 88, 3457 (1966).
600. Hogben, M. G., and W. A. G. Graham, J. Am. Chem. Soc., 91, 283 (1969).
601. Holliman, F. G., and F. G. Mann, J. Chem. Soc., 1943, 547.
602. Holliman, F. G., and F. G. Mann, J. Chem. Soc., 1947, 1634.
603. Hooge, F. N., and P. J. Christen, Rec. Trav. Chim., 77, 911 (1958).
604. Horizons, Inc., Neth. 6,409,367 (1965); C. A., 63, 181e (1965).
605. Horner, L., and W. Balzer, Tetrahedron Lett., 1965, 1157.

606. Horner, L., W. D. Balzer, and D. J. Peterson, Tetrahedron Lett., 1966, 3315.
607. Horner, L., and W. D. Balzer, Chem. Ber., 102, 3542 (1969).
608. Horner, L., and P. Beck, Chem. Ber., 93, 1371 (1960).
609. Horner, L., P. Beck, and H. Hoffmann, Chem. Ber., 92, 2088 (1959).
610. Horner, L., P. Beck, and R. Luckenbach, Chem. Ber., 101, 2899 (1968).
610a. Horner, L., P. Beck, and V. G. Toscano, Chem. Ber., 94, 1317 (1961).
611. Horner, L., J. P. Bercz, and C. V. Bercz, Tetrahedron Lett., 1966, 5783.
612. Horner, L., and J. Dörges, Tetrahedron Lett., 1965, 763.
613. Horner, L., and M. Ernst, Chem. Ber., 103, 318 (1970).
613a. Horner, L., I. Ertel, H. D. Ruprecht, and O. Belovsky, Chem. Ber., 103, 1582 (1970).
614. Horner, L., H. Fuchs, H. Winkler, and A. Rapp, Tetrahedron Lett., 1963, 965.
615. Horner, L., and J. Haufe, Chem. Ber., 101, 2903 (1968).
616. Horner, L., and J. Haufe, Chem. Ber., 101, 2921 (1968).
617. Horner, L., and H. Hoffmann, Angew. Chem., 68, 483 (1956).
618. Horner, L., and H. Hoffmann, Neuere Methoden der Präparativen Organischen Chemie, Vol. II, Verlag Chemie, 1960, p. 108.
619. Horner, L., H. Hoffmann, and P. Beck, Chem. Ber., 91, 1583 (1958).
620. Horner, L., and W. Hofer, Tetrahedron Lett., 1966, 3321.
621. Horner, L., W. Jurgeleit, and K. Klüpfel, Ann. Chem., 591, 108 (1955).
622. Horner, L., R. Luckenbach, and W. D. Balzer, Tetrahedron Lett., 1968, 3157.
623. Horner, L., and A. Mentrup, Ger. 1,114,190 (1961); C. A., 57, 2256f (1962).
624. Horner, L., and A. Mentrup, Ann. Chem., 646, 65 (1961).
625. Horner, L., and H. Nickel, Chem. Ber., 89, 1681 (1956).
626. Horner, L., H. Reuter, and E. Herrmann, Ann. Chem., 660, 1 (1962).
627. Horner, L., and F. Roettger, Korrosion, 16, 57 (1963); C. A., 61, 14266g (1964).
628. Horn, P. E., and E. Rothstein, J. Chem. Soc., 1963, 1036.
629. Horner, L., F. Schedlbauer, and P. Beck, Tetrahedron Lett., 1964, 1421.

630. Horner, L., and H. Winkler, Tetrahedron Lett., 1964, 175.
631. Horner, L., and H. Winkler, Tetrahedron Lett., 1964, 455.
632. Horner, L., and H. Winkler, Tetrahedron Lett., 1964, 461.
633. Horner, L., and H. Winkler, Tetrahedron Lett., 1964, 3275.
634. Horner, L., and H. Winkler, Ann. Chem., 685, 1 (1965).
635. Horner, L., H. Winkler, A. Rapp, A. Mentrup, H. Hoffmann, and P. Beck, Tetrahedron Lett., 1961, 161.
636. Horvat, R. J., and A. Furst, J. Am. Chem. Soc., 74, 562 (1952).
637. Horward, E., Jr., and M. Braid, A.C.S., 140th Meeting, Chicago, Sept. 1961, Abstr., p. 40Q.
638. Houalla, D., R. Miquel, and R. Wolf, Bull. Soc. Chim. Fr., 1963, 1152.
638a. Houalla, D., R. Marty, and R. Wolf, Z. Naturforsch., B, 25, 451 (1970).
639. Howells, E. R., F. M. Lovell, D. Rogers, and A. J. C. Wilson, Acta Crystallogr., 7, 298 (1954).
640. Howell, T. E. W., S. A. J. Pratt, and L. M. Venanzi, J. Chem. Soc., 1961, 3167.
641. Huff, T., and E. Perry, J. Polym. Sci., Part A, 1, 1553 (1963).
642. Huff, T., and E. Perry, U.S. 3,262,995 (1966); C. A., 65, 13844a (1966).
643. Hughes, L. J., and E. Perry, J. Polym. Sci., Part A, 3, 1527 (1965).
644. Hughes, A. N., and S. Uaboonkul, Tetrahedron, 24, 3437 (1968).
645. Imperial Chemical Industries, Ltd., Fr. 1,410,590 (1965); C. A., 64, 11133c (1966).
646. Ingold, C. K., Structure and Mechanism in Organic Chemistry, Cornell University Press, 1953, pp. 90, 691.
647. Inst. f. Silikon and Fluorocarbon Chemie, Fr. 1,536,120 (1968); C. A., 71, 4078v (1969).
648. Interrante, L. V., M. A. Bennett, and R. S. Nyholm, Inorg. Chem., 5, 2212 (1966).
649. Issleib, K., and H. Anhöck, Z. Naturforsch., B, 16, 837 (1961).
650. Issleib, K., and L. Baldauf, Pharm. Zentralh. Deut., 99, 329 (1960).
651. Issleib, K., and A. Balszuweit, Chem. Ber., 99, 1316 (1966).
652. Issleib, K., and R. D. Bleck, Z. Anorg. Allg. Chem., 336, 234 (1965).
653. Issleib, K., and A. Brack, Z. Anorg. Allg. Chem., 277, 258 (1954).
654. Issleib, K., and A. Brack, Z. Anorg. Allg. Chem., 292, 245 (1957).

654a. Issleib, K., and H. Bruchlos, Z. Anorg. Allg. Chem., 316, 1 (1962).
655. Issleib, K., and L. Brüsehaber, Z. Naturforsch., B, 20, 181 (1965).
656. Issleib, K., and G. Döll, Chem. Ber., 94, 2664 (1961).
657. Issleib, K., and G. Döll, Z. Anorg. Allg. Chem., 324, 259 (1963).
658. Issleib, K., and G. Döll, Chem. Ber., 96, 1544 (1963).
659. Issleib, K., and H. O. Fröhlich, Z. Naturforsch., B, 14, 349 (1959).
659a. Issleib, K., and H. O. Fröhlich, Chem. Ber., 95, 375 (1962).
660. Issleib, K., and G. Grams, Z. Anorg. Allg. Chem., 299, 58 (1959).
661. Issleib, K., and D. Haferburg, Z. Naturforsch., B, 20, 916 (1965).
662. Issleib, K., and R. M. Haftendorn, Z. Anorg. Allg. Chem., 351, 9 (1967).
663. Issleib, K., and B. Hamann, Z. Anorg. Allg. Chem., 339, 289 (1965).
664. Issleib, K., and G. Harzfeld, Chem. Ber., 95, 268 (1962).
665. Issleib, K., and G. Harzfeld, Chem. Ber., 97, 3430 (1964).
666. Issleib, K., and G. Harzfeld, Z. Anorg. Allg. Chem., 351, 18 (1967).
667. Issleib, K., and S. Häusler, Chem. Ber., 94, 113 (1961).
668. Issleib, K., and M. Hoffmann, Chem. Ber., 99, 1320 (1966).
669. Issleib, K., and D. Jacob, Chem. Ber., 94, 107 (1961).
670. Issleib, K., and K. Jasche, Chem. Ber., 100, 412 (1967).
671. Issleib, K., and K. Jasche, Chem. Ber., 100, 3343 (1967).
672. Issleib, K., and F. Krech, Chem. Ber., 94, 2656 (1961).
673. Issleib, K., and F. Krech, Z. Anorg. Allg. Chem., 328, 21 (1964).
674. Issleib, K., and K. Krech, Chem. Ber., 98, 1093 (1965).
675. Issleib, K., and F. Krech, J. Organometal. Chem., 13, 283 (1968).
676. Issleib, K., K. Krech, and K. Gruber, Chem. Ber., 96, 2186 (1963).
677. Issleib, K., and R. Kümmel, J. Organometal. Chem., 3, 84 (1965).
678. Issleib, K., and R. Kümmel, Z. Naturforsch., B, 22, 784 (1967).

679. Issleib, K., and R. Kümmel, Z. Chem., 7, 235 (1967).
680. Issleib, K., and R. Kümmel, Chem. Ber., 100, 3331 (1967).
681. Issleib, K., R. Kümmel, H. Oehme, and I. Meissner, Chem. Ber., 101, 3612 (1968).
682. Issleib, K., R. Kümmel, and H. Zimmermann, Angew. Chem., 77, 172 (1965).
683. Issleib, K., and R. Lindner, Ann. Chem., 699, 40 (1966).
684. Issleib, K., and R. Lindner, Ann. Chem., 707, 120 (1967).
685. Issleib, K., and R. Lindner, Ann. Chem., 713, 12 (1968).
686. Issleib, K., and M. Lischewski, J. Prakt. Chem., 311, 857 (1969).
687. Issleib, K., and O. Löw, Z. Anorg. Allg. Chem., 346, 241 (1966).
688. Issleib, K., and H. Matschiner, Z. Anorg. Allg. Chem., 340, 34 (1965).
689. Issleib, K., H. Matschiner, and M. Hoppe, Z. Anorg. Allg. Chem., 351, 251 (1967).
690. Issleib, K., H. Matschiner, and S. Naumann, J. Electroanal. Chem., 16, 563 (1968).
691. Issleib, K., and H. M. Möbius, Chem. Ber., 94, 102 (1961).
692. Issleib, K., and D. W. Müller, Chem. Ber., 92, 3175 (1959).
693. Issleib, K., and H. Oehme, Tetrahedron Lett., 1967, 1489.
694. Issleib, K., and H. Oehme, Chem. Ber., 100, 2685 (1967).
695. Issleib, K., H. Oehme, R. Kümmel, and E. Leissring, Chem. Ber., 101, 3619 (1968).
696. Issleib, K., and E. Priebe, Chem. Ber., 92, 3183 (1959).
697. Issleib, K., and R. Rieschel, Chem. Ber., 98, 2086 (1965).
698. Issleib, K., and K. Rockstroh, Chem. Ber., 96, 407 (1963).
699. Issleib, K., and H. R. Roloff, Z. Anorg. Allg. Chem., 324, 250 (1963).
700. Issleib, K., and H. R. Roloff, Chem. Ber., 98, 2091 (1965).
701. Issleib, K., and W. Seidel, Chem. Ber., 92, 2681 (1959).
702. Issleib, K., and K. Standtke, Chem. Ber., 96, 279 (1963).
703. Issleib, K., and G. Thomas, Chem. Ber., 93, 803 (1960).
704. Issleib, K., and G. Thomas, Chem. Ber., 94, 2244 (1961).

705. Issleib, K., and G. Thomas, Z. Anorg. Allg. Chem., 330, 295 (1964).
706. Issleib, K., and A. Tzschach, Chem. Ber., 92, 704 (1959).
707. Issleib, K., and A. Tzschach, Chem. Ber., 92, 1118 (1959).
708. Issleib, K., and A. Tzschach, Chem. Ber., 92, 1397 (1959).
709. Issleib, K., and A. Tzschach, Chem. Ber., 93, 1852 (1960).
710. Issleib, K., A. Tzschach, and H. U. Block, Chem. Ber., 101, 2931 (1968).
711. Issleib, K., A. Tzschach, and R. Schwarzer, Z. Anorg. Allg. Chem., 338, 141 (1965).
712. Issleib, K., and F. Ungváry, Z. Naturforsch., B, 22, 1238 (1967).
713. Issleib, K., and H. Völker, Chem. Ber., 94, 392 (1961).
714. Issleib, K., and B. Walter, Chem. Ber., 97, 3424 (1964).
715. Issleib, K., and H. Weichmann, Chem. Ber., 97, 721 (1964).
716. Issleib, K., and H. Weichmann, Chem. Ber., 101, 2197 (1968).
717. Issleib, K., and H. Zimmermann, Z. Anorg. Allg. Chem., 353, 197 (1967).
718. Itoh, K., M. Fukui, and Y. Ishii, J. Chem. Soc., C, 1969, 2002.
719. Ivin, S. Z., K. V. Karavanov, and V. V. Lysenko, Zh. Obshch. Khim., 34, 852 (1964); C. A., 60, 15902g (1964).
720. Jackson, I. K., W. C. Davies, and W. J. Jones, J. Chem. Soc., 1931, 2109.
720a. Jackson, I. K., W. C. Davies, and W. J. Jones, J. Chem. Soc., 1930, 2298.
721. Jackson, I. K., and W. C. Jones, J. Chem. Soc., 1931, 575.
722. Jaffé, H. H., J. Chem. Phys., 22, 1430 (1954).
723. Jaffé, H. H., and L. D. Freedman, J. Am. Chem. Soc., 74, 1069 (1952).
724. Jain, S. R., and H. H. Sisler, Inorg. Chem., 8, 1243 (1969).
725. Jakobsen, H. J., and J. A. Nielsen, J. Mol. Spectrosc., 31, 230 (1969).
726. Jakobsen, H. J., and J. Nielsen, Acta Chem. Scand., 23, 1070 (1969).
726a. Jakobsen, H. J., and J. A. Nielsen, J. Mol. Spectrosc., 33, 474 (1970).
727. Jaro, M., P. Knowlton, J. E. Bissey, H. Goldwhite, and W. R. Carper, Mol. Phys., 13, 165 (1967); C. A., 68, 68338a (1968).

728. Jenkins, A. D., and L. J. Wolfram, Brit. 959,356 (1964); C. A., 61, 5812b (1964); U.S. 3,248,376 (1966); C. A., 65, 826f (1966).
729. Jenkins, A. D., and L. J. Wolfram, U.S. 3,256,154 (1966); C. A., 65, 5299d (1966).
730. Jennings, B. E., Brit. 1,077,958 (1967); C. A., 67, 82542y (1967).
730a. Jensen, K. A., J. Prakt. Chem., 148, 101 (1937).
731. Jensen, K. A., and P. H. Nielsen, Acta Chem. Scand., 17, 549 (1963).
732. Jensen, K. A., and P. H. Nielsen, Acta Chem. Scand., 17, 1875 (1963).
733. Jerchel, D., Chem. Ber., 76, 600 (1943).
734. Jerchel, D., and J. Kimmig, Chem. Ber., 83, 277 (1950).
735. Jl'yasov, A. V., Yu. M. Kargin, Ya. A. Levin, B. V. Melnikov, and V. S. Galeev, Izv. Akad. Nauk SSSR, Ser. Khim., 1968, 2841; C. A., 70, 92195p (1969).
736. Joannis, A., C. R. Acad. Sci., Paris, 119, 557 (1894); Ann. Chim. Phys., (8), 7, 105 (1906).
737. Job, A., and G. Dussolier, C. R. Acad. Sci., Paris, 184, 1454 (1927).
738. Johnson, A. W., and H. L. Jones, J. Am. Chem. Soc., 90, 5232 (1968).
739. Johnson, F., R. S. Gohlke, and W. A. Nasutavicus, J. Organometal. Chem., 3, 233 (1965).
740. Jolly, W. L., Inorg. Syn., 11, 124 (1968).
741. Jolly, W. L., Inorg. Syn., 11, 126 (1968).
742. Jones, E. R. H., and F. G. Mann, J. Chem. Soc., 1955, 4472.
743. Jones, W. J., W. C. Davies, T. S. Bowden, C. Edwards, V. E. Davis, and L. H. Thomas, J. Chem. Soc., 1947, 1446.
744. Jurgeleit, H. W., U.S. 3,027,359 (1962); C. A., 57, 2424d (1962).
745. Kaabak, L. V., M. I. Kabachnik, A. P. Tomilov, and S. L. Varshavskii, Zh. Obshch. Khim., 36, 2060 (1966); C. A., 66, 85828m (1967).
746. Kabachnik, M. I., Dokl. Akad. Nauk SSSR, 110, 393 (1956).
746a. Kabachnik, M. I., and G. A. Balueva, Izv. Acad. Nauk SSSR, 1962, 536.
747. Kabachnik, M. I., Chang Chung-Yu, and E. N. Tsvetkov, Dokl. Akad. Nauk SSSR, 135, 603 (1960); C. A., 55, 12272a (1961).
748. Kabachnik, M. I., and E. S. Shepeleva, Izv. Akad. Nauk SSSR, 1949, 56.
749. Kabachnik, M. I., and E. N. Tsvetkov, Dokl. Akad. Nauk SSSR, 117, 817 (1957); C. A., 52, 8070b (1958).
750. Kabachnik, M. I., and E. N. Tsvetkov, Dokl. Akad. Nauk SSSR, 143, 592 (1962).

751. Kabachnik, M. I., E. N. Tsvetkov, and Chang Chung-Yu, Tetrahedron Lett., 1962, 5.
752. Kabachnik, M. I., V. V. Voevodskii, T. A. Mastryukova, S. P. Solodovnikov, and T. A. Melent'eva, Zh. Obshch. Khim., 34, 3234 (1964); C. A., 62, 3906e (1964).
752a. Kaesz, H. D., and F. G. A. Stone, J. Am. Chem. Soc., 82, 6213 (1960).
753. Kaesz, H. D., and F. G. A. Stone, J. Org. Chem., 24, 635 (1959).
754. Kaesz, H. D., and F. G. A. Stone, Spectrochim. Acta, 15, 360 (1959).
755. Kaesz, H. D., and F. G. A. Stone, in Organometallic Chemistry, H. Zeiss, Ed., Reinhold, 1960, p. 88.
756. Kali-Chemie Akt. Ges. Brit. 823,483 (1959); C. A., 54, 14125c (1960).
757. Kamai, G., Zh. Obshch. Khim., 2, 524 (1932); C. A., 27, 966 (1933).
758. Kamai, G., and L. A. Khismatullina, Dokl. Akad. Nauk SSSR, 92, 69 (1953); C. A., 48, 9945f (1954).
759. Kamai, G., and L. A. Khismatullina, Zh. Obshch. Khim., 26, 3426 (1956); C. A., 51, 9512h (1957).
760. Kamai, G., and L. A. Khismatullina, Izv. Kazan. Filiala Akad. Nauk SSSR, Ser. Khim. Nauk, 1957, No. 4, 79; C. A., 54, 6601a (1960).
761. Kamai, G., and G. M. Rusetskaya, Dokl. Akad. Nauk SSSR, 143, 596 (1962); C. A., 57, 3476i (1962).
762. Kamai, I. G., and G. M. Rusetskaya, Zh. Obshch. Khim., 32, 2848 (1962); C. A., 58, 7965d (1963).
763. Kamai, I. G., and G. M. Rusetskaya, Zh. Obshch. Khim., 32, 2854 (1962); C. A., 58, 7965f (1963).
764. Kaplan, E. D., and E. R. Thornton, J. Am. Chem. Soc., 89, 6644 (1967).
765. Kastening, B., Naturwissenschaften, 49, 130 (1962).
766. Katz, T. J., J. C. Carnaham, Jr., G. M. Clarke, and N. Acton, J. Am. Chem. Soc., 92, 734 (1970).
767. Katz, T. J., C. R. Nicholson, and C. A. Reilly, J. Am. Chem. Soc., 88, 3832 (1966).
768. Kaufman, M. L., and C. E. Griffin, Tetrahedron Lett., 1965, 769.
769. Kawada, I., and R. Allmann, Angew. Chem., 80, 40 (1968).
770. Kawasumi, S., K. Maemoto, and M. Ouishi, Jap. 29,366 (1964); C. A., 62, 13274e (1964).
771. Keat, R., Chem. Ind. (London), 1968, 1362.
772. Keiter, R. L., and J. G. Verkade, Inorg. Chem., 8, 2115 (1969).
773. Kelbe, W., Chem. Ber., 11, 1499 (1878).
774. Kemmitt, R. D. W., D. I. Nichols, and R. D. Peacock, J. Chem. Soc., A, 1968, 2149.
775. Kennedy, J., E. S. Lane, and J. L. Williams, J. Chem. Soc., 1956, 4670.

776. Kiso, Y., M. Kobayashi, Y. Kitaoka, K. Kawamoto, and J. Takada, Bull. Chem. Soc. Jap., 40, 2779 (1967).
777. King, R. B., and P. N. Kapoor, J. Am. Chem. Soc., 91, 5191 (1969); 93, 4158 (1971); Angew. Chem., 83, 766 (1971).
777a. King, R. B., and A. Efraty, Inorg. Acta, 4, 123 (1970).
778. Klages, G., and R. Langpape, Z. Electrochem., 63, 533 (1959).
779. Klamann, D., and P. Weyerstahl, Chem. Ber., 97, 2534 (1964).
780. Knobloch, P., and M. Stockhausen, Angew. Chem., 76, 186 (1964).
781. Knowles, W. S., and M. J. Sabacky, Chem. Commun., 1968, 1445.
782. Knunyants, I. L., and R. N. Sterlin, C. R. Acad. Sci., U.S.S.R., 56, 49 (1947); C. A., 42, 519g (1948).
783. Kochetov, E. V., A. A. Berlin, and N. S. Enikolopyan, Vysokomolekul. Soedin., 8, 1018 (1966); C. A., 65, 13829b (1966).
784. Kodama, G., J. R. Weaver, J. LaRochelle, and R. W. Parry, Inorg. Chem., 5, 710 (1966).
785. Köhler, H., and A. Michaelis, Chem. Ber., 10, 807 (1877).
786. Köhler, H., and A. Michaelis, Chem. Ber., 10, 807 (1877).
787. Koenigs, E., and H. Friedrich, Ann. Chem., 509, 138 (1934).
788. Koeppl, G. W., D. S. Sagatys, G. S. Krishnamurthy, and S. I. Miller, J. Am. Chem. Soc., 89, 3396 (1967).
789. Köster, R., and Y. Morita, Angew. Chem., 77, 589 (1965).
790. Kojima, T., E. L. Breig, and C. C. Lin, J. Chem. Phys., 35, 2139 (1961).
791. Kolitowska, J. H., Rocz., Chem., 8, 568 (1928).
792. Kolling, O. W., and D. A. Garber, Anal. Chem., 39, 1562 (1967); C. A., 67, 113513w (1967).
793. Kolninov, O. V., and Z. V. Zvonkova, Zh. Fiz. Khim., 36, 2228 (1962); C. A., 58, 5158g (1963).
794. Koppers Company, Ger. 1,113,827 (1961).
795. Koral, J. N., Makromol. Chem., 62, 148 (1963).
796. Koral, J. N., U.S. 3,163,622 (1964); C. A., 62, 6594b (1965).
797. Koral, J. N., U.S. 3,214,410 (1965); C. A., 63, 18300g (1965).
798. Koral, J. N., and B. W. Song, J. Polym. Sci., 54, S34 (1961).
799. Koral, J. N., and B. W. Song, U.S. 3,122,524 (1964); C. A., 61, 3227b (1964).
800. Korshak, V. V., V. A. Zamyatina, A. I. Solomatina, E. I. Fedin, and P. V. Petrovskii, J. Organometal. Chem., 17, 201 (1969).

801. Kosolapoff, G. M., Organophosphorus Compounds, Wiley, 1950.
802. Kostyanovskii, R. G., I. I. Chervin, V. V. Yakshin, and A. U. Stepanyants, Izv. Akad. Nauk SSSR, Ser. Khim., 1967, 1629, 2128; C. A., 68, 25353j (1968).
803. Kostyanovskii, R. G., and V. V. Yakshin, Izv. Akad. Nauk USSR, Ser. Khim., 1967, 2363; C. A., 68, 77526n (1968).
804. Kostyanovskii, R. G., and V. V. Yakshin, Izv. Akad. Nauk SSSR, Ser. Khim., 1969, 478.
805. Kostyanovakii, R. G., V. V. Yakshin, and I. I. Chervin, Dokl. Akad. Nauk SSSR, 188, 366 (1969); C. A., 72, 11915q (1970).
806. Kostyanovskii, R. G., V. V. Yakshin, and S. L. Zimont, Izv. Akad. Nauk SSSR, Ser. Khim., 1967, 1398.
807. Kostyanovskii, R. G., V. V. Yakshin, and S. L. Zimont, Izv. Akad. Nauk SSSR, Ser. Khim., 1968, 651; C. A., 69, 14069x (1968).
808. Kostyanovskii, R. G., V. V. Yakshin, S. L. Zimont, and I. I. Chervin, Izv. Akad. Nauk SSSR, Ser. Khim., 1968, 188; C. A., 69, 43988v (1968).
809. Kostyanovskii, R. G., V. V. Yakshin, S. L. Zimont, and I. I. Chervin, Izv. Akad. Nauk SSSR, Ser. Khim., 1968, 190; C. A., 69, 43987u (1968).
810. Kostyanovskii, R. G., V. V. Yakshin, S. L. Zimont, and I. I. Chervin, Izv. Akad. Nauk SSSR, Ser. Khim., 1968, 391.
811. Kozlow, E. S., A. I. Sedlov, and A. V. Kirsanov, Zh. Obshch. Khim., 38, 1881 (1968); C. A., 70, 4215s (1969).
812. Krafft, F., and R. Neumann, Chem. Ber., 34, 565 (1901).
812a. Kraft, M. Ya., and V. P. Parini, Sb. Statei Obshch. Khim. Akad. Nauk SSSR, 1, 729 (1953); C. A., 49, 909e (1955).
813. Kramolowsky, R., Angew Chem., 81, 182 (1969).
814. Krasil'nikova, E. A., N. A. Moskva, and A. I. Razumov, Zh. Obshch. Khim., 39, 216 (1969); C. A., 70, 87910p (1969).
815. Kratzer, R. H., and K. L. Paciorek, J. Org. Chem., 32, 853 (1967).
816. Krespan, C. G., J. Am. Chem. Soc., 83, 3432 (1961).
817. Krespan, C. G., B. C. McKusik, and T. L. Cairns, J. Am. Chem. Soc., 82, 1515 (1960).
818. Kreutzkampf, N., Chem. Ber., 87, 919 (1954).
819. Kuchen, W., and H. Buchwald, Angew. Chem., 68, 791 (1956).
820. Kuchen, W., and H. Buchwald, Angew. Chem., 71, 162 (1959); Ann. Chem., 652, 28 (1962).
821. Kuchen, W., and H. Buchwald, Chem. Ber., 91, 2296 (1958).

822. Kuchen, W., and H. Buchwald, Chem. Ber., 91, 2871 (1958).
823. Kuchen, W., and H. Buchwald, Chem. Ber., 92, 227 (1959); Angew. Chem., 69, 307 (1957).
824. Künzel, E., A. Giefer, and W. Kern, Makromol. Chem., 96, 17 (1966).
825. Kuhn, H. J., K. Plieth, Naturwissenschaften, 53, 359 (1966); 52, 12 (1965).
826. Kumada, M., and K. Noda, Mem. Fac. Eng. Osake City Univ., 4, 173 (1962); C. A., 59, 8782c (1963).
827. Kupchik, E., and V. A. Perciaccante, J. Organometal. Chem., 10, 181 (1967).
828. Kuwajima, I., and T. Mukaiyama, J. Org. Chem., 29, 1385 (1964); 28, 2024 (1963).
829. Kuzmin, K. I., Z. U. Panfilovich, and L. A. Pavlova, Tr. Kazan. Khim. Tekhnol. Inst., No. 34, 392 (1965); C. A., 68, 69088f (1968).
830. Kuznetsov, E. V., D. A. Faizullina, and R. P. Tyurikova, Vysokomolekul. Soedin., 7, 761 (1965); C. A., 63, 7121b (1965).
831. Kuznetsov, E. V., and E. K. Ignat'eva, U.S.S.R. 203,215 (1967); C. A., 69, 11097n (1968).
832. Kuznetsov, E. V., T. V. Sorokina, and R. K. Valetdinov, Zh. Obshch. Khim., 33, 2631 (1963); C. A., 60, 542f (1964).
833. Labarre, M. C., J. Chim. Phys., 65, 549 (1968).
834. Labarre, M. C., D. Voigt, and F. Gallais, Bull. Soc. Chim. Fr., 1967, 3328.
835. Lagowski, J. J., Quart. Rev. (London), 13, 233 (1959).
836. Lal, J., U.S. 2,833,741 (1958); C. A., 52, 14219e (1958).
837. Lambert, J. B., W. L. Oliver, Jr., and G. F. Jackson, Tetrahedron Lett., 1969, 2027.
838. Lamza, L., J. Prakt. Chem., 25, 294 (1964).
839. Lannon, J. A., and E. R. Nixon, Spectrochim. Acta, 23A, 2713 (1967).
840. Lautsch, W. F., Chem. Tech. (Berlin), 10, 419 (1958); C. A., 53, 43c (1959).
841. Laskorin, B. N., and V. F. Smirnov, Zh. Prikl. Khim., 38, 2232 (1965); C. A., 64, 2801h (1966).
842. Laszkiewicz, B., Wiadomosei Chem., 19, 629 (1965); C. A., 64, 828d (1966).
843. Latscha, H. P., P. B. Hormuth, and H. Vollmer, Z. Naturforsch., B, 24, 1237 (1969).
844. Lautenschlager, H., and D. Wittenberg, Ger. 1,131,669 (1962); C. A., 58, 1490f (1963).
845. Leavitt, F. C., T. A. Manuel, and F. Johnson, J. Am. Chem. Soc., 81, 3163 (1959); 82, 5099 (1960); U.S. 3,412,119 (1969).
846. Lecog, H., Bull. Soc. Chim. Belges, 42, 199 (1933).
846a. Lee, M. M., U.S. 2,520,601 (1950); C. A., 44, 11175b (1950).

847. Leffler, A. J., and E. G. Teach, A.C.S., 133rd Meeting, San Francisco, April 1958, Abstr., p. 29N.
848. Leqoux, C., C. R. Acad. Sci., Paris, 209, 47 (1939).
848a. Leto, J. R., and M. F. Leto, J. Am. Chem. Soc., 83, 2944 (1961).
849. Letsinger, R. L., J. R. Nazy, and A. S. Hussey, J. Org. Chem., 23, 1806 (1958).
850. Letts, E. A., and R. F. Blake, Trans. Roy. Soc. Edinburgh, 35, 527 (1889); J. Chem. Soc., 58, 766 (1890).
851. Letts, E. A., and N. Collie, Phil. Mag. (5), 22, 183 (1886).
852. Levison, M. E., A. S. Josephson, and D. M. Kirschenbaum, Experientia, 25, 126 (1969).
853. Levy, J. B., G. O. Doak, and L. D. Freedman, J. Org. Chem., 30, 660 (1965).
854. Levy, J. B., L. D. Freedman, and G. O. Doak, J. Org. Chem., 33, 474 (1968).
855. Lewis, A. F., and L. J. Forrestal, Am. Soc. Testing Mater., Spec. Tech. Publ. No. 360 59-75, 76-7 (1963); C. A., 62, 5389b (1964).
856. Lewis, R. A., K. Naumann, K. E. DeBruin, and K. Mislow, Chem. Commun., 1969, 1010.
857. Lide, D. R., Jr., and D. E. Mann, J. Chem. Phys., 29, 914 (1958).
857a. Lindner, E., H. D. Ebert, and A. Haag, Chem. Ber., 103, 1872 (1970).
858. Lindner, E., and H. Kranz, Z. Naturforsch., B, 22, 675 (1967).
859. Lindner, E., and H. Kranz, Chem. Ber., 101, 3438 (1968).
860. Linton, H. R., and E. R. Nixon, Spectrochem. Acta, 15, 146 (1959).
861. Linville, R. G., U.S. 3,053,809 (1962); C. A., 58, 616h (1963).
862. Lippincott, E. R., G. Nagarajan, and J. M. Stutman, J. Phys. Chem., 70, 78 (1966).
863. Livingstone, S. E., and T. N. Lockyer, Inorg. Nucl. Chem. Lett., 3, 35 (1967).
864. Lobanov, D. I., E. N. Tsvetkov, and M. I. Kabachnik, Zh. Obshch. Khim., 39, 841 (1969).
865. Lobanov, D. I., E. N. Tsvetkov, E. V. Saltanova, E. A. Yakovleva, A. I. Shatenshtein, and M. I. Kabachnik, Izv. Akad. Nauk SSSR, Ser. Khim., 1968, 2050; C. A., 70, 20163d (1969); Tetrahedron, 25, 1165 (1969).
865a. Loewenthal, M., Schweiz. Z. Pathol. Bakteriol., 12, 313 (1949); C. A., 44, 5006e (1950).
866. Long, D. A., and R. B. Gravenor, Spectrochim. Acta, 19, 961 (1963).
867. Long, L. H., and J. F. Sackmann, Trans. Faraday Soc., 53, 1606 (1957).

868. Lorenz, H. J., and V. Franzen, D.A.S. 1,265,745 (1968).
869. Lorrenz, W. J., and H. Fischer, Fr. Mezhdunar. Kongr. Korroz. Metal., 3rd, 2, 98 (1966); C. A., 71, 108 288e (1969).
870. Low, H., and P. Tavs, Tetrahedron Lett., 1966, 1357.
871. Lucken, E. A. C., J. Chem. Soc., A, 1966, 1357.
872. Luckenbach, R., Ph.D. Thesis, University of Mainz (1969).
873. Lundberg, K. L., R. J. Rowatt, and N. E. Miller, Inorg. Chem., 8, 1336 (1969).
874. McAllister, T., and F. P. Lossing, J. Phys. Chem., 73, 2996 (1969).
875. McClure, J. D., U.S. 3,225,083 (1965); C. A., 64, 19427e (1966).
876. McClure, J. D., U.S. 3,227,745 (1966).
877. McClure, J. D., Tetrahedron Lett., 1967, 2401
878. McClure, J. D., U.S. 3,444,208 (1969); C. A., 71, 49284a (1969).
879. McCoy, C. R., and A. L. Allred, J. Inorg. Nucl. Chem., 25, 1219 (1963).
880. McDonald, R. N., U.S. 3,347,829 (1967); C. A., 67, 117 551e (1967); U.S. 2,828,286 (1958); C. A., 52, 10640h (1958).
881. McEwen, W. E., A. Bladé-Font, and C. A. Vander Werf, J. Am. Chem. Soc., 84, 677 (1962).
882. McEwen, W. E., K. F. Kumli, A. Bladé-Font, M. Zanger, and C. A. Vander Werf, J. Am. Chem. Soc., 86, 2378 (1964).
883. McEwen, W. E., and A. P. Wolf, J. Am. Chem. Soc., 84, 676 (1962).
884. McEwen, W. E., C. A. Vander Werf, A. Bladé-Font, C. B. Parisek, G. Keldsen, D. C. Velez, D. P. Young, K. Kumli, and G. Axelrad, A.C.S., 140th Meeting, Chicago, Sept. 1961, Abstr., p. 96Q.
885. McFarlane, W., Chem. Commun., 1967, 58.
886. McFarlane, W., Chem. Commun., 1968, 229.
886a. McFarlane, W., Org. Magnetic Resonance, 1, 3 (1969).
887. McFarlane, W., and J. A. Nash, Chem. Commun., 1969, 127.
888. McWhinnie, W. R., and R. C. Poller, Spectrochim. Acta, 22, 501 (1966).
889. Märkl, G., Z. Naturforsch., B, 18, 84 (1963).
890. Märkl, G., Angew. Chem., 75, 168 (1963).
891. Märkl, G., Angew. Chem., 78, 907 (1966).
892. Märkl, G., D. E. Fischer, and H. Olbrich, Tetrahedron Lett., 1970, 645.
893. Märkl, G., and F. Lieb, Tetrahedron Lett., 1967, 3489.
894. Märkl, G., and F. Lieb, Angew.Chem., 80, 702 (1968); Angew. Chem. Int. Ed., 7, 733 (1968).

895. Märkl, G., F. Lieb, and A. Merz, Angew, Chem., 79, 59 (1967); Angew. Chem. Int. Ed., 6, 87 (1966).
896. Märkl, G., F. Lieb, and A. Merz, Angew. Chem., 79, 475 (1967); Angew. Chem. Int. Ed., 6, 458 (1967).
897. Märkl, G., F. Lieb, and A. Merz, Angew Chem., 79, 947 (1967); Angew. Chem. Int. Ed., 6, 944 (1967).
898. Märkl, G., and A. Merz, Tetrahedron Lett., 1968, 3611.
899. Märkl, G., and A. Merz, Tetrahedron Lett., 1969, 1231.
900. Märkl, G., and H. Olbrich, Angew. Chem., 78, 598 (1966); Angew. Chem. Int. Ed., 5, 589 (1966).
901. Märkl, G., and H. Olbrich, Angew. Chem., 78, 598 (1966); Angew. Chem. Int. Ed., 5, 588 (1966).
902. Märkl, G., and H. Olbrich, Tetrahedron Lett., 1968, 3813.
903. Märkl, G., and R. Potthast, Angew. Chem., 79, 58 (1967).
904. Märkl, G., and R. Potthast, Tetrahedron Lett., 1968, 1755.
905. Mahler, W., and A. B. Burg, J. Am. Chem. Soc., 80, 6161 (1958).
906. Maier, L., Angew. Chem., 71, 574 (1959).
907. Maier, L., Technical Report, Contr. AF 33(616)-6950 (1960).
908. Maier, L., Chem. Ber., 94, 3043 (1961).
909. Maier, L., Chem. Ber., 94, 3056 (1961).
910. Maier, L., J. Inorg. Nuclear Chem., 24, 1073 (1962).
911. Maier, L., "Preparation and Properties of Primary, Secondary, and Tertiary Phosphines," in Progress in Inorganic Chemistry, F. A. Cotton, Ed., Vol. V, Interscience, 1963, p. 27.
912. Maier, L., Helv. Chim. Acta, 46, 1812 (1963).
913. Maier, L., Helv. Chim. Acta, 46, 2026 (1963).
914. Maier, L., Helv. Chim. Acta, 46, 2667 (1963).
915. Maier, L., Helv. Chim. Acta, 47, 2129 (1964).
916. Maier, L., Helv. Chim. Acta, 47, 2137 (1964).
917. Maier, L., Angew. Chem., 77, 549 (1965).
918. Maier, L., Helv. Chim. Acta, 48, 1034 (1965).
919. Maier, L., Helv. Chim. Acta, 48, 1190 (1965).
920. Maier, L., U.S. 3,253,033 (1966); C. A., 65, 5488f (1966).
921. Maier, L., Helv. Chim. Acta, 49, 842 (1966); D.A.S. 1,299,639 (1969).
922. Maier, L., Helv. Chim. Acta, 49, 1000 (1966).
923. Maier, L., Helv. Chim. Acta, 49, 1119 (1966).
924. Maier, L., Helv. Chim. Acta, 49, 1718 (1966).
925. Maier, L., Helv. Chim. Acta, 49, 2458 (1966).
926. Maier, L., U.S. 3,321,557 (1967); C. A., 67, 73684u (1967).
927. Maier, L., Helv. Chim. Acta, 50, 1723 (1967).

928. Maier, L., Helv. Chim. Acta, 50, 1747 (1967).
929. Maier, L., Fortsch. Chem. Forsch., 8, 1 (1967).
930. Maier, L., Helv. Chim. Acta, 51, 1608 (1968).
931. Maier, L., Helv. Chim. Acta, 52, 858 (1969).
932. Maier, L., in preparation.
933. Maier, L., unpublished; and Helv. Chim. Acta, 54, 1651 (1971)
934. Maier, L., D. Seyferth, F. G. A. Stone, and E. G. Rochow, J. Am. Chem. Soc., 79, 5884 (1957); Z. Naturforsch., B, 12, 263 (1957).
935. Malatesta, L., Gazz. Chim. Ital., 77, 509, 518 (1947); C. A., 42, 5411h (1948).
936. Malhotra, S. C., Inorg. Chem., 3, 902 (1964).
937. Mallion, K. B., and F. G. Mann, Chem. Ind. (London), 1963, 654.
938. Mallion, K. B., and F. G. Mann, Chem. Ind. (London), 1963, 1558.
939. Mallion, K. B., and F. G. Mann, J. Chem. Soc., 1964, 5716.
940. Mallion, K. B., and F. G. Mann, J. Chem. Soc., 1964, 6121.
941. Mallion, K. B., and F. G. Mann, J. Chem. Soc., 1965, 4115.
942. Mallion, K. B., F. G. Mann, B. P. Tong, and V. P. Wystrach, J. Chem. Soc., 1963, 1327.
943. Manatt, S. L., E. A. Cohen, and A. H. Cowley, J. Am. Chem. Soc., 91, 5919 (1969).
944. Manatt, S. L., D. D. Elleman, A. H. Cowley, and A. B. Burg, J. Am. Chem. Soc., 89, 4544 (1967).
945. Manatt, S. L., G. L. Juvinall, and D. D. Elleman, J. Am. Chem. Soc., 85, 2664 (1963).
946. Manatt, S. L., G. L. Juvinall, R. I. Wagner, and D. D. Elleman, J. Am. Chem. Soc., 88, 2689 (1966).
947. Mann, F. G., and E. J. Chaplin, J. Chem. Soc., 1937, 527.
948. Mann, F. G., and I. T. Millar, J. Chem. Soc., 1951, 2205.
949. Mann, F. G., and I. T. Millar, J. Chem. Soc., 1952, 3039.
950. Mann, F. G., and I. T. Millar, J. Chem. Soc., 1952, 4453.
951. Mann, F. G., I. T. Millar, and B. B. Smith, J. Chem. Soc., 1953, 1130.
952. Mann, F. G., I. T. Millar, and F. H. C. Stewart, J. Chem. Soc., 1954, 2832.
953. Mann, F. G., I. T. Millar, and H. R. Watson, J. Chem. Soc., 1958, 2516.
954. Mann, F. G., and M. J. Pragnell, J. Chem. Soc., 1965, 4120; Chem. Ind. (London), 1964, 1386.
955. Mann, F. G., and D. Purdie, J. Chem. Soc., 1935, 1549; 1936, 873.

956. Mann, F. G., B. P. Tong, and V. P. Wystrach, J. Chem. Soc., 1963, 1155.
957. Mann, F. G., and H. J. Watson, J. Org. Chem., 13, 502 (1948).
958. Mann, F. G., and H. R. Watson, Chem. Ind. (London), 1958, 1264.
959. Mann, F. G., and H. R. Watson, J. Chem. Soc., 1957, 3945, 3950.
960. Mann, F. G., A. F. Wells, and D. Purdie, J. Chem. Soc., 1936, 1503; 1937, 1828.
961. Mao, T. J., Brit. 1,054,533 (1967); C. A., 66, 76540j (1967).
962. Mao, T. J., and R. J. Eldred, Polym. Preprints, 6, 267 (1965); J. Polym. Sci., Part A, 5, 1741 (1967).
963. Maretina, I. A., and A. A. Petrov, Zh. Obshch. Khim., 34, 1685 (1964); C. A., 61, 5686a (1964).
963a. Margulis, T. N., and D. H. Templeton, J. Am. Chem. Soc., 83, 995 (1961).
964. Markevich, M. A., L. K. Pakhomova, and N. S. Enikolopyan, Dokl. Acad. Nauk SSSR, 187, 609 (1969); C. A., 71, 113 321q (1969).
965. Marshall, M. D., and H. J. Harwood, U.S. 3,031,509 (1962); C. A., 57, 8618i (1962).
966. Marsi, K. L., Chem. Commun., 1968, 846.
967. Marsi, K. L., J. Am. Chem. Soc., 91, 4724 (1969).
968. Martz, M. D., and L. D. Quin, J. Org. Chem., 34, 3195 (1969).
969. Mason, R. F., U.S. 3,435,076 (1969); C. A., 71, 22187j (1969).
969a. Matchiner, H., L. Krause, and F. Krech, Z. Anorg. Allg. Chem., 373, 1 (1970).
970. Mathey, F., C. R. Acad. Sci., Paris, Ser. C., 269, 1066 (1969).
971. Mathey, F., and G. Muller, C. R. Acad. Sci., Paris, Ser. C, 269, 158 (1969).
972. Mathis, R., M. Barthelat, C. Charrier, and F. Mathis, J. Mol. Strukt., 1, 481 (1968).
973. Mauret, P., J. P. Fayet, D. Voigt, M. C. Labarre, and J. F. Labarre, J. Chim. Phys., 65, 549 (1968).
974. Mavel, G., C. R. Acad. Sci., Paris, 248, 3699 (1959).
974a. Mavel, G., and G. Martin, C. R. Acad. Sci., Paris, Ser. C., 252, 110 (1961).
975. Mayer, C., W. J. Lorenz, and H. Fischer, Z. Phys. Chem., 52, 193 (1967); C. A., 66, 110 955k (1967).
976. Mayo, F. R., and C. Walling, Chem. Rev., 27, 351 (1940).
977. Meisenheimer, J., Ann. Chem., 449, 213 (1906).
978. Meisenheimer, J., J. Casper, M. Höring, W. Lauter, L. Lichtenstadt, and W. Samuel, Ann. Chem., 449, 213 (1926).
979. Meriwether, L. S., and J. R. Leto, J. Am. Chem. Soc., 83, 3192 (1961).

979a. Meriwether, L. S., E. C. Colthup, G. W. Kennerly, and R. N. Reusch, J. Org. Chem., 26, 5155 (1961).
979b. Meriwether, L. S., and M. L. Fiene, J. Am. Chem. Soc., 81, 4200 (1959).
980. Messinger, J., and C. Engels, Chem. Ber., 21, 326 (1888); Chem. Ber., 21, 2919 (1888).
981. Metzger, S. H., O. H. Basedow, and A. F. Isbell, J. Org. Chem., 29, 627 (1964).
982. Michaelis, A., Chem. Ber., 7, 6 (1874).
983. Michaelis, A., Ann. Chem., 181, 265 (1876).
984. Michaelis, A., Ann. Chem., 293, 193 (1896).
985. Michaelis, A., Ann. Chem., 294, 1 (1896).
986. Michaelis, A., Ann. Chem., 315, 43 (1901).
987. Michaelis, A., and J. Ananoff, Chem. Ber., 8, 493 (1875).
988. Michaelis, A., and F. Dittler, Chem. Ber., 12, 338 (1879).
989. Michaelis, A., and L. Gleichmann, Chem. Ber., 15, 801, 1961 (1882).
990. Michaelis, A., and A. Link, Ann. Chem., 207, 193 (1881).
991. Michaelis, A., and C. Panek, Ann. Chem., 212, 203 (1882).
992. Michaelis, A., and A. Reese, Chem. Ber., 15, 1610 (1882).
993. Michaelis, A., and A. Schenk, Chem. Ber., 21, 1497 (1888), Ann. Chem., 260, 1 (1890).
993a. Michaelis, A., and H. v. Soden, Chem. Ber., 17, 921 (1884).
994. Michaelis, A., and H. v. Soden, Ann. Chem., 229, 295 (1885); Chem. Ber., 17, 921 (1884).
995. Midland Silicones Ltd., Ger. 1,175,672 (1964).
996. Mikhailov, B. M., and N. F. Kucherova, Dokl. Akad. Nauk SSSR, 74, 501 (1950); C. A., 45, 3343c (1951).
997. Mikhailov, B. M., and N. F. Kucherova, Zh. Obshch. Khim., 22, 792 (1952); C. A., 47, 5388g (1953).
998. Militskova, E. A., and L. A. Zavyalina, U.S.S.R. 142,427 (1961); C. A., 56, 14483c (1962).
999. Miller, F. A., and D. H. Lemmon, Spectrochim. Acta, 23A, 1099 (1967).
1000. Miller, G. R., A. W. Yankowsky, and O. S. Grim, J. Chem. Phys., 51, 3185 (1969).
1001. Miller, J. M., J. Chem. Soc., A, 1967, 828.
1002. Miller, R. C., J. Org. Chem., 24, 2013 (1959).
1003. Mingoia, A., Gazz. Chim. Ital., 60, 144 (1930).
1004. Mironova, Z. N., E. N. Tsvetkov, A. V. Nikolaev, and M. I. Kabachnik, Zh. Obshch. Khim., 37, 2747 (1967); C. A., 69, 19262h (1968).
1005. Mironova, Z. N., E. N. Tsvetkov, A. V. Nikolaev, and M. I. Kabachnik, U.S.S.R. 247,296 (1969); C. A., 71, 124 650x (1969).

1006. Mitchell, R. W., L. J. Kuzma, R. J. Pirkle, and J. A. Merritt, Spectrochim. Acta, 25A, 819 (1969).
1006a. Mitchener, J. P., and A. M. Aguiar, Org. Preparative Procedures, 1, 259 (1969).
1007. Mitterhofer, F., and H. Schindlbauer, Monatsh. Chem., 98, 206 (1967).
1008. Moedritzer, K., L. Maier, and L. D. C. Groenweghe, J. Chem. Eng. Data, 7, 307 (1962); J. Phys. Chem., 66, 901 (1962).
1009. Moedritzer, K., Technical Report, Contr. AF 33(616)-6950 (1960).
1010. Möslinger, W., Ann. Chem., 185, 49 (1877); Chem. Ber., 9, 1005 (1876).
1011. Monagle, J. J., J. V. Mangenhauser, and D. A. Jones, Jr., J. Org. Chem., 32, 2477 (1967).
1012. Monsanto Co., Fr. 1,347,066 (1963); C. A., 60, 12055c (1964).
1013. Moore, C. G., and B. R. Trego, J. Appl. Polym. Sci., 5, 299 (1961); 8, 723, 1824 (1964).
1014. Mootz, D., H. Altenburg, and D. Lücke, Z. Kristallogr., 130, 239 (1969).
1015. Mootz, D., and H. Brinkel, Naturwissenschaften, 48, 402 (1961).
1016. Mootz, D., and G. Sassmannshausen, Z. Anorg. Allg. Chem., 355, 200 (1967); Z. Kristallogr., 117, 233 (1962).
1017. Morgan, P. W., and B. C. Herr, J. Am. Chem. Soc., 74, 4526 (1952).
1018. Morita, K., Z. Suzuki, and H. Hirose, Bull. Chem. Soc. Jap., 41, 2815 (1968).
1019. Moritz, A. G., Spectrochim. Acta, 22, 1015 (1966).
1020. Moritz, A. G., J. D. Sayby, and S. Sternhell, Aust. J. Chem., 21, 2565 (1968).
1021. Mortimer, G. A., U.S. 3,377,300 (1968); C. A., 69, 3317u (1968).
1022. Moulds, G. M., and R. E. Walck, U.S. 3,230,193 (1966); C. A., 64, 8399b (1966).
1023. M. & T. Chemical Inc., Neth. 6,501,268 (1965); C. A., 64, 2127h (1966).
1024. M. & T. Chemical Inc., Neth. 6,514,669 (1966); C. A., 65, 12238f (1966).
1025. Mukaiyama, T., R. Yoda, and I. Kuwajima, Tetrahedron Lett., 1969, 23.
1026. Muller, N., P. C. Lauterbur, and J. Goldendon, J. Am. Chem. Soc., 78, 3557 (1956).
1027. Murahashi, S., S. Nozakura, and K. Hatada, Bull. Chem. Soc. Jap., 34, 939 (1961); C. A., 56, 1589h (1962).
1028. Murin, A. N., and V. D. Nefedov, Primenenie Mechenykh Atomov v Anal. Khim., Akad. Nauk SSSR, Inst. Geokhim. Anal. Khim., 1955, 75; C. A., 50, 3915i (1956).

1029. Nakaguchi, K., and M. Kirooka, Jap. 20,735 (1963); C. A., 60, 3123f (1964).
1030. Narasimhan, P. T., and M. T. Rogers, J. Chem. Phys., 34, 1049 (1961).
1031. National Distiller and Chem. Corp., Neth. 6,602,627; 6,602,495 (1966); C. A., 66, 37462r, 37463s (1967).
1032. Natta, G., G. F. Pregaglia, G. Mazzanti, V. Zamboni, and M. Binaghi, European Polym. J., 1, 25 (1965); C. A., 63, 8512d (1965).
1033. Naumann, K., G. Zon, and K. Mislow, J. Am. Chem. Soc., 91, 7012 (1969).
1034. Naumann, K., G. Zon, and K. Mislow, J. Am. Chem. Soc., 91, 2788 (1969).
1034a. Nelson, A. J., J. C. Clardy, and T. J. Barton, A.C.S., 159th Meeting, Houston, Texas, Feb. 1970, Abstr., ORGN 92.
1035. Nelson, R., J. Chem. Phys., 39, 2382 (1963).
1036. Nesmeyanov, N. A., V. M. Novikov, and O. A. Reutov, Zh. Organ. Khim., 2, 942 (1966); C. A., 65, 15420h (1966).
1037. Neunhoeffer, O., and L. Lamza, Chem. Ber., 94, 2514 (1961).
1038. Neunhoeffer, O., and L. Lamza, Chem. Ber., 94, 2519 (1961).
1039. Neunhoeffer, O., L. Lamza, and G. Tomaschewski, Naturwissenschaften, 48, 477 (1961).
1040. Neuse, E. W., and G. J. Chris, J. Makromol. Sci., A, 1, 371 (1967).
1041. Nicco, A., and B. Boucheron, Ger. 1,814,640 (1969); C. A., 71, 81929a (1969).
1042. Nicpon, P., and W. Meek, Inorg. Chem., 6, 145 (1967).
1043. Niebergall, H., Ger. 1,086,896 (1958).
1044. Niebergall, H., Ger. 1,116,411 (1958); C. A., 56, 7523d (1962).
1045. Niebergall, H., Ger. 1,118,781 (1959); C. A., 56, 11622a (1962).
1046. Niebergall, H., Angew. Chem., 72, 210 (1960).
1047. Niebergall, H., Ger. 1,113,827 (1961); C. A., 56, 14475g (1962).
1048. Niebergall, H., Makromol. Chem., 52, 218 (1962); Ger. 1,103,590 (1961).
1049. Niebergall, H., and B. Langenfeld, Ger. 1,083,262 (1960); U.S. 2,959,621 (1960); C. A., 55, 7289h (1961).
1050. Niebergall, H., and B. Langenfeld, Chem. Ber., 95, 64 (1962).
1051. Nielsen, J. R., and J. D. Walker, Spectrochim. Acta, 21, 1163 (1965).
1052. Nikolaev, A. V., Yu. A. Afanas'ev, and A. D. Starostin, Izv. Sib. Otd. Akad. Nauk SSSR, Ser. Khim. Nauk, 1968, 3; C. A., 71, 7169m (1969).

1053. Nishimura, S., Kogyo Kagaku Zasshi, 65, 2065 (1962); C. A., 58, 12766e (1963).
1054. Niwa, E., H. Aoki, H. Tanaka, and K. Munakata, Chem. Ber., 99, 712 (1966).
1055. Nobis, J. F., L. F. Moormeier, and R. E. Robinson, Advan. Chem. Ser., 23, 63 (1959).
1055a. Noble, M. L., Brit. 713,325 (1954); C. A., 50, 6500f (1956).
1056. Nöth, H., Z. Naturforsch., B, 15, 327 (1960).
1057. Noltes, J. G., Rec. Trav. Chim., 84, 782 (1965).
1058. Normant, H., C. R. Acad. Sci., Paris, 239, 1510 (1954).
1059. Normant, H., T. Cuvigny, J. Normant, and B. Angelo, Bull. Soc. Chim., Fr., 1965, 3446.
1059a. Novikova, E. N., Vestsi Akad. Navuk Belarusk, SSR, Ser. Fiz-Tekhn. Navuk, No. 1, 47 (1957); C. A., 52, 2805i (1958).
1060. Novikova, Z. S., E. A. Efimova, and I. F. Lutsenko, Zh. Obshch. Khim., 38, 2345 (1968).
1061. Novikova, Z. S., M. V. Proskurina, L. I. Petrovskaya, I. V. Bogdanova, N. P. Galitskova, and I. F. Lutsenko, Zh. Obshch. Khim., 37, 2080 (1967); C. A., 68, 78392c (1968).
1061a. N. V. de Bataatsche Petroleum Maatschappij, Neth. 90,245 (1959); C. A., 54, 11568e (1960).
1062. Oda, R., T. Kawabata, and S. Tanimoto, Tetrahedron Lett., 1964, 1653.
1063. Ojima, I., K. Akiba, and N. Inamoto, Bull. Chem. Soc. Jap., 42, 2975 (1969).
1064. Okhlobystin, O. Yu., and L. I. Zakharkin, Izv. Akad. Nauk SSSR, Otdel Khim. Nauk, 1958, 1006; C. A., 53, 1122f (1959).
1065. Olah, G. A., and C. W. McFarland, J. Org. Chem., 34, 1832 (1969).
1066. Onken, H., and J. Lottermoser, Naturwissenschaften, 54, 560 (1967).
1067. Oppegard, A. L., U.S. 2,687,437 (1954); C. A., 49, 11000h (1955).
1067a. Otsuka, S., and S. Murahashi, Nippon Kagaku Zasshi, 75, 884 (1954); C. A., 51, 14614e (1957).
1068. Owen, W. J., and F. C. Saunders, Brit. 1,007,333 (1965); C. A., 63, 18154g (1965).
1069. Paciorek, K. L., and R. H. Kratzer, J. Org. Chem., 31, 2426 (1966).
1070. Packer, K. J., J. Chem. Soc., 1963, 960.
1071. Pailer, M., and H. Huemer, Monatsh. Chem., 95, 373 (1964).
1072. Paisley, D. M., and C. S. Marvel, J. Polym. Sci., 56, 533 (1962).
1073. Paolillo, L., and J. A. Ferretti, Congr. Conv. Simp. Sci., 11, 103 (1967).

1074. Papp, G. P., and S. A. Buckler, Fr. 1,405,747 (1965); C. A., 63, 16385c (1965).
1075. Papp, G. P., and S. A. Buckler, J. Org. Chem., 31, 588 (1966); Brit. 1,021,820 (1966); U.S. 3,314,993 (1967).
1076. Park, P. J. D., and P. J. Hendra, Spectrochim. Acta, 24A, 2081 (1968).
1077. Parshall, G. W., J. Inorg. Nucl. Chem., 14, 291 (1960).
1078. Parshall, G. W., Inorg. Chem., 4, 52 (1965).
1079. Parshall, G. W., Inorg. Syn., 11, 157 (1968).
1080. Parshall, G. W., D. C. England, and R. V. Lindsey, Jr., J. Am. Chem. Soc., 81, 4801 (1959).
1081. Parshall, G. W., W. H. Knoth, and R. A. Schunn, J. Am. Chem. Soc., 91, 4990 (1969).
1082. Parshall, G. W., and R. V. Lindsey, Jr., J. Am. Chem. Soc., 81, 6273 (1959).
1083. Pass, F., and H. Schindlbauer, Monatsh. Chem., 90, 148 (1959).
1084. Pass, F., E. Steininger, and H. Schindlbauer, Monatsh. Chem., 90, 792 (1959).
1085. Pass, F., E. Steininger, and H. Zorn, Monatsh. Chem., 93, 230 (1962); Can. 618,333 (1961); Ger. 1,126,867 (1962); C. A., 57, 8619f (1962).
1086. Partheil, A., and A. Gronover, Arch. Pharm., 241, 411 (1903).
1087. Pellon, J., J. Am. Chem. Soc., 83, 1915 (1961).
1087a. Pellon, J., J. Polym. Sci., 43, 537 (1960).
1088. Pellon, J., J. Polym. Sci., Part A, 1, 3561 (1963).
1089. Pellon, J., and W. G. Carpenter, J. Polym. Sci., Part A, 1, 863 (1963); Brit. 897,680 (1962).
1089a. Perry, E., J. Polym. Sci., 54, 546 (1961).
1090. Perry, E., J. Appl. Polym. Sci., 8, 2605 (1964); C. A., 62, 6584d (1965).
1091. Perry, E., U.S. 3,248,362 (1966); C. A., 65, 877f (1966).
1092. Perry, E., U.S. 3,272,786 (1966); C. A., 66, 11294u (1967).
1093. Perry, E., U.S. 3,322,742 (1967); C. A., 67, 54593x (1967).
1094. Perveev, F. Ya., and K. Rikhter, Zh. Obshch. Khim., 30, 784 (1960); C. A., 55, 1580h (1961).
1095. Peters, G., J. Am. Chem. Soc., 82, 4751 (1960).
1096. Peters, G., J. Org. Chem., 27, 2198 (1962).
1097. Peterson, D. J., J. Org. Chem., 31, 950 (1966).
1098. Peterson, D. J., J. Org. Chem., 32, 1717 (1967).
1099. Peterson, D. J., J. Organometal. Chem., 8, 199 (1967).
1100. Peterson, D. J., and J. H. Collins, J. Org. Chem., 31, 2373 (1966).

1101. Peterson, D. J., and H. R. Hays, J. Org. Chem., 30, 1939 (1965); U.S. 3,414,624 (1968); C. A., 70, 58017x (1969).
1102. Peterson, D. J., and T. J. Logan, J. Inorg. Nucl. Chem., 28, 53 (1966).
1103. Peterson, L. K., and G. L. Wilson, Can. J. Chem., 46, 685 (1968).
1104. Petrov, A. A., and V. A. Kormer, Dokl. Akad. Nauk SSSR, 132, 1095 (1960); C. A., 54, 22327g (1960).
1105. Petrov, K. A., V. A. Parshina, and V. A. Gaidamak, Zh. Obshch. Khim., 31, 3411 (1961); C. A., 57, 4692d (1962).
1106. Petrov, K. A., and V. A. Parshina, Zh. Obshch. Khim., 31, 2729 (1961); C. A., 56, 11612g (1962).
1107. Petrov, K. A., and V. A. Parshina, Zh. Obshch. Khim., 31, 3417 (1961); C. A., 57, 4692g (1962).
1108. Petrov, K. A., and V. A. Parshina, Zh. Obshch. Khim., 31, 3421 (1961); C. A., 57, 4693b (1962).
1109. Petrov, K. A., and V. A. Parshina, Usp. Khim., 37, 1218 (1968).
1110. Petrov, K. A., V. A. Parshina, and M. B. Luzanova, Zh. Obshch. Khim., 32, 553 (1962); C. A., 58, 5714h (1963).
1111. Petrov, K. A., V. A. Parshina, and A. F. Manuilov, Zh. Obshch. Khim., 35, 2062 (1965).
1112. Petrov, K. A., V. A. Parshina, B. A. Orlov, and G. M. Tsypina, Zh. Obshch. Khim., 32, 4017 (1962); C. A., 59, 657c (1963).
1113. Petrov, K. A., V. A. Parshina, and G. M. Petrova, Zh. Obshch. Khim., 39, 1247 (1969); C. A., 71, 70693t (1969).
1114. Petrov, K. A., V. A. Parshina, and G. M. Tsypina, Plasticheskie Massy, 1963, 11; C. A., 60, 5651f (1964).
1115. Petrovskaya, L. I., M. V. Proskurina, Z. S. Novikova, and I. F. Lutsenko, Izv. Akad. Nauk SSSR, Ser. Khim., 1968, 1277.
1116. Pettit, L. D., and H. M. N. H. Irving, J. Chem. Soc., 1964, 5336.
1117. Pfeiffer, P., I. Heller, and H. Pietsch, Chem. Ber., 37, 4620 (1904).
1118. Philips Petroleum Co., Belg. 713,185 (1968).
1119. Philips, G. M., J. S. Hunter, and L. E. Sutton, J. Chem. Soc., 1945, 146.
1120. Pike, R. A., J. Org. Chem., 27, 2186 (1962); U.S. 3,057,902 (1962); 3,153,662 (1964).
1121. Pike, R. A., U.S. 3,122,581 (1964); C. A., 60, 12053f (1964).
1121a. Plazek, E., and R. Tyka, Rocz. Chem., 33, 549 (1959); C. A., 53, 21750c (1959).

1122. Plazek, E., and R. Tyka, Zeszyty Nauk. Politech., Wroclaw., Chem., No. 4, 79 (1957); C. A., 52, 20156c (1958).
1123. Plets, V. M., Dissertation, Kazan, 1938. (Cited in Ref. 801).
1124. Popov, E. M., E. N. Tsvetkov, J. Yu. Chang, and T. Ya. Medved, Zh. Obshch. Khim., 32, 3255 (1962); C. A., 58, 5165c (1963).
1125. Potenza, J. A., E. H. Poindexter, P. J. Caplan, and R. A. Dwek, J. Am. Chem. Soc., 91, 4356 (1969).
1126. Pregaglia, G., and M. Binaghi, Ital. 694,415 (1965); C. A., 67, 22352b (1967).
1127. Price, A. H., J. Phys. Chem., 62, 773 (1958).
1128. Price, C. C., T. Parasaran, and T. V. Lakshminarayan, J. Am. Chem. Soc., 88, 1034 (1966).
1129. Prikoszovich, W., and H. Schindlbauer, Chem. Ber., 102, 2922 (1969).
1130. Proskurina, M. V., I. F. Lutsenko, Z. S. Novikova, and N. P. Voronova, Khim. Org. Soedin. Fosfora, Akad. Nauk SSSR, Otd. Obshch. Tekh. Khim., 1967, 8; C. A., 69, 52217v (1968).
1131. Proskurina, M. V., Z. S. Novikova, and I. F. Lutsenko, Dokl. Akad. Nauk SSSR, 159, 619 (1964); C. A., 62, 6508h (1964).
1132. Protopopov, I. S., and M. Ya. Kraft, Med. Prom. SSSR, 13, No. 12, 5 (1959); C. A., 54, 10914c (1960).
1133. Protopopov, I. S., and M. Ya. Kraft, Zh. Obshch. Khim., 33, 3050 (1963); C. A., 60, 1789g (1964).
1134. Protopopov, I. S., and M. Ya. Kraft, Zh. Obshch. Khim., 34, 1446 (1964); C. A., 61, 5685f (1964).
1135. Ptitsyna, O. A., M. E. Pudeeva, and O. A. Reutov, Dokl. Akad. Nauk SSSR, 165, 838 (1965); C. A., 64, 5129a (1966).
1136. Pudovik, A. N., and M. A. Pudovik, Zh. Obshch. Khim., 33, 3353 (1963); C. A., 60, 9308a (1964).
1137. Pudovik, A. N., and M. A. Pudovik, Vysokomolekul. Soedin., Ser. A, 9, 2241 (1967); C. A., 68, 13649x (1968).
1138. Pullman, B. J., and B. O. West, Aust. J. Chem., 17, 30 (1964).
1139. Quin, L. D., and J. G. Bryson, J. Am. Chem. Soc., 89, 5984 (1967).
1140. Quin, L. D., J. G. Bryson, and C. G. Moreland, J. Am. Chem. Soc., 91, 3308 (1969).
1141. Quin, L. D., J. P. Gratz, and T. P. Barket, J. Org. Chem., 33, 1034 (1968).
1142. Quin, L. D., J. P. Gratz, and R. E. Montgomery, Tetrahedron Lett., 1965, 2187.
1143. Quin, L. D., and D. A. Mathews, Chem. Ind. (London), 1963, 210.

1144. Quin, L. D., and D. A. Mathews, J. Org. Chem., 29, 836 (1964).
1145. Quin, L. D., J. A. Peters, C. E. Griffin, and M. Gordon, Tetrahedron Lett., 1964, 3689.
1146. Quin, L. D., and H. E. Shook, Jr., Tetrahedron Lett., 1965, 2193.
1147. Quin, L. D., and H. E. Shook, J. Org. Chem., 32, 1604 (1967).
1148 Quin, L. D., J. H. Somers, and R. H. Prince, J. Org. Chem., 34, 3700 (1969).
1149. Rabinowitz, R., A. C. Henry, and R. Marcus, J. Polym. Sci., Part A, 3, 2055 (1965).
1150. Rabinowitz, R., and R. Marcus, J. Org. Chem., 26, 4157 (1961).
1151. Rabinowitz, R., and R. W. Marcus, Fr. 1,314,883 (1963); C. A., 58, 11401d (1963).
1152. Rabinowitz, R., R. Marcus, and J. Pellon, J. Polym. Sci., Part A, 2, 1233 (1964).
1153. Rabinowitz, R., R. Marcus, and J. Pellon, J. Polym. Sci., Part A, 2, 1241 (1964).
1154. Rabinowitz, R., and J. Pellon, J. Org. Chem., 26, 4623 (1961).
1155. Radcliffe, L. G., and W. H. Brindley, Chem. Ind. (London), 42, 64 (1923).
1156. Rakshys, J. W., R. W. Taft, and W. A. Sheppard, J. Am. Chem. Soc., 90, 5236 (1968).
1157. Ramirez, F., and D. Rhum, J. Org. Chem., 24, 894 (1959).
1158. Ramsden, H. E., U.S. 2,912,465 (1959); C. A., 54, 2170g (1960).
1159. Ramsden, H. E., U.S. 2,913,498 (1959); C. A., 54, 3315c (1960).
1160. Rao, C. N. R., J. Ramachandran, M. S. C. Iah, S. Somasekhara, and T. V. Rajakumar, Nature, 183, 1475 (1959).
1161. Raudnitz, H., Chem. Ber., 60, 743 (1927).
1162. Rauhut, M. M., U.S. 3,116,316 (1963); C. A., 60, 12054f (1964).
1163. Rauhut, M. M., U.S. 3,116,316 (1963); C. A., 60, 12054g (1964).
1164. Rauhut, M. M. U.S. 3,206,496 (1965); C. A., 63, 14906c (1965).
1164a. Rauhut, M. M., Bernheimer, and A. A. Semsel, J. Org. Chem., 28, 478 (1963).
1165. Rauhut, M. M., G. B. Borowitz, and H. C. Gillham, J. Org. Chem., 28, 2565 (1963).
1166. Rauhut, M. M., and H. A. Currier, U.S. 2,953,595 (1960); C. A., 55, 11309f (1961).
1167. Rauhut, M. M., and H. A. Currier, J. Org. Chem., 26, 4626 (1961).
1168. Rauhut, M. M., and H. Currier, U.S. 3,074,999 (1963).
1169. Rauhut, M. M., H. A. Currier, and V. P. Wystrach, J. Org. Chem., 26, 5133 (1961).

1170. Rauhut, M. M., H. A. Currier, G. A. Peters, F. C. Schaefer, and V. P. Wystrach, J. Org. Chem., 26, 5135 (1961).
1171. Rauhut, M. M., H. A. Currier, A. M. Semsel, and V. P. Wystrach, J. Org. Chem., 26, 5138 (1961).
1172. Rauhut, M. M., I. Hechenbleikner, and H. A. Currier, U.S. 2,953,596 (1960).
1173. Rauhut, M. M., I. Hechenbleikner, H. A. Currier, F. C. Schaefer, and V. P. Wystrach, J. Am. Chem. Soc., 81, 1103 (1959).
1174. Rauhut, M. M., I. Hechenbleikner, H. A. Currier, and V. P. Wystrach, J. Am. Chem. Soc., 80, 6690 (1958).
1175. Rauhut, M. M., and A. M. Semsel, U.S. 3,060,241 (1962); C. A., 58, 6862c (1963).
1176. Rauhut, M. M., and A. M. Semsel, J. Org. Chem., 28, 471 (1963); U.S. 3,060,241.
1177. Rauhut, M. M., and A. M. Semsel, J. Org. Chem., 28, 473 (1963); U.S. 3,099,691 (1963); C. A., 60, 555a (1964).
1178. Rausch, M. D., F. E. Tibbets, and H. B. Gordon, J. Organometal. Chem., 5, 493 (1966).
1179. Raver, Ch. R., A. B. Bruker, and L. Z. Soborovskii, Zh. Obshch. Khim., 32, 588 (1962); C. A., 58, 6857c (1963).
1180. Razuvaev, G. A., and N. A. Osanova, Dokl. Akad. Nauk SSSR, 104, 552 (1955); C. A., 50, 11268b (1956).
1181. Reiff, H. F., and B. C. Pant, J. Organometal. Chem., 17, 165 (1969).
1181a. Reppe, W., and W. J. Schweckendiek, Ann. Chem., 560, 104 (1948).
1182. Retcofsky, H. L., and C. E. Griffin, Tetrahedron Lett., 1966, 1975.
1183. Reuter, M., and W. Duersch, Ger. 1,235,910 (1967); C. A., 67, 32777h (1967).
1184. Reuter, M., and L. Orthner, Ger. 1,035,135 (1958); C. A., 54, 14125a (1960).
1185. Reuter, M., and L. Orthner, Ger. 1,075,610 (1960); 55, 13316h (1961).
1186. Reuter, M., and E. Wolf, Ger. 1,078,574 (1960); C. A., 55, 16427c (1961).
1187. Reuter, M., and E. Wolf, Ger. 1,082,910 (1960).
1188. Reuter, M., E. Wolf, and L. Orthner, D.A.S. 1,061,513 (1958).
1189. Rhomberg, A., and P. Tavs, Monatsh. Chem., 98, 105 (1967).
1190. Rhone-Poulenc S. A., Brit. 1,003,656 (1965); C. A., 64, 605a (1966).
1191. Richards, E. M., J. C. Tebby, and R. S. Wards, J. Chem. Soc., C, 1969, 1542.

1192. Robins, R. K., and B. E. Christensen, J. Org. Chem., 16, 324 (1951).
1193. Röhrscheid, F., and R. H. Holm, J. Organometal. Chem., 4, 335 (1965).
1194. Rosenbaum, E. J., D. J. Rubin, and C. R. Sandberg, J. Chem. Phys., 8, 366 (1940).
1195. Rosenbaum, E. J., and C. R. Sandberg, J. Am. Chem. Soc., 62, 1622 (1940).
1196. Ross, E. P., and G. R. Dobson, J. Inorg. Nucl. Chem., 30, 2363 (1968).
1197. Ross, E. P., and G. R. Dobson, Inorg. Chem., 6, 1256 (1967).
1198. Rothstein, E., R. W. Saville, and P. E. Horn, J. Chem. Soc., 1953, 3994.
1199. Roy, F., Jr., J. Gas Chromatogr., 6, 245 (1968).
1200. Rozinov, V. G., E. F. Grechkin, and A. V. Kalabina, Zh. Obshch. Khim., 39, 712 (1969).
1201. Rüdorff, W., and W. Müller, Doctoral thesis, W. Müller, Tübingen, 1957.
1201a. Rümpel, W., Mitt. Chem. Forschungsinst. Oesterr., 4, 113 (1950); C. A., 45, 2603h (1951).
1201b. Ruff, J. K., Inorg. Chem., 2, 813 (1963).
1202. Ryan, M. T., and W. L. Lehn, J. Organometal. Chem., 4, 455 (1965).
1203. Sacconi, L., and I. Bertini, J. Am. Chem. Soc., 90, 5443 (1968).
1204. Sacconi, L., and J. Gelsomini, Inorg. Chem., 7, 291 (1968).
1205. Sacconi, L., and R. Morassi, J. Chem. Soc., A, 1968, 2997.
1206. Sacconi, L., and R. Morassi, Inorg. Nucl. Chem. Lett., 4, 449 (1968).
1207. Sacconi, L., G. P. Speroni, and R. Morassi, Inorg. Chem., 7, 1521 (1968).
1208. Sachs, H., Chem. Ber., 25, 1514 (1892).
1209. Saegusa, T., Y. Ito, and S. Kobayashi, Tetrahedron Lett., 1968, 935.
1210. Salvesen, B., and J. Bjerrum, Acta Chem. Scand., 16, 735 (1962).
1211. Sander, M., Chem. Ber., 93, 1220 (1960).
1212. Sander, M., Chem. Ber., 95, 473 (1962).
1213. Santhanam, K. S. V., and A. J. Bard, J. Am. Chem. Soc., 90, 1118 (1968).
1214. Satgé, J., and C. Couret, C. R. Acad. Sci., Paris, Ser. C, 264, 2169 (1967).
1215. Satgé, J., and C. Couret, C. R. Acad. Sci., Paris, Ser. C, 267, 173 (1968).
1216. Satgé, J., and C. Couret, Bull. Soc. Chim. Fr., 1969, 333
1216a. Satgé, J., C. Couret, and J. Escudié, C. R. Acad. Sci., Paris, Ser. C, 270, 351 (1970).

1217. Saunders, M., and G. Burchman, Tetrahedron Lett., 1959, 8.
1218. Savage, M. P., and S. Trippett, J. Chem. Soc., C, 1967, 1998.
1219. Savage, M. P., and S. Trippett, J. Chem. Soc., C, 1968, 591.
1220. Savides, C., U.S. 3,324,091 (1967); C. A., 67, 44391r (1967).
1221. Sayre, R., J. Am. Chem. Soc., 80, 5483 (1958).
1222. Screttas, C., and A. F. Isbell, J. Org. Chem., 27, 2573 (1962).
1223. Schenker, E., Angew. Chem., 73, 81 (1961).
1224. Schiemenz, G. P., Tetrahedron Lett., 1964, 2729.
1225. Schiemenz, G. P., Chem. Ber., 98, 65 (1965).
1226. Schiemenz, G. P., Angew. Chem., 78, 145 (1966).
1227. Schiemenz, G. P., Angew. Chem., 78, 777 (1966).
1228. Schiemenz, G. P., Chem. Ber., 99, 504 (1966).
1229. Schiemenz, G. P., Chem. Ber., 99, 514 (1966).
1230. Schiemenz, G. P., Naturwissenschaften, 53, 476 (1966).
1231. Schiemenz, G. P., Tetrahedron Lett., 1966, 3023.
1232. Schiemenz, G. P., Angew. Chem., 80, 558 (1968).
1233. Schiemenz, G. P., Angew. Chem., 80, 559 (1968).
1233a. Schiemenz, G. P., Org. Syn., 49, 66 (1969).
1234. Schiemenz, G. P., and H. U. Siebeneick, Chem. Ber., 102, 1883 (1969).
1235. Schiemenz, G. P., and J. Thobe, Chem. Ber., 99, 2663 (1966).
1235a. Schimuzu, K., J. Chem. Soc. Jap. Pure Chem. Sect., 77, 1103 (1956).
1236. Schindlbauer, H., Monatsh. Chem., 94, 99 (1963).
1237. Schindlbauer, H., Monatsh. Chem., 96, 1021 (1965).
1238. Schindlbauer, H., Monatsh. Chem., 96, 1793 (1965).
1239. Schindlbauer, H., Monatsh. Chem., 96, 2012 (1965).
1240. Schindlbauer, H., Monatsh. Chem., 96, 2051 (1965).
1241. Schindlbauer, H., Monatsh. Chem., 96, 2058 (1965).
1242. Schindlbauer, H., Allg. Prakt. Chemie (Wien), 18, 242 (1967).
1243. Schindlbauer, H., Chem. Ber., 100, 3432 (1967).
1244. Schindlbauer, H., L. Golser, and V. Hilzensauer, Chem. Ber., 97, 1150 (1964).
1245. Schindlbauer, H., and H. Hagen, Monatsh. Chem., 96, 285 (1965).
1246. Schindlbauer, H., and G. Hajek, Chem. Ber., 96, 2601 (1963).
1247. Schindlbauer, H., and V. Hilzensauer, Monatsh. Chem., 96, 961 (1965).
1248. Schindlbauer, H., and V. Hilzensauer, Monatsh. Chem., 98, 1196 (1967).
1249. Schindlbauer, H., K. Kirsch, and L. Lalla, Ger. 1,150,981 (1963); C. A., 60, 557a (1964).

1250. Schindlbauer, H., and F. Mitterhofer, Z. Anal. Chem., 221, 394 (1966).
1251. Schindlbauer, H., and W. Prikoszovich, Chem. Ber., 102, 2914 (1969).
1252. Schindlbauer, H., and E. Steininger, Monatsh. Chem., 92, 868 (1961).
1253. Schlosser, M., Chem. Ber., 97, 3219 (1964).
1254. Schmerling, L., U.S. 2,902,517 (1959); C. A., 54, 2254h (1960).
1255. Schmidbauer, H., and W. Malisch, Chem. Ber., 102, 83 (1969).
1256. Schmidt, M., and H. Bipp, Sitzber. Ges. Beförder. Naturw. Marburg/Lahn, 83/84, 523 (1961/62); C. A., 59, 6436a (1963).
1257. Schmidt, U., and I. Boie, Angew. Chem., 78, 1061 (1966).
1258. Schmidt, U., I. Boie, C. Osterroht, R. Schroer, and H. F. Grützmacher, Chem. Ber., 101, 1381 (1968).
1259. Schmidt, U., F. Geiger, A. Müller, and K. Markau, Angew. Chem., 75, 640 (1963).
1260. Schmidt, U., K. Kabitzke, K. Markau, and A. Müller, Chem. Ber., 99, 1497 (1966).
1261. Smitz-DuMont, O., and H. Klieber, Z. Anorg. Allg. Chem., 371, 115 (1969).
1262. Schmitz-DuMont, O., B. Ross, H. Klieber, and W. Jansen, Angew. Chem., 79, 869 (1967).
1263. Schönberg, A., and R. Michaelis, Chem. Ber., 69, 1080 (1936).
1264. Schroeder, H., U.S. 3,155,630 (1964); C. A., 63, 704h (1965).
1265. Schulz, R. C., G. Wegner, and W. Kern, J. Polym. Sci., Part C, 16, 989 (1967).
1266. Schumann, H., Angew. Chem., 81, 970 (1969).
1267. Schumann, H., P. Jutzi, and M. Schmidt, Angew. Chem., 77, 812 (1965); Chem. Ber., 101, 24 (1968).
1268. Schumann, H., P. Jutzi, and M. Schmidt, Angew Chem., 77, 912 (1965).
1269. Schumann, H., H. Köpf, and M. Schmidt, Z. Anorg. Allg. Chem., 331, 200 (1964).
1270. Schumann, H., O. Stelzer, and U. Niederreuther, J. Organometal. Chem., 16 P64 (1969).
1271. Schwarzenbach, G., Chem. Zvesti, 19, 200 (1965).
1272. Schweizer, E. E., J. G. Thompson, and T. A. Ulrich, J. Org. Chem., 33, 3082 (1968).
1273. Senear, A. E., W. Valient, and J. Wirth, J. Org. Chem., 25, 2001 (1960).
1274. Sennewald, K., H. Baeder, K. Gehrmann, L. Lugosy, and W. Vogt, Ger. 1,245,360 (1967); C. A., 68, 21498u (1968).
1275. Seyferth, D., J. Am. Chem. Soc., 80, 1336 (1958).

1276. Seyferth, D., in Progress in Inorganic Chemistry, Vol. III, F. A. Cotton, Ed., Interscience, 1962, p. 129.
1277. Seyferth, D., and K. A. Brändle, J. Am. Chem. Soc., 83, 2055 (1961).
1278. Seyferth, D., and J. M. Burlitch, J. Org. Chem., 28, 2463 (1963).
1279. Seyferth, D., W. B. Hughes, and J. K. Heeren, J. Am. Chem. Soc., 87, 3467 (1965).
1280. Seel, F., and K. H. Rudolph, Z. Anorg. Allg. Chem., 363, 233 (1968).
1281. Seyferth, D., Y. Sato, and M. Takamizawa, J. Organometal. Chem., 2, 367 (1964).
1282. Seyferth, D., and T. Wada, Inorg. Chem., 1, 78 (1962).
1283. Sharts, C. M., U.S. 3,396,197 (1968); C. A., 69, 59367c (1968).
1284. Shaw, G., J. K. Becconsall, R. M. Canadine, and R. Murray, Chem. Commun., 1966, 425.
1285. Shaw, R. A., B. C. Smith, and C. P. Thakur, Chem. Commun., 1966, 228.
1286. Sheline, R. K., J. Chem. Phys., 18, 602 (1950).
1287. Shell Internat. Res. Maatschappij, N. V., Belg. 635,518 (1964); C. A., 61, 13347a (1964).
1288. Shell Int. Maatschappij N. V., Neth. 294,714 (1965); C. A., 63, 13442f (1965).
1289. Shell Int. Res. Maatschappij, N. V., Fr. 1,411,003 (1965); C. A., 64, 1972e (1966).
1290. Shell Internat. Res. Maatschappij, N. V., Belg. 654,329 (1965); C. A., 65, 8960h (1966).
1291. Shell Internat. Res. Maatschappij, N. V., Neth. 6,604,094 (1966); C. A., 66, 65101r (1967).
1291a. Shokal, E. C., U.S. 2,840,617 (1958); C. A., 52, 16795b (1958).
1291b. Shokal, E. C., U.S. 2,768,153 (1956); C. A., 51, 4057a (1957).
1292. Shook, E., Jr., and L. D. Quin, J. Am. Chem. Soc., 89, 1841 (1967).
1293. Shutt, J. R., and S. Trippett, J. Chem. Soc., C, 1969, 2038.
1294. Siebert, H., Z. Anorg. Allg. Chem., 273, 161 (1953).
1295. Siebert, W., W. E. Davidsohn, and M. C. Henry, J. Organometal. Chem., 17, 65 (1969).
1296. Siebert, W., W. E. Davidsohn, and M. C. Henry, J. Organometal. Chem., 15, 69 (1968).
1297. Sieckhaus, J. F., and T. Layloff, Inorg. Chem., 6, 2185 (1967).
1298. Simonnin, M. P., J. Organometal. Chem., 5, 155 (1966).
1299. Simonnin, M. P., Bull. Soc. Chim., Fr., 1966, 1774.

1300. Simonnin, M. P., and B. Borecka, Bull. Soc. Chim., Fr., 1966, 3842.
1301. Simonnin, M. P., and C. Charrier, C. R. Acad. Sci., Paris, Ser. C, 267, 550 (1968).
1302. Simonnin, M. P., and C. Charrier, Org. Magnetic Resonance, 1, 27 (1969).
1303. Singh, S., and C. N. R. Rao, Can. J. Chem., 44, 2611 (1966).
1304. Siuda, A., Nukleonika, 10, 459 (1965); C. A., 64, 16600h (1966).
1305. Sladkov, A. M., L. Yu. Ukhin, and V. V. Korshak, Izv. Akad. Nauk SSSR, Ser. Khim., 1963, 2213; C. A., 60, 9187g (1964).
1306. Slota, P. J., Jr., Ph.D. thesis, Temple University, 1954.
1307. Smith, D. C., and A. H. Soloway, J. Med. Chem., 11, 1060 (1968).
1308. Smith, D. C., A. H. Soloway, and R. W. Turner, J. Med. Chem., 9, 360 (1966).
1309. Smith, D. J. H., and S. Trippett, Chem. Commun., 1969, 855.
1309a. Snyder, E. H., U.S. 2,825,458 (1958); C. A., 52, 11718h (1958).
1310. Soborovskii, L. Z., A. B. Bruker, and Kh. R. Raver, U.S.S.R. 140,058 (1960); C. A., 56, 10192e (1962).
1310a. Societa Appl. Ricerche Chimiche, Ital. 510,939 (1955); C. A., 52, 19109g (1958).
1311. Soc. Ital. Resine, SpA., Neth. 6,609,104 (1967); C. A., 67, 12139u (1967).
1312. Sollott, G. P., and E. Howard, Jr., J. Org. Chem., 27, 4034 (1962).
1313. Sollott, G. P., H. E. Mertwoy, S. Portnoy, and J. L. Snead, J. Org. Chem., 28, 1090 (1963).
1314. Sollott, G. P., and W. R. Peterson, J. Organometal. Chem., 4, 491 (1965).
1315. Sollott, G. P., and W. R. Peterson, J. Organometal. Chem., 19, 143 (1969).
1316. Sollott, G. P., J. L. Snead, S. Portnoy, W. R. Petersen, Jr., and H. E. Merthwoy, U.S. Dept. Commerce, Office Technical Services, A. D. 611869, Vol. II, 411 (1965); C. A., 63, 18147b (1965).
1317. Solvay et Cie., Belg. 617,902 (1962); C. A., 58, 9294h (1963).
1317a. Springall, H. D., and L. O. Brockway, J. Am. Chem. Soc., 60, 996 (1938).
1318. Stafford, S. L., and J. D. Baldeschwieler, J. Am. Chem. Soc., 83, 4473 (1961).
1318a. Stallmann, O., U.S. 2,671,082 (1954); C. A., 49, 2510g (1955).
1319. Staudinger, H., and E. Hauser, Helv. Chim. Acta, 4, 861, 1887 (1921).

1320. Stepanov, B. I., and A. I. Bokanov, Zh. Obshch. Khim., 34, 3849 (1964); C. A., 62, 6509b (1965).
1321. Stepanov, B. I., A. I. Bokanov, and B. A. Korolev, Zh. Obshch. Khim., 37, 2139 (1967); C. A., 68, 68328x (1968); and Teor. Eksp. Khim., 4, 354 (1968); C. A., 69, 63127w (1968).
1322. Stepanov, B. I., A. I. Bokanov, and B. A. Korolev, Teor. Eksp. Khim., 4, 354 (1968); C. A., 69, 63217w (1968).
1323. Stepanov, B. I., A. I. Bokanov, B. A. Korolev, and V. A. Plakhov, Khim. Org. Soedin. Fosfora, Akad. Nauk SSSR, Otd. Obshch. Tekh. Khim., 1967, 162; C. A., 69, 36224q (1968).
1324. Stepanov, B. I., E. N. Karpova, and A. I. Bokanov, Zh. Obshch. Khim., 39, 1544 (1969).
1325. Steger, E., Abh. Deut. Akad. Wiss. Berlin, Kl. Math., Physik Techn., 1962, 19; C. A., 59, 2295f (1967).
1326. Steger, E., and K. Stopperka, Chem. Ber., 94, 3023 (1961).
1327. Stein, R. A., and V. Slawson, Anal. Chem., 35, 1008 (1963).
1328. Steiner, A., Chem. Ber., 8, 1177 (1875).
1329. Steininger, E., Chem. Ber., 95, 2541 (1962).
1330. Steininger, E., Chem. Ber., 96, 3184 (1963).
1331. Steininger, E., Ger. 1,194,579 (1965); C. A., 63, 5848h (1965).
1332. Steininger, E., and M. Sander, Angew. Chem., 75, 88 (1963).
1333. Steininger, E., and M. Sander, Kunststoffe, 54, 507 (1964).
1334. Steinkopf, W., and K. Buchheim, Chem. Ber., 54, 1024 (1921).
1335. Sterlin, R. N., and S. S. Dubov, Zh. Vses. Khim., Obshchest., 7, 117 (1962); C. A., 57, 294e (1962).
1336. Sterlin, R. N., R. D. Yatsenko, L. N. Pinkina, and I. L. Khunyants, Khim. Nauka i Promy., 4, 810 (1959); C. A., 54, 10838c (1960).
1337. Stetter, H., and W. D. Last, Chem. Ber., 102, 3364 (1969).
1337a. Stewart, A. P., and S. Trippett, J. Chem. Soc., C, 1970, 1263.
1338. Stiles, A. R., F. F. Rust, and W. E. Vaughan, J. Am. Chem. Soc., 74, 3282 (1952); and A. R. Stiles, F. F. Rust, and W. E. Vaughan, U.S. 2,803,597 (1957); C. A., 52, 2049a (1958).
1339. Stockel, R. F., Can. J. Chem., 46, 2625 (1968).
1340. Stockel, R. F., Can. J. Chem., 47, 867 (1969).
1341. Stone, F. G. A., Chem. Rev., 58, 101 (1958).
1342. Streuli, C. A., Anal. Chem., 32, 985 (1960).
1343. Strubel, W., J. Prakt. Chem., 18, 113 (1962).

1344. Struck, R. F., and Y. F. Shealy, J. Med. Chem., 9, 414 (1966).
1345. Sugimoto, S., K. Kuwata, S. Ohnishi, I. Nitta, Nippon Hoshasen Kob. Kenk. Ky. Namp., 7, 199 (1965); C. A., 67, 77760p (1967).
1346. Sumito Chem. Co., Brit. 1,000,673 (1965); C. A., 63, 16505a (1965).
1346a. Taft, R. W., Jr., in Steric Effects in Organic Chemistry, M. S. Newman, Ed., Wiley, 1956, p. 556.
1347. Taft, R. W., Jr., R. H. Martin, and E. W. Lampe, J. Am. Chem. Soc., 87, 2490 (1965).
1348. Takashina, N., and C. C. Price, J. Am. Chem. Soc., 84, 489 (1962).
1349. Tamborski, C., U.S. 3,354,214 (1967); C. A., 69, 27526n (1968).
1349a. Tamborski, C., U.S. 3,499,041 (1970); C. A., 72, 100879y (1970).
1350. Tamborski, C., F. E. Ford, W. L. Lehn, G. J. Moore, and E. J. Soloski, J. Org. Chem., 27, 619 (1962).
1351. Tanaka, M., Japan, Pat. 1277 (1953); C. A., 48, 2097h (1954).
1352. Tavs, P., Angew. Chem., 81, 742 (1969).
1353. Taylor, R. C., G. R. Dobson, and R. A. Kolodny, Inorg. Chem., 7, 1886 (1968).
1354. Technochemie GmbH, Brit. 1,087,311 (1967); C. A., 68, 114742c (1968).
1355. Tefteller, W., Jr., R. A. Zingaro, and A. F. Isbell, J. Chem. Eng. Data, 10, 301 (1965).
1356. Thomas, R., and K. Eriks, Inorg. Syn., 9, 59 (1967).
1356a. Thomson, C., and D. Kilcast, Angew. Chem., 82, 325 (1970).
1357. Tokareva, L. G., N. N. Mikhailov, Z. I. Potemkina, and P. A. Kirpichnikov, U.S.S.R. 142,423 (1962); C. A., 56, 14474d (1962).
1358. Tolmachev, A. I., and E. S. Kozlov, Zh. Obshch. Khim., 37, 1922 (1967); C. A., 68, 105298k (1968).
1358a. Tolman, Ch. A., J. Am. Chem. Soc., 92, 2956 (1970).
1359. Tomaschewski, G., J. Prakt, Chem., 305, 168 (1966).
1359a. Toy, A. D. F., and T. M. Beck, J. Am. Chem. Soc., 72, 3191 (1950).
1360. Trippett, S., J. Chem. Soc., 1961, 2813.
1361. Trippett, S., and B. J. Walker, J. Chem. Soc., C, 1966, 887.
1361a. Tsvetkov, E. N., G. Borisov, Kh. Sivriev, R. A. Malevannaya, and M. I. Kabachnik, Zh. Obshch. Khim., 40, 285 (1970); C. A., 72, 121634e (1970).
1362. Tsvetkov, E. N., D. I. Lobanov, and M. I. Kabachnik, Zh. Obshch. Khim., 38, 2285 (1968); C. A., 71, 13176c (1969).
1363. Tsvetkov, E. N., P. I. Lobanov, M. M. Makhamatkhanov, and M. I. Kabachnik, Tetrahedron, 25, 5623 (1969).

1364. Turner, R. W., and A. H. Soloway, J. Org. Chem., 30, 4031 (1965).
1365. Tyka, R., and E. Plazek, Bull. Acad. Pol. Sci., Ser. Sci. Chim., 9, 577 (1961); C. A., 60, 4182a (1964).
1366. Tyka, R., and E. Plazek, Rocz. Chem., 37, 283 (1963); C. A., 59, 7555e (1963).
1367. Tyler, J. K., J. Chem. Phys., 40, 1170 (1964).
1368. Tzschach, A., and W. Fischer, Z. Chem., 7, 196 (1967).
1369. Tzschach, A., and W. Lange, Chem. Ber., 95, 1360 (1962).
1370. Uhlig, E., and M. Maaser, Z. Anorg. Allg. Chem., 344, 205 (1966).
1371. Vahrenkamp, H., and H. Nöth, J. Organometal. Chem., 12, 281 (1968).
1372. Valetdinov, R. K., E. V. Kuznetsov, R. R. Belova, R. K. Mukhaeva, T. I. Malykina, and M. Kh. Khasanov, Zh. Obshch. Khim., 37, 2269 (1967); C. A., 68, 87350q (1968).
1373. Valetdinov, R. K., E. V. Kuznetsov, and S. L. Komissarova, Zh. Obshch. Khim., 39, 1744 (1969).
1374. Valetdinov, R. K., E. V. Kuznetsov, M. R. Rakhimova, and M. Kh. Khasanov, Zh. Obshch. Khim., 37, 2522 (1967); C. A., 68, 87353t (1968).
1375. Van Wazer, J. R., Phosphorus and Its Compounds, Vol. II, Interscience, Chapter 19 by D. H. Chadwick and R. S. Watt, 1961, p. 1278.
1376. Van Wazer, J. R., C. F. Callis, J. N. Schoolery, and R. C. Jones, J. Am. Chem. Soc., 78, 5715 (1956).
1377. Vaughan, L. G., and R. V. Lindsey, Jr., J. Org. Chem., 33, 3088 (1968).
1377a. Vereinigte Glanzstoff-Fabriken AG., Brit. 763,950 (1956); C. A., 51, 11761d (1957); Fr. 1,134,751 (1957); C. A., 51, 9213b (1957); Brit. 794,169 (1958); C. A., 52, 19252c (1958).
1378. Vilesov, F. I., and V. M. Zaitsev, Dokl. Akad. Nauk SSSR, Ser. Khim., 154, 886 (1964); C. A., 60, 11870h (1964).
1379. Vinokurova, G. M., Zh. Obshch. Khim., 37, 1652 (1967); C. A., 68, 29798f (1968).
1380. Vinokurova, G. M., and I. A. Aleksandrova, Izv. Akad. Nauk SSSR, Ser. Khim., 1967, 362.
1381. Vinokurova, G. M., and S. G. Fattaklov, Zh. Obshch. Khim., 36, 67 (1966); C. A., 64, 14209g (1966).
1382. Vinokurova, G. M., and S. G. Fattakhov, Izv. Akad. Nauk SSSR, Ser. Khim., 1969, 1762.
1383. Vinokurova, G. M., and Kh. Kh. Nagaeva, Izv. Akad. Nauk SSSR, Ser. Khim., 1967, 414; C. A., 67, 21977x (1967).
1384. Vobecky, M. , V. D. Nefedov, and E. N. Sinotova, Zh. Obshch. Khim., 33, 4023 (1963); C. A., 60, 8672e (1964).

1385. Voigt, D., R. Turpin, M. C. Labarre, and J. F. Labarre, J. Chim. Phys., 66, 906 (1969).
1386. Voigt, D., R. Turpin, and M. Torres, C. R. Acad. Sci., Paris, Ser. C, 265, 884 (1967).
1387. Volans, P., Brit. 975,970 (1964); C. A., 64, 3800d (1966).
1388. Vol'fkovich, S. I., V. K. Kuskov, and K. F. Koroteeva, Izv. Akad. Nauk SSSR, Otd. Khim. Nauk, 1954, 5; C. A., 49, 6859c (1955).
1389. Voskuil, W., and J. F. Arens, Rec. Trav. Chim., 81, 993 (1962).
1390. Voskuil, W., and F. Arens, Rec. Trav. Chim., 82, 302 (1963).
1391. Voskuil, W., and J. F. Arens, Rec. Trav. Chim., 83, 1301 (1964).
1392. Vullo, W. J., Ind. Eng. Chem., Prod. Res. Develop., 5, 346 (1966).
1393. Vullo, W. J., J. Org. Chem., 33, 3665 (1968).
1393a. Vullo, W. J., U.S. 3,475,479 (1969); C. A., 72, 55649k (1970).
1394. Wada, J., and R. W. Kiser, J. Phys. Chem., 68, 2290 (1964).
1395. Wagner, F. L., and M. Grayson, Fr. 1,344,698 (1963); C. A., 60, 14542c (1964).
1396. Wagner, F. L., and M. Grayson, U.S. 3,380,837 (1968); C. A., 69, 44018r (1968).
1397. Wagner, P. L., U.S. 3,476,715 (1969); C. A., 72, 13202d (1970).
1398. Wagner, I., and C. O. Wilson, Inorg. Chem., 5, 1009 (1966).
1399. Wagner, R. I., U.S. 3,086,053 (1963); C. A., 59, 10124d (1963).
1400. Wagner, R. I., U.S. 3,086,056 (1963); C. A., 60, 559e (1964).
1401. Wagner, R. I., and A. B. Burg, J. Am. Chem. Soc., 75, 3869 (1953).
1402. Wagner, R. I., L. V. D. Freeman, H. Goldwhite, and D. G. Rowsell, J. Am. Chem. Soc., 89, 1102 (1967).
1403. Wagstaffe, F. J., and H. W. Thompson, Trans. Faraday Soc., 40, 41 (1944).
1404. Waite, N. E., and J. C. Tebby, J. Chem. Soc., C, 1970, 386.
1404a. Walden, P., and R. Swinne, Z. Phys. Chem., 79, 714 (1912).
1405. Wall, L. A., R. E. Donadio, and W. J. Pummer, J. Am. Chem. Soc., 82, 4846 (1960); U.S. 3,046,313 (1962); C. A., 57, 15003b (1962).
1406. Walling, C., U.S. 2,437,795 (1948); C. A., 42, 4198h (1948).
1407. Walling, C., U.S. 2,437,796; 2,437,798; C. A., 42, 4199a (1948).

1408. Walling, C., and W. Helmreich, J. Am. Chem. Soc., 81, 1144 (1959).
1409. Wang, I. Y. M., C. O. Britt, A. H. Cowley, and J. E. Boggs, J. Chem. Phys., 48, 812 (1968).
1410. Watson, H. R., cited by J. Chatt and F. A. Hart, J. Chem. Soc., 1960, 1385.
1411. Watt, G. W., and R. C. Thomson, J. Am. Chem. Soc., 70, 2295 (1948).
1412. Watt, G. W., and R. C. Thompson, Jr., J. Am. Chem. Soc., 70, 2295 (1949).
1413. Wawzonek, S., and J. H. Wagenknecht, Angew. Chem., 76, 927 (1964).
1414. Weaver, J. R., and R. W. Parry, Inorg. Chem., 5, 718 (1966).
1415. Webb, R. L., U.S. 2,293,286 (1966); C. A., 66, 76194z (1967).
1416. Webb, R. L., Proc. U. N. Int. Conf. Peaceful Uses At. Energy, 2nd, Geneva, 1958, 29, 331; C. A., 54, 21951e (1960).
1417. Wedekind, E., Chem. Ber., 45, 2933 (1912).
1418. Weil, T., B. Prijs, and H. Erlenmeyer, Helv. Chim. Acta, 35, 616 (1952).
1419. Weil, T., B. Prijs, and H. Erlenmeyer, Helv. Chim. Acta, 36, 142 (1953).
1420. Weiner, M. A., and G. Pasternack, J. Org. Chem., 32, 3707 (1967).
1421. Weiner, M. A., and G. Pasternack, J. Org. Chem., 34, 1130 (1969).
1422. Weiss, P., Pure Appl. Chem., 15, 587 (1967).
1423. Welcher, R. P., U.S. 3,105,096 (1963); C. A., 60, 5553d (1964).
1424. Welcher, R. P., and N. E. Day, J. Org. Chem., 27, 1824 (1962); Brit. 969,129 (1964); Brit. 971,669 (1964).
1425. Welcher, R. P., G. A. Johnson, and V. P. Wystrach, J. Am. Chem. Soc., 82, 4437 (1960); J. Org. Chem., 27, 1824 (1962).
1425a. Weston, R. E., Jr., and J. Bigeleisen, J. Am. Chem. Soc., 76, 3074 (1954).
1426. Weyer, K., Doctoral thesis, T.-H., Aachen (1956).
1427. Whistler, R. L., and C. C. Wang, J. Org. Chem., 33, 4455 (1968).
1428. Whitesides, G. M., J. L. Beauchamp, and J. D. Roberts, J. Am. Chem. Soc., 85, 2665 (1963).
1429. Whitesides, G. M., J. P. Sevenair, and R. W. Goeth, J. Am. Chem. Soc., 89, 1135 (1967).
1430. Wibant, J. P., Rec. Trav. Chim., 44, 239 (1924).
1431. Wiberg, E., and H. Nöth, Z. Naturforsch., B, 12, 125 (1957).
1432. Wichelhaus, H., Ann. Chem. Suppl., 6, 257 (1868).

1433. Wiley, R. A., and J. H. Collins, J. Med. Chem., 12, 146 (1969).
1434. Wiley, R. A., and H. N. Godwin, J. Pharm. Sci., 54, 1063 (1965); C. A., 63, 13311c (1965); U.S. 3,442,948 (1969).
1435. Williams, D. H., R. S. Ward, and R. G. Cooks, J. Am. Chem. Soc., 90, 966 (1968).
1436. Willans, J. L., Chem. Ind. (London), 1957, 235.
1437. Wittenberg, D., and H. Müller, Fr. 1,385,883 (1965); C. A., 62, 14508e (1965).
1438. Wittenberg, D., and H. Gilman, J. Org. Chem., 23, 1063 (1958).
1439. Wittig, G., and E. Benz, Chem. Ber., 92, 1999 (1959).
1440. Wittig, G., and H. J. Cristau, Bull. Soc. Chim., Fr., 1969, 1293.
1441. Wittig, G., H. J. Cristau, and H. Braun, Angew. Chem., 79, 721 (1967).
1442. Wittig, G., H. Eggers, and P. Duffner, Ann. Chem., 619, 10 (1958).
1443. Wittig, G., and G. Geissler, Ann. Chem., 580, 44 (1953).
1444. Wittig, G., and A. Maercker, Chem. Ber., 97, 747 (1964).
1445. Wittig, G., and H. Matzura, Ann. Chem., 732, 97 (1970).
1446. Workman, M. O., G. Dyer, and D. W. Meek, Inorg. Chem., 6, 1543 (1967).
1447. Worrall, D. E., J. Am. Chem. Soc., 52, 2933 (1930).
1448. Worrall, D. E., J. Am. Chem. Soc., 62, 2514 (1940).
1448a. Wright, G., and J. Bjerrum, Acta Chem. Scand., 16, 1262 (1962).
1449. Wu, C., and F. J. Welch, J. Org. Chem., 30, 1229 (1965).
1450. Wunsch, G., K. Wintersberger, and H. Geierhaas, Z. Anorg. Allg. Chem., 369, 33 (1969); Ger. 1,247,310 (1967); C. A., 68, 105363c (1968).
1451. Wymore, C. E., and J. C. Bailar, Jr., J. Inorg. Nucl. Chem., 14, 42 (1960).
1452. Wymore, C. E., Ph.D. thesis, University of Illinois, 1957.
1453. Wystrach, V. P., and G. A. Peters, U.S. 3,188,605 (1964); C. A., 61, 7045d (1964).
1454. Yakovleva, E. A., E. N. Tsvetkov, D. I. Lobanov, M. I. Kabachnik, and A. I. Shatenshtein, Tetrahedron Lett., 1966, 4161; and Dokl. Akad. Nauk SSSR, 170, 1103 (1966); C. A., 66, 37038g (1967).
1454a. Yamashita, Y., and T. Shimamura, Kôgyô Kagaku Zasshi, 60, 423 (1957); C. A., 53, 9025h (1959).
1454b. Yoneda, H., Bull. Chem. Soc. Jap., 31, 708 (1958); C. A., 53, 9752i (1959).

1455. Yoshida, T., M. Iwamoto, and S. Yuguchi, Jap. 11,934 (1967); C. A., 68, 105358e (1968).
1456. Young, D. P., W. E. McEwen, D. C. Velez, J. W. Johnson, and C. A. VanderWerf, Tetrahedron Lett., 1964, 359.
1456a. Yust, V. E., and J. L. Bame, U.S. 2,828,195 (1958); C. A., 52, 13244a (1958).
1457. Zabolotny, E. R., and H. Gesser, J. Am. Chem. Soc., 81, 6091 (1959).
1458. Zakharkin, L. I., V. I. Bregadze, and O. Yu. Okhlobystin, J. Organometal. Chem., 4, 211 (1965).
1459. Zakharkin, L. I., O. Yu. Okhlobystin, and B. N. Strunin, Izv. Akad. Nauk SSSR, Otd. Khim. Nauk, 1962, 2002; C. A., 58, 9131h (1963).
1460. Zakharkin, L. I., and G. G. Zhigareva, Zh. Obshch. Khim., 37, 1791 (1967).
1461. Zbiral, E., Monatsh. Chem., 95, 1759 (1964).
1462. Zbiral, E., Tetrahedron Lett., 1964, 1649.
1463. Zecchini, F., Gazz. Chim. Ital., 23, I, 97 (1893).
1464. Zeeh, B., and J. B. Thomson, Tetrahedron Lett., 1969, 111.
1465. Zhmurova, I. N., and A. V. Kirsanov, Zh. Obshch. Khim., 36, 1248 (1966); C. A., 65, 16826g (1966).
1466. Zingaro, R. A., and R. E. McGlothlin, J. Chem. Eng. Data, 8, 226 (1963).
1467. Zletz, A., and D. R. Carmody, U.S. 2,893,202 (1959); C. A., 53, 22955f (1959).
1467a. Zletz, A., and D. R. Carmody, U.S. 2,892,305 (1959); C. A., 53, 22955d (1959).
1468. Zon, G., K. E. DeBruin, K. Naumann, and K. Mislow, J. Am. Chem. Soc., 91, 7023 (1969).
1469. Zorn, H., Neth. 6,502,108 (1965); C. A., 64, 3598d (1966).
1470. Zorn, H., F. Pass, and E. Steininger, Ger. 1,126,867 (1962); C. A., 57, 8619f (1962).
1471. Zorn, H., H. Schindlbauer, and H. Hagen, Monatsh. Chem., 95, 422 (1964).
1472. Zorn, H., H. Schindlbauer, and H. Hagen, Chem. Ber., 98, 2431 (1965).
1473. Zorn, H., H. Schindlbauer, and D. Hammer, Monatsh. Chem., 98, 73 (1967).
1474. Zorn, H., and E. Steininger, U.S. 3,382,173 (1968).

Chapter 2. Organophosphorus-Metal Compounds, Biphosphines, Triphosphines, Tetraphosphines, Cyclopolyphosphines, and Corresponding Oxides, Sulfides, and Selenides

LUDWIG MAIER

Monsanto Research S. A., Zürich, Switzerland

Nearly all the compounds discussed in this chapter are sensitive to oxygen and water. In spite of this a great amount of work has been done in this area in the past few years, mainly because of the importance of organosubstituted phosphides for preparative organic chemistry and for the coordination chemistry of metals. Additional inputs came from the unusual thermal and hydrolytic stability of phosphinoborines. This led to the preparation and investigation of a large number of other organosubstituted phosphorus metal compounds.

A. ORGANOSUBSTITUTED ALKALI PHOSPHIDES

A.1. Methods of Preparation

I. FROM PHOSPHINES AND ALKALI METAL OR ORGANOALKALI COMPOUNDS

The hydrogen atoms in PH_3 can be selectively replaced by an alkali metal to give metal phosphides of the type MPH_2 or M_3P.[168,333] M_2PH has not been prepared in a pure state by this procedure. The monosodium and potassium[333,555,564,570] phosphides can be prepared from the metals and PH_3 in liquid ammonia with appropriate precautions to repress substitution beyond the desired limit.[165] Addition of Fe_2O_3 and THF usually increases the rate of reaction.[304] Other solvents such as toluene-glycol ether can be used.[376] $NaPH_2$ was also prepared from Na_4P_2, sodium, and NH_4Br in liquid ammonia,[168] and from sodium, white phosphorus, and NH_4Br in liquid NH_3.[550] A more clean-cut reaction is obtained when triphenylmethylsodium,[12] phenyllithium,[12,319,357] or butyllithium[460] is made the source of the metal. In this procedure triphenylsodium or phenyllithium is slowly added to an ethereal,[12,357] or ether-pentane[319] solution of phosphine through which a rapid stream of phosphine is passed so that PH_3 is always present in excess. Even then, the Li/P ratio is always greater than 1:1, usually 1.35:1. This indicates that higher substituted products, such as Li_2PH and Li_3P, are also formed.[319] Exhaustive metalation of phosphine with phenyllithium in diethyl ether apparently leads not to pure Li_3P[319] but mainly to Li_2PH.[303] Na_3P is best prepared from white phosphorus and a sodium dispersion in an inert organic media (toluene) at 80 to 145°.[467]

Primary and secondary phosphines react in much the same way with alkali metals or organoalkali compounds to give RPHM and RPM_2, and R_2PM, respectively. Monoorganosubstituted sodium and potassium phosphides can be prepared from the metals and a primary phosphine in liquid ammonia,[258,304,564,567] or more conveniently by refluxing a primary phosphine with finally divided sodium metal in an inert solvent[463] such as hydrocarbon at temperatures from 50 to 80°. Only above 90° is the second hydrogen replaced by sodium.[463] In order to increase the rate of

$$RPH_2 + M \longrightarrow RPHM + 1/2H_2$$

$$RPH_2 + 2M \longrightarrow RPM_2 + H_2$$

$$R_2PH + M(MR) \quad R_2PM + 1/2H_2(RH)$$

reaction, utilization has been made of the disubstituted product by carrying out the reaction above 90°, since it has been found that it interacts with a primary phosphine to give the monosubstituted product.[463] Ethylphosphine and cyclohexylphosphine apparently do not react with sodium even with variation in the solvent.[279]

Potassium has also been used in several cases.[212,279,292,567] In this case an excess of primary phosphine had to be used, otherwise disubstitution occurred.[292] By observing this condition phenylphosphine,[292] butylphosphine,[568] cyclohexylphosphine, and ethylphosphine[279] gave with potassium in hydrocarbon or with isodipropyl ether as solvent the monosubstituted product RPHK.

Phenyllithium yields with primary phosphines only the disubstituted product $RPLi_2$[319] which on treatment with RPH_2 can be converted to the monoderivative RPHLi.[398]

Secondary phosphines are also readily metalated by alkali metals, either by refluxing the secondary phosphine with sodium or potassium in an inert solvent such as ether, hydrocarbons,[361,463,567] or liquid ammonia,[250,259,555,564] or by reaction with methyl- or phenyllithium.[279,292,319] Some phosphides, such as $LiP(C_6H_{11})_2$ and $LiP(Bu\text{-}t)_2$ are ammonolyzed in liquid ammonia and therefore cannot be prepared in this solvent.[303] Because of the high acidity of $PhPH_2$ and Ph_2PH, they can also be metalated with $NaNH_2$.[303]

II. FROM PHOSPHONOUS, PHOSPHONIC, PHOSPHINOUS, OR PHOSPHINIC ACID DERIVATIVES AND ALKALI METALS

The reaction of PCl_3 with sodium and potassium dispersions in toluene affords the phosphides Na_3P and K_3P in good yields.[268] In much the same way, phenylphosphonous dichloride was reduced to the disubstituted product by lithium hydride,[270] by lithium in THF at -40°,[47,415] or by a finely divided dispersion of sodium in toluene.[462] The reduction of cyclohexylphosphonic dichloride, $C_6H_{11}POCl_2$, and of quasi-phosphonium compounds of the type $[(PhO)_3RP]^+I^-$ and $[(R_2N)_3RP]^+I^-$ was also effected with a sodium suspension in boiling toluene.[268] Usually not the alkali phosphide was isolated in these reactions but rather the primary phosphine formed by hydrolysis.[268,270,462]

$$PhPCl_2 + 4Na \longrightarrow PhPNa_2 + 2NaCl$$

Phosphinous and phosphinic acid derivatives such as R_2PCl,[250,268,336,342,361,439,440,552] R_2PNR_2,[269] R_2PS_2H,[268] and R_2POCl,[268] have also been successfully reduced with sodium, potassium, lithium, or sodium hydride in boiling dibutyl ether, THF, or a hydrocarbon. The reduction proceeds stepwise, and Ph_2PCl with sodium first yields tetraphenylbiphosphine which reacts with excess sodium to

$$2R_2PCl + 2Na \longrightarrow R_2P\text{-}PR_2 + 2NaCl$$

$$R_2P\text{-}PR_2 + 2Na \longrightarrow 2R_2PNa$$

give sodium diphenylphosphide.[361] Dialkylphosphinous chlorides with sodium in boiling dipropyl or dibutyl ether,[313] or with lithium in THF,[250] also yield tetraalkylbiphosphines. Further reduction occurs only slowly (see Section A.1.III). A similar stepwise reduction has been observed in the reaction of Ph_2POCl with sodium. A 1:2 ratio of the two components gave the sodium derivative of diphenylphosphine oxide which then underwent further reaction with excess sodium to yield sodium diphenylphosphide.[268]

$$Ph_2POCl + 2Na \longrightarrow Ph_2P(O)Na + NaCl$$

$$Ph_2P(O)Na + 2Na \longrightarrow Ph_2PNa + Na_2O$$

III. FROM CYCLOPOLYPHOSPHINES, BIPHOSPHINES, OR BI-
PHOSPHINE DISULFIDES AND ALKALI METALS OR ORGANO-
ALKALI COMPOUNDS

The first identifiable cleavage products of pentamethyl-cyclopentaphosphine,[311] ethylcyclopolyphosphine $(EtP)_{4,5}$,[300,311] and pentaphenylcyclopentaphosphine[300] with alkali metals seem to be the 1,4-, and 1,5-dialkali tetra- and pentaphosphides (1a and 1b) (for discussion see Ref. 406). The ^{31}P NMR spectrum indicates, however, that these compounds are not obtained in a pure state.[188,311] Excess

$$(RP)_n + 2M \quad\begin{cases} \begin{array}{c} R\ \ R\ \ R\ \ R \\ |\ \ \ |\ \ \ |\ \ \ | \\ M\text{-}P\text{-}P\text{-}P\text{-}P\text{-}M \\ (\underline{1a}) \end{array} \\[2em] \begin{array}{c} M\text{-}P\text{-}P\text{-}P\text{-}P\text{-}P\text{-}M \\ |\ \ \ |\ \ \ |\ \ \ |\ \ \ | \\ R\ \ R\ \ R\ \ R\ \ R \\ (\underline{1b}) \end{array} \end{cases}$$

(R = Me, Et, Ph)

(M = Li, Na, K)

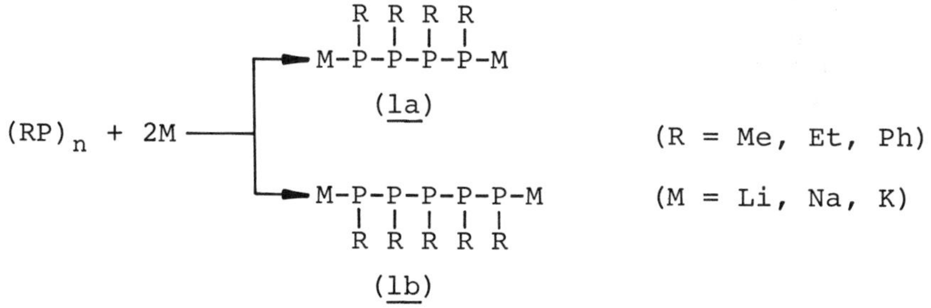

potassium or sodium cleaves $(EtP)_{4,5}$ in THF mainly to 1,2-dipotassium-1,2-diethylbiphosphide[299] (2) and small amounts of the triphosphide (3).[188] A further cleavage to K_2PEt does not occur.

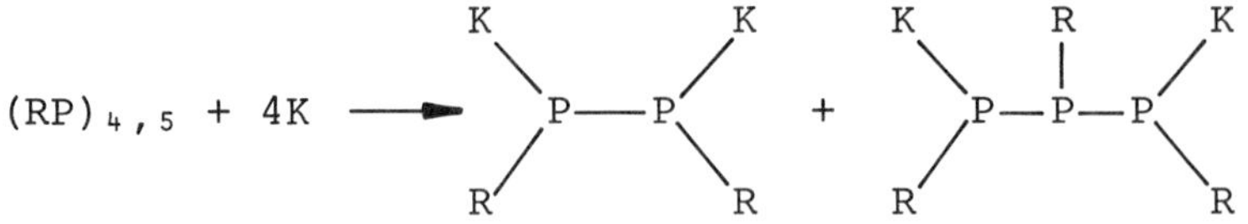

(2) (3)

(R = Et, Ph)

Depending upon the conditions used, $(PhP)_5$ yields with alkali metals a tetraphosphide (1), triphosphide[300] (3), biphosphide (2),[47,300,359,476] and also a dialkali phenylphosphide $PhPM_2$ (M = Li, Na, K).[47,300,359] The triphosphide may have a cyclic structure similar to that of the cyclopropenylium cation (4) when the alkali metal is potassium,[282] cesium, or lithium.[302]

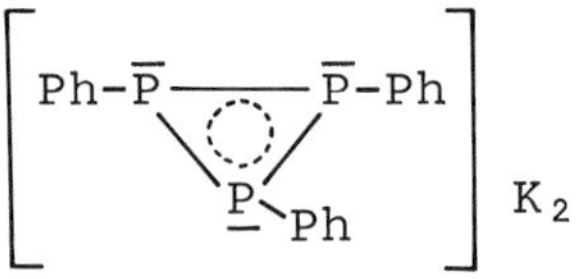

(4)

In contrast to $(EtP)_{4,5}$ and $(PhP)_5$, the cyclohexyl- and t-butyl-substituted cyclotetraphosphines, $(c\text{-}C_6H_{11}P)_4$[299] and $(t\text{-}BuP)_4$,[291] are cleaved only with potassium (and not with Na and Li) in THF to the tetraphosphides (1, R = t-Bu, $c\text{-}C_6H_{11}$). A further cleavage does not occur.[291]

Alkali phosphides have also been obtained in the interaction of phenyllithium,[188,300,302a] or lithium phosphides[302] with cyclopolyphosphines. Thus with phenyllithium $(PhP)_5$ gave 1-lithium-1,2,3,4,4'-pentaphenyltetraphosphide (5),[300,302a] whereas $(EtP)_{4,5}$ and phenyllithium produced lithium phenylethylphosphide (6) and lithium P-ethyl-P'-phenylethylbiphosphide (7).[188]

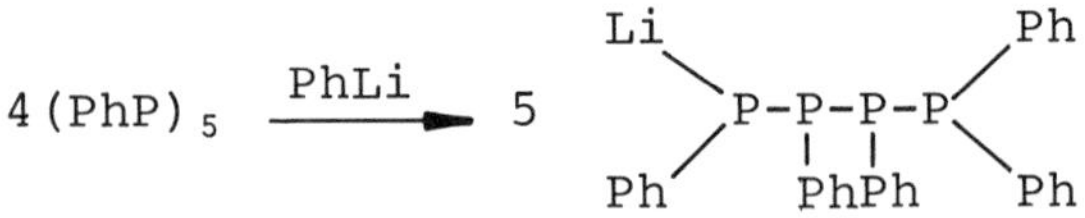

(5)

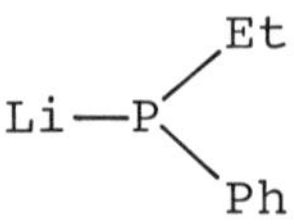

(6)

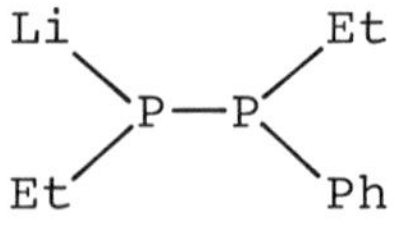

(7)

Complex phosphides of the type Li-$(PPh)_n$-Li (n = 2,3,4) were formed when $(PhP)_n$ was treated with Li_2PPh.[302] The failure to isolate lithium phosphides of composition Li-$(PPh)_n$-PPh_2 (n > 1) in the interaction of $(PhP)_n$ with $LiPPh_2$ was attributed to the instability of these products. As a result, only their decomposition products, that is, Li-$(PPh)_4$-Li, Ph_2P-PPh_2, and LiPPh-PPh_2 could be isolated.[302]

Homolytic cleavage of P-P bonds in biphosphines and biphosphine disulfides is also a useful source of alkali dialkyl- or diarylphosphides since the starting material is readily available (see Section E). The ease of cleavage varies with the substituents and decreases in the order:

$$Ph > Me > Et > Pr \sim Bu$$

Thus tetraphenylbiphosphine was cleaved with sodium, potassium,[361] or lithium,[552] in boiling ether or THF to give Ph_2PM. In the aliphatic series this method had only limited success. While tetramethyl- and tetraethylbiphos-

$$R_2P\text{-}PR_2 + 2M \longrightarrow 2R_2PM$$

phine were cleaved with sodium in boiling dioxan,[321] or with lithium in boiling THF[250] (not in ether), no such cleavage was observed with Pr_2P-PPr_2 and Bu_2P-PBu_2.[321] In decalin, however, these compounds were cleaved with sodium or potassium[438,440] at temperatures ranging from 120 to 150°. Alternatively, the more easily accessible biphosphine disulfides $[R_2P(S)]_2$ were also reduced with excess sodium in dioxan to R_2PNa when R = Me or Et, but if R = Pr or Bu reduction proceeded only to the first stage and gave biphosphines.[321] In other solvents, such as decalin, and at higher temperatures, cleavage of the P-P bond in these compounds should also occur. Certain organometallic reagents also cause cleavage of biphosphines,[296] for example,

$$R_2P\text{-}PR_2 + PhLi \longrightarrow R_2PLi + PhPR_2$$

The relative ease of biphosphine cleavage with PhLi closely resembles that for the alkali metals: Ph > Me > Et > Pr > Bu.[301] The cyclohexyl,[296] isopropyl, and t-butyl[301] derivatives could not be cleaved. The reactivity of the organometallic reagent follws the sequence: PhLi > EtLi > Et_2Mg > Et_3Al >> Et_4Pb.[296] Cleavage of the unsymmetrical biphosphine Et_2P-$P(C_6H_{11})_2$ with PhLi yields Et_2PhP and $LiP(C_6H_{11})_2$.[298] This result indicates that Et_2P^- is less nucleophilic than $(C_6H_{11})_2P^-$.

The interaction of Ph_2P-PPh_2 and $LiPR_2$ (R = Et, C_6H_{11}) or Li_2PPh produces in the primary step unsymmetrical biphosphines Ph_2P-PR_2 and lithium biphosphide, which react with further lithium phosphides to form symmetrical biphosphines

$$Ph_2P\text{-}PPh_2 + LiPR_2 \longrightarrow LiPPh_2 + [Ph_2P\text{-}PR_2 \xrightarrow{LiPR_2} Ph_2PLi + R_2P\text{-}PR_2]$$

$$Ph_2P\text{-}PPh_2 + Li_2PPh \longrightarrow LiPPh_2 + [Ph_2P\text{-}PLiPh \xrightarrow{Li_2PPh} Ph_2PLi + PhLiP\text{-}PLiPh]$$

and dilithium-1,2-diphenylbiphosphide,[302] respectively.

IV. FROM TERTIARY PHOSPHINES AND ALKALI METALS

In the presence of at least one aromatic group,[324] tertiary phosphines are cleaved with Li in THF,[8,575] or in liquid ammonia,[250,478] and with sodium or potassium in dioxan[283,324] or THF[57,58] to give alkali phosphides R_2PM (M = Li, Na, K). With excess alkali metal further reduction to the radical anion takes place, which has been identified by ESR.[57] On the basis of reaction rates of various aromatic tertiary phosphines with potassium in

$$Ph_3P + 2M \longrightarrow Ph_2PM + MPh$$

$$Ph_2PM + M \longrightarrow Ph_2PM^{\overset{\bullet}{-}} + M^+$$

dioxan under comparable conditions, the following cleavage series was proposed.[324]

α-Naphthyl > phenyl > p-tolyl > 2,5-dimethylphenyl > Et > c-C_6H_{11}

This order agrees with the cleavage series of tetraarylphosphonium hydroxides[251,260] and tertiary phosphine oxides[269] in which the most electronegative substituent is cleaved first. In a few cases exceptions have been noted. Thus Et_2PhP gave EtPhPK and (c-C_6H_{11})$_2$PhP gave (c-C_6H_{11})-PhPK, but $EtPh_2P$ yielded the expected product EtPhPK.[324]

Since tertiary aliphatic and cycloaliphatic phosphines are not cleaved by alkali metals, the presence of at least one resonance-stabilized aromatic group is apparently essential for the formation of alkali phosphides from tertiary phosphines. The usefulness of lithium diphenylphosphide prepared by this method has been extended considerably by selectively eliminating the by-product phenyllithium through reaction with t-butyl chloride.[8]

A.2. General Chemistry

All the alkali phosphides are extremely sensitive to water and other protic solvents, and many of them inflame in contact with air. Most of the alkali phosphides are soluble in ether, THF, and dioxan, and insoluble in benzene and petroleum ether.[275] They normally crystallize with 1 to 2

$$RPNa_2 + 2H_2O \longrightarrow RPH_2 + 2NaOH$$

moles of dioxan, which is an indication of the unsaturated character of the alkali metal in alkali phosphides. They also give adducts with tertiary phosphine oxides.[297] The reactivity of alkali phosphides is determined by the polarity of the alkali metal-phosphorus bond, by the organic substituents, and by the degree of association.

$$>\ddot{P}-M; \quad >P\blacktriangleright M; \quad [>P:]^-M^+$$

From this it follows that with the same organosubstituents on phosphorus the potassium derivative is more reactive than the lithium compound, and $LiPEt_2$ is more reactive than $LiPPh_2$ because in the latter compound the negative charge is distributed through the whole phosphide anion owing to the mesomeric effect of the phenyl group. The degree of association in dioxan increases from 1.20 for $LiPPh_2$ to 1.24 for $LiP(C_6H_{11}\text{-}c)_2$ to 2.30 for $LiPEt_2$.[276] Therefore $LiP(C_6H_{11}\text{-}c)_2$, which is less associated than $LiPEt_2$, is more reactive than $LiPEt_2$. The following reactivity series may be given.[276]

$$Ph_2P^- < Et_2P^- < (c\text{-}C_6H_{11})_2P^-$$

Numerous reactions with alkali phosphides have been described.[275,276,398] Alkali phosphides are important starting materials for the preparation of phosphorus-containing compounds. They undergo a great number of substitution, exchange, and addition reactions (see Chapter 1). Furthermore, most of the compounds having a metal-phosphorus bond have been prepared by reaction of alkali phosphides with metal halides (see the following discussion).

A.3. General Physical Properties

The dipole moments of alkali phosphides indicate that the bond polarity of, for example, potassium diphenylphosphide, is comparable with that of triphenylmethylsodium.[276] In accord with this, alkali phosphides of the type $MPPh_2$ (M = Li, Na, K, Rb, Cs) and KPHPh represent 1:1 electrolytes in dimethyl sulfoxide. With the exception of the lithium derivatives and KPHPh, which are very weak electrolytes, the compounds of type $MPPh_2$ generally show a degree of dissociation of about 0.8.[305] An equivalent conductance of 14.5 was estimated for the diphenylphosphide ion Ph_2P^-.[305] ^{31}P NMR measurements of $MPPh_2$ show that in THF the existence of free diphenylphosphide anions can be excluded.[186] Although one would expect that the chemical shift would be found at higher field strength as the ionic character increases, the contrary has been observed:[186] ^{31}P chemical shift of Ph_2PM: M = Li +23.0 ppm; Na +24.4 ppm; K +12.4 ppm; Rb +7.8 ppm; Cs ±0.0 ppm. This has been ascribed to a change in the asymmetry of the occupation of p functions and to a change in the hybridization (going from p^3 to sp^3).

The ^1H and ^{31}P chemical shifts of the unsubstituted alkali phosphides MPH_2 (M = Li, Na, K, Rb, Cs) exhibit strong solvent, temperature, and concentration dependence.[350,537] Both the ^1H and ^{31}P chemical shifts as well as the activation energies and lifetimes for the hydrogen exchange are quite different for sodium as compared to potassium, rubidium, and cesium, but the lithium value was found to drop back not far from the values of potassium, rubidium, and cesium. To explain these results ionization and aggregation with the solvent have been postulated.[350]

B. ORGANOSUBSTITUTED ALKALINE EARTH PHOSPHIDES

B.1. Methods of Preparation

I. FROM PHOSPHINES AND ALKALINE EARTH METALS OR ORGANOALKALINE EARTH COMPOUNDS

The acidity of the hydrogen atoms in phosphines is such as to react with Grignard reagents[12,334] and organocalcium compounds[418] to form organomagnesium and calcium phosphides, for example,

$$PH_3 + EtMgBr \longrightarrow BrMgPH_2 + C_2H_6$$

$$PhPH_2 + 2RMgX \longrightarrow PhP(MgX)_2 + 2C_6H_6$$

$$Ph_2PH + Ph_3CCaCl \longrightarrow Ph_2PCaCl + Ph_3CH$$

The reaction with phenylphosphine proceeds more smoothly when PhMgBr is used[414] instead of EtMgBr.[334] A calcium phosphide has been obtained in the reaction of PH_3 with calcium metal in liquid ammonia.[564] This monophosphide when heated to 50° under vacuum loses PH_3 and gives a

$$2PH_3 + Ca \xrightarrow{NH_3} Ca(PH_2)_2{\cdot}6NH_3 \xrightarrow[\text{in vacuo}]{50°} CaPH + PH_3 + 6NH_3$$

secondary phosphide.[564] Heating of dimethylberyllium with
Me_2PH to 190° for 5 hr resulted in the formation of about
0.5 mole methane and nonvolatile, evidently polymeric
material.[112]

Two other reactions have been found useful for preparing magnesium phosphides. One method involves reaction of
phenylphosphine, diphenylphosphine, or secondary alkylenediphosphines with diethylmagnesium in ether--aliphatic and

$$PhPH_2 + Et_2Mg \longrightarrow PhHPMgEt$$

$$Ph_2PH + Et_2Mg \left[\begin{array}{l} 1:1 \longrightarrow Ph_2PMgEt \\ 2:1 \longrightarrow (Ph_2P)_2Mg \end{array} \right.$$

$$PhHP(CH_2)_3PHPh + MgEt_2 \xrightarrow{-2C_2H_6} PhP \overset{\displaystyle CH_2}{\underset{\displaystyle Mg}{\langle \begin{array}{c} CH_2 \quad CH_2 \\ \end{array} \rangle}} PPh$$

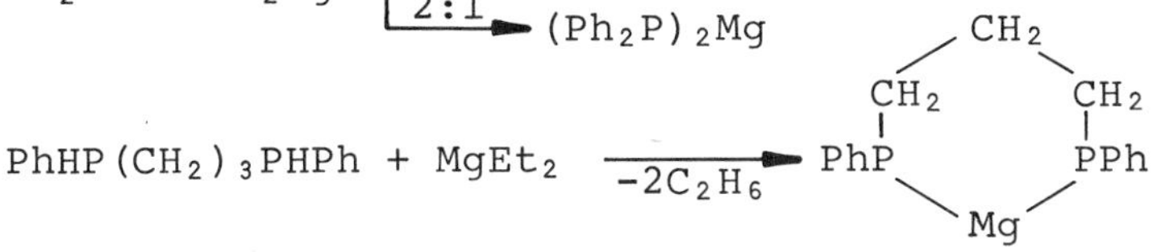

cycloaliphatic substituted phosphines do not react under
these conditions[278]--and the other involves interaction of
alkali phosphides with $MgBr_2 \cdot 4THF$. In the latter case
aliphatic substituted magnesium phosphides are also obtainable.[278]

$$MgBr_2 + 2KPPh_2 \longrightarrow Mg(PPh_2)_2 + 2KBr$$

In this connection it is interesting to note that the addition of THF to bromo- and iodomagnesium diphenylphosphide,
prepared in ether, causes a dismutation similar to that
observed with EtMgI.[278]

$$2Ph_2PMgX \xrightarrow{THF} (Ph_2P)_2Mg + MgX_2 \quad (X = Br, J)$$

Cyclic magnesium phosphides (8 and 9) are formed in
the reaction of 1,4- and 1,5-dialkali tetra- and pentaphosphides, respectively, with $MgBr_2$.[311]

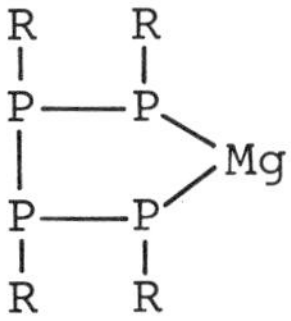

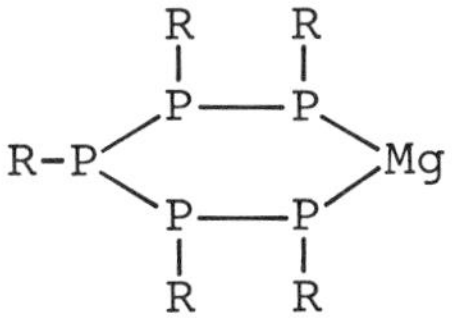

II. FROM BIPHOSPHINES AND DIETHYLMAGNESIUM

Similar to phenyllithium, diethylmagnesium also cleaves the P-P bond in tetraphenylbiphosphine and yields, depending upon the ratio used, a monophosphide or a diphosphide.[296] In contrast to phenyllithium, however, this cleavage is incomplete in ether or benzene and requires refluxing for several hours in diisobutyl ether for completion.[296]

$$MgEt_2 + Ph_2P\text{-}PPh_2 \longrightarrow EtMgPPh_2 + EtPPh_2$$

$$MgEt_2 + 2Ph_2P\text{-}PPh_2 \longrightarrow Mg(PPh_2)_2 + 2EtPPh_2$$

B.2. General Chemistry

The properties of alkaline earth phosphides are similar to those of alkali phosphides. Thus they are also very sensitive to water and other protic solvents, and many of them inflame in contact with air. They undergo reaction with alkyl halides and metal halides to give tertiary phosphines and compounds containing a phosphorus-metal bond. As a

$$RP(MgBr)_2 + 2R'X \longrightarrow RPR'_2 + 2MgXBr$$

$$Ph_2PMgBr + ClSnR_3 \longrightarrow Ph_2P\text{-}SnR_3 + MgClBr$$

class, the alkaline earth phosphides have not been investigated in detail yet.

C. ORGANOPHOSPHORUS COMPOUNDS OF THE GROUP III ELEMENTS

The known chemistry of Group III element phosphorus organic compounds largely concerns derivatives of the elements in their trivalent state. Bonding occurs through donation of a pair of electrons by P(III) and through acception of an electron pair by the Group III element. The compounds formed have been divided into complexes, for example, $Me_3P \cdot BMe_3$ and into phosphinoboranes, -alanes, etc. Only the last-mentioned class of compounds is discussed here. Depending on the groups attached to the elements, monomeric, dimeric, trimeric, and polymeric compounds are formed. However, there appears to be no general rule that can be used to predict the degree of association of these compounds. There seem to be four factors that affect the degree of association: steric and electronic interference, valency angle strain, entropy, and the nature of the intermediates involved in their formation.[33] Thus when the electron deficiency of Group III elements is ameliorated by attachment of electron-donating groups, they become too weak acids to coordinate with another $R_2PMR'_2$ (M = B, Al) molecule.[115 447]

Bulky substituents also stabilize monomeric compounds.[115] These and other aspects are discussed in detail in a recent review on boron-phosphorus compounds.[458]

C.1. Methods of Preparation

I. BY PYROLYSIS OF THE ADDUCTS OF PHOSPHINES WITH GROUP III COMPOUNDS

The most versatile method for preparation of cyclic phosphine Group III compounds is the pyrolysis of the appropriate phosphine complex, such as $R_2PH{\cdot}BX_2H$, $R_2PH{\cdot}GaMe_3$, or $R_4P_2{\cdot}2BH_3$. Thus $Me_2PH{\cdot}BH_3$ smoothly loses 1 mole of hydrogen on pyrolysis at 150°,[88] while the adduct

$$R_2PH{\cdot}BHX_2 \xrightarrow{\Delta} [R_2BX_2]_x + H_2$$

$$R_2PH{\cdot}GaMe_3 \xrightarrow{\Delta} [R_2PGaMe_2]_x + CH_4$$

$Me_3Al{\cdot}HPMe_2$ requires heating to 215° for several hours to produce $[Me_2AlPMe_2]_3$.[147] The more acidic hydrogen of diphenylphosphine reacts with Et_3Al already in boiling benzene to give the dimeric $[Ph_2P\text{-}AlEt_2]_2$,[336] which is cleaved with ether to the monomeric etherate.[277] Pyrolysis of the gallium and indium adducts, $Me_3Ga{\cdot}HPMe_2$ and $Me_3In{\cdot}HPMe_2$, to $[Me_2GaPMe_2]_x$ and $(Me_2InPMe_2]_x$, respectively, proceeds at lower temperature, that is, at 150 to 155° and 25 to 40°, respectively.[33] The latter two compounds are polymeric glasses in the condensed state but cyclic trimers in benzene solution.[33] The biphosphine-diborane adduct $[Me_2P]_2{\cdot}2BH_3$ requires heating to 220° to produce a mixture of trimeric and tetrameric phosphinoborane,[65,442] while the adduct $Me_2P\text{-}P(CF_3)_2{\cdot}BH_3$ decomposes into $(CF_3)_2PH$ and $(Me_2PBH_2)_3$.[228] The adducts need not be isolated in every case since pyrolysis of a mixture of the two components often gives similar results.[69] In several cases pyrolysis of silyl-phosphine-BX_3 adducts also produced phosphinoboranes, usually dimers.[4,448] Without handling the air-sensitive starting materials, large quantities of the

$$2Me_3SiPEt_2{\cdot}BCl_3 \xrightarrow{120°} 2Me_3SiCl + [Et_2P\text{-}BCl_2]_2$$

phosphinoboranes have been prepared by reaction of $Me_2P(O)Cl$ with $NaBH_4$ in boiling diglyme,[86] and by fusion of tetramethylbiphosphine disulfide and $LiBH_4$ or $NaBH_4$ under

$$3Me_2P(S)\text{-}(S)PMe_2 + 6LiBH_4 \longrightarrow 2[Me_2P\text{-}BH_2]_3 + 3H_2S + 3Li_2S + 3H_2$$

reduced pressure.[44,352] A similar reaction of biphosphines with $LiAlH_4$ proceeded only to the phosphide-substituted alanate stage, $Li[AlH(PR_2)_3]$.[322] Reduction of Ph_2PCl and $PhPCl_2$ with $LiAlH_4$ apparently also gives a phosphide-substituted alanate.[184]

II. BY DEHYDROHALOGENATION OF THE ADDUCTS

The simplest phosphinoborane monomer, Me_2BPH_2, was prepared by reaction of phosphine with bromodimethylborane in the presence of triethylamine.[88] More complex phosphinoboranes[115,116,218] and B-halogenated phosphinoboranes (usually dimers)[218] have likewise been prepared by this method. This procedure, however, was not successful for preparing similar phosphinoaluminum compounds.[82] In this case an adduct with dimethylphosphine, $Me_2P{\cdot}AlH_2{\cdot}HPMe_2$, was apparently formed.[82]

$$R_2PH + BrBH_2 \xrightarrow{R_3N} R_2PBH_2 + R_3NHBr$$

$$2Ph_2PH{\cdot}BBr_3 \xrightarrow{R_3N} [Ph_2P\text{-}BBr_2]_2 + 2R_3NHBr$$

III. FROM ALKALI PHOSPHIDES AND HALOGEN-CONTAINING GROUP III COMPOUNDS OR FROM PHOSPHORUS HALIDES AND ALKALI GROUP III COMPOUNDS

Metathesis of alkali phosphides with halogen-containing compounds of Group III elements is a generally applicable method for preparing this type of compound.[115] With chlorodiaryl- and diaminoboranes monomers are usually obtained,[115,445-447] but boron trihalides or the aluminum compounds[195,277] normally yield dimers[277,448] or trimers.[204,207] The phosphino derivatives of pentaborane[73] and decaborane[428,429,499,500] have been made either by this

$$Ph_2PLi + ClBPh_2 \longrightarrow Ph_2P\text{-}BPh_2 + LiCl$$

$$3AlCl_3 + 3LiPEt_2 \longrightarrow [Et_2P\text{-}AlCl_2]_3 + 3LiCl$$

method or from the corresponding lithium boride and chlorophosphine. Excess alkali phosphide causes further reaction and leads to ionic complexes, for example,[199]

$$AlCl_3 + 4LiPEt_2 \longrightarrow Li[Al(PEt_2)_4] + 3LiCl_3$$

It is significant that most of the cyclic phosphorus Group III compounds obtained by the dehydrohalogenation or alkali phosphide route are dimers, in contrast to the trimers obtained by pyrolytic syntheses. Whether the ring size is governed thermodynamically or kinetically is not clear yet.

IV. BY SUBSTITUTION REACTIONS

This type of reaction is limited to the phosphinoboranes since when substitution was attempted with the corresponding alanes only cleavage reactions were observed, for example,[207]

$$[Cl_2Al\text{-}PEt_2]_3 + LiPh \longrightarrow LiAlPh_4 \text{ and other products}$$

The hydrogens attached to boron atoms in cyclic phosphinoboranes are readily replaced by halogen to give B-halophosphinoboranes. The halogenation is achieved either directly with halogenes[92,264] or with hydrogen halides,[92,226] alkyl halides,[44,226] $CHCl_3$, CCl_4, CH_2Cl_2, and N-bromo- or N-iodosuccinimide.[226,565] By metathetical reactions of the monoiodo derivative of P-hexamethylphosphinoborane, several functional groups such as alkoxy, cyano, isocyanato, thiocyanato, thio, methylthio, and ethyl have been attached to a boron atom.[565]

C.2. General Chemistry

Cyclic phosphinoboranes have good chemical and thermodynamic stability,[88] but the corresponding aluminum,[195,207] gallium,[33,113] and indium[33] compounds are much less stable and often inflame in air and are hydrolyzed by water. The cyclic phosphinoborane trimer $[Me_2P\text{-}BH_2]_3$ is completely stable at 250° and decomposes only very slowly at 360°. The corresponding tetramer $[Me_2P\text{-}BH_2]_4$ depolymerizes slowly at 300° to the trimer and other products.[88] The trimer $[Me_2P\text{-}BH_2]_3$ ignites spontaneously in air at 240°, but the halogen-substituted derivatives have a much higher ignition temperature, for example, $[Me_2P\text{-}BCl_2]_3$ at 395° and $[Me_2P\text{-}BBr_2]_3$ at >420°.[226]

The hydrolytic stability of the B-H bonds in phosphinoboranes is also remarkable, especially when compared to the explosive hydrolysis of the lower boranes. Thus 11 days were required for complete degradation of the trimer $(Me_2P\text{-}BH_2)_3$ to hydrogen, boric acid, and dimethylphosphinic acid by concentrated HCl at 300°.[88] However, the monosubstituted cyano trimer $Me_6P_3B_3H_5CN$, which was also stable in concentrated HCl when refluxed for 5 hr, was completely decomposed by a solution of H_2SO_4 in aqueous acetic acid.[565] The difference in the two experiments is that the trimer was completely soluble in acetic acid solution but only slightly soluble in HCl solution. As expected, hexahalogen-substituted phosphinoborane derivatives are much less stable to hydrolysis.[226]

When the pyrolysis of phosphine-borane adducts is carried out in the presence of bases such as phosphine[65] or amine,[566] white thermoplastic poly(phosphinoboranes) have been obtained. It has been postulated[65,566] that the base initially forms a complex of the type base$\cdot H_2B-PR_2$ which successively adds R_2P-BH_2 units to it. Cyclization to trimer or tetramer is prevented by coordination of the base to the terminal BH_2 group. The polymers obtained by this method are thermoplastic and are soluble in hot benzene.[566] The properties suggest that the polymer chains are not cross-linked. At 330° the polymer seems to break down to trimer and tetramer.[65]

Cross-linked or thermosetting phosphinoborane polymers have been obtained by pyrolysis of disecondary alkylene-diphosphine diborane complexes[354] or by copyrolysis of dialkylphosphine-borane complexes with pentaborane, aminophosphines,[87] and alkylenediphosphines,[90,265] respectively.

While several substitution reactions have been successfully carried out with the P-B system[44,92,226,264,565] (see Section C.1.IV), all attempts to substitute chlorine in $[Et_2P-AlCl_2]$ by other groups were unsuccessful. In each case a cleavage of the P-Al bond was observed.[204] Phosphinoboranes have been claimed to be useful as dielectrics,[94,224] hydraulic fluids,[94] metal coatings, and laminating resins.[224]

Several trimeric phosphinoboranes catalyze the polycondensation of alkane diisocyanates.[468]

C.3. General Physical Properties

X-ray analysis shows that the dimer $[Ph_2P-BJ_2]_2$,[64] the trimer $[Me_2P-BH_2]_3$,[237] and the tetramer $[Me_2P-BH_2]_4$[223] have simple monocyclic rings of alternating boron and phosphorus atoms. Burg and Wagner[88] suggested that the unusual stability of the boron-phosphorus bond in cyclic phosphinoboranes arises in large part from the availability of d orbitals on the phosphorus atoms. The deviation of the B-P-B angles from a tetrahedral value in the structures of the trimer[237] $(Me_2P-BH_2)_3$ (B-P-B 118.1°) and the tetramer $(Me_2P-BH_2)_4$[223] (125°) seems to support such a view, although the bond order of the P-B bond is not far from one in the trimer[237] (P-B 1.94, calculated from the sum of the radii 1.95 Å) and possibly even less than one in the tetramer[223] (P-B 2.08 Å) and the dimer $[Ph_2P-BI_2]_2$[64] (P-B 2.03 Å).

The ring stretching vibrations of the phosphinoborane trimer and tetramer in the IR and Raman spectra in the region of 500 to 700 cm^{-1} also indicate single-bonded ring structures in solution and in the vapor phase[106] but provide no evidence for weakening of the B-H bonds by donation of B-H bonding electrons to the ring bonds as suggested by Burg and Wagner.[88] On the basis of dipole moment studies of monomeric phosphinoboranes, the effect of internal B-P π bonding was also discounted in this class;[115] the π bonding is less extensive than in the analogous aminoboranes.

The IR spectra of several aluminum-[33,207] gallium-[33] and indium-[33] substituted organophosphorus derivatives have also been reported.

Thermal decomposition and mass spectral data indicate that $[Me_2P\text{-}BH_2]_3$ does not depolymerize into its monomer but decomposes in a more complicated manner to give hydrogen, phosphorus, and methylboranes.[179]

D. ORGANOPHOSPHORUS COMPOUNDS OF GROUP IV ELEMENTS

The first organophosphorus compound with a Group IV element was described in 1957. Since then many articles and reviews (Si-P,[62,109,158a,195,382] Ge-P,[158a,221,267,501] Sn-P,[501,518] Pb-P,[501] derivatives) have appeared on this subject. As a class, all the organophosphorus Group IV element compounds are sensitive to water and oxygen and therefore must be handled in a dry nitrogen box or in a vacuum line.

D.1. Methods of Preparation

I. BY REACTION OF ALKALI PHOSPHIDES WITH GROUP IV HALIDES

The reaction of alkali phosphides with Group IV halides seems to be a general method for making organophosphorus derivatives of silicon,[109,195,361,376,460] germanium,[221,267,501] and tin.[60,98,361] Replacement of alkali phosphides by bromomagnesium phosphide[98] or $LiAl(PH_2)_4$[450a] produces similar results.

$$R_3MX + LiPR'_2 \longrightarrow R_3M\text{-}PR'_2 + LiX \qquad (M = Si, Ge, Sn)$$

$$4R_3MX + LiAl(PH_4)_4 \longrightarrow 4R_3M\text{-}PH_2 + LiAlX_4$$

Phosphinosubstituted lead compounds have not been prepared by this route. Silicon and tin di-, tri-, and tetrahalides react in the same way, but germanium tri-, and tetrahalides produce in this reaction colored polymeric materials and tetraphenylbiphosphine.[59] These probably

$$MCl_4 + 4LiPR_2 \longrightarrow [R_2P]_4M + 4LiCl \qquad (M = Si, Sn)$$

result from halogen-metal exchange reactions, and possibly reduction of Ge(IV) to Ge(II). When the starting materials contained hydrogen bonds, products were obtained that were evidently the result of a series of consecutive reactions, for example, KPH_2 or $LiPH_2$ and Me_3SiCl also gave in addition to the monosubstituted product Me_3SiPH_2 the trisubstituted derivative $[Me_3Si]_3P$.[60,460] Similarly, MeHPLi[24] and PhPHNa[98] gave with R_3SnCl the disubstituted product. In an extensive study, Fritz and co-workers[197,198] showed recently that $LiPEt_2$ interacts with Si-H bonds and yields, depending on the starting material, either a phosphorylated product or a metalated silicon compound, for example,

$$H_3Si\text{-}PEt_2 + LiPEt_2 \longrightarrow H_2Si(PEt_2)_2 + LiH$$

and

$$HSi(PEt_2)_3 + LiPEt_2 \longrightarrow LiSi(PEt_2)_3 + HPEt_2$$

Methyl- and butyllithium cleave the Si-P.[197] Furthermore, chlorine- and hydrogen-containing phosphinosilanes readily undergo disproportionation.[198]

$$2H_3Si\text{-}PEt_2 \longrightarrow SiH_4 + H_2Si[PEt_2]_2$$

Cyclic phosphinosilanes have been synthesized by reaction of dialkali phosphides with diethyldichlorosilane.[460] Cyclic phosphinosubstituted germanes and stannanes were similarly obtained.[504,505] In several instances the interaction of chlorophosphines with alkali[399,514,518] or mercury[234] derivatives of Group IV elements also produced phosphinosubstituted Group IV compounds.

$$2Ph_3SiLi + Br_2PMe \longrightarrow [Ph_3Si]_2PMe + 2LiBr$$

$$[Me_3Si]_2Hg + 2IP(CF_3)_2 \longrightarrow 2Me_3Si\text{-}P(CF_3)_2 + HgI_2$$

The reported formation of a Si-P bond in the reaction of sodium dialkylphosphites with R_3SiCl,[23,346,435] and $(RO)_3SiCl$,[23,171] or from Ph_3SiLi and $ClP(O)(OR)_2$[435] needs further confirmation, particularly in view of the reported formation of a Si-O-P bond in the analogous reaction of

$$(RO)_2P(O)Na + R_3^1SiCl \longrightarrow (RO)_2P(O)SiR_3^1 + NaCl$$

the sodium salts of secondary phosphine oxides with trialkylchlorosilanes.[325]

$$R_2P(O)Na + ClSiR_3 \longrightarrow R_2POSiR_3 + NaCl$$

II. FROM PHOSPHINES AND GROUP IV HALIDES BY DEHYDROHALOGENATION

Most of the phosphinosubstituted germanium[520,522] and tin[503,512,520,522] and all of the lead[517,519,520,522] derivatives have been made by reaction of phosphines with Group IV halides in the presence of a tertiary amine. The reaction is usually run in benzene and gives clean products in good yields.

$$R_3SnCl + HPPh_2 \xrightarrow{Et_3N} R_3SnPPh_2 + Et_3NHCl$$

A six-membered P-Sn ring[503] (10) and what is believed to have a cubanelike structure (11)[502] have also been made by this route. Surprisingly, the six-membered ring was also obtained in the interaction of Ph_3SnLi with Ph_2PCl.[514] This means that one Ph-Sn and one Ph-P bond must have been broken.

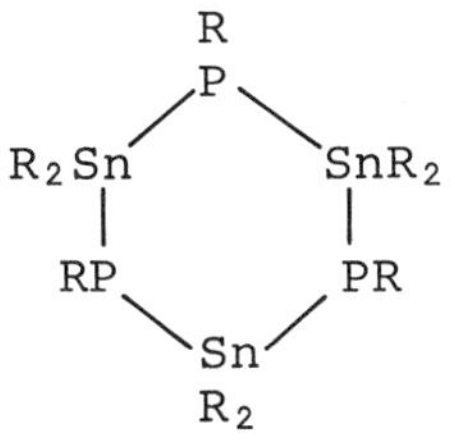

(10)

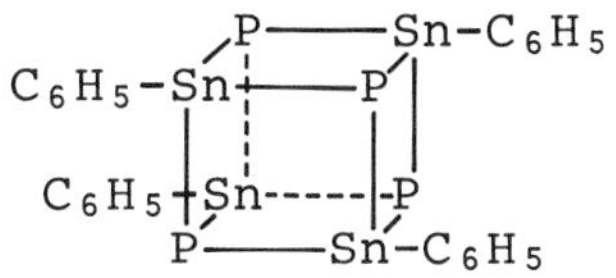

(11)

No phosphinosilane, however, has been prepared by this procedure. Whether this method does not work in this case is not known.

III. FROM PHOSPHINES AND AMINOSUBSTITUTED DERIVATIVES OF GROUP IV ELEMENTS

Phosphines cleave Ge-N[480,485,506,527] and Sn-N[249,337] bonds under mild conditions and give the corresponding phosphinosubstituted derivatives in high yield.

$$R_3SnNMe_2 + HPPh_2 \longrightarrow R_3SnPPh_2 + HNMe_2$$

$$3Me_3GeNMe_2 + PH_3 \longrightarrow (Me_3Ge)_3P + 3HNMe_2$$

The reactions are usually carried out in ether or benzene and the products are isolated by fractional distillation. Only in the reaction of diphenylphosphine with aminogermanes was it necessary to heat the mixture to 200° to effect this cleavage.[480,485] Attempts to substitute in primary phosphines only one P-H bond by a phosphorus-metal bond were only partially successful. Thus $MePH_2$ gave in the reaction with Me_3GeNMe_2 only the disubstituted product $MeP[GeMe_3]_2$, but $PhPH_2$ yielded both the mono- and the disubstituted derivative.[527] The reaction has one further restriction. Only class "b" acceptors (Ge and Sn--with lead the reaction has not been reported yet) seem to give this exchange reaction, whereas in aminosilanes (silicon appears to be a class "a" acceptor) the amino group is not displaced upon treatment with Ph_2PH.[337]

IV. BY MISCELLANEOUS METHODS

Organostannyl-substituted phosphines, such as (<u>10</u>) and (<u>12</u>) [structure of (<u>12</u>) uncertain] have been obtained in the direct reaction of tetraphenyltin with white phosphorus at 235 to 250°C.

$$[Ph_3Sn]_2P\text{-}P[SnPh_3]_2$$

(<u>12</u>)

The presence of other tin-phosphorus compounds has been indirectly shown through identification of their oxidative degradation products.[513,518] Above 300° Ph_3P is the only product isolated.[518]

Partial hydrolysis of $(Me_3Si)_3P$ with H_2O or D_2O under homogeneous conditions in diglyme is a convenient method for the preparation of mono- and disilylphosphine.[63]

$$(Me_3Si)_3P + H_2O \longrightarrow [Me_3Si]_2O + Me_3SiPH_2$$

Hydrolysis under heterogeneous conditions yields PH_3 and siloxane only. A partial cleavage of the Si-P and Sn-P bond has also been effected by reaction with disulfides at ∿50°.[61] This reaction seems not to be easily controlled,

$$MeP[SnEt_3]_2 + Et_2S_2 \xrightarrow{50°} Et_3SnSEt + Me(EtS)PSnEt_3$$

and the yields were rather low. Biphosphines cleave the Si-Si bond on heating and produce a phosphinosilane.[382a]

$$Cl_3Si\text{-}SiCl_3 + Me_2P\text{-}PMe_2 \longrightarrow 2Cl_3SiPMe_2$$

The originally reported formation of Si-P[23] and Sn-P[20,22] bonds in an Arbuzov-type reaction from trialkylphosphites and halogen derivatives of silicon or tin could not be confirmed later.[410,411] It was shown that in each case compounds with an M-O-P bond were formed.[21,410,411]

D.2. General Chemistry

All phosphinosubstituted compounds of the Group IV elements are sensitive to water and oxygen and therefore must be handled under conditions in which air and humidity are excluded. A report claiming the isolation of the primary oxidation product $Bu_3SnP(O)Ph_2$ in the reaction of Bu_3SnPPh_2 with oxygen[488] could not be verified.[510,512] In all studies involving Si-P,[6,376,460] Ge-P,[59,221] Sn-P,[98,510,512-514] and Pb-P[519] bonds, oxidation with oxygen or oxygen transfer agents[460,510] produced the corresponding phosphates, phosphonates, or phosphinates, respectively.

$$R_3MPR_2 + O_2 \longrightarrow R_3MOP(=O)R_2 \qquad (M = Si, Ge, Sn, Pb)$$

Sulfur gave a similar reaction with Ph_3SnPPh_2, giving $Ph_3SnSP(S)Ph_2$,[510] but with $(Me_3Si)_3P$ the formation of the phosphine sulfide $(Me_3Si)_3PS$ has been described.[60] With $(Ph_3Sn)_2PPh$ or $(Ph_3Sn)_3P$ and sulfur, only the cleavage products $Ph_3SnSSnPh_3$ and $[PhPS_2]_2$ or P_2S_5, respectively, were isolated.[510] On treatment with MeI, the unstable phosphonium salts $[Me_3SiPEt_3]^+I^-$ [201,444] and $[Ph_3SnPMePh_2]^+I^-$ [512] (structure uncertain[526]) were obtained. Compounds of the type $[Me_3SiPMe_3]^+[Co(CO)_4]^-$, obtained from the reaction of tertiary phosphines with $Me_3Si\text{-}Co(CO)_4$, were stable, however.[25a]

This contrasts with the behavior of the corresponding germanium compound and of trialkylstannyldiphenylphosphines,[98] which gave only cleavage products.[59] Cleavage of the Si-P,[197] Ge-P,[59] and Sn-P[98,521] bonds by alkali or

$$Et_3GePPh_2 + 2MeI \longrightarrow Me_3GeI + [Ph_2PMe_2]^{\oplus}I^{\ominus}$$

organoalkali derivatives seems to be a general reaction. Halogens,[4,59,444] fluorophosphoranes,[430] chlorophosphines,[516] haloarsines,[4,516] chloroamines,[257] hydrogen halides,[444,481] boron halides,[4,444] and several other metal halides[4,376] also effect this cleavage. One of the more interesting reactions of phosphinosubstituted Group IV compounds is the insertion reaction with 1,2-dipolar compounds. The reactions investigated are summarized in the scheme on p. 312. Carbon dioxide inserted only into the Si-P bond[6] but not into Ge-P[482] and Sn-P[508] bonds. Also remarkable is the different mode of addition of acetylenes, olefins, and ketene to the various Group IV-substituted phosphines. Insertion of $(CF_3)_2CO$ and SO_2 into the Ge-P[484a] or Si-P bond,[6] of COS, $CSCl_2$, or $CS(NH_2)_2$[507] into the Sn-P bond, and of phenylazide into the Sn-P and Pb-P[515] bonds has also been described. Phosphino-germanes give with α,β-unsaturated ketones the 1,4-addition adducts.[484a]

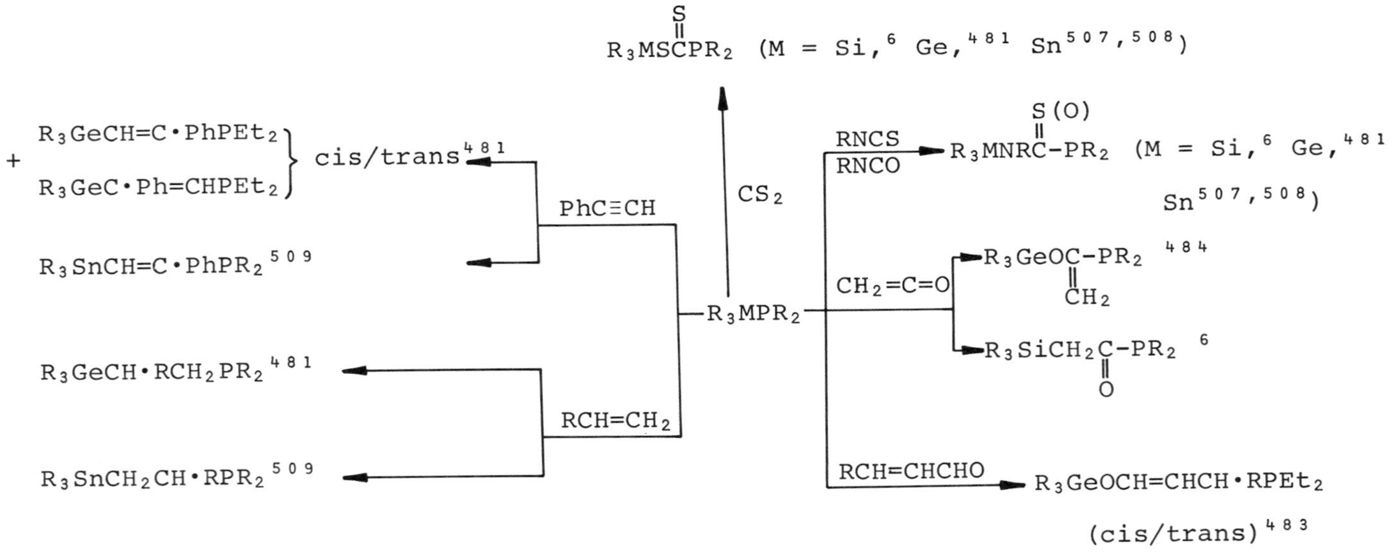
R3MSCPR2 (M = Si,6 Ge,481 Sn507,508)
S
CS2
R3MPR2
PhC≡CH
R3GeCH=C•PhPEt2
R3GeC•Ph=CHPEt2
+
cis/trans481
R3SnCH=C•PhPR2 509
RCH=CH2
R3GeCH•RCH2PR2 481
R3SnCH2CH•RPR2 509
RNCS
RNCO
R3MNRC-PR2 (M = Si,6 Ge,481 Sn507,508)
S(O)
CH2=C=O
R3GeOC-PR2 484
CH2
R3SiCH2C-PR2 6
O
RCH=CHCHO
R3GeOCH=CHCH•RPEt2
(cis/trans)483

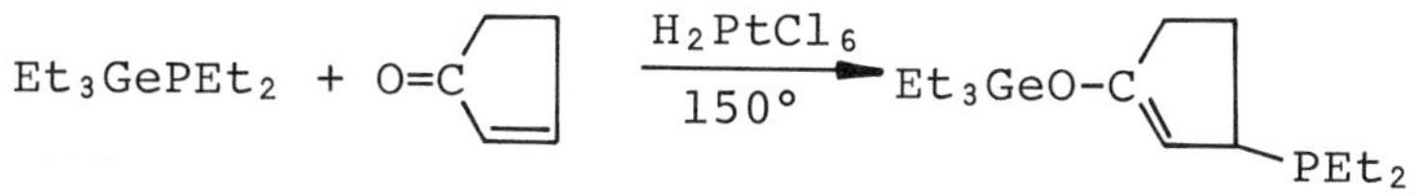

The interesting heterocyclic system, substituted 1,4-dihydrophosphorin-4-one, was obtained from the radical initiated reaction of $PhP(SiMe_3)_2$ with bis(phenylethynyl) ketone.[387] Substituted phosphorins were isolated from the interaction of $(Me_3Si)_3P$ and pyrrylium salts[386] (see Chapter 1). Tris(trimethylsilyl)phosphine was also found to be a useful starting material for the synthesis of phosphamethin-cyanines.[385] And, finally, the Sn-P bond is broken in a hydrostannolysis reaction with Ph_3SnH.[132]

D.3. General Physical Properties

With the exception of phosphinosubstituted lead compounds which all decompose at or before 110°,[517,520] the thermal stability of the other Group IV phosphino derivatives is good. Thus $(Me_3Si)_3P$ can be refluxed at its boiling point at 242° without noticeable decomposition.[460] Diphenylphosphinotriethylgermane shows no sign of decomposition at 180° in a nitrogen atmosphere,[59,221] and several stannylphosphines have been prepared at 235 to 250° in the direct reaction.[518] IR, Raman, and ^{31}P NMR investigations indicate that all compounds of the type $(Me_3M)_3P$ (M = Si, Ge, Sn, Pb) have a pyramidal configuration with C_{3v} symmetry. These conclusions are in agreement with an electron diffraction study which proved for the molecule $(H_3Si)_3P$ a pyramidal structure (Si-P-Si 95° ± 2°).[34] The symmetric and asymmetric stretching modes of the M_3P groups have been assigned at: for Si_3P ν_{as} 461, ν_s 380;[62,63] for Ge_3P ν_{as} 397, ν_s 320;[166,506] for Sn_3P ν_{as} 351, ν_s 290;[166,249] and for Pb_3P ν_{as} 313, ν_s 286[517] [cm^{-1}]. IR and Raman assignments for the M_2P and MP symmetric and asymmetric stretching modes have also been given in the literature (M = Si,[62,63] Ge,[481,520,527] Sn,[98,166,502,520] and Pb[517,520]).

Additional evidence for the pyramidal structure of the M_3P molecule comes from ^{31}P NMR investigations.[166,185] With the exception of white phosphorus (+462 ppm),[166] the ^{31}P chemical shifts of these molecules lie at the highest magnetic field of all three coordinated phosphorus compounds: ^{31}P chemical shift for $(Me_3Si)_3P$ +251.2 ppm;[185] $(Me_3Ge)_3P$ +228 ppm;[166] $(Me_3Sn)_3P$ +330 ppm.[166] These high chemical shifts have been attributed to considerable p-σp^3 bonding in the phosphorus-metal bond.[166,185] This makes the s-electron density at the phosphorus atom high and causes a high chemical shift. In other words, this then means that the molecules must have a pyramidal structure.

A $p\pi$-$d\pi$ bonding of the phosphorus-metal bonding in

Et_3SnPEt_2 has also been deduced from UV and dipole moment measurements.[98]

The phosphorus atom, however, retains in these molecules its donor character to an appreciable degree since it forms coordination complexes with silver iodide,[59] BH_3,[460] and several metal carbonyls.[523-525]

E. ORGANOPHOSPHORUS COMPOUNDS OF THE GROUP V ELEMENTS (EXCLUDING P-N COMPOUNDS)

Phosphorus, similar to silicon and sulfur, which are situated to the left and right of phosphorus in the periodic table, forms numerous chain and cyclic structures. Interest in the study of these compounds came (a) from the importance of determining the relationship of P-P bonds to other compounds with M-M bonds, (b) from the type of bonding, that is, the extent to which the lone electron pairs participate in the bonding between the phosphorus atom; and (c) from the ready availability of these compounds and thus for use in further syntheses.

E.1. Organosubstituted Biphosphines

E.1.1. Methods of Preparation

I. FROM SECONDARY PHOSPHINES AND PHOSPHINOUS DERIVATIVES

The first organosubstituted biphosphine was synthesized over 80 years ago by Dörken[156] by the reaction of diphenylphosphine with diphenylphosphinous chloride in boiling petroleum ether.[156,360]

$$Ph_2PH + ClPPh_2 \longrightarrow Ph_2P\text{-}PPh_2 + HCl$$

More recently, this method has been extended to the synthesis of tetraalkylbiphosphines[66,77] and the unsymmetrical biphosphines $Me_2P\text{-}P(CF_3)_2$,[228] $MeHP\text{-}P(CF_3)_2$,[74] $Me(CF_3)P\text{-}P(CF_3)Me$,[75] and $Ph_2P\text{-}P(C_6H_{11}\text{-}c)_2$.[298] The latter reactions are facilitated by the presence of tertiary amines which act as HCl scavengers. Biphosphine formation is also favored in the interaction of Me_2PH and Me_2PNMe_2 because the P-P bond can form two π bonds as opposed to only one π bond in the aminophosphine.[66,77] The phosphinous derivative need not be present in every case but may be formed in situ in reactions with secondary phosphines. Thus treatment of Ph_2PH with chloroamine resulted in the isolation of $Ph_2P\text{-}PPh_2$. Excess chloroamine must be avoided since biphosphines are cleaved by this reagent.[256]

II. BY REDUCTION OF PHOSPHINOUS HALIDES

The Wurtz-Fittig synthesis is a well-known method of coupling carbon atoms. The first application of this approach to phosphorus chemistry was the coupling of PI_3 with mercury to diphosphorus tetraiodide.[42] This procedure was also successfully used for the synthesis of tetrakis-(trifluoromethyl)biphosphine,[38,76] 1,2-dimethyl-1,2-bis-(trifluoromethyl)biphosphine,[75] $Me(CF_3)P-P(CF_3)Me$, tetraphenylbiphosphine,[558,574] and tetrakis(pentafluorophenyl)-biphosphine.[18] The reaction probably proceeds via an

$$2(CF_3)_2PI + Hg \longrightarrow (CF_3)_2P\text{-}P(CF_2)_2 + HgI_2$$

intermediate phosphorus-mercury compound.[76] The activity of the halogen toward mercury increases in the order Cl < Br < I. Thus $(CF_3)_2PI$ reacts at room temperature, $(CF_3)_2$-PBr requires 100°,[72] $(CF_3)_2PCl$ does not attack mercury, and $Me(CF_2)PCl$ reacts very slowly and incompletely.[75] Other reducing agents such as lithium,[250] LiH,[262] sodium,[291,313,358,360,438,439] potassium,[438,439] magnesium,[177,240,439] CaC_2,[543] or tertiary phosphines[193] have been successfully used for the coupling of the less reactive but more accessible phosphinous chlorides. N-Substituted aminoorganochlorophosphines are reduced with alkali metals in the same way as diaryl- and dialkylphosphinous chlorides. Difficulties attributable to disproportionation of the so obtained aminoorganobiphosphines are minimized by lowering the reaction temperature to -10 to -20° and by using the more reactive sodium-potassium alloy.[536] The alkali metal coupling probably proceeds via the alkali metal phosphide intermediate,[313,440] since biphosphines have been produced

$$R_2PCl + 2M \longrightarrow R_2PM + MCl$$

$$R_2PM + ClPR_2 \longrightarrow R_2P\text{-}PR_2 + MCl$$

from the second step of the above reaction.[107,560] As discussed in Section A.1.III, biphosphines are cleaved by alkali metals; therefore excess alkali metal must be avoided in these coupling reactions. The formation of tetramesitylbiphosphine from PX_3 and mesitylmagnesium bromide in THF is unique and indicates that a reductive coupling step must be involved here.[549] Excess magnesium has been made responsible for this reduction step.[549]

III. BY DESULFURIZATION OF ORGANOSUBSTITUTED BIPHOSPHINE DISULFIDES

One of the most suitable methods for the preparation

of organosubstituted biphosphines is the reduction of substituted biphosphine disulfides, which are easily accessible (Section E). Among the reducing agents that effect this desulfurization are zinc dust,[362,393,400,442] alkali metal,[321,441] powdered iron, copper, lead, cadmium,[372a,438,440] and many other metals,[441] triethylphosphite,[456] and tertiary phosphines[393,400] such as tributylphosphine [397,400,456] or triphenylphosphine.[398,400] Desulfurization with tributylphosphine proceeds stepwise, and tetraalkylbiphosphine monosulfides are formed initially.[397] Excess Bu_3P causes further reduction and gives the biphosphines in high yield.[397] It was by this method that the first aliphatic substituted unsymmetrical biphosphines, for

$$R_2P(S)\text{-}P(S)R_2 + Bu_3P \longrightarrow R_2P\text{-}P(S)R_2 + Bu_3PS$$

$$R_2P(S)\text{-}P(S)R_2 + 2Bu_3P \longrightarrow R_2P\text{-}PR_2 + 2Bu_3PS$$

example, MeEtP-PEtMe, were obtained in stereoisomeric forms.[397] Reduction of thiophosphinic anhydrides $[R_2P(S)]_2S$ with Bu_3P also produces biphosphines.[181]

In a preliminary report[37] it has been noted that complexes of biphosphine disulfides with metal salts decompose on heating or on boiling with water or alkali hydroxide to give a biphosphine and other products.

IV. FROM ALKALI ORGANOPHOSPHIDES AND HALOGEN SOURCES

Whereas several organic dihalides react normally with organophosphides to give ditertiary phosphines[398] (see Chapter 1), lithium dialkylphosphides, $LiPR_2$ (R = Et, c-C_6H_{11}), react differently with CH_2Cl_2, $ClCH_2CH_2Cl$, $BrCH_2$-CH_2Br,[307] 1-chloro-2-phenylacetylene,[289] halocarboxylic esters,[317] and $COCl_2$[309] in ether or dioxan. Here a lithium-halogen interconversion takes place and biphosphines are formed, for example,

$$R_2PLi + XCH_2CH_2X \longrightarrow XCH_2CH_2Li + R_2PX$$

$$XCH_2CH_2Li \longrightarrow LiX + CH_2{=}CH_2$$

$$R_2PX + R_2PLi \longrightarrow R_2P\text{-}PR_2 + LiX$$

A report describing the formation of a disecondary biphosphine, $[Ph(H)P]_2$,[292] from PhPHNa and Br_2 could not be confirmed.[300] In this case a cyclopolyphosphine, $(PhP)_5$, was produced.[300] An interesting application of the metal-halogen exchange reaction has made available the novel heterocyclic biphosphine (13),[295] which is also obtained in the interaction of the triphosphide with 1,3-dichloropropane.[300]

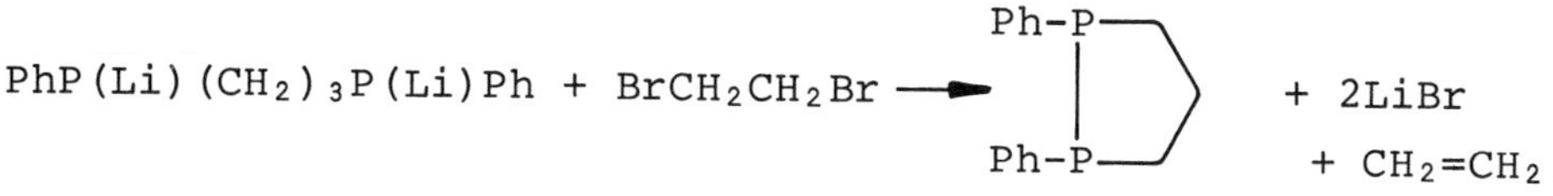

$$PhP(Li)(CH_2)_3P(Li)Ph + BrCH_2CH_2Br \longrightarrow \underset{(\underline{13})}{\overset{Ph-P}{\underset{Ph-P}{\Big|}}} + 2LiBr + CH_2=CH_2$$

$$KPPh-PPh-PPhK + Cl(CH_2)_3Cl \longrightarrow \left[\overset{Ph-P}{\underset{Ph-P}{\Big|}} \underset{\underset{Ph}{P}}{} \right] + 2KCl$$

$$\longrightarrow \underset{(\underline{13})}{\overset{Ph-P}{\underset{Ph-P}{\Big|}}} + 1/5 \ (PhP)_5$$

It should be noted that the halogen-metal exchange reaction is sensitive to change in solvent. Thus when LiPEt$_2$ was treated with ClCH$_2$CH$_2$Cl in THF the ditertiary phosphine was formed in high yield.[250]

The interaction of 1,2-dialkali-1,2-diorganobiphosphide and organo halides[299,300,476] or 1,4-dichlorobutane[299,300] leads to the formation of unsymmetrical biphosphines[476] and a heterocyclic biphosphine, 1,2-diorgano-1,2-diphosphacyclohexane ($\underline{14}$),[299,300] respectively.

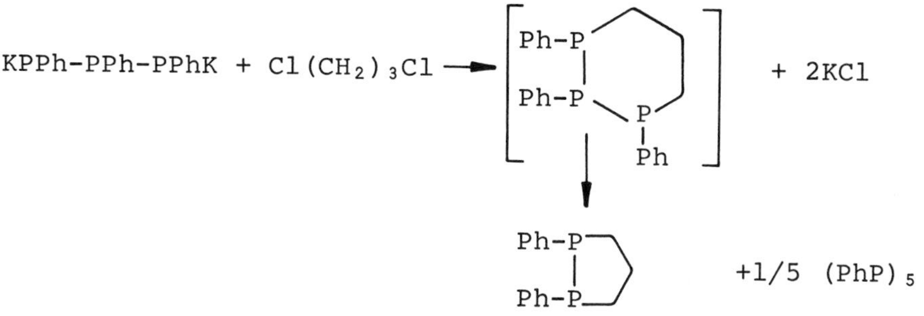

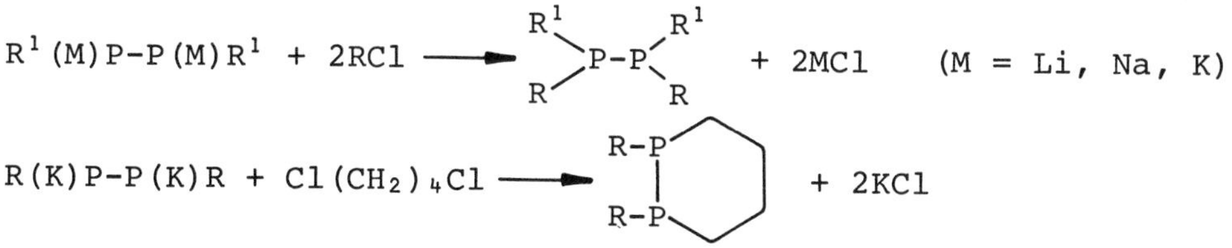

$$R^1(M)P-P(M)R^1 + 2RCl \longrightarrow \underset{R}{\overset{R^1}{>}}P-P\underset{R}{\overset{R^1}{<}} + 2MCl \quad (M = Li, Na, K)$$

$$R(K)P-P(K)R + Cl(CH_2)_4Cl \longrightarrow \overset{R-P}{\underset{R-P}{\Big|}} + 2KCl$$

$$(\underline{14})$$

Attempts to prepare straight-chain tetraphosphines by reaction of 1,4-dialkali tetraphosphides with alkyl halides failed.[299,300] Although the reaction proceeded, the reaction products were unstable and disproportionated into a biphosphine and a cyclopolyphosphine.[299,300]

$$MRP\text{-}PR\text{-}PR\text{-}PRM + 2R^1X \longrightarrow [R^1RP\text{-}PR\text{-}PR\text{-}PRR^1] + 2MX$$

$$\downarrow$$

$$R^1RP\text{-}PRR^1 + 2/n[RP]_n$$

$$Li(PEt)_4Li + 2ClPEt_2 \longrightarrow Et_2P\text{-}PEt_2 + (EtP)_4 + 2LiCl$$

Simple organophosphides, however, give high yields of biphosphines when treated with chlorophosphines.[107,560]

$$R_2PM + ClPR^1_2 \longrightarrow R_2P\text{-}PR^1_2 + MCl$$

V. MY MISCELLANEOUS METHODS

Among the miscellaneous methods that have been used to obtain biphosphines are the coupling of phosphino radicals produced either from secondary phosphines and azoisobutyronitrile[379] or from Ph_2PCl and $AgClO_4$,[209] the simultaneous

$$2Ph_2PCl + 2AgClO_4 \xrightarrow[-2AgCl]{} [2Ph_2PClO_4] \longrightarrow 2[Ph_2P\cdot + ClO_4\cdot] \longrightarrow$$

$$Ph_2P\text{-}PPh_2 + 2ClO_2 + 2O_2$$

desulfurization and coupling of $Et_2P(S)Cl$ with copper bronze,[440] the reaction of Ph_2PNa with benzophenone,[320] the interaction of Me_3SiPPh_2 and Ph_2PCl [4] or fluorophosphoranes,[431] and the decompositions of long-chain polyphosphines of the type $R_2P(PR)_xPR_2$.[74,83,299,300,360,574] For completeness, the formation of the unsymmetrical biphosphine $(CF_3)_2P\text{-}PMe_2$ from $(CF_3)_4P_2$ and Me_2PH,[103,232] of Me_4P_2 from $(CF_3P)_4$ and Me_2PH,[121a] and of Ph_4P_2 from $Me_4P_2S_2$ and Ph_2PCl[395] should be mentioned. Biphosphines were also obtained as by-products in several reactions of alkali phosphides with metal halides, particularly when the metal halide was reduced to a lower valence state.[4,59,153,284-286,288,329] The heterocyclic biphosphine 9,9'-bi-(9-phosphafluorenyl) (15) is isolated as a by-product in the alkali metal cleavage of alkyl-[152] or aryl-substituted[58] biphenylenephosphines. It has been suggested that because of the steric protection of the phosphorus atoms by the biphenylene groups this biphosphine (15) is very stable[152] and the equilibrium between metal biphenylenephosphide and its corresponding biphosphine lies in favor of the biphosphine (15).[58]

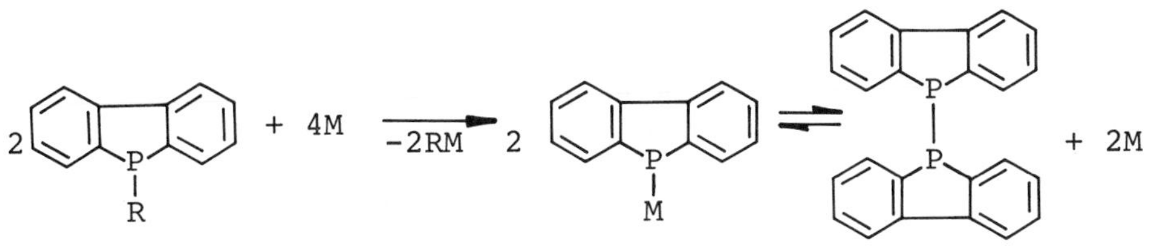

(15)

The dehydration-cyclization reaction of acetylenic and hydroxy groups containing primary phosphines has been claimed to give biphosphines (15a).[466] Additional physical data are needed to confirm the structure of (15a).

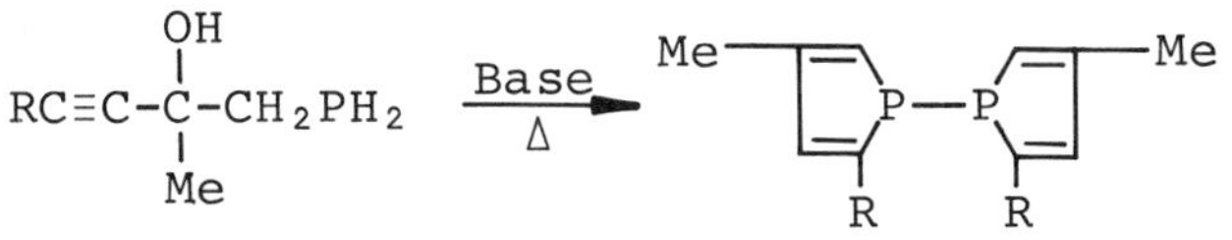

(15a)

Similar to phosphonous difluorides, phosphinous fluorides also disproportionate to give a biphosphine and a trifluorodialkylphosphorane.[59a,431,529]

$$3Me_2PF \longrightarrow Me_2P\text{-}PMe_2 + Me_2PF_3$$

The claimed formation of a triphenyltin-substituted biphosphine in the high-temperature reaction of white phosphorus with Ph_4Sn[513] could not be verified.[526] However, in several other cases breakdown of more complex structures that involve several P-P bonds has been successfully used for the synthesis of biphosphines. Thus fission of the P-P bonds in white phosphorus with sodium metal in liquid ammonia gives, in addition to other products, a biphosphide $Na_2P\text{-}PNa_2$ which forms biphosphines when treated with alkyl halides.[54,55,168] Electrolysis of a suspension of white phosphorus in Me_2NCHO in the presence of butyl bromide produced a low yield of $Bu_2P\text{-}PBu_2$ in addition to other products.[340] When solutions of white phosphorus in CCl_3Br were exposed to ionizing radiation (^{60}Co γ source) or maintained at 100° for 1 hr, the biphosphines $CCl_3(Br)P\text{-}P(Br)(CCl_3)$ and $CCl_3(Br)P\text{-}PBr_2$ were formed.[10] This is one of the simplest biphosphine syntheses known.

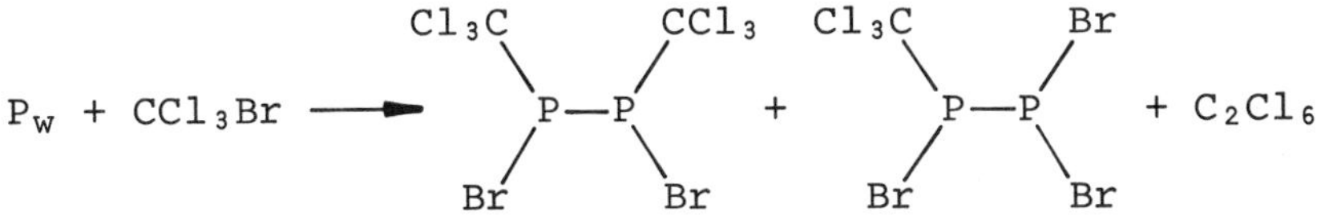

Mahler and Burg isolated the biphosphine $CF_3(H)P$-$P(H)CF_3$ as one of the products of plain water hydrolysis of $(CF_3P)_4$ or $(CF_3P)_5$.[391] Reaction of these cyclopolyphosphines with tertiary phosphines gives polar isomers of biphosphines, R_3P^+-$\bar{P}CF_3$, which consist of phosphonium and phosphide ions in the same molecule and may be considered analogs of ylids.[81] The reaction of cyclopolyphosphines with acetylenes[388] or dienes,[493,494] initiated thermally [388,494] or by irradiation with UV light,[493,494] has produced several new heterocyclic biphosphines (16 and 17) and triphosphines (18).

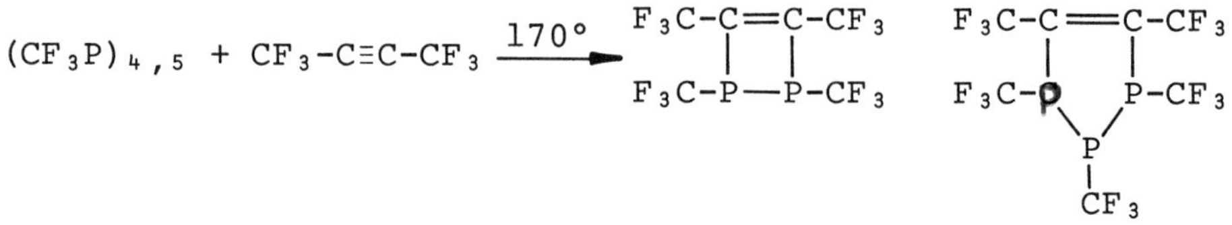

(16) (18)

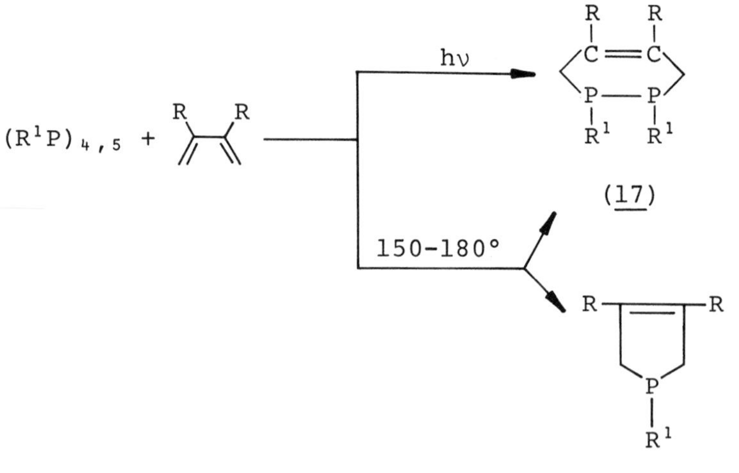

(17)

Reaction of $(PhP)_5$ with MeI, EtI, iodine,[261] $PhPBr_2$, or bromine[29] leads to ring cleavage with the formation of Ph(X)P-P(X)Ph (X = I, Br) and other products.[29,261] Interestingly, the same biphosphine is formed on dissolving $PhPI_2$ in ether and on treatment of $PhPCl_2$ with LiI in ether (X = I).[172] Other diiododiarylbiphosphines are obtained similarly.[172a,172c]

E.1.2. General Chemistry

(1) Oxidation

All the known tetraalkyl- and tetraarylbiphosphines are sensitive to oxygen, and the lower members ignite spontaneously in air. Controlled oxidation with dry air in benzene solution yields biphosphine dioxides,[313,325,360,449]

$$R_2P\text{-}PR_2 + O_2 \longrightarrow R_2P(O)\text{-}P(O)R_2$$

but stronger oxidizing agents such as nitric acid[360] or H_2O_2[240] produce phosphinic acids.

(2) Sulfurization

Biphosphines add 2 equiv of sulfur to give biphosphine disulfides.[296,298-301,313,360,364,449,532,533] The reaction is usually carried out in boiling benzene or boiling CS_2. The addition of only 1 equiv of sulfur to give

$$R_2P\text{-}PR_2 + 2S \longrightarrow R_2P(S)\text{-}P(S)R_2$$

biphosphine monosulfides $R_2P\text{-}P(S)R_2$ has also been achieved.[397,397a]

(3) Metalation

As discussed previously (see Sections A.1.III, B.1.II, and C.1.I), biphosphines are readily cleaved by alkali metals, organometallic compounds, and $LiAlH_4$ to give metal-substituted diorganophosphides, for example,

$$R_2P\text{-}PR_2 + 2M \longrightarrow 2R_2PM$$

(4) Reaction with Halogen

Treatment of biphosphines with equimolar amounts of halogens in an inert solvent such as CCl_4, ether, or benzene causes quantitative P-P bond cleavage with the formation of phosphinous halides.[72,152,313,360,397,535,560] Excess of halogen must be avoided; otherwise dialkyltrihalophosphoranes are formed.[72,313,314,449,560] As an exception,

$$R_2P\text{-}PR_2 + X_2 \longrightarrow 2R_2PX \xrightarrow{X_2} 2R_2PX_3$$

the heterocyclic biphosphine $C_2P_2(CF_3)_4$ (16) could not be cleaved with iodine,[388] and in $(CF_3)_2P\text{-}P(CF_3)_2$[72] and CCl_3-(Br)P-P(Br)CCl_3,[10] simultaneous cleavage of P-C bonds occurred when the bromination was carried out at 80°[72] or at room temperature, respectively.[10]

(5) Hydrolyses

In general, substituted biphosphines do not react with water at room temperature. Hydrolytic cleavage of the P-P bond is possible, however, under more forcing conditions. Thus hydrogen chloride cleaves the P-P bond in symmetrical and unsymmetrical biphosphines.[74,75,442,531,560] Interaction of $[(CF_3)_2P]_2$ and HI in the presence of mercury gives $(CF_3)_2PH$ in high yield.[101]

Basic hydrolysis of CF_3-substituted biphosphines gives 3 moles of CHF_3, while plain water or dilute HCl hydrolysis at 100° yields only 2 moles.[39] The observation that bis-(trifluoromethyl)phosphine, $(CF_3)_2PH$, one of the initially formed cleavage products, hydrolyzes with plain water or dilute acid without fluoroform evolution explains this result.

$$(CF_3)_2P-P(CF_3)_2 \xrightarrow{\text{H}_2\text{O or OH}^-} (CF_3)_2PH + (CF_3)_2POH$$

$$\text{H}_2\text{O} \diagdown \quad \diagdown \text{OH}^- \qquad\qquad \downarrow \text{H}_2\text{O or OH}^-$$

$$CF_3PH_2, F^-, CO_3^{2-} \qquad CHF_3, F^- \qquad 2CHF_3, H_3PO_3$$

$$CO_3^{2-} \text{ and } CF_3P \text{ acid}$$

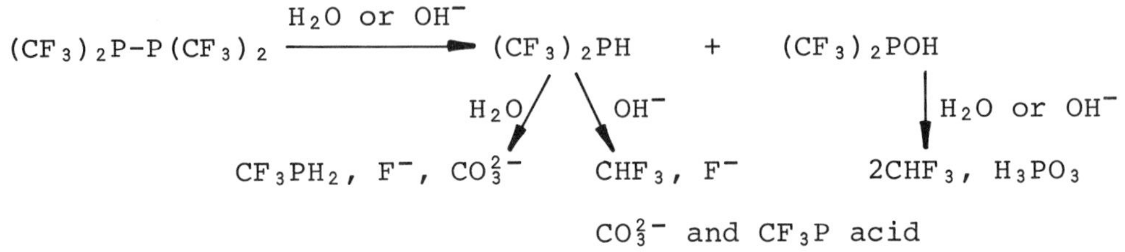

Other CF_3-, and CCl_3-substituted biphosphines such as $[CF_3PH]_2$,[391] $Me_2P-P(CF_3)_2$,[228] and $CCl_3(Br)P-P(Br)CCl_3$,[10] hydrolyze similarly. Because of better solubility, alcoholysis of the P-P bond in biphosphines is more easily achieved and has been observed to occur at room temperature[10,232] or slightly above.[254,290,407] Secondary phosphines and phosphinites are produced as initial products. The same products were observed in the reaction of diphenyl-

$$R_2P-PR_2 + ROH \longrightarrow R_2PH + R_2POR$$

phosphino radicals $Ph_2P\cdot$, produced by the action of heat (∿180°) or UV light on Ph_4P_2, with alcohols.[148] Initial attack of the $Ph_2P\cdot$ radical on the oxygen of the alcohol was postulated in this case.[148] Furthermore, it was observed that aldehydes,[150] aliphatic,[150] and aromatic carboxylic acids[149] also cleaved the P-P bond in Ph_4P_2 when the mixtures were heated to 180°. The initially formed products then underwent further reaction to produce a mixture of phosphine oxides and/or phosphines.

$$Ph_2P-PPh_2 + RCOOH \longrightarrow RCOPPh_2 + Ph_2P(O)H$$

$$Ph_2P-PPh_2 + RCHO \longrightarrow RCOPPh_2 + Ph_2PH$$

Disulfides,[538] diselenides, and tetraphenylhydrazine[539] also effect cleavage of P-P bonds in biphosphines. The reactions are initiated by heating or irradiation with UV light and are believed to proceed by a radical mechanism.[538]

$$R'S-SR' \longrightarrow 2R'S\cdot$$

$$R_2P-PR_2 + R'S\cdot \longrightarrow R_2PSR' + R_2P\cdot$$

(6) Phosphonium Salt Formation

Simple monophosphonium salt formation takes place when tetraalkylbiphosphines are treated with alkyl halides.[313,442] It is not possible to quaternize the second phosphorus atom also. Instead, heating of tetraalkylbiphosphines with excess alkyl iodide to 200°[442] or with perfluorocyclo-

$$R_2P\text{-}PR_2 + R'I \longrightarrow [R_2P\text{-}PR_2R']^{\oplus}I^{\ominus}$$

$$(R = Me, Et, Bu;\ R' = Me, Et)$$

butene or 1,2-dichlorotetrafluorocyclobutene to 130°[139] causes cleavage of the P-P bonds. Also, cleavage rather than quaternization is observed when the P-P bond is weakened by electron-withdrawing substituents such as CF_3[137] or phenyl groups,[262,320] or when the alkyl groups are replaced by the bulkier t-butyl[291] or c-C_6H_{11} groups.[313] Aminosubstituted biphosphines are similarly cleaved by alkyl iodide.[531,532,534] An interesting application of

$$Ph_2P\text{-}PPh_2 + R'I \longrightarrow Ph_2PI + R'PPh_2$$

this cleavage reaction consists in the synthesis of cyclic phosphonium salts (19) starting from tetraphenylbiphosphine.[384]

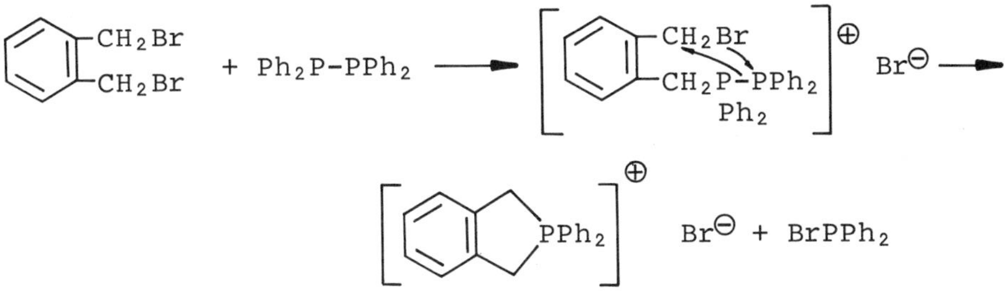

(19)

Interaction of chloramine with biphosphines also produces no simple addition compound. Instead, cleavage of the P-P bond has been observed, apparently giving an amminophosphine and a phosphinous chloride.[194] These products are then presumed to undergo further chloramination and ammonolysis to produce diamminodiorganophosphonium chlorides.[194]

$$R_2P\text{-}PR_2 + NH_2Cl \longrightarrow [R_2PNH_2 + ClPR_2] \xrightarrow{NH_2Cl} 2[R_2(NH_2)_2P]^{\oplus}Cl^{\ominus}$$

(7) Reaction with Ethylene and Benzyne
The biphosphines Me_4P_2, $(CF_3)_4P_2$,[66,71] and Ph_2P-PPh_2[138] add across ethylenic or acetylenic multiple bonds in much the same way as a halogen molecule (Burg's "pseudohalogen" concept of the R_2P moiety).[66,71] o-Phenylenebis(diethylphosphine) has been obtained by adding Et_4P_2 across the triple bond of benzyne.[108]

$$R_2P\text{-}PR_2 + CH_2{=}CH_2 \longrightarrow R_2PCH_2CH_2PR_2$$

(8) Donor and Acceptor Properties
By reason of their lone pairs of electrons, biphosphines form complexes with several Lewis acids such as BH_3, BF_3, Et_3Al, and transition metal halides. These reactions are discussed in Chapter 3A. The $(CF_3)_2P$ group in Me_2P-$P(CF_3)_2$ is acidic enough to form a complex with the donor molecule Me_3N.[228]

(9) Miscellaneous Reactions
When heated with biphosphine disulfides at 190°, biphosphines produce a quantitative yield of biphosphine monosulfides.[397a]

$$R_2P\text{-}PR_2 + R_2P(S)\text{-}P(S)R_2 \longrightarrow 2R_2P\text{-}P(S)R_2$$

A similar reaction was observed when the double borane adduct of Me_2P-PMe_2 was heated with Me_2P-PMe_2. The monoborane adduct[78] Me_2P-$PMe_2 \cdot BH_3$ was the sole product formed. Burg's pseudohalogen concept of biphosphines is further manifested in the reaction of Me_2P-PMe_2 with $(CF_3P)_4$, which produced a triphosphine, $CF_3P[PMe_2]_2$.[121a]

E.1.3. General Physical Properties

The absorption due to vibration of the P-P bond in organosubstituted biphosphines is shifted toward longer wavelength as compared to the unsubstituted biphosphine H_2P-PH_2. Thus an absorption band due to the P-P bond in Et_2P-PEt_2 and Bu_2P-PBu_2 was observed at 424 and 419 cm^{-1}, respectively,[313,546] whereas that in H_2P-PH_2 was found at 437 cm^{-1}. Evidence for extensive electron delocalization comes from the intensive UV spectra of biphosphines. For example, $(CF_3)_2P$-$P(CF_3)_2$ shows an intensive band at 2160 Å.[38] Replacement of the CF_3 by the CH_3 group moves the maximum to higher wavelength.[228] The intensity of the band is simultaneously diminished, indicating that there is less delocalization of the lone electron pairs in Me_4P_2 than in $(CF_3)_4P_2$.

The P-P bond energy in Et_4P_2 has recently been found by electron impact studies to be 86 kcal/mole[233] and is thus much higher than in elemental phosphorus (44 to 48 kcal/mole) or $H_2P\text{-}PH_2$ (61.2 kcal/mole). It is therefore understandable that organosubstituted biphosphines are thermally much more stable than $H_2P\text{-}PH_2$. While $H_2P\text{-}PH_2$ decomposes rapidly above 0°, organosubstituted biphosphines are stable up to 300°. Slow disproportionation takes place above this temperature,[66,360,391] giving a tertiary phosphine, a cyclopolyphosphine, or red phosphorus.

$$3Ph_2P\text{-}PPh_2 \longrightarrow 4Ph_3P + 2P$$

The unsymmetrical biphosphines of type $R^1_2P\text{-}PR_2$ are noticeably less thermally stable.[228,298,560] In these cases decomposition is probably facilitated by the formation of intermolecular P-P bonds between the relatively basic and acidic phosphorus atoms. A similar argument may account for the low thermal stability of $(Me_2N)_4P_2$.[449]

Unsymmetrical substituted biphosphines of the type $R^1RP\text{-}PRR^1$ have been obtained in stereoisomeric forms.[397] They show in the ^{31}P NMR spectra two separate signals due to the meso and dl forms.[187,397]

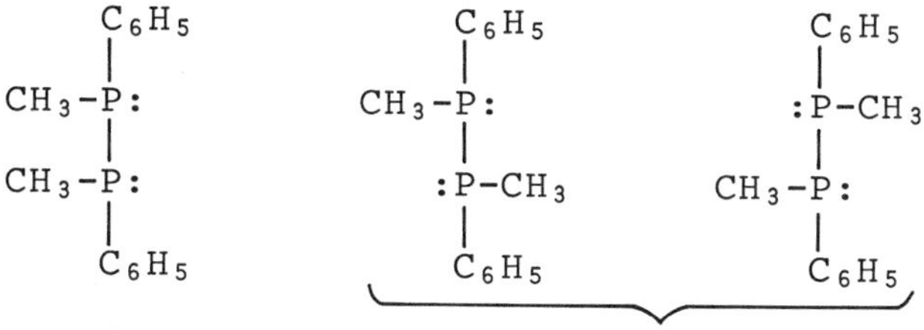

Meso form d,l forms

The Arrhenius activation energy attributable to an interconversion between the meso and the dl forms in 1,2-dimethyl-1,2-diphenylbiphosphine through inversion about phosphorus has been found by ^1H NMR to be 26 ± 2 kcal/mole,[372] (23.6 kcal/mole[372a]). Other unsymmetrically substituted biphosphines, MeRP-PRMe (R = $PhCH_2$, $G^\ddagger$ 24 kcal/mole; R = $4\text{-}MeC_6H_4$, $\Delta G^\ddagger$ 22.5 kcal/mole; R = $4\text{-}CF_3C_6H_4$, $\Delta G^\ddagger$ 22.6 kcal/mole), show a similar barrier to inversion.[372a] The lower barrier for inversion of biphosphines as compared to monophosphines (see Chapter 1, $\Delta G^\ddagger \geq 30$ kcal/mole) has been attributed to stabilization of the transition state through $p\pi\text{-}d\pi$ bonding between the phosphorus atoms.[372,372a]

The coupling constant between directly bonded phosphorus atoms varies with valence state and substitutents similar to other coupling constants. Thus J_{PP} is -108.2 Hz for P_2H_4,[381] between -179.7 and -396 Hz for organosubstituted biphosphines,[174,187,188] between -220 and -243 Hz for tetraalkylbiphosphine monosulfides,[239,397] and between 18.7

and 69 Hz for tetraalkylbiphosphine disulfides.[187,239] The coupling constants may be positive or negative. The variations in J_{PP} can be explained by the dependence of the resonance integral between the outer-shell s electrons of the two nuclei on the electronegativity and bulk of the substituents.[175]

E.2. Organosubstituted Biphosphine Disulfides

E.2.1. Methods of Preparation

I. FROM $P(S)Cl_3$, $RP(S)Cl_2$, $R_2P(S)Cl$, AND GRIGNARD REAGENTS

In 1949, Kabachnik and Shepeleva[341] found that the reaction of MeMgI with $P(S)Cl_3$ gives tetramethylbiphosphine disulfide rather than the expected tertiary phosphine sulfide as reported previously.[551] Since then several biphosphine disulfides have been prepared by what has been called the "anomalous" Grignard reaction. Several generalizations may be made concerning either biphosphine disulfide or tertiary phosphine sulfide formation (see also Chapter 7).[41,121,398,403,440] With thiophosphoryl chloride, biphosphine disulfide formation is restricted to

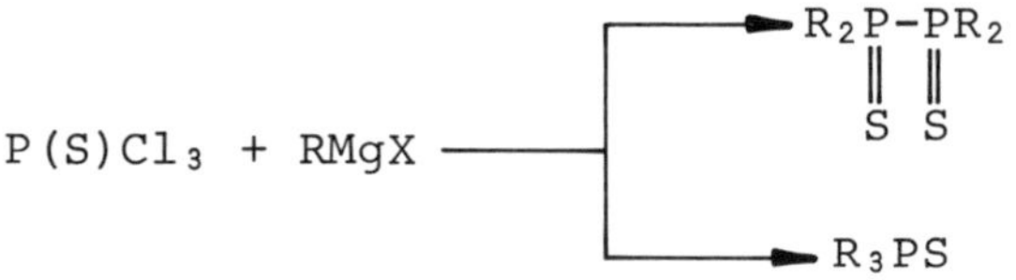

reactions with lower alkylmagnesium halides, RMgX (R = Me, Et, CH_2=$CHCH_2$, Bu, Am)[110,318,362,440] and α,ω-alkylene bis-(magnesium bromide),[495] while acetylenic, vinyl, styryl, hindered alkyl, long-chain aliphatic, and aromatic Grignard reagents produce tertiary phosphine sulfides (see Chapter 7).

In certain cases, such as benzyl[318] and amyl,[362] the biphosphine disulfide has been isolated by one group but not by others, who obtained tertiary phosphine sulfides or less substituted products (benzyl,[133,440] amyl[110]). The yield of biphosphine disulfide depends on several factors,[110,318,362,440] such as the nature of the Grignard, the phosphorus thiotrihalide ($P(S)Cl_3$ or $P(S)Br_3$), the solvent, the rate of reaction, and the reaction temperature. Optimum yields appear to be obtained with a 1:3.2 ratio of $P(S)Cl_3$ to RMgBr in ether or THF at the 5 to 20° range.[110,318,362,398,440] Under the same conditions $P(S)Br_3$ has been said to give mainly tertiary phosphine sulfides.[440] Also, RMgCl and higher temperatures (∿70°) favor tertiary

phosphine sulfide formation,[440] while Grignard reagents
prepared from isopropyl, s-butyl, t-butyl, and cyclohexyl
halides and temperatures below zero seem to produce less
substituted products.[110,296] It is interesting to note
that biphosphine disulfides were also obtained from the
reaction of $(EtO)_2P(S)H$ with BuMgBr or PrMgBr and sulfur.[420]

   In reactions involving phosphonothioic dihalides, some
of the above restrictions no longer apply. Phenyl groups
may be present either in the phosphonothioic dihalide or
in the Grignard reagent since 1,2-dimethyl-1,2-diphenyl-
biphosphine disulfide is produced in both reactions shown
below.[394,400] From these reactions racemic and meso forms
have been isolated by Maier.[394] Biphosphine disulfides
were also obtained from $MeP(S)Br_2$ and alkyl-,[394] ben-

$PhP(S)Cl_2$ + MeMgBr

$MeP(S)Br_2$ + PhMgBr

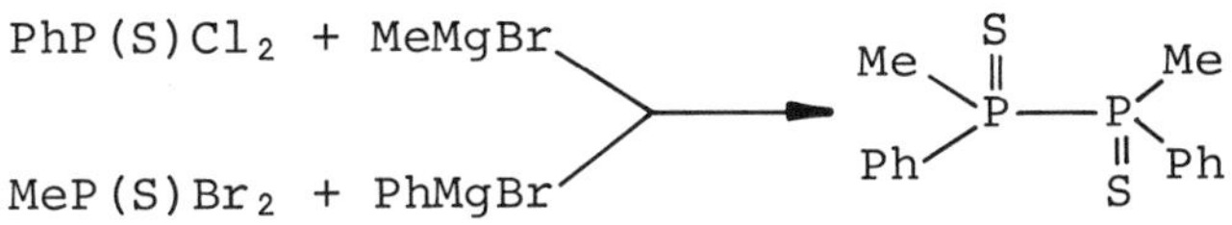

zyl-,[372a,394] substituted aryl-,[372a] or styrylmagnesium
bromide,[53] and from $PhP(S)Cl_2$ and ethyl-[469] or benzylmag-
nesium halide.[134] However, formation of tertiary phosphine sulfides from $PhP(S)Cl_2$ and tolyl-,[427] alkynyl-,[51]
or styrylmagnesium halide,[53] and from $RP(S)Cl_2$ and cyclo-
hexyl-,[394] or alkynylmagnesium halide,[50,52] parallels the
reactions of $P(S)Cl_3$. Depending upon the substituent on
the Grignard reagent, phosphinothioic halides containing
an aryl group gave both biphosphine disulfides and tertiary phosphine sulfides,[464] but dialkylphosphinothioic
halides produced in the same reaction only tertiary phosphine sulfides.[133,241,464]

$$Ph_2P(S)Cl \begin{cases} \xrightarrow{PhMgBr} Ph_3PS \\ \\ \xrightarrow{EtMgBr} \underset{\|}{\overset{S}{Ph_2P}}-\underset{\|}{\overset{S}{PPh_2}} \end{cases}$$

   A report claiming the formation of biphosphine disulfides in the reaction of $Et_2P(S)Cl$ with EtMgBr and of
$Pr_2P(S)Cl$ with PrMgBr[48] could not be confirmed.[464] It is
worthwhile to mention that MePhP(S)Cl gave with EtMgBr or
PhMgBr only meso-1,2-dimethyl-1,2-diphenylbiphosphine
disulfide.[464] Steric interference in the transition state
has been invoked to explain this result.[464]

   Two mechanisms have been proposed that could account
for the products observed. While Crofts and Fox[133] favor
a "phosphinidene sulfide" intermediate which then inserts
into a phosphorus-halogen bond to produce a compound with

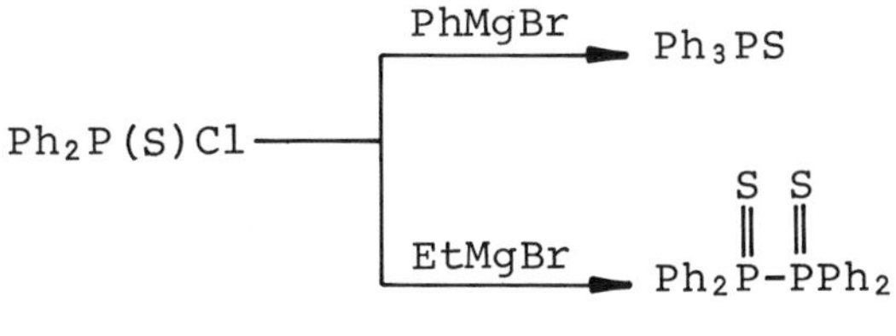

a P-P bond, Patel and Harwood[464] suggest a thiophosphoryl Grignard reagent, arising from a metal-halogen exchange process, as an intermediate which then couples with a P-X bond to give biphosphine disulfides.

$$RP(S)Cl_2 + RMgX \longrightarrow RCl + MgXCl + [RP=S \xrightarrow{>P(S)Cl}] \underset{R}{\overset{\displaystyle S \ \ S}{\underset{}{>P-P}}}\overset{Cl}{\underset{R}{}}$$

$$>P(S)Cl + RMgX \longrightarrow RCl + \overset{S}{\overset{\|}{>PMgX}} \xrightarrow{+ClP<} MgXCl + \overset{S \ \ S}{\overset{\| \ \|}{P-P}}$$

$$\longrightarrow \overset{S}{\overset{\|}{>PR}} + MgXCl; \text{ nucleophilic displacement}$$
leads to tertiary phosphines

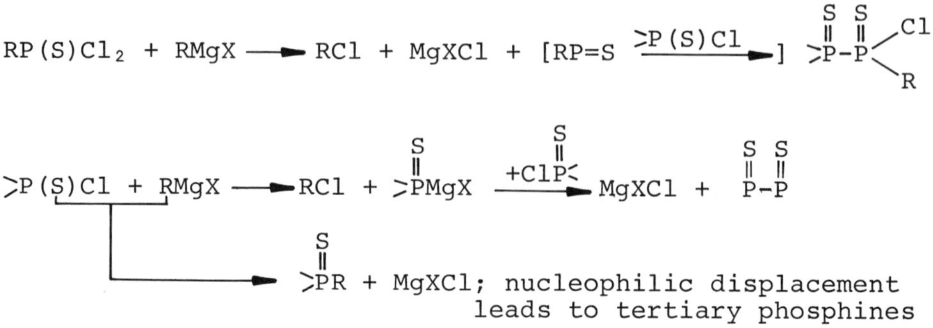

There is some indication that a metal-halogen exchange process precedes the coupling step since several phosphino-thioic halides were noted to yield biphosphine disulfides when treated with Grignard reagents.[464]

II. FROM BIPHOSPHINES AND SULFUR

When biphosphine disulfides are not available from the "anomalous" Grignard reaction, they are usually prepared by the addition of 2 equivalents of sulfur to a biphos-phine. Since this reaction is highly exothermic, it is usually contacted in an inert organic solvent.[55,296,298,313,360,449] Only one case is known in which this reaction

$$R_2P-PR_2 + 2S \longrightarrow R_2P(S)-P(S)R_2$$

failed to give a biphosphine disulfide. When tetra-t-butylbiphosphine was treated with sulfur, insertion into the P-P bond occurred and a thiophosphinic anhydride was produced.[291]

III. BY MISCELLANEOUS METHODS

Biphosphine disulfides have also been obtained in the reaction of $R_2P(S)H$ with $R_2P(S)Cl$,[440] and in low yield in a Wurtz-type coupling of $Ph_2P(S)Cl$[422] or $Et_2P(S)Cl$ with sodium.[440] Attempts to synthesize unsymmetrical biphos-phine monosulfides by the interaction of $Me_4P_2S_2$ and Ph_2PCl gave instead $[Ph_2P]_2$, $[Ph_2P(S)]_2$, Me_2PCl, and $Me_2P(S)Cl$[395] (disproportionation and sulfurization prod-ucts). Reaction of some transition metal phosphides such as $(R_2P)_3V$ or $(R_2P)_2Ti$ with sulfur also produced biphosphine disulfides.[331]

E.2.2. General Chemistry

Under certain circumstances it is possible for the P-P and P=S bonds of biphosphine disulfides to undergo reactions independently. It is more unusual, however, for the P-P and P=S bonds to react simultaneously.

(1) Reduction
Mention has already been made of the desulfurization of biphosphine disulfides with certain metals[321,362,393,440] and tertiary phosphines[393] without P-P bond cleavage (see Section E.1.1.III). Other reducing agents such as $LiAlH_4$ [318,393,396,400,417,459] and hydrogenation with Raney copper[438,440,441] lead directly to secondary phosphines, while reduction with metal hydrides[440] results in mixtures of biphosphines and secondary phosphines. The dry reaction of $Me_4P_2S_2$ with $NaBH_4$ or $LiBH_4$ at 250° gave a useful yield of phosphinoborane, $(H_2BPMe_2)_{3,4}$.[44,352]

(2) Halogenation
Halogenation with theoretical quantities of Cl_2 or Br_2,[118,134,241,362,393,395,400,464,469] PCl_5,[362] SO_2Cl_2,[395,489] Hg_2Cl_2,[395] or SCl_2[395] are examples of P-P bond cleavage without attack on the P-S bonds. An excess of Cl_2[362,477] or Br_2[362] causes halogenation of both the P-P and P=S bonds, as does fluorination with SbF_3 or AsF_3.[495,497] Under more

$$R_2\overset{\overset{S}{\|}}{P}-\overset{\overset{S}{\|}}{P}R_2 + X_2 \longrightarrow 2R_2P(S)X$$

$$R_2P(S)-P(S)R_2 + 5Cl_2 \longrightarrow 2R_2PCl_3 + 2SCl_2$$

vigorous conditions the C-H bonds are also chlorinated.[477]

$$Me_4P_2S_2 + 17Cl_2 \longrightarrow 2(CCl_3)_2PCl_3 + 2SCl_2 + 12HCl$$

A sulfur-oxygen exchange plus P-P bond cleavage results from the reaction of $SOCl_2$[393,396,469] or excess SO_2Cl_2[177] with biphosphine disulfides. It has been shown by ^{31}P NMR spectroscopy[396] that P-P bond cleavage precedes sulfur-oxygen exchange in the reaction with $SOCl_2$. Chlorination

$$R_2P(S)-P(S)R_2 + SOCl_2 \xrightarrow[-SO]{} 2R_2P(S)Cl \xrightarrow{2SOCl_2} 2R_2P(O)Cl + S_x + S_2Cl_2$$

plus desulfurization of the P-P bond takes place when $Me_4P_2S_2$

is heated with Ph_2PCl[395] or $PhPCl_2$.[455]

$$Me_2P(S)\text{-}P(S)Me_2 + PhPCl_2 \longrightarrow 2Me_2PCl + (PhPS_2)_x$$

(x probably 2)

(3) Oxidation
Biphosphine disulfides are readily oxidized to phosphinic acids by a variety of oxidizing agents such as HNO_3,[341,355] H_2O_2,[355,362,393] organic peroxides,[393] HgO,[362,368,393] and PbO.[368] Mercury oxide oxidation produced the phosphinic anhydride as intermediate, which was isolated under anhydrous conditions,[368] and peroxybenzoic acid yields as an

$$R_2P(S)\text{-}P(S)R_2 + 3HgO \longrightarrow R_2P(O)OP(O)R_2 \xrightarrow{H_2O} 2R_2P(O)OH$$

intermediate the insertion product of oxygen into the P-P bond,[274] which reacted further to give finally a phosphinic acid. Although biphosphine disulfides have been found to

$$Et_2P(S)\text{-}P(S)Et_2 + [O] \longrightarrow Et_2P(S)OP(S)Et_2$$

be unreactive toward sulfur,[368] ready cleavage of the P-P bond takes place when they are heated with metals (Zn, Cd) or metal sulfides (ZnS, CdS, Na_2S) and sulfur to temperatures ranging from 130 to 180°.[365-367,369] This procedure gives the salts of dialkylphosphinodithioic acids in high yield.

$$R_2P(S)\text{-}P(S)R_2 + M + 2S \longrightarrow (R_2PS_2)_2M \quad (M = Zn, Cd)$$

$$R_2P(S)\text{-}P(S)R_2 + Na_2S + S \longrightarrow 2R_2PS_2Na$$

(4) Hydrolysis
Hydrolysis of biphosphine disulfides with aqueous alkaline (∿10%) produces equimolar amounts of phosphinothioic acid and secondary phosphine sulfide.[119] This cleavage reaction seems to be limited to aliphatic substituted biphosphine disulfides, since it was reported previously that $Ph_4P_2S_2$

$$R_2P(S)\text{-}P(S)R_2 + NaOH \longrightarrow R_2P(S)ONa + R_2P(S)H$$

does not react with 10% NaOH solution even on boiling.[360]

(5) Ethylene Insertion
Another example in which the P=S bonds in biphosphine disulfides remain intact is the ethylene insertion reaction which takes place at ∿270° in the presence of iodine as a catalyst.[456,495]

$$R_2P(S)\text{-}P(S)R_2 + CH_2{=}CH_2 \longrightarrow R_2P(S)CH_2CH_2P(S)R_2$$

(6) Miscellaneous Reactions
Similar to tertiary phosphine sulfides, biphosphine disulfides act as complexing ligands for some metal halides.[421,553] In a redistribution-type reaction, biphosphine disulfides interact with cacodyl to give a phosphorus-arsenic compound.[239]

E.2.3. Application of Biphosphine Disulfides

Biphosphine disulfides have been claimed to be useful as insecticides and additives in lubricating oil and gasoline.[400,422] It is interesting to note that the high-melting form of MeEtP(S)-P(S)EtMe showed a higher insecticidal activity than the low-melting form.[400]

E.2.4. General Physical Properties

As a class, biphosphine disulfides are thermally very stable compounds. While unsymmetrical biphosphines of the type $R_2P\text{-}PR^1_2$ are unstable, the corresponding disulfides show no tendency to disproportionate.[298]

We[393,394,400] and others[300,469,535] have found that several asymmetric biphosphine disulfides containing two asymmetric phosphorus atoms can be separated into optically inactive meso (20) and racemic forms (21). It has been

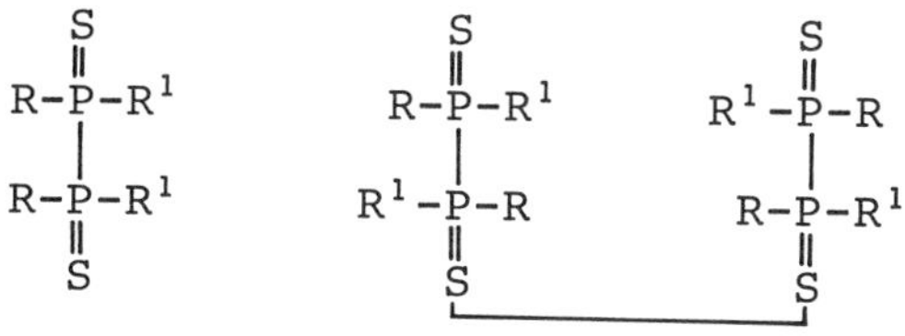

(20) (21)

shown by x-ray structural analysis[573] that the high-melting form (B) of 1,2-dimethyl-1,2-diphenylbiphosphine disulfide is meso (20, R = Me, R^1 = Ph) and the low-melting form (A) is probably racemic (21, R = Me, R^1 = Ph). X-ray crystallographic studies have furthermore shown that $Me_4P_2S_2$,[465] $[(CH_2)_4P(S)]_2$,[375] $[(CH_2)_5P(S)]_2$,[375a] $Et_4P_2S_2$,[160] and the meso form of $Me_2Ph_2P_2S_2$[573] exist in the trans conformation in the solid state. The same conformation has been suggested for other biphosphine disulfides not only in the solid state but also in solution on the basis of their dipole moments[220] and IR[110,130,131,227,461] and Raman spectra.[110,130,131,220,227,461] Depending on the substituents, the P=S antisymmetric stretching vibration was found in the 540 to 640 cm^{-1} range and the P-P stretching mode in the 440 to 510 cm^{-1} range.

The ^{31}P chemical shifts of biphosphine disulfides are found in the same range as those of tertiary phosphine sulfides (see Chapter 7). In fact, the close relationship is remarkable, for example, Me_3PS -30.9 ppm, $Me_4P_2S_2$ -34.7 ppm; Et_3PS -51.9 ppm; $Et_4P_2S_2$ -49.4 ppm.

The half-wave reduction potentials of biphosphine disulfides follow the Taft σ^* constants, with the exception of $[i\text{-}Pr_2P(S)]_2$ and $[(c\text{-}C_6H_{11})_2P(S)]_2$.[420a]

E.3. Organosubstituted Biphosphine Monosulfides

Biphosphine monosulfides can be prepared by removal of one sulfur atom from a biphosphine disulfide with tertiary phosphines[397] by heating a 1:1 mixture of $R_4P_2S_2$ and R_4P_2,[239,397a] or by addition of 1 equiv of sulfur to a biphosphine.[397a]

$$R_2P(S)\text{-}P(S)R_2 + Bu_3P \longrightarrow R_2P(S)\text{-}PR_2 + Bu_3PS$$

Similar to biphosphine disulfides, the P-P bond in biphosphine monosulfides is cleaved by halogens.[397] Because of the trivalent phosphorus atom, biphosphine monosulfides are sensitive to oxidation. The primary oxidation product $R_2P(O)\text{-}P(S)R_2$, however, could not be isolated.[397] The structure of biphosphine monosulfides was established by ^1H[239] and ^{31}P NMR[397] spectroscopy. In contrast to the trifluoromethyl-substituted compound, $(CF_3)_2\text{-}PSP(CF_3)_2$,[70,105] aliphatic and aromatic substituted biphosphine monosulfides have the unsymmetrical structure (22).[397]

$$R_2\overset{\overset{\displaystyle S}{\|}}{P}\text{—}PR_2$$

(22)

E.4. Organosubstituted Biphosphine Monoxides

Biphosphine monoxides were obtained in the reaction of phosphinous halides with water in the presence of triethylamine as hydrogen halide acceptor.[383] That this reaction probably proceeds through the initial formation of secondary phosphine oxides[383] was demonstrated by the finding

$$Ph_2PCl + H_2O \xrightarrow{\text{base}} Ph_2\overset{\overset{\displaystyle O}{\|}}{P}H \xrightarrow{+\ Ph_2PCl/\text{base}} Ph_2\overset{\overset{\displaystyle O}{\|}}{P}\text{-}PPh_2$$

that $Bu_2P(O)H$ interacts with Bu_2PCl in the presence of a tertiary amine to give the monoxide $Bu_2P(O)\text{-}PBu_2$.[325,326a] A variation of this reaction consists in the formation of biphosphine monoxides from aliphatic substituted secondary phosphine oxides and $SiCl_4$[325] or $Ph_2P(O)Cl$[473] in the presence of a base. Apparently, part of the secondary phosphine oxide is converted by $SiCl_4$ or $Ph_2P(O)Cl$ to phosphinous chloride. The aromatic substituted secondary phosphine oxide $Ph_2P(O)H$, however, reacts differently with

$$4R_2P(O)H + SiCl_4 \xrightarrow{NEt_3} 2R_2P\text{-}P(O)R_2 + 4HCl + SiO_2$$

$$2(PhCH_2)_2P(O)H + 2Ph_2P(O)Cl \xrightarrow{2R_3N} (PhCH_2)_2P\text{-}P(O)(CH_2Ph)_2 + [Ph_2P(O)]_2O + 2R_3NHCl$$

$Ph_2P(O)Cl$, giving diphenylphosphinic diphenylphosphinous anhydride, $Ph_2POP(O)Ph_2$.[473] Useful yields of biphosphine monoxide were also obtained in the partial oxidation of Ph_4P_2 with air in benzene.[443]

Application of the Arbuzov reaction seems to be a generally applicable method for the synthesis of biphosphine monoxides. Interaction of phosphonous or phosphinous chlorides with trivalent phosphorus esters in refluxing benzene produces biphosphine monoxides in high yield.[183]

$$Ph_2PCl + ROPPh_2 \longrightarrow Ph_2P\text{-}P(O)Ph_2 + RCl$$

$$PhPCl_2 + (RO)_3P \longrightarrow PhClP\text{-}P(O)(OR)_2 + RCl$$

The observation[471] that the residue of a Ph_2POEt preparation from Ph_2PCl and EtOH contained $Ph_2P(O)\text{-}P(O)Ph_2$ also points to an Arbuzov-type reaction. The initially formed biphosphine monoxide (crude, m. 147 to 150°[471]) was probably oxidized to a biphosphine dioxide during the course of purification.

The remaining lone electron pair makes biphosphine monoxides sensitive to oxidation either by oxygen[325,383] or sulfur,[383] and, as in other biphosphine derivatives,

$$Ph_2P\text{-}P(O)Ph_2 + S \longrightarrow Ph_2P(S)\text{-}P(O)Ph_2$$

the P-P bond can be cleaved by halogens.[383] Attempts to prepare quaternary salts of $Ph_2P\text{-}P(O)Ph_2$, however, were not successful.[383] It is not yet clear whether reduction of $Ph_2P\text{-}P(O)Ph_2$ with $LiAlH_4$ led to $Ph_2P\text{-}PPh_2$.[383] The structure of biphosphine monoxides was established by ^{31}P NMR[183,325] spectroscopy. In contrast to the trifluoromethyl-substituted compound $(CF_3)_2POP(CF_3)_2$,[232] aliphatic[325] and aromatic substituted biphosphine monoxides have the unsymmetrical structure (<u>23</u>).

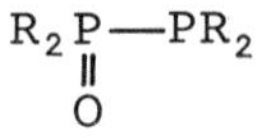

(23)

E.5. Organosubstituted Biphosphine Dioxides

Biphosphine dioxides are generally prepared by the controlled oxidation of biphosphines[283,360,449,543] or biphosphine monoxides.[325,383] Tetramethylbiphosphine dioxide was isolated in the reaction of $Me_2P-P(CF_3)_2$ with $(CF_3)_2POH$ and from Me_2PH and $(CF_3)_2POP(CF_3)_2$.[232] Several aryl-substituted biphosphine dioxides have been prepared by exposure to air of an ethereal, THF, or benzene solution containing equimolar amounts of a diarylphosphinous chloride and a tertiary amine in the presence of small quantities of water.[473]

$$2Ph_2PCl + 2R_3N + H_2O + 1/2O_2 \longrightarrow Ph_2P(O)-P(O)Ph_2 + 2R_3NHCl$$

The mechanism of this reaction is not clear, but it seems that here also a secondary phosphine oxide is formed initially. Finally, biphosphine dioxides were also obtained from the reaction of $R_2P(O)Cl$ with $LiSiR_3$.[325] A metal-

$$2R_2P(O)Cl + 2LiSiR_3 \longrightarrow R_2P(O)-P(O)R_2 + R_3SiSiR_3 + 2LiCl$$

halogen exchange reaction could account for the observed products. Related to this reaction is a Wurtz-type coupling of $Ph_2P(O)Cl$ with sodium/naphthalene in THF which gave $Ph_4P_2O_2$.[422] The same biphosphine dioxide was formed as a by-product in the reduction of the phosphinate $Ph_2P(O)OEt$ with $LiAlH_4$.[412] Probably, the biphosphine is formed initially and then oxidized during work-up.[412]

Biphosphine dioxides are readily cleaved in alkaline solution.[360] Cleavage of $Ph_2P(O)-P(O)Ph_2$ in a neutral medium allowed recovery of $Ph_2P(O)H$ which in an alkaline solution is rapidly disproportionated.[471] Oxidation of $Ph_4P_2O_2$

$$Ph_2P(O)-P(O)Ph_2 + H_2O \longrightarrow Ph_2P(O)H + Ph_2P(O)OH$$

with peroxybenzoic acid produced diphenylphosphinic anhydride, $Ph_2P(O)OP(O)Ph_2$.[274]

E.6. Miscellaneous Compounds Containing a P-P Bond

One of the simplest methods for producing compounds that contain a P-P bond probably consists in the interaction of tertiary phosphines with phosphonous or phosphinous halides.[531,542]

$R_3P + R_2PX \longrightarrow [R_3P-PR_2]X \qquad (X = Cl, Br, I)$

$Me_3P=PCF_3 + MeI \longrightarrow [Me_3P-PMeCF_3]I$

The same type of phosphonium salts are obtained in the reaction of the "ylid" $Me_3P=PCF_3$ with alkyl iodides[75] and in the quaternization of certain biphosphines with alkyl halides (see Section E.1.2). Treatment of dithiophosphonic anhydrides with equimolar amounts of Bu_3P results in the formation of phosphonium salts with a "zwitterion" structure.[181]

$$RP{\overset{\displaystyle S}{\underset{\displaystyle S}{\diagup}}}{\overset{\displaystyle S}{\underset{\displaystyle S}{\diagdown}}}PR + 2Bu_3P \longrightarrow 2R\overset{\displaystyle S}{\underset{\displaystyle S^{\ominus}}{\overset{\displaystyle \|}{P}}}{}^{\oplus}-PBu_3$$

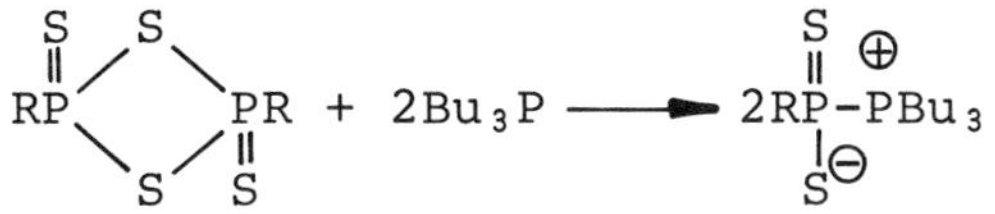

Reduction of the same anhydride with potassium in $(Me_2N)_3PO$ yields a dimercaptosubstituted biphosphine disulfide,[182] while phenylphosphonic anhydride gives with

$[PhPS_2]_2 + 2K \longrightarrow (KS)PhP(S)-P(S)Ph(SK)$

$[PhPO_2]_2 + 2K \longrightarrow (KO)PhP(O)-P(O)Ph(OK)$

potassium in liquid ammonia a dihydroxy-substituted biphosphine dioxide.[182]

Complex compounds containing P-P bonds were isolated from the reactions of hexamethyldisilazane and Ph_2PCl,[443] or from $[Ph_3PNHSiMe_3]Br$ and Ph_2PCl.[416] The proposal that bis(diphenylphosphino)amine, the primary reaction product, reacts with excess Ph_2PCl at the more nucleophilic phosphorus atoms and not at the NH group seems to account for the product isolated.[443]

$(Me_3Si)_2NH + 2Ph_2PCl \longrightarrow 2Me_3SiCl + Ph_2PNHPPh_2$

$+ Ph_2PCl \quad | \quad + 2Ph_2PCl$

$[Ph_2P-PPh_2-NH-PPh_2]Cl \qquad [Ph_2P-PPh_2=N=PPh_2-PPh_2]Cl$

deprotonation

with Et_3N

$Ph_2P-PPh_2=N-PPh_2$

E.7. Organosubstituted Triphosphines, Tetraphosphines, and Corresponding Oxides and Sulfides

E.7.1. Methods of Preparation

In general, the same type of reactions that led to biphosphines have also been successfully applied to the synthesis of triphosphines and tetraphosphines.

I. FROM PRIMARY OR SECONDARY PHOSPHINES AND PHOSPHONOUS OR PHOSPHINOUS DERIVATIVES

A general approach that leads to the formation of fully substituted triphosphines is the reaction of a monofunctional with a difunctional reagent in a 2:1 ratio. These reagents can be a primary phosphine and a phosphinous halide,[74,83] using a tertiary amine as hydrogen halide acceptor, a phosphonous dihalide and a secondary phosphine,[559,574] or a phosphonous diamide and a secondary

$$MePH_2 + 2(CF_3)_2PI + 2R_3N \longrightarrow MeP[P(CF_3)_2]_2 + 2R_3N{\cdot}HI$$

phosphine.[404,405] The analogous reaction of Ph_2PCl with

$$PhPBr_2 + 2Ph_2PH + 2R_3N \longrightarrow PhP[PPh_2]_2 + 2R_3N{\cdot}HBr$$

$PhPH_2$ failed to give a triphosphine. Because of the high reaction temperature, only the disproportionation product, that is, a biphosphine and a cyclopolyphosphine could be isolated.[360] In one case the phosphine was not added as

$$MeP(NMe_2)_2 + 2HPPh_2 \longrightarrow MeP[PPh_2]_2 + 2Me_2NH$$

such but was produced in situ by reduction of part of the $PhPBr_2$ with LiH, which also functions as the HBr acceptor.[574]

II. FROM ALKALI PHOSPHIDES AND PHOSPHONOUS OR PHOSPHINOUS HALIDES

The synthesis of triphosphines by this method was less successful than in the case of biphosphines.[574] This is mainly due to the fact that fully aliphatic substituted triphosphines have a great tendency to disproportionate.

$$EtPCl_2 + 2LiPEt_2 \longrightarrow EtP[PEt_2]_2 + 2LiCl$$

The method failed completely for the synthesis of branched tetraphosphines. Only the disproportionation products, biphosphine and elemental phosphorus, could be

$$2PX_3 + 6MPR_2 \xrightarrow[-6MX]{} 2[P(PR_2)_3] \longrightarrow 2P + 3R_2P\text{-}PR_2$$

isolated.[574]

III. FROM BIPHOSPHINES OR CYCLOPOLYPHOSPHINES

Triphosphines have also been obtained by coupling of a monofunctional biphosphine with a phosphinous halide,[74] by a disproportionation reaction of $(Me_2N)_2P\text{-}P(NMe_2)_2$[449]

$$(CF_3)_2P\text{-}PHMe + (CF_3)_2PI + R_3N \longrightarrow MeP[P(CF_3)_2]_2 + R_3N{\cdot}HI$$

and by an insertion reaction of "trifluoromethyl phosphinidene," CF_3P, into the P-P bond of tetramethylbiphosphine.[121a]

$$4Me_2P\text{-}PMe_2 + (CF_3P)_4 \longrightarrow 4CF_3P[PMe_2]_2$$

The first triphosphine, $H(PCF_3)_3H$,[84,392] and tetraphosphine, $H(PCF_3)_4H$,[84] to be isolated were two of the hydrolysis products of $(CF_3P)_{4/5}$. That Burg's pseudohalogen concept for biphosphines is also valid for cyclopolyphosphines was demonstrated in the reaction of $(CF_3P)_{4/5}$ with $CF_3C{\equiv}CCF_3$, which gave a heterocyclic biphosphine (16) and a heterocyclic triphosphine (18).[388]

Another heterocyclic triphosphine, 1,2,3-triphenyl-1,2,3-triphosphaindane (24), in addition to a phosphanthren (25), was obtained in the interaction of Li_2PPh and o-bromoiodobenzene or o-diiodobenzene.[415] The reaction was formulated as involving addition of "phenylphosphinidene," PhP, or $(PhP)_5$ to benzyne.[415]

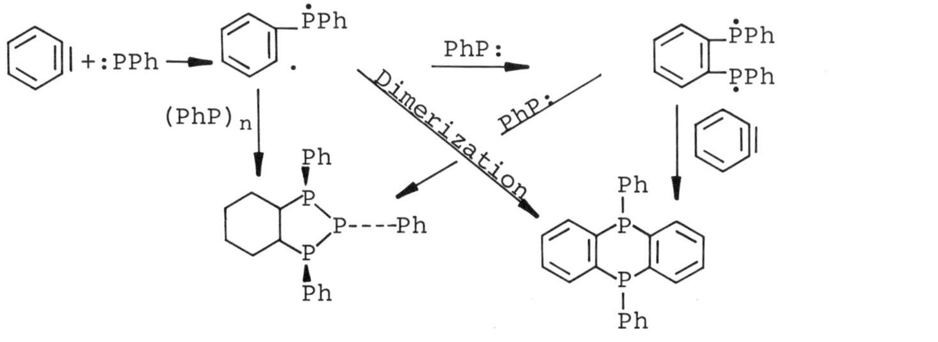

(24) (25)

IV. BY MISCELLANEOUS METHODS

The Arbuzov reaction of phosphonous dichlorides with

trivalent phosphorus esters in refluxing benzene seems to be a general method for preparing compounds that contain two P-P bonds. In these compounds the phosphorus atoms are in different valence states.[183] The phosphorus esters need not be

$$PhPCl_2 + 2P(OR)_3 \longrightarrow (RO)_2\overset{O}{\overset{\|}{P}}\text{-}PPh\text{-}\overset{O}{\overset{\|}{P}}(OR)_2 + 2RCl$$

prepared separately since a compound with two P-P bonds is also obtained in the interaction of $PhPCl_2$ and alcohols in the presence of pyridine.[183] If the reaction of $RPCl_2$ with $RP(OR)_2$ is carried out at lower temperature (0 to 5°), redistribution products [RP(OR)Cl] are formed (see Chapter 10).

The only branched tetraphosphine known was obtained in the reaction of tris(trimethylstannyl)phosphine with diphenylphosphinous chloride.[516]

$$(Me_3Sn)_3P + 3ClPPh_2 \longrightarrow 3Me_3SnCl + P[PPh_2]_3$$

Other attempts to prepare open-chain or branched polyphosphines involving more than three phosphorus atoms have been frustrated by disproportionation.[299,300,574] However, some alkali phosphides of tetraphosphines and pentaphosphines are known (see Section A).

E.7.2. Chemical and Physical Properties

Triphosphines[74,121a,388] show UV absorption in the same region as diphosphines,[38,228] suggesting that there is some delocalization of the lone-pair electrons by interaction with adjacent phosphorus 3d orbitals. Support for this concept comes from the observation that the heterocyclic bi- and triphosphines $C_2P_2(CF_3)_4$ (16) and $C_2P_3(CF_3)_5$ (18) absorb UV light at longer wavelength than the corresponding open-chain di- and triphosphines.[388] This suggests that the orbitals of the polyphosphine portion of the ring overlap with the carbon-carbon π orbitals.

Chemical evidence for this suggestion includes the observations that only the terminal atoms of $PhP[PPh_2]_2$ are quaternized with HBr,[574] and that only a disulfide of probably structure $MeP[P(S)Ph_2]_2$ is obtained from $MeP[PPh_2]_2$ and sulfur.[404] The cyclic triphosphine 1,2,3-triphenyl-1,2,3-triphosphaindene (24) gave two isomeric disulfides in which the sulfur atoms were assigned in one form the cis-1,3- (26), and in the other the trans-1,2-position (27).[415] It was postulated that the failure to form a trisulfide is caused by steric hinderance, for the 1,2,3-triethyl-1,2,3-triphosphaindane readily forms a crystalline trisulfide.[415] X-ray analysis confirms the structure of 1,2,3-triphenyl-1,2,3-triphosphaindane and

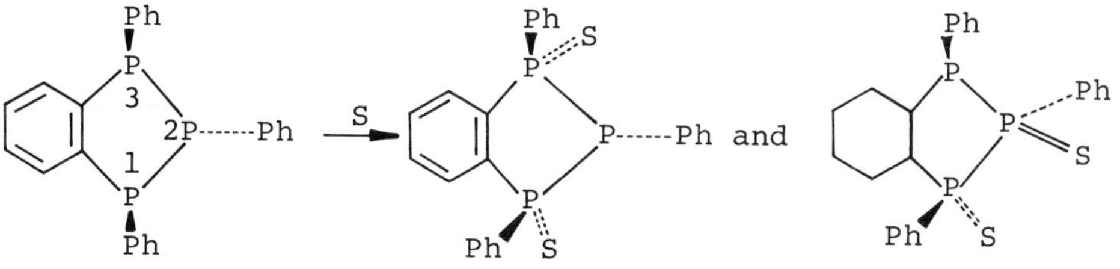

(24) (26) (27)

shows that the bicyclic phosphaindane nucleus is roughly planar. The phenyl group in the 2-position is trans to those in the 1- and 3-positions and also lies in the molecular mirror plane.[114] As in biphosphines, the P-P bonds in triphosphines are readily cleaved by oxidizing agents such as HNO_3, H_2SO_4,[574] and halogens.[404] Iodine, however, converts the heterocyclic triphosphine $P_3C_2(CF_3)_5$ (18) to the biphosphine $P_2C_2(CF_3)_4$ (16) which is resistant to further action of iodine.[388]

E.8. Cyclopolyphosphines

"Phosphobenzol," the first cyclopolyphosphine, was prepared in 1877 by Köhler and Michaelis.[351] Then this field lay dormant until 1952 when the second report appeared on this subject.[571] However, intensive investigation in this area started only in 1957 with the appearance of three papers by Ressor and Wright,[476] Mahler and Burg,[388,389] and Kuchen and Buchwald.[359] Since then more than 100 publications have appeared on this subject, among them several reviews dealing with different aspects of cyclopolyphosphine chemistry.[27,67,121,271,272,398,453,479,574] In 1967, two reviews appeared, one by Cowley and Pinel[124] and the other by Maier,[406] which covered the subject completely up to 1966. We therefore keep this chapter rather short. In doing so we refer frequently to the two reviews mentioned.

E.8.1. Methods of Preparation

I. BY REACTION OF PRIMARY PHOSPHINES WITH PHOSPHONOUS HALIDES

Perhaps the most generally applicable route to cyclopolyphosphines is that originally used by Köhler and Michaelis[351] for the preparation of "phosphobenzol," that is, reaction of a primary phosphine with the corresponding phosphonous dihalide (see Ref. 406, Table 4, p. 30). Whereas aromatic substituted primary phosphines react already at room temperature with arylphosphonous dichlorides,

$$nRPH_2 + nRPCl_2 \longrightarrow (RP)_{2n} + 2nHCl$$

aliphatic substituted derivatives need higher temperatures. In this case the reaction is usually carried out in refluxing benzene or toluene. In the presence of bulky groups, higher-boiling solvents such as xylene are necessary for completion of the reaction. The yields are normally high by this route, but the reaction product is not always uniform. Thus in the reaction of $EtPH_2$ with $EtPCl_2$, the four-membered ring $(EtP)_4$ and the five-membered ring $(EtP)_5$ were formed simultaneously.[30] Compounds that contain longer aliphatic groups apparently also form higher molecular compounds in addition to the four-membered ring.[125,129] The ring size of aromatic substituted cyclopolyphosphines seems to be mainly determined by the reactant quantities, reactant concentration, temperature, and solvent. Running the reaction (a) in ether favors the five-membered ring, (b) in benzene gives the six-membered ring, and (c) without a solvent yields a cyclopolyphosphine whose ring size is not known yet.[146,245,404,409]

II. FROM PHOSPHONOUS DIHALIDES AND REDUCING AGENTS

In order to avoid working with primary phosphines that are very toxic and inflame in air[398] (see Chapter 1), cyclopolyphosphines have recently been prepared directly by reduction of the corresponding alkyl- or arylphosphonous dihalides. Among the reducing agents that have been used are LiH,[30,270,462] $LiAlH_4$,[35,245,270,408,409,462,476,487] lithium,[47,244,245] sodium,[291,462] magnesium,[213,245,363,364,404] zinc,[491] mercury,[38,123a,127,389,392,558,574] antimony,[67] and tertiary phosphines.[193,261,542] Tertiary phosphines also reduced $(CF_3P)_4S$[68] and dithiophosphonic anhydrides, $[RPS_2]_2$,[180,181] to cyclopolyphosphines. As intermediates in the latter reaction, inner phosphonium salts are formed which can be isolated.[181] Since cyclopolyphosphines are cleaved by many of the reducing agents

$$[RPS_2]_2 + 2Bu_3P \longrightarrow 2R\text{-}\overset{\overset{\displaystyle S}{\|}}{\underset{\underset{\displaystyle S^{\ominus}}{|}}{P}}\text{-}\overset{\oplus}{P}Bu_3 \xrightarrow{2Bu_3P} 2/n(RP)_n + 2Bu_3PS$$

with the formation of primary phosphines or metal phosphides (see Section A.1.III), excess of the reducing agent must be avoided. The isolation of $PhP(SEt)_2$ in the reaction of $PhPCl_2$ with zinc in the presence of diethyl disulfide

$$PhPCl_2 + 2M \longrightarrow (PhP) + 2MCl$$

$$n(PhP) \longrightarrow [PhP]_n \xrightarrow{+nM} n/2\,PhP(M)\text{-}PPh(M) \xrightarrow{+nM} nPhPM_2$$

$$(M = Li,\ Na,\ K)$$

seems to indicate the intermediate formation of "phenylphosphinidene," PhP,[491] but direct reaction of $(PhP)_5$ with Et_2S_2 cannot be excluded.

Other attempts to trap "phosphinidene" in reductions with lithium or magnesium, however, were unsuccessful.[245]

Less than equivalent amounts of reducing agent sometimes allow the isolation of intermediate products, for example, in the reduction of phenylphosphonous dibromide with LiH in benzene solution at 5°C, 1,2,3-triphenyltriphosphine was isolated.[574]

$$3PhPBr_2 + 6LiH \longrightarrow PhHP\text{-}PPh\text{-}PPhH + 6LiBr + 2H_2$$

As in the reaction of primary phosphines with phosphonous dihalides, the reduction of phosphonous dihalides with metals and metal hydrides does not always result in uniform products. Thus in the coupling of CF_3PI_2[389,392] or CF_3PBr_2[72] with mercury or antimony,[67] a 3:2 mixture of the tetramer $(CF_3P)_4$ and the pentamer $(CF_3P)_5$ was formed. Similarly, the reduction of $EtPCl_2$ with LiH and of $EtPBr_2$ with magnesium gave not only the tetramer $(EtP)_4$[245,306] but the pentamer $(EtP)_5$[30,364] as well. Reductions of other alkylphosphonous dichlorides with magnesium seem to yield generally the four-membered ring and higher molecular compounds.[125,129] Phenylphosphonous dichloride gave in the reduction with magnesium in THF the five-membered ring $(PhP)_5$[245,404] and the six-membered ring $(PhP)_6$ simultaneously.[404] And recently, it was claimed that reduction of $C_2F_5PI_2$ with mercury gave a 2:3 mixture of the three-membered ring $(C_2F_5P)_3$ and the four-membered ring compound $(C_2F_5P)_4$.[123a] Other workers failed to confirm the existence of the trimer $(C_2F_5P)_3$.[18a]

III. BY DEHYDRATION OF PRIMARY PHOSPHINE OXIDES

The conversion of primary phosphine oxides into cyclotetraphosphines constitutes the first example of the formation of P-P bonds by dehydration[245,246] (see also ref. 466). The dehydration is effected by heating the crude primary phosphine oxides to 60° at a pressure of 1 mm Hg.

$$4RP(O)H_2 \longrightarrow (RP)_4 + 4H_2O$$

$$[R = Et_2CH\ (15\%);\ R = c\text{-}C_6H_{11}\ (22\%)]$$

The yields are rather low, probably because of the disproportionation of primary phosphine oxides into primary phosphines and phosphonic acids.

IV. FROM PHOSPHONOUS DIFLUORIDES BY DISPROPORTIONATION

Methylphosphonous difluoride,[370,528] $ClCH_2PF_2$,[441a] CF_3PF_2,[498] and $C_6H_5PF_2$[498,173] disproportionate on standing at room temperature and faster on slight heating into alkyl- or aryltetrafluorophosphorane and cyclopolyphosphine.

$$2RPF_2 \longrightarrow RPF_4 + 1/n[RP]_n$$

(R = Me, n = 5;[370] R = CF_3, n = 4/5; R = Ph, n = 5,[498] n = unknown;[173] R = $ClCH_2$, n unknown[441a])

The ring size of the formed cyclopolyphosphine depends upon the substituent on phosphorus. $C_6F_5PF_2$ and $(C_6F_5)_2PF$ are stable, however, under the above conditions and only at higher temperatures and upon prolonged heating redox-disproportionation does take place.[178]

V. FROM ELEMENTAL PHOSPHORUS

In contrast to the reactions of white phosphorus with PhLi or PhNa and alkyl halides, which yield secondary and tertiary phosphines, alkylmagnesium bromide and alkyl bromide interact with white phosphorus (ratio 2:2:1) in refluxing THF with the formation of mainly cyclotetraphosphine (42%) in addition to small amounts of secondary (6%) and tertiary phosphines.[474,475] This reaction seems to be generally applicable to the preparation of distillable cyclopolyphosphines.[126,129] Electrolysis of a suspension

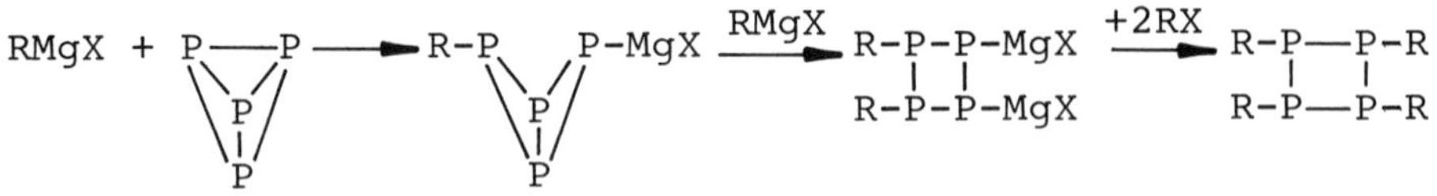

of white phosphorus in Me_2NCHO in the presence of BuBr gave $(BuP)_4$ and $MeBu_3P_4$ (low yields) in addition to other products.[340]

Tetrakis(trifluoromethyl)cyclotetraphosphine was isolated from the gas-phase reaction of trifluoromethyl radicals with white phosphorus.[569] Similar results were obtained by passing fluoroform into a solution of benzoyl peroxide and phosphorus in carbon disulfide-dioxan mixtures.

VI. BY THERMAL DECOMPOSITION OF VARIOUS MONOMERS

Heating of the biphosphines $[(CF_3)_2P]_2$ and $[CF_3PH]_2$ or the secondary phosphine $(CF_3)_2PH$ in sealed tubes between 300 and 350° resulted in the formation of $(CF_3)_3P$, white phosphorus, and a mixture of cyclopolyphosphines consisting of the tetramer $(CF_3P)_4$, pentamer $(CF_3P)_5$, and higher polymers.[80,392] Catalytic decomposition of the triphosphine $[(CF_3)HP]_2PCF_3$ and the tetraphosphine $[CF_3(H)P\text{-}PCF_3\text{-}]_2$ on nickel,[392] mercury, or with a tertiary phosphine[84] produced CF_3PH_2, $(CF_3PH)_2$, and a mixture of cyclopolyphosphines $(CF_3P)_{4,5}$ and higher polymers. Decomposition of the unsymmetrical biphosphine $CF_3(H)P\text{-}P(CF_3)_2$ and of the triphosphine $CF_3P[P(CF_3)_2]_2$ gave also a mixture of cyclopolyphosphines, $(CF_3P)_{4,5}$.[83] Similar decomposition reactions were observed with aliphatic and aromatic substituted triphosphines and biphosphines[74,299,300,574] as well. Interestingly, in these decomposition reactions aliphatic substituted cyclopolyphosphines are formed, such as $(EtP)_n$[574] and $(MeP)_n$,[74] whose molecular weights seem to be much higher than that corresponding to a four- or five-membered ring. The thermal decomposition of disecondary diphosphines $MeHPCH_2PHMe$[563] and $PhHPCH_2CH_2PHPh$[316] also yields

$$EtP(PEt_2)_2 \longrightarrow 1/n(EtP)_n + Et_2P\text{-}PEt_2$$

$$Me(H)P\text{-}P(CF_3)_2 \longrightarrow 1/n(MeP)_n + (CF_3)_2PH$$

cyclopolyphosphines, in addition to primary and secondary phosphines. In all attempts to prepare aliphatic or aromatic substituted straight-chain tetra- or hexaphosphines, only their decomposition products, that is, a primary phosphine or biphosphine and a cyclopolyphosphine, could be isolated.[299,300,574] Heating of bis(alkylamino)phenylphosphines $PhP(NHR)_2$ at 140 to 160° did not result in deamination but caused rearrangement to yield free amine,

$$RHP\text{-}PR\text{-}PR\text{-}PHR \longrightarrow RPH_2 + (RP)_n$$

$$RR'P\text{-}PR\text{-}PR\text{-}PRR' \longrightarrow RR'P\text{-}PRR' + (RP)_n$$

some condensed material, a phosphazene $RN{=}PPh(NHR)_2$, and pentaphenylcyclopentaphosphine, $(PhP)_5$.[373,374] Similar results were obtained by heating bis(phenylcarbamoyl)-phenylphosphine, $PhP(CONHPh)_2$, for 2 hr at 180°.[208] The products in this case were CO, diphenylurea, and $(PhP)_5$ (40%). The existence of "phenylphosphinidene," PhP, was postulated in the above reaction, but attempts to trap it were not successful. In an experiment designed to yield phosphoarsenobenzene, PhPAsPh, from $PhPH_2$ and $PhAsCl_2$, only the decomposition products $(PhP)_5$ and a cyclic arsine could be isolated.[548]

VII. BY MISCELLANEOUS METHODS

Several miscellaneous preparations of the cyclopolyphosphines have been described in the literature, but these preparations have been applied only to one or two compounds in the series.

In attempts to prepare bis(diethylamino)diorganobiphosphines by reductive dehalogenation of $R_2N(R)PCl$ with sodium[535,560] or mercury,[560] only the decomposition products, that is, bis(diethylamino)organophosphine and a cyclic polyphosphine, could be isolated. Similarly, reaction of the biphosphine Ph(I)P-P(I)Ph with Me_2NH did not produce the expected aminosubstituted biphosphine but the decomposition products aminophosphine and cyclopolyphosphine.[560] Aminosubstituted biphosphines, obtained when

$$\text{PhIP-PIPh} + 4\text{Me}_2\text{NH} \longrightarrow \text{PhP(NMe}_2)_2 + 1/n\,(\text{PhP})_n + 2\text{Me}_2\text{NH}\cdot\text{HI}$$

the reduction of $R_2N(R)PCl$ was carried out at -10 to -20° with a Na/K alloy, are cleaved by HCl to give phosphonous dichlorides and a cyclopolyphosphine.[531] Pentaphenylcyclopentaphosphine, $(PhP)_5$, is formed when bis(dimethylamino)phenylphosphine is heated with phenylphosphine (1 hr,

$$\text{Et}_2\text{N(R)P-P(R)NEt}_2 + 4\text{HCl} \longrightarrow 1/n\,(\text{RP})_n + \text{RPCl}_2 + 2\text{Et}_2\text{NH}\cdot\text{HCl}$$

170°, yield 83%)[404,405] or phenylphosphonous acid, $PhP(OH)_2$ (yield 10%).[210]

$$5\text{PhP(NMe}_2)_2 + 5\text{PhPH}_2 \longrightarrow 2\,(\text{PhP})_5 + 10\text{Me}_2\text{NH}$$

The same cyclopolyphosphine, $(PhP)_5$, is produced in the interaction of $PhPCl_2$ and $PhP(OCH_2CH_2OEt)_2$ at room temperature.[433] Cyclopolyphosphines can also result from certain halogen-metal exchange reactions, such as the reactions of alkali phosphides with bromine, CH_2Cl_2, or 1,2-dibromoethane.[279,281,300] An earlier report[292] describing the formation of a disecondary biphosphine, $[Ph(H)P]_2$,

$$\text{R(H)PK} + \text{Br}_2 \xrightarrow[-\text{KBr}]{} [\text{R(H)PBr}] \longrightarrow 1/n\,(\text{RP})_n + \text{HBr}$$

$$\text{R(H)PK} + \text{CH}_2\text{Cl}_2 \xrightarrow[-\text{ClCH}_2\text{K}]{} [\text{R(H)PCl}] \longrightarrow 1/n\,(\text{RP})_n + \text{HCl}$$

$$(\text{R} = \text{Et, c-C}_6\text{H}_{11}, \text{Ph})$$

in these reactions was found to be in error.[300]

An interesting application of the metal-halogen exchange

reaction resulted in the preparation of pure tetraethylcyclotetraphosphine and pure pentaethylcyclopentaphosphine.[311] To achieve this the tetra- and pentaphosphides were caused to react with 1,2-dibromoethane according to:

$$\text{KEtP-EtP-EtP-PEtK} + \text{BrCH}_2\text{CH}_2\text{Br} \xrightarrow[-2\text{KBr}]{-\text{CH}_2=\text{CH}_2} (\text{EtP})_4$$

$$\text{KEtP-EtP-EtP-EtP-PEtK} + \text{BrCH}_2\text{CH}_2\text{Br} \xrightarrow[-2\text{KBr}]{-\text{CH}_2=\text{CH}_2} (\text{EtP})_5$$

As intermediates, bromophosphides of the type $KEtP(PEt)_n$-PEtBr (n = 2,3) were postulated. The tetramer $(EtP)_4$ was found to be unstable, even at room temperature, converting to the stable pentamer $(EtP)_5$.[311]

A variety of organomercury compounds have been found to react with phenylphosphine, forming pentaphenylcyclopentaphosphine, $(PhP)_5$, mercury, and a hydrocarbon.[470] The same cyclopolyphosphine is obtained in the reactions

$$5\text{PhPH}_2 + 5\text{R}_2\text{Hg} \longrightarrow (\text{PhP})_5 + 5\text{Hg} + 10\text{RH}$$

of $PhPH_2$ with tetraphenylcyclopentadienone,[211] with chloramine,[256] or with iodine in the presence of triethylamine.[491] Unsymmetrically substituted cyclopolyphosphines were postulated to be formed on heating a mixture of $(MeP)_5$ and $(EtP)_5$ to 100 to 125° at a pressure of 0.1 mm Hg. The mixed products were identified by mass spectroscopic analysis.[492]

$$(\text{MeP})_5 + (\text{EtP})_5 \longrightarrow (\text{MeP})_5 + \text{Me}_4\text{EtP}_5 + \text{Me}_3\text{Et}_2\text{P}_5 + \text{Me}_2\text{Et}_3\text{P}_5 + \text{MeEt}_4\text{P}_5 + (\text{EtP})_5$$

VIII. POLYCYCLIC POLYPHOSPHINES

A polycyclic polyphosphine, probably with adamantane-like structure (28), was obtained in the reaction of PCl_3 or $POCl_3$ with dilithium phenylphosphide.[574] Both products (28 and its oxide) are soluble in benzene, toluene, xylene,

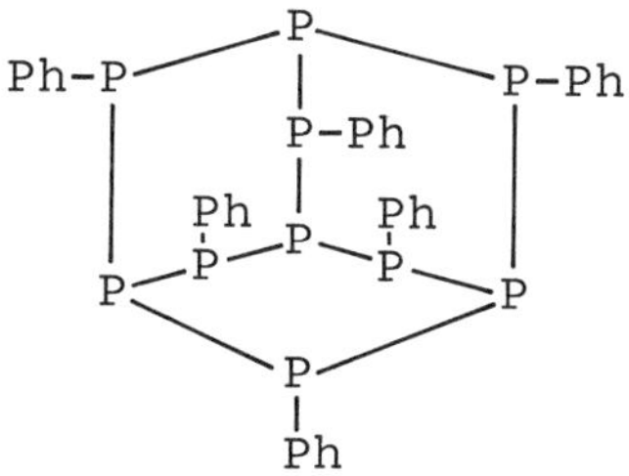

(28)

THF, $CHCl_3$, and dioxan. Above 60° the compounds begin to

$$4PCl_3[POCl_3] + 6Li_2PPh \longrightarrow P_4(PPh)_6[P_4O_4(PPh)_6] + 12LiCl$$

soften. When heated to above 200°, the compounds no longer dissolve completely in benzene. This has been ascribed to the formation of higher polymers.

The interaction of sodium with $Me_2N(Ph)PCl$ produced in addition to the biphosphine a compound of composition $P_{10}Ph_6(NMe_2)_2$ for which a decalinlike structure (29) was proposed.[560] And finally, decomposition of tetrakis(dimethylamino)biphosphine gave, in addition to other products,

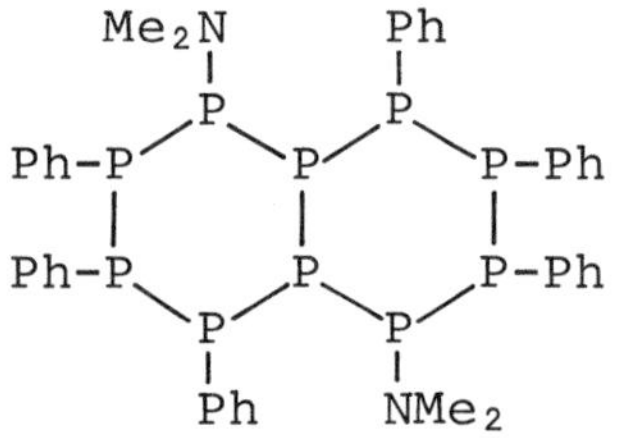

(position of ligands uncertain)

(29)

a polyphosphine of composition $P_{18}(NMe_2)_{12}$, which might have a structure (30) analogous to tetracene.[449] It should

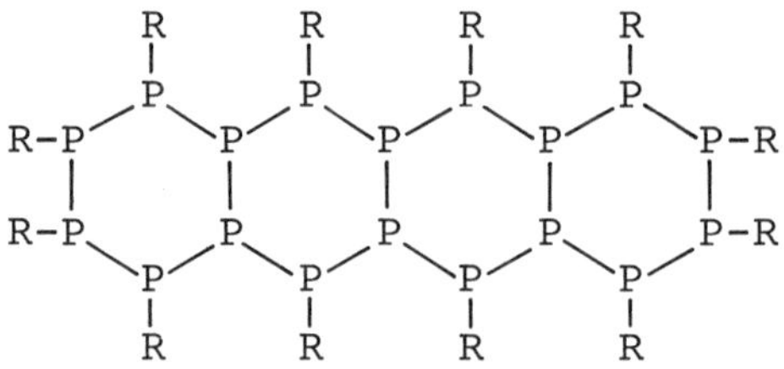

$(R = NMe_2)$

(30)

be pointed out that none of these polycyclic ring structures has been ascertained.

E.8.2. General Chemistry of Cyclopolyphosphines

(1) Hydrolysis

Hydrolysis of $(CF_3P)_4$ and $(CF_3P)_5$ with water yields 1,2-bis(trifluoromethyl)biphosphine and 1,2,3-tris(trifluoromethyl)triphosphine in addition to CF_3PH_2, H_3PO_3, and CHF_3.[392] Aqueous alkali hydrolyzes $(CF_3P)_4$ at room temperature and yields one-half the trifluoromethyl groups as

$$(CF_3P)_4 \xrightarrow{\text{H}_2\text{O, 140°}} (CF_3PH)_2, \ CF_3PH_2, \ H_3PO_3, \ CHF_3$$

$$(CF_3P)_5 \xrightarrow[50°]{\text{H}_2\text{O/diglyme}} H(CF_3P)_3H, \ (CF_3PH)_2, \ CF_3PH_2, \ H_3PO_3,$$

$$CHF_3$$

fluoroform[389,392] (basic hydrolysis rule). Likewise,
hydrolysis of the pentamer yielded 2.5 moles of fluoroform
per mole of cyclopolyphosphine. Alcoholysis of $(CF_3)_{4,5}$

$$(CF_3P)_4 + xNaOH \ aq \longrightarrow 2CF_3PH_2 + 2H_3PO_3 + 2CHF_3$$

begins at room temperature[84,392] and it was by this route
that the first triphosphine, $CF_3(H)P\text{-}PCF_3P(H)CF_3$,[84,392]
and tetraphosphine, $CF_3(H)P\text{-}PCF_3\text{-}PCF_3\text{-}P(H)CF_3$,[84] were prepared. Cleavage of $(CF_3P)_{4,5}$ with HI in the presence of
mercury results in the formation of CF_3PH_2 in high yield.[101]
   Aliphatic and aromatic substituted cyclopolyphosphines
appear to be hydrolytically more stable than trifluoromethyl-substituted cyclopolyphosphines. Thus $(PhP)_5$ was
stable toward water at 95°,[351,358,462] and only heating
at 95° for 3.5 hr in aqueous HCl caused hydrolytic cleavage.[351,462,476] The hexamer $(PhP)_6$ is cleaved in benzene
solution by dry HCl to give $PhPH_2$ and $PhPCl_2$, which condense to give $(PhP)_5$ in about 10% yield.[404] Pentaphenyl-

$$(PhP)n + nH_2O \longrightarrow n/2 \ PhPH_2 + n/2 \ PhP(OH)_2$$

cyclopentaphosphine, $(PhP)_5$, is also cleaved by $NH_3\text{-}NH_2Cl$
in benzene solution, as are biphosphines, to yield a phosphonium salt (30a) and a linear phosphonitrilic derivative
(30b).[194] Alcohol and alkali do not seem to attack $(PhP)_5$.[351]

$$[Ph(NH_2)_3P]^{\oplus}Cl^{\ominus}$$

$$Cl\left[\begin{array}{c} NH_2 \\ | \\ -P=N \\ | \\ Ph \end{array}\right]_5 \begin{array}{c} NH_2 \\ | \\ -P=NH \\ | \\ Ph \end{array}$$

(30a)   (30b)

The hydrolytic cleavage of aliphatic substituted cyclopolyphosphines has not yet been reported.

(2) Oxidation
Nearly all the cyclopolyphosphines are extremely sensitive
to oxidizing agents. Thus $(CF_3P)_4$ inflames in air,[389,392]
and the water-clear liquids $(MeP)_5$ and $(EtP)_{4,5}$ immediately
become milky on contact with air.[30,364] The aromatic substituted pentamer $(PhP)_5$ is relatively stable to air oxidation in the solid state. In solution, however, air attacks

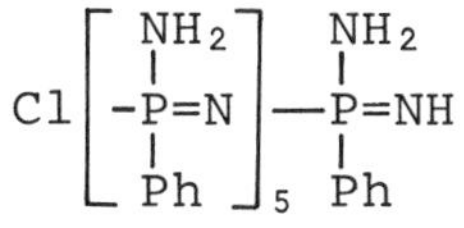

the ring easily.[359] Tetracyclohexylcyclotetraphosphine seems to be stable against air oxidation in the solid state as well as in solution.[315] Controlled oxidation of cyclopolyphosphines, $(RP)_n$ (R = CF_3, C_6H_5, n = 4, 5), in inert solvents also destroys the ring, leading to polymeric phosphonic anhydrides $(RPO_2)_n$.[359,392]

(3) Sulfurization and Selenation

Sulfurization of $(PhP)_5$ with equivalent amounts of sulfur results in ring cleavage to give a product of composition $(PhPS)_3$ (m. 150.5°).[146,404]

The structure of this product is not yet known with certainty. Out of the six possible structures (excluding S-S bonds), a six-membered ring structure (31) has been proposed.[146,404,406] IR and mass spectroscopic studies seem to indicate a three-membered ring structure (32).[32]

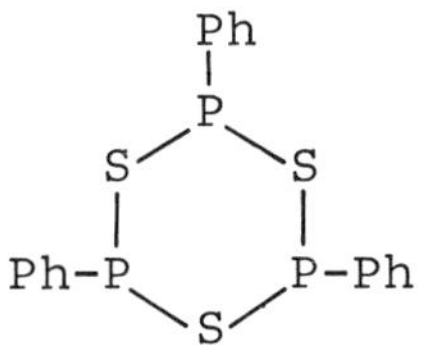

(31)

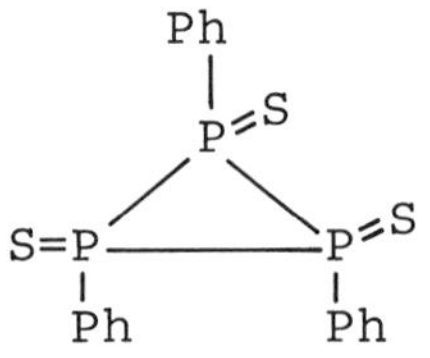

(32)

However, we found in the IR an absorption band at 578 cm^{-1}, which indicates the presence of P-S-P bonds.[408] The same product apparently was also formed in the reaction of $PhPCl_2$ with $(Me_2SiS)_{2,3}$.[2]

Two equivalents of sulfur per PhP unit lead to dimeric phenyldithiophosphonic anhydride (33), for which a trans structure has been established.

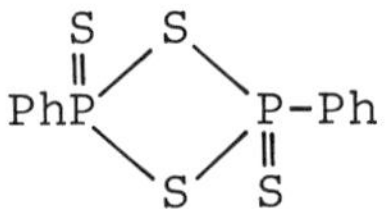

(33)

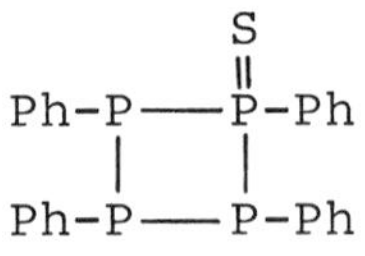

(34)

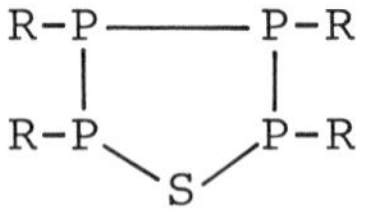

(R = CF_3, Ph)

(35)

$(PhPS)_3$ and $[PhPS_2]_2$ were also produced in the reaction of $PhPH_2$ with sulfur; in addition, a third product of composition $(PhP)_4S$ was isolated and structure (34) or (35) was proposed.[401]

On heating equivalent amounts of $(PhP)_5$ with N_4S_4, a yellow crystalline compound of composition $(PhPNS)_x$ (m. 200 to 201°) is formed.[339] The structure of the product is unknown. Interaction of the tetramer $(CF_3P)_4$ with sulfur at 170 to 250° yields a variety of products among which tetrakis(trifluoromethyl)cyclotetraphosphathiolan, [(35), R = CF_3, b. 183° (760 mm Hg)] was isolated in pure form.[68] With excess sulfur at 180 to 200°, the thiotetraphosphine (35) is converted to the very stable dithiotriphosphorane heterocycle (35a).[83a]

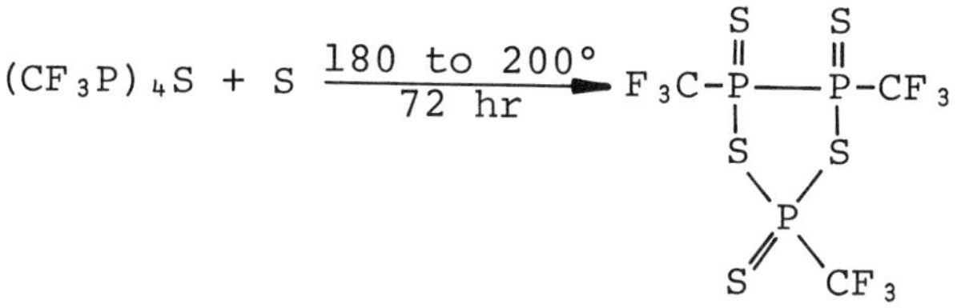

(m. 133°)

(35a)

Whereas $(EtP)_{4,5}$[306] and $(t\text{-}BuP)_4$[291] do not interact with sulfur, $(c\text{-}C_6H_{11}P)_4$ reacts with this reagent but without giving a definitive product.[315] However, tetra-alkylcyclotetraphosphine tetrasulfides seem to be obtainable from the reaction of primary phosphines and sulfur.[402,406,426] The hexamer $(PhP)_6$ does not react with sulfur.[404]

$$4RPH_2 + S_8 \longrightarrow (RPS)_4 + 4H_2S$$

Treatment of $(PhP)_5$ with red selenium leads to a compound of composition $(PhPSe)_n$ (m. 71 to 72°).[356] The same compound is obtained from $PhPH_2$ and selenium.[402] A four-membered ring structure with exocyclic selenium has been proposed for this adduct.[356] In view of the complexities in the various sulfurized products, this structure must be considered uncertain and warrants further investigation.[404] In contrast to the sulfur adduct $(PhPS)_3$, the selenium adduct does not react with more selenium.[356] Whether aliphatic substituted cyclopolyphosphines also react with selenium is not known.

(4) Metallation

Metallation of cyclopolyphosphines leads to tetraphosphides, triphosphides, biphosphides, and monophosphides. These reactions are discussed in Section A.1.III.

(5) Halogenation

All the cyclopolyphosphines undergo ring cleavage with halogens and form with stoichiometric amounts of halogen alkyl- or arylphosphonous dihalides.[127,306,315,351,358,359,364,392,476]

$$(RP)_n + nX_2 \longrightarrow nRPX_2$$

(n = 4, 5; R = CF_3, Et, $c\text{-}C_6H_{11}$, Ph, C_6F_5; X = Cl, Br, I)

This reaction probably follows an ionic mechanism.[424] Cleavage occurs stepwise, and in the case of iodination of $(PhP)_5$ it is possible to isolate 1,2-diiodo-1,2-diphenylbiphosphine, Ph(I)P-P(I)Ph, if a ratio of 2.5:1 is used.[261] The same biphosphine is formed when $PhPI_2$ is dissolved in ether.[172]

$$2PhPI_2 \xrightarrow{Et_2O} Ph(I)P\text{-}P(I)Ph + I_2$$

With the exception of $(C_6F_5P)_5$, all other cyclopolyphosphines are converted into tetrahalophosphoranes when excess halogen is used.[306,315,364,392] In the case of the pentafluorophenyl-substituted derivative it was shown that $C_6F_5PCl_4$ is unstable and decomposes into Cl_2 and $C_6F_5PCl_2$.[127] The cleavage of cyclopolyphosphines by iodine has recently been used in the determination of the oxidation state of phosphorus in cyclopolyphosphines.[364]

(6) Quaternization

Alkyl halides react with cyclopolyphosphines in several different ways, depending upon the nature of the alkyl group on the alkyl halide and upon the electronegativity of the substituents on the cyclopolyphosphines. Thus methyl iodide undergoes a simple addition reaction with $(EtP)_4$, $(EtP)_5$, $(MeP)_5$,[311] and $(t\text{-}BuP)_4$[291] to give a monoquaternary salt.

$$(RP)_{4,5} + MeI \longrightarrow [(RP)_{4,5}\ Me]I$$

Reaction of $(PhP)_5$ with MeI or EtI, however, leads to ring cleavage with formation of Ph(I)P-P(I)Ph, a phosphonium salt, and $PhPI_2$.[261]

$$0.8(PhP)_5 + 5RI \longrightarrow [Ph(I)P]_2 + [PhR_3P]I + PhR_2PI_2$$

Complete ring cleavage results in the reaction of $(PhP)_5$ with benzyl halides[261] and CF_3I,[35] giving in the first case a phosphonium salt, a diiodophosphorane, and $PhPI_2$, and with CF_3I a tertiary phosphine $Ph(CF_3)_2P$, a phosphinous iodide $PhCF_3PI$, and hexafluoroethane.[35]

(7) Complex Formation

Cyclopolyphosphines form complexes with boron halides,[126,359] metal carbonyls,[16,17,78,140,191,192,282,293,404,408] and transition metal salts.[140,191,252,306] Several conflicting reports have been published, and the picture has not been clarified by any means during the past 3 years. The conclusion drawn in a recent review[406] still holds, that is, although there are signs that in some cases ring expansion or ring contraction of cyclopolyphosphines may occur on complex formation, not a single example has yet been rigorously proved. Careful future investigations, in which the cyclopolyphosphine used as a starting material is exactly defined, will have to clarify the validity of these reports.

(8) Reactions with Phosphines, Arsines, and Amines (Ring Contraction and Ring Expansion)

The trimer $(C_2F_5P)_3$,[123a] tetramer $(CF_3P)_4$, and pentamer $(CF_3P)_5$ react reversibly with tertiary phosphines, aminophosphines, or amines to form monomer complexes of possible ylid structure.[81,123] The reversibility of the reaction requires that tertiary phosphines catalyze the intercon-

$$R_3P{=}CF_3;\ (Me_2N)Me_2P{=}PCF_3;\ R_3N{=}PCF_3;\ Me_3P{=}PC_2F_5$$

version of $(CF_3P)_4$ and $(CF_3P)_5$, and this has been observed. In solution in ether or hexane, Me_3P converts the tetramer mostly to the pentamer, but without solvent, the tetramer is strongly favored by its solid-state energy.[81] Tetramethylhydrazine also effects this interconversion.[123] On heating $(PhP)_5$ at 80° for 5-1/2 hr in piperidine, the insoluble form C (m. 260°) $(PhP)_n$, is obtained.[404] Forms B, $(C_6H_5P)_6$, C, and D of phenylcyclopolyphosphine are converted at their melting points into the pentamer, $(PhP)_5$ (form A), which is apparently the most stable form of phenylcyclopolyphosphine.[404] A similar interconversion has been observed between the A and B forms of pentafluorophenylcyclopentaphosphine, $(C_6F_5P)_5$.[127] The tetramer $(CF_3P)_4$ is cleaved in reactions with primary phosphines, but not with PH_3, to give trifluoromethylphosphine and cyclopentaphosphines.[121a] The extent of this reaction

$$1/4(CF_3P)_4 + RPH_2 \longrightarrow CF_3PH_2 + 1/5(RP)_5$$

$$(R = CH_3,\ Ph)$$

appears to depend on the basicity of the primary phosphine. In the reaction of Me_2PH with the tetramer $(CF_3P)_4$, trifluoromethylphosphine, the biphosphine $Me_2P{-}PMe_2$, and the triphosphine $CF_3P[PMe_2]_2$ are formed, the latter arising from the interaction of $Me_2P{-}PMe_2$ with $(CF_3P)_4$ as shown by an independent experiment.[121a] The tetramer $(CF_3P)_4$ reacts also with Me_4As_2 and Me_2S_2 to insert a CF_3P group into the As-As or S-S bond,[123] but no reaction takes place between $(CF_3P)_4$ and Me_6Si_2 up to 138°.[123]

(9) Miscellaneous Reactions with Phosphinidene as an Intermediate

Phenylphosphinidene has been postulated as an intermediate in the reaction of zinc with $PhPCl_2$ at 25°,[491] in the oxidation of $PhPH_2$ with iodine in the presence of Et_3N,[491] and on heating $(PhP)_5$ to above 160°.[189,491] It was trapped by reaction with disulfides, benzil,[491] di-1(2-butanon-yl)-mercury[470] and acrylonitrile.[46] The products can also be explained, however, without invoking the intermediate formation of phosphinidene.[406,494] Pyrolysis at 400° seems to produce PhP, however.[235]

$(PhP)_5$

$PhPCl_2 + Zn \longrightarrow PhP: + EtSSEt \longrightarrow PhP(SEt)_2$

$PhPH_2 + I_2$

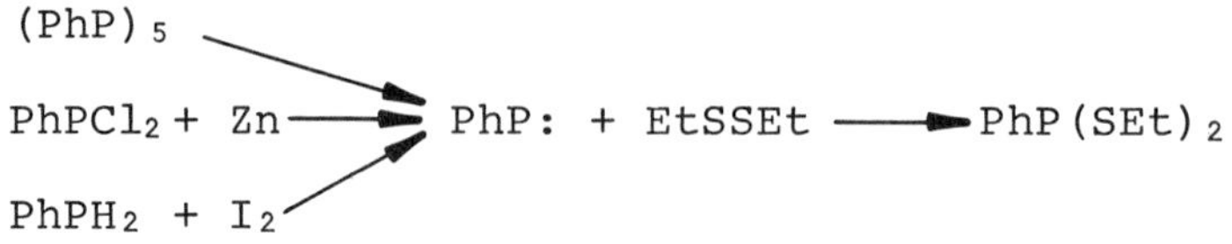

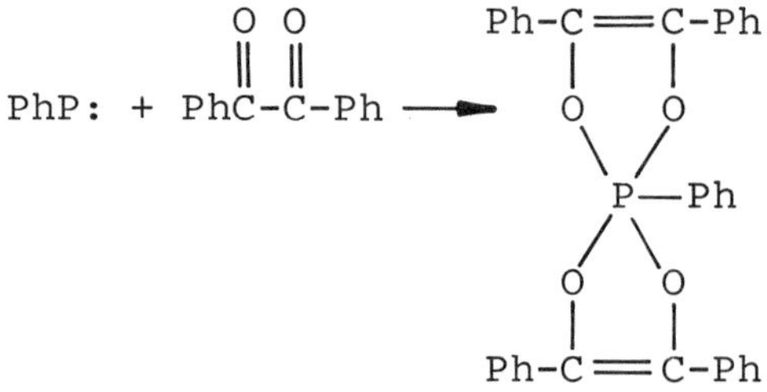

$PhP: + Hg(CH_2COEt)_2 \longrightarrow PhP(OC(Et)=CH_2)_2 + Hg$

Treatment of $(CF_3P)_4$ or $(CF_3P)_5$ with excess hexafluorobut-2-yne at 170° for 70 hr yields two new heterocyclic systems (16 and 18)[388] (see Section E.1.1.V).

The formation of 1,2,3-triphenyl-1,2,3-triphosphaindane (24) and 5,10-dihydro-5,10-diphenylphosphanthrene (25) from Li_2PPh_2 and o-bromoiodobenzene or o-diiodobenzene was formulated as involving reaction of phenylphosphinidene, PhP, with benzyne[415] (see Section E.7.1.III).

In the catalytic cleavage of Bu_3P with iodine at 250°, "butylphosphinidene," BuP, is possibly formed as an intermediate, since in the presence of diphenylacetylene 1-butyl-2,3,4,5-tetraphenylphosphol (36) was isolated in 3% yield.[335] Red phosphorus or Ph_3P, however, did not undergo this reaction.

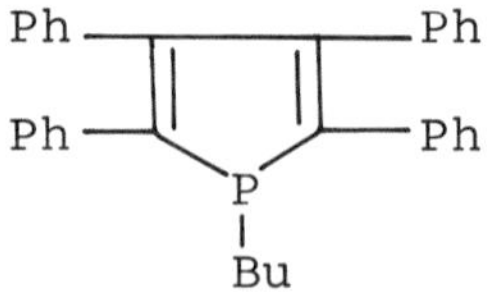

(36)

The scrambling reaction of $(MeP)_5$ with $(EtP)_5$ at 130°, which gave all possible mixed substituted cyclopolyphosphines, $Me_nEt_{5-n}P_5$ (n = 1 to 5),[493] may involve phosphinidene as an intermediate, but a biradical mechanism $\cdot PR(PR)_3$-$PR\cdot$ also seems possible.[494] And finally, heating a cyclopolyphosphine with a diene at 150 to 180° for 2 hr results in the formation of phospholenes (37) and tetrahydro-1,2-diphosphorines (17) in 40 to 60% yield.[493,494] The reaction may also be initiated by irradiation with UV light.[493,494] In this case the tetrahydrodiphosphorine (17) was formed exclusively.

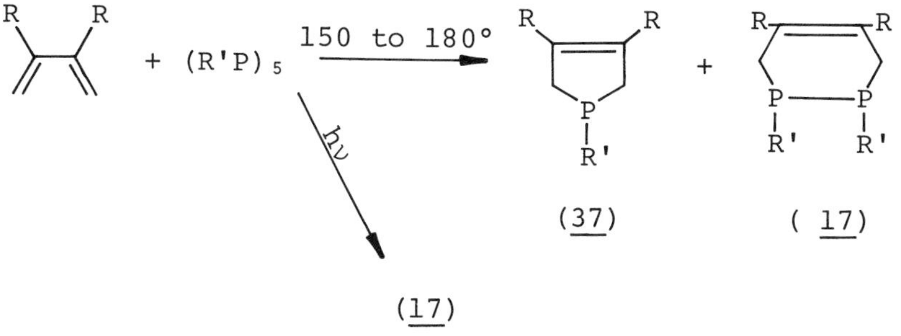

(R = H, Me; R' = Me, Et, Ph)

E.8.3. General Physical Properties

X-ray crystallographic analysis confirmed the structures of $(CF_3P)_4$[454] and $(CF_3P)_5$[157,544,545] first deduced from molecular weight determinations and chemical reactions,[389,392] Crystals of $(CF_3P)_4$ are tetragonal and consist of a puckered four-membered phosphorus ring with bond angles of 84.7° and ring torsion angles of 34°. The average P-P bond distance is 2.213 Å.[454]

The five-membered phosphorus ring in $(CF_3P)_5$ is also nonplanar, with bond angles of 93.9 to 108.3° and ring torsion angles of 18 to 58°. The P-P distance varies from 2.202 to 2.252 Å.[544,545] The molecular size of the polymeric $(CF_3P)_n$ material obtained as by-products is unknown.[72,83,389,392] The yellow-green color may indicate the presence of a six-membered ring, $(CF_3P)_6$.[392]

With the exception of the dianion $(PhP)_3^{2-}$,[282,302] tris(pentafluoroethyl)cyclotriphosphine, $(C_2F_5P)_3$, is the only three-membered phosphorus ring structure known.[123a] Its P_3 ring structure was deduced from molecular weight determination, the ^{19}F NMR spectrum, and mass spectroscopic analysis.[123a] More recently the existence of $(C_2F_5P)_3$ was questioned.[18a]

The methyl-substituted cyclopentaphosphine possesses, according to moecular weight determinations,[30,245,364,370] the UV spectrum,[245] the mass spectrum,[125] and the ^{31}P[245] and ^{1}H NMR[128,528] spectra, a nonplanar five-membered phosphorus ring structure. The structure of the also described polymeric, yellow $(MeP)_n$ is unknown.[74,364]

On the basis of molecular weight determinations in

solution,[245,291,306,315,474] the UV spectra,[245] Raman spectra,[14] and ^{31}P NMR spectra, four-membered ring structures were assigned to the cyclopolyphosphines substituted by higher alkyl groups, [$(RP)_4$, R = Et, Pr, i-Pr, Bu, i-Bu, c-C_6H_{11}, n-C_8H_{17}, $NCCH_2CH_2$, and Et_2CH]. The phosphorus four-membered ring structure for the cyclohexyl derivative (c-$C_6H_{11}P)_4$ was confirmed by an x-ray crystallographic analysis.[28,146] Crystals of (c-$C_6H_{11}P)_4$ are tetragonal (two different forms[28,146]) and consist of a puckered four-membered phosphorus ring structure with bond angles of 85.5° and ring torsion angles of 31.4°. The average P-P distance is 2.224 Å.[28]

Mass spectrometric analyses of the ethyl-, propyl-, and butylcyclopolyphosphines indicate, however, that pentamers are also present besides tetramers.[125] Gas chromatographic analysis[364] and molecular weight determinations[30,364] confirm for the ethyl derivative the presence of a mixture, consisting of the tetramer $(EtP)_4$ and the pentamer.[406]

Four different forms of phenyl-substituted cyclopolyphosphines have been reported (for a detailed discussion see ref. 406) in the literature: A, m. 149 to 156°;[245,270,351,358,359,364,404,462,476,574] B, m. 188 to 193°;[245,404,476,571] C, m. 252 to 255°;[245,404,476] and D, m. 260 to 285°.[245] The molecular weight determinations, UV spectra, ^1H NMR spectra, mass spectra,[245] and Raman spectra[13] of A and B have been interpreted in terms of a four-membered phosphorus ring structure. In contrast to this, x-ray analysis of form A established a five-membered nonplanar phosphorus ring structure with bond angles of 94.0 to 107.2° and ring torsion angles of 2.2 to 60.6°. The average P-P distance is 2.217 Å.[141,145] Phosphobenzene form B was found to be polymorphic, crystallizing in at least four different forms[145,404] (rhombohedric, monoclinic, triclinic, and tetragonal). All four forms possess a six-membered phosphorus ring structure. The rhombohedric form crystallizes with one molecule of benzene.[145,404] The trigonal modification consists of a six-membered phosphorus ring in the chair configuration with the phenyl groups occupying the equatorial positions.[142,146] The triclinic form of B is very similar to that of the trigonal form, differing mainly in the orientation of the phenyl rings.[143] The higher-melting forms C and D were formulated as polymers.[245,476] They might have the same structure.[406] X-ray powder photographs show that they are highly crystalline. Therefore the degree of polymerization cannot be very high.[404]

Other aromatic substituted cyclopolyphosphines also exist in different ring sizes, for example, $(RP)_{4,5,6}$ (R = 4-ClC_6H_4, 1-MeC_6H_4,[409] C_6F_5,[127,176] 4-FC_6H_4,[487] 4-$MeOC_6H_4$, 4-$EtOC_6H_4$, 2-$C_{10}H_7$[181]).

The intense UV spectra[245,392] and the nonbasic properties of aliphatic substituted cyclopolyphosphines suggest electron delocalization is taking place between the lone pair on one phosphorus atom and the vacant 3d orbital of the neighboring phosphorus atom. The following types of resonance structures can be written.

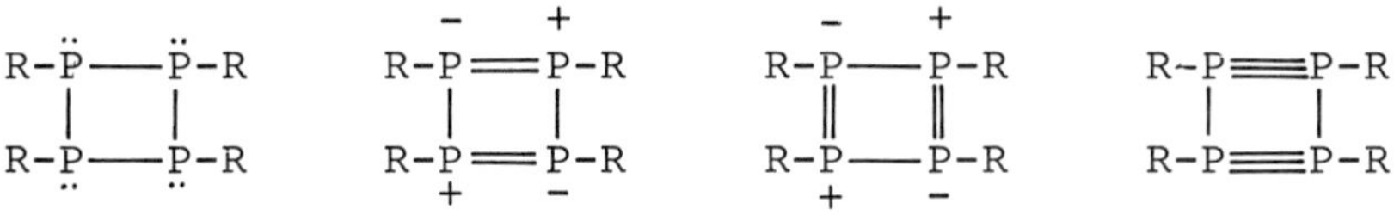

IR spectroscopic studies of several cyclopolyphosphines have been described in the literature.[30,245,359,409,476] In the Raman spectra of aliphatic substituted cyclopolyphosphines,[14,30] a symmetric ring stretching vibration was found between 390 and 410 cm^{-1} and an asymmetric ring stretching vibration between 465 and 490 cm^{-1}.[14] The thermal stability of cyclopolyphosphines is very dependent upon the substituents and increases in the following series.

$$(PhP)_6 < (PhP)_5 < (CF_3P)_5 < (CF_3P)_4 < (c\text{-}C_6H_{11}P)_4 < (EtP)_{4,5}$$

Thus the phenyl-substituted cyclopolyphosphines decompose at their melting points or a little above,[351,359,402] whereas the ethyl derivative can be distilled undecomposed at normal pressure.[306] As decomposition products of $(PhP)_5$, the phosphine Ph_3P, the biphosphine $Ph_2P\text{-}PPh_2$, and phosphorus have been identified.[359] The conclusion from a ^{31}P NMR study that in a melt of $(PhP)_5$ monomeric PhP and dimeric PhP=PPh are present[189] does not agree with a mass spectroscopic study.[494] Forms B, C, and D of phenylcyclopolyphosphine decompose at their melting points to the five-membered ring $(PhP)_5$.[404]

E.9. Arsinosubstituted Phosphines

Early attempts to prepare "phosphoarsenobenzene," PhP=AsPh, by reaction of phenylphosphine with phenyldichloroarsine were not successful. Only the disproportionation products could be isolated.[548] In the meantime, two different series of phosphorus-arsenic compounds were described. In one series both elements are in the trivalent state, and in the other phosphorus is quadrivalent with a positive charge, thus belonging to the phosphonium-type compounds.

E.9.1. Methods of Preparation

I. ARSINOSUBSTITUTED PHOSPHINES

Three methods have been reported for preparing arsinosubstituted phosphines. One involves reaction of stannylphosphines with arsenic halides[219,516] or stannylarsines with phosphorus halides;[516] the second involves cleavage of biphosphines with secondary arsines or of biarsines with secondary phosphines;[103]

$$3Me_3SnPPh_2 + AsF_3 \longrightarrow 3Me_3SnF + As(PPh_2)_3$$

$$(Me_3Sn)_3As + 3ClPPh_2 \longrightarrow 3Me_3SnCl + As(PPh_2)_3$$

$$(CF_3)_2P\text{-}P(CF_3)_2 + Me_2AsH \longrightarrow (CF_3)_2P\text{-}AsMe_2 + (CF_3)_2PH$$

$$(CF_3)_2As\text{-}As(CF_3)_2 + Me_2PH \longrightarrow (CF_3)_2As\text{-}PMe_2 + (CF_3)_2AsH$$

and the third method uses the insertion of "CF_3P," generated from $(CF_3P)_4$ or from Me_3PPCF_3, into the As-As bond of cacodyl.[122,123] In a redistribution reaction of bi-

$$4Me_2As\text{-}AsMe_2 + (CF_3P)_4 \longrightarrow 4CF_3P[AsMe_2]_2$$

phosphine disulfides with cacodyl at 220° for 16 hr, an arsinophosphine sulfide was produced.[238] Similarly, heating a mixture of biphosphine with biarsine apparently yielded an arsinophosphine, but the product could not be

$$Me_2P(S)\text{-}P(S)Me_2 + Me_2As\text{-}AsMe_2 \longrightarrow 2Me_2P(S)\text{-}AsMe_2$$

isolated.[103] In an experiment designed to produce dimethylarsinodiphenylphosphine by reaction of Me_3SiPPh_2 with

$$(CF_3)_2P\text{-}P(CF_3)_2 + (CF_3)_2As\text{-}As(CF_3)_2 \longrightarrow 2(CF_3)_2P\text{-}As(CF_3)_2$$

Me_2AsI, only the disproportionation products cacodyl and tetraphenylbiphosphine were obtained.[4]

$$Me_3SiPPh_2 + Me_2AsI \longrightarrow Me_3SiI + [Ph_2PAsMe_2 \longrightarrow (Me_2As)_2 + (Ph_2P)_2]$$

II. ARSINOSUBSTITUTED PHOSPHONIUM SALTS

Arsinosubstituted phosphonium salts are obtained in the reaction of tertiary phosphines with haloarsines.[56,114,332] The reaction can also be carried out in a solvent such as ether.[56]

$$Me_2AsI + R_3P \longrightarrow [Me_2AsPR_3]^{\oplus}I^{\ominus}$$

The stability of arsinophosphonium salts depends upon the tertiary phosphine and the haloarsine. Thus $[Me_2AsPMe_3]I$

was stable to air and water, but $[Ph_2AsPEt_3]I$ was unstable and evidence for the formation of the corresponding chloride, which could not be isolated in a pure state, was obtained only from a conductimetric titration.[56] With tributylphosphine a different type of reaction occurred. Both diphenylhaloarsines and $AsCl_3$ were reduced to give tetraphenylbiarsine or elemental arsenic, respectively, but phenyldihaloarsines gave arsinophosphonium salts.[332] No explanation is apparent for this different behavior of haloarsines.

$$Bu_3P + 2Ph_2AsCl \longrightarrow Bu_3PCl_2 + Ph_2As\text{-}AsPh_2$$

Attempts to prepare antimony- or bismuth-substituted phosphonium salts failed. When Ph_2SbI and Ph_2BiI were treated with Et_3P, unstable salts were formed, but in neither instance could a pure product be isolated.[56] An attempt to obtain an addition compound of Me_3Sb with PCl_3 was also unsuccessful. While Me_3As formed a 1:1 complex with PCl_3, the reaction of trimethylstibine with PCl_3 led to complete reduction to elemental phosphorus and the formation of Me_3SbCl_2.[266]

III. ARSINOSUBSTITUTED PHOSPHONATES

Compounds with an As-P bond have also been claimed to be formed in the reactions of trialkylphosphites with haloarsines,[343-345] and of alkali phosphites with haloarsines.[343,344] However, since analogously obtained "$R_3MP(O)(OR)_2$" (M = metal) compounds have been shown to contain all M-O-P

$$Me_2AsBr + (RO)_3P \longrightarrow Me_2As\text{-}P(O)(OR)_2 + RBr$$

units, the arsenic compounds need confirmation.

E.9.2. General Chemistry

All the arsinophosphines are very sensitive to oxidizing agents and decompose on contact with air.[516] The As-P bond is also cleaved by HCl.[123] Of the arsinophosphonium salts studied, only $[Me_2AsPMe_3]I$ was stable to air and water.[56] All the other salts decomposed in air by hydrolysis and oxidation but were stable when stored under nitrogen.[56] The arsinophosphonium salts are hydrolyzed to tertiary phosphines and arsinous anhydride.[56] This result was attributed to the reversible dissociation of the

$$2[R_2AsPR'_3]^{\oplus} + 2OH^{\ominus} \longrightarrow (R_2As)_2O + 2R'_3P + H_2O$$

arsinophosphinium salt, followed by rapid hydrolysis of

$$[R_2AsPR_3]^{\oplus}X^{\ominus} \rightleftharpoons R_2AsX + PR_3$$

the haloarsine.[56]

E.9.3. General Physical Properties

Arsinophosphines of the type $(CF_3)_2AsPMe_2$ and $Me_2AsP(CF_3)_2$ can withstand temperatures of 185° for a few minutes,[103] and the diphenyl-substituted arsinophosphines also seem to be stable up to 170°.[516] The stretching frequency of the P-As bond has been found in the IR spectrum at 353 cm^{-1}.[516] Other assignments made were for the groups: ν_{as} P_3As 357 cm^{-1}, ν_s P_3As 280 cm^{-1}, and ν_{as} As_3P 311 cm^{-1}, ν_s As_3P at 274 cm^{-1}; ν_{as} PAs_2 354 cm^{-1}, ν_s PAs_2 282 cm^{-1}, and ν_{as} P_2As 311 cm^{-1}, ν_s P_2As 293 cm^{-1}.[516]

Arsinophosphonium salts are 1:1 electrolytes in nitrobenzene[56] and nitromethane,[332] but they seem not to be ionic in deuterium-chloroform solution.[332] The IR spectra of arsinophosphonium salts show little or no change in frequencies when compared to the starting materials.[332]

F. ORGANOPHOSPHORUS COMPOUNDS OF TRANSITION METALS

This class of compounds has been investigated only briefly and not much is known about their structure and chemical and physical properties. The few articles concerned with this area have been restricted mainly to the preparation of individual compounds. Often the compounds obtained are polymeric, probably with a phosphide-bridged structure. We limit the discussion in this section to metal phosphides, whereas Chapter 3A deals with true phosphine complexes. The latter contain some phosphide-bridged types defined as being directly derived from a neutral phosphine ligand. A few examples will make this point clear. Compounds of the type $[Ph_2PCu]$, R_2PZnR, $(R_2P)_2Co$, $(R_2P)_3V$, and $[R_2PNi(CO)_3]K$ are discussed in this section, whereas complexes of the type $Ph_2PCu(HPPh_2)$, $(Ph_2P)_2Ni(HPPh_2)_2$ or $(CO)_nM(PR_2)_2M(CO)_n$ appear in Chapter 3A.

F.1. Methods of Preparation

I. FROM ALKALI PHOSPHIDES AND TRANSITION METAL COMPOUNDS

Metathesis of alkali phosphides with transition metal halides is a general method for preparing this type of compound.[284,286,329,330]

$$ZnCl_2 + 2KPPh_2 \longrightarrow Zn(PPh_2)_2 + 2KCl$$

$$CuBr + KPPh_2 \longrightarrow Cu(PPh_2) + KBr$$

Excess alkali phosphide causes further reaction and leads to ionic complexes.[153,284,286,328,330] In several

$$CuBr + 2KPPh_2 \longrightarrow KBr + KCu(PPh_2)_2$$

$$NiBr_2 + 3LiP(C_6H_{11})_2 \longrightarrow LiNi[P(C_6H_{11}\text{-}c)_2]_3 + 2LiBr$$

cases the reaction of transition metal halides with alkali phosphides is accompanied by a redox reaction. Thus $CuBr_2$ was reduced to $Cu(I)PR_2$,[284] and treatment of $TiCl_3$ or $TiCl_4$ with $LiPR_2$ led only to Ti(II) phosphide and a biphosphine.[329] Similarly, bis-π-cyclopentadienyltitanium

$$TiCl_3 + 3LiPR_2 \longrightarrow Ti(PR_2)_2 + 0.5\ R_2P\text{-}PR_2 + 3LiCl$$

dichloride and zirconium dibromide gave on treatment with a lithium dialkylphosphide a Ti(III) and a Zr(III) phosphide,[153,288] respectively. In some cases reduction of the metal to a lower valence state can be avoided when the reaction is run under homogeneous conditions.[286]

Metal carbonyls[310] and ammonia complexes of metal carbonyls[294] also give transition metal phosphides on treatment with alkali phosphides. While interaction of $FeBr_2$ with $LiP(C_6H_{11}\text{-}c)_2$ led to the formation of the expected

$$Ni(CO)_4 + KPPh_2 \longrightarrow K[Ni(CO)_3PPh_2] + CO$$

$$M(CO)_3(NH_3)_3 + 3KPPh_2 \longrightarrow K_3[M(CO)_3(PPh_2)_3] + 3NH_3$$

$$(M = Cr, Mo, W)$$

Fe(II) phosphide $Fe(PR_2)_2$,[330] an unusual product of composition $P[FePPh_2]_3$ was isolated when $FeBr_2$ was treated with $KPPh_2$.[285] The following reaction sequence was proposed.[285]

$$FeBr_2 + 2KPPh_2 \longrightarrow Fe(PPh_2)_2 + 2KBr$$

$$2Fe(PPh_2)_2 \longrightarrow 2FePPh_2 + Ph_2P\text{-}PPh_2$$

$$3Ph_2P\text{-}PPh_2 \longrightarrow 4Ph_3P + 2P$$

$$3FePPh_2 + P \longrightarrow P[FePPh_2]_3$$

The reverse reaction, that is, treatment of a transition metal alkali compound with a chlorophosphine in some cases also led to transition metal phosphides.[230,253]

$$(CO)_4CoNa + ClPPh_2 \longrightarrow (CO)_4CoPPh_2 + NaCl$$

Finally, other metal phosphides that have properties similar to alkali phosphides have also been used successfully for the formation of transition metal phosphides. Thus treatment of trimethylsilyldiphenylphosphine with CuCl, $PdCl_2$, or $NiCl_2$ produced the corresponding metal phosphide and trimethylchlorosilane.[4]

$$NiCl_2 + 2Ph_2PSiMe_3 \longrightarrow [(Ph_2P)_2Ni]_n + 2Me_3SiCl$$

Similar to alkali phosphides, the silylphosphine in some instances also gave a redox reaction with transition metal halides. For example, interaction of Me_3SiPPh_2 with CuBr, CuI, HgX_2, or AgX (X = Cl, Br, I) produced Me_3SiX, Cu, or Ag, and $Ph_2P\text{-}PPh_2$.[4]

II. FROM PHOSPHINES AND TRANSITION METAL COMPOUNDS

Most of the known zinc phosphides have been made by the reaction of secondary phosphines with diorganozinc compounds.[284,450]

$$ZnEt_2 + HPPh_2 \longrightarrow EtZnPPh_2 + C_2H_6$$

$$ZnEt_2 + 2HPPh_2 \longrightarrow Zn(PPh_2)_2 + 2C_2H_6$$

The reaction proceeds stepwise, and the zinc monophosphide as well as the zinc diphosphides have been made in this way.[450] $(c\text{-}C_6H_{11})_2PH$, whose acidity is too low, does not react with $ZnEt_2$.[330] In a few cases transition metal phosphides have been obtained directly from the interaction of secondary phosphines and cyclopentadienyl metal compounds[155] or metal halides, usually in the presence of amines as HCl scavengers.[4]

$$Ph_2PH + CuCl \xrightarrow{Et_3N} [Ph_2PCu]_n + Et_3NHCl$$

$$Ph_2PH + AgCl \xrightarrow{NH_3} [Ph_2PAg]_n + NH_4Cl$$

The preparation of mercury phosphides from HgX_2 and aliphatic substituted secondary phosphines requires no tertiary amine.[43,203]

$$HgCl_2 + Et_2PH \longrightarrow Et_2PHgCl + HCl$$

This reaction proceeds quantitatively and has been used for the estimation of secondary phosphines. In one case a mercury phosphide was obtained by cleavage of a biphosphine with HgI_2.[534] Phenylphosphine, however, forms with mercuric chloride in ethanol an unstable complex which reacts further with the solvent over a period of days at room temperature to give phenylphosphonous acid, ethyl chloride, HCl, and metallic mercury.[151]

$$PhPH_2 + 2HgCl_2 + 2EtOH \longrightarrow PhP(OH)_2 + 2Hg + 2HCl + 2EtCl$$

Other metal halides such as $NiBr_2$[327] and $PdCl_2$[242,243,327] also interact with secondary phosphines but form phosphine complexes of metal phosphides. These compounds are discussed in Chapter 3A, as are those obtained from carbonyls and biphosphines. There are some indications that $V(CO)_5$ initially formed vanadium tetracarbonyl phosphide, $V(CO)_4PR_2$, in the reaction with secondary phosphines.[255] The product, however, could not be isolated in a pure state.

Bis(trifluoromethyl)phosphide-substituted transition metal compounds were obtained from the reactions of $Mn_2(CO)_{10}$ and $Fe(CO)_5$ with $(CF_3)_2PI$.[164] The products formed, $(CF_3)_2PMn_2(CO)_8I$ and $[(CF_3)_2PFe(CO)_3I]_2$, are believed to contain a phosphorus and/or iodine bridge.[164]

F.2. General Chemistry

Phosphides of iron, cobalt, nickel, and copper possess properties of complex compounds and give with phosphines phosphide-phosphine complexes,[284,330] but elements having d^5 or lower electron configuration,

$$LiFe[P(C_6H_{11})_2]_3 + MeOH \longrightarrow (C_6H_{11})_2PH \cdot Fe[P(C_6H_{11})_2]_2 + MeOLi$$

$$LiCu[P(C_6H_{11})_2]_2 + H_2O \longrightarrow (C_6H_{11})_2PH \cdot CuP(C_6H_{11})_2 + LiOH$$

$$CuP(C_6H_{11})_2 + HP(C_6H_{11})_2 \longrightarrow (C_6H_{11})_2PH \cdot CuP(C_6H_{11})_2$$

are solvolytically decomposed by H_2O or ROH and give no phosphine complexes. Zinc phosphides behave very similarly to alkali phosphides.[284,432,450] They are cleaved by H_2O and alcohols,[284] and they give insertion reactions with isocyanates,[432,450] isothiocyanates, and carbon disulfide.[450] Contrary to the nitrogen analogs, which are inert toward

$$RZnPR_2 + RNCO\,(RNCS) \longrightarrow RZnNR\overset{O}{\overset{\|}{C}}PR_2\;(RZnNR\overset{S}{\overset{\|}{C}}PR_2)$$

$$RZnPR_2 + CS_2 \longrightarrow RZnS\underset{S}{\underset{\|}{C}}PR_2$$

methyl iodide, $EtZnPPh_2$ affords the quaternary salt $[EtZnPPh_2Me]^+I^-$. The Zn-P bond, however, is cleaved in refluxing methyl iodide with the formation of EtZnI and $[Ph_2PMe_2]I$.[450]

The Ti-P, Zr-P,[288,329] Fe-P, and Cu-P[330] bonds are also cleaved by methyl iodide, acetic acid[329] and HCl.[288] Iodine oxidizes $Ti(PR_2)_2$ first to I_2TiPR_2. Excess iodine, however, cleaves the Ti-P and V-P bonds.[329] Sulfur also oxidizes Ti(II) and V(III) with concomitant cleavage of

$$Ti(PR_2)_2 + I_2 \longrightarrow TiI_2(PR_2)_2$$

$$Ti(PR_2)_2 + 5I_2 \longrightarrow TiI_4 + 2R_2PI_3$$

the Ti-P and V-P bonds.[331]

$$2(R_2P)_2Ti + 9S \longrightarrow [R_2P(S)STiS]_2S + R_2P(S){-}P(S)R_2$$

F.3. General Physical Properties

Very little is known about the physical properties of transition metal phosphides. In addition to melting points, decomposition points, and a few dipole moments, the stability of the transition metal-phosphorus bond has been deduced from the rate of reaction of the transition metal phosphides with methyl iodide. The following stability series was given for metal-phosphorus bonds.[329]

$$Ti > V > Cr > Mn < Fe < Co < Cu$$

G. LIST OF COMPOUNDS

The list of compounds is divided into 14 major types (the subsections correspond to the sections in the text discussion), very unequal in size, which are subdivided into 51 structural classes, arranged so as to bring together compounds with similar structure.

Within each structural class the compounds are listed in order of increasing complexity. Compounds containing functionally modified substituents are listed after the unmodified compound, for example, EtPHNa, EtPHK, $H_2NCH_2CH_2$-PHNa, $EtNHCH_2CH_2PHNa$, $Et_2NCH_2CH_2PHNa$.

The entry for each compound indicates the method by which it has been prepared (Roman numerals; in classes in which only one method has been used to prepare the compound, the Roman numeral is omitted), records values of common physical constants, and gives literature references to these and more complex physical methods. The listing is as comprehensive as possible.

Organosubstituted Alkali Phosphides

Since alkali phosphides containing no organic group have been used very often for the synthesis of organophosphines and phosphorus-metal-containing compounds, this type of compound is also listed. Roman numerals refer to method of preparation described in Section A.1 of the text.

TYPE: Alkali Phosphides Containing No Organic Group

H_2PLi. I. Yellow,[319,357,377,460] ^1H NMR,[350] ^{31}P +251 ppm (0.9 M in Me_2NCHO).[350]
H_2PNa. I. White to pale-yellow solid,[12,165,168,333,376,377,564] ^1H NMR,[45,350,537] ^{31}P +279 ppm (0.9 M in Me_2-NCHO),[350] J_{PH} 138.7 Hz.[537]
H_2PK. I. White solid,[60,377,555,570] monoclinic crystals, distorted NaCl structure;[40] reacts with white P in Me_2NCHO to give KP_5H_2 and other products;[349] ^1H NMR,[349,350,537] ^{31}P +252.5 ppm (0.9 M in Me_2NCHO),[349,350] +272 ± 2 ppm (in Me_3N),[190] +280 ppm (5% in NH_3),[190,537] J_{PH} 139 Hz.[190]
H_2PRb. I. Monoclinic crystals, distorted NaCl structure,[40] ^1H NMR, ^{31}P +250 ppm (0.9 M in Me_2NCHO).[350]
H_2PCs. I. Solid, ^1H NMR, ^{31}P +243 ppm (0.9 M in Me_2NCHO).[350]
$HPLi_2$. I. Not obtained pure; contained PLi_3.[319,377,460]
$HPNa_2$. I. Not obtained pure.[377]
HPK_2. I. Not obtained pure.[377]
PLi_3. I. Yellow-brown, not pure.[63,319,460]
PNa_3. II.[268] Also obtained from white P and Na in an organic solvent.[467]
$PNaK_2$. Obtained from Na/K and white P in liquid NH_3; yellow.[168]
PK_3. II.[268]
Li_2P-PLi_2. Obtained from white P and Li in liquid NH_3, yellow.[168]
Na_2P-PNa_2. Obtained from white P and Na in liquid NH_3; yellow.[167]

TYPE: RPHM

MePHK. I.[60]
EtPHNa. I.[567,575a]
EtPHK. I.[279] Colorless crystals.[281]
$H_2NCH_2CH_2PHNa$. I.[304]
$EtNHCH_2CH_2PHNa$. I.[304]
$Et_2NCH_2CH_2PHNa$. I.[304]
BuPHNa. I. Yellow to dark-colored solid.[463,567]
BuPHK. I. White solid.[567]
n-$C_6H_{13}PHNa$. I.[463]
c-$C_6H_{11}PHLi$. I. Colorless solid with 1 mole of dioxan.[312]
c-$C_6H_{11}PHK$. I. Pale yellow.[279]
n-$C_8H_{17}PHNa$. I.[463]

PhPHLi. I. Colorless solid with 1 mole of dioxan.[312]
PhPHNa. I. Orange in NH_3,[258] yellow in Et_2O.[463]
PhPHK. I. Pale yellow.[292]
1,4-$(NaHP)_2C_6H_4$. I. Scarlet in NH_3 solution.[169]

TYPE: RPM_2

$MePLi_2$. I.[25]
$MePNa_2$. II.[268]
$EtPNa_2$. II.[268]
$BuPNa_2$. I. Dark brown.[463]
$(c\text{-}C_6H_{11})PLi_2$. I. Pale yellow; gives chemoluminescence with traces of O_2.[319]
$(c\text{-}C_6H_{11})PNa_2$. II.[268]
$PhPLi_2$. I.[319,460] II.[47] III.[47] Pale yellow with 1 mole of THF;[300,319] useful as catalyst for the polymerization of methacrylates and styrene.[248]
$PhPNa_2$. II. Greenish color when admixed with NaCl.[1,268,462] III.[359]
$PhPK_2$. II. Pale yellow.[359]
4-$EtC_6H_4PNa_2$. II.[462]
Me_3SiPLi_2. I.[460]

TYPE: R_2PM

Me_2PLi. I. Colorless solid with 0.5 mole of ether.[321] II.[380] III.[296]
Me_2PNa. I. Orange in NH_3.[564] III.[348,440] Yellow-white powder with 0.5 mole of dioxan.[321]
Me_2PK. I. Golden yellow in NH_3, white solid when dry.[555] II.[439] III.[440]
Et_2PLi. I.[319] II.[250] III.[250,296] Crystallizes with 1 mole of dioxan; colorless plates, white solid without dioxan,[319] catalyst for butadiene polymerization.[153]
Et_2PNa. II.[439] III. Colorless crystals with 0.5 mole of dioxan.[321]
Et_2PK. I.[567] II.[440] III. Yellow powder.[440]
$(Et_2NCH_2CH_2)_2PLi$. I.[304]
EtPhPLi. III.[188] I.[316] Pale-yellow solid with 1 mole of dioxan.[316]
EtPhPNa. I. Orange.[258] IV. Pale-yellow needles with 1 mole of dioxan.[324]
EtPhPK. IV. Lemon-yellow needles with 1 mole of dioxan.[324]
$(EtOCH_2CH_2)$PhPNa. I.[258]
$(H_2NCH_2CH_2)$PhPLi. I. Pale-yellow, powdery solid.[308]
$(H_2NCH_2CH_2)$PhPNa. I. Yellow in liquid NH_3.[308]
$(EtNHCH_2CH_2)$PhPNa. I.[304]
$(Et_2NCH_2CH_2)$PhPNa. I.[304]
$(EtSCH_2CH_2)$PhPLi. I. White powder.[323]
$(EtSCHMe{\cdot}CH_2)$PhPLi. I. Yellow crystals with 1 mole of dioxan.[323]

Pr_2PLi. I. III. Colorless needles with 0.5 mole of Et_2O; loses Et_2O at 100°.[296]
i-Pr_2PLi. I. Colorless solid.[301]
$Pr(n\text{-}C_6H_{13})PNa$. I.[463]
Bu_2PLi. III. Solid.[296]
Bu_2PNa. I.[463]
t-Bu_2PLi. I. Pale-yellow powder with 0.5 mole of Et_2O or dioxan.[301]
$(c\text{-}C_6H_{11})_2PLi$. I. Pale yellow.[319] III.[298]
$(c\text{-}C_6H_{11})PhPK$. IV. Orange-yellow solution in dioxan.[324]
$(n\text{-}C_8H_{17})_2PNa$. I.[463]
Ph_2PLi. I.[57,319] II.[336,552] III.[296,302,552] IV.[8,57,250,478,575] Yellow crystals with 1 mole of dioxan; is associated in solution;[319] orange in THF solution,[336] UV,[319] ^{31}P +23 ppm (in THF);[186] below -50° a radical anion is formed ($Ph_2PLi^{\cdot -}$ + Li^+) with excess Li.[57]
Ph_2PNa. I.[57,361] II.[319,361] III.[361] IV.[250] Yellow crystals with 1 mole of dioxan,[319] UV,[319] ^{31}P +24.4 ppm (in THF);[186] below -50° a radical anion ($Ph_2PNa^{\cdot -}$ + Na^+) is formed with excess Na.[57]
Ph_2PK. I.[57,361] II.[319,361] III.[361] IV.[250] Orange-red crystals with 2 moles of dioxan,[319] UV,[319] ^{31}P +12.4 ppm (in THF);[186] with excess K a radical anion is formed ($Ph_2PK^{\cdot -}$ + K);[57] useful as a catalyst for the polymerization of methacrylates, styrene,[248] and dienes.[170]
Ph_2PRb. I. Orange-red crystals with 2 moles of dioxan,[186] ^{31}P +7.8 ppm.[186]
Ph_2PCs. I. Orange-red crystals with 2 moles of dioxan,[186] ^{31}P 0.0 ppm.[186]
$Ph(1\text{-}C_{10}H_7)PK$. IV. Violet-red crystals with 2 moles of dioxan.[324]
$(4\text{-}MeC_6H_4)_2PK$. IV. Orange-red crystals with 1 mole of dioxan.[324]
$(2,5\text{-}Me_2C_6H_3)_2PK$. IV. Orange-red solution in dioxan; not completely pure.[324]
$(1\text{-}C_{10}H_7)_2PK$. IV. Violet solution in dioxan.[324]

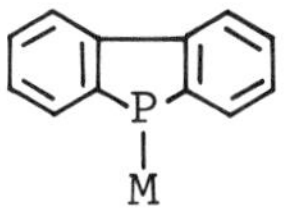

IV. (M = Li, Na, K, Cs).[58] Yellow in THF; with excess alkali metal a radical anion is formed at -50°;[58] the phosphide is in equilibrium with the biphosphine.[58]
$(Me_3Si)_2PLi$. I.[63,460]
$(Ph_3Sn)_2PLi$. From cleavage of $(Ph_3Sn)_3P$ with BuLi.[521]
$Ph_3Sn(Ph)PLi$. From cleavage of $(Ph_3Sn)_2PPh$ with BuLi[521]

TYPE: Alkylenediphosphides

$Ph_2P(CH_2)_3PPhLi$. III. Pale-yellow solid with 1 mole of dioxan.[296]

$LiRP(CH_2)_nPRLi$. I. (n = 2; R = Ph). Yellow, crystalline solid with 1 or 2 moles of dioxan.[316]
(n = 2; R = Et). Colorless crystals with 2 moles of dioxan.[280]
(n = 2; R = c-C_6H_{11}). Colorless crystals.[280]
(n = 3; R = c-C_6H_{11}). Slightly yellow crystals with 2 moles of dioxan.[279]
(n = 3; R = Ph). Slightly yellow solid with 2 moles of dioxan.[295]
(n = 3; R = Et). Colorless solid with 2 moles of dioxan.[281]
(n = 4; R = c-C_6H_{11}). Pale yellow.[279]
(n = 4; R = Ph). Pale-yellow solid with 2 moles of dioxan.[295]
(n = 4; R = Et). Colorless solid with 2 moles of dioxan.[281]
(n = 5; R = c-C_6H_{11}). Pale yellow.[279]
(n = 5; R = Et). Colorless solid with 2 moles of dioxan.[281]
(n = 5; R = Ph). Slightly yellow solid with 4 moles of dioxan; loses 3 moles of dioxan on heating.[295]
(n = 6; R = c-C_6H_{11}). Pale-yellow solid with 2 moles of dioxan.[279]
(n = 6; R = Ph). Pale-yellow solid with 2 moles of dioxan.[295]
(n = 6; R = Et). Colorless solid with 2 moles of dioxan.[281]
All these phosphides show green chemoluminescence with traces of oxygen.

$C[CH_2PPhNa]_4$. IV. Orange-yellow solid, nonelectrolyte; forms an adduct with THF; IR.[162]

$C[CH_2PPhK]_4$. IV. Orange-yellow solid, nonelectrolyte; forms an adduct with THF; IR.[162]

$PhCH[SCH_2CH_2PPhLi]_2$. I. Yellow crystals with 1 mole of dioxan.

$PhCH[SCH{\cdot}MeCH_2PPhLi]_2$. I. Solid.[323]

TYPE: Bi-, Tri-, Tetra-, and Pentaphosphinediphosphides

$EtPhP_\alpha{-}P_\beta EtLi$. III.[188] ^{31}P: P_α +17.2, P_β +112.8 ppm; J_{PP} 396 Hz.[188]

$M(R)P{-}PM(R)$. III. (R = Et; M = K). Crystals with 1 mole of THF,[299] ^{31}P +79.6 ppm.[188]
(R = Ph; M = Li). Crystals with 1 mole THF.[47,300,302]
(R = Ph; M = Na). Orange-yellow.[359,476]

(R = Ph; M = K). Slightly yellow with 1 mole of THF.[300]

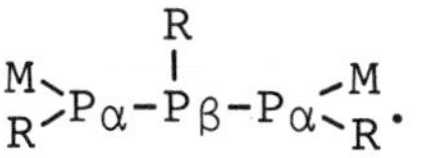 III. (R = Et; M = K). ^{31}P: P_{α} +78.5, P_{β} +23.9; J_{PP} 306 Hz.[188]

(R = Ph; M = K). Crystals with 2 moles of THF;[300] may have cyclic structure;[282] ^{31}P +49.8 ppm.[282]

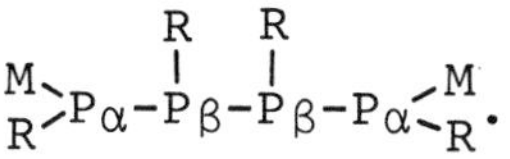 III. (R = Me; M = Li). Yellow solid with 2 moles of THF,[311] ^{31}P: P_{α} +149.9 and +161.2, P_{β} +31.7 and +42.7 ppm.[311,188]

(R = Me; M = K). Yellow crystals with 2 moles of THF.[311]

(R = Et; M = Li). Yellow crystals with 3 moles of THF.[299]

(R = Et; M = K). Yellow crystals.[299,311]

(R = t-Bu; M = K). Yellow crystals with 1 mole of THF.[291]

(R = c-C_6H_{11}; M = K). Yellow.[299]

(R = Ph; M = Li). Slightly yellow with 4 moles of dioxan,[300,302] ^{31}P: P_{α} +86, P_{β} +8.2 ppm/ J_{PP} 216 Hz.[188]

(R = Ph; M = Na). Yellow solid, with 2 moles of dioxan.[300]

(R = Ph; M = K). Slightly yellow with 4 moles of dioxan.[300]

$Ph_2P(PPh)_3Li$. III.[300]

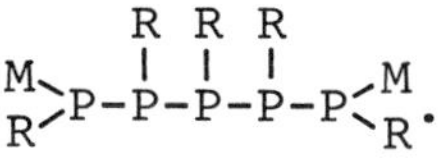 III. (R = Me; M = Li). Solid with 2 moles of THF.[311]

(R = Me; M = Na). Solid with 2 moles of THF,[311] ^{31}P many peaks.[31]

(R = Me; M = K). Yellow crystals with 2 moles of THF.[311]

(R = Et; M = K). Yellow crystals.[311]

Organosubstituted Alkaline Earth Phosphides

Roman numerals refer to methods of preparation described in Section B.1 of the text.

H_2PMgBr. I. Not isolated.[12]

$(H_2P)_2Ca \cdot 6NH_3$. I.[377,564]

HPCa. From above by pyrolysis.[564]

PhHPMgEt. I. Crystals;[278] loses ethane on heating to 160° under vacuum and yields polymeric $(PhP \cdot Mg_x)$.[278]

(PhHP)$_2$Mg.   I.[278]
PhP(MgBr)$_2$.   I.[258,334,414]
(Me$_2$PBeMe)$_n$.   I. Polymeric.[112]
Ph$_2$PMgBr.   I.[278,334]
Ph$_2$PMgI.   I.[278]
Ph$_2$PMgEt.   I.   II.   Slightly gray solid,[296] colorless
     crystals with 1 mole of THF.[278]
(Ph$_2$P)$_2$Mg.   I.   II.   Nearly colorless solid;[296] crystal-
     lizes with 4 moles of THF;[278] loses 3.5 moles of THF
     on heating; is cleaved by MeI; has reactivity similar
     to alkali phosphides.[278]
Ph$_2$PCaCl.   I.   Yellow solid.[278]

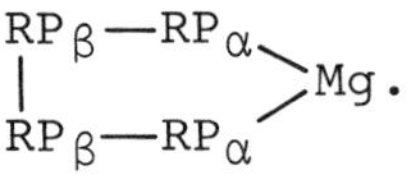

$$\begin{matrix} RP_\beta\!-\!RP_\alpha \\ | \\ RP_\beta\!-\!RP_\alpha \end{matrix}\!\!\Big\rangle Mg.$$ Obtained from the corresponding di-K phos-

     phide and MgBr$_2$.
     (R = Et).   Crystals with 2 moles of THF,[311] [31]P: P$_\alpha$ +120.
     and +130.8 ppm, P$_\beta$ +5.7 and +15.6 ppm,[311] (main signals).
     (R = c-C$_6$H$_{11}$).   Crystals with 2 moles of THF.[311]
     (R = Ph).   Yellow crystals with 2 moles of THF,[311]
     loses THF at 180°.

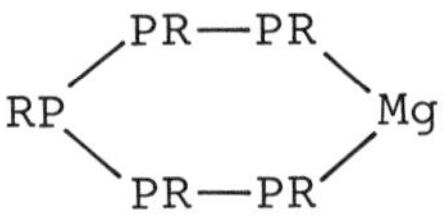

$$RP\overset{\displaystyle PR\!-\!PR}{\underset{\displaystyle PR\!-\!PR}{<\quad>}}Mg$$ Obtained from the corresponding di-K

     phosphide and MgBr$_2$.
     (R = Me).   Solid with 2 moles of THF.[311]
     (R = Et).   Solid with 1 mole of THF.[311]

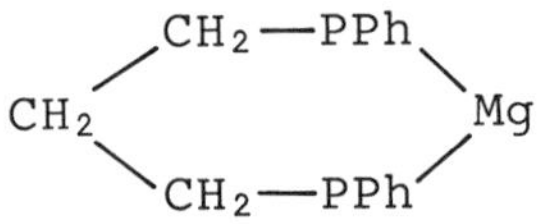

$$CH_2\overset{\displaystyle CH_2\!-\!PPh}{\underset{\displaystyle CH_2\!-\!PPh}{<\quad>}}Mg$$ I.   Crystals.[278]

Organophosphorus Compounds of Group III Elements

Roman numerals refer to methods of preparation described
in Section C.1 of the text.

Organophosphorus-Boron Compounds

TYPE: Monomeric Derivatives

Li[B(PEt$_2$)$_4$].   III.   Colorless crystals, m. 264 to 266°
     (dec.), unstable in air.[199]
Li[H$_3$BPEt$_2$BH$_3$].   III.   Colorless crystals, m. 152 to 156°
     (from benzene) (dec.).[199]

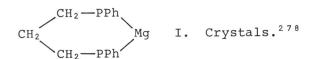

$Na[BH_3PMe_2]$. I. Crystallizes with 0.5 mole of dioxan.[7]
$K[Me_3BPMe_2BMe_3]$. III. White solid, m. 180°, sensitive to H_2O and air; at 215° yields small amounts of $(Me_2PBMe_2)_3$.[556]
H_2P-BMe_2. II. Solid, m. 45 to 50°;[88] could not be prepared by methods I and III; sensitive to air and H_2O; changes gradually into a polymer.[88]
Et_2P-BPh_2. III. Solid, m. 192° (from benzene), IR, μ 1.1 D, stable to hydrolysis.[115]
Et_2P-$B(C_6H_4Br$-$4)_2$. III. Solid, m. 202° (from benzene/hexane), IR, μ 0.7 D, methiodide, m. 310 to 312° (dec.).[115]
Et_2P-$B(NMe_2)_2$. III. Liquid, b_{53} 134 to 136°,[445,446] n_D^{20} 1.4872, IR, ^1H NMR, sensitive to air and protic solvents, methiodide unstable; cleaved by sulfur.[446]
Et_2P-$BBu(NMe_2)$. III. Liquid, b_{12} 109 to 112°.[447]
Et_2P-$BH(NEt_2)$. III. Liquid, $b_{0.001}$ 36°, not very stable; gives polymer.[447]
Et_2P-$BCl(NEt_2)$. III. Liquid, $b_{0.002}$ 46 to 47°,[445,447] n_D^{20} 1.4803.
Et_2P-$BBu(NEt_2)$. III. Colorless liquid, $b_{0.001}$ 74 to 79°.[447]
Et_2P-$BNEt_2(OC_6H_{11}$-c). III. From bis(phosphino) compound below and HOR, yellow liquid, $b_{0.001}$ 89 to 94°, n_D^{20} 1.4831, sensitive to air and H_2O.[447]
$(Et_2P)_2\cdot BNEt_2$. III. Lemon-yellow oil, $b_{0.001}$ 79 to 81°,[445,447] n_D^{20} 1.5183, IR, sensitive to air and protic reagents, methiodide unstable.[447]
Bu_2P-$B(NMe_2)_2$. III. Liquid, b_{11} 136 to 139°.[446]
PhHP-BClPh. II. Liquid, $b_{0.001}$ 98 to 100°[116] (prepared without amine).
Ph_2P-BPh_2. II. III. Solid; sublimes at 240°/10^{-3} torr;[115] m. 234°,[218] IR,[115] relatively resistant to hydrolysis and oxidation,[115] mass spect.[4]
$(Ph_2P)_2BPh$. II. Colorless crystals, m. 143 to 144° (from benzene/hexane);[116] on heating yields a yellow polymer, m. 120 to 132°.
$PhP[BPEt_2(NEt_2)]_2$. III. Oil, decomposes on dist.[447]
$PhP[B(NMe_2)_2]_2$. III. Colorless crystals, m. 60 to 64°.[447]
$PhP[BPh_2]_2$. II. Colorless crystals, m. 148 to 150° (from benzene/hexane), sensitive to air and moisture.[116]
$(3$-$MeC_6H_4)_2P$-BPh_2. III. Colorless solid, m. 123 to 124°, IR, μ 2.2 D.[115]
$(3$-$MeC_6H_4)_2P$-$B(C_6H_4Me$-$3)_2$. III. Crystals, m. 257 to 258° (from benzene/hexane), IR, μ 2.9 D.[115]
$(3$-$MeC_6H_4)_2P$-$B(C_6H_4Br$-$4)_2$. III. Crystals, m. 281° (from benzene/hexane), IR, μ 0.6 D.[115]
$(3$-$MeC_6H_4)_2P$-$B(C_6H_2Me_3$-$2,4,6)_2$. III. White solid, m. 264 to 265°, IR.[116]
$(3$-$MeC_6H_4)_2P$-$B(C_6H_4Ph$-$4)_2$. II. Solid, m. 77 to 78°, IR.[116]
$Et_2P(Et_2N)B$-$B(NEt_2)(PEt_2)$. From $Et_2P(Et_2N)BCl$ and K, liquid, $b_{0.001}$ 160 to 170°.[445,447]

TYPE: Dimeric Derivatives

$[Et_2P\text{-}BCl_2]_2$. I. Solid, m. 136 to 138° (from petroleum ether),[445,448] IR.[448]
$[Et_2P\text{-}BBr_2]_2$. I. Solid, m. 172 to 175° (sublimes), IR, μ 2.54 D.[448]
$[Et_2P\text{-}BPr_2]_2$. I. Solid, m. 72 to 74°, IR.[448]
$[Et_2P\text{-}B(OBu)_2]_2$. I. Liquid, $b_{0.001}$ 120° (not obtained pure).[448]
$[Bu_2P\text{-}BH_2]_2$. I. Liquid.[458]
$[Bu_2P\text{-}BBr_2]_2$. I. Solid, m. 111 to 113°, $b_{0.001}$ 185°.[444,448]
$[Ph_2P\text{-}BBr_2]_2$. II. Colorless monoclinic laths, m. 183 to 184° (from benzene),[216,218,444] not stable in moist air, IR;[217] cleaved by nucleophiles.[217]
$[Ph_2P\text{-}BI_2]_2$. II. Colorless monoclinic crystals, m. 194 to 195° (dec.) (from petroleum ether/$CHCl_3$),[216,444] x-ray analysis gives[64] nonplanar, four-membered P-B ring, P-B 2.03 Å, <P-B-P 88°,[64] stable in moist air, IR.[217]
$[PhP\text{-}BPh]_2$. II. Colorless solid, m. 89 to 91°.[116]

TYPE: Trimeric Derivatives

$[Me_2P\text{-}BH_2]_3$. I. Orthorhombic crystals, m. 85 to 86° (from MeOH),[65,85,86,88,228,442] m. 87 to 88°,[44,66,100,228,444] b_{760} 235° (est.),[88] log P_{mm} (solid) = 13.1 - (4515/T) and log P_{mm} (liq.) = 8.557 - (2887/T),[88] d_4^{20} 0.94 to 0.96,[88] d_4^{20} (calc. 0.959);[237] hydrolyzes slowly at 300°; stable to air,[88] IR,[44,106] Raman,[106] ^{31}P +47 ppm;[425] mass spect. indicates that dec. begins at 360°,[179] x-ray analysis[237] shows cyclohexanelike ring with P-B 1.93, P-C 1.83 Å, P-B-P 112.2°, B-P-B 118.1°, P-B-C 109°, and C-P-C 100.4°,[237] spontaneous ignition temp. 240°.[226]
$[Me_2P\text{-}BMe_2]_3$. II. Solid, m. 102 to 104° (?),[159] m. 334° (from MeOH), b. nearly the same,[88] log P_{mm} = 9.883 - (4184/T), stable in air,[88] IR, Raman,[106] spontaneous ignition temp. 302°.[226]
$Me_5(H_2N)P_3B_3H_6$. IV. Solid, m. 35.5 to 35.7°.[88]
$Me_6P_3B_3H_5X$. IV. (X = Br). m. 76 to 78°;[224] (X = I). m. 80 to 82°;[565] (X = Et). m. 36.5 to 37.0°; (X = HCOO), m. 45.5 to 46.5°; (X = CH_3COO), m. 44 to 45°; (X = C_6H_5COO), m. 81.0 to 81.5°;[565] (X = CN), m. 108 to 110°; (X = NCO), m. 71 to 72°,[224,565] (X = SCN), m. 63 to 64°; (X = SH), m. 51 to 52°; (X = SMe), m. 60 to 61°;[225,565] (X = $NHCONEt_2$), m. 101 to 102°;[224,565] (X = NCO) plus ethylenediamine gives polymer.[224]
$[Me_2P\text{-}BPr_2]_3$. I. Liquid, b_2 165 to 167°.[44]
$[Me_2P\text{-}BF_2]_3$. IV. Needles, m. 129 to 130° (from methylcyclohexane),[226] m. 127 to 128°,[92] d_4^{25} 1.24,[226] IR.[106]
$[Me_2P\text{-}BCl_2]_3$. IV. Solid, m. 384 to 385°,[226] m. 377 to

378°,[92] m. 393 to 394° (from benzene),[44] d_4^{25} 1.44, IR,[44,106] spontaneous ignition temp. 395°.[226]

$[Me_2P\text{-}BBr_2]_3$. IV. Solid, m. above 411° (dec.),[92,226] d_4^{25} 2.26,[226] IR,[106] spontaneous ignition tem. >420°.[226]

$[Me_2P\text{-}BI_2]_3$. IV. Solid, m. above 411° (dec.),[92,226] d_4^{23} 2.75, spontaneous ignition temp. >430°.[226]

$[(CF_3)_2P\text{-}BH_2]_3$. I. Solid, m. 30.5°, b_{760} 177° (est.), dec. at 200°, hydrolyzes in NaOH.[69]

$[MeEtP\text{-}BH_2]_3$. I. Liquid.[91]

$[MeEtP\text{-}BCl_2]_3$. IV. Solid, m. 210 to 211°.[226]

$Me_3Et_3P_3B_3HCl_5$. IV. Low-melting wax.[92]

$[Me(Me_2NCH_2CH_2CH_2)P\text{-}BH_2]_3$. I. Liquid; purified by molecular dist.; IR.[562]

$[Me(Me_2N(\rightarrow O)CH_2CH_2CH_2)P\text{-}BH_2]_3$. From amino compound by oxidation. Glass, IR.[562]

$[Me(CH_2{=}CHCH_2)P\text{-}BH_2]_3$. By pyrolysis of the aminoxide above, liquid, $b_{0.005}$ 80 to 88°, IR.[562]

$[MePhP\text{-}BH_2]_3$. I. Solid, m. 125 to 126°.[85,226]

$[Et_2P\text{-}BH_2]_3$. I.[44,100,236,446,448] III.[199] [Solid, m. 67 to 68° [263] (?)], $b_{1.5}$ 133 to 134°,[86] b_{12} 185°,[100] $b_{0.1}$ 106 to 108°,[44] $b_{0.005}$ 112°,[199] n_D^{20} 1.5312,[44] IR,[446,448] μ 2.66 D,[448] stable in air.[86]

$Et_5HP_3B_3H_6$. I. Liquid, not pure, b_3 158 to 162°, n_D^{25} 1.521.[86]

$[Et_2P\text{-}BF_2]_3$. IV. Colorless needles, m. 69 to 70° (from methylcyclohexane).[226]

$[Et_2P\text{-}BCl_2]_3$. IV. Rods, m. 181.5° (from petroleum ether), IR.[44]

$[Et_2P\text{-}BBr_2]_3$. IV. Rods, m. 304 to 305° (from $CHCl_3$/petroleum ether), IR.[44]

$[Et_2P\text{-}BI_2]_3$. IV. Rhombic crystals, m. 325 to 330° (from CCl_4/petroleum ether), IR.[44]

$[Bu_2P\text{-}BH_2]_3$. I. Feathery crystals, $b_{0.001}$ 160°, m. not given.[448]

$[C_4H_8P\text{-}BH_2]_3$. I. Long needles, m. 169.3°.[87,89]

$[(i\text{-}C_5H_{11})_2P\text{-}BH_2]_3$. I. Liquid.[85]

$[C_5H_{10}P\text{-}BH_2]_3$. I. Liquid.[85]

$[(c\text{-}C_6H_{11})_2P\text{-}BH_2]_3$. I.[85,89]

$[PhHP\text{-}BH_2]_3$. I. Orange-yellow solid.[236]

$[PhHP\text{-}BClPh]_3$. II. Waxy solid; prepared without amine; sublimed under vacuum; IR;[554] also obtained as monomer.[116]

$[Ph_2P\text{-}BH_2]_3$. I. Solid, m. 160°,[565] m. 161° (from benzene/EtOH);[85,214] and when tried at 140°, m. 179°,[215] m. 177°,[263] IR.[214]

$Ph_6P_3B_3H_{6-n}Cl_n$. IV. (n = 4). Solid, m. 207 to 208° (not pure).[264]

(n = 5). Solid, m. 249°.[264]

(n = 6). Colorless crystals, m. 258 to 259° (from

benzene/EtOH),[214] m. 243°,[264] IR.[214]
$Ph_6P_3B_3H_{6-n}Br_n$. IV. (n = 1) m. 183°; (n = 2) m. 210 to 212°; (n = 3) m. >260° (dec.) (form A), m. >200° (dec.) (form B);[215] (n = 4) m. 288° (dec.); (n = 5) m. 270° (dec.);[215] (n = 6) colorless needles, m. 274° (from benzene),[214,215] IR,[214] $CHCl_3$ adduct, m. 160° (dec.).[214]
$Ph_6P_3B_3H_{6-n}I_n$. IV. (n = 1) m. 185°; (n = 2) m. 217°; (n = 3) m. 285° (dec.).[215]
$Ph_6P_3B_3H_2Br_3I$. IV. Solid, m. 200°.[89]
$[(4\text{-}MeC_6H_4)_2P\text{-}BH_2]_3$. I. Solid, m. 200°.[89]

TYPE: Tetrameric Derivatives

$[Me_2P\text{-}BH_2]_4$. I. Twinned crystals, m. 160 to 161° (from benzene),[66,88] b_{760} 310° (est.),[88] log P_{mm} (solid) = 12.28 - (4900/T) and log P_{mm} (liquid) = 8.364 - (3196/T),[88] IR, Raman,[106] x-ray analysis[223] gives eight-membered puckered ring with P-B 2.08, P-C 1.84 Å, ∠B-P-B 125°, P-B-P 104°, C-P-C 103°;[223] hydrolyzes slowly at 300°; on heating converts to trimer and other products.[88]
$[(CF_3)_2P\text{-}BH_2]_4$. I. Solid, m. 116°, log P_{mm} (solid) = 12.683 - (4489.2/T).[69]

TYPE: Polymeric Derivatives

$[MeHP\text{-}BH_2]_n$. I. Nonvolatile oil.[88]
$[Me_2P\text{-}BHMe]_n$. I. Solid, not obtained pure, m. 120 to 132°.[88]
$[Me_2P\text{-}BH_2]_{22}\cdot Me_2PH$. I. Solid, m. 180 to 200°.[66]
$[Me_2P\text{-}BH_2]_n$. I.[65,66,88,94,565,566] (n = 80) m. 170 to 172°;[94,566] (n = 213) m. 169 to 172°.[94]
$[Me_2PBH_2]_4\cdot MePBH$. I. Solid, with naphthalenelike structure, m. 98 to 99°.[93]
$[MeEtP\text{-}BH_2]_{21}$. I. Solid, m. 118 to 126°;[94,566] has interesting plastic properties.[566]
$[MePrP\text{-}BH_2]_{34}$. I. Liquid.[94,565]
$[Me(CH_2{=}CHCH_2)P\text{-}BH_2]_n$. I. Solid, m. 76 to 78°.[94,562]
$[MeBuP\text{-}BH_2]_{26}$. I. Liquid.[565]
$[MeAmP\text{-}BH_2]_{14}$. I. Liquid.[565]
$[MeC_6H_{13}P\text{-}BH_2]_8$. I. Liquid.[565]
$[MeC_{12}H_{25}P\text{-}BH_2]$. I. Liquid, n_D^{25} 1.4926 (mol. wt. 870, n ∿ 4).[94]
$[MePhP\text{-}BH_2]_6$. I. Liquid.[92,565]
$[Et_2P\text{-}BH_2]_n$. I. Solid, m. 263 to 267°.[94]
$[(Et_2P)_2BH]_n$. III. Solid, m. 135 to 145°, sensitive to air and H_2O.[447]
$[\overline{(CH_2)_4P}\text{-}BH_2]_{15}$. I. Glass.[94,565]
$[PhHP\text{-}BH_2]_n$. I. Eliminates $PhPH_2$ at 250°.[353]
$[Ph_2PBH_2]_n$. I. Thermoplastic; prepared at 164 to 170° for 3 to 4 hr.[371]

$[PhP\text{-}BPh]_n$. I.[444] II.[167] Yellow, crystalline powder,[444] white powder, m. 142 to 144°.[167]
$[\text{-}N(CH_2)_3PhP\text{-}BH_2\text{-}]_n$. I. From $H_2N(CH_2)_3PhPH$ and BH_3NEt_3 at 100°.[578]
$[Me_3SiP\text{-}BH_2]_n$. I. Solid, m. 140 to 205°.[376]

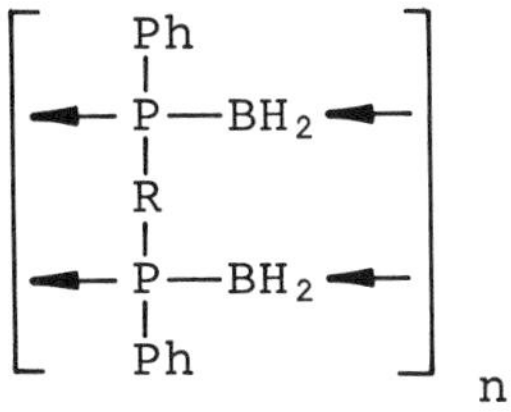

I.[354] By heating a disecondary diphosphine with a small excess of $Et_3N \cdot BH_3$ at increasing temp. of 300°; linear, glassy, insoluble polymers are formed which are stable up to 350 to 400° and, depending on the bridging group, have good chemical and hydrolytic stability.[354]
$[R = \text{-}(CH_2)_{10}\text{-}]$. Transparent solid, IR,[354] thermal stability.[354]
$[R = 1,4\text{-}CH_2CH_2C_6H_4CH_2CH_2\text{-}]$. Glassy substance, IR, thermal stability.[354]
$[R = \text{-}(CH_2)_3O(CH_2)_3\text{-}]$. Opaque solid, IR, thermal stability.[354]
$[R = \text{-}(CH_3)_3O(CH_2)_3\text{-}HB(CH_2)_3O(CH_2)_3\text{-}]$. Glassy polymer, m. 110 to 113°, IR.[354]
$[R = \text{-}CH_2CH_2SiMe_2OSiMe_2CH_2CH_2\text{-}]$. Transparent solid, IR, thermal stability. Si-O-Si bonds hydrolyzed in 2.5 N NaOH.[354]
$[R = \text{-}(CH_2)_3OBHO(CH_2)_3\text{-}]$. Tough, opaque mass which on heating at 260° gave a white solid, IR.[354]

TYPE: Other Borane Derivatives

2,3-μ-$Me_2PB_5H_8$. III. Liquid, m. 17°, b_{760} 191° (est.), IR, ^1H and ^{11}B NMR, ^{30}P +85 ppm,[73] log P_{mm} = 7.0157 +1.75 log T - 0.005T - (3010/T),[73] dec. at 85° or at 25° over 1 week; is hydrolyzed by HCl at 100° for 24 hr.[73]
2,3-μ-$MeCF_3PB_5H_8$(A). III. Liquid, m. 14°, b_1 25°; isomerizes irreversibly to form B at room temp. in 48 hr; ^1H ^{19}F, and ^{11}B NMR, ^{31}P +33.5 ppm, IR;[73] is associated in solution; mass spect.[73]
2,3-μ-$MeCF_3PB_5H_8$(B). From form A above. m. -28°, b_{760} 137 (est.), log P_{mm} = 6.3335 + 1.75 log T - 0.005T - (2450/T), IR, ^1H, ^{19}F, and ^{11}B NMR, ^{31}P +47 ppm,[73] mass spect.; hydrolyzed in H_2O at 100° for 48 hr.
1-$(CF_3)_2PB_5H_8$. III. Liquid, m. -42°, b_{760} 184° (est.), IR, ^1H, ^{19}F, and ^{11}B NMR, ^{31}P +25.1 ppm, log P_{mm} = 6.6114 + 1.75 log T - 0.005T - (2790/T).[73]

$B_{10}H_{13}\cdot PPh_2$. III. Solid, m. 149°,[499,500] m. 136 to 137° (from cyclohexane);[428] behaves as a strong acid in aqueous media;[428] forms salts.[428,429,499,500]

$B_9H_{14}\cdot PPh_2$. III. Solid, m. 170° (dec.) (from EtOH/$CHCl_3$),[428] m. 169°.[500]

$B_9H_{13}Br\cdot PPh_2$. III. Solid, m. 143° (from EtOH/$CHCl_3$).[428]

1,2-$B_{10}H_{10}$CHP. III.[557] Solid, m. 353.5 to 354.5°, IR, UV, ^{1}H and ^{11}B NMR, ^{31}P +57 ppm,[557] mass spect.; does not react with Br or MeI; behaves as Lewis acid; properties similar to the isoelectronic 1,2-$B_{10}H_{11}C_2H_2$; EtONa removes phosphorus atom, and treatment with piperidine gives [7,8-$B_9H_{10}CHP^-$][$C_5H_{10}NH_2^+$], m. 307 to 308°;[557] this anion gives with $M(CO)_6$ compounds of the type [7,8-$B_9H_{10}CHP\cdot M(CO)_5$]$NMe_4$,[541] (M = Mo, W, Cr);[541] deprotonation of 7,8-$B_9H_{10}CHP^-$ with NaH or Et_3N generates the phosphacarbollyl anion 7,8-$B_9H_9CHP^{2-}$ which forms sandwich-bonded transition metal compounds.[378a]

1,2-$B_{10}H_{10-x}Br_x$CHP. From above by treatment with $AlCl_3/Br_2$.
(x = 1). Crystals, m. 244.5 to 245.5°.
(x = 2). Crystals, m. 255 to 256°.
(x = 3). Crystals, m. 352 to 353°.[557]

1,7-$B_{10}H_{10}$CHP. From 1,2-$B_{10}H_{10}$CHP by heating to 550°,[557] solid, m. 319 to 320°,[577] m. 325 to 327°, IR, UV, ^{1}H and ^{11}B NMR, ^{31}P +71 ppm, mass spect.;[557] treatment with piperidine gives the anion [7,9-$B_9H_{10}CHP^-$][C_5H_{10}-NH_2^+], m. 296 to 297°;[557] this anion gives with $M(CO)_6$ compounds of the type [7,9-$B_9H_{10}CHP\cdot M(CO)_5$]$NMe_4$ (M = Mo, W, Cr);[541] further deprotonation of the anion with NaH or Et_3N yields the phosphacarbollyl anion 7,9-$B_9H_9CHP^{2-}$ which forms sandwich-bonded transition metal compounds.[378a]

1,7-$B_{10}H_{10-x}Br_x$CHP. From above by treatment with $AlCl_3/Br_2$.
(x = 1). Crystals, m. 226 to 227°.
(x = 2). Crystals, m. 234 to 235°.[557]

1,12-$B_{10}H_{10}$CHP. From 1,7-$B_{10}H_{10}$CHP by heating to 650°,[557] or 700 to 800°.[577] Crystals, m. 309 to 310° (from hexane),[577] m. 314 to 315.5°, IR, ^{1}H and ^{11}B NMR, mass spect.; is stable to base.[557]

1,7-$B_{10}H_{10}C(CO_2H)$P. From m-$B_{10}H_{10}$CHP, LiBu, and CO_2. Crystals, m. 177 to 178°, pK_a 3.6.[576]

1,7-$B_{10}H_{10}$C(HgMe)P. Crystals, m. 214 to 215°.[576]

1,12-$B_{10}H_{10}C(CO_2H)$P. From p-$B_{10}H_{10}$CHP, LiBu, and CO_2. Crystals, m. 162 to 163°.[577]

7,8-B_9H_{10}CHPMe. From 7,8-$B_9H_{10}CHP^-$ and MeI. Crystals, m. 108 to 109°, IR, ^{1}H NMR, mass spect.;[557] deprotonation with NaH or Et_3N generates the anion 7,8-$B_9H_9CHPMe^-$ which forms sandwich-bonded transition metal compounds.[378a]

7,9-B_9H_{10}CHPMe. From 7,9-$B_9H_{10}CHP^-$ and MeI. Crystals, m. 112 to 113°, IR, ^{1}H NMR, ^{31}P +95 ppm, mass spect.;[557] deprotonation with NaH or Et_3N gives the anion 7,9-$B_9H_9CHPMe^-$ which forms sandwich-bonded transition metal compounds.[378a,571a] New numbering system designates these compounds as 1,7 derivatives.

7,9-B_9H_{10}CHPH. From 7,9-$B_9H_{10}CHP^-$ and acid, solid, m. 206 to 210° (dec.), IR, mass spect.[577]

Organophosphorus-Aluminum Compounds

TYPE: Monomeric Derivatives

Li[$AlH_3(PEt_2)$]. III. White crystals; inflames in air.[207]

Li[$AlH_2(PEt_2)_2$]. III. Colorless crystals; inflames in air;[206,207] IR.[207]

Li[$AlH(PMe_2)_3$]. II. White solid; hydrolyzes in H_2O.[82]

Li[$AlH(PEt_2)_3$]. III. Not obtained completely pure; inflames in air; IR.[82]

Li[AlH{$P(C_6H_{11}$-c$)_2$}$_3$]. I. Yellow, amorphous solid; hydrolyzes in water.[322]

Li[$AlH(PPh_2)_3$]. I. Yellow, amorphous solid; hydrolyzes in water.[322]

Li[$Al(PEt_2)_4$]. III. Colorless crystals, m. 224 to 226° (from ether/pentane);[205,207] inflames in air.[207]

Me_2P-AlH_2·Me_2PH. II. Solid, not obtained pure, stable to 100°, reactive to air and moisture.[82]

Me_2P-$AlCl_2$·Et_2O. II. Crystalline solid, ether complex uncertain.[82]

$(Me_2P)_2AlCl$·Et_2O. II. Transparent crystals which on standing become opaque, stable at 150°, ether complex uncertain.[82]

Et_2P-$AlCl_2$·$OPCl_3$. From $(Et_2P$-$AlCl_2)_3$ and $POCl_3$. Orange-red, viscous liquid.[204]

Ph_2P-$AlEt_2$·Et_2O. I. III. Colorless needles, m. 148°,[336] m. 151°,[277] m. 145 to 150°,[296] sensitive to air and H_2O.[277]

TYPE: Dimeric Derivatives

[Et_2P-$AlEt_2$]$_2$. III. Colorless crystals, m. 15°, b_{2-3} 148° (slight dec.).[277]

[(c-$C_6H_{11})_2P$-$AlEt_2$]$_2$. III. Crystals, m. 123°.[277]

[Ph_2P-$AlMe_2$]$_2$. I. Crystals, m. 230°; inflames in air.[113]

[Ph_2P-$AlEt_2$]$_2$. I. III. Colorless needles, m. 124 to 126° (from benzene/petroleum ether); is cleaved to monomeric complex by Lewis bases.[336]

TYPE: Trimeric Derivatives

[Me_2P-$AlMe_2$]$_3$. I. White solid, m. >200°,[147] m. 238 to 244°;[33] cleaved by HCl;[147] glassy polymer at room temperature, trimer in benzene solution or vapor phase, IR.[33]

[Me_2P-$AlEt_2$]$_3$. I. Solid, m. 35°;[120] is decomposed by Me_2NH.[120]

$[Et_2P\text{-}AlH_2]_3$. III. Liquid, m. 1 to 2°, sensitive to air and water,[206,207] IR.[207]
$[Et_2P\text{-}AlMe_2]_3$. I. Solid; sublimes at 120 to 130°; glassy polymer at room temp., trimer in benzene solution, IR.[33]
$[Et_2P\text{-}AlCl_2]_3$. III. White crystals, m. above 80° (dec.),[204] sensitive to H_2O.

TYPE: Polymeric Derivatives

$[(Me_2P)_3Al]_n$. I. Stable to high temp., low polymer.[82]
$Me_2PAl_5H_4$. I. Nonvolatile.[82]
$[(Et_2P)_3Al]_n$. III. Dark, viscous oil; inflames in air; not obtained completely pure.[207]
$[(Et_2P)_2AlH]_n$. III. Highly viscous, yellow liquid, IR.[207]
$(Ph_2Al\text{-}PEt_2)_n$. III.[207]

Organophosphorus-Gallium Compounds

$Ph_2P\text{-}GaMe_2$. III. Liquid, $b_{0.001}$ 125 to 150°; gives dimer slowly.[113]
$[Ph_2P\text{-}GaMe]_2$. I. III. Crystals, m. 194° (from benzene), sensitive to air and H_2O.[113]
$[Me_2P\text{-}GaH_2]_n$. I. ^1H NMR, unstable; decomposes at room temp. to Me_2PH, Ga, and hydrogen.[231]
$[Me_2P\text{-}GaMe_2]_3$. I. Solid, m. 265 to 267° (dec.), IR,[33] glassy polymer at room temp., trimer in benzene solution.[33]
$[Et_2P\text{-}GaMe_2]_3$. I. Solid, subl. temp. 120 to 130°, IR,[33] glassy polymer at room temp., trimer in benzene solution.[33]
$[PhP\text{-}GaMe]_n$. I. Colorless below 200°, yellow between 200 and 280°, not well characterized.[33]

Organophosphorus-Indium Compounds

$[Me_2P\text{-}InMe_2]_3$. I. Solid, m. 115 to 120° (dec.), glassy polymer at room temp., trimer in benzene solution, IR.[33]
$[Et_3P\text{-}InMe_2]_3$. I. Solid, m. 100 to 104° (dec.), glassy polymer at room temp., trimer in benzene solution, IR.[33]
$[PhP\text{-}InMe]_n$. I. Yellow-orange, not well characterized.[33]
$[HP\text{-}InMe]_n$. I. Yellow amorphous polymer, decomposes at 120°.[154]

Organophosphorus Compounds of Group IV Elements

Roman numerals refer to methods of preparation described in Section D.1 of the text.

TYPE: P-Si Compounds

H_2P-$SiMe_3$. I.[460] V.[63] Liquid, b_{760} 77.5 to 78°,[63,460] m. -75°,[63] n_D^{25} 1.4638,[63] IR, Raman,[63,63a] ^{31}P +239.0 ppm,[185] inflammable in air; IR and Raman also of D_2P-$SiMe_3$.[63a]
H_2P-$SiHMe_2$. I. ^{31}P +244 ± 3 ppm, J_{PH} 182 Hz.[450a]
H_2P-$SiMe(NEt_2)_2$. I. Liquid, b_5 80°, n_D^{20} 1.4606, d_4^{20} 0.8756.[25]
MeHP-SiH_3. I. Liquid, vapor pressure at 0° is 113 ± 1 mm Hg, IR, ^1H NMR.[135]
MeHP-$SiMe_3$. I. Liquid, b_{153} 54 to 55°, d_4^{20} 0.7969.[60]
MeHP-$SiMe_2NEt_2$. I. Liquid, b_{19} 75 to 76°, n_D^{20} 1.4702, d_4^{20} 0.8711.[25]
MeHP-$SiMe(NEt_2)_2$. I. Liquid, $b_{2.5}$ 82 to 83°, n_D^{20} 1.4800, d_4^{20} 0.8962.[25]
Me_2P-SiH_3. I. Liquid, vapor pressure at 0° is 61 ± 1 mm Hg, IR, ^1H NMR.[135]
Me_2P-$SiMe_3$. Cleaved by NOCl to give Me_3SiCl, $Me_3SiOSiMe_3$, $Me_2P(O)Cl$, N_2O, NO, N_2.[97a]
Me_2P-$SiCl_3$. I.[195,200] IV.[382a] Liquid, b_{25} ∿50°,[200] b_3 83°.[195]
$(CF_3)_2P$-$SiMe_3$. I. Liquid, b_{760} 112°, log P_{mm} = 7.795 - (1892/T), IR, ^1H and ^{19}F NMR.[234]
Et_2P-$SiCl_3$. I. Liquid, $b_{0.06}$ 32 to 33°.[196]
Et_2P-SiH_3. I. Liquid, b_{79} 53.5°, ^1H NMR;[196] decomposes at room temperature to SiH_4 and $(Et_2P)_2SiH_2$.[198]
Et_2P-SiH_2Me. I. Liquid, b_{760} 139 to 140°, ^1H NMR.[196]
Et_2P-$SiHMe_2$. I. Liquid, b_{25} 58.5°, ^1H NMR.[196]
Et_2P-$SiClMe_2$. I. Liquid, b_5 53 to 54°,[196,200] ^1H NMR.[196]
Et_2P-$SiMe_3$. I. Liquid, b_{20} 71 to 72°,[196,200] b_{11} 56 to 57°,[448] ^1H NMR;[196] gives a salt with EtI, m. 122 to 123°;[201] excess EtI cleaves P-Si bond.[201]
Bu_2P-$SiMe_3$. I. Liquid, b_{10} 106°, n_D^{20} 1.4683.[448]
$Ph(H_2NCH_2CH_2)P$-$SiMe_3$. I. Liquid, b_5 122 to 124°,[308] IR; hydrolyzes in H_2O.[308]
Ph_2P-$SiMe_3$. I. Colorless liquid, b_1 126 to 127°,[361] $b_{0.05}$ 90°,[5] n_D^{20} 1.6000,[5,361] sensitive to H_2O,[361] ^1H NMR,[163] magnitude and sign of coupl. const.[163]
$HP[SiMe_3]_2$. I.[60,460] V.[63] Liquid, b_{760} 170 to 172°,[460] b_{20} 60 to 61°,[60] b_{16} 60°,[63] m. -38°,[63] d_4^{20} 0.8188,[60] n_D^{25} 1.4637,[460] IR, Raman,[63,63a] ^{31}P +237.4 ppm,[185] IR and Raman also of $DP[SiMe_3]_2$.[63a]
$MeP[SiH_3]_2$. I. Liquid, vapor pressure at 0° is 34 ± 1 mm Hg, IR, ^1H NMR.[135]
$MeP[SiMe_3]_2$. I. Liquid, b_{20} 75 to 76°, d_4^{20} 0.8433.[60]
$MeP[SiMe_2(OMe)]_2$. I. Liquid, b_{10} 85 to 86°, n_D^{20} 1.4670, d_4^{20} 0.9600;[25] ignites in air.
$MeP[SiMe_2(OBu\text{-}i)]_2$. I. Liquid, b_4 117 to 118°, n_D^{20} 1.4603, d_4^{20} 0.8991;[25] ignites in air.
$MeP[SiMe_2NEt_2]_2$. I. Liquid, b_1 106 to 108°, n_D^{20} 1.5020,

d_4^{20} 0.9184;[25] cleaved by acetone.
$MeP[SiPh_3]_2$. I. Solid, m. 80°.[399]
$PhP[SiMe_3]_2$. I. Liquid, $b_{0.05}$ 76°, ^1H NMR.[338]
$P[SiMe_3]_3$. I.[60,63,376,460] b_{760} 242 to 243°,[460] b_{16} 102 to 105°,[63,376] m. 24°,[63] d_4^{20} 0.8670,[60] n_D^{25} 1.5077, IR, Raman, ^1H NMR,[63,63a] ^{31}P +251.2 ppm,[185,540] inflammable in air,[460] stable at 243°; oxidized with N_2O_4 to phosphate ester;[460] adducts: $L{\cdot}BH_3$, m. 100 to 107° (dec.);[460] $L{\cdot}CoNO(CO)_2$, rubin-red crystals dec. 100°;[524,525b] $L{\cdot}Fe(CO)_4$, yellow crystals dec. 110°;[525] $L{\cdot}Ni(CO)_3$, ^{31}P +334 ppm,[525a] dec. 30°;[523] $L{\cdot}Cr(CO)_5$, yellow crystals, dec. 100°.[524]
$[H_2P]_2SiMe_2$. I. Liquid, b_{760} 117.9° (calc.), log P_{mm} = 8.3384 - (2134.1/T), m. -108.0°,[450a] ^{31}P +220 ± 3 ppm J_{PH} 187 Hz,[450a] change of geminal coupl. const. with temp.[436]
$[Me_2P]_2SiCl_2$. I. Liquid, b_5 47°.[200]
$[Et_2P]_2SiH_2$. I. Liquid, $b_{0.005}$ 45 to 46°, ^1H NMR.[196]
$[Et_2P]_2SiCl_2$. I. Liquid, $b_{0.001}$ 82 to 89°.[196]
$[Et_2P]_2SiHMe$. I. Liquid, $b_{0.02}$ 54 to 55°, ^1H NMR.[196]
$[Et_2P]_2SiMe_2$. I. Liquid, b_1 60 to 65°,[200] b_5 65 to 69°,[196] ^1H NMR.[196]
$[Me_2P]_3SiCl$. I. Liquid, $b_{0.001}$ 120°.[196]
$[Et_2P]_3SiH$. I. Liquid, $b_{0.005}$ 93 to 105°, ^1H NMR;[196] disproportionates readily.[196]
$[Et_2P]_3SiCl$. I. Liquid, b_1 138 to 142°.[196]
$[Et_2P]_3SiMe$. I. Liquid, $b_{0.005}$ 94 to 96°,[196] $b_{0.02}$ 101 to 102°,[197] ^1H NMR.[196]
$[Et_2P]_4Si$. I. Liquid, b_1 ∿160°,[202] ^1H NMR,[196] sensitive to air and H_2O.[202]
$[Et_2P]_3Si\text{-}SiH_3$. I. Liquid, not isolated pure, ^1H NMR.[197]
$[Et_2P]_3Si\text{-}SiHMe_2$. I. Liquid, $b_{0.005}$ 129 to 130°, ^1H NMR.[197]
$(Et_2P)_3Si\text{-}SiMe_3$. I. Liquid, $b_{0.02}$ 115 to 117°, ^1H NMR.[197]
$RPh_2Si[P{\cdot}SiMe_3{\cdot}SiPh_2]_2{\cdot}R$. From $P[SiMe_3]_3$ and Ph_2SiCl_2 (R = OH or Me), m. 265 to 269° (from benzene).[376]

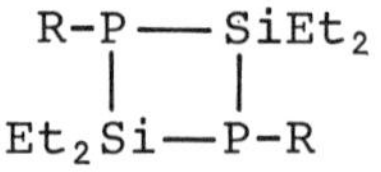
I. (R = H). Liquid, $b_{0.06}$ 107 to 110°, n_D^{25} 1.5829.[460]
(R = Ph). Solid, m. 43 to 47° (from ligroin), $b_{0.02}$ 151 to 153°.[460]
(R = $SiMe_3$). Colorless viscous liquid, $b_{0.2}$ 96 to 98°, n_D^{20} 1.5522.[460]

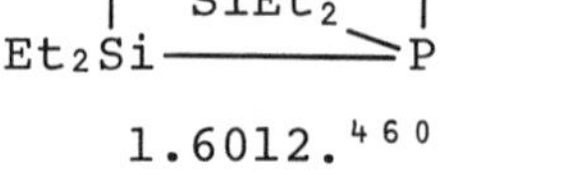
I. Liquid, $b_{0.005}$ 130 to 134°, n_D^{25} 1.6012.[460]

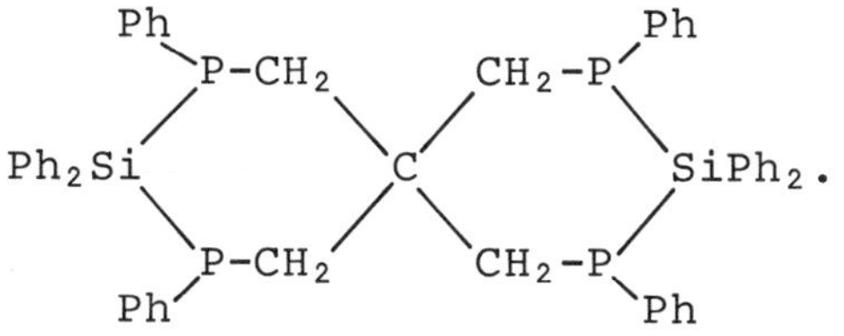

. I. Solid, m. 126°.[162]

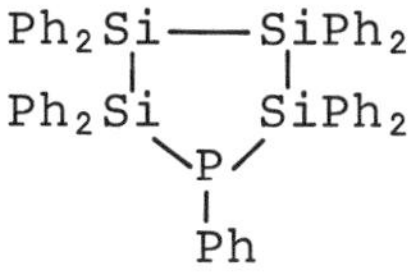

I. Solid, m. 203 to 206° (from benzene/ligroin).[247]

$(Ph_2Si\text{-}PPh)_3$. Only complex of type L•Mo$(CO)_3$, bright yellow crystals, dec. >165°, described.[504a]
$(PhSiP)_7$. I. Solid, dec. 300°;[505] a cage structure consisting of Si_2P_2 four-membered rings and Si_3P_3 six-membered rings was proposed.[505]
$(PhP)_6(SiPh)_4$. I. Colorless solid, dec. >125°, IR;[504] adamantanelike structure proposed.[504]
$(EtS)_2P\text{-}SiMe_3$. V. Liquid, b_3 69 to 72°, n_D^{20} 1.5357, d_4^{20} 1.0263.[61]
$(MeO)_2P(O)\text{-}SiMe_3$. I. Liquid, b_7 61°,[435] [may have a phosphite structure $(MeO)_2POSiMe_3$].
$(EtO)_2P(O)\text{-}SiEt_3$. I. Liquid, b_{10} 158 to 159°, d_4^{20} 0.9659, n_D^{20} 1.4390,[23] (may have a phosphite structure).
$(EtO)_2P(O)\text{-}Si(OMe)_3$. I. Liquid, b_{12} 100 to 103°, Raman,[171] (phosphite structure possible).
$(EtO)_2P(O)\text{-}Si(OEt)_3$. I. Liquid, b_{12} 60 to 61°, b_{12} 113°,[23] b_{12} 114 to 118°,[171] d_4^{20} 0.9282, n_D^{20} 1.4080, n_D^{20} 1.3860,[23] Raman,[171] (may have a phosphite structure).
$(BuO)_2P(O)\text{-}SiMe_3$. I. Liquid, $b_{4.5}$ 93 to 94°,[346] n_D^{20} 1.4150, d_4^{25} 1.0531,[346] [may have a phosphite structure $(BuO)_2POSiMe_3$].
$(PhCH_2O)_2P(O)\text{-}SiPh_3$. I. Solid, m. 222°,[435] [may have a phosphite structure, $(PhCH_2O)_2POSiPh_3$].
$[R_3^1P\text{-}SiR_3]^+[Co(CO)_4]^-$. From $R_3Si\text{-}Co(CO)_4$ and R_3^1P. ($R = R^1$ = Me). Solid, IR, ^1H NMR.[25a]
(R = Me; R^1 = Et). Solid, IR.[25a]
($R = R^1$ = Et). Solid, IR.[25a]

TYPE: P-Ge Compounds

$H_2P\text{-}GeMeH_2$. From H_2PSiH_3 and $MeGeClH_2$; condenses at -95° under vacuum; IR, ^1H NMR.[158]
$H_2P\text{-}GeMeHCl$. From H_2PSiH_3 and $MeGeCl_2H$, ^1H NMR.[158]
$H_2P\text{-}GeMe_2H$. I. Liquid, b_{760} 74.2° (calc.), log $P_{,,}$ = 8.2458 - (1863.3/T), m. -106.5°, ^{31}P +243 ± 3 ppm,

J_{PH} 176 Hz.[450a]
H_2P-$GeMe_3$. I. Liquid, b_{760} 95.8° (calc.), log P_{mm} = 7.4784 - (1696.2/T), m. -97.2°, ^{31}P +236 ± 3 ppm, J_{PH} 177 Hz.[450a]
Et_2P-$GeEt_3$. I. Liquid, b_{15} 120°, n_D^{20} 1.4845, IR, ^1H NMR,[481] P-Ge bond cleavage by NH_3, HCl; insertion between the P-Ge bond takes place with CO_2, CS_2, RNCS, RNCO, and RC≡CH.[481]
PhHP-$GeMe_3$. I. III. Liquid, b_2 60°, ^1H NMR, IR.[527]
Ph_2P-$GeMe_3$. III. Liquid, b_{12} 185 to 187°, n_D^{20} 1.6089.[480]
Ph_2P-$GeEt_3$. I.[59,222] III.[485] Liquid, $b_{0.003}$ 146°, easily oxidized. P-Ge bond cleaved by MeI, H_2O;[59,222] adduct: [L·AgI]$_4$, m. 183° (dec.).[59]
Ph_2P-$GePh_3$. I.[59] II.[522] Colorless needles, m. 154 to 156° (from methyl cyclohexane),[59] m. 159 to 161°,[522] IR.[522]
MeP[$GeMe_3$]$_2$. III. Liquid, b_1 53°, ^1H NMR, IR.[527]
PhP[$GeMe_3$]$_2$. III. Liquid, b_2 107°, ^1H NMR, IR.[527]
PhP[$GePh_3$]$_2$. II. Solid, m. 110°; IR indicates pyramidal structure;[520,522] P-Ge bond cleaved by KOH/EtOH.
PhP[$GeClPh_2$]$_2$. I. Solid, m. 118 to 120°.[501]
PhP($GePh_3$)$SnPh_3$. I. Solid, m. 115 to 119°; cleaved by KOH/EtOH, IR.[521,522]
P[$GeMe_3$]$_3$. I. III.[506] II.[522] Liquid, $b_{0.1}$ 62 to 63°,[506,522] IR,[166,506] Raman,[166,522] ^1H NMR,[506] ^{31}P +228 ppm;[166] from IR and NMR data pyramidal structure deduced; adducts: L·Ni(CO)$_3$, ^{31}P +204.7 ppm;[525a] L·Cr(CO)$_5$ yellow crystals;[524] L·Fe(CO)$_4$, yellow crystals, dec. 120°;[525] L·CoNO(CO)$_2$, rubin-red crystals, dec. 132°;[524,525b] L·Ni(CO)$_3$, colorless solid, dec. 100°.[523]
P[$GePh_3$]$_3$. II. Solid, m. 128°, IR,[166,520,522] Raman.[522]
P($GePh_3$)($SnPh_3$)$_2$. I. Solid, m. 160°, IR.[521,522]
[H_2P]$_2GeMe_2$. I. Liquid, m. -100.2°, ^{31}P +222 ± 3 ppm, J_{PH} 180 Hz.[450a] Change of geminal coupl. const. with temp.[436]
[Ph_2P]$_2GePh_2$. I. Colorless needles, m. 182 to 185° (from methylcyclohexane).[59]
[Ph_2P-GePh]$_2$. I. Structure uncertain, pale-yellow powder, m. 108 to 112°.[59]
[PhP-$GePh_2$]$_3$. I. Solid, m. 112 to 114°.[501]
[PhP-$GePh_2$]$_2$. I. Solid, m. 40 to 42°.[501]
(PhP)$_3$(GePh)$_2$. I. Solid, m. 126° (dec.);[501,504] dimeric, adamantanelike structure proposed; IR.[504]
(PhGeP)$_7$. I. Solid, m. 140 to 142°,[501,505] structure unknown; cage structure consisting of Ge_2P_2 four-membered rings and Ge_3P_3 six-membered rings proposed.[505]

TYPE: P-Sn Compounds

Me_2P-$SnMe_3$. I. Decomposes at 10°.[501]
Me(EtS)P-$SnEt_3$. V. [From MeP($SnEt_3$)$_2$ and Et_2S_2]. Liquid,

b$_2$ 86 to 88°, n_D^{20} 1.5485, d_4^{20} 1.2498.[61]
$Et_2P-SnEt_3$. I. Liquid, $b_{0.3}$ 70°, IR, UV, μ 1.05 D.[98]
$EtPhP-SnEt_3$. Liquid, $b_{0.2}$ 99 to 100°.[98]
$EtPhP-SnPr_3$. I. Liquid, $b_{0.3}$ 125 to 126°.[98]
$t-Bu_2P-SnMe_3$. I and III. Liquid, $b_{0.1}$ 73°,[501] IR, ^1H NMR, ^{31}P -20.3 ppm;[514a] adducts: L•Ni(CO)$_3$, dec. 130°; L•Fe(CO)$_4$, dec. 165°; L•Mo(CO)$_5$, dec. 98°.[514a,525a]
$Bu_2P-SnBu_3$. I. Liquid, $b_{0.15}$ 122 to 124°.[434]
$Ph_2P-SnMe_3$. I.[98] III.[337] Liquid, $b_{0.7}$ 141 to 142°,[98] $b_{0.8}$ 150°,[337] ^1H NMR and sign of coupl. const.;[163] adducts: L•Ni(CO)$_3$, liquid,[3] solid,[525a] L_2•Ni(CO)$_2$, crystals.[3]
$Ph_2P-SnEt_3$. I.[98,361] III.[337] Liquid, $b_{0.7}$ 167 to 168°,[98] $b_{0.7}$ 170°,[337,361] μ 1.05 D;[98] did not react with sulfur; is cleaved by MeI[98] and H_2O.[361]
$Ph_2P-SnPr_3$. I. Liquid, $b_{0.6}$ 176 to 177°.[98]
$Ph_2P-SnBu_3$. I.[98] II.[511,512] Solid, m. 90 to 96°;[512] the also reported m. 60°[511] is incorrect;[526] liquid(?) $b_{0.6}$ 192°,[98] μ 0.96 D;[98] with O_2 yields phosphinate ester.[98,512]
$Ph_2P-SnPh_3$. I.[98,511,514] II.[512,522] Solid, m. 127 to 130°,[512,514,522] m. 126°,[511] m. 103 to 105° (from benzene),[98] IR, Raman,[166,522] ^{31}P +56 ppm;[166] the reported formation of a phosphonium salt with MeI, dec. 40°,[512] is incorrect.[526]
$MeP[SnMe_3]_2$. I.[42] II.[501a] Liquid, $b_{0.1}$ 68 to 71°,[501a] b_3 89 to 90°, n_D^{20} 1.5778, d_4^{20} 1.5601,[42] IR;[501a] adducts: L•Ni(CO)$_3$, dec. -10°; L•Cr(CO)$_5$, m. 75 to 78°; L•Mo(CO)$_5$, m. 95 to 98°.[501a]
$MeP[SnEt_3]_2$. I. Liquid, b_1 143 to 145°, n_D^{20} 1.5621, d_4^{20} 1.3725.[42]
$PhP[SnMe_3]_2$. II. Solid, m. 35°, $b_{0.01}$ 132 to 136°;[515] L•Ni(CO)$_3$, dec. 50°.[525a]
$PhP[SnEt_3]_2$. I. Liquid, $b_{0.3}$ 150 to 151°, IR.[98]
$PhP[SnBu_3]_2$. I. Liquid, $b_{0.01}$ 178 to 181°.[326]
$PhP[SnPh_3]_2$. I.[514] II.[520,522] Solid, m. 150°,[520] m. 146 to 150°,[514] m. 180°(?),[522] IR,[166,520] Raman,[166,522] ^{31}P +163 ppm,[166] (pyramidal structure).
$PhP(SnPh_3)(PbPh_3)$. I. Solid, dec. 110°, IR.[521]
$P[SnMe_3]_3$. I.[60] II.[522] III.[249] Liquid, $b_{1.5}$ 116 to 117°,[522] b_3 136 to 137°, d_{20}^{20} 1.6769, n_D^{20} 1.5970;[60] does not ignite in air, IR, Raman,[166,249,522] ^1H NMR,[522] ^{31}P +330 ppm;[166] adducts: L•Cr(CO)$_5$, yellow crystals, dec. 190°;[524] L•Fe(CO)$_4$, yellow crystals, dec. 142°;[525] L•CoNO(CO)$_2$, rubin-red crystals, dec. 190°;[524,525b] L•Ni(CO)$_3$ colorless solid, dec. 90°,[523] ^{31}P +307.3 ppm.[525a]
$P[SnPh_3]_3$. I.[514] II.[520] Solid, m. 197 to 201°,[514] m. 201,[520,522] IR,[166] Raman,[166,522] ^{31}P +323 ppm.[166]
$P(SnPh_3)_2PbPh_3$. I. Solid, dec. 171 to 172°, IR.[521,522]
$[Ph_2P]_2SnMe_2$. II. Solid, m. 110 to 114°;[512] adduct:

L·Ni(CO)$_2$, orange solid.[3]
[Ph_2P]$_2$$SnBu_2$. II. Solid, m. 98 to 102°.[512]
[Ph_2P]$_2$$SnPh_2$. I.[98] II.[512] Solid, m. 78 to 80°,[512] orange powder, m. 144 to 147°.[98]
[Ph_2P]$_3$SnPh. II. Solid, m. 115 to 117°.[512]
[Ph_2P]$_4$Sn. II. Solid, m. 106 to 107°;[512] adducts:[161] L·4Fe(NO)$_2$CO, red solid; L·4CoNO(CO)$_2$, red solid;[161] spirocyclic adducts: L·2Cr(CO)$_4$, pale green, dec. 170°; L·2Mo(CO)$_4$, red-brown, dec. 160°; L·2W(CO)$_4$, bright red, m. 160° (dec.);[161] L·2Fe(NO)$_2$, dark red, m. 312° (dec.); L·2Co(NO)(CO), dark red, dec. 305°; L·2Ni(CO)$_2$, pale-green, dec. 160°.[161]
[Ph_3Sn]$_2$P-P[$SnPh_3$]$_2$. IV. Yellow, crystalline powder, m. 95 to 100°;[513] very likely does not have a biphosphine structure.[526]
[PhP-SnR_2]$_3$. I.[514] II.[503] III.[513] Six-membered ring.
(R = Me). Solid, m. 134°, IR, ^{1}H NMR;[503] cleaved by HCl; decomposes above 175°.
(R = Bu). Oil, IR.[503]
(R = Ph). Solid, m. 155 to 160°,[503] m. 161 to 163°;[501] yellow crystals, the also reported m. 55 to 60°,[513,504] and m. 64°,[511] are incorrect;[526] IR.[503]
[Ph_3SnP-$SnPh_2$]$_3$. I. Six-membered ring, yellow crystals, m. 98 to 101°,[514] m. 99°.[511]
[P-$SnPh_2$]$_4$. II.[502] IV.[513] Cubanelike structure proposed; yellow solid, dec. 160°, IR.[502]
$P_4(SnPh_2)_6$. I. Adamantane structure proposed; solid.[501]

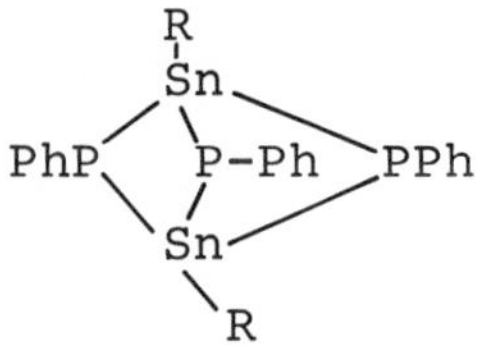

I. (R = Ph). Solid, m. 133 to 135°.[501,504]

(R = Bu). Colorless solid, m. 89 to 91° (dec.), ^{31}P +5.2 ppm, IR.[504]

TYPE: P-Pb Compounds

Ph_2P-$PbPh_3$. II. Solid, dec. 100°, not stable at room temperature, very sensitive to air and H_2O,[519,522] IR.[522]
PhP[$PbPh_3$]$_2$. II. Solid, dec. 110°; IR indicates pyramidal structure.[520,522]
P[$PbMe_3$]$_3$. II. Solid, m. 46 to 47°,[517] m. 48 to 49°,[522] dec. at 50°, IR, ^{1}H NMR;[517,522] adduct: L·Ni(CO)$_3$, colorless solid, dec. 85°.[523,525a]
P[$PbPh_3$]$_3$. II. Solid, dec. 110°, IR.[517,520,522]
(Ph_2P)$_2$$PbPh_2$. II. Solid, dec. at 10°.[519]

Organosubstituted Biphosphines

Roman numerals refer to methods of preparation described in Section E.1 of the text.

TYPE: R_2P-PR_2

Me_2P-PMe_2. I.[65,66,442] II.[438,440] III.[321,397,438,440,441,442,456] V.[121a,529] Liquid, b_{16} 38 to 40°,[397] b. 140.2° (est.),[66] n_D^{20} 1.5252, d_4^{20} 0.9024,[424a] log P_{mm} = 6.5244 - 0.005588T + 1.75 log T - (2444/T),[66] m. -2.25 to -2.15°,[66] m. -5 to -1°,[456] UV,[228] ^1H NMR,[174,239] mass spect.,[49] μ 1.75 D,[424a] J_{PP} -179.7 Hz,[174] ^{31}P +59.5,[397] +60.2 ppm,[372a] stable up to 300°;[66] adducts: L•MeI, solid, m. 128 to 130°,[65,321] m. 135 to 136°;[442] $L{\cdot}BMe_3$, m. 24 to 27°;[442] $L{\cdot}BH_3$;[65,66,442] $L{\cdot}2BH_3$;[65,66,442] x-ray structure analysis of $L{\cdot}2BH_3$;[99] $L{\cdot}CS_2$, red oil;[321] $L{\cdot}2AlEt_3$.[111]

$(CF_3)_2P-P(CF_3)_2$. II. Liquid, 84°,[38,76] UV,[38] mass spect.;[102] adduct: $L{\cdot}NMe_3$;[81] flash photolysis gives CF_2 in the singlet ground state.[104]

Et_2P-PEt_2. II.[240,250,313] III.[321,397,438,440,561] IV.[299] Liquid, $b_{0.5}$ 42 to 60°,[240] b_1 55 to 60°,[298] b_4 75°,[299] b_{740} 220 to 223°,[313,321,397,440] n_D^{20} 1.5180, d_4^{20} 0.8951,[424a] n_D^{20} 1.507, d_4^{20} 0.8885,[561] UV,[55a] mass spect.,[49,233] $\bar{D}$(P-P) 86 kcal/mole,[233] μ 2.76 D[424a] mag. susceptibility,[561] Raman,[313] ^{31}P +34.3 ppm;[397] adducts: L•EtI, m. 98 to 103°;[313,321] $L{\cdot}CS_2$, red oil.[313]

Pr_2P-PPr_2. III.[321,362,438,440,561] Liquid, b_5 112 to 113°, b_{16} 144 to 145°,[321] b_{20} 148°,[561] n_D^{20} 1.492, d_4^{20} 0.8797,[561] mag. susceptibility.[561]

$i\text{-}Pr_2P-PPr_2\text{-}i$. IV. Liquid, b_{10} 110°.[301]

Bu_2P-PBu_2. II.[313] III.[321,438,440] V.[340] Liquid, b_2 150 to 152°,[340] b_{14} 180 to 181°,[313,321,440] b_{18} 185°,[561] n_D^{20} 1.504, d_4^{20} 0.8869,[561] Raman,[313] mag. susceptibility,[561] mass spect.,[49] adducts: $L{\cdot}CS_2$, red oil;[313] L•EtI, solid.[313]

$t\text{-}Bu_2P-PBu_2\text{-}t$. II.[291] IV.[301] Solid, m. 48°, $b_{0.01}$ 92 to 98°,[301] b_1 124 to 127°,[291] ^{31}P -40 ppm;[291] adducts: $L{\cdot}NiBr_2$, red-brown crystals, dec. 205 to 210°;[291] $L{\cdot}CoBr_2$, blue crystals, dec. 327 to 332°.[291]

$(c\text{-}C_6H_{11})_2P-P(C_6H_{11}\text{-}c)_2$. II.[313] IV.[299] Solid, m. 173°,[313] ^{31}P +21.5 ppm (in benzene);[187,298] cleaved by MeI;[313] adducts: L•2AgI, solid, m. 129 to 130°;[530] $L{\cdot}NiBr_2$, m. 192°;[313] $L{\cdot}HgI_2$, m. 97 to 99°.[530]

Ph_2P-PPh_2. I.[156,256,360] II.[107,193,360,363,364,441,543] III.[181,438,440] V.[59a,209,379,395] Solid, m. 120.5°,[360] m. 119 to 120°,[59a,193] m. 121 to 122°,[364,395,440] m. 124 to 126°,[256] b_1 258 to 260°,[360] IR,[360] $E_{1/2}$ -2.05 V,[420a] x-ray powder diagram;[360] yields $Ph_2P{\cdot}$ radicals on irradiation with UV light;[490] ^{31}P +15.2 ppm (in

C_6H_6);[187,298] +16 ppm;[59a] adducts: L•$AlEt_3$, solid, m. 82 to 83°;[111,296] L•BF_3, oil;[360] gives with N_3SiPh_3 the diimine, m. 236 to 238°;[452] is cleaved by MeI;[320] gives with pyridine a yellow color.[59a]
(4-MeC_6H_4)$_2$P-P(C_6H_4Me-4)$_2$. I.[440] Solid.
(2,4,6-$Me_3C_6H_2$)$_2$P-P($C_6H_2Me_3$-2,4,6). II. Solid, m. 200 to 215° (dec.), IR, UV.[549]
(C_6F_5)$_2$P-P(C_6F_5)$_2$. II.[15,176] Solid, m. 159 to 163°,[176] m. 165 to 170°,[15] IR,[15,176] mass spect.[15]
(Ph_3Sn)$_2$P-P($SnPh_3$)$_2$. V. Solid, m. 95 to 110°,[513] (this compound very likely does not have a biphosphine structure).[526]
(Me_2N)$_2$P-P(NMe_2)$_2$. II. Solid, m. 48° $b_{0.01}$ 50°,[449] IR; cleaved by MeI and HCl;[449] adducts: L•2BH_3, dec. 132°; L•CS_2, brown-red oil.[449]

TYPE: $R^1R^2P-PR^2R^1$

CF_3(H)P-P(H)CF_3. V.[390,391] Liquid, b_{760} 69.5° (est.), log P_{mm} = 6.4475 - 0.006115T + 1.75 log T - (2024/T),[391] UV, IR,[391] dec. 225°.[391]
MeCF_3P-PCF_3Me. I. II.[75] Liquid, b_{760} 120.4° (est.), log P_{mm} = 6.2094 - 0.005T + 1.75 log T - (2323/T),[75] IR; is cleaved by HCl; adducts: L•BCl_3; L•BH_3.[75]
MeEtP-PEtMe. III.[393,397,400] Liquid, b_{740} 188 to 190°,[397] ^{31}P +44.7 and +46.2 ppm (two peaks are due to stereoisomers),[397] +45.9 and +48.1 ppm, ^1H NMR.[372a]
MePrP-PPrMe. III. Liquid, $b_{0.18}$ 95°.[400]
MeBuP-PBuMe. III. Liquid, $b_{0.01}$ 51 to 52°.[398,400]
Me(t-Bu)P-P(Bu-t)Me. II. Liquid, $b_{0.1}$ 73 to 75°; becomes glassy at low temp.; ignites in air, ^1H NMR.[486]
MePhP-PPhMe. III.[372,397] IV. Solid, m. 75 to 77° (from ligroin),[476] m. 73 to 76°,[397] $b_{0.5}$ 128 to 130°,[397] ^1H NMR,[372,372a] activation energy for the inversion through phosphorus is 26 ± 2 kcal/mole (23.6 kcal/mole[372a]) as determined from ^1H NMR,[372] ^{31}P +38.2 and +41.7 ppm[372a,397] (two peaks are due to stereoisomers).[397]
Me($PhCH_2$)P-P(CH_2Ph)Me. III. ^1H NMR, ^{31}P +39.4 and +40.7 ppm (stereoisomers); activation energy for inversion through phosphorus $\Delta G^{\ddagger}_{155}$ 24 kcal/mole.[372a]
Me(4-ClC_6H_4)P-P(C_6H_4Cl-4)Me. III. ^1H NMR, ^{31}P +36.4 and +38.3 ppm (stereoisomers).[372a]
Me(4-MeC_6H_4)P-P(C_6H_4Me-4)Me. III. ^1H NMR, ^{31}P +67.6 and +73.3 ppm (stereoisomers, structure uncertain), $\Delta G^{\ddagger}_{155}$ 22.5 kcal/mole.[372a]
Me(4-$CF_3C_6H_4$)P-P($C_6H_4CF_3$-4)Me. III. ^1H NMR, ^{31}P +38.3 and +41.9 ppm (stereoisomers), $\Delta G^{\ddagger}_{155}$ 22.6 kcal/mole.[372a]
Et(Et_2N)P-P(NEt_2)Et. II. Liquid, b_{12} 142 to 145°.[535]
EtBuP-PBuEt. IV. Liquid, b_3 106°,[299] ^{31}P +37.5 ppm[187,299] (method of synthesis yields only one form).
Et(c-C_6H_{11})P-P(C_6H_{11}-c)Et. IV. Plates, m. 115° (from ether).[299]

EtPhP-PPhEt. IV. Liquid, b_4 160 to 162°,[300] m. 63 to 68°,[302] $E_{1/2}$ -2.295 V,[420a] ^{31}P +21.5 and +28.3 ppm (in benzene)[187] (two peaks are due to stereoisomers).
BuPhP-PPhBu. IV. Liquid, b_4 188 to 190°.[300]
$(c\text{-}C_6H_{11})(Et_2N)P\text{-}P(NEt_2)(C_6H_{11}\text{-}c)$. II. Liquid, b_{15} 234 to 235°, b_{1-2} 180 to 182°.[535]
$(c\text{-}C_6H_{11})(C_5H_{10}N)P\text{-}P(NC_5H_{10})(C_6H_{11}\text{-}c)$. II. Colorless needles, m. 121 to 122° (from benzene);[530,532] adducts: $L\cdot HgI_2$, m. 230 to 232°; L·2AgI, m. 180 to 184° (dec.);[530,532] L·2CuBr, m. 155 to 156°; $L\cdot CdI_2$, m. 108 to 112°; $L\cdot ZnI_2$.[532]
$Ph(PhCH_2)P\text{-}P(CH_2Ph)Ph$. I. II. Solid, m. 134° (from ligroin), b_2 240 to 265°.[547]
Ph(PhCO)P-P(COPh)Ph. IV. Bright-yellow crystals, m. 117 to 117.5° (from MeOH).[476]
$Ph(Me_2N)P\text{-}P(NMe_2)Ph$. II. Solid, m. 85° (from ligroin), $b_{0.5}$ 150°.[560]
$Ph(Et_2N)P\text{-}P(NEt_2)Ph$. II. Liquid, b_{15} 236 to 238°, b_{1-2} 188 to 190°.[535]
$Ph(C_5H_{10}N)P\text{-}P(NC_5H_{10})Ph$. II. Solid, m. 82° (from ether);[533] adducts: L·2CuBr; L·2AgI; $L\cdot ZnI_2$, m. 122° (dec.); $L\cdot CdI_2$, m. 92° (dec.).[533]
$Ph(C_6F_5)P\text{-}P(C_6F_5)Ph$. II. Solid, m. 122 to 125°.[176]
$CCl_3(Br)P\text{-}P(Br)CCl_3$. V. Liquid, $b_{0.4}$ 77°, n_D^{25} 1.628, d_4^{23} 2.00; IR, UV.[10]
$(c\text{-}C_6H_{11})(I)P\text{-}P(I)(C_6H_{11}\text{-}c)$. V. Solid, m. 78° (dec.);[172c] gives with $C_6H_{11}I$ the adduct $[(C_6H_{11})_3P\text{-}P(C_6H_{11})_3]I^+I_5^-$.
Ph(Br)P-P(Br)Ph. V. Solid, m. 119°.[29]
Ph(I)P-P(I)Ph. V. Solid, m. 178 to 180°;[172,172a,261] yields with $C_6H_{11}I$ an alkylated product, $[Ph(C_6H_{11})_2P]_2 I^+I_5^-$, m. 172 to 173°.[172b]
$(4\text{-}FC_6H_4)(I)P\text{-}P(I)(C_6H_4F\text{-}4)$. V. Solid, m. 138 to 140°.[172a]
$(4\text{-}ClC_6H_4)(I)P\text{-}P(I)(C_6H_4Cl\text{-}4)$. V. Solid, m. 167 to 168°.[172a]
$(4\text{-}BrC_6H_4)(I)P\text{-}P(I)(C_6H_4Br\text{-}4)$. V. Solid, m. 175 to 177°.[172a]
$(4\text{-}MeC_6H_4)(I)P\text{-}P(I)(C_6H_4Me\text{-}4)$. V. Solid, m. 154 to 156°.[172a]

TYPE: $R_2P\text{-}PR_2^1$

$Me_2P_\alpha\text{-}P_\beta(CF_3)_2$. I.[103,228] Liquid, b_{760} 120° (est.), m. -79.2 to -79.1°[228] log P_{mm} = 8.0938 - (2022/T);[228] mass spect.,[103] ^1H and ^{19}F NMR,[103,413] ^{31}P +55.5 (P_α) and -12.6 (P_β) ppm,[103] $J_{P_\alpha P_\beta}$ 252 Hz; adducts: $L\cdot BH_3$; $L\cdot NMe_3$.[228]
$(Me_2N)_2P\text{-}PPh_2$. IV. Pale-yellow liquid, $b_{0.001}$ 137 to 140°, n_D^{21} 1.6221;[560] adduct: $L\cdot 2BH_3$, m. 123°.[560]
$Et_2P_\alpha\text{-}P_\beta(C_6H_{11}\text{-}c)_2$. IV. Liquid, b_2 155 to 160°,[298] ^{31}P +42.2 (P_α) and +13.8 (P_β) ppm; $J_{P_\alpha P_\beta}$ 282 Hz;[187] disproportionates readily on heating to the symmetrical biphosphines $R_2P\text{-}PR_2$.
$Et_2P\text{-}PPh_2$. IV. Colorless oil; disproportionates readily

to the symmetrical biphosphines R_2P-PR_2.[298]

$Ph_2P_\alpha-P_\beta(C_6H_{11}-c)_2$. I. IV. Solid, dec. 92°,[298] ^{31}P +28.8 (P_α) and +8.2 (P_β) ppm; J_{PP} 224 Hz.[187]

TYPE: $R_2P-PR^1R^2$

$Me_2P-P(CF_3)NMe_2$. I. From $(Me_2P)_2PCF_3$ and $(Me_2N)_2PCF_3$ at 132°, ^1H and ^{19}F NMR.[123]

$(CF_3)_2P-P(H)Me$. I. Liquid, b_{760} 91.6° (est.),[74] $\log P_{mm} = 6.1043 - 0.005T + 1.75 \log T - (2146/T)$, UV, IR;[74] is decomposed by sunlight and heat; forms no BH_3 adduct.[74]

$(CF_3)_2P-P(H)CF_3$. I. Liquid, vapor pressure is 61 mm Hg at 0°,[83] compound is unstable.

$(c-C_6H_{11})_2P-P(C_6H_{11}-c)(NC_5H_{10})$. Preparation not described; adducts: L·2CuBr, m. 240 to 245°;[534] $[L\cdot CuBr]_2$, solid; $L\cdot 2AgI$, solid; $L\cdot CdI_2$, solid.[534]

$Br_2P-P(Br)(CCl_3)$. V. Solid, m. 45°, $b_{0.4}$ 46°, d_4^{23} 2.26; ignites when brought in contact with paper tissue; UV.[10]

TYPE: Cyclic Biphosphines

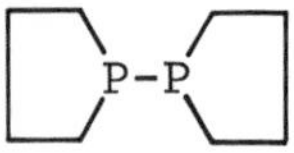

III. Liquid, $b_{0.05}$ 50°.[496]

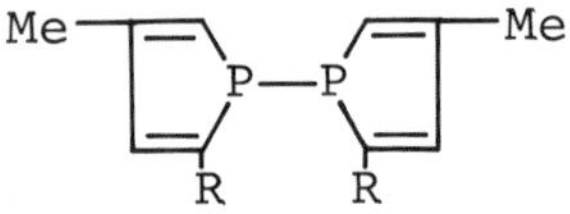

V. (R = Me). Liquid, $b_{0.5}$ 78°, n_D^{20} 1.5832, d_4^{20} 1.0724[466] (structure uncertain).
(R = Bu). Liquid, b_1 138 to 140°, n_D^{20} 1.5448, d_4^{20} 1.0010[466] (structure uncertain).

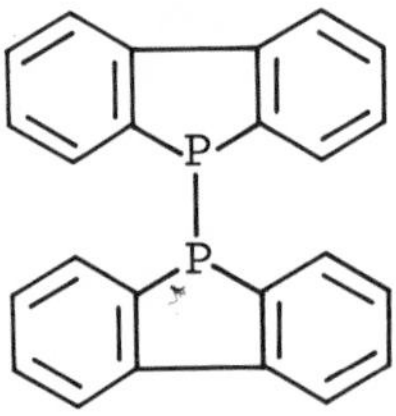

V.[58,152] Yellow crystals, m. 242 to 243° (from pyridine),[152] m. 230° (crude);[58] is cleaved by H_2O_2, I_2, and $PhCH_2Br$;[152] forms a radical anion on treatment with alkali metal.[58]

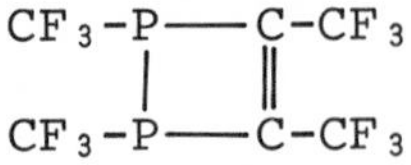

V. Liquid, b. 110° (est.), vapor pressure

is 6 mm at 0°, 26 mm at 25°;[388] inflammable in air, stable at 200°;[388] IR, UV, ^{19}F NMR, ^{31}P +40 ppm (J_{PP} 55 Hz),[388] resistant to iodine cleavage.[388]

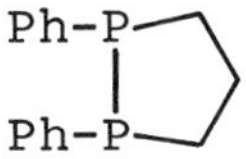

IV. Liquid, b_4 184 to 190°,[295] $b_{1.5}$ 180°.[300]

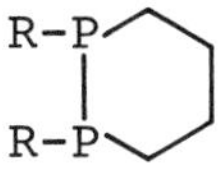

IV. (R = Ph). Crystals, m. 90° (from THF).[300]

(R = Et). Liquid, b_2 70 to 71°,[299] ^{31}P +46.0 ppm.[187,299]

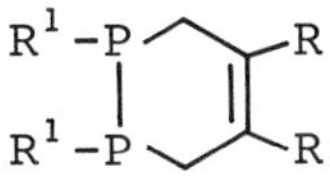

V. (R = H; R^1 = Me). Liquid, b_{11} 82 to 83°.[492,494]

(R = H; R^1 = Et). Liquid, b_9 114°.[494]
(R = Me; R^1 = Me). Liquid, b_8 100°,[494] ^1H NMR,[492] mass spect.[494]
(R = Me; R^1 = Et). Liquid, $b_{0.4}$ 78 to 80°.[494]
(R = Me; R^1 = Ph). Liquid, $b_{0.005}$ 150°.[494]

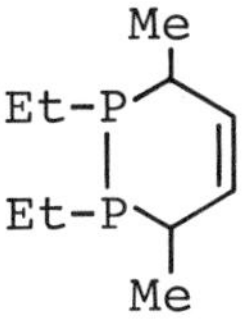

V. Liquid, $b_{0.2}$ 68 to 70°.[494]

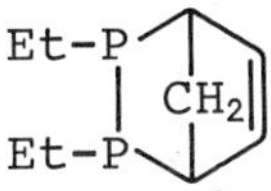

V. Liquid, $b_{0.4}$ 74 to 78°.[494]

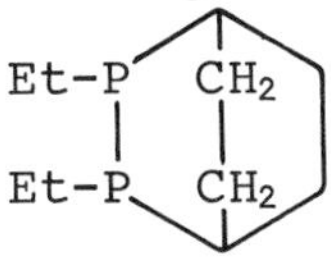

V. Liquid, $b_{0.5}$ 73 to 77°.[494]

Organosubstituted Biphosphine Disulfides

Roman numerals refer to methods of preparation described in Section E.2.1 of the text.

TYPE: $R_2P(S)-P(S)R_2$

$Me_2P(S)-P(S)Me_2$. I.[110,133,341,372a,394,457,469,477] II.[55] Crystals, m. 215 to 221°,[110] m. 220 to 223°,[133,469] m. 227° (from toluene/EtOH),[372a,394,440,477] m. 233.5°,[239] UV,[130] IR,[110,130,131,227] Raman,[110,130,131,220,227] space group 2/mC,[130] ^1H NMR, J_{PP} 18.7 Hz,[239]

$E_{1/2}$ -2.48 V,[420a] ^{31}P -34.7 ppm (in $CHCl_3$),[394] -34.7 ppm (in benzene);[187] x-ray analysis[465] shows P-P 2.18, P-C 1.74, P=S 1.96 Å, ∠P-P-S 113°; sulfur atoms are trans to each other;[465] adducts: $L_2 \cdot CuClO_4$, solid;[421] $[L_2Cu][CuCl_2]$, solid;[421] $L \cdot SnCl_4$, m. 145 to 147°;[36,553] $L \cdot SnBr_4$, m. 152 to 154°;[553] $L \cdot CuCl_2$;[36] $L \cdot CuBr_2$;[36] $L \cdot ZnCl_2$;[36] $L \cdot 2CdCl_2$;[37] $L \cdot HgX_2$ (X = Cl, Br, I);[36] $L \cdot SbCl_3$;[36] $L \cdot BiCl_3$.[36]

$(Me_2N)_2P(S)-P(S)(NMe_2)_2$. II. Plates, m. 227° (from ligroin).[449]

$Et_2P(S)-P(S)Et_2$. I.[48,110,318,362,419,469,561] II.[55,250,313] Crystals, m. 76 to 77° (from acetone/H_2O),[110,250,313,318,362,420,561] m. 77 to 78°,[440] UV,[55a,130] IR,[110,130,131,461] Raman,[110,131,220,461] μ 1.57 D,[131,220] d_4^{20} 1.25,[160] d_4^{80} 1.0698, n_D^{80} 1.581,[561] $E_{1/2}$ -2.45 V,[420a] μ 1.63 D,[424a] mag. susceptibility,[561] ^{31}P -49.4 ppm (in benzene),[187] x-ray analysis[160] shows space group P $\bar{1}$, P-P 2.22 P=S 1.94, P-C 1.83 Å, ∠P-P-S 112.8°; sulfur atoms are trans to each other;[160] adducts: $L \cdot CuClO_4$, solid;[421] $L \cdot SnCl_4$, m. 146 to 149°;[553] $L \cdot SnBr_4$, m. 138 to 139°;[553] $L \cdot CuX$ (X = Cl, Br), solid, ionic structure, space group.[240a]

$Pr_2P(S)-P(S)Pr_2$. I.[48,110,133,362] Crystals, m. 144 to 145° (from EtOH),[110,133,420] m. 145.6 to 146°,[48,362] m. 147 to 148°,[440] d_4^{16} 0.9356,[561] IR,[110,131] Raman,[220] μ 1.58 D,[131] $E_{1/2}$ -2.482 V,[420a] mag. susceptibility;[561] useful for separating Pd from Pt;[273] adduct: $L \cdot PdCl_2$, red-brown solid, m. 73 to 74°.[273]

$i\text{-}Pr_2P(S)-P(S)Pr_2\text{-}i$. II. Solid, m. 96° (from MeOH),[301] $E_{1/2}$ -2.315 V.[420a]

$(CH_2{=}CHCH_2)_2P(S)-P(S)(CH_2CH{=}CH_2)_2$. I. Crystals, m. 48 to 49°,[438] m. 58 to 59° (from MeOH),[440] IR.[131]

$Bu_2P(S)-P(S)Bu_2$. I.[110,318,420,469] II.[313] Crystals, m. 73 to 73.5° (from MeOH),[110,420] m. 74 to 75° (from benzene or MeOH),[313,318,362,440,469] d_4^{80} 0.9571, n_D^{80} 1.527,[561] IR,[110,131] Raman,[131,220] μ 1.45 D,[220] mag. susceptibility,[561] $E_{1/2}$ -2.517 V.[420a]

$i\text{-}Bu_2P(S)-P(S)Bu_2\text{-}i$. I. Crystals, m. 92 to 93° (from MeOH), IR.[110]

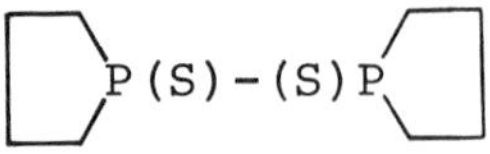

I. White prisms, m. 185° (from toluene/EtOH),[495] IR,[131,495] ^{31}P -62.2 ppm (in $CHCl_3$),[495] x-ray analysis[375] shows triclinic, space group P $\bar{1}$, P-P 2.20, P=S 1.95 Å,[375] ∠C-P-C 97°.[375]

$Am_2P(S)-P(S)Am_2$. I. Crystals, m. 43.5° (from MeOH).[362]

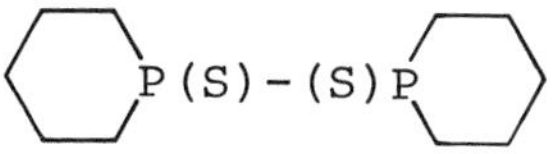

I. Solid, m. 185 to 225° (from toluene/EtOH),[495] IR,[131,495] ^{31}P -37.5 ppm (in $CHCl_3$);[495] x-ray-analysis[375a] shows triclinic, P-P 2.21, P=S 1.95 Å, ∠C-P-C 101.8°.[375a]

$(c\text{-}C_6H_{11})_2P(S)\text{-}P(S)(C_6H_{11}\text{-}c)_2$. II.[313] III.[331,440] Crystals, m. 205° (from CS_2);[313,440] two crystalline modifications: rhombohedric and triclinic;[287] IR,[131] ^{31}P -56.1 ppm (in benzene),[187] $E_{1/2}$ -2.335 V.[420a]

$(PhCH_2)_2P(S)\text{-}P(S)(CH_2Ph)_2$. Crystals, m. 145 to 150° (from MeOH),[318] $E_{1/2}$ 2.11 V.[420a]

$Ph_2P(S)\text{-}P(S)Ph_2$. I.[469] II.[296,360,364,379,440] III.[395,422,440] Crystals, m. 167 to 168°,[296,362,364,395,440] m. 170 to 172° (from EtOH),[422,469,543] m. 174 to 176°,[379] IR,[131,360] x-ray powder diagram,[360] ^{31}P -37.9 ppm (in toluene),[187] $E_{1/2}$ -1.535 V.[420a]

$(4\text{-}MeC_6H_4)_2P(S)\text{-}P(S)(C_6H_4Me\text{-}4)_2$. II.[451] III.[440,543] Crystals, m. 183 to 184° (from acetic acid),[440,543] m. 184 to 185°.[451]

TYPE: $R^1R^2P(S)\text{-}P(S)R^2R^1$

Most of the compounds were obtained in two forms: A and B; one form must be the racemate and the other the meso form.

MeEtP(S)-P(S)EtMe. I.[393,394,400] II.[55] Crystals, (A) m. 103 to 104° (from acetone/H_2O)[394] m. 96 to 103°,[372a] and (B) m. 159 to 160° (from EtOH),[394] m. 150 to 153°,[372a] ^{31}P -42.1 ppm (in $CHCl_3$),[394] -44.5 ppm (in toluene),[187] mass spect.;[372a] form B shows a higher insecticidal activity than form A.[400]

MePrP(S)-P(S)PrMe. I.[393,394,400] Crystals, (A) m. 92 to 94° (from H_2O) and (B) m. 155 to 156° (from EtOH),[394] ^{31}P -40.3 ppm (in $CHCl_3$ or toluene).[187,394]

MeBuP(S)-P(S)BuMe. I.[393,394,400] Crystals, (A) m. 47 to 50° (from ligroin) and (B) m. 126 to 128° (from EtOH).[394]

$Me(PhCH_2)P(S)\text{-}P(S)(CH_2Ph)Me$. I.[394,400] Crystals, (A) m. 120 to 123° (from EtOH) and (B) m. 188 to 189° (from acetone),[394] m. 183 to 184°,[372a] mass spect.

Me(PhCH=CH)P(S)-P(S)(CH=CHPh)Me. I. Crystals, m. 213.5 to 214.0°,[53] (only one form isolated), IR, UV.[53]

MePhP(S)-P(S)PhMe. I.[393,394,400,469,489] Crystals, (A) m. 145 to 146° (from EtOH)[372a,394,469] and (B) m. 206 to 208° (from acetone),[372a,394] m. 216 to 218°,[469] ^{31}P -37.0 ppm (in $CHCl_3$ or toluene),[187,394] mass spect.;[372a] x-ray analysis[573] established that the high-melting form B is the meso compound: P-P 2.21, P-C(Me) 1.82, P-C(Ph) 1.88, P=S 1.98 Å, ∠P-P-S 111.8°.[573]

$Me(4-ClC_6H_4)P(S)-P(S)(C_6H_4Cl-4)Me$. I. Crystals, (A) m. 45 to 50° (from EtOH), and (B) m. 209 to 211° (from acetone),[372a] mass spect.

$Me(4-MeC_6H_4)P(S)-P(S)(C_6H_4Me-4)Me$. I. Crystals, (A) m. 132 to 138° (from EtOH) and (B) m. 208 to 210° (from acetone), mass spect.,[372a] structure uncertain.

$Me(4-CF_3C_6H_4)P(S)-P(S)(C_6H_4CF_3-4)Me$. I. Crystals, (A) m. 137 to 140° (from EtOH), and (B) m. 187 to 192° (from acetone), mass spect.[372a]

EtBuP(S)-P(S)BuEt. II. Liquid, not obtained pure, b_4 165° (dec.),[299] ^{31}P -47.6 ppm (in toluene).[187,299]

EtPhP(S)-P(S)PhEt. I. Crystals, (A) m. 85 to 87° (from MeOH) and (B) m. 156 to 157° (from EtOH);[469] $E_{1/2}$ -1.777 V.[420a]

$Et_2N(c-C_6H_{11})P(S)-P(S)(C_6H_{11}-c)NEt_2$. II. Crystals, (A) m. 158 to 160° (from MeOH/benzene), and (B) m. 191° (from benzene).[535]

$Et_2NPhP(S)-P(S)PhNEt_2$. II. Crystals, (A) m. 123° (from MeOH), and (B) m. 128° (from benzene).[535]

BuPhP(S)-P(S)PhBu. II. Crystals, (A) m. 88° (from MeOH/acetone), $E_{1/2}$ -1.91 V,[420a] and (B) m. 107° (from MeOH).[300]

$(c-C_6H_{11})(C_5H_{10}N)P(S)-P(S)(NC_5H_{10})(C_6H_{11}-c)$. II. Solid, m. 244° (only one form isolated, from benzene).[532]

$Ph(C_5H_{10}N)P(S)-P(S)(NC_5H_{10})Ph$. II. Quadratic crystals, m. 183° (only one form isolated, from benzene, CH_3OH).[533]

$Ph(PhCH_2)P(S)-P(S)(CH_2Ph)Ph$. I. II. Crystals, m. 189.5 to 190.5° (from ligroin/EtOH), only one form isolated,[134] space group $P2_1/c$.[134]

TYPE: $R_2P(S)-P(S)R^1_2$

$Et_2P(S)-P(S)(C_6H_{11}-c)_2$. II. Crystals, m. 117 to 118° (from CS_2),[298] ^{31}P -57.1 and -47.6 ppm (in toluene), J_{PP} 69 Hz.[187,298]

$Et_2P_\beta(S)-P_\alpha(S)Ph_2$. II. Crystals, m. 110 to 112° (from MeOH),[298] ^{31}P -26.0 (P_α) and -57.7 ppm (P_β) (in toluene).[187,298]

$(c-C_6H_{11})_2P_\beta(S)-P_\alpha(S)Ph_2$. II. Crystals, m. 193 to 194° (from acetone),[298] ^{31}P -20.9 (P_α) and -64.0 ppm (P_β) (in toluene).[187,298]

TYPE: Cyclic Biphosphine Disulfides

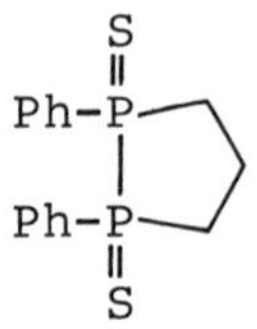

II. Crystals, m. 178 to 180°.[295]

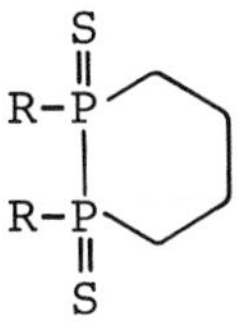

II. (R = Ph). Crystals, m. 171° (from benzene),[300] $E_{1/2}$ -1.88 V.[420a]
(R = Et). Crystals, m. 115° (from acetone),[299] ^{31}P -39.4 ppm (in benzene).[187]

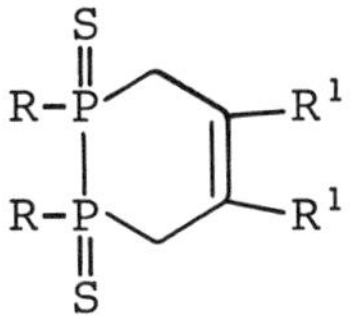

II. (^{1}R = H; R = Me). Solid, m. 176 to 177°.[494]
(^{1}R = H; R = Et). Solid, m. 94°.[494]
(^{1}R = Me; R = Me). Solid, m. 193°.[494]
(^{1}R = Me; R = Et). Solid, m. 115 to 116°.[494]
(^{1}R = Me; R = Ph). Solid, m. 178 to 179°.[494]

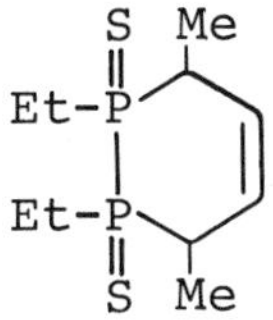

II. Solid, m. 100°.[494]

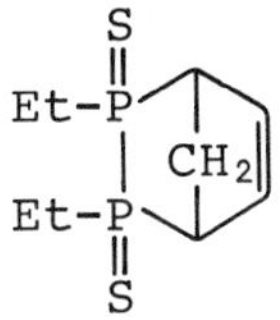

II. Solid, m. 100°.[494]

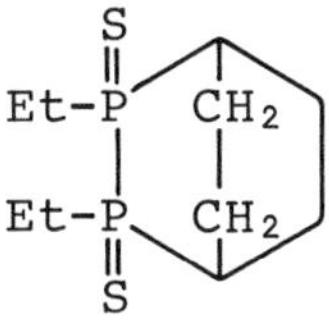

II. Solid, m. 118°.[494]

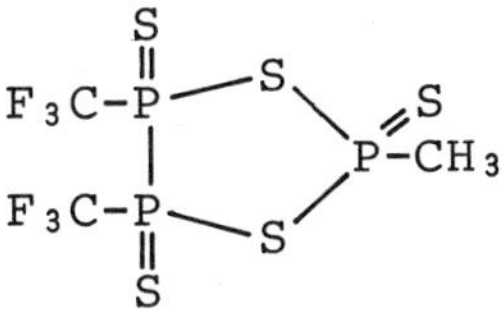

III. From $(CF_3P)_4S$ + S at 180 to

200°.[83a] Crystals, m. 133°, log P_{mm} (solid) = 14.327 - (5044/T), d 2.45; IR, ^{19}F NMR, stable at 185°,[83a] desulfuration by Hg restores $(CF_3P)_4S$.[83a]

Organosubstituted Biphosphine Monosulfides

Roman numerals refer to methods of preparation described in Section E.3.1 of the text.

TYPE: $R_2P(S)-PR_2$

$Me_2P_{\alpha}(S)-P_{\beta}Me_2$. Solid at room temp.,[239,397] $b_{0.5}$ 65 to 70°,[397] ^1H NMR,[239] ^{31}P -37.1 (P_{α}) and +62.1 (P_{β}) ppm;[397] J_{PP} 220.3, 239,[397] 220 Hz.[239] (The coupling constants of biphosphine monosulfides given in a recent monograph[136] are wrong; there they have been calculated by assuming that the spectra were obtained at 24.3 MHz; this, however, was not so; they were all obtained at 16.2 MHz[397].) Could not be reduced electrochemically.[420a]

$Et_2P_{\alpha}(S)-P_{\beta}Et_2$. Liquid, ^{31}P -55.3 (P_{α}) and +37 (P_{β}) ppm; J_{PP} 243, 244 Hz.[397]

$MeEtP_{\alpha}(S)-P_{\beta}EtMe$. Liquid, $b_{0.025}$ 85 to 86°,[397] $b_{0.5}$ 92°,[397a] ^{31}P -47.0 (P_{α}) and +48.3 (P_{β}) ppm; J_{PP} 222, 241 Hz.[397] (The different coupling constants measured on the negative and positive part of the spectrum are due to experimental errors.)

$MePhP_{\alpha}(S)-P_{\beta}PhMe$. Only obtained as a solution in Bu_3PS,[397] ^{31}P -38.8 (P_{α}) and +42.4 (P_{β}) ppm; J_{PP} 159, 160 Hz,[397] -40.8 (P_{α}) and +36.7 (P_{β}) ppm,[372a] ^1H NMR[372a] (shows no change over a temperature of 100 to 200°).

$Ph_2P(S)-PPh_2$. Crystals, m. 138° (from ligroin),[420a] $E_{1/2}$ -1.795 V.[420a]

Organosubstituted Biphosphine Monoxides

Roman numerals refer to methods of preparation described in Section E.4.1 of the text.

TYPE: $R_2P(O)-PR_2$

$Bu_2P_{\alpha}(O)-P_{\beta}Bu_2$. Liquid, $b_{0.01}$ 120 to 125°,[325,326a] ^{31}P -50 (P_{α}) and +51.1 (P_{β}) ppm, J_{PP} 216 Hz.[325,326a]

$(PhCH_2)_2P(O)-P(CH_2Ph)_2$. Solid, m. 153 to 156°, not obtained pure.[473]

$Ph_2P_{\alpha}(O)-P_{\beta}Ph_2$. Solid, m. 155 to 158° (from acetone/H_2O),[383] m. 153 to 157° (from benzene),[443] m. 158 to 161°,[183] IR;[383] forms no quaternary salt; is cleaved by Br_2;[383] ^{31}P -36.9 (P_{α}) and +21.6 (P_{β}) ppm, J_{PP} 224 Hz.[183]

$(2,4,6\text{-}Me_3C_6H_2)_2P(O)-P(C_6H_2Me_3\text{-}2,4,6)$. Crystals, m. 205 to 208° (from acetone),[549] IR, UV;[549] forms no salt

with MeI.[549]
$Ph_2P_\alpha(O)-P_\beta(Cl)Ph$. ^{31}P -50.3 (P_α) and +26.6 (P_β) ppm, J_{PP} 202 Hz.[183]
$(MeO)_2P_\alpha(O)-P_\beta Ph_2$. Liquid, decomposes on attempted distillation, ^{31}P -35.7 (P_α) and +34.1 (P_β) ppm, J_{PP} 192 Hz.[183]
$(MeO)_2P_\alpha(O)-P_\beta(OMe)Ph$. ^{31}P -46.4 (P_α) and +50.6 (P_β) ppm, J_{PP} 202 Hz.[183]
$(EtO)_2P_\alpha(O)-P_\beta Ph_2$. ^{31}P -31.8 (P_α) and +33.5 (P_β) ppm, J_{PP} 178 Hz.[183]
$(EtO)_2P_\alpha(O)-P_\beta(OEt)Ph$. ^{31}P -43.3 (P_α) and +51.1 (P_β) ppm, J_{PP} 201 Hz.[183]
$(EtO)PhP_\alpha(O)-P_\beta Ph_2$. ^{31}P -47.0 (P_α) and +26.8 (P_β) ppm, J_{PP} 205 Hz.[183]

Biphosphine Oxide Sulfides

TYPE: $R_2P(O)-P(S)R_2$

$Ph_2P(O)-P(S)Ph_2$. From $Ph_2P(O)-PPh_2$ and sulfur, white needles, m. 166 to 170° (from acetone/H_2O), IR.[383]

Organosubstituted Biphosphine Dioxides

Roman numerals refer to methods of preparation described in Section E.5.1 of the text.

TYPE: $R_2P(O)-P(O)R_2$

$Me_2P(O)-P(O)Me_2$. Solid, m. 132.5 to 132.7°, IR.[232]
$(Me_2N)_2P(O)-P(O)(NMe_2)_2$. Liquid, $b_{1.5}$ 145 to 150°.[449]
$Bu_2P(O)-P(O)Bu_2$. Liquid, $b_{0.01}$ 150 to 152°.[325]
$(c\text{-}C_6H_{11})_2P(O)-P(O)(C_6H_{11}\text{-}c)_2$. Crystals, m. 205° (from benzene/ligroin).[313]
$(PhCH)_2P(O)-P(O)(CH_2Ph)_2$. Crystals, m. 158 to 159° (from benzene/ether).[473]
$Ph_2P(O)-P(O)Ph_2$. Crystals, m. 167°,[360] m. 167 to 169°,[422,471-473,543] m. 163 to 166°,[451] m. 179.5 to 181.5° (from acetone),[412] m. 169 to 173° (from toluene),[383] IR,[360,383,472] x-ray powder diagram.[360]
$Ph(4\text{-}ClC_6H_4)P(O)-P(O)(C_6H_4Cl\text{-}4)Ph$. Solid, m. 155 to 156°.[472,473]
$Ph(4\text{-}NCC_6H_4)P(O)-P(O)(C_6H_4CN\text{-}4)Ph$. Solid, m. 180 to 182°.[473]
$(4\text{-}MeC_6H_4)_2P(O)-P(O)(C_6H_4Me\text{-}4)_2$. Solid, m. 165 to 168°,[473] m. 159 to 163;[451] recrystallization from Et_2O/THF raised the m. to 183 to 185° with no change in the IR spectrum.[473]
$(2,4,6\text{-}Me_3C_6H_2)_2P(O)-P(O)(C_6H_2Me_3\text{-}2,4,6)$. Solid, m. 188 to 190°, IR, UV.[549]

Miscellaneous Compounds Containing a P-P Bond

Roman numerals refer to methods of preparation described in Section E.6.1 of the text.

$Me_2(Me_2N)\overset{+}{P}-\overset{-}{P}CF_3$. From $Me_2(Me_2N)P$ and $(CF_3P)_4$ at room temp. ^1H and ^{19}F NMR.[123]

$Me_3\overset{+}{P}-\overset{-}{P}CF_3$. From $(CF_3P)_{4/5}$ and Me_3P. ^1H NMR,[81] signs of coupl. const.[413]

$Bu_3\overset{+}{P}-\overset{-}{P}CF_3$. From $(CF_3P)_{4/5}$ and Bu_3P, ^1H NMR.[81]

$Me_3\overset{+}{P}-\overset{-}{P}C_2F_5$. From $(C_2F_5P)_3$ and Me_3P, ^1H and ^{19}F NMR.[123a]

$[Me_3P-PMeCF_3]I$. From above and MeI. White solid; cleaved by HCl to $MeCF_3PCl$ + MePHCl.[75]

$[Et_3P_\alpha-P_\beta Me_2]Cl$. From R_3P and R_2PCl. Solid, m. 87 to 90°, ^1H NMR, ^{31}P -33.7 (P_α) and +42.2 (P_β) ppm.[542]

$[Pr_3P_\alpha-P_\beta Me\ ]Cl$. Liquid, ^1H NMR, ^{31}P -22.5 (P_α) and +47.9 (P_β) ppm.[542]

$[Bu_3P_\alpha-P_\beta Me_2]Cl$. Liquid, ^1H NMR, ^{31}P -21.2 (P_α) and +51.0 (P_β) ppm.[542]

$[(C_8H_{17})_3P_\alpha-P_\beta Me_2]Cl$. Viscous oil, ^1H NMR, ^{31}P -14.5 (P_α) and +25.9 (P_β) ppm.[542]

$[Et_3P-PMeCl]Cl$. Solid, m. 76 to 78°, ^1H NMR.[542]

$[Bu_3P_\alpha-P_\beta MeCl]Cl$. Liquid, ^{31}P -16.5 (P_α) and +14.7 (P_β) ppm.[542]

$[Et_3P-PPhCl]Cl$. White crystals, ^1H NMR;[542] decomposes readily to $(PhP)_5$ and Et_3PCl_2.[542]

$[Ph_3P-P(O)Cl_2]Cl$. From Ph_3P and $P(O)Cl_3$. Colorless solid, m. 87° (dec.), IR.[378]

$[Ph_3P-P(O)Br_2]Br$. From Ph_3P and $P(O)Br_3$. Light-brown microcrystalline solid, m. 197° (dec.), IR.[378]

$Et_3P-PF_2Cl_3$. From Et_3P and PF_2Cl_3[347] (tentatively suggested). Decomposes easily to Et_3PCl_2, $[Et_3PCl]PF_6$, PF_2Cl, and other products.[347]

$[Ph_2P_\alpha-P_\beta Ph_2-NH-P_\gamma Ph_2]Cl$. From $(Me_3Si)_2NH$ and Ph_2PCl. Solid, m. 105 to 110°, IR;[433] ^{31}P +10.6 (P_α), -17.2 (P_β) and -40.2 ppm (P) $J_{P_\alpha P_\beta}$, 249 $J_{P_\beta NP_\gamma}$ 93 Hz.[421a]

$Ph_2P-PPh_2=N-PPh_2$. From above and Et_3N. Solid, m. 124 to 127°, IR;[443] yields with sulfur $Ph_2P-PPh_2=NP(S)Ph_2$,[421a] m. 138 to 142°, ^{31}P.[421a]

$[Ph_2P_\alpha-P_\beta Ph_2{\cdots}N{\cdots}P_\beta Ph_2-P_\alpha Ph_2]Cl$. From $(Me_3Si)_2NH$ and Ph_2PCl. Solid, m. 115° (dec.), IR,[443] ^{31}P +19.2 (P_α) and -24.8 (P_β) ppm, J_{PP} 260 Hz;[187,443] also obtained from above and Ph_2PCl.[421a]

$[Ph_2P-PPh_2{\cdots}N{\cdots}PPh_2-PPh_2]I$. From above and MeI, crystals, m. 175° (dec.), IR, ^1H NMR.[421a]

$S=PPh_2-PPh_2=N-PPh_2=S$. From above and sulfur. Solid, m. 166 to 168°, IR;[421a,443] ^{31}P -43.3 ppm.[421a]

$[Ph_3P_x{\cdots}N{\cdots}P_BPh_2-P_APh_2]^+Cl^-$. From $Ph_3P=N-PPh_2$ and Ph_2PCl or from $[Ph_3PNHSiMe_3]Br$ and Ph_2PCl. Solid, m. 166° (dec.), ^{31}P: P_X -20.2, P_B -26.1, P_A +19.2 ppm.[416]

$PhP_{\alpha}(=S)(S^{\ominus})-P_{\beta}^{\oplus}Bu_3$. From $[RPS_2]_2$ + Bu_3P. Solid, m. 97 to 98°, ^{31}P -60 (P_{α}) and -7.0 (P_{β}) ppm,[181] J_{PP} 118 Hz.[181]

$4\text{-}MeC_6H_4P_{\alpha}(=S)(S^{\ominus})-P_{\beta}^{\oplus}Bu_3$. As above. Solid, m. 87°, ^{31}P -60 (P_{α}) and -7.4 (P_{β}) ppm, J_{PP} 109 Hz.[181]

$4\text{-}EtOC_6H_4P_{\alpha}(=S)(S^{\ominus})-P_{\beta}^{\oplus}Bu_3$. As above. Solid, m. 85.7°, ^{31}P -60.9 (P_{α}) and -7.0 (P_{β}) ppm, J_{PP} 108 Hz.[181]

$1\text{-}C_{10}H_7P_{\alpha}(=S)(S^{\ominus})-P_{\beta}^{\oplus}Bu_3$. As above, Solid, m. 140°, ^{31}P -61.6 (P_{α}) and -9.5 (P_{β}) ppm, J_{PP} 96 Hz.[181]

$PhP(=O)(OK)-P(=O)(OK)Ph$. From $(PhPO_2)_2$ + K. Colorless needles, ^{31}P -5.2 ppm.[182]

$PhP(=S)(SK)-P(=S)(SK)Ph$. From $(PhPS_2)_2$ + K. Pale-yellow solid, ^{31}P -95.2 ppm.[182] The reported m. of the free acid $Ph(HS_2)P\text{-}P(S_2H)Ph$, m. 225° (dec.),[292] seems to be in error since in a later report[300] it was said that the biphosphine PhHP-PHPh used to synthesize this acid was in fact a cyclopolyphosphine which had included water.[300]

Organosubstituted Triphosphines, Tetraphosphines, and Corresponding Oxides and Sulfides

Roman numerals refer to methods of preparation described in Section E.7.1 of the text.

TYPE: Triphosphines, Oxides, and Sulfides

$CF_3P[P(H)CF_3]_2$. III.[84 391] Liquid, b 187.7° (est.), log P_{mm} = 6.4627 + 1.75 log T - 0.0054 T - (2375/T),[84] UV, IR;[391] is decomposed by Hg, stopcock grease, and tertiary phosphines.[84]

$CF_3P[P(CF_3)_2]_2$. I. Liquid, b. 140° (est.), log P_{mm} = 7.1080 + 1.75 log T - 0.006 T - (2616/T),[83] UV, IR.[83]

$CF_3P[PMe_2]_2$. III. Liquid, b. 220° (est.), log P_{mm} = 7.3510 - (2176/T), IR, UV, ^{19}F NMR,[121a,123] ^1H NMR.[123]

$MeP[P(CF_3)_2]_2$. I. III. Liquid, b. 151° (est.), m. -66°, log P_{mm} = 7.5474 + 1.75 log T - 0.00635 T - (2786/T),[74] UV, IR.[74]

$Me_2NP[P(NMe_2)_2]_2$. III. Solid, m. 18 to 20°, $b_{0.01}$ 85 to 90°, IR.[449]

$EtP[PEt_2]_2$. II. Liquid; disproportionates readily.[574]

$MeP[PPh_2]_2$. I. Crystals, m. 130 to 134° (from $CHCl_3$);[404] is cleaved with Br_2 and MeI and gives with sulfur a disulfide of probable structure $MeP[P(S)Ph_2]_2$, m. 163.5 to 166°.[404]

$PhP[P(H)Ph]_2$. I. Crystals, m. 113.5° (from benzene); is oxidized with HNO_3.[574]

$PhP[PPh_2]_2$. II. Colorless crystals, m. 70° (dec.),[574] IR; is decomposed by HNO_3, H_2SO_4, $AgNO_3$, and MeI.[574]

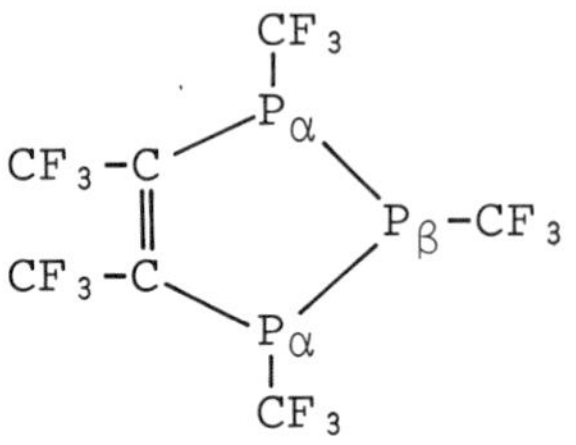

III. Liquid, b. 160° (est.), vapor pressure is 1.5 mm at 25°, inflammable in air,[388] UV, IR, ^{19}F NMR, ^{31}P +41 ppm (P_α) and -55 ppm (P_β); converted by I_2 to $P_2C_2(CF_3)_4$.[388]

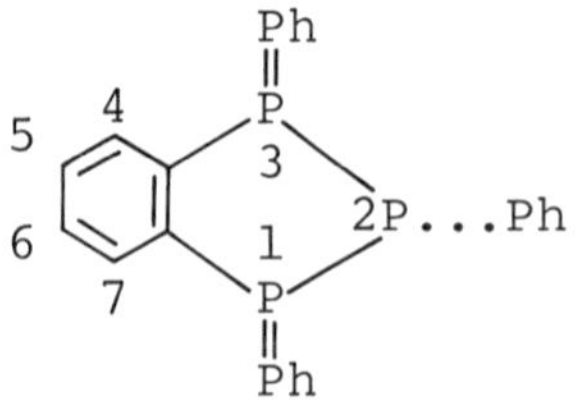

II. III. Crystals, m. 184 to 186° (from acetone);[415] monomethyl iodide, m. 175°; but also cleaved by MeI, $KMnO_4$, Br_2;[415] gave with sulfur a monosulfide, m. 122 to 123°, two isomeric disulfides (cis-1,3 and trans-1,2), m. 182 to 183° and 203 to 204°;[415] x-ray analysis[144] shows bicyclic phosphaindane nucleus is roughly planar and the phenyl group in the 2-position is trans to those in the 1- and 3-positions, P-P 2.209, P-C 1.825 Å, valence angles at phosphorus vary from 97.3 to 103.6°.[144]

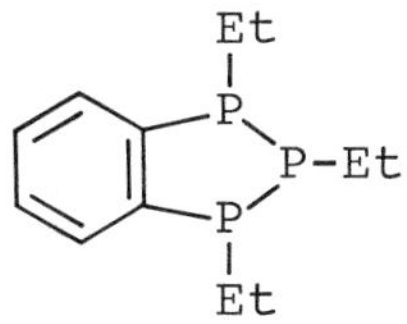
II. III.[415] Liquid, not obtained pure; contained $(EtP)_4$, $b_{0.2}$ 124 to 135°. Cleaved by MeI, gave with sulfur a trisulfide, m. 120.2° (from MeOH).[415]

$PhP_\alpha[P_\beta(O)(OMe)_2]_2$. IV. ^{31}P +64.9 ppm (P_α) and -33.7 ppm (P_β), J_{PP} 168 Hz.[183]
$PhP_\alpha[P_\beta(O)(OEt)_2]_2$. IV. ^{31}P +63.3 ppm (P_α) and -26.7 ppm (P_β), J_{PP} 157 Hz.[183]
$PhP_\alpha[P_\beta(O)(OMe)Ph]_2$. IV. ^{31}P +45.6 ppm (P_α) and -45.7 ppm (P_β), J_{PP} 210 Hz.[183]
$PhP_\alpha[P_\beta(O)(OEt)Ph]_2$. IV. ^{31}P +45.5 ppm (P_α) and -43.5 ppm (P_β), J_{PP} 204 Hz.[183]

TYPE: Tetraphosphines

$H(PCF_3)_4H$. III. Liquid, vapor pressure at -45° is 5 mm;[84] decomposes at room temperature.
$P[PPh_2]_3$. IV. Crystals, m. 118 to 120° (ether/pentane), IR;[516] ^{31}P +16.9 and +27.0 ppm.[501]

Cyclopolyphosphines

Roman numerals refer to methods of preparation described in Section E.8.1 of the text.

TYPE: $(RP)_x$

$(ClCH_2P)_n$. IV. Structure unknown.[441a]
$(MeP)_5$. II.[30,244,245,311,364,542] IV.[370,528] VI.[563] Liquid, $b_{0.0002}$ 86°,[30] $b_{0.2}$ 99°,[364] b_2 123°,[370] b_1 110 to 112°;[245,311] IR,[30,491] Raman,[30] UV,[245,494] mass spect.;[125,235,492] pyrolysis at 900° seems to produce mainly $(MeP)_3$ and small amounts of "MeP";[235] ^1H NMR,[128,528] ^{31}P -21.0 ppm (in CS_2);[245] +92.8 ppm;[18a] adducts: L•MeI, white crystals;[311] $L\cdot Cr(CO)_4$, yellow solid, dec. >200°;[17] $L\cdot Mo(CO)_4$, yellow crystals, dec. 200°;[17] $L\cdot W(CO)_4$, yellow crystals, dec. 205°;[17] $Me_4P_4[Fe(CO)_3]_2$, orange prisms, m. >300°;[15] L•CuBr, yellow solid, dec. >250°;[140] $(MeP)_4CuCl$, yellow solid, dec. >250°.[140]
$(MeP)_n$. II.[364] VI.[74,563] Highly viscous liquid[74] or pale-yellow solid.[364,563]

$(CF_3P)_4$. II.[68,79,309,392] IV.[498] VI.[80,83,84,392] White crystals, m. 66.4°,[389,392] b_{760} 135° (est.), log P_{mm} = 8.3935 - (2251.6/T) (liquid), and log P_{mm} = 11.7239 - (3384.6/T) (solid),[392] d_4^{20} 2.0,[454] $d_4^{66.4}$ 1.54 (liquid), UV,[392] mass spect.,[102,498] ^{19}F NMR,[572] ^{31}P +74.3 ppm;[572] x-ray analysis[454] shows four-membered puckered ring with bond angles of 84.7°, ring torsion angles of 34°, P-P 2.213, P-C 1.867 Å;[454] adducts: with sulfur $(CF_3P)_4S$, five-membered ring, liquid, b_{760} 183°;[68,83a] with $Ni(CO)_4$ polymeric product,[78] $[(CF_3P)_4Ni_{1.77}(CO)_{4.45}]$; reacts with Me_4As_2 and Me_2S_2 to give $(Me_2As)_2PCF_3$ and $(MeS)_2$-PCF_3,[123] respectively; does not react with Me_6Si_2.[123]

$(CF_3P)_5$. II.[68,80,392] IV.[498] VI.[80,83,84,392] Liquid, b_{760} 190° (est.),[392] log P_{mm} = 6.9302 - 0.004913 T + 1.75 log T - (2982.6/T),[392] d_4^{20} 1.60, d_4^{-100} 2.12 (solid),[544] UV,[392] mass spect.,[498] ^{19}F NMR,[572] ^{31}P -8.7 ppm;[572] x-ray analysis[157,544] shows puckered five-membered phosphorus ring with bond angles of 101.3 ± 4.9°, torsion angles of 18 to 58°, P-P 2.202 to 2.252, P-C 1.873 to 1.925 Å.[544]

$(CF_3P)_n$. II.[72,389,392] VI.[83] Structure unknown; may contain $(CF_3P)_6$.[392]

$(EtP)_{4,5}$. I.[306] II.[30,245,364] VI.[299] V.[125] VII.[281,311] Liquid, $b_{0.0002}$ 112 to 114°,[30] $b_{0.05}$ 124 to 129°,[245] $b_{0.2}$ 123°,[364] b_{3-4} 160,[299,311] b_{15} 168 to 170;[306] UV, IR,[30] Raman,[14,30] mass spect.,[125,492] $E_{1/2}$ -2.38 V,[420a] ^{31}P -17 ppm (CS_2),[245] -16 ppm,[125] -15.7 ppm[188,189] -15.8 ppm;[187] adducts: $(EtP)_4$· MeI, m. 84°;[311] $(EtP)_5$·MeI, m. 105°,[311] $(EtP)_2$·BX_3 (X = Cl, Br, F), polymeric;[126] $(EtP)_5Cr(CO)_5$, yellow crystals, m. 70°;[17] $(EtP)_5Mo$-$(CO)_5$, yellow crystals, m. 71°; $(EtP)_5$-$Mo(CO)_4$, pale yellow, m. 124°;[140,191] x-ray analysis established in this complex a five-membered phosphorus ring;[96,97] $(EtP)_4Mo(CO)_2$, yellow polymer, dec. 250°;[140] $(EtP)_5W(CO)_5$, yellow crystals, m. 69°;[17] $(EtP)_4W(CO)_4$, pale-yellow crystals, m. 148.5°;[140,191] preliminary x-ray analysis of this complex indicates a four-membered phosphorus ring;[96] $Et_4P_4[Fe(CO)_3]_2$, orange-red plates (structure unknown), m. 196°;[17] $(EtP)_4$·$2Fe(CO)_4$;[293] $(EtP)_4$-$PdCl_2$, orange-yellow, dec. 250°;[140] $(EtP)_4CuX$ (X = Cl, Br), pale yellow, dec. 150°.[140,191]

$(EtP)_n$. VI. Red-brown solid; inflames in air; structure unknown.[574]

$(C_2F_5P)_3$. II. Colorless liquid, b_{760} 151.4° (calc.), log P = 7.831 - (2101.9/T),[123a] UV, ^{19}F NMR, mass spect.;[123a] yields with Me_3P the ylid Me_3P=PC_2F_5;[123a] structure could not be confirmed.[18a]

$(C_2F_5P)_4$. II. White crystals, m. 23.5°;[123a] b_{22} 80°, ^{19}F, ^{31}P +71.9 ppm.[18a]

$(C_2F_5P)_5$. II. Liquid, b_{51} 72 to 74°, ^{19}F, ^{31}P +144.2 ppm, converts to the tetramer, mass spect.[18a]

$(PrP)_4$. I.[245] III.[246] V.[125] Liquid, $b_{0.03}$ 140 to 145°,[245] $b_{1.0}$ 120 to 124°,[246] Raman,[14] mass spect.,[125] ^{31}P -13 ppm (in CS_2),[245] -53, -16, -12 ppm[125] (the various peaks attributed to different ring sizes, for discussion see ref. 406); adduct: $(PrP)_2BX_3$ (X = F, Cl, Br), structure unknown.[126]

$(i\text{-}PrP)_4$. II. Liquid, $b_{0.04}$ 110 to 114°, m. 23 to 24°,[245] Raman,[14] ^{31}P +66 ppm (in CS_2).[245]

$(C_3F_7P)_4$. II. Liquid, b_{16} 118 to 120°, ^{19}F, ^{31}P +66.4 and +141.4 ppm; mass spect.[18a]

$(C_3F_7P)_5$. II. Liquid, b_{30} 108 to 110° (?), ^{19}F NMR, mass spect.[18a]

$(NCCH_2CH_2P)_4$. I. Solid, m. 87 to 89° (from acetone/H_2O),[245] ^{31}P -2 ppm (in CS_2).[245]

$MeBu_3P_4$. V. Liquid, b_8 214°.[340]

$(BuP)_4$. I.[245] V.[125,340,474,475] Liquid, $b_{0.007}$ 136 to 140°,[474] $b_{0.02}$ 170°,[245] $b_{0.2}$ 182 to 185°,[340] IR, UV,[474] mass spect.,[125] ^{31}P -14 ppm (in CS_2),[245] -13, -17 ppm[125] (attributed to different ring sizes, for discussion see ref. 406); adduct: $(BuP)_2{\cdot}BX_3$ (X = F, Cl, Br), structure unknown.[126]

$(i\text{-}BuP)_4$. I.[245] II.[245,246] Liquid, $b_{0.1}$ 140°,[245] $b_{1.0}$ 145 to 148°,[246] UV,[245] Raman,[14] ^{31}P -13 ppm (in CS_2);[245] adducts: $[i\text{-}BuPS]_4$, solid, m. 78 to 82°, structure uncertain;[402,406,426] L·CuX (X = halogen) yellow solids;[252] $L{\cdot}[CuX]_2$.[252]

$(t\text{-}BuP)_4$. II. Solid, m. 167 to 196°, ^{31}P +57.8 ppm,[291] adduct: L·MeI, solid, m. 183 to 187°.[291]

$(Et_2CHP)_4$. III.[245,246] Solid, m. 91 to 91.4° (from ligroin),[245] m. 92 to 93°,[246] UV,[245] ^{31}P +70 ppm (in CS_2).[245]

$(c\text{-}C_6H_{11}P)_4$. I.[245,315] III.[245] VI.[299] VII.[279] Crystalline solid, m. 222 to 224° (from toluene),[245] m. 219 to 220°,[299,315] d_4^{20} 1.157[28] UV,[245] ^{31}P +70 ppm (in CS_2),[245] from benzene tetragonal crystals;[28,146] x-ray analysis[28] gives puckered four-membered phosphorus ring with bond angles of 85.5°, torsion angles of 31.4° and P-P 2.224 Å, P-C 1.874 Å;[28] adducts: $L{\cdot}[Fe(CO)_4]_2$, yellow crystals, m. 190°;[293] $L{\cdot}Ni_3(CO)_8$, polymeric, m. 120° (dec.);[293] $L{\cdot}Ni_{1.8}(CO)_x$ polymeric, white solid, m. 138°;[293] L·CuX (X = halogen), colorless solid;[252] $L{\cdot}(CuX)_2$.[252]

$(n\text{-}C_8H_{17}P)_4$. I. II.[245,246] Liquid, $b_{0.1}$ 230°,[245] UV,[245] ^{31}P -13 ppm (in CS_2);[245] adducts: $[n\text{-}C_8H_{17}PS]_4$, liquid, $b_{0.2-0.1}$ 245 to 250°;[402,426] L·CuX (X = halogen), yellow solids;[252] $L{\cdot}[CuX]_2$.[252]

$(PhP)_3K_2$. From $(PhP)_5$ and K in THF; cyclopropenylium cation analogous structure proposed;[282] orange-red crystals with 2 moles of THF, ^{31}P +49.8 ppm (in THF); hydrolyzes with H_2O to $(PhP)_5$ and $PhPH_2$; adducts: $[(PhP)_3{\cdot}Ni(CO)_3]K_2{\cdot}$THF, orange-red crystals, dec. 95°; $[(PhP)_3Fe(CO)_4]K{\cdot}$THF, brown crystals, dec. 195°.[282]

$(PhP)_3Li_2$. From $(PhP)_5$ and Li_2PPh in ether, unstable in THF, ^{31}P +56.7 ppm.[302]

$(PhP)_5$ (Form A). I.[145,245,351,358,359,404,462,476] II.[35,47,126,181,193,213,245,270,363,364,404,462,491,558,574] IV.[498] VI.[208,316,373,374,548] VII.[210,211,300,404,405,433,470,491,560] Crystalline solid, m.

150°,[351] m. 151 to 152°,[358,359] m. 154 to 156°,[245,404,476] (from benzene, or acetonitrile),[404] d_4^{20} 1.324;[405] UV,[245,571] IR,[245,359,476] Raman,[13] mass spect.,[31,245,494,498] $E_{1/2}$ -1.725 V;[420a] pyrolysis at 400° seems to produce "PhP";[235] ^{1}H NMR,[245] ^{31}P +9 ppm (in CS_2),[245] +2.3 and +3.1 ppm (in CS_2),[498] in benzene: +4.4 ppm,[187] +4.6 ppm,[437] +4.3 ppm;[498] x-ray powder diagram;[406] x-ray analysis[141,145] shows puckered five-membered phosphorus ring with bond angles of 94 to 107.2°, ring torsion angles of 2.2 to 60.6°, P-P 2.217, P-C 1.843 Å;[141,145] adducts: $(PhP)_4S$, m. 154 to 156°;[401,404] $(PhPS)_3$, probably six-membered ring, m. 148°;[146,359,364,401,406] $[PhPSe]_4$, m. 71 to 72°;[356,402,406] $(PhPNS)_x$ m. 200 to 201°;[339,406] $(PhP)_3 \cdot BF_3$, m. 130 to 140°;[126,359] $(PhP)_3 \cdot BCl_3$, m. 185 to 195°;[126] $(PhP)_3 \cdot 2BBr_3$;[126] L·Cr(CO)$_4$, yellow crystals, m. 300°;[17] $(PhP)_4Cr(CO)_4$, brown solid;[192] L·Mo(CO)$_5$, cream crystals, m. 183°;[16,17] $(PhP)_4Mo(CO)_4$, brown solid;[192] L·W(CO)$_5$, cream crystals, m. 190°;[16,17] $(PhP)_4W(CO)_4$, brown solid;[192] L·Ni(CO)$_3$, yellow crystals,[125,406] dec. 80°;[406] $(PhP)_4$-$Ni(CO)_3$, yellow solid;[16,17] $(PhP)_4[Fe(CO)_4]_2$;[16] $(PhP)_4$-$[Fe(CO)_3]_2$, orange plates, m. 225°.[17]

$(PhP)_6$ (Form B). I.[146,245,476,571] II.[404] Crystalline solid, m. 188 to 192°,[476,571] m. 195 to 199°,[245] UV,[245,476] IR,[245,476] Raman,[13] mass spect.,[245] ^{1}H NMR;[245] four different crystalline modifications:[406] rhombohedric, m. 183 to 186° (with 1 mole benzene), monoclinic, m. 189 to 193°, triclinic, m. 184.5 to 189°, trigonal, m. 190 to 195°;[145,146,402] all four forms have a six-membered phosphorus ring structure; full x-ray analysis of the trigonal[142,146] and the triclinic[143] form establishes for the P_6 ring chair configuration with the phenyl groups in equatorial positions; trigonal: bond angles 94.6°, torsion angles 85°, P-P 2.237, P-C 1.843 Å;[142,146] triclinic: bond angles 95.5°, torsion angles 83.6°, P-P 2.23, P-C 1.843 Å;[143] d_4^{20} 1.257 to 1.275 (depending on crystal structure);[404] forms no adduct with sulfur or $Ni(CO)_4$.[404,406]

$(PhP)_n$ (Form C). I.[245,404,476] Solid, m. 252 to 256°,[476] m. 260 to 280,[245] m. 289 to 305°,[404] (m. depends on rate of heating[404]), IR,[245] x-ray powder diagram,[404,406] (ring size unknown).

$(PhP)_n$ (Form D). I.[245] IV.[173] Solid, m. 280°,[173] m. 260 to 285°,[245] IR,[245] probably identical with form C.[245]

$(1\text{-}MeC_6H_4P)_5$ (Form A). I. II. Solid, m. 281 to 287°, m. 290 to 299°,[409] IR,[409] x-ray powder diagram.[409]

$(1\text{-}MeC_6H_4P)_4$ (Form C). I. Tetragonal crystals, m. 198 to 202°, d_4^{20} 1.283,[409] IR, x-ray powder diagram.[409]

$(4\text{-}ClC_6H_4P)_5$ (Form A). Yellow, crystalline powder, m. 191 to 193° (from benzene),[409] sensitive to O_2 and H_2O, IR, x-ray powder diagram.[409]

$(4\text{-}ClC_6H_4P)_6$ (Form B). I. Microcrystalline powder, m. 194 to 198° (from EtOH/THF),[409] IR.[409]
$(4\text{-}ClC_6H_4P)_4$ (Form C). I. Yellow, crystalline solid, m. 161 to 165°,[409] (ring size uncertain), IR, x-ray powder diagram.[409]
$(4\text{-}FC_6H_4P)_n$. I. Solid, m. 167 to 169° (from THF/H_2O), and m. 195 to 197° (from benzene/AcEt).[487]
$(4\text{-}MeOC_6H_4P)_5$. II. Solid, m. 188 to 192°, ^{31}P +11.8 ppm.[181]
$(4\text{-}EtOC_6H_4P)_5$. II. Solid, m. 188°, ^{31}P +12.0 ppm.[181]
$(C_6F_5P)_4$. I. Solid, m. 151°, ^{31}P +67.0 ppm;[176] x-ray analysis.[478a]
$(C_6F_5P)_5$ (Form A). II. Solid, m. 156 to 161°,[127] ^{19}F NMR, mass spect., x-ray powder diagram.[127]
$(C_6F_5P)_5$ (Form B). By sublimation of form A, or by dissolution of form A in ether; solid, m. 145°, x-ray powder diagram.[127]
$(2\text{-}C_{10}H_7P)_n$. II. Glassy product, ^{31}P -3.3 ppm.[181]

TYPE: Polycyclic Compounds.

$P_{10}Ph_6$. Prepared from PCl_3 + Li_2PPh, yellow solid, softening point ~60°, IR; adamantanelike structure proposed; oxide, $P_{10}Ph_6O_4$, from $POCl_3$ + Li_2PPh, yellow solid, softening point ~60°, IR.[574]
$P_{10}Ph_6(NMe_2)_2$. II.[560] Brown yellow solid, not well characterized; decalinelike structure proposed.
$P_{18}(NMe_2)_{12}$. Obtained from the decomposition of $[(Me_2N)_2P]_2$, solid, not well characterized; tetracenelike structure proposed.[449]

Arsinosubstituted Phosphines

Roman numerals refer to methods of preparation described in Section E.9.1 of the text.

TYPE: $R_2As\text{-}PR_2$

$Me_2P\text{-}As(CF_3)_2$. I. Liquid, b_{760} 123°, log P_{mm} = (-2000/T) + 7.93,[103] mass spect., ^1H and ^{19}F NMR, ^{31}P +43.5 ppm.[103]
$(CF_3)_2P\text{-}AsMe_2$. I. Liquid, b_{760} 125°, log P_{mm} = (-1981/T) + 7.85,[103] mass spect., ^1H and ^{19}F NMR, ^{31}P -10.7 ppm.[103]
$(CF_3)_2P\text{-}As(CF_3)_2$. I. ^{19}F NMR, mass spect.[103]
$Me_2P(S)\text{-}AsMe_2$. I. ^1H NMR.[238]
$Ph_2P\text{-}AsPh_2$. I. Solid, m. 115 to 117°, IR.[516]
$[Ph_2P]_2AsPh$. Solid, m. 125 to 129°, IR,[516] ^{31}P +15.3 ppm;[501] $L\cdot Ni(CO)_3$, dec. 112°.[525a]
$CF_3P[(AsMe_2)(PMe_2)]$. I. From $(Me_2P)_2PCF_3$ and $(Me_2As)_2PCF_3$ by heating to 130°, not isolated pure, ^1H and ^{19}F NMR.[123]
$CF_3P[AsMe_2]_2$. I. From $(CF_3P)_4$ + $(Me_2As)_2$ at room temp., vapor pressure at 24° is 0.5 mm Hg, IR, UV,[123] ^1H

NMR;[122,123] adduct with $2BH_3$;[123] is cleaved by HCl to give CF_3PH_2 and Me_2AsCl.[123]
$PhP[AsPh_2]_2$. I. Solid, m. 155 to 158°, IR,[516] ^{31}P -36.1 ppm.[501]
$[Ph_2P]_3As$. I.[219,516] Solid, m. 120 to 123°, IR,[516] ^{31}P +15.2 ppm.[501]
$P[AsPh]_3$. I. Solid, m. 169 to 172°, IR,[516] ^{31}P +59.1 ppm.[501]

TYPE: $[R_3P\text{-}AsR_2]X$

$[Me_3P\text{-}AsMe_2]I$. II. Solid, m. 270 to 273° (from $EtOH/H_2O$).[56,114]
$[Et_3P\text{-}AsMe_2]X$. II. (X = Cl). m. 73 to 75° (from EtOH-Et_2O).[56,114]
(X = Br). m. 142 to 145° (from EtOH).[56]
(X = I). m. 132 to 135° (from EtOH).[56,114]
$[Et_3P\text{-}AsPh_2]I$. II. Solid, m. 85 to 87° (from EtOH).[56]
$[Pr_3P\text{-}AsMe_2]I$. II. Solid, m. 100° (from PrOH).[56]
$[Bu_3P\text{-}AsMe_2]I$. II. White crystals, m. 81 to 82° (from CH_3CN), IR, ^1H NMR.[332]
$[Bu_3P\text{-}AsMeCl]Cl$. II. White crystals, m. 69 to 70° (from ligroin), IR, ^1H NMR.[332]
$[Bu_3P\text{-}AsMeI]I$. II. Yellow crystals, m. 89 to 90° (from CH_3CN), IR, ^1H NMR.[332]
$[Bu_3P\text{-}AsPhCl]Cl$. II. White crystals, m. 80 to 81° (from ligroin), IR, ^1H NMR.[332]
$[Bu_3P\text{-}AsPhI]I$. II. Orange crystals, m. 94 to 96° (from CH_3CN), IR, ^1H NMR.[332]
$[Me_2PhP\text{-}AsMe_2]X$. II. (X = Cl). Solid, m. 115 to 116° (from acetone).[56,114]
(X = Br). Solid, m. 178° (from EtOH).[56]
(X = I). Solid, m. 147° (from EtOH).[56,114]
$[Me_2PhP\text{-}AsEt_2]I$. II. Solid, m. 112 to 118° (from EtOH).[56]
$[Et_2PhP\text{-}AsMe_2]Br$. II. Solid, m. 142 to 148° (from acetone).[56]
$[MePh_2P\text{-}AsMe_2]I$. II. Solid, m. 115 to 117° (from EtOH).[56]

TYPE: $(RO)_2P(O)\text{-}AsR_2$

Structure of all these compounds is uncertain; they may all have a phosphite structure $(RO)_2POAsR_2$.

$(MeO)_2P(O)\text{-}AsMe_2$. III. Liquid, b_1 76.5°, d_0^0 1.4011.[343]
$(MeO)_2P(O)\text{-}AsEt_2$. III. Liquid, b_1 98.5 to 99.5°, d_0^0 1.3205.[343]
$(EtO)_2P(O)\text{-}AsMe_2$. III. Liquid, b_1 83°, d_0^0 1.3036, n_D^{10} 1.2932.[343]
$(EtO)_2P(O)\text{-}AsEt_2$. III. Liquid, b_1 105.5 to 106.5°, d_0^0 1.2276, d_0^{15} 1.2129, n_D^{15} 1.4778.[343]
$(EtO)P(O)\text{-}AsEt_2BuI$. III. Crystals, m. 182 to 183° (from

$EtOH/Et_2O$).[344]
$(EtO)_2P(O)$-AsEtBu. III. Liquid, b_1 112 to 113°, d_0^0 1.2054, d_0^{14} 1.1865.[344]
$(EtO)_2P(O)$-AsEtAm-i. III. Liquid, b_1 118 to 120°, d_0^0 1.2858, d_0^{14} 1.2718.[344]
$(EtO)_2P(O)$-AsEtPh. III. Liquid, b_1 144 to 145°, d_0^0 1.2869, d_0^{14} 1.2734.[344]
$(EtO)_2P(O)$-As$(CH_2CH{=}CH_2)$Ph. III. Liquid, b_1 142 to 143°, d_0^0 1.2568.[344]
$(EtO)_2P(O)$-AsBuPh. III. Liquid, b_1 162 to 163°, d_0^0 1.2411, d_0^{17} 1.2345.[344]
$(EtO)_2P(O)$-$AsPh_2$. III. Liquid, b_1 176 to 177°, d_0^0 1.2971, d_0^{16} 1.2845.[344]
$(PrO)_2P(O)$-$AsMe_2$. III. Liquid, b_1 101.5°, d_0^0 1.2343, n_D^{10} 1.2242.[343]
$(PrO)_2P(O)$-$AsEt_2$. III. Liquid, b_1 124 to 125°, d_0^0 1.1817, d_0^{15} 1.1675, n_D^{15} 1.4901.[343]
$(PrO)_2P(O)$-AsEtPh. III. Liquid, b_2 165 to 166°, d_0^0 1.2620, d_4^{24} 1.2427, n_D^{24} 1.5375.[345]
$(i\text{-}PrO)_2P(O)$-$AsMe_2$. III. Liquid, b_1 82 to 83°, d_0^0 1.2112, n_D^{20} 1.2015.[343]
$(i\text{-}PrO)_2P(O)$-$AsEt_2$. III. Liquid, b_1 106 to 107°, d_0^0 1.1678, d_0^{15} 1.1529, n_D^{15} 1.4782.[343]
$(i\text{-}PrO)_2P(O)$-AsMeEt. III. Liquid, b_2 99.0 to 99.5°, b_{4-5} 110 to 112°, d_0^0 1.1932, d_0^{18} 1.1733, n_D^{18} 1.4761.[345]
$(BuO)_2P(O)$-$AsMe_2$. III. Liquid, b_1 122 to 123°, d_0^0 1.1933.[343]
$(BuO)_2P(O)$-$AsEt_2$. III. Liquid, b_1 144 to 145°, d_0^0 1.0522.[343]
$(BuO)_2P(O)$-AsMeEt. III. Liquid, b_3 127 to 128°, d_0^0 1.0884, d_0^{20} 1.0710.[345]
$(BuO)_2P(O)$-AsEtBu. III. Liquid, b_2 146 to 147°, d_0^0 1.1226, d_0^{20} 1.1048, n_D^{20} 1.4775.[345]
$(BuO)_2P(O)$-AsEtBu-i. III. Liquid, b_2 138.5 to 140°, d_0^0 1.1268, d_0^{20} 1.1087, n_D^{20} 1.4738.[345]
$(BuO)_2P(O)$-AsEtPh. III. Liquid, b_2 176 to 176.5°, d_0^0 1.2031, d_0^{13} 1.1875, n_D^{18} 1.5304.[345]

Organophosphorus Compounds of Transition Metals

Roman numerals refer to methods of preparation described in Section F.1 of the text.

TYPE: Group IB Phosphorus Compounds

$(c\text{-}C_6H_{11})_2PCu$. I. Colorless diamagnetic crystals, dec. 195°, probably polymeric structure, sensitive to air.[330]
$[(c\text{-}C_6H_{11})_2P]_2CuLi\cdot 2THF$. From above and $LiP(C_6H_{11}\text{-}c)_2$, solid, dec. 135°; gives on treatment with H_2O a phosphine complex $(c\text{-}C_6H_{11})_2PCu\cdot HP(C_6H_{11}\text{-}c)_2$, pale-yellow crystals, m. 158°.[330]
Ph_2PCu. I.[284] II.[4] Red-violet solid, dec. >300°,[284] bright-red solid, m. 190 to 192° (dec.),[4] stable in H_2O,

polymeric structure; gives with $HPPh_2$ an adduct, dec. 93°.[284]

$[Ph_2P]_2CuK$. From above and $KPPh_2$, orange-red, diamag. solid, dec. 177 to 181°, trimeric in benzene;[284] gives on treatment with ROH a phosphine complex, $Ph_2PCu \cdot HPPh_2$, orange-yellow, diamag. solid, dec. 93°, trimeric in dioxan solution.[284]

$[Ph_2PAg]_n$. II. White, pale-yellow powder, air-sensitive.[4]

TYPE: Group IIB Phosphorus Compounds

Bu_2PZnEt. II. Solid, m. 43 to 45°, associated in benzene solution.[450]

Bu_2PZnBu. II. Liquid, associated in benzene solution.[450]

$[Bu_2P]_2Zn$. II. Solid, dec. begins at ∿260°, m. 310°, polymeric structure.[450]

$[c\text{-}C_6H_{11})_2P]_5Br_2Zn_3Li \cdot 2THF$. I. Pale-yellow crystals, dec. 360°, sensitive to air and H_2O.[330]

Ph_2PZnEt. II. Solid, m. >300°, polymeric structure,[450] MeI adduct, m. 158 to 160°; P-Zn bond cleaved in refluxing MeI.[450]

Ph_2PZnPh. II. Solid, m. 179 to 180° (from benzene/pentane),[450] m. 165 to 170°,[432] degree of polymerization ∿7.[450]

$[Ph_2P]_2Zn$. I.[284] II.[284,450] Infusible and insoluble white,[450] to pale yellow,[284] polymeric solid, sensitive to air and H_2O.

$[Ph_2P]_3Cl_2Zn_2Li \cdot 3THF$. I. Pale-yellow crystalline solid.[284]

Et_2PHgCl. II. White solid,[203] $HgCl_2$ is used to determine quantitatively PH_3 and primary and secondary phosphines.

$(c\text{-}C_6H_{11})_2PHgI$. II. Colorless solid, dec. 150°.[534]

$P(HgCl)_3$. II. Pale-yellow solid.[43]

TYPE: Group IVB Phosphorus Compounds

$[(c\text{-}C_6H_{11})_2P]_2Ti$. I. Black-brown pyrophoric crystals, mag. moment 0.60 μB, sensitive to H_2O;[329] gives on treatment with iodine $[c\text{-}C_6H_{11})_2P]_2TiI_2$ and $(c\text{-}C_6H_{11})_2\text{-}PTiI_3$.[329]

Me_2PTiCp_2. I. ESR spect.;[346a] yields with excess $NaPMe_2$ an ionic complex $\{[Me_2P]_2TiCp_2\}Na$, ESR spect.[346a]

Et_2PTiCp_2. I. Dark violet, diamag. crystals, dec. 203 to 205°, probably dimeric.[288]

$Et_2PTi(NMe_2)_3$. I. Yellow liquid, $b_{0.001}$ 51 to 52°, IR, ^1H NMR.[63b]

$Et_2PTi(NEt_2)_3$. I. Blood-red crystals, $subl._{0.0001}$ 90°, IR, ^1H NMR.[63b]

$(Me_3Si)_2PTi(NMe_2)_3$. I. Yellow liquid, $b_{0.01}$ 70 to 72°, IR, ^1H NMR.[63b]

Bu_2PTiCp_2. I. Dark violet, diamag. crystals, dec. 158 to

160°, probably dimeric.[288]
Ph_2PTiCp_2. I. ESR spect.;[346a] yields with excess $NaPPh_2$ an ionic complex, $\{[Ph_2P]_2TiCp_2\}Na$, ESR spect.[346a]
Et_2PZrCp_2. I. Brown-red, diamag. crystals, dec. 280 to 282°, probably dimeric.[288]
Bu_2PZrCp_2. I. Brown-red, diamag. crystals, dec. 238 to 240°, probably dimeric.[288]
$[(Ph_2P)_2ZrCp_2]Na$. I. ESR spect.[346a]

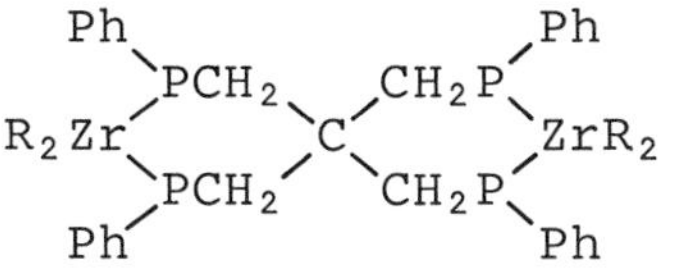

I. (R = Cp). Yellowish-white solid, m. 270° (dec.).[162]
(R = Cl). Pale-brown solid.[162]

TYPE: Group VB Phosphorus Compounds

$[(c\text{-}C_6H_{11})_2P]_3V$. I. Black-brown solid, m. 215°, mag. moment 1.0 μB;[329] gives on treatment with iodine $[(c\text{-}C_6H_{11})_2P]_3VI$, solid.[329]
$Ph_2PV(CO)_4$. II. Green in hexane, not obtained pure.[255]

TYPE: Group VIB Phosphorus Compounds

$[(c\text{-}C_6H_{11})_2P]_3Cr$. I. Grey-green, pyrophoric solid, m. 160°, mag. moment 2.93 μB, sensitive to H_2O;[329] gives $CrAc_3$ on treatment with acetic acid.
$[(Ph_2P)_2NH_3(CO)_3Cr]K_2$. I. Dark-red crystals; decomposes at room temp.[294]
$[(Ph_2P)_3Cr(CO)_3]K_3 \cdot 6$ dioxan. I. Dark-red crystals, dec. 78 to 79°;[294] gives on treatment with $Ba(SCN)_2$ the Ba salt, yellow crystals, dec. 87 to 88°.
$[(Ph_2P)_3Mo(CO)_3]K_3$. Red crystals, dec. 84 to 85°;[294] gives with $Ba(SCN)_2$ the Ba salt, pale-yellow crystals, dec. 88 to 89°.[294]
$(C_6F_5)_2PMo(CO)_2Cp$. II. Yellow crystals, m. 117°, IR, ^1H and ^{19}F NMR, Mass spect.[120a]
$[(Ph_2P)_3W(CO)_3]K_3$. I. Yellow crystals, dec. 87 to 88°;[294] gives with $Ba(SCN)_2$ the Ba salt, pale-yellow crystals, dec. 93 to 95°.[294]

TYPE: Group VIIB Phosphorus Compounds

$[Ph_2P]_2Mn$. I. Not obtained pure; contained KBr.[328]
$[Ph_2P]_3MnK \cdot 2THF$. I. Yellow-orange crystals, dec. 195 to 200°, decomposed by H_2O and EtOH; loses THF at 110°.[328]
$[Ph_2P]_4MnK_2 \cdot 2$ dioxan. I. Orange-red, pyrophoric crystals, dec. 217 to 221°, mag. moment 5.65 μB, sensitive

to H_2O and O_2; loses dioxan at 145°.[328]
$[(c\text{-}C_6H_{11})_2P]_3MnLi$. I. Red-brown, pyrophoric crystals, dec. 253 to 255°.[328]
$(CF_3)_2PMn_2(CO)_8I$. II. Orange-red crystals; sublimes under vacuum at 40 to 50°; IR.[164]
$[Ph_2PMn(CO)_4]_2$. I. Yellow crystals, m. 242 to 243°, IR.[230]
$Ph_2PMn_2(CO)_8H$. By sublimation of above compound. Yellow crystals, m. 154 to 155°,[230] IR, ^1H NMR.[230]

TYPE: Group VIIIB Phosphorus Compounds

$\{[(CF_3)_2P]_2NiCp\}_n$. II. Black, air stable, crystalline solid, m. 297 to 299°, dimeric, mass spect., ^1H NMR.[155]
$[(Et_2P)_2Ni(CO)_2]Li_2\cdot 3$ THF. I. Pale-yellow crystals, m. 85 to 88°, dec. 100°.[310]
$[(c\text{-}C_6H_{11})_2P]_3NiLi$. I. Not obtained pure; isolated after hydrolysis as the phosphine complex $[(R_2P)_2Ni\cdot HPR_2]_2$, black-brown solid, dec. 115°.[286]
$[PhHPNi(CO)_3]K\cdot$THF. I. Pale-orange crystals, dec. 86°.[310]
$[(PhHP)_2Ni(CO)_2]K_2\cdot$2THF. I. Dark-brown crystals, m. 115 to 118°, dec. 135°.[310]
$[(Ph_2P)_2Ni]_n$. I.[4,286] Green crystals, mag. moment 1.13 μB,[286] dimeric,[286] pentameric,[4] sensitive to air.[4,286]
$[Ph_2PNi(CO)_3]K\cdot$THF. I. Slightly red crystals, m. 70 to 75°.[310]
$[(Ph_2P)_2Ni(CO)_2]K_2\cdot$2THF. I. Red-brown crystals, m. 102 to 108°, dec. 115°.[310]
$[(Ph_2P)_2Ni(CO)_2]Rb_2$. I. Brown-orange crystals, m. 80.5 to 83.5°, dec. 99°.[310]
$[PhP\text{-}PPhNi(CO)_2]K_2\cdot$2THF. I. Dark-brown crystals, m. 167 to 172°.[310]
$\{[(c\text{-}C_6H_{11})_2P]_2Co\}_n$. I. Black-brown crystals, probably dimeric, mag. moment 4.71 μB.[286]
$[(Ph_2P)_2Co]_n$. I. Air-sensitive solid, mag. moment 2.77 μB, dimeric.[286]
$[Ph_2PCo(CO)_3]_n$. Orange, amorphous solid, dec. >110°.[253]
$[(c\text{-}C_6H_{11})_2P]_2Fe$. I. Brown, pyrophoric crystals, m. 145° (from benzene/MeOH), monomeric in benzene;[330] gives a phosphine complex of type $(R_2P)_2Fe\cdot HPR_2$, dark-brown solid, m. 155°.[330]
$[Ph_2PFe]_3P$. I. Black, pyrophoric solid, mag. moment 5.78 μB, monomeric in THF,[285] sensitive to water.
$[(CF_3)_2PFe(CO)_3I]_2$. II. Brick-red crystals; sublimes under vacuum at 80 to 90°; IR, UV.[164]
$(C_6F_5)_2PFe(CO)_2Cp$. II. Red crystals, m. 132°, IR, ^1H and ^{19}F NMR, mass spect.[120a]
$[(Ph_2P)_2Pd]_n$. I. Brown-red solid, tetrameric in benzene.[4]

(received July 9, 1970)

BIBLIOGRAPHY

Review Articles on Phosphorus-Metal and Metalloid Compounds

Alkali Phosphides

Maier, L., Progr. Inorg. Chem., 5, 27 (1963).
Issleib, K., Z. Chem., 2, 163 (1962).
Issleib, K., Pure Appl. Chem., 9, 205 (1964).

Organophosphorus Group III Compounds

Parshall, G. S., "Boron-Phosphorus Compounds," in Chemistry of Boron and Its Compounds, E. L. Muetterties, Ed., Wiley, 1967.
Bregadze, V. I., and O. Yu. Okhlobystin, "Organoelement Derivatives of Barenes," Organometal. Chem. Rev., Sect. A, 4, 345 (1969).

Organophosphorus Group IV Compounds

Si-P Compounds

Abel, E. W., and S. M. Illingworth, Organometal. Chem. Rev., Sect. A, 5, 143 (1970).
Fritz, G., Angew. Chem., 78, 80 (1966).
MacDiarmid, A. G., Advan. Inorg. Chem., Radiochem., 3, 246 (1961).
Chernyshev, E. A., and E. F. Bugerenko, Organometal. Chem. Rev., 3, 469 (1968).
Bürger, H., Organometal. Chem. Rev., 3, 425 (1968).
Drake, J. E., and C. Riddle, Quart. Rev., 24, 263 (1970).

Ge-P Compounds

Glockling, F., The Chemistry of Germanium, Academic, 1969.
Hooton, K. A., Preparative Inorg. Reactions, 4, 147 (1968).
Schumann, H., Angew. Chem., 81, 970 (1969).

Sn-P and Pb-P Compounds

Schumann, H., and M. Schmidt, Angew. Chem., 77, 1049 (1965).
Schumann, H., Angew. Chem., 81, 970 (1969).

Organophosphorus Group V Compounds

Huheey, J. E., J. Chem. Educ., 40, 153 (1963).
Maier, L., Progr. Inorg. Chem., 5, 27 (1963).
Maier, L., Fortschr. Chem. Forsch., 8, 1 (1967).
Banks, R. E., and R. N. Haszeldine, Advan. Inorg. Chem. Radiochem., 1, 337 (1961).
Paddock, N. L., Roy. Inst. Chem. (London). Lectures, Monographs, Rept. No. 2 (1962).
Sasse, K., Methoden der Organischen Chemie, Thieme Verlag, Vol. XII, Part 1 (1963).

Cowley, A. H., Chem. Rev., 65, 617 (1965).
Cowley, A. H., and R. P. Pinnell, in Topics in Phosphorus Chemistry, M. Grayson and E. J. Griffith, Eds., Vol. 4, Interscience, 1967, p. 1.
Cullen, W. R., Advan. Organometal. Chem., 4, 145 (1966).
Burg, A. B., Accounts Chem. Res., 2, 353 (1969).

REFERENCES

1. Abbiss, T. P., A. H. Soloway, and V. H. Mark, J. Med. Chem., 7, 763 (1964).
2. Abel, E. W., D. A. Armitage, and R. P. Bush, J. Chem. Soc., 1964, 5584.
3. Abel, E. W., J. P. Crow, and S. M. Illingworth, J. Chem. Soc., A, 1969, 1631.
4. Abel, E. W., R. A. N. McLean, and I. H. Sabherwal, J. Chem. Soc., A, 1968, 2371.
5. Abel, E. W., and I. H. Sabherwal, J. Organometal. Chem., 10, 491 (1967).
6. Abel, E. W., and I. H. Sabherwal, J. Chem. Soc., A, 1968, 1105.
7. Aftandilian, V. D., H. C. Miller, and E. L. Muetterties, J. Am. Chem. Soc., 83, 2471 (1961).
8. Aguiar, A. M., J. Beisler, and A. Miller, J. Org. Chem., 27, 1001 (1962).
9. Ahrland, S., J. Chatt, and N. R. Davies, Quart. Rev. (London), 12, 265 (1958).
10. Airey, P. L., Z. Naturforsch., B, 24, 1393 (1969).
11. Albers, H., W. Künzel, and W. Schuler, Chem. Ber., 85, 239 (1952).
12. Albers, H., and W. Schuler, Chem. Ber., 76, 23 (1943).
13. Amster, R. L., W. A. Henderson, and N. B. Colthup, Can. J. Chem., 42, 2577 (1964).
14. Amster, R. L., N. B. Colthup, and W. A. Henderson, Spectrochim. Acta, 19, 1841 (1963).
15. Ang, H. G., and J. M. Miller, Chem. Ind. (London), 1966, 944.
16. Ang, H. G., J. S. Shannon, and B. O. West, Chem. Commun., 1965, 10.
17. Ang, H. G., and B. O. West, Aust. J. Chem., 20, 1133 (1967
18. Ang, H. G., and J. M. Miller, Chem. Ind. (London), 1966, 944.
18a. Ang, H. G., M. E. Redwood, and B. O. West, Aust. J. Chem., 25, 493 (1972).
19. Announcement, Chem. Eng. News, 30, 4515 (1952).
20. Arbuzov, B. A., and N. P. Grechkin, Zh. Obshch. Khim., 17, 2166 (1947); C. A., 42, 4522g (1948).
21. Arbuzov, B. A., and N. P. Grechkin, Izv. Akad. Nauk SSSR, Ser. Khim., 1956, 440; C. A., 50, 16661h (1956).
22. Arbuzov, B. A., and A. N. Pudovik, Zh. Obshch. Khim., 17, 2158 (1947); C. A., 42, 4522a (1948).

23. Arbuzov, B. A., and A. N. Pudovik, Dokl. Akad. Nauk SSSR, 59, 1433 (1948); C. A., 47, 4281g (1953).
24. Balashova, L. D., A. B. Bruker, and L. Z. Soborovskii, Zh. Obshch. Khim., 35, 2207 (1965); C. A., 64, 12718g (1966).
25. Balashova, L. D., A. B. Bruker, and L. Z. Soborovskii, Zh. Obshch. Khim., 36, 73 (1966); C. A., 64, 14211e (1966).
25a. Bald, J. F., and A. G. MacDiarmid, J. Organometal. Chem., 22, C22 (1970).
26. Balon, W. J., U. S. 2,853,518 (1958); C. A., 53, 5202b (1959).
27. Banks, R. E., and R. N. Haszeldine, Advan. Inorg. Chem. Radiochem., 3, 337 (1961).
28. Bart, J. C. J., Acta Crystallogr., B, 25, 762 (1969).
29. Baudler, M., O. Gehlen, K. Kipker, and P. Backes, Z. Naturforsch., B, 22, 1354 (1967).
30. Baudler, M., U. K. Hammerström, Z. Naturforsch., B, 20, 810 (1965).
31. Baudler, M., K. Kipker, and H. W. Valpertz, Naturwissenschaften, 53, 612 (1966).
32. Baudler, M., K. Kipker, and H. W. Valpertz, Naturwissenschaften, 54, 43 (1967).
33. Beachley, O. T., and G. E. Coates, J. Chem. Soc., 1965, 3241.
34. Beagley, B., A. G. Robiette, and G. M. Sheldrick, Chem. Commun., 1967, 601.
35. Beg, M. A. A., and H. C. Clark, Can. J. Chem., 39, 564 (1961).
36. Beg, M. A. A., and K. S. Hussain, Chem. Ind. (London), 1966, 1181.
37. Beg, M. A. A., and S. H. Khawaja, Spectrochim. Acta, 24A, 1031 (1968).
38. Bennett, F. W., H. J. Emeléus, and R. N. Haszeldine, J. Chem. Soc., 1953, 1565.
39. Bennett, F. W., H. J. Emeléus, and R. N. Haszeldine, J. Chem. Soc., 1954, 3896.
40. Bergerhoff, G., Acta Crystallogr., 15, 420 (1962).
41. Berlin, K. D., T. H. Austin, M. Peterson, and M. Nagabhushnam, in Topics in Phosphorus Chemistry, M. Grayson and E. J. Griffith, Eds., Vol. 1, Interscience, 1964, p. 17.
42. Besson, A., C. R. Acad. Sci., Paris, 122, 140, 814, 1200 (1896).
43. Beyer, K., Z. Anorg. Allg. Chem., 250, 312 (1942).
44. Biddulph, R. H., M. P. Brown, R. C. Cass, R. Long, and H. B. Silver, J. Chem. Soc., 1961, 1822; Brit. 882,532 (1962); C. A., 56, 12948d (1962).
45. Birchall, T., and W. L. Jolly, Inorg. Chem., 5, 2177 (1966).
46. Bloch, B., and Y. Gounelle, C. R. Acad. Sci., Paris, Ser. C, 266, 220 (1968).

47. Bloomfield, P. R., and K. Parvin, Chem. Ind. (London), 1959, 541.
48. Bogolyubov, G. M., Zh. Obshch. Khim., 35, 754 (1965); C. A., 63, 4327 (1965).
49. Bogolyubov, G. M., N. N. Grishin, and A. A. Petrov, Zh. Obshch. Khim., 39, 1808 (1969).
50. Bogolyubov, G. M., K. S. Mingaleva, and A. A. Petrov, Zh. Obshch. Khim., 35, 1566 (1965); C. A., 63, 17860c (1965).
51. Bogolyubov, G. M., and A. A. Petrov, Zh. Obshch. Khim., 35, 704 (1965); C. A., 63, 4430 (1965).
52. Bogolyubov, G. M., and A. A. Petrov, Zh. Obshch. Khim., 35, 988 (1965); C. A., 63, 9981h (1965).
53. Bogolyubov, G. M., and A. A. Petrov, Zh. Obshch. Khim., 36, 724 (1966); C. A., 65, 8954a (1966).
54. Bogolyubov, G. M., and A. A. Petrov, Zh. Obshch. Khim., 36, 1505 (1966).
55. Bogolyubov, G. M., and A. A. Petrov, Dokl. Akad. Nauk SSSR, 173, 1076 (1967); C. A., 67, 90887e (1967).
55a. Bogolyubov, G. M., and Yu. N. Shlyk, Zh. Obshch. Khim., 39, 1759 (1969).
56. Braddock, J. M. F., and G. E. Coates, J. Chem. Soc., 1961, 3208.
57. Britt, A. D., and E. T. Kaiser, J. Phys. Chem., 69, 2775 (1965).
58. Britt, A. D., and E. T. Kaiser, J. Org. Chem., 31, 112 (1966).
59. Brooks, E. H., F. Glockling, and K. A. Hooton, J. Chem. Soc., 1965, 4283.
59a. Brown, C., M. Murray, and R. Schmutzler, J. Chem. Soc., C, 1970, 878.
60. Bruker, A. A., L. D. Balashova, and L. Z. Soborovskii, Dokl. Akad. Nauk SSSR, 135, 843 (1960); C. A., 55, 13301a (1961).
61. Bruker, A. B., L. D. Balashova, and L. Z. Soborovskii, Zh. Obshch. Khim., 36, 75 (1966); C. A., 64, 14211g (1966).
62. Bürger, H., Organometal. Chem. Rev., 3, 425 (1968).
63. Bürger, H., and U. Goetze, J. Organometal. Chem., 12, 451 (1968).
63a. Bürger, H., U. Goetze, and W. Sawodny, Spectrochim. Acta, 26A, 671 (1970).
63b. Bürger, H., and H. J. Neese, Inorg. Nucl. Chem. Lett., 6, 299 (1970).
64. Bullen, G. J., and P. R. Mallinson, Chem. Commun., 1969, 132.
65. Burg, A. B., J. Inorg. Nucl. Chem., 11, 258 (1959).
66. Burg, A. B., J. Am. Chem. Soc., 83, 2226 (1961).
67. Burg, A. B., in Topics in Modern Inorganic Chemistry, Proceedings of the Robert A. Welch Foundation Conferences on Chemical Research, Vol. VI, Houston Texas, 1962, S. 133.

68. Burg, A. B., J. Am. Chem. Soc., 88, 4298 (1966).
69. Burg, A. B., and G. Brendel, J. Am. Chem. Soc., 80, 3198 (1958).
70. Burg, A. B., and K. Gosling, J. Am. Chem. Soc., 87, 2113 (1965).
71. Burg, A. B., and L. R. Grant, U. S. 3,118,951 (1964); C. A., 60, 10718b (1964).
72. Burg, A. B., and J. E. Griffiths, J. Am. Chem. Soc., 82, 3514 (1960).
73. Burg, A. B., and H. Heinen, Inorg. Chem., 7, 1021 (1968).
74. Burg, A. B., and K. K. Joshi, J. Am. Chem. Soc., 86, 353 (1964).
75. Burg, A. B., K. K. Joshi, and J. F. Nixon, J. Am. Chem. Soc., 88, 31 (1966).
76. Burg, A. B., and W. Mahler, J. Am. Chem. Soc., 79, 4242 (1957).
77. Burg, A. B., P. J. Slota, Jr., and W. Mahler, A.C.S., 134th Meeting, Chicago, Abstr., 1958, p. 78P.
78. Burg, A. B., and W. Mahler, J. Am. Chem. Soc., 80, 2334 (1958).
79. Burg, A. B., and W. Mahler, U. S. 2,923,742 (1960); C. A., 54, 9765b (1960).
80. Burg, A. B., and W. Mahler, U. S. 2,923,741 (1960); C. A., 54, 9767b (1960).
81. Burg, A. B., and W. Mahler, J. Am. Chem. Soc., 83, 2388 (1961).
82. Burg, A. B., and K. Moedritzer, J. Inorg. Nucl. Chem., 13, 318 (1960).
83. Burg, A. B., and J. F. Nixon, J. Am. Chem. Soc., 86, 356 (1964).
83a. Burg, A. B., and D. M. Parker, J. Am. Chem. Soc., 92, 1898 (1970).
84. Burg, A. B., and L. K. Peterson, Inorg. Chem., 5, 943 (1966).
85. Burg, A. B., and P. J. Slota, Jr., U. S. 2,877,272 (1959); C. A., 53, 16062g (1959).
86. Burg, A. B., and P. J. Slota, Jr., J. Am. Chem. Soc., 82, 2145 (1960).
87. Burg, A. B., and P. J. Slota, Jr., J. Am. Chem. Soc., 82, 2148 (1960).
88. Burg, A. B., and R. I. Wagner, J. Am. Chem. Soc., 75, 3872 (1953).
89. Burg, A. B., and R. I. Wagner, U. S. 2,925,440 (1960); C. A., 54, 15408g (1960).
90. Burg, A. B., and R. I. Wagner, U. S. 2,948,689 (1960); C. A., 55, 7355h (1961).
91. Burg, A. B., and R. I. Wagner, Brit. 852,970 (1960); C. A., 56, 505e (1962).
92. Burg, A. B., and R. I. Wagner, U. S. 3,025,326 (1962); C. A., 57, 8619b (1962).

93. Burg, A. B., and R. I. Wagner, U. S. 3,065,271 (1963); C. A., 58, 9140h (1963).
94. Burg, A. B., and R. I. Wagner, U. S. 3,071,553 (1963), C. A., 59, 5198b (1963).
95. Burrus, C. A., J. Chem. Phys., 28, 427 (1958).
96. Bush, M. A., V. R. Cook, and P. Woodward, Chem. Commun., 1967, 630.
97. Bush, M. A., and P. Woodward, J. Chem. Soc., A, 1968, 1221.
97a. Byrne, J. E., and C. R. Russ, J. Organometal. Chem., 22, 357 (1970).
98. Campbell, I. G. M., G. W. A. Fowles, and L. A. Nixon, J. Chem. Soc., 1964, 1389
99. Carrell, H. L., and J. Donohue, Acta Crystallogr., B24, 699 (1968).
100. Cass, R. C., R. Long, and M. P. Brown, Brit. 882,532 (1961); C. A., 56, 12948d (1962).
101. Cavell, R. G., and R. C. Dobbie, J. Chem. Soc., A, 1967, 1308.
102. Cavell, R. G., and R. C. Dobbie, Inorg. Chem., 7, 690 (1968).
103. Cavell, R. G., and R. C. Dobbie, J. Chem. Soc., A, 1968, 1406.
104. Cavell, R. G., R. C. Dobbie, and W. J. R. Tyerman, Can. J. Chem., 45, 2849 (1967).
105. Cavell, R. G., and H. J. Emeléus, J. Chem. Soc., 1964, 5825.
106. Chapman, A. C., Trans. Faraday Soc., 59, 806 (1963).
107. Chatt, J., and D. A. Thornton, J. Chem. Soc., 1964, 1005.
108. Chatt, J., F. A. Hart, and H. C. Fielding, U. S. 2,922,819 (1960); C. A., 54, 9847b (1960).
109. Chernyshev, E. A., and E. F. Bugerenko, Organometal. Chem. Rev., 3, 469 (1968).
110. Christen, P. J., L. M. van der Linde, and F. N. Hooge, Rec. Trav. Chim., 78, 161 (1959).
111. Clemens, D. F., H. H. Sisler, and W. S. Brey, Inorg. Chem., 5, 527 (1966).
112. Coates, G. E., F. Glockling, and N. D. Huck, J. Chem. Soc., 1952, 4512.
113. Coates, G. E., and J. Graham, J. Chem. Soc., 1963, 233.
114. Coates, G. E., and J. G. Livingstone, Chem. Ind. (London), 1958, 1366.
115. Coates, G. E., and J. G. Livingstone, J. Chem. Soc., 1961, 1000.
116. Coates, G. E., and J. G. Livingstone, J. Chem. Soc., 1961, 5053.
117. Cölln, R., and G. Schrader, D.A.S. 1,056,606 (1959).
118. Cölln, R., and G. Schrader, D.A.S. 1,054,453 (1959); C. A., 55, 6375b (1961).

119. Cölln, R., and G. Schrader, Ger. 1,138,771 (1962); C. A., 58, 12601a (1963).
120. Cohen, B. M., A. R. Cullingworth, and J. D. Smith, J. Chem. Soc., A, 1969, 2193.
120a. Cooke, M., M. Green, and D. Kirkpatrick, J. Chem. Soc., A, 1968, 1507.
121. Cowley, A. H., Chem. Rev., 65, 617 (1965).
121a. Cowley, A. H., J. Am. Chem. Soc., 89, 5990 (1967).
122. Cowley, A. H., 4th International Conference on Organometallic Chemistry, University of Bristol, August 1969, Abstracts of Papers, Z 4.
123. Cowley, A. H., and D. S. Dierdorf, J. Am. Chem. Soc., 91, 6609 (1969).
123a. Cowley, A. H., T. A. Furtsch, and D. S. Dierdorf, Chem. Commun., 1970, 523.
124. Cowley, A. H., and R. P. Pinnell, in Topics in Phosphorus Chemistry, M. Grayson and E. J. Griffith, Eds., Vol. 4, Interscience, 1967, p. 1.
125. Cowley, A. H., and R. P. Pinnell, Inorg. Chem., 5, 1459 (1966).
126. Cowley, A. H., and R. P. Pinnell, Inorg. Chem., 5, 1463 (1966).
127. Cowley, A. H., and R. P. Pinnell, J. Am. Chem. Soc., 88, 4533 (1966).
128. Cowley, A. H., and R. P. Pinnell, private communication.
129. Cowley, A. H., and R. P. Pinnell, A.C.S., 150th Meeting, Atlantic City, Sept. 1965, Abstr., 25-0.
130. Cowley, A. H., and H. Steinfink, Inorg. Chem., 4, 1827 (1965).
131. Cowley, A. H., and W. D. White, Spectrochim. Acta, 22, 1431 (1966).
132. Creemers, H. M. J. C., F. Verbeek, and J. G. Noltes, J. Organometal. Chem., 8, 469 (1967).
133. Crofts, P. C., and I. S. Fox, J. Chem. Soc., B, 1968, 1416.
134. Crofts, P. C., and K. Gosling, J. Chem. Soc., 1964, 2486.
135. Crosbie, K. D., C. Glidewell, and G. M. Sheldrick, J. Chem. Soc., A, 1969, 1861.
136. Crutchfield, M. M., C. H. Dungan, J. H. Letcher, V. Mark, and J. R. Van Wazer, in Topics in Phosphorus Chemistry, M. Grayson and E. J. Griffith, Eds., Vol. 6, Interscience, 1967, p. 361.
137. Cullen, W. R., Can. J. Chem., 38, 439 (1960).
138. Cullen, W. R., and D. S. Dawson, Can. J. Chem., 45, 2887 (1967).
139. Cullen, W. R., D. S. Dawson, and P. S. Dhaliwal, Can. J. Chem., 45, 683 (1967).
140. Cundy, C. S., M. Green, F. G. A. Stone, and A. Taunton-Rigby, J. Chem. Soc., A, 1968, 1776.

141. Daly, J. J., J. Chem. Soc., 1964, 6147.
142. Daly, J. J., J. Chem. Soc., 1965, 4789.
143. Daly, J. J., J. Chem. Soc., A, 1966, 428.
144. Daly, J. J., J. Chem. Soc., A, 1966, 1020.
145. Daly, J. J., and L. Maier, Nature, 203, 1167 (1964).
146. Daly, J. J., and L. Maier, Nature, 208, 383 (1965).
147. Davidson, N., and H. C. Brown, J. Am. Chem. Soc., 64, 316 (1942).
148. Davidson, R. S., R. A. Sheldon, and S. Trippett, J. Chem. Soc., C, 1966, 722.
149. Davidson, R. S., R. A. Sheldon, and S. Trippett, J. Chem. Soc., C, 1967, 1547.
150. Davidson, R. S., R. A. Sheldon, and S. Trippett, J. Chem. Soc., C, 1968, 1700.
151. Davies, T. H., Chem. Ind. (London), 1964, 1755.
152. Davis, M., and F. G. Mann, J. Chem. Soc., 1964, 3770.
153. DelaMare, H. E., and F. E. Neumann, Ger. 1,938,289 (1970); C. A., 72, 90999w (1970).
154. Didchenko, R., J. E. Alix, and R. H. Toeniskoetter, J. Inorg. Nucl. Chem., 14, 35 (1960).
155. Dobbie, R. C., M. Green, and F. G. A. Stone, J. Chem. Soc., A, 1969, 1881.
156. Dörken, C., Chem. Ber., 21, 1505 (1888).
157. Donohue, J., Acta Crystallogr., 15, 708 (1962).
158. Drake, J. E., N. Goddard, and C. Riddle, J. Chem. Soc., C, 1969, 2704.
158a. Drake, J. E., and C. Riddle, Quart. Rev., 24, 263 (1970).
159. Dumont, E., and H. Reinhart, Ger. 1,044,811 (1958); C. A., 55, 1444f (1961).
160. Dutta, S. N., and M. M. Woolson, Acta Crystallogr., 14, 178 (1961).
161. Ellermann, J., and K. Dorn, Z. Naturforsch., B, 23, 420 (1968).
162. Ellermann, J., and F. Poersch, Angew. Chem., 79, 380 (1967).
163. Elser, H., and H. Dreeskamp, Ber. Bunsenges. Phys. Chem., 73, 619 (1969).
164. Emeléus, H. J., and J. Grobe, Angew. Chem., 74, 467 (1962).
165. Emeléus, H. J., and K. M. Mackay, J. Chem. Soc., 1961, 2676.
166. Engelhardt, G., R. Reich, and H. Schumann, Z. Naturforsch., B, 22, 352 (1967).
167. English, W. D., Ger. 1,098,210 (1961); Brit. 848,656 (1960); C. A., 55, 7906g (1961).
168. Evers, E. C., E. H. Street, Jr., and S. L. Jung, J. Am. Chem. Soc., 73, 5088, 2038 (1951).
169. Evleth, E. M., Jr., L. V. D. Freedman, and F. I. Wagner, J. Org. Chem., 27, 2192 (1962).

170. Farbenfabriken Bayer A.G., Neth. 6,500,606 (1965); C. A., 64, 5276a (1966).
171. Fehér, F. G., G. Kuhlbörsch, A. Blümcke, H. Keller, and K. Lippert, Chem. Ber., 90, 134 (1957).
172. Feshchenko, N. G., and A. V. Kirsanov, Zh. Obshch. Khim., 31, 1399 (1961); C. A., 55, 27169b (1961).
172a. Feshchenko, N. G., T. V. Kovaleva, and A. V. Kirsanov, Zh. Obshch. Khim., 39, 2184 (1969); C. A., 72, 43796t (1970).
172b. Feshchenko, N. G., T. V. Kovaleva, and A. V. Kirsanov, Zh. Obshch. Khim., 39, 2188 (1969); C. A., 72, 43798v (1970).
172c. Feshchenko, N. G., E. A. Mel'nichuk, and A. V. Kirsanov, Zh. Obshch. Khim., 39, 2139 (1969).
173. Finch, A., P. J. Gardner, A. Hameed, and K. K. S. Gupta, Chem. Commun., 1969, 854.
174. Finer, E. G., and R. K. Harris, Mol. Phys., 13, 65 (1967).
175. Finer, E. G., and R. K. Harris, Chem. Commun., 1968, 110.
176. Fild, M., I. Hollenberg, and O. Glemser, Z. Naturforsch., B, 22, 248 (1967).
177. Fild, M., I. Hollenberg, and O. Glemser, Naturwissenschaften, 54, 89 (1967).
178. Fild, M., and R. Schmutzler, J. Chem. Soc., A, 1969, 840.
179. Florin, R. E., L. A. Wall, F. L. Mohler, and E. Quinn, J. Am. Chem. Soc., 76, 3344 (1954).
180. Fluck, E., and H. Binder, Angew. Chem., 78, 677 (1966).
181. Fluck, E., and H. Binder, Z. Anorg. Allg. Chem., 354, 113 (1967).
182. Fluck, E., and H. Binder, Angew. Chem., 79, 903 (1967).
183. Fluck, E., and H. Binder, Inorg. Nucl. Chem. Lett., 3, 307 (1967).
184. Fluck, E., and H. Binder, Z. Naturforsch., B, 22, 1001 (1967).
185. Fluck, E., H. Bürger, and U. Goetze, Z. Naturforsch., B, 22, 912 (1967).
186. Fluck, E., and K. Issleib, Z. Naturforsch., B, 20, 1123 (1965).
187. Fluck, E., and K. Issleib, Chem. Ber., 98, 2674 (1965).
188. Fluck, E., and K. Issleib, Z. Anorg. Allg. Chem., 339, 274 (1965).
189. Fluck, E., and K. Issleib, Z. Naturforsch., B, 21, 736 (1966).
190. Fluck, E., and V. Novobilshy, Forstchr. Chem. Forsch., 13, 125 (1969).
191. Forster, A., C. S. Cundy, M. Green, and F. G. A. Stone, Inorg. Nucl. Chem. Lett., 2, 233 (1966).

192. Fowles, G. W. A., and D. K. Jenkins, Chem. Commun., 1965, 61.
193. Frazier, S. E., R. P. Nielsen, and H. H. Sisler, Inorg. Chem., 3, 292 (1964).
194. Frazier, S. E., and H. H. Sisler, Inorg. Chem., 5, 925 (1966).
195. Fritz, G., Angew. Chem., 78, 80 (1966).
196. Fritz, G., G. Becker, and D. Kummer, Z. Anorg. Allg. Chem., 372, 171 (1970).
197. Fritz, G., and G. Becker, Z. Anorg. Allg. Chem., 372, 180 (1970).
198. Fritz, G., and G. Becker, Z. Anorg. Allg. Chem., 372, 196 (1970).
199. Fritz, G., and F. Pfannerer, Z. Anorg. Allg. Chem., 373, 30 (1970).
200. Fritz, G., and G. Poppenburg, Angew. Chem., 72, 208 (1960).
201. Fritz, G., and G. Poppenburg, Naturwissenschaften, 49, 449 (1962).
202. Fritz, G., and G. Poppenburg, Angew. Chem., 75, 297 (1963).
203. Fritz, G., and G. Poppenburg, Z. Anorg. Allg. Chem., 331, 147 (1964).
204. Fritz, G., and G. Trenczek, Z. Anorg. Allg. Chem., 313, 236 (1961).
205. Fritz, G., and G. Trenczek, Angew. Chem., 74, 942 (1962); Angew. Chem. Int. Ed., 1, 663 (1962).
206. Fritz, G., and G. Trenczek, Angew. Chem., 75, 723 (1963); Angew. Chem. Int. Ed., 2, 482 (1963).
207. Fritz, G., and G. Trenczek, Z. Anorg. Allg. Chem., 331, 206 (1964).
208. Fritzsche, H., U. Hasserodt, and F. Korte, Angew. Chem., 75, 1205 (1963).
209. Gallagher, M. J., J. L. Garnett, and W. Sollich-Baumgartner, Tetrahedron Lett., 1966, 4465.
210. Gallagher, M. J., and I. D. Jenkins, Chem. Commun., 1965, 587; J. Chem. Soc., C, 1966, 2176.
211. Gallagher, M. J., and J. D. Jenkins, J. Chem. Soc., C, 1969, 2605.
212. Garner, A. Y., U. S. 3,010,999 (1961); C. A., 56, 5002e (1962).
213. Garner, A. Y., U. S. 3,271,460 (1966); C. A., 65, 15427f (1966).
214. Gee, W., J. B. Holden, R. A. Shaw, and B. C. Smith, J. Chem. Soc., 1965, 3171.
215. Gee, W., J. B. Holden, R. A. Shaw, and B. C. Smith, J. Chem. Soc., A, 1967, 1545.
216. Gee, W., R. A. Shaw, B. C. Smith, and G. J. Bullen, Proc. Chem. Soc., 1961, 432.
217. Gee, W., R. A. Shaw, and B. C. Smith, J. Chem. Soc., 1964, 4180.

218. Gee, W., R. A. Shaw, and B. C. Smith, J. Chem. Soc., 1964, 4180.
219. George, T. A., and M. F. Lappert, Chem. Commun., 1966, 463.
220. Gerding, H., D. H. Zijp, F. N. Hooge, G. Blasse, and P. J. Christen, Rec. Trav. Chim., 84, 1274 (1965).
221. Glockling, F., The Chemistry of Germanium, Academic, 1969.
222. Glockling, F., and K. A. Hooton, Proc. Chem. Soc., 1963, 146.
223. Goldstein, P., and R. A. Jacobson, J. Am. Chem. Soc., 84, 2457 (1962).
224. Goodrow, M. H., U. S. 3,272,781 (1966); C. A., 65, 20252b (1966).
225. Goodrow, M. H., R. I. Wagner, and F. F. Caserio, U. S. 3,347,800 (1967); C. A., 68, 22058f (1968).
226. Goodrow, M. H., R. I. Wagner, and R. D. Stewart, Inorg. Chem., 3, 1212 (1964).
227. Goubeau, J., H. Reinhardt, and D. Bianchi, Z. Physik. Chem., N. F., 12, 387 (1957).
228. Grant, L. R., and A. B. Burg, J. Am. Chem. Soc., 84, 1834 (1962).
229. Grayson, M., P. T. Keough, and G. A. Johnson, J. Am. Chem. Soc., 81, 4803 (1959); U. S. 3,005,013 (1958); C. A., 56, 4798b (1962); U. S. 3,148,206 (1964).
230. Green, M. L. H., and J. T. Moelwyn-Hughes, Z. Naturforsch., B, 17, 783 (1962).
231. Greenwood, N. N., J. E. Roos, and A. Storr, J. Chem. Soc., A, 1966, 706.
232. Griffith, J. E., and A. B. Burg, J. Am. Chem. Soc., 84, 3442 (1962).
233. Grishin, N. N., G. M. Bogolyubov, and A. A. Petrov, Zh. Obshch. Khim., 38, 2683 (1968).
234. Grobe, J., Z. Naturforsch., B, 23, 1609 (1968).
235. Grützmacher, H. F., W. Silhan, and U. Schmidt, Chem. Ber., 102, 3230 (1969).
236. Haber, C. P., and C. O. Wilson, Jr., U. S. 2,892,873 (1959); C. A., 54, 3203i (1960).
237. Hamilton, W. C., Acta Crystallogr., 8, 199 (1955).
239. Harris, R. K., and R. G. Hayter, Can. J. Chem., 42, 2282 (1964).
240. Hart, F. A., and F. G. Mann, J. Chem. Soc., 1957, 3939.
240a. Hartung, H., Z. Chem., 10, 153 (1970); Z. Anorg. Allg. Chem., 372, 150 (1970).
241. Harwood, H. J., and K. A. Pollart, J. Org. Chem., 28, 3430 (1963).
242. Hayter, R. G., J. Am. Chem. Soc., 84, 3046 (1962).
243. Hayter, R. G., Nature, 193, 872 (1962).
244. Henderson, Wm. A., Jr., U. S. 3,029,289 (1962); C. A., 57, 8618e (1962).

245. Henderson, Wm. A., Jr., M. Epstein, and F. S. Seichter, J. Am. Chem. Soc., 85, 2462 (1963).
246. Henderson, Wm. A., Jr., S. A. Buckler, and M. Epstein, U. S. 3,032,591 (1962); C. A., 57, 11239f (1962).
247. Hengge, E., and U. Brychcy, Monatsh. Chem., 97, 1309 (1966).
248. Henrici-Olivé, G., and S. Olivé, J. Polym. Sci., 62, 58 (1962).
249. Hester, R. E., and K. Jones, Chem. Commun., 1966, 317.
250. Hewertson, W., and H. R. Watson, J. Chem. Soc., 1962, 1490.
251. Hey, L., and C. K. Ingold, J. Chem. Soc., 1933, 531.
252. Hicks, D. G., and J. A. Dean, Chem. Commun., 1965, 172.
253. Hieber, W., and H. Duchatsch, Z. Naturforsch., B, 18, 1132 (1963).
254. Hieber, W., and R. Kummer, Z. Anorg. Allg. Chem., 344, 292 (1966).
255. Hieber, W., and E. Winter, Chem. Ber., 97, 1037 (1964).
256. Highsmith, R. E., and H. H. Sisler, Inorg. Chem., 7, 1740 (1968).
257. Highsmith, R. E., and H. H. Sisler, Inorg. Chem., 8, 1029 (1969).
258. Hitchcock, C. H. S., and F. G. Mann, J. Chem. Soc., 1958, 2081.
259. Hoffmann, A., J. Am. Chem. Soc., 52, 2995 (1930).
260. Hoffmann, H., Ann. Chem., 634, 1 (1960).
261. Hoffmann, H., and R. Grünewald, Chem. Ber., 94, 186 (1961).
262. Hoffmann, H., R. Grünewald, and L. Horner, Chem. Ber., 93, 861 (1960).
263. Hofmann, E., Brit. 908,106 (1962); C. A., 58, 6862e (1963).
264. Hofmann, E., Brit. 928,350 (1963); C. A., 59, 12843h (1963).
265. Hofmann, E., Brit. 941,556 (1963).
266. Holmes, R. R., and E. F. Bertat, J. Am. Chem. Soc., 80, 2983 (1958).
267. Hooton, K. A., Preparative Inorg. Reactions, 4, 147 (1968).
268. Horner, L., P. Beck, and H. Hoffmann, Chem. Ber., 92, 2088 (1959).
269. Horner, L., and H. Hoffmann, Neuere Methoden der Präparativen Organischen Chemie, Vol. 11, Verlag Chemie, 1960, p. 108.
270. Horner, L., H. Hoffmann, and P. Beck, Chem. Ber., 91, 1583 (1958).
271. Hudson, R. F., "Structure and Mechanism in Organo-Phosphorus Chemistry," in Organic Chemistry, A. T.

Bloomquist, Ed., Vol. 6, Academic, 1965, p. 34.
272. Huheey, J. E., J. Chem. Educ., 40, 153 (1963).
273. Il'ina, L. A., and S. V. Larionov, Izv. Akad. Nauk SSSR, Ser. Khim., 1968, 1895.
274. Inamoto, N., T. Emoto, and R. Okazaki, Chem. Ind. (London), 1969, 832.
275. Issleib, K., Z. Chemie, 2, 163 (1962).
276. Issleib, K., Pure Appl. Chem., 9, 205 (1964).
277. Issleib, K., and H. J. Deyling, Z. Naturforsch., B, 17, 198 (1962).
278. Issleib, K., and H. J. Deyling, Chem. Ber., 97, 946 (1964).
279. Issleib, K., and G. Döll, Chem. Ber., 94, 2664 (1961).
280. Issleib, K., and G. Döll, Chem. Ber., 96, 1544 (1963).
281. Issleib, K., and G. Döll, Z. Anorg. Allg. Chem., 324, 259 (1963).
282. Issleib, K., and E. Fluck, Angew. Chem., 78, 597 (1966).
283. Issleib, K., and H. O. Fröhlich, Z. Naturforsch., B, 14, 349 (1959).
284. Issleib, K., and H. O. Fröhlich, Chem. Ber., 95, 375 (1962).
285. Issleib, K., and H. O. Fröhlich, Chem. Ber., 97, 1659 (1964).
286. Issleib, K., H. O. Fröhlich, and E. Wenschuh, Chem. Ber., 95, 2742 (1962).
287. Issleib, K., and W. Gründler, Z. Kristallogr., 119, 472 (1964).
288. Issleib, K., and H. Häckert, Z. Naturforsch., B, 21, 519 (1966).
289. Issleib, K., and G. Harzfeld, Chem. Ber., 95, 268 (1962).
290. Issleib, K., and G. Harzfeld, Z. Anorg. Allg. Chem., 351, 18 (1967).
291. Issleib, K., and M. Hoffmann, Chem. Ber., 99, 1320 (1966).
292. Issleib, K., and D. Jacob, Chem. Ber., 94, 107 (1961).
293. Issleib, K., and M. Keil, Z. Anorg. Allg. Chem., 333, 10 (1964).
294. Issleib, K., and W. Kratz, Z. Naturforsch., B, 20, 1303 (1965).
295. Issleib, K., and F. Krech, Chem. Ber., 94, 2656 (1961).
296. Issleib, K., and F. Krech, Z. Anorg. Allg. Chem., 328, 21 (1964).
297. Issleib, K., and K. Krech, Z. Anorg. Allg. Chem., 328, 69 (1964).
298. Issleib, K., and K. Krech, Chem. Ber., 98, 1093 (1965).
299. Issleib, K., and K. Krech, Chem. Ber., 98, 2545 (1965).

300. Issleib, K., and K. Krech, Chem. Ber., 99, 1310 (1966).
301. Issleib, K., and F. Krech, J. Organometal. Chem., 13, 283 (1968).
302. Issleib, K., and F. Krech, J. Prakt. Chem., 311, 463 (1969).
302a. Issleib, K., and F. Krech, Z. Anorg. Allg. Chem., 372, 65 (1970).
303. Issleib, K., and R. Kümmel, J. Organometal. Chem., 3, 84 (1965).
304. Issleib, K. R. Kümmel, H. Oehme, and I. Meissner, Chem. Ber., 101, 3612 (1968).
305. Issleib, K., G. Lux, and R. Stolz, Z. Anorg. Allg. Chem., 350, 44 (1967).
306. Issleib, K., and B. Mitcherling, Z. Naturforsch., B, 15, 267 (1960).
307. Issleib, K., and D. W. Müller, Chem. Ber., 92, 3175 (1959).
308. Issleib, K., and H. Oehme, Chem. Ber., 100, 2685 (1967).
309. Issleib, K., and E. Priebe, Chem. Ber., 92, 3183 (1959).
310. Issleib, K., and W. Rettkowski, Z. Naturforsch., B, 21, 999 (1966).
311. Issleib, K., Ch. Rockstroh, I. Duchek, and E. Fluck, Z. Anorg. Allg. Chem., 360, 77 (1968).
312. Issleib, K., and H. R. Roloff, Chem. Ber., 98, 2091 (1965).
313. Issleib, K., and W. Seidel, Chem. Ber., 92, 2681 (1959).
314. Issleib, K., and W. Seidel, Z. Anorg. Allg. Chem., 303, 155 (1960).
315. Issleib, K., and W. Seidel, Z. Anorg. Allg. Chem., 303, 155 (1960).
316. Issleib, K., and K. Standtke, Chem. Ber., 96, 279 (1963).
317. Issleib, K., and G. Thomas, Chem. Ber., 93, 803 (1960).
318. Issleib, K., and A. Tzschach, Chem. Ber., 92, 704 (1959).
319. Issleib, K., and A. Tzschach, Chem. Ber., 92, 1118 (1959).
320. Issleib, K., and A. Tzschach, Chem. Ber., 92, 1397 (1959).
321. Issleib, K., and A. Tzschach, Chem. Ber., 93, 1852 (1960).
322. Issleib, K., A. Tzschach, and R. Schwarzer, Z. Anorg. Allg. Chem., 338, 141 (1965).
323. Issleib, K., and F. Ungváry, Z. Naturforsch., B, 22, 1238 (1967).
324. Issleib, K., and R. Völker, Chem. Ber., 94, 392 (1961).

325. Issleib, K., and B. Walther, Angew. Chem., 79, 59 (1967).
326. Issleib, K., and B. Walther, J. Organometal. Chem., 10, 177 (1967).
326a. Issleib, K., and B. Walther, J. Organometal. Chem., 22, 375 (1970).
327. Issleib, K., and E. Wenschuh, Z. Anorg. Allg. Chem., 305, 15 (1960).
328. Issleib, K., and E. Wenschuh, Z. Naturforsch., B, 17, 778 (1962).
329. Issleib, K., and E. Wenschuh, Chem. Ber., 97, 715 (1964).
330. Issleib, K., and E. Wenschuh, Z. Naturforsch., B, 19, 199 (1964).
331. Issleib, K., E. Wenschuh, and B. Fritsche, Z. Chem., 5, 143 (1965).
332. Jain, S. R., and H. H. Sisler, Inorg. Chem., 7, 2204 (1968).
333. Joannis, A., C. R. Acad. Sci., Paris, 119, 557 (1894); Ann. Chim. Phys., 8, 7, 105 (1906).
334. Job, A., and G. Dussolier, C. R. Acad. Sci., Paris, 184, 1454 (1927).
335. Johnson, F., R. S. Gohlke, and W. A. Nasutavicus, J. Organometal. Chem., 3, 233 (1965).
336. Johnson, A. W., W. D. Larson, and G. H. Dahl, Can. J. Chem., 43, 1338 (1965).
337. Jones, J., and M. F. Lappert, Proc. Chem. Soc., 1964, 22; J. Organometal. Chem., 3, 295 (1965).
338. Jtoh, K., M. Fukin, and Y. Ishii, J. Chem. Soc., C, 1969, 2002.
339. Jung, H., Dissertation, Technische Hochschule München, 1962.
340. Kaabak, L. V., M. I. Kabachnik, A. P. Tomilov, and S. L. Varsharskii, Zh. Obshch. Khim., 36, 2060 (1966); C. A., 66, 85828m (1967).
341. Kabachnik, M. I., and E. S. Shepeleva, Izv. Akad. Nauk SSSR, 1949, 56; C. A., 43, 5739e (1949).
342. Kali Chemie Akt. Ges., Brit. 823,483 (1959); C. A., 54, 14125c (1960).
343. Kamai, G., and E. M. Sh. Bastanov, Dokl. Akad. Nauk SSSR, 89, 693 (1953); C. A., 48, 6374e (1954).
344. Kamai, G., and O. N. Belorossova, Izv. Akad. Nauk SSSR, 1947, 191; C. A., 42, 4133f (1948).
345. Kamai, G., and O. N. Belorossova, Izv. Akad. Nauk SSSR, Ser. Khim., 1950, 198; C. A., 44, 8880g (1950).
346. Keeber, W. H., and H. W. Post, J. Org. Chem., 21, 509 (1956).
346a. Kenworthy, J. G., J. Myatt, and P. F. Todd, Chem. Commun., 1969, 263; J. Chem. Soc., B, 1970, 791.
347. Kesavadas, T., and D. S. Payne, J. Chem. Soc., A, 1967, 1001.

348. King, R. B., and R. N. Kapoor, J. Inorg. Nucl. Chem., 31, 2169 (1969).
349. Knoll, F., and G. Bergerhoff, Monatsh. Chem., 97, 808 (1966).
350. Knoll, F., and J. R. Van Wazer, J. Inorg. Nucl. Chem., 31, 2620 (1969).
351. Köhler, H., and A. Michaelis, Chem. Ber., 10, 807 (1877).
352. Korshak, V. V., A. I. Solomatina, N. I. Bekasova, and V. A. Zamyatina, Izv. Akad. Nauk SSSR, Ser. Khim., 1963, 1856.
353. Korshak, V. V., V. A. Zamyatina, and A. I. Solomatina, Izv. Akad. Nauk SSSR, Ser. Khim., 1964, 1541.
354. Korshak, V. V., V. A. Zamyatina, A. I. Solomatina, E. I. Fedin, and P. V. Petrovskii, J. Organometal. Chem., 17, 201 (1969); Zh. Neorg. Khim., 14, 1589 (1969); Izv. Akad. Nauk SSSR, Neorg. Mater., 5, 894 (1969).
355. Kosolapoff, G. M., and R. M. Watson, J. Am. Chem. Soc., 73, 5466 (1951).
356. Krauss, H. L., and H. Jung, Z. Naturforsch., B, 15, 545 (1960).
357. Kreutzkampf, N., Chem. Ber., 87, 919 (1954).
358. Kuchen, W., and H. Buchwald, Angew. Chem., 68, 791 (1956).
359. Kuchen, W., and H. Buchwald, Chem. Ber., 91, 2296 (1958).
360. Kuchen, W., and H. Buchwald, Chem. Ber., 91, 2871 (1958).
361. Kuchen, W., and H. Buchwald, Chem. Ber., 92, 227 (1959); Angew. Chem., 69, 307 (1957).
362. Kuchen, W., H. Buchwald, K. Strolenberg, and J. Metten, Ann. Chem., 652, 28 (1962).
363. Kuchen, W., and W. Grünewald, Angew. Chem., 75, 576 (1963).
364. Kuchen, W., and W. Grünewald, Chem. Ber., 98, 480 (1965).
365. Kuchen, W., and J. Metten, Angew. Chem., 72, 584 (1960).
366. Kuchen, W., and J. Metten, D.A.S. 1,137,732 (1962); C. A., 58, 6863b (1963).
367. Kuchen, W., J. Metten, and A. Judat, Chem. Ber., 97, 2306 (1964).
368. Kuchen, W., K. Strolenberg, and H. Buchwald, Chem. Ber., 95, 1703 (1962).
369. Kuchen, W., K. Strolenberg, and J. Metten, Chem. Ber., 96, 1733 (1963).
370. Kulakova, V. N., Y. M. Zinov'ev, and L. Z. Soborovskii, Zh. Obshch. Khim., 29, 3957 (1969); C. A., 54, 20846e (1960).
371. Kuznetsov, N. L., Zh. Neorg. Khim., 9, 1817 (1964).

372. Lambert, J. B., G. F. Jackson, and D. C. Mueller, J. Am. Chem. Soc., 90, 6401 (1968).
373. Lane, A. P., P. A. Morton-Blake, and D. S. Payne, J. Chem. Soc., A, 1967, 1492.
374. Lane, A. P., and D. S. Payne, Proc. Chem. Soc., 1964, 403.
375. Lee, J. D., and G. W. Goodacre, Naturwissenschaften, 1968, 543.
375a. Lee, J. D., and G. W. Goodacre, Acta Crystallogr., B26, 507 (1970).
376. Leffler, A. J., and E. G. Teach, J. Am. Chem. Soc., 82, 2710 (1960).
377. Leqoux, C., C. R. Acad. Sci., Paris, 209, 47 (1939).
378. Lindner, E., and H. Schless, Chem. Ber., 99, 3331 (1966).
378a. Little, J. L., P. S. Welcker, N. J. Loy, and L. J. Todd, Inorg. Chem., 9, 63 (1970).
379. Low, H., and P. Tavs, Tetrahedron Lett., 1966, 1357.
380. Lundberg, K. L., R. J. Rowatt, and N. E. Miller, Inorg. Chem., 8, 1336 (1969).
381. Lynden-Bell, R. M., Trans. Faraday Soc., 57, 888 (1961).
382. MacDiarmid, A. G., Advan. Inorg. Chem. Radiochem., 3, 246 (1961).
382a. MacDiarmid, A. G., and T. A. Banford, A.C.S., 158th Meeting, New York, Sept. 1969, Abstr., INOR. 122; Inorg. Nucl. Chem. Lett., 8, 733 (1972).
383. McKechnie, J., D. S. Payne, and W. Sim, J. Chem. Soc., 1965, 3500.
384. Märkl, G., Angew Chem., 75, 859 (1963).
385. Märkl, G., and F. Lieb, Tetrahedron Lett., 1967, 3489.
386. Märkl, G., F. Lieb, and A. Merz, Angew. Chem., 79, 475 (1967).
387. Märkl, G., and H. Olbrich, Tetrahedron Lett., 1968, 3813.
388. Mahler, W., J. Am. Chem. Soc., 86, 2306 (1964).
389. Mahler, W., and A. B. Burg, J. Am. Chem. Soc., 79, 251 (1957).
390. Mahler, W., and A. B. Burg, J. Am. Chem. Soc., 79, 251 (1957).
391. Mahler, W., and A. B. Burg, J. Am. Chem. Soc., 80, 6161 (1958).
392. Mahler, W., and A. B. Burg, J. Am. Chem. Soc., 80, 6161 (1958).
393. Maier, L., Angew. Chem., 71, 575 (1959).
394. Maier, L., Chem. Ber., 94, 3043 (1961).
395. Maier, L., Chem. Ber., 94, 3051 (1961).
396. Maier, L., Chem. Ber., 94, 3056 (1961).
397. Maier, L., J. Inorg. Nucl. Chem., 24, 275 (1962).
397a. Maier, L., Helv. Chim. Acta, 45, 2381 (1962).

398. Maier, L., in Progress in Inorganic Chemistry, F. A. Cotton, Ed., Vol. 5, Interscience, 1963, p. 27.
399. Maier, L., Helv. Chim. Acta, 46, 2667 (1963).
400. Maier, L., U. S. 3,075,017 (1963); C. A., 58, 13995c (1963).
401. Maier, L., Helv. Chim. Acta, 46, 1812 (1963).
402. Maier, L., Helv, Chim. Acta, 48, 1190 (1965).
403. Maier, L., Topics in Phosphorus Chemistry, M. Grayson and J. J. Griffith, Eds., Vol. 2, Interscience, 1965, p. 43.
404. Maier, L., Helv. Chim. Acta, 49, 1119 (1966).
405. Maier, L., U. S. 3,242,216 (1966); C. A., 64, 15922f (1966).
406. Maier, L., Fortschr. Chem. Forsch., 8, 1 (1967).
407. Maier, L., Helv, Chim. Acta, 51, 1608 (1968).
408. Maier, L., unpublished results.
409. Maier, L., and J. J. Daly, Helv. Chim. Acta, 50, 1747 (1967).
410. Malatesta, L., Gazz. Chim. Ital., 80, 527 (1950); C. A., 46, 4472i (1952).
411. Malatesta, L., A. Sacco, and L. Ormezzano, Gazz. Chim. Ital., 80, 658 (1950); C. A., 46, 4473d (1952).
412. Mallion, K. B., and F. G. Mann, J. Chem. Soc., 1964, 6121.
413. Manatt, S. L., D. D. Elleman, and A. H. Cowley, and B. Burg, J. Am. Chem. Soc., 89, 4544 (1967).
414. Mann, F. G., and I. T. Millar, J. Chem. Soc., 1952, 3039.
415. Mann, F. G., and M. J. Pragnell, J. Chem. Soc., C, 1966, 916.
416. Mardersteig, H. G., L. Meinel, and H. Nöth, Z. Anorg. Allg. Chem., 368, 254 (1969).
417. Marshall, M. D., and W. J. Harwood, U. S. 3,031,509 (1962).
418. Masthoff, R., G. Krieg, and C. Vieroth, Z. Anorg. Allg. Chem., 364, 316 (1969).
419. Mastryukowa, T. A., A. E. Shipov, and M. I. Kabachnik, Zh. Obshch. Khim., 29, 1450 (1959); C. A., 54, 9729f (1960).
420. Mastryukova, T. A., A. E. Shipov, and M. I. Kabachnik, Zh. Obshch. Khim., 31, 507 (1961); C. A., 55, 22101f (1961).
420a. Matschiner, H., F. Krech, and A. Steinert, Z. Anorg. Allg. Chem., 371, 256 (1969).
421. Meek, D. W., and P. Nicpon, J. Am. Chem. Soc., 87, 4951 (1965).
421a. Meinel, L., and H. Nöth, Z. Anorg. Allg. Chem., 373, 36 (1970).
422. Meinhardt, N. A., U. S. 3,065,273 (1962); C. A., 58, 9139a (1963).

423. Meriwether, L. S., E. C. Colthup, G. W. Kennerly, and R. N. Reusch, J. Org. Chem., 26, 5155 (1961).
424. Miller, B., in Topics in Phosphorus Chemistry, M. Grayson and J. J. Griffith, Eds., Vol. 2, Interscience, New York, 1965, p. 142.
424a. Mingaleva, K. S., G. M. Bogolyubov, Yu. N. Shlyk, and A. A. Petrov, Zh. Obshch. Khim., 39, 2679 (1969).
425. Moedritzer, K., L. Maier, and L. D. C. Groenweghe, J. Chem. Eng. Data, 7, 307 (1962).
426. Monsanto Co., Belg. 639,193 (1964); C. A., 62, 9175a (1965).
427. Morgan, P. W., and B. C. Herr, J. Am. Chem. Soc., 74, 4526 (1952).
428. Muetterties, E. L., and V. D. Aftandilian, Inorg. Chem., 1, 731 (1962).
429. Muetterties, E. L., U.S. 3,118,932 (1964); C. A., 60, 9312g (1964).
430,
431. Murray, M., and R. Schmutzler, Chem. Ind. (London), 1968, 1730.
432. Nedelandse Centrale Organisatie voor Toegepast-Natuurwetenschappelijk Onderzoek, Neth. 6,500,454 (1966); C. A., 66, 2641q (1967).
433. Nesterov, L. V., and N. A. Aleksendrova, Zh. Obshch. Khim., 39, 931 (1969).
434. Neumann, W. P., B. Schneider, and R. Sommer, Ann. Chem., 692, 1 (1966).
435. Newlands, M. J., Proc. Chem. Soc., 1960, 123.
436. Newmark, R. A., A. D. Norman, and R. W. Rudolph, Chem. Commun., 1969, 893.
437. Nielsen, M. L., J. V. Pustinger, Jr., and J. Strobel, J. Chem. Eng. Data, 9, 167 (1964).
438. Niebergall, H., Angew. Chem., 72, 210 (1960).
439. Niebergall, H., and B. Langenfeld, Ger. 1,083,262 (1960); U.S. 2,959,621 (1960); C. A., 55, 7289h (1961).
440. Niebergall, H., and B. Langenfeld, Chem. Ber., 95, 64 (1962).
441. Niebergall, H., and B. Langenfeld, D.A.S. 1,149,355 (1963).
441a. Nixon, J. F., personal communication; and cited by R. Schmutzler, Angew. Chem., 77, 530 (1965), ref. 29.
442. Nöth, H., Z. Naturforsch., B, 15, 327 (1960).
443. Nöth, H., and L. Meinel, Z. Anorg. Allg. Chem., 349, 225 (1967).
444. Nöth, H., and W. Schrägle, Z. Naturforsch., B, 16, 473 (1961).
445. Nöth, H., and W. Schrägle, Angew. Chem., 74, 587 (1962).
446. Nöth, H., and W. Schrägle, Chem. Ber., 97, 2218 (1964).
447. Nöth, H., and W. Schrägle, Chem. Ber., 97, 2374 (1964).

448. Nöth, H., and W. Schrägle, Chem. Ber., 98, 352 (1965).
449. Nöth, H., and H. J. Vetter, Chem. Ber., 94, 1505 (1961).
450. Noltes, J. G., Rec. Trav. Chim., 84, 782 (1965).
450a. Norman, H. D., Inorg. Chem., 9, 870 (1970).
451. Okon, K., J. Sobczynski, J. Sowinski, and K. Niewielski, Biul. Wojskowa, Akad. Tech., 13, 109 (1964); C. A., 62, 4050f (1965).
452. Paciore, K. L., and R. H. Kratzer, J. Org. Chem., 31, 2426 (1966).
453. Paddock, N. L., Roy. Inst. Chem. (London), Lecture Series, No. 2 (1962).
454. Palenik, G. J., and J. Donohue, Acta Crystallogr., 15, 564 (1962).
455. Parshall, G. W., J. Inorg. Nucl. Chem., 12, 372 (1960).
456. Parshall, G. W., J. Inorg. Nucl. Chem., 14, 291 (1960).
457. Parshall, G. W., Org. Synth., 45, 102 (1965).
458. Parshall, G. W., "Boron-Phosphorus Compounds" in Chemistry of Boron and Its Compounds, E. L. Muetterties, Ed., Wiley, 1967.
459. Parshall, G. W., Inorg. Synth., 11, 157 (1968).
460. Parshall, G. W., and R. V. Lindsey, Jr., J. Am. Chem. Soc., 81, 6273 (1959).
461. Park, P. J. D., G. Chambers, E. Wyn-Jones, and P. J. Hendra, J. Chem. Soc., A, 1967, 646.
462. Pass, F., and H. Schindlbauer, Monatsh. Chem., 90, 148 (1959).
463. Pass, F., E. Steininger, and H. Schindlbauer, Monatsh. Chem., 90, 792 (1959).
464. Patel, N. K., and H. J. Harwood, J. Org. Chem., 32, 2999 (1967).
465. Pedone, C., and A. Sirigu, J. Chem. Phys., 47, 339 (1967).
466. Perveev, F. Ya., and K. Rikhter, Zh. Obshch. Khim., 30, 784 (1960); C. A., 55, 1580h (1961).
467. Peterson, D. J., and T. J. Logan, J. Inorg. Nucl. Chem., 28, 53 (1966).
468. Phelisse, J. A., and M. Vilanneau, Fr. 1,337,271 (1963).
469. Pollart, K. A., and H. Y. Harwood, J. Org. Chem., 27, 4444 (1962).
470. Postnikova, G. B., and I. F. Lutsenko, Zh. Obshch. Khim., 33, 4029 (1963); C. A., 60, 9309a (1964).
471. Quin, L. D., and H. G. Anderson, J. Org. Chem., 29, 1859 (1964).
472. Quin, L. D., and H. G. Anderson, J. Am. Chem. Soc., 86, 2090 (1964).
473. Quin, L. D., and H. G. Anderson, J. Org. Chem., 31, 1206 (1966).

474. Rauhut, M. M., and A. M. Semsel, J. Org. Chem., 28, 473 (1963).
475. Rauhut, M. M., and A. M. Semsel, U.S. 3,099,690 (1963); C. A., 60, 556a (1964).
476. Reeser, J. W. B., and G. F. Wright, J. Org. Chem., 22, 385 (1957).
477. Reinhardt, H., D. Bianchi, and D. Mölle, Chem. Ber., 90, 1656 (1957).
478. Rüdorff, W., and W. Müller, Doctoral thesis, W. Müller, Tübingen, 1957.
478a. Sanz, F., and J. J. Daly, J. Chem. Soc., A, 1971, 1083.
479. Sasse, K., in Houben-Weyl, Methoden der Organischen Chemie, E. Müller, Ed., Vol. XII, Organische Phosphorverbindungen, Part 1, 1963; Verlag Thieme.
480. Satgé, J., and M. Baudet, C. R. Acad. Sci., Paris, Ser. C, 263, 435 (1966).
481. Satgé, J., and C. Couret, C. R. Acad. Sci., Paris, Ser. C, 264, 2169 (1967).
482. Satgé, J., C. Couret, and M. Lesbre, Bull. Soc. Chim. Fr., 1967, 774.
483. Satgé, J., and C. Couret, C. R. Acad. Sci., Paris, Ser. C, 267, 173 (1968).
484. Satgé, J., and C. Couret, Bull. Soc. Chim. Fr., 1969, 333.
484a. Satgé, J., C. Couret, and J. Escudié, C. R. Acad. Sci., Paris, Ser. C, 270, 351 (1970).
485. Satgé, J., M. Lesbre, and M. Baudet, C. R. Acad. Sci., Paris, Ser. C, 259, 4733 (1964).
486. Scherer, O. J., and W. Gick, Chem. Ber., 103, 71 (1970).
487. Schindlbauer, H., personal communication.
488. Schindlbauer, H., and D. Hammer, Monatsh. Chem., 94, 644 (1963).
489. Schlör, H., and G. Schrader, D.A.S. 1,067,021 (1959).
490. Schmidt, U., F. Geiger, A. Müller, and K. Markau, Angew. Chem., 75, 640 (1963).
491. Schmidt, U., and Ch. Osterroht, Angew. Chem., 77, 455 (1965).
492. Schmidt, U., R. Schröer, and H. Achenbach, Angew. Chem., 78, 307 (1966).
493. Schmidt, U., and I. Boie, Angew. Chem., 78, 1061 (1966).
494. Schmidt, U., I. Boie, C. Osterroht, R. Schröer, and H. F. Grützmacher, Chem. Ber., 101, 1381 (1968).
495,
496. Schmutzler, R., Inorg. Chem., 3, 421 (1964).
497. Schmutzler, R., U.S. 3,246,032 (1966); C. A., 64, 19678f (1966).
498. Schmutzler, R., personal communication; see also Ang., H. G., and R. Schmutzler, J. Chem. Soc., A, 1969, 702.

499. Schroeder, H. J., Inorg. Chem., 2, 390 (1963).
500. Schroeder, H. J., U. S. Dept. Commerce, Office of Technical Services, AD 267,991 (1961); C. A., 58, 7595e (1963).
501. Schumann, H., Angew. Chem., 81, 970 (1969).
501a. Schumann, H., and U. Arbenz, J. Organometal. Chem., 22, 411 (1970).
502. Schumann, H., and H. Benda, Angew. Chem., 80, 846 (1968).
503. Schumann, H., and H. Benda, Angew. Chem., 80, 845 (1968).
504. Schumann, H., and H. Benda, Angew. Chem., 81, 1049 (1969).
504a. Schumann, H., and H. Benda, Angew Chem., 82, 46 (1970).
505. Schumann, H., and H. Benda, J. Organometal. Chem., 21, P12 (1970).
506. Schumann, H., and H. Blass, Z. Naturforsch., B, 21, 1105 (1966).
507. Schumann, H., and P. Jutzi, Chem. Ber., 101, 24 (1968).
508. Schumann, H., P. Jutzi, and M. Schmidt, Angew. Chem., 77, 812 (1965).
509. Schumann, H., P. Jutzi, and M. Schmidt, Angew. Chem., 77, 912 (1965).
510. Schumann, H., P. Jutzi, A. Roth, P. Schwabe, and E. Schauer, J. Organometal. Chem., 10, 71 (1967).
511. Schumann, H., H. Köpf, and M. Schmidt, Angew. Chem., 75, 672 (1963).
512. Schumann, H., H. Köpf, and M. Schmidt, J. Organometal. Chem., 2, 159 (1964).
513. Schumann, H., H. Köpf, and M. Schmidt, Chem. Ber., 97, 1458 (1964).
514. Schumann, H., H. Köpf, and M. Schmidt, Chem. Ber., 97, 2395 (1964).
514a. Schumann, H., L. Rösch, and O. Stelzer, J. Organometal. Chem., 21, 351 (1970).
515. Schumann, H., and A. Roth, J. Organometal. Chem., 11, 125 (1968).
516. Schumann, H., A. Roth, and O. Stelzer, Angew. Chem., 80, 240 (1968); J. Organometal. Chem., 24, 183 (1970).
517. Schumann, H., A. Roth, O. Stelzer, and M. Schmidt, Inorg. Nucl. Chem. Lett., 2, 311 (1966).
518. Schumann, H., and M. Schmidt, Angew. Chem., 77, 1049 (1965).
519. Schumann, H., P. Schwabe, and M. Schmidt, J. Organometal. Chem., 1, 366 (1964).
520. Schumann, H., P. Schwabe, and M. Schmidt, Inorg. Nucl. Chem., Lett., 2, 309 (1966).
521. Schumann, H., P. Schwabe, and M. Schmidt, Inorg. Nucl. Chem. Lett., 2, 313 (1966).

522. Schumann, H., P. Schwabe, and O. Stelzer, Chem. Ber., 102, 2900 (1969).
523. Schumann, H., and O. Stelzer, Angew. Chem., 79, 692 (1967).
524. Schumann, H., and O. Stelzer, Angew. Chem., 80, 318 (1968).
525. Schumann, H., and O. Stelzer, J. Organometal. Chem., 13, P25 (1968).
525a. Schumann, H., O. Stelzer, U. Niederreuther, and L. Rösch, Chem. Ber., 103, 1383 (1970).
525b. Schumann, H., O. Stelzer, and U. Niederreuther, Chem. Ber., 103, 1391 (1970).
526. Schumann, H., private communication.
527. Schumann-Ruidisch, I., and J. Kuhlmey, J. Organometal. Chem., 16, P26 (1969).
528. Seel, F., K. Rudolph, and R. Budenz, Z. Anorg. Allg. Chem., 341, 196 (1966).
529. Seel, F., K. Rudolph, and W. Gombler, Angew. Chem. Int. Ed., 6, 708 (1967); Z. Anorg. Allg. Chem., 363, 233 (1968).
530. Seidel, W., Z. Chem., 3, 429 (1963).
531. Seidel, W., Z. Anorg. Allg. Chem., 330, 141 (1964).
532. Seidel, W., Z. Anorg. Allg. Chem., 335, 316 (1965).
533. Seidel, W., Z. Anorg. Allg. Chem., 341, 70 (1965).
534. Seidel, W., Z. Anorg. Allg. Chem., 342, 165 (1966).
535. Seidel, W., and K. Issleib, Z. Anorg. Allg. Chem., 325, 113 (1963).
536. Seidel, W., and K. Issleib, Z. Anorg. Allg. Chem., 325, 113 (1963).
537. Sheldrick, G. M., Trans. Faraday Soc., 63, 1065 (1967).
538. Shlyk, Yu. N., G. M. Bogolyubov, and A. A. Petrov, Dokl. Akad. Nauk SSSR, 176, 1327 (1967).
539. Shlyk, Yu. N., G. M. Bogolyubov, and A. A. Petrov, Zh. Obshch. Khim., 38, 193 (1968).
540. Siebert, H., J. Eints, and E. Fluck, Z. Naturforsch., B, 23, 1006 (1968).
541. Silverstein, H. T., D. C. Beer, and L. J. Todd, J. Organometal. Chem., 21, 139 (1970).
542. Spangenberg, S. F., and H. H. Sisler, Inorg. Chem., 8, 1006 (1969).
543. Spanier, E. J., and F. E. Caropreso, Tetrahedron Lett., 1969, 199.
544. Spencer, C. J., and W. N. Lipscomb, Acta Crystallogr., 14, 250 (1961).
545. Spencer, C. J., P. G. Simpson, and W. N. Lipscomb, Acta Crystallogr., 15, 509 (1962).
546. Steger, E., and K. Stopperka, Chem. Ber., 94, 3029 (1961).
547. Steininger, E., Chem. Ber., 96, 3184 (1963).
548. Steinkopf, W., and H. Dudek, Chem. Ber., 62, 2494 (1929).

549. Stepanov, B. I., E. N. Karpova, and A. I. Bokanova, Zh. Obshch. Khim., 39, 1544 (1969).
550. Stewart, R. D., and R. I. Wagner, U. S. 2,900,416 (1959); C. A., 54, 2171c (1960).
551. Strecker, W., and Ch. Grossmann, Chem. Ber., 49, 63 (1916).
552. Tamborski, C. F., E. Ford, W. L. Lehn, G. J. Moore, and E. J. Soloski, J. Org. Chem., 27, 619 (1962).
553. Teichmann, H., Angew. Chem., 77, 809 (1965).
554. Tevebaugh, A. D., Inorg. Chem., 3, 302 (1964).
555. Thompson, N. R., J. Chem. Soc., 1965, 6288.
556. Thompson, N. R., J. Chem. Soc., 1965, 6290.
557. Todd, L. J., J. L. Little, and H. T. Silverstein, Inorg. Chem., 8, 1698 (1969).
558. Van Ghemen, M., and E. Wiberg, U. S. 3,361,830 (1968); C. A., 69, 10539w (1968).
559. Van Ghemen, M., and E. Wiberg, U. S. 3,390,189 (1968); C. A., 69, 77473p (1968).
560. Vetter, H. J., and H. Nöth, Chem. Ber., 96, 1816 (1963).
561. Voigt, D., R. Turpin, and M. C. Labarre, Bull. Soc. Chim. Fr., 1968, 3561.
562. Wagner, I., and C. O. Wilson, Inorg. Chem., 5, 100g (1966).
563. Wagner, R. I., cited in (74), ref. 3.
564. Wagner, R. I., and A. B. Burg, J. Am. Chem. Soc., 75, 3869 (1953).
565. Wagner, R. I., and A. B. Burg, WADC Technical Report, 57-126, Part VI (1962).
566. Wagner, R. I., and F. F. Caserio, Jr., J. Inorg. Nucl. Chem., 11, 259 (1959).
567. Walling, C., U. S. 2,437,797; 2.437,795 (1948); C. A., 42, 419f (1948).
568. Walling, C., U. S. 2,437,796; 2,437,798; C. A., 42, 4199a (1948).
569. Watson, W. H., Texas J. Sci., 11, 471 (1959); C. A., 54, 13928c (1960).
570. Watt, G. W., and R. C. Thompson, Jr., J. Am. Chem. Soc., 70, 2295 (1949).
571. Weil, Th., B. Prijs, and H. Erlenmeyer, Helv. Chim. Acta, 35, 616 (1952).
571a. Welcker, P. S., and L. J. Todd, Inorg. Chem., 9, 286 (1970).
572. Wells, E. J., H. P. K. Lee, and L. K. Peterson, Chem. Commun., 1967, 894.
573. Wheatley, P. J., J. Chem. Soc., 1960, 523.
574. Wiberg, E., M. Van Ghemen, and G. Müller-Schiedmayer, Angew. Chem., 75, 814 (1963); U. S. 3,471,568 (1969); C. A., 71, 124660a (1969).
575. Wittenberg, D., and H. Gilman, J. Org. Chem., 23, 1063 (1958).

575a. Wymore, C. E., and J. C. Bailar, Jr., J. Inorg. Nucl. Chem., 14, 42 (1960).
576. Zakharkin, L. I., and V. I. Kyskin, Zh. Obshch. Khim., 39, 928 (1969); C. A., 71, 50096x (1969).
577. Zakharkin, L. I., and V. I. Kyskin, Izv. Akad. Nauk SSSR, Ser. Khim., 1969, 1167; C. A., 71, 50103x (1969).
578. Zamyatina, V. A., A. I. Solomatina, V. V. Korshak, and B. V. Lokshin, Izv. Akad. Nauk SSSR, Neorg. Mater., 3, 524 (1967); C. A., 67, 33127h (1967).

Chapter 3A. Phosphine Complexes with Metals

G. BOOTH

Imperial Chemical Industries Limited, Blackley,
Manchester, England

Abbreviations

X (or Y) = Cl, Br, I or pseudohalogen anions

M = metal	Bu = butyl, Oct = octyl
diph = $Ph_2PCH_2CH_2PPh_2$	All = allyl
R = alkyl or aryl	Am = amyl, Pe = pentyl
Me = methyl	Cy = cyclohexyl
Et = ethyl	Bz = benzyl
Pr = propyl	Ph = phenyl

The importance of this class of complex is reflected by the 1500 or so publications summarized in this chapter. Over a 1000 of these references have been published in the last 6 years, including, significantly, ultimate x-ray crystal structure determinations for over 100 complexes. The potential of the early work of Hofmann[761] and Cahours and Gal[238,239] was not recognized until developed in the schools of Mann and Jensen 60 years later. Advances in the field over the subsequent 30 years to the present (i.e., the end of 1969) have been remarkable and stimulated by discoveries that phosphine ligands can stabilize many unusual oxidation states of transition metals and form isolatable and characterizable organo (σ- and π-bonded) hydrido, carbonyl, and nitrosyl derivatives[192] and, more recently still, nitrogen, acyl,[272,373] nitrene,[63] and carbene[1003,1336] complexes.

Recent reviews compare the properties of transition metal complexes of phosphines with those of arsines and stibines[192] and their relative affinities.[18] Other reviews emphasize the importance of hydride complexes,[607] metal carbonyl complexes[966] and the application of complexes in catalysis.[403,1259]

Complexes have found important outlets as catalysts in many different types of reactions and as model compounds in kinetic (Section B.3) and metal surface simulation[1287] studies, often with wide-reaching significance. Nitrogen complexes, for example, are important as possible models for nitrogenase.[333]

Most of the newer physical techniques have found application in the many investigations into structure and bonding.

A. SYNTHETIC ROUTES

A.1. Metal(O) Phosphine Complexes (with No Other Ligands)

I. From free metal, usually in activated form, by analogy with nickel tetracarbonyl.

$$Ni + PCl_2Me \xrightarrow{\text{reflux}} [Ni(PCl_2Me)_4]$$ [1133]

II. From the metal carbonyl, as for the preparation of carbonyl complexes (see Route V) but under more vigorous conditions, especially with tertiary diphosphines.[301]

$$[Ni(CO)_4] + \text{diphosphine} \longrightarrow [Ni(\text{diphosphine})_2]$$

III. By replacement of ligands other than carbonyl, for example, cyano,[134] π-allyl,[1347] π-cyclopentadienyl,[1302] or π-aryl.[356]

$$K_4[Ni(CN)_4] \xrightarrow[\text{liq. } NH_3]{PPh_3} [Ni(PPh_3)_4]$$

$$[Mo(\pi\text{-}C_6H_6)_2] \longrightarrow [Mo(Me_2PCH_2CH_2PMe_2)_3]$$

$$[Pd(\pi\text{-}C_3H_5)(\pi\text{-}C_5H_5)] \xrightarrow[\text{petrol}]{PMePh_2} [Pd(PMePh_2)_4]$$ [1029]

IV. Reduction of a suitable, usually halide, complex if necessary in the presence of the phosphine. Typical reducing agents are sodium borohydride,[307] hydrazine,[944] sodium in liquid ammonia and sodium naphthalene.[357]

$$[MX_2(\text{diphosphine})_2] \longrightarrow [M(\text{diphosphine})_2]$$

$$CrCl_3 + Me_2PCH_2CH_2PMe_2 \xrightarrow[\text{in THF}]{Na^+(C_{10}H_8)^-} [Cr(Me_2PCH_2CH_2PMe_2)_3]$$

A.2. Carbonyl Complexes

V. Direct replacement of carbon monoxide by a phosphine ligand, reaction being either neat or in a suitable solvent. Substitution can be enhanced by temperature control or by exposure to UV light.[918,1075] Metals with more than one carbonyl derivative, for example, iron and cobalt, usually give a corresponding number of series of phosphine-carbonyl complexes. The solvent may be sufficiently important to determine the constitution of the product, for example, with vanadium and cobalt. The product is often not a pure species and is purified by crystallization or by more refined techniques of sublimation,[453] chromatography,[465,486] or TLC.[269]

$$[V(CO)_6] \xrightarrow[\text{ether}]{PPh_3} [V(Et_2O)_6][V(CO)_5(PPh_3)]_2$$ [747]

$$[Cr(CO)_6] \xrightarrow[\text{diglyme/160°}]{PR_3} [Cr(CO)_4(PR_3)_2]$$ [932]

This method applied to carbonyl-halide complexes is a standard procedure for the production of phosphine-carbonyl-halide complexes, in some cases by a straight-through process from the metal (e.g., rhodium[344]) halide.

$$[RhCl(CO)_2]_2 + PPh_3 \xrightarrow{C_6H_6/60°} [RhCl(CO)(PPh_3)_2]$$ [732]

VI. Indirect replacement by a phosphine ligand by reaction of a carbonyl derivative containing a ligand more readily displaced than carbonyl.

$$[Mo(\text{norbornadiene})(CO)_4] \xrightarrow[\text{benzene}]{PR_3} [Mo(CO)_4(PR_3)_2]$$ [1236]

$$[Cr(\pi\text{-}C_6H_6)(CO)_3] \xrightarrow{PR_3} [Cr(CO)_3(PR_3)_3]$$ [932]

VII. Reaction of a metal carbonyl with a base to form a reactive intermediate prior to substitution with the phosphine.

$$[Mo_2(CO)_6(OH)_3H_3] \longrightarrow [Mo(CO)_3(PPh_3)_3]$$ [719]

VIII. Ligand displacement with carbon monoxide (possibly in the presence of a reducing agent).

$$[Ru(\text{2-methylallyl})_2(PPh_3)_2] \xrightarrow{CO} [Ru(CO)_3(PPh_3)_2]$$ [1123]

$$[OsX_2(CO)_2(PPh_3)_2] \xrightarrow[\text{Zn/DMF}]{CO} [Os(CO)_3(PPh_3)_2]$$ [419]

IX. Direct synthesis from the metal by simultaneous reaction with a phosphine and carbon monoxide.

$$Ni + PPh_3 \xrightarrow[\text{MeOH}]{\text{CO/20 atm/150°}} [Ni(CO)_2(PPh_3)_2]$$ [1017]

IXA. In certain cases metal halide phosphine complexes (possibly prepared in situ) can be reacted with carbon monoxide to form derived carbonyl halide complexes.

$$RhCl_3 + PEt_3 + CO \xrightarrow{\text{boiling EtOH}} [RhCl_3(CO)(PEt_3)_2]$$ [344]

The carbon monoxide for this reaction may be abstracted

from the reaction medium, which is usually an alcohol[340,341] but possibly other oxygen-containing solvents such as DMF, THF, dioxan, and cyclohexanone.[1169]

$$RhCl_3 + PPh_3 \xrightarrow[DMF]{boiling} [RhCl(CO)(PPh_3)_2]^{1169}$$

$$[RhCl_3(PEt_3)_3] \xrightarrow{EtOH/KOH} [RhCl(CO)(PEt_3)_2]^{781}$$

This abstraction reaction has been extended to prepare the first thiocarbonyl complexes of transition metals.[83]

$$[RhCl(PPh_3)_3] \xrightarrow{CS_2} [RhCl(CS)(PPh_3)_2] \quad (\text{structure}^{484})$$

A.3. Halide Complexes

X. Mixing of stoichiometric quantities of metal salt and phosphine ligand produces a complex which with the appropriately chosen solvent, frequently ethanol, may crystallize out. The stability of the derivatives from non-transition metals is generally low, and although products are usually classified as simple addition compounds they are nevertheless included in this chapter for completeness.

$$MX_2 + nPR_3 \longrightarrow [MX_2(PR_3)n]$$

As with carbonyl complexes (Route V), the solvent may determine the nature of the product, for example, MnI_2 forms $[MnI_2(PPh_3)_2]$ in THF,[1036] but $[MnI_4][PPh_3H]_2$ in acetone.[1038] One variation is to mix in the absence of solvent, for example, by using molten triphenylphosphine with a zinc or copper salt.[110] Purification is often possible by recrystallization or by more refined techniques such as chromatography.[862]

The products are usually four-, five-, or six-coordinated with varying stereochemistry. Various techniques have evolved for separation of isomers and adjustment of conditions to favor one isomer, for example, complexes of platinum[359] and palladium.[361] Photochemical isomerization is now established with, for example, complexes of platinum[655] and iridium;[216] some of the earlier experimental work may therefore have been unknowingly affected by light.

Complexes are also formed with a variety of derivatives other than halides, for example, metal thiophenolates[187,1060] and dithioketone[1193,1194] compounds.

Primary and secondary phosphines often behave similarly to tertiary phosphines, although there is a tendency toward unusual coordination numbers, but with certain metals, notably cobalt, nickel, and copper, they may alternatively form phosphidophosphine complexes (listed in Section D as pseudohalogen compounds). For example,

$$NiBr_2 + PHPh_2 \longrightarrow [Ni(PPh_2)_2(PHPh_2)_2]$$ [817]

Primary phosphines may also form dimeric phosphido-bridged complexes (see Route XXVIII) under the conditions usually employed for simple halide complex formation, especially with palladium.[817]

XA. Certain metals, for example, cadmium, mercury, palladium, and platinum, also form halogen-bridged dimeric species under appropriate conditions. The general method of preparation for halogen-bridged complexes involves reaction of further metal halide with the monomeric complex.[352,354,623,955]

$$[PdX_2(PR_3)_2] \xrightarrow[(NH_4)_2PdCl_4]{} [Pd_2X_4(PR_3)_2]$$

It is therefore feasible to prepare bridged complexes of this type containing two different metal atoms.[956] Thiocyanatobridged complexes of platinum[285,300,303,1081] and analogous thiobridged[302,328,1234] and oxalatobridged[330] complexes of platinum and palladium have been examined in detail.

XB. The bridged complexes are split by reaction with phosphine so this provides a convenient method for the preparation of "mixed" phosphine complexes such as $[PtCl_2(PR_3)(PR'_2Cl)]$.[315]

XI. Displacement of a ligand from an existing halogeno complex by a phosphine.

$$[UCl_5(\text{trichloroacryloyl chloride})] \xrightarrow{PPh_3} [UCl_5PPh_3]$$ [1201]

$$[WCl_4(MeCN)_2] \longrightarrow [WCl_3(\text{diphosphine})]$$ [189]

$$[FeI_2(CO)_4] \longrightarrow [FeI_2(PPh_3)_2]$$ [721]

$$[RhCl(C_2H_4)_2]_2 \longrightarrow [RhCl(PHPh_2)_3]$$ [685]

$$[Ru_2Cl_3(PEt_2Ph)_6]Cl \longrightarrow cis\text{-}[RuCl_2(\text{diphosphine})]$$ [310]

Calcium[676] and Th(IV)[1235] are reported not to form complexes with specific ligands. Some of the earlier references to simple complexes have been shown to be erroneous, for example, not $[UX_4(PR_3)_2]$[34] but $(R_3PH)_2[UO_2X_4]$.[479,596] Phosphine oxide complexes of uranium[1203] and rhenium[851,925] exemplify the ease with which phosphines can be oxidized by metals in higher oxidation states, thus hindering or preventing preparation of a phosphine complex.

In work on halide complexes, there is also a history of formulations omitting other essential ligand atoms, now mainly characterized by IR spectroscopic methods. For example,

$[ReOCl_3(PPh_3)_2]$[925] was originally formulated $[ReCl_3(PPh_3)_2]$,[591] $[ReO(OEt)Br_2(PPh_3)_2]$[335] was originally formulated $[ReBr_2(PPh_3)_2]$,[591] $[ReNCl_2(PPh_3)_2]$[298] was originally formulated $[ReCl_2(PPh_3)_2]$,[335] and $[MHCl(CO)(PPh_3)_3]$ (M = Ru or Os)[341,1313] were originally formulated as $[RuCl(PPh_3)_3]$[1318] and $[OsCl(PPh_3)_3]$.[1303]

Other cases of "mistaken identity" (e.g., complexes of copper[910] and platinum[369]) stress the need for careful and unambiguous characterization in preparation of new complexes.

XII. Reaction with halogen or hydrohalic acid of a complex such as a hydride or a halogen complex of a lower oxidation state.

$$[ReH_3(PPh_3)_2] \xrightarrow[\text{or } Cl_2]{HCl} [ReCl_4(PPh_3)_2]^{945}$$

$$[ReCl_3(PEt_2Ph)_2] \xrightarrow{Cl_2/CCl_4} [ReCl_4(PEt_2Ph)_2]^{296}$$

$$[AuClPEt_3] \xrightarrow{Cl_2/CHCl_3} [AuCl_3PEt_3]^{957}$$

XIIA. This reaction can be extended to halogenation of phosphine carbonyl complexes to produce carbonyl-halide complexes.

$$[Ru(CO)_3(PPh_3)_2] \xrightarrow{I_2} [RuI_2(CO)_2(PPh_3)_2]^{416}$$

$$[RhCl(CO)(PPh_3)_2] \xrightarrow{Cl_2/CCl_4} [RhCl_3(CO)(PPh_3)_2]^{1298}$$

XIII. Employment of a phosphine as a reducing agent in reaction, for example, with a halogen complex of a higher oxidation state.

$$[ReOCl_3(PR_3)_2] \xrightarrow[\text{in } C_6H_6]{PR_3} [ReCl_3(PR_3)_3]^{324}$$

$$OsO_4 \xrightarrow[EtOH/HCl]{PEt_3} [OsCl_3(PEt_3)_3]^{324}$$

XIV. Certain halide (mainly iodide) and most pseudohalide complexes are formed by double-decomposition reactions with the chloride complex.

$$[RhCl_3(PMe_2Ph)_3] \xrightarrow[EtOH]{NaI} [RhI_3(PMe_2Ph)_3]$$ [215]

A convenient modification is the use of the acetato complex as starting material.

$$[Au(OCOCH_3)PPh_3] \xrightarrow{HX} [AuXPPh_3]$$ [1050]

A.4. Nitrosyl Complexes

XV. Reaction of a phosphine with a complex already containing a nitrosyl ligand. With halide complexes this may involve either direct addition, often by scission of halogen bridges, or reduction in the theoretical oxidation state of the metal (as in Route XIII).

$$[Ru(NO)Cl_3] \xrightarrow[EtOH]{PR_3} [RuCl_3(NO)(PR_3)_2]$$ [555]

$$[Co_2Cl_2(NO)_4] \xrightarrow{PPh_3} [CoCl(NO)_2PPh_3]$$ [165]

$$[Ni(NO)Cl_2] \xrightarrow{PPh_3/100°} [NiCl(NO)(PPh_3)_2]$$ [16]

With carbonyl complexes this usually involves displacement of carbon monoxide.

$$[Co(CO)_3NO] \xrightarrow[diglyme]{diph} [Co(NO)(CO)diph]$$ [1301]

$$[Fe(NO)_2(CO)_2] \xrightarrow[toluene]{PPh_3} [Fe(NO)_2(PPh_3)_2]$$ [941]

XVI. Reaction of a suitable complex with nitric oxide, or organic N-nitroso compound, involving either displacement or addition.

$$[Pt(PPh_3)_4] \xrightarrow[C_6H_6]{NO} [Pt(NO)(PPh_3)_2]_2$$ [1115]

$$[V(CO)_4(PPh_3)_2] \xrightarrow[n\text{-hexane}]{NO} [V(CO)_4(NO)PPh_3]$$ [1338]

$$[Mn(CO)_4PPh_3] \xrightarrow{NO} [Mn(NO)_3PPh_3]$$ [713]

$$[CoCl_2(PEt_3)_2] \xrightarrow[C_6H_6]{NO} [CoCl_2(NO)(PEt_3)_2]$$ [194]

XVIA. In certain cases a nitrosyl halide can be used.

$$[Fe(CO)_3(PPh_3)_2] \xrightarrow[MeCN/-10°]{NOCl} [FeCl(CO)_2(NO)(PPh_3)_2]$$ [460]

XVII. Oxidation reactions may form the nitrosyl ligand in situ with reagents such as the nitrite anion or nitric acid. Reduction of the nitro or nitrato ligand with carbon monoxide can also generate a nitrosyl group indirectly.

$$[NiBr_2(PPh_3)_2] \xrightarrow[THF/reflux]{NaNO_2} [NiBr(NO)(PPh_3)_2]$$ [562]

$$[Ni(NO_2)_2(PPh_3)_2] \xrightarrow[C_6H_6]{CO} [Ni(NO_2)(NO)(PPh_3)_2]$$ [194]

Derivatives of nitrosyl complexes are available also by metathetical replacement reactions (Route XIV). Typical classes of compound are listed in Section D.7.

A.5. Hydrido Complexes

XVIII. Treatment of a halide complex with a reducing agent such as N_2H_4, $LiAlH_4$, $NaBH_4$, aluminum (hydrido) alkyls[926] or $GeHMe_3$.[218] It may be possible to simultaneously react the metal salt, phosphine, and reducing agent in a through-process variation.

$$[PtX_2(PR_3)_2] \xrightarrow[EtOH]{N_2H_4} [PtHX(PEt_3)_2]$$ [343]

$$[OsCl_3(PMe_2Ph)_3] \xrightarrow[EtOH]{NaBH_4} [OsH_4(PMe_2Ph)_3]$$ [514]

$$[FeCl_2(diphosphine)_2] \xrightarrow[THF]{LiAlH_4} [FeH_2(diphosphine)_2]$$ [311]

In the special cases of nickel, copper, and silver (see Section D.8), BH_4 is retained in the complex following borohydride reduction.

XVIIIA. Under certain conditions nitrogen is simultaneously abstracted from the environment during reduction with a reagent such as an aluminum alkyl. This leads to the production of nitrogen or hydridonitrogen complexes (see Section D.10). Hydrogen and nitrogen in these complexes are often interchangeable.

$$[RuHCl(PPh_3)_3] \xrightarrow{AlEt_3/N_2} [Ru(N_2)H_2(PPh_3)_2] \underset{N_2}{\overset{H_2}{\rightleftharpoons}} [RuH_4(PPh_3)_3]$$ [887]

XIX. Addition of gaseous hydrogen, or sometimes a hydrohalic acid, to a complex. Synthesis from the components is possible in rare cases.

$$[Os(CO)_4PPh_3] \xrightarrow{H_2} [OsH_2(CO)_3PPh_3]$$ [913]

$$Fe + H_2 + \text{diphosphine} \longrightarrow [FeH_2(\text{diphosphine})_2]$$ [306]

XXA. Hydrogen abstraction from water or an alcohol (usually employed as solvent in reaction of a metal halide with a tertiary phosphine).

XXB. Intramolecular hydride transfer from a ligand.

$$[IrCl_3(PEt_3)_3] \xrightarrow{EtOH/KOH} [IrHCl_2(PEt_3)_3] (+CH_3CHO)$$ [340]

$$[RuCl_2(Me_2PCH_2CH_2PMe_2)_2] \xrightarrow{Na^+C_{10}H_8^-} [Ru(C_{10}H_8)(Me_2PCH_2CH_2PMe_2)_2] \rightleftharpoons [RuH(\beta\text{-}C_{10}H_7)(Me_2PCH_2CH_2PMe_2)_2]$$ [280]

XXI. Protonation of certain phosphine carbonyl complexes by treatment with acid.

$$[Fe(CO)_4PPh_3] \xrightarrow{H_2SO_4} [FeH(CO)_4PPh_3]^+$$ [476]

Metathetical replacement reactions apply as with halide complexes (Route XIV) and are extended to formation of hydridohalide complexes by interaction of a di- or polyhydride with a halogen or hydrohalic acid. Ligand displacement (Route XI), by reaction of a phosphine with a hydrido complex containing an alternative ligand such as carbonyl, is also applicable.

A.6. Hydrocarbyl and Related Complexes

XXII. Treatment of a halide complex with an alkylating or arylating agent such as a Grignard compound or an alkyl or aryl lithium or aluminum derivative.

$$[PtCl_2(PR_3)_2] \xrightarrow[\text{or MeLi}]{\text{MeMgBr}} [PtMe_2(PR_3)_2]$$ [338,339]

$$[NiX_2(PR_3)_2] \xrightarrow[\text{liq. } NH_3]{PhC{\equiv}CNa} [Ni(C{\equiv}CPh)_2(PR_3)_2]$$ [337]

XXIII. Oxidative addition of, for example, an alkyl halide to a coordinatively unsaturated complex or precursor.

$$[RhCl(PPh_3)_3] \xrightarrow{\text{MeI}} [RhMeICl(PPh_3)_2]$$ [908]

$$[Pd(PPh_3)_4] \xrightarrow{\text{MeI}} [PdMeI(PPh_3)_2]$$ [572]

$$Na[Mn(CO)_3(PPh_3)_2] \xrightarrow{\text{MeI}} [MnMe(CO)_3(PPh_3)_2]$$ [733]

XXIV. Insertion reaction between a hydride complex and an olefin, in the case of tetracoordinate hydride complexes proceeding via a pentacoordinate π complex.[385]

$$[PtHCl(PEt_3)_2] \xrightarrow[\text{THF}/100°]{C_2H_4} [PtEtCl(PEt_3)_2]$$ [278]

XXV. Intramolecular cleavage from a ligand.

$$Pd(PPh_3)_4 + PdCl_2 \xrightarrow[\text{DMSO}]{130°} [PdCl_2(PPh_3)_2] + [PdPhCl(PPh_3)_2]$$ [456]

Methods applicable to simple complexes, for example, Routes X and XI, also apply in formation of hydrocarbyl complexes from existing alkyl, aryl, or acetylide derivatives, for example, a palladium-σ-azobenzene complex.[1225]

Points of note implicit from the list of typical hydrocarbyl complexes (Section D.9) are the increased stability of aryl and perfluoroalkyl over simple alkyl complexes (cf. Ni series) and the general increased stability on descending a subgroup.

XXVI. Olefin and acetylene (and oxygen) complexes are prepared by addition of the olefin or acetylene (or oxygen) to a coordinatively unsaturated complex often formed in situ by reaction or disproportionation.

$$[Ni(PPh_3)_2] \xrightarrow{C_2H_4} [Ni(C_2H_4)(PPh_3)_2]^{1348,1350}$$

$$[PtCl_2(PPh_3)_2] \xrightarrow[2.\ \text{stilbene}]{1.\ N_2H_4} [Pt(\text{stilbene})(PPh_3)_2]^{348}$$

$$[IrCl(CO)(PPh_3)_2] \xrightarrow[25°/3\ \text{atm}]{C_2F_4} [Ir(C_2F_4)Cl(CO)(PPh_3)_2]^{1088}$$

$$[M(PPh_3)_n] \xrightarrow{O_2} [M(O_2)(PPh_3)_2] \quad (M = Ni,\ Pd,\ Pt)^{1350}$$

$$[Pt(PPh_3)_4] \xrightarrow{O_2/CO_2} [Pt(OCO_3)(PPh_3)_2]^{693}$$

XXVII. Unsaturated hydrocarbon ligands may displace other complexed ligands such as CO and more especially, for example, styrene displaces ethylene[1348] and acetylenes displace olefins.[348]

$$[Pt(PhC{\equiv}CH)(PPh_3)_2] \longrightarrow [Pt(CF_3C{\equiv}CCF_3)(PPh_3)_2]^{200}$$

π-Allyl complexes can be prepared by a related displacement type of reaction.

$$[Ni(CO)_3PPh_3] \xrightarrow{CH_2{=}CHCH_2Cl} [(\pi\text{-}C_3H_5)NiClPPh_3]^{705}$$

A.7. Phosphidobridged Complexes

XXVIII. Reaction of primary phosphines with certain metal halides can produce phosphidobridged complexes (cf. Route X), but a general method for the production of dinuclear palladium and platinum phosphidobridged complexes involves treatment of a halogene-bridged complex with a secondary phosphine.

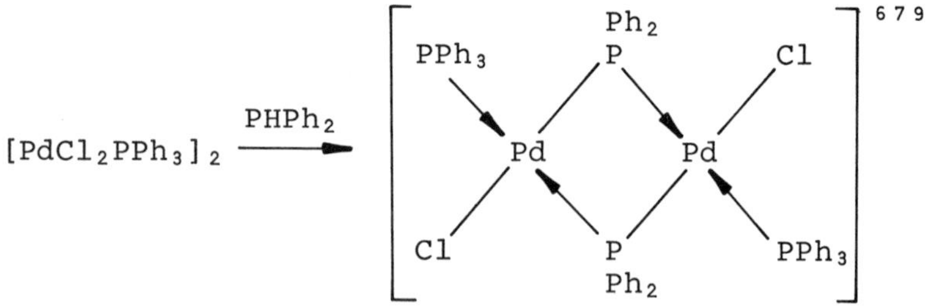

Only rarely do metal carbonyls form phosphidobridged complexes on interaction with secondary phosphines.

$$[V(CO)_6] \xrightarrow{PHPh_2} \left[(CO)_4V \underset{\underset{Ph_2}{P}}{\overset{\overset{Ph_2}{P}}{\diamond}} V(CO)_4 \right]^{754}$$

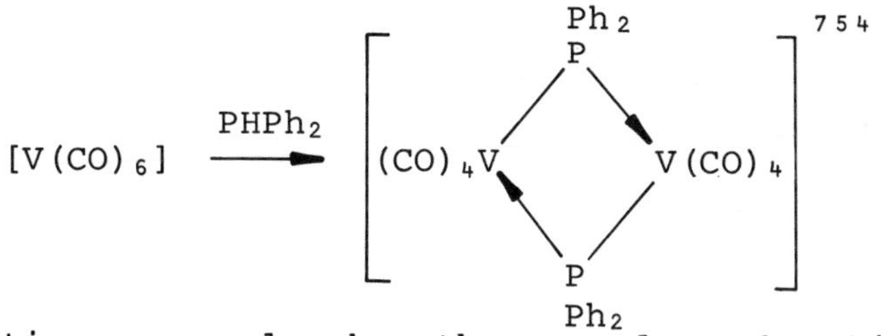

The reaction may apply when the secondary phosphine is already coordinated to a metal, in which case complexes with phosphidobridged metal-metal heterobonds can be produced,[156] for example, di-μ-chlorobis(μ-diphenylphosphido-tetracarbonyl iron palladium).[873]

XXIX. The general method for the preparation of phosphidobridged dinuclear metal carbonyl derivatives involves heating the metal carbonyl with a diphosphine (R_2PPR_2) at high temperature.

$$[Cr(CO)_6] \xrightarrow[260°/20\ hr]{Me_2P\cdot PMe_2} \left[(CO)_4Cr \underset{\underset{Me_2}{P}}{\overset{\overset{Me_2}{P}}{\diamond}} Cr(CO)_4 \right]^{350}$$

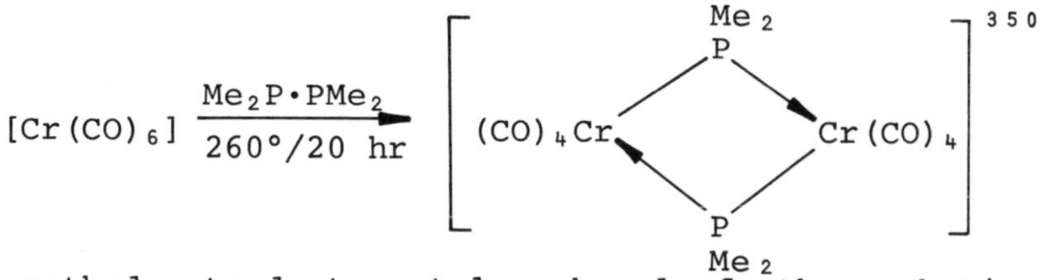

This method extends to metal carbonyls further substituted by, for example, nitrosyl[691] or π-cyclopentadienyl[681,683,690] groups.

XXX. A further method for the preparation of phosphidobridged carbonyl derivatives involves either reaction of R_2PX with an anionic carbonyl, or reaction of KPR_2 or $Me_3Si\cdot PPh_2^4$ with a carbonyl or nitrosyl halide derivative of a metal.

$$NaMn(CO)_5 + Ph_2PCl \longrightarrow \left[(CO)_4Mn \underset{\underset{Ph_2}{P}}{\overset{\overset{Ph_2}{P}}{\diamond}} Mn(CO)_4 \right]^{637}$$

$$[ReCl(CO)_5] \xrightarrow[THF]{KPPh_2} \left[(CO)_4Re \underset{\underset{Ph_2}{P}}{\overset{\overset{Ph_2}{P}}{\diamond}} Re(CO)_4 \right]^{745}$$

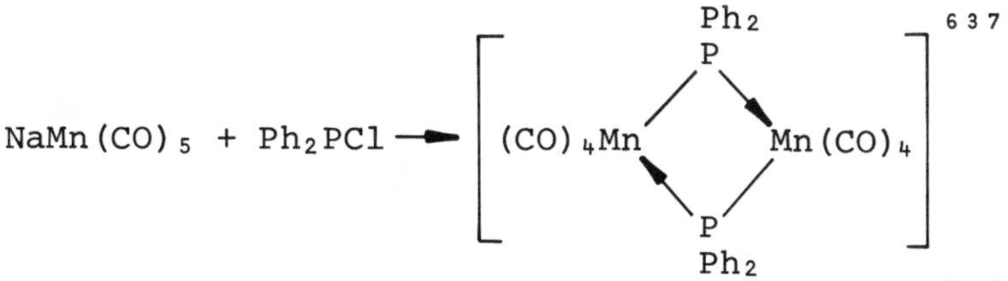

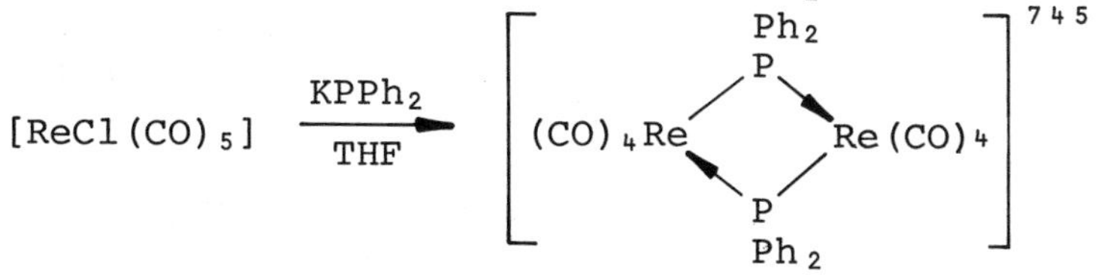

When R_2PX is reacted directly with the carbonyl, then

binuclear derivatives having phosphido and X bridges where X may be halogen[538,650,651] or sulfur[845] are formed (see also Chapter 2).

B. REACTIONS

B.1. General

Many of the general reactions of halide and carbonyl complexes have already been covered in Section A, and further references are given in Section D. Reduction and oxidation reactions, for example, may involve changes in oxidation state and the former, alternatively, formation of hydrido complexes. Ligand displacement/replacement reactions also cover a large number of different classes, and hydrogen and CO abstraction reactions have been stressed as important. Halogenation, alkylation, and related reactions are described in Section A.

Special reactions of halide complexes involve reagents such as p-toluenesulfinic acid (Pt),[332] aryl diazo compounds (Pt),[1086] thiourea (Pt),[648] ammonia (Pt),[885] amines (Ru, Os),[327 604] (Pt-bridged),[353] ethylenediamine (Pt),[649] β-diketones (Ru),[604] (Pt, Pd),[849] (Au),[603] thiols (Au),[394,864] nitriles (Ru),[604] carboxylic acids (Re),[1165] and isocyanides (Re),[590] (Pt).[71] A reaction apparently specific to nickel complexes is the formation of quaternary phosphonium salts,[192] originally formulated as addition compounds,[106,114,972,1368] by reaction with alkyl halides (see Section B.2 for catalytic properties), for example, $[NiBr_2(PPh_3)_2] + BuBr \rightarrow [NiBr_3PPh_3][PPh_3Bu]$. The corresponding ammonium salts, for example, $[NiBr_3PPh_3]$-$[NEt_4]$[443,446] and related salts, for example, $[NiCl_4][PMePh_3]_2$,[625] have also been studied.

The ease with which rhenium forms multiple bonds with nitrogen[211,212,508] is demonstrated by the very ready preparation of nitrido and arylimido complexes such as $[ReNX_2(PR_3)_{2\text{ or }3}]$.[293,297,314,436]

Decomposition reactions are sometimes important as methods of preparation, for example, synthesis of metal aryl derivatives by either elimination of SO_2 from S-sulfinates[417] or elimination of nitrogen from aryl azo derivatives.[1086]

Insertion reactions are typified by reaction of carbon monoxide with a phosphine-stabilized metal-alkyl complex to form a metal-acyl complex.

$$[PtXMe(PEt_3)_2] \xrightarrow{CO} [PtX(COMe)(PEt_3)_2]$$ [193]

Similar reactions have been reported with alkyl derivatives of molybdenum,[94] manganese,[90,240,890,891] iron,[166] cobalt,[699-701,704,712,1033] rhodium,[513,867] palladium and platinum.[193,1357] The reverse reaction may take place on heating or by decarbonylation using a CO acceptor such as $[RhCl(PPh_3)_3]$.[24] Analogous insertion reactions with hydrido complexes to yield formyl complexes have not been achieved,[77] but reaction of $[PtHCl(PEt_3)_2]$ with bis(trifluoromethyl) diazomethane results in formation of $[Pt[CH(CF_3)_2]Cl(PEt_3)_2]$, defined as an insertion reaction.[432] Diazoacetonitrile reacts similarly with $[PdCl_2PPh_3]_2$.[971] Olefin insertion reactions have already been referred to as a preparative method for alkyl derivatives. This is particularly important with cobalt hydride derivatives because of the relevance to the OXO synthesis. For example,

$$RCH{=}CH_2 + [HCo(CO)_4] + PPh_3 \longrightarrow [RCH_2CH_2Co(CO)_3PPh_3] \downarrow [RCH_2CH_2COCo(CO)_3PPh_3]$$ [703]

Epoxides behave similarly to olefins.[697]

The addition reactions already covered under the section on synthetic routes, namely, addition of olefins, acetylenes, CO, NO, N_2, or O_2 are particularly important because of the analogy with gas adsorption on metal surfaces. Further related addition reactions involving coordinatively unsaturated complexes such as $[IrCl(CO)(PPh_3)_2]$, or dissociative species such as $[RhCl(PPh_3)_2]$ or $[Pt(PPh_3)_2]$, are known with reactants such as carbon disulfide,[80,81,83,491,1362] carbon subsulfide (C_3S_2),[610] and sulfur dioxide.[235,428,914,1311] For example,

$$[IrCl(CO)(PPh_3)_2] \underset{>100°}{\overset{25°}{\rightleftharpoons}} [Ir(SO_2)Cl(CO)(PPh_3)_2]$$ [1311]

The precise nature of the bonding involved in all these complexes is not established in spite of many x-ray structural determinations, for example, oxygen,[989] SO_2,[907,1028] and fumaronitrile[1027] complexes. The possibility of the complexed molecule behaving as an acceptor must be taken into account since related adducts have been prepared by using BX_3[1128,1199] and trinitrobenzene.[79]

The category of oxidative-addition reactions is reserved here for the addition, to the same types of coordinatively unsaturated complexes as those described above, of a molecule A-B with cleavage of the A-B bond. The B atom is usually hydrogen or halogen, and the formal oxidation number of the complexed metal is simultaneously increased by two units, for example, Ir(I) to Ir(III) as in

$$[IrX(CO)(PR_3)_2] \xrightarrow{A-B} [IrXAB(CO)(PR_3)_2]$$ [698]

Alkyl halides are widely exemplified[82,429,490,698] A-B molecules as in the reactions (cf. Route XXIII).

$$[RhX(CO)(PPh_3)_2] \xrightarrow{MeI} [RhXIMe(CO)(PPh_3)_2]^{513}$$

$$[Pt(PPh_3)_2] \xrightarrow{MeI} [PtMeI(PPh_3)_2]^{1289}$$

Halogens,[420] hydrogen halides,[82,420,1310] and HCN[1230] are simple inorganic adding molecules, extended to $HgCl_2$,[15,418,529,909,977,1061] $SnCl_2/HCl$[502,1268,1300] and $ClSF_5$.[871] An extensive range of organic adding molecules includes allyl and acyl halides,[487,491,1205,1279,1323] chloroolefins,[410,573] chloroacetylenes,[410] picryl chloride,[1057] arylsulfonyl chlorides,[417,491] chloroformic esters,[489] thiols,[1230] imides,[1161] diketene,[888] carboxylic acids,[492] acetylacetone,[59] diazo compounds,[430,605,950] azides,[412,415] diethylazodicarboxylate,[635] polyfluoroacetone,[387] trifluoroacetonitrile,[185] trifluoroacetyl cyanide,[633] R_3SiH,[270,485,668,672] and BPh_2X.[1188]

Certain substituted acetylenes[413,1043] differentiate the oxidative addition reaction from the normal acetylene complex formation by reacting to produce hydrides, for example,

$$[IrCl(CO)(PPh_3)_2] \xrightarrow{HC\equiv CCOOEt} [IrHCl(C\equiv CCOOEt)(CO)(PPh_3)_2]^{413}$$

The reverse reaction (reductive subtraction?) can be carried out in certain cases, for example, dehydrochlorination.

$$[IrH_2Cl(CO)(PPh_3)_2] \xrightarrow{base} [IrH(CO)(PPh_3)_2]^{667}$$

Interaction of $[AuClPPh_3]$ with coordinatively unsaturated complexes is a form of oxidative addition,[202,510,909] leading to metal-metal bond formation.

$$[IrCl(CO)(PPh_3)_2] \xrightarrow{[AuClPPh_3]} [IrCl_2(AuPPh_3)(CO)(PPh_3)_2]^{510}$$

$[PtCl(SnPh_3)(PPh_3)_2]$ is similarly prepared from $[Pt(PPh_3)_n]$.[175]

Oxidative addition has thus provided one synthetic route for the preparation of stabilized complexes containing metal-metal bonds, a study of which is of fundamental importance in catalysis and related fields. Very often, only x-ray structural analysis, for example, with rhenium[442,445] or rhodium[268] complexes can determine the extent of metal-metal interaction.

Other methods of preparation of this class include simple displacement reactions such as

$$Na[Co(CO)_3PPh_3] \xrightarrow{HgX_2} Hg[Co(CO)_3PPh_3]_2$$ [714]

$$Na[Mn(CO)_5] \xrightarrow{AuClPPh_3} [Ph_3PAuMn(CO)_5]$$ [400,858]

$$Na[Mn(CO)_4PPh_3] \xrightarrow{Ph_3SnCl} [Ph_3SnMn(CO)_4PPh_3]$$ [625]

(x-ray[227])

providing also examples of substituted carbonyl derivatives of vanadium,[859] tungsten,[1346] iron,[367] rhodium, and iridium.[421]

The reaction

$$[AuClPPh_3] \xrightarrow{Ph_3GeLi} [Ph_3PAuGePPh_3]$$ [615,618]

typifies the reactions of certain phosphine metal halide complexes with other compounds of Group IV elements such as $LiSiMePh_2$,[291] Ph_3SiLi,[78] SiH_3X,[158] R_3SiH,[292,785] Hg-$(SiMe_3)_2$,[616] Me_3SnNMe_2,[249] and $LiPbPh_3$[493] to give complexes containing Pt-Si,[78,158,291,292,616,617] Pd-Ge,[217] Pt-Ge,[461,617,1354] Pt-Sn,[78,249] and Pt-Pb[493,494] bonds with elimination of MX, HX, H_2, or $HNMe_2$. Tin compounds can also form bridges in, for example, iron complexes,[771,881] and stannous halides interact with cobalt carbonyl derivatives by insertion into existing metal-metal bonds,[96,188,1091]

$$[Co(CO)_3PBu_3]_2 \xrightarrow[\text{ether/CO/100 atm}]{SnX_2} [\{Bu_3PCo(CO)_3\}_2SnX_2]$$

and with halides of certain metals in another form of insertion reaction. For example,

$$[(\pi\text{-}C_3H_5)PdClPPh_3] \xrightarrow{SnCl_2} [(\pi\text{-}C_3H_5)Pd(SnCl_3)PPh_3]$$ [969]

Related complexes, for example, $[(\pi\text{-}C_5H_5)NiPPh_3]SnCl_3$,[1300] may, however, be ionic.

Intramolecular hydrogen transfer from a phosphine ligand has already been mentioned (Route XXB); other related hydrogen abstraction reactions are known involving rhodium[150,866] and platinum[207] complexes. A frequently occurring, and usually undesirable, reaction of complexed phosphine ligands is oxidation to the oxide. Occasionally, for example, with cobalt and nickel complexes, the derived phosphine oxide complex can be isolated.

$$[NiX_2(PR_3)_2] \text{ or } \xrightarrow[H_2O_2^{468}]{NO^{194}} [NiX_2(OPR_3)_2]$$

Reactions of other complexed ligands in phosphine complexes are most important in catalytic processes. Specific examples involving isolated species in addition to those mentioned in Section A include protonation of complexed (Pt) olefins and acetylenes[91] and decarbonylation of acetylenic ketones (Rh).[1030]

B.2. Catalysis

The many articles and patents describing the use of phosphine complexes as catalysts and as models for studies on the mechanisms of catalytic reactions are summarized below. The major areas exemplified are oligomerization, polymerization, hydrogenation, and carbonylation, the last being of industrial importance. Group VIII metals, especially nickel, predominate.

B.2.1. Oligomerization and Polymerization

B.2.1.1. Monoalkenes

(a) Dimerization and codimerization

$Ni(II)/EtAlCl_2/PCl_2Ph$.[790]
$[NiX_2(PR_3)_2]/EtAlCl_2$.[792,1042]

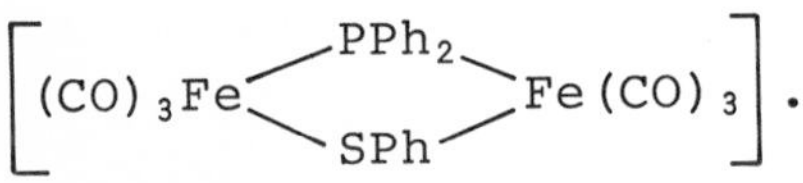

.[793]

$[NiI(NO)(PPh_3)]$.[1103]

(b) Polymerization

$[Ni(CO)_2(PPh_3)_2]$.[98,794]
$[TiMeCl_3PPh_3]$.[128]
$[NiCl_2(PR_3)_2]/BF_3$.[1294]
$[Ni(CO)_x(PR_3)_y]$/benzoquinone.[108]
$[Ni(CO)_2(PR_3)_2]/Cl_2C{=}CHCl$.[797]

B.2.1.2. Dienes

(a) Formation of cycloolefins, for example,

1,3-butadiene ⟶ 1,5-cyclooctadiene and/or 1,5,9-cyclododecatriene.

$Ni(II)/Al/PPh_3$.[104,242]
$[NiPPh_3\{P(OPh)_3\}]$.[378]
$[Ni(CO)(PR_2)_3]$.[426]

$[Ni(CH_2=CHCN)_2(PR_3)_2]$.[1040]
$[Fe(NO)_2(PPh_3)_2]$.[1142]
$[Ni(diene)(PPh_3)_2]$.[1349]

(b) Trimerization, for example,

$$\text{Butadiene} \xrightarrow{[Pd(PPh_3)_4]} \text{1,3,7-octatriene linear dimer[1260]}$$

(c) Polymerization
$[Ni(CO)_x(PR_3)_y]/BF_3$.[99]
$[Pd(PPh_3)_n]$.[102]
$RuCl_3/PBu_3$.[1216]

(d) Copolymerization, for example,

$$\text{Butadiene + ethylene} \xrightarrow{Co(II)/diphosphine/AlEt_2Cl} \text{hexa-1,4-}$$

diene[522,820]

$[Ni(CN)_2(PR_3)_2]$.[1192]

B.2.1.3. Acetylenes

Acetylenes can be polymerized to give (a) aromatic trimers, for example,

$$RC{\equiv}CR \xrightarrow{\text{catalyst}}$$

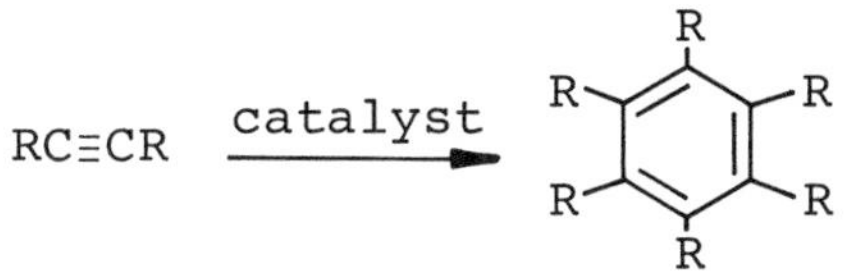

(b) linear oligomers or polymers, for example, $PhC{\equiv}CH \rightarrow PhC{\equiv}C-CH=CH-Ph$, or (c) unresolvable mixtures. Acetylene itself can give either (c) or a mixture of styrene and benzene. The most commonly used catalysts are nickel carbonyl derivatives which have been separately reviewed.[994,997]

(a) Cyclic trimerization
$[Ni(CO)_3PPh_3]$. $ROCH_2C{\equiv}CCH_2OR$.[599]
$[Ni(CO)_2(PPh_3)_2]$. C_2H_2,[1366] $PhC{\equiv}CH$,[1156] $Me_2C(OH){\equiv}CMe$.[364]
$[Ni(CO)_2\{P(C_2H_4CN)_3\}_2]$. $RC{\equiv}CH$.[39]
$[Ni(CO)_2(diphosphine)]$. $RC{\equiv}CH$.[39,547,787]
$[Ni(CO)_3]_2Ph_2P{\cdot}PPh_2$. C_2H_2.[112]
$[Ni(CN)_2]_3PPh_3$[1015] and $[Ni(CN)_4(PPh_3)_2]$.[97] C_2H_2.
$[NiX_2(PR_3)_2]/NaBH_4$. $RC{\equiv}CH$.[511,639,930,1295]
$[Ni(CH_2=CHCN)_2PPh_3]$. $RC{\equiv}CH$.[1191]
$[Mo(CO)_5PPh_3]$.[37]
$[Fe(CO)_x(PPh_3)_y]$.[1144]
$[RhPh(diene)PPh_3]$.[1263]

$[IrCl(C_5H_5)(PPh_3)_2]$.[414]
$[Pd(PPh_3)_n]$.[102]
$[CuI(PEt_3)_2]$.[1011]

(b) Linear dimerization
$[RhCl(PPh_3)_3]$.[872,1229]

B.2.1.4. Polymerization of Other Compounds

Propylene $\xrightarrow{[Ni(CO)_2(PPh_3)_2]/AlCl_3}$ dimer.[674]

Acrylonitrile $\xrightarrow{[Ni(CO)_x(PR_3)_y]}$ dimer.[1041]

Allene $\xrightarrow{[Ni(CO)_2(PPh_3)_2]}$ cyclic trimers + tetramer.[157]

Ketene $\xrightarrow{[PdCl_2(PR_3)_2]}$ product of polyester structure.[1210]

Allylic alcohol $\xrightarrow{[NiCl_2(PPh_3)_2]/PrMgBr}$ mixed olefins.[371,560]

Propylene oxide $\xrightarrow{[NiX_2(PR_3)_2]}$ polymer.[1296]

Methyl methacrylate $\xrightarrow{[Re_2(CO)_{10}]/PPh_3/CCl_4}$ polymer.[88]

Vinyl compounds polymerized on fibers to improve their dyeability: $[Ni_2(CO)_6(Ph_2PPPh_2)]/CCl_4$.[792]

B.2.2. Hydrogenation

B.2.2.1. Olefins

$$>C=C< \xrightarrow{H_2/catalyst} >CH-CH<$$

$[RuHCl(PPh_3)_3]$. Alk-1-enes.[663,664]
$[RuCl_2(PPh_3)_2]$. Selective hydrogenation, for example, hexene-1 preferentially reduced to cyclohexene.[784]
$[RuH(OCOR)(PPh_3)_2]$. Alk-1-enes.[1155]
$[CoH(CO)(PR_3)_3]$. Alk-1-enes.[1021,1023]
$[Co(CO)_2PR_3]_3$. Olefin hydrogenation and isomerization.[1129]
$Co(II)/PR_3$. General hydrogenation catalyst mixtures.[1209,1214]
$[RhCl(PPh_3)_3]$. n-1-Dodecene,[167] cyclopropylalkenes[694] unsaturated aldehydes.[825]
$[RhX(PR_3)_3]$. Various alkenes;[173,783,823,1024,1072,1228,1373] dicyanobutene $\rightarrow$ adiponitrile.[523]
$[RhCl_3(PR_3)_3]$. Hex-1-ene and other alkenes.[472,666]
$[RhX(CO)(PR_3)_2]$. Hept-1-ene.[1258]
$[RhH(CO)(PPh_3)_3]$. Selective hydrogenation of terminal olefins.[1064,1065]
$[Rh(diene)(PPh_3)_2]^+$. General hydrogenation.[1204]
$[IrH_3(PPh_3)_2]$. Ethylene and hex-1-ene.[612]
$[IrH_3(PPh_3)_3]$. Oct-1-ene $\rightarrow$ n-octane.[788]
$[PtHCl(PEt_3)_2]/HCl/EtOH$. Olefins.[613]

$[PtCl_2(PPh_3)_2]/SnCl_2$. Polyolefins.[1266]
$[MX_2(PR_3)_2]/SnCl_2$ (M = Ni, Pd, or Pt). Methyl linoleate mixture of partially hydrogenated esters.[73]

B.2.2.2. Acetylenes

$$\text{-C}\equiv\text{C-} \xrightarrow{H_2/\text{catalyst}} \text{-CH}_2\text{-CH}_2\text{-}$$

$RhCl(PPh_3)_3$. Hex-1-yne $\rightarrow$ n-hexane.[824,1072]
$[MX(CO)(PPh_3)_2]$ (M = Rh or Ir). Acetylene (and ethylene).[1317]

B.2.3. Olefin Isomerization and Disproportionation

Conditions may be adjusted so that isomerization (e.g., hex-1-ene $\rightarrow$ cis-hex-2-ene + trans-hex-2-ene) is the principal reaction when olefins are in contact with selected metal complex catalysts. It is therefore always a possible side reaction in other olefin reactions, such as catalytic hydrogenation or hydroformylation.
$[RhCl(PPh_3)_3]$.[69,174]
$[IrX_3(PR_3)_3]$.[401]
$[NiX_2(PR_3)_2]$.[1351]
$[PtCl_2(PPh_3)_2]/SnCl_2$.[1265]
Cyclohexa-1,4-diene is converted to a mixture of benzene and cyclohexene under mild conditions with $[IrX(CO)(PPh_3)_2]$/hydride as catalyst.[931]

B.2.4. Hydroformylation of Olefins (OXO Reaction)

$$RCH{=}CH_2 \xrightarrow{CO/H_2/\text{catalyst}} RCH_2CH_2CHO \left[+ \text{ some } \underset{\displaystyle RCHCH_3}{\overset{CHO}{|}}\right]$$

$$RCH_2CH_2CHO \xrightarrow{H_2} RCH_2CH_2CH_2OH$$

Catalysts, especially the most commonly used cobalt carbonyl types, can be modified by the use of phosphine ligands so that the product contains an increased proportion of the more commercially desirable linear, rather than branched, product. It has been established with certain catalysts that this is due to stereoselectivity rather than the alternative double-bond isomerization.
$[Ru(CO)_3(PPh_3)_2]$.[550]
$[Co_2(CO)_6(PBu_3)_2]$.[544-546]
$[Co_2(CO)_8]/PBu_3$.[707,1215,1218]
Co compound/PR_3.[886,1212]
$[CoH(CO)_3PR_3]$.[1280]
$[Co(\pi\text{-crotyl})(CO)_2PBu_3]$.[998]
$[Rh(CO)_x(PR_3)_y]$.[1217]

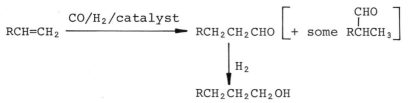

$[RhCl_3(PPh_3)_3]$.[1073]
$[RhCl(CO)(PEt_3)_3]$.[781]
$[RhX(CO)(PPh_3)_2]$.[214,551]
$[RhH(CO)(PPh_3)_3]$.[219,1360]
$[Ir(CO)_x(PR_3)_y]$.[160] Ir compound/CO/PR_3.[1219]

Dienes form dialdehydes with $[Rh(CO)_x(PR_3)_y]$ as catalyst.[561,1296]

B.2.5. Carboxylation

$$RCH{=}CH_2 \xrightarrow[\text{catalyst}]{CO/H_2O/ROH} RCH_2CH_2COOR$$

with $[PdO(PPh_3)_2]$[100] or $[PdX_2(PR_3)_2]$.[101]

$$CH_2{=}CH{-}CH{=}CH_2 \xrightarrow[[PdX_2(PBu_3)_2]]{CO/MeOH} CH_3CH{=}CHCH_2COOMe^{208}$$

$$CH{\equiv}CH \xrightarrow{CO/\text{catalyst}} CH_2{=}CHCOOH$$

$$CH{\equiv}CH \xrightarrow{CO/ROH/\text{catalyst}} CH_2{=}CHCOOR$$

with $[Ni(CO)_x(PPh_3)_y]$,[1365] $[Ni(CN)_2]_3PPh_3$,[1013] or $[NiBr_2(PPh_3)_2]$/BuBr.[105-107,780,795,1012,1014,1016,1365,1367]

$$MeOH \xrightarrow[300°/250\ atm]{CO/[NiI_2(PPh_3)_2]/I_2} MeCOOMe.^{1018}$$

B.2.6. Carbonylation

Cyclic dienes[209] and α,ω-dienes[210] give mixed cyclic carbonyl products on reaction with $CO/[PdI_2(PR_3)_2]$.

$$MeOH \xrightarrow{CO/[RhCl(CO)(PPh_3)_2]} MeCOOH.^{1092}$$

$$PhNO_2 \xrightarrow{CO/[IrCl(CO)(PPh_3)_2]} [PhNCO] \xrightarrow{MeOH} PhNHCOOMe.^{791}$$

B.2.7. Decarbonylation

$$RCH_2CHO \xrightarrow[\text{or } [RhCl(PPh_3)_3]^{84,1067,1278}]{[Ru_2Cl_3(PEt_2Ph)_6]^{+\ 1130}} RCH_3\ (+\ \text{some } RCH{=}CH_2).$$

$$RCOCl \xrightarrow[-CO]{RhCl(PPH_3)_3} RCl\ (\text{if no } \beta\text{-H atom}).^{1067}$$

$$RCH_2CH_2COCl \xrightarrow[-CO/-HCl]{} RCH{=}CH_2.$$ [1067]

B.2.8. Decomposition Reactions

$$HCOOH \xrightarrow{\text{catalyst}} H_2 + CO_2.$$

$[IrH_2Cl(PPh_3)_3]$ and related complexes.[402,789]
Hydroperoxides: $[RhCl(PPh_3)_3]$.[168]

B.2.9. Condensation of Primary Alcohols (Guerbet Reaction)

$$RCH_2OH + R^1CH_2CH_2OH \longrightarrow \begin{array}{l} R\text{-}CH_2 \\ \quad | \\ R^1CHCH_2OH \end{array}$$

For example, BuOH $\xrightarrow{RhCl_3/PBu_3}$ 2-ethylhexanol.[1022]

B.2.10. Alkylation

Secondary amine $\xrightarrow[[Co_2(CO)_8]/PR_3]{\text{olefin}/H_2}$ tertiary amine.[1213]

B.2.11. Oxygenation

Cyclohexylisocyanide $\xrightarrow[O_2]{[Pt(PPh_3)_4]}$ cyclohexylisocyanate.[1261]

B.2.12. Additives and Auxiliary Products

Antiknocks: $[Mn(CO)_4PPh_3]$,[103] $[Cu(C_5H_5)PR_3]$.[1211]
Antioxidant: $[Fe(CO)_3(PPh_3)_2]$.[1020]
Flame retardant: $[CuX(R_2PC_2H_4PR_2)]$.[38]

B.3. Reaction Mechanisms and Kinetic Studies

Many reactions of phosphine-stabilized complexes have been studied in detail to obtain fundamental kinetic and mechanistic data.

In many-substitution reactions of metal carbonyls with phosphine ligands, the rate-determining step, usually CO dissociation, has been determined, for example, complexes of Group VI metals,[47,48,1332,1333,1335,1377,1379] manganese,[46,627,657,763,975,1055] rhenium,[658,1380] iron,[51,404,1244] ruthenium,[247] cobalt,[696,1272] and rhodium.[628,1197] The kinetics of CO exchange reactions, using ^{14}CO, are relevant to this,[116,204,206] and in certain examples CO insertion is involved.[237,975] Other exchange reactions studied involve hydrides and deuterides,[1089,1308] ^{15}NO,[796] and phosphine ligands.[363,630,765,1112,1125]

Nucleophilic substitution reactions of halide complexes are usually studied with relatively stable platinum complexes as models. Typically, reactions of trans-[$PtCl_2$-$(PEt_3)_2$] with halide and pseudohalide ions[137-139,142,557,1269] or thiourea[267] show the importance of ligand exchange and solvent interaction. Related substitution reactions involve amine complexes,[1066] bridged,[1095] and five-coordinate[1096] complexes. Similar studies with alkyl or aryl platinum complexes, for example, trans-[$PtPh_2(PEt_3)_2$], involve nucleophiles[141,143,558] or electrophilic agents[62,140] acting through a nucleophilic addition and rearrangement process. Nickel, palladium, and platinum have been compared in reactions of [$MXR(PEt_3)_2$] with pyridine.[117]

Other examples of work in this area include stabilities of [$Ni(CN)_2(PR_3)_2$][1149] and [Pt(acetylene)$(PR_3)_2$],[28] dissociation of [$Pt(PPh_3)_3$],[175,176] hydrolysis of [$Pt(NO_2)_2$-$(PEt_3)_2$],[1248] and addition reactions of [IrX(CO)$(PPh_3)_2$].[368] Studies on the isomerization of cis- and trans-[$PtCl_2$-$(PBu_3)_2$][656] and [Pt(π-Cl_2C=CCl_2)$(PPh_3)_2$] to [Pt(σ-CCl=CCl_2)-Cl$(PPh_3)_2$][180-182] again highlight the general importance of solvent effects.

C. PHYSICAL PROPERTIES

Inspection of the other sections of this chapter shows the large number of measured physical parameters appropriate to phosphine complexes. Magnetic properties, dipole moments, and spectral measurements are interpreted to study factors such as stereochemistry and electronic state of the metal, the nature of the bonding in the complex, and ligand field effects. It has already been mentioned that a large number of phosphine complexes have also been considered worthy of x-ray structural analysis. The fundamental property of phosphine complexes that promotes their wide use as model compounds is their stable, nonionic, and crystalline nature leading frequently to solubility in organic solvents. The analogy with wholly organic compounds often extends to a sharp melting point improved by recrystallization.

This all-important stability is influenced by many factors; for example, the metal and its oxidation state, the nature of the complexed ligands, and the chemical and physical environment. All metals form complexes with a tendency for increased stability on descending a periodic series and on crossing to the right of the transition metals; platinum and gold thus form the most stable complexes. Instability is due to dissociation or to oxidation of either metal or ligand, in the latter case the metal itself may be the oxidizing agent. The use of bidentate phosphine ligands is one way to overcome instability, for example, with complexes of metals in low oxidation states. One comparison[675] of simple phosphine ligands suggests the order $R_2PH > R_3P > RPH_2$ for stabilizing influence, but generalizations such as this should be viewed with care. Group VIII metals usually give more stable halide complexes with PPh_3 than with PEt_3 but this is almost certainly due to a dissociative effect with stability related to phosphine reactivity.

Examples of the applications of spectral measurements to phosphine complexes are summarized here to demonstrate the wide range of problems covered.

Space does not permit a detailed consideration of the variation in the measured parameter with constitution or with the interpretation of results.

(1) UV and Visible

Ligand (crystal) field effects:
trans-[$PtCl_2$(piperidine)PR_3].[295]
[$MX_2(PPh_3)_2$] (M = Co or Ni).[624]
trans-[RuXY($Me_2PCH_2CH_2PMe_2$]$_2$.[309]

Low-temperature polarized:
[$CoCl_2(PPh_3)_2$].[1226]

Charge transfer:
[$CoX_2(PEt_3)_2$].[842]

(2) IR

Carbonyl complexes (γ_{CO})[9,49,170,1231] and comparison with nitrogen complexes.[331]
Hydrides (γ_{M-H}).[7,287,1309]
Alkyl and acyl complexes (platinum).[8,10,14]
Thiocyanato[283,1097] and nitro[284] complexes.
Amine complexes (γ_{N-H}).[288,289,290,521]

(3) Far-IR

Metal-halogen vibrations:
[$PtX_2(PR_3)_2$].[13,1084]
[$PtX_4(PR_3)_2$].[11]
[$PdX_2(PR_3)]_2$ and [$PtX_2(PR_3)]_2$.[12,622]
[$RhX(PPh_3)_3$], [$RhX_3(PR_3)_3$], and [RhX(CO)$(PR_3)_2$].[145]
[$IrX_3(PR_3)_3$].[832]
[$MX_4(PR_3)_2$] (M = Re, Os, or Ir).[321]
M-X and M-P vibrations in [$MX_2(PPh_3)_2$] (M = Zn, Cd, or Hg).[480]

(4) Raman

[$PtX_2(PR_3)_2$] and [$PtX_4(PR_3)_2$].[322,520]
[$PtHX(PEt_3)_2$].[120]
Complexes of $Ph_2PC{\equiv}CPPh_2$.[258]

(5) ESR

[$CoAr_2(PR_3)_2$].[973]

$[CoR_2diph]$.[706]
$[RuX_3(PR_3)_3]$ and $[OsX_3(PR_3)_3]$.[771]

(6) Mössbauer
$[FeX_2(tetraphosphine)]$.[575]

(7) Mass Spectra
Polynuclear carbonyl complexes.[917]
Phosphorus bridged complexes.[850]
$[Mn(C_5H_5)(CO)_2PPh_3]$.[1031]

(8) NMR
This is a rapidly expanding field of study with particular application to hydrides and their identification, and studies of stereochemistry, isomerization, and metal-ligand bonding. The literature up to 1966 has been reviewed,[968] and special mention should be made of the use[1321] of NMR data to support an equivocal argument against conventional theories of dative π bonding from phosphorus to metal (a subject that justifies consideration separate from this chapter).

Hydrides, for example, $[PtHX(PR_3)_2]$[229,1122] have been studied through ^1H spectroscopy,[556,631,640,857] ^{31}P spectroscopy,[483,1238] and studies of ^{13}C-^1H coupling constants,[1343] and ^{195}Pt-^1H coupling constants.[68,483,987]

Halide, carbonyl, and related complexes have also responded to ^1H NMR[897-900,1106,1111] and ^{31}P NMR[478,644,646,695,868,1135] spectroscopic studies, with complexes of PMe_2Ph being especially valuable since the ^1H NMR pattern of the Me usually indicates the stereochemistry of the complex, for example, Group VIII halide complexes.[325,831,833,834,1025,1207] Other work has specifically involved ^{195}Pt,[1109] ^{103}Rh,[220] and ^{19}F (in fluorophenyl[764,1087] and fluorophosphine[93,762] derivatives) and ^{195}Pt-^{31}P coupling.[33,987,1107,1108]

Other organic ligands such as dienes[1324] and π-allyl types[122,1241] have also been studied in their complexed state.

D. LIST OF COMPOUNDS

All carbonyl and halide complexes from primary, second ary, and tertiary phosphines PH_2R, PHR_1R_2, and $PR_1R_2R_3$, where R (R_1 and R_2) is Me, Et, Pr, Bu, All, Cy, Ph, tolyl, are included in Sections D.1 through D.3. Related complexes from less common phosphine ligands are not individually listed but the appropriate reference is given in Sections D.3.5 and D.4. Complexes containing no ligands other than phosphines are listed in Sections D.3.1 and D.5.1.

With multidentate ligands, 1,2-bis(diphenylphosphino)-

ethane (diph) has been taken as representative, and derived carbonyl and halide complexes are fully shown in Sections D.5.2 and D.5.4. Again, related complexes from less common multidentate phosphines and from unsymmetrical phosphine ligands are not individually listed but the appropriate references are given by listing the ligands in Sections D.5.3, 5.5, 5.6, 6.1, and 6.2.

With carbonyl and halide complexes thus summarized as the important primary complexes, derivatives such as carbonyl halide, nitrosyl, hydrido, hydrocarbyl, nitrogen, oxygen, and phosphidobridged complexes are listed, in the remaining sections, by references to types rather than individual complexes.

D.1. Primary Phosphine Complexes

$[TiCl_4PH_2ME]$. X. Orange, IR.[119]
$[ZnCl_2PH_2Et]$. X. Di- or polymeric.[675]
$[CoCl_2(PH_2Et)_2]$. X. Mag. moment 4.3 μB.[675]
$[CrCl_3(PH_2Cy)_2]_2$. XI. (THF). m. 188°, mag. moment 3.78 μB.[811]
$[FeCl_2(PH_2Cy)_2]$. X. Red, m. 135°.[811]
$[FeBr_3(PH_2Cy)_3]$. X. Red, m. 126°, mag. moment 3.48 μB.[811]
$[NiBr_2(PH_2Cy)_2]$. X. (C_6H_6). Red, m. 134°, diamag.[811]
$[CuBr(PH_2Cy)]_2$. X. (C_6H_6). Colorless, m. 28°, diamag.[811]
$[CrCl_3(PH_2Ph)_3]$. X. (Toluene). Blue, m. 123°.[819]
$[Fe(PH_2Ph)_4]Cl_2$. X. (EtOH). Orange.[819]
$[Fe(PH_2Ph)_4]Br_2$. X. (EtOH). Red, dec. 155°.[819]
$[CoCl_2(PH_2Ph)_2]$. X. (EtOH). Black.[819]
$[CoBr_2(PH_2Ph)_2]$. X. (EtOH). Black, m. 153 to 157°, mag. moment 2.5 μB.[819]
$[Co(NO_3)_2(PH_2Ph)_2]$. X. (Toluene). Brown.[819]
$[CoI_2(PH_2Ph)_4]$. X. (EtOH). Yellow, mag. moment, dipole.[819]
$[CoI(PHPh)(PH_2Ph)_3]$. (From above). Phosphido deriv., black, dec. 135 to 140°.[819]
$[NiBr_2(PH_2Ph)_4]$. X. Black, dec. 64°.[819]
$[CuCl(PH_2Ph)]_4$. X. (C_6H_6). White, m. 75 to 76°.[819]
$[V(CO)_4(PH_2Ph)_2]$. V.[754]

D.2. Secondary Phosphine Complexes

D.2.1. Monodentate

$[TiCl_4PHMe_2]$. X. Orange-red, IR.[119]
$[CrCl_2(PHMe_2)_n]$. X. (0°).[163]
$[PdCl_2(PHMe_2)_2]$. XI. Colorless, dec. 230°.[685]
$[ZnX_2(PHEt_2)_2]$. X. Colorless oil.[675]
$[CrCl_3(PHEt_2)_3]$. X. Violet, dec. 69 to 71°, mag. moment 3.84 μB.[804]

$[FeCl_2(PHEt_2)_2]$. X. Red, dec. 75 to 80°, mag. moment 3.61 μB.[804]
$[RuCl_2(PHEt_2)_4]$. X. (EtOH). Yellow, m. 175 to 190°.[685]
$[CoCl_2(PHEt_2)_2]$. X. Black-green.[675]
$[CoCl_2(PHEt_2)_4]$. X. Green, unstable in air,[675] dec. 78 to 82°.[804]
$[CoBr_2(PHEt_2)_4]$. X. Green, unstable in air,[675] dec. 108 to 112°, mag. moment, dipole.[804]
$[RhCl_3(PHEt_2)_3]$. X. (EtOH). Orange, dec. 212 to 215°.[685]
$[NiCl_2(PHEt_2)_4]$. X. Red.[804]
$[NiBr_2(PHEt_2)_4]$. X. Red, dec. 110 to 115°, paramag., μ 7.88 D.[804]
$[PdCl_2(PHEt_2)_2]$. X. (EtOH). Colorless, dec. 170°.[688]
$[PdBr_2(PHEt_2)_2]$. X. Pale yellow, dec. 175°.[688]
$[PdI_2(PHEt_2)_2]$. X. Yellow, dec. 147 to 149°.[688]
$[FeCl_2(PHCy_2)_2]$. X. Colorless, dec. 95 to 98°, mag. moment 5.12 μB.[804]
$[Fe(PCy_2)_2PHCy_2]$. X.[818]
$[CoBr_2(PHCy_2)_2]$. X. Blue, dec. 112 to 114°, mag. moment, dipole.[804]
$[NiBr_2(PHCy_2)_2]$. X. Red, m. 195 to 198°.[815]
$[Cu(PCy_2)PHCy_2]$. Yellow, m. 158°.[818]
$[RuCl_2(PHPh_2)_4]$. X. Yellow-orange, dec. 245 to 255°.[685]
$[CoBr_2(PHPh_2)_3]$. X. (MeOH). Brown, dec. 163 to 165°, mag. moment 2.01 μB;[817] x-ray.[164]
$[CoBr(PHPh_2)_3]Br$. X. Green, dec. 141 to 143°.[817]
$[CoBr(PHPh_2)_4]Br$. X. Yellow, dec. 135 to 136°.[817]
$[RhCl(PHPh_2)_3]$. XI. Yellow-brown, m. 173 to 175°.[685]
$[RhCl_3(PHPh_2)_3]$. X. Cis and trans.[685]
$[Ni(PPh_2)_2(PHPh_2)_2]$. X. Yellow, dec. 172 to 173°.[817]
$[NiI_2(PHPh_2)_2]$. X. (CH_2Cl_2). Brown, m. 170°.[682]
$[NiCl_2(PHPh_2)_3]$. X. (C_6H_6). Brown, m. 98°, diamag.[682]
$[NiBr_2(PHPh_2)_3]$. X. (C_6H_6). Brown, m. 95°, diamag.[682]
$[NiI_2(PHPh_2)_3]$. X. (C_6H_6). Blue, m. 115°.[682] Mag. moment 0.6 μB.[500]
$[PdX_2(PHPh_2)_2]$. X.[679]
$[PdBr(PHPh_2)_3]Br$. X. Bronze, dec. 140 to 160°.[679]
$[Pd(PPh_2)_2(PHPh_2)_2]$. X. Pink, dec. 110 to 120°.[679]
$[PtCl_2(PPr_3)(PHPh_2)]$. XI.[281]
$[CuBr(PHPh_2)]_4$. X. Colorless, dec. 160 to 162°.[817]
$[CuX(PHPh_2)]$. X. (EtOH). White, IR, UV, NMR.[3]
$[CuBr(PHPh_2)_2]$. X. (EtOH). White, IR, UV, NMR.[3]
$[CuI(PHPh_2)_2]$. X. (EtOH). White, IR, UV, NMR.[3]
$[CuX(PHPh_2)_3]$. X. (EtOH). White, IR, UV, NMR.[3]
$[Cu_2Cl_3(PHPh_2)_3]$. X. (EtOH). White, IR, UV, NMR.[3]
$[Cu(PPh_2)(PHPh_2)]$.[806]
$[V(CO)_4(PHPh_2)_2]$. V.[754]
$[Cr(CO)_5PHPh_2]$. V. (140°). Yellow, m. 55 to 65°, IR, NMR.[1236]
$[Mo(CO)_5PHPh_2]$. V. (160°). m. 73 to 75°, IR, NMR, μ 4.88 D.[1236]

$[Mo(CO)_4(PHPh_2)_2]$. VI. m. 95 to 98°, IR, NMR, μ 6.22 D.[1236]
$[Mo(CO)_3(PHPh_2)_3]$. VI. Purple, m. 140 to 150°.[1236]
$[W(CO)_5PHPh_2]$. V. (140°). m. 90 to 92°, IR, NMR.[1236]
$[Fe(CO)_4PHPh_2]$. V. (170°). Yellow, m. 73 to 76°, NMR,[1236] x-ray.[875]
$[Fe(CO)_3(PHPh_2)_2]$. V. Yellow, dec. 160°.[1236]
$[M(CO)_5PH(C_6F_5)_2]$. V. (M = Cr, Mo, W). IR.[636]
$[Mo(CO)_4\{PH(C_6F_5)_2\}_2]$. VI. White, m. 131 to 132°, IR.[636]
$[Mo(CO)_3\{PH(C_6F_5)_2\}_3]$. VI. Buff, dec. 150°, IR.[636]

D.2.2. Bidentate

$[CoX_2(RHPCH_2CH_2PHR)]$. X. (R = Et or Ph).[816]
$[NiX(RHPCH_2CH_2PHR)]X$. X. (R = Et or Ph).[816]

D.3. Monodentate Tertiary Phosphine Complexes

D.3.1. With No Other Ligands

$[Ni(PEt_3)_4]$. III.[1347]
$[Ni(PF_2Me)_4]$. III. IR.[1200]
$[Ni(PCl_2Me)_4]$. I. (Reflux). Yellow, dec. 170°.[1133]
$[Ni(PBr_2Me)_4]$. I. (-70°). Yellow-orange, dec. 110°.[934]
$[Ni\{(CF_3)_2PF\}_4]$. II and III. m. 57 to 58°, b. 218°,[232] NMR.[1054]
$[Ni\{(CF_3)PF_2\}_4]$. II and III. Colorless liq., m. -84°, b. 160°;[232] IR, NMR.[1054]
$[Ni(PPh_3)_4]$. III.[134]
$[Pd(PPh_3)_4]$. III. Yellow, dec. 104 to 106°, dissoc. in solution.[571,938]
$[Pt(PPh_3)_2]$. Yellow, m. 157 to 160°, unstable.[1289]
$[Pt(PPh_3)_3]$. IV. Yellow, m. 125 to 130°,[944] x-ray structure.[19]
$[Pt(PPh_3)_4]$. IV. Yellow, m. 118°.[944]
$[Pd(PF_3)_2(PPh_3)_2]$. III.[892]
$[Pt(PPh_3)]_4$. III. Cluster compound.[606]
$[Pt(PPh_3)_2]_3$. III. Red trimer.[606]
$[Pd(PMePh_2)_4]$. III. Yellow, m. 80 to 83°.[1029]
$[Ni(PF_2Ph)_4]$. III. m. 63.5°, IR.[1200]
$[Ni(PCl_2Ph)_4]$. II. Yellow, m. 86.5°.[948] III. m. 93 to 94°.[1133]
$[Pd\{P(p\text{-}MeC_6H_4)_3\}_3]$. IV. Yellow, dec. 110°.[938]
$[Pd\{P(p\text{-}FC_6H_4)_3\}_3]$. IV. Yellow, dec. 136 to 139°, IR, NMR.[27]
$[Pd\{P(p\text{-}ClC_6H_4)_3\}_3]$. IV. Yellow, m. 90 to 100°.[938]
$[Pd\{P(p\text{-}ClC_6H_4)_3\}_4]$. IV. Yellow, dec. 120°.[938]
$[Pt\{P(p\text{-}ClC_6H_4)_3\}_3]$. IV. Yellow, m. 186°.[944]

D.3.2. Carbonyl Complexes

$[Cr(CO)_4(PMe_3)_2]$. VI. trans, Yellow, m. 79 to 84°, IR,

NMR;[830] cis, colorless, unstable.[835]

fac-$[Cr(CO)_3(PMe_3)_3]$. VI. Yellow, dec. 132°, IR, NMR.[830]

cis-$[Mo(CO)_4(PMe_3)_2]$. VI. White, m. 93 to 95°, IR, NMR.[830,83

fac-$[Mo(CO)_3(PMe_3)_3]$. VI. dec. 155°, IR, NMR.[830]

cis-$[W(CO)_4(PMe_3)_2]$. VI. Yellow, m. 106 to 108°, IR, NMR.[830,835]

$[Co_2(CO)_5(PMe_3)_3]$. V. Red, m. 96°.[1098]

$[Co_2(CO)_4(PMe_3)_4]$. VI. Red, dec. 152°.[1098]

$[Ni(CO)_3PMe_3]$. V. (60°). Colorless liq.,[169] IR.[170,1257]

$[Ni(CO)(PMe_3)_3]$. V. IR.[172,256]

$[V(CO)_4(PEt_3)_2]$. V. Red-brown, mag. moment 1.8 μB.[754]

$[Mo(CO)_5PEt_3]$. V. (70°). Liq., b. 110° (5×10^{-4} mm), IR.[1116]

cis-$[Mo(CO)_4(PEt_3)_2]$. V. (125°). White, m. 73°, IR.[1116]

trans-$[Mo(CO)_4(PEt_3)_2]$. V. (125°). Yellow, m. 58°, IR,[1116] NMR.[996]

$[Mo(CO)_3(PEt_3)_3]$. V. Unstable liq.,[1116] IR.[1117]

$[W(CO)_4(PEt_3)_2]$. V. Mixture of isomers,[1116] x-ray structure.[770]

$[Mn(CO)_4PEt_3]$. V. Orange, IR, mag. moment 1.88 μB.[726]

$[Mn(CO)_4PEt_3]_2$. V. (UV). Yellow, m. 153 to 154°,[1075] IR,[916,1085] x-ray structure.[155]

$[Fe(CO)_4PEt_3]$. V. IR.[1138]

trans-$[Fe(CO)_3(PEt_3)_2]$. V. IR.[1138]

$[Co_2(CO)_5(PEt_3)_3]$. V. Red, m. 78°.[1098]

$[Co(CO)_3PEt_3]_2$. V. (60°). Red,[727] IR.[248]

$[Co(CO)_3(PEt_3)_2]^+X^-$. V.[727] Hg deriv., x-ray structure.[228]

$[Ni(CO)_3PEt_3]$. V. IR.[1257]

$[Ni(CO)_2(PEt_3)_2]$. V. (40°). Dec. 100°,[172] IR,[170] dipole,[171] NMR.[996]

$[Pt_3(CO)_3(PEt_3)_4]$. VIII. Red, m. 107 to 108°, IR, dipole.[196]

$[V(CO)_4(PPr_3)_2]$. V. Red-brown, mag. moment 1.8 μB.[754]

$[Mn(CO)_4(PPr_3)]_2$. V. m. 125 to 127°, IR.[916]

$[W(CO)_5(PPr^i_3)]$. V. (diglyme). White, dec. 150 to 160°.[458]

$[W(CO)_5(PPr^c_3)]$. V. (UV). White, m. 100 to 102°, NMR.[458]

$[Cr(CO)_5PBu_3]$. V. Liq., IR,[932] NMR.[647]

trans-$[Cr(CO)_4(PBu_3)_2]$. V. Yellow, m. 40 to 41°, IR, NMR.[645]

$[Mo(CO)_5PBu_3]$. V. Brown liq., NMR.[647]

cis[$(Mo(CO)_4(PBu_3)_2]$. V. Colorless, m. 32°, IR, NMR.[645]

trans-$[Mo(CO)_4(PBu_3)_2]$. V. Yellow, m. 44 to 45°, IR, NMR.[645]

cis-$[W(CO)_4(PBu_3)_2]$. V. Colorless, m. 37 to 38°, IR, NMR.[645]

trans-$[W(CO)_4(PBu_3)_2]$. V. Yellow, m. 46 to 47°, IR, NMR.[645]

$[Mn(CO)_4PBu_3]_2$. V. m. 70°, IR.[916]

$[Ru(CO)_3PBu_3]_3$. V.[1105]

$[Co(CO)_3PBu_3]_2$. V. Red, x-ray structure.[228,779]

$[Ni(CO)_2(PBu_3)_2]$. V. Orange liq., IR,[995] NMR.[996]

$[V(CO)_4(PCy_3)_2]_2$. V. Yellow, diamag.[754]

$[Cr(CO)_5PCy_3]$. V. (Bu_2O). Yellow, m. 135 to 137°, IR.[1334]

$[Mo(CO)_5PCy_3]$. V. Colorless, m. 146 to 147°, IR.[1334]

$[Mn(CO)_4PCy_3]_2$. V. Orange, m. 206 to 207°, IR.[726,916]

$[Co(CO)_3PCy_3]_2$. V. (40°). Brown.[727]
$[Ni(CO)_3PCy_3]$. V. IR.[1257]
$[Cr(CO)_4(PMe_2Ph)_2]$. V. Yellow, m. 101 to 103°, IR, NMR.[830]
$[Mo(CO)_5PMe_2Ph]$. V. White, m. 26 to 28°, IR, NMR.[830]
cis-$[Mo(CO)_4(PMe_2Ph)_2]$. V. White, m. 110 to 112°, IR, NMR.[830]
fac-$[Mo(CO)_3(PMe_2Ph)_3]$. V. White, m. 156 to 157°, IR, NMR.[830]
mer-$[Mo(CO)_3(PMe_2Ph)_3]$. V. Yellow, m. 95 to 97°, IR, NMR.[830]
cis-$[W(CO)_4(PMe_2Ph)_2]$. V. Yellow, m. 115 to 117°, IR, NMR.[830]
$[Fe_3(CO)_9(PMe_2Ph)_3]$. V. (THF). Black, IR, NMR, x-ray structure.[986]
$[Pt_4(CO)_5(PMe_2Ph)_4]$. VI. Brown, m. 101 to 104°,[197] x-ray structure.[1326]
$[Cr(CO)_4(PEt_2Ph)_2]$. V. Yellow, m. 95°, μ 5.9 D.[358]
$[Mo(CO)_4(PEt_2Ph)_2]$. V. Buff, m. 98 to 99°, μ 6.95 D.[358]
$[W(CO)_4(PEt_2Ph)_2]$. V. Yellow, m. 104°, μ 7.25 D.[358]
$[Mn(CO)_4PEt_2Ph]_2$. V. m. 176 to 177°, IR.[916]
$[Re(CO)_3(PEt_2Ph)_2]$. V. Diamag. solid, paramag. in solution.[1063]
$[Ni(CO)_2(PEt_2Ph)_2]$. V. NMR.[996]
$[Mo(CO)_4(PAll_2Ph)_2]$. V. Yellow, m. 101 to 103°.[830]
trans-$[W(CO)_4(PAll_2Ph)_2]$. V. Yellow, m. 101 to 105°.[830]
$[Cr(CO)_5PBu_2Ph]$. V. Yellow, m. 67 to 68°, IR, NMR.[647]
trans-$[Cr(CO)_4(PBu_2Ph)_2]$. V. Yellow, m. 108 to 109°, IR, NMR.[645]
$[Mo(CO)_5PBu_2Ph]$. V. Brown liq., IR, NMR.[647]
cis-$[Mo(CO)_4(PBu_2Ph)_2]$. V. Colorless, m. 99 to 100°, IR, NMR.[645]
trans-$[(Mo(CO)_4(PBu_2Ph)_2]$. V. Yellow, m. 103 to 104°, IR, NMR.[645]
$[W(CO)_5PBu_2Ph]$. V. Yellow, m. 39 to 41°, IR, NMR.[647]
cis-$[W(CO)_4(PBu_2Ph)_2]$. V. Colorless, m. 107 to 108°, IR, NMR.[645]
trans-$[W(CO)_4(PBu_2Ph)_2]$. V. Yellow, m. 103 to 104°, IR, NMR.[645]
$[Cr(CO)_5PMePh_2]$. V. Yellow, m. 90 to 92°, IR, NMR.[647]
$[Mo(CO)_5PMePh_2]$. V. Colorless, m. 87 to 88°, IR, NMR.[647]
$[W(CO)_5PMePh_2]$. V. Colorless, m. 98 to 99°, IR, NMR.[647]
$[Pt(CO)(PMePh_2)_3]$. V. Cream, m. 108 to 110°, IR.[197]
$[Pt_3(CO)_3(PMePh_2)_4]$. V. Red-violet, dec. 139 to 142°, IR.[197]
$[Cr(CO)_5PEtPh_2]$. V. Yellow, m. 75 to 76°, IR, NMR.[647]
$[Mo(CO)_5PEtPh_2]$. V. Colorless, m. 52 to 53°, IR, NMR.[647]
$[W(CO)_5PEtPh_2]$. V. Colorless, m. 59 to 60°, IR, NMR.[647]
$[Mn(CO)_4PEtPh_2]_2$. V. m. 165 to 167°, IR.[916]
$[Fe(CO)_3(PEtPh_2)_2]$. V. Yellow, m. 151 to 154°, IR.[965]
$[Ni(CO)_2(PEtPh_2)_2]$. V. NMR.[996]
$[Pt_3(CO)_3(PEtPh_2)_4]$. V. Red, dec. 142 to 146°, IR.[197]

$[Cr(CO)_5PPr^iPh_2]$. V. Yellow, m. 101 to 102°, IR, NMR.[647]
$[Mo(CO)_5PPr^iPh_2]$. V. Colorless, m. 92 to 93°, IR, NMR.[647]
$[W(CO)_5PPr^iPh_2]$. V. Colorless, m. 98 to 100°, IR, NMR.[647]
$[Cr(CO)_5PBuPh_2]$. V. Yellow, m. 100 to 102°, IR, NMR.[647]
trans-$[Cr(CO)_4(PBuPh_2)_2]$. V. Yellow, dec. 155 to 157°, IR, NMR.[645]
$[Mo(CO)_5PBuPh_2]$. V. Colorless, m. 82 to 83°, IR, NMR.[647]
cis-$[Mo(CO)_4(PBuPh_2)_2]$. V. Colorless, dec. 154 to 156°, IR, NMR.[645]
trans-$[Mo(CO)_4(PBuPh_2)_2]$. V. Yellow, dec. 160 to 163°, IR, NMR.[645]
$[W(CO)_5PBuPh_2]$. V. Colorless, m. 94 to 95°, IR, NMR.[647]
cis-$[W(CO)_4(PBuPh_2)_2]$. V. Colorless, dec. 157 to 159°, IR, NMR.[645]
trans-$[W(CO)_4(PBuPh_2)_2]$. V. Yellow, m. 175 to 176°, IR, NMR.[645]
$[Ni(CO)_2(PBuPh_2)_2]$. V. White, IR.[1301]
$[V(CO)_4(PPh_3)_2]$. V. Orange, dec. 142°, unstable in air, mag. moment 1.78 μB, IR.[1337]
$[Cr(CO)_5PPh_3]$. V. Yellow, m. 127 to 128°,[932,974] IR, NMR,[647] x-ray structure.[1113]
$[Cr(CO)_4(PPh_3)_2]$. V. Yellow, m. 250 to 252°.[746,932] VI.[133]
$[Cr(CO)_3(PPh_3)_3]$. VI. m. 175 to 177°.[1049]
$[Cr(CO)_4PH_3(PPh_3)]$. V. Yellow, dec. 131 to 133°.[568]
$[Cr(CO)_3NH_3(PPh_3)_2]$. VI.[710]
$[Mo(CO)_5PPh_3]$. V. White, m. 138 to 139°.[932,974]
$[Mo(CO)_4(PPh_3)_2]$. V. Yellow, dec. 273 to 275°, diamag., IR.[746] VI.[1236] Halogenation.[920]
$[Mo(CO)_3(PPh_3)_3]$. VI. Yellow, dec. 170°, IR.[2] VII.[719] Halogenation.[920]
$[Mo(CO)_3(bipy)PPh_3]$. V.[1254] Reaction with SO_2.[774]
$[Mo(CO)_2(MeCN)_2(PPh_3)_2]$. VI.[501]
$[W(CO)_5PPh_3]$. V. Yellow, m. 146 to 147°.[932,974]
$[W(CO)_4(PPh_3)_2]$. V. Yellow, dec. 292 to 294°, diamag., IR.[746]
$[Mn(CO)_4PPh_3]$. V. (120°). Red, dec. 205°,[103,726] mag. moment 1.84 μB.[753]
$[Mn(CO)_4PPh_3]_2$. V. (UV). Orange, m. 189 to 190°,[1075] IR.[916,1085]
$[Mn_2(CO)_9PPh_3]$. V. (UV). Orange-red, dec. 145°.[1328,1375]
$[Re(CO)_4PPh_3]_2$. V. White, m. 232°, mag. moment 3.47 μB.[582]
$[Re(CO)_3(PPh_3)_2]$. V. Cream, m. 214°, mag. moment 1.5 μB.[582,588,1063,1077]
$[Fe(CO)_4PPh_3]$. V. (110°). Yellow, dec. 201 to 202°,[109,391,476] IR.[453]
$[Fe(CO)_3(PPh_3)_2]$. V. Yellow, dec. 260°,[391,476] IR.[453] VI.[967]
$[Fe(CO)_2(PPh_3)_3]$. VI. Yellow-orange.[967]
$[Fe_3(CO)_{11}PPh_3]$. V. Green-black,[50] x-ray structure.[467]
$[Ru(CO)_4PPh_3]$. V. (UV).[913] Yellow, dec. 120°.[1104]
$[Ru(CO)_3(PPh_3)_2]$. VIII.[416,1105,1123]

$[Ru(CO)_3PPh_3]_3$. V. Violet, m. 174 to 176°, IR.[246,847]
$[Os(CO)_4PPh_3]$. V. (UV).[913]
$[Os(CO)_3(PPh_3)_2]$. VIII. Colorless, IR,[419] x-ray structure.[1244]
$[Os(CO)_3PPh_3]_3$. V.[201]
$[Co(CO)_4PPh_3]^+X^-$.[716]
$[Co(CO)_3PPh_3]_2$. V. (0°). Brown, diamag.[725,1006,1170]
$[Co(CO)_3(PPh_3)_2]^+X^-$. V.[580,725,750,1174] IR.[1322]
$[Co_2(CO)_7PPh_3]$. V. IR.[198]
$[Co_2(CO)_2(PPh_3)_6]$. VIII. Yellow, IR, NMR.[1227]
$[Co_4(CO)_{11}PPh_3]$. V. IR.[269]
$[Rh_2(CO)_4(PPh_3)_4]$. V. Yellow, dec. 104°, IR.[821]
$[Rh(CO)_2(PPh_3)_2]^+$. V.[723]
$[Ir_4(CO)_{10}(PPh_3)_2]$. VII. Orange, IR,[942] x-ray structure.[20]
$[Ir_4(CO)_9(PPh_3)_3]$. VII. IR,[942] x-ray structure.[20]
$[Ir(CO)_2(PPh_3)_2]^+$. V.[723]
$[Ni(CO)_3PPh_3]$. V. (20°). m. 123°,[109,1144,1365] IR, Raman.[530]
$[Ni(CO)_2(PPh_3)_2]$. V. (50°). White, dec. 215°,[1156,1365] NMR,[996] IR.[256] IX. (150°).[1017]
$[Pd(CO)(PPh_3)_3]$. VIII. Cream-yellow, dec. 110°, unstable in solution.[1005]
$[Pd(CO)PF_3(PPh_3)_2]$. VIII. IR.[892]
$[Pd(CO)PPh_3]_n$. VIII. Red, dec. 73°, IR.[1005]
$[Pd_3(CO)_3(PPh_3)_4]$. VIII. Orange, dec. 70°, IR.[1005]
$[Pt(CO)_2(PPh_3)_2]$. V. White, dec. 105°, IR.[197] VIII. m. 118°, dissoc. to mixture,[944] IR.[256]
$[Pt(CO)(PPh_3)_2]$. VIII. IR.[256]
$[Pt(CO)(PPh_3)_3]$. VIII. Orange-red, m. 130°,[944] IR,[256] x-ray.[21] Brown, m. 100 to 110°,[196] x-ray.[22]
$[Pt_3(CO)_4(PPh_3)_3]$. V. Black, dec. 177 to 179°, IR.[197]
$[Pt_3(CO)_3(PPh_3)_4]$. VIII. Orange, m. 145 to 148°, IR.[196,197]
$[Ni(CO)_3P(p\text{-}MeC_6H_4)_3]$. IX. m. 158°,[1017,1144] IR.[1257]

D.3.3. Halide and Pseudohalide Complexes

$[CdBr_2PMe_3]_2$. X. m. 195 to 198°.[554]
$[CdI_2PMe_3]_2$. X. dec. 174 to 176°.[554]
$[AlCl_3(PMe_3)_2]$. X. Unstable adduct, IR, Raman.[121]
$[GaCl_3PMe_3]$. X. Unstable adduct,[121] IR.[87]
$[InCl_3(PMe_3)_2]$. X. Unstable adduct, IR, Raman.[121]
$[InI_3(PMe_3)_2]$. X. Unstable adduct, IR, Raman.[121]
$[TiCl_4PMe_3]$. X. (Benzene). Red, monomeric, IR.[119]
$[CrCl_2PMe_3]$. X. Green, unstable in air, mag. moment 4.94 μB.[163]
$[CoBr_2(PMe_3)_2]$. X. Green.[841]
$[NiCl_2(PMe_3)_2]$. X. Crimson, m. 199 to 200°, diamag.[130,840]
$[NiBr_2(PMe_3)_2]$. X. Crimson, m. 178 to 181°, diamag.[130,840]
$[NiI_2(PMe_3)_2]$. X. Brown, UV spect.[130]
$[Ni(NO_3)_2(PMe_3)_2]$. X. Red, mag. moment 3.17 μB.[130]
$[NiBr_2(PMe_3)_3]$. X. Blue, dec. 140 to 150°, diamag.[273,840]

$[NiI_2(PMe_3)_3]$. X. Black, dec. 160 to 170°.[840]
$[Ni(CN)_2(PMe_3)_3]$. Orange-red, m. 165 to 168°.[840]
$[Ni(NO_3)_2(PMe_3)_4]$. X. Red, dec. 104 to 110°.[840]
$[PdCl_2(PMe_3)_2]$. X. Yellow, dec. 282°,[961] IR,[621] NMR.[620]
$[PdBr_2(PMe_3)_2]$. X. Yellow, m. 228 to 230°,[553] IR.[621]
$[PdI_2(PMe_3)_2]$. X. Orange, m. 181 to 182°,[553] IR.[621]
$[PdCl_2PMe_3]_2$. XA. Red, dec. 285°,[961] IR.[621,622]
$[PdBr_2PMe_3]_2$. XA. Red, m. 300 to 305°, IR.[621]
$[PdI_2PMe_3]_2$. XA. Brown, dec. 280 to 284°, IR.[621]
cis-$[PtCl_2(PMe_3)_2]$. X. White, m. 324 to 326°, μ 13.1 D,[129] IR,[621] x-ray.[250,1000]
cis-$[PtBr_2(PMe_3)_2]$. XIV. White, dec. 311°,[619] IR.[621]
trans-$[PtI_2(PMe_3)_2]$. XIV. Yellow, m. 194°,[619] IR,[621] Raman.[520]
cis-$[PtCl_2(PPh_2Cl)(PMe_3)]$. XB. Colorless, m. 213 to 215°, hydrolysis.[315]
$[PtCl_2PMe_3]_2$. XA. Yellow-orange, dec. 217 to 220°,[352,623] IR.[621,622]
$[PtBr_2PMe_3]_2$. XA. Yellow, m. 255 to 260°, IR.[621]
$[PtI_2PMe_3]_2$. XA. Orange, dec. 296 to 299°, IR.[621]
$[AgIPMe_3]$. XIV. White, m. 131 to 133°.[553]
$[AgNO_3PMe_3]$. X. m. 101 to 102°, useful source of PMe_3, liberated from aq. soln. by thiourea.[553]
$[AuBrPMe_3]$. XIII. Dec. 225°.[957]
$[AuBr_3PMe_3]$. XII. Orange-red, m. 162°,[957] x-ray (planar).[1100]
$[ZnCl_2(PEt_3)_2]$. X. m. 97 to 98°,[398] IR.[676]
$[ZnBr_2(PEt_3)_2]$. X. m. 130°,[398] IR.[676]
$[CdBr_2(PEt_3)_2]$. X. Dec. 103 to 104°.[554]
$[CdI_2(PEt_3)_2]$. X. m. 132 to 134°.[554]
$[CdBr_2PEt_3]_2$. X. m. 163 to 164°, x-ray structure.[554]
$[CdI_2PEt_3]_2$. X. m. 141°.[554]
$[HgI_2(PEt_3)_2]$. X. (EtOH). Colorless, m. 156 to 157°.[265]
$[HgBr_2PEt_3]_2$. X. m. 106°.[554]
$[HgI_2PEt_3]_2$. X. m. 121 to 123°.[554]
$[(HgBr_2)_3(PEt_3)_2]$. X. m. 130°.[554]
$[(HgI_2)_3(PEt_3)_2]$. X. m. 109 to 110°.[554]
$[(HgCl_2)_4(PEt_3)_2]$. X. m. 163 to 164°.[554]
$[(HgBr_2)_4(PEt_3)_2]$. X. m. 149 to 151°.[554]
$[SnCl_4(PEt_3)_2]$. X. (EtOH). White, m. 145 to 150°.[34]
$[SnBr_4(PEt_3)_2]$. X. m. 170 to 171°.[34]
$[TiCl_4(PEt_3)_2]$. X. Red.[312]
$[VCl_3(PEt_3)_2]$. X. Red, dec. 160 to 163°, mag. moment 2.83 μB.[801]
$[Cr(SCN)_4(PEt_3)_2]^-$. X. Analog of Reinecke's salt.[814,1270]
$[CrCl_2PEt_3]_n$. X. Grey polymer.[806]
$[CrCl_2(PEt_3)_2]$. X. Blue, unstable, forms polymer.[806]
$[CrCl_3(PEt_3)_2]_2$. X. Olive-green, m. 167°, mag. moment 3.8 μB.[806]
$[CrBr_3(PEt_3)_2]$. X. Red-brown, dec. 168°.[806]
$[OsCl_3(PEt_3)_3]$. XIII. (EtOH). Red, m. 99 to 100°, mag. moment 1.9 μB.[324]

$[CoCl_2(PEt_3)_2]$. X. (EtOH). Blue, m. 101 to 102°, μ 8.7 D,[839] mag. moment.[676]
$[CoBr_2(PEt_3)_2]$. X. Blue-green, dec. 134 to 136°.[342]
$[CoI_2(PEt_3)_2]$. XIV. Green, m. 128 to 135°,[192] mag. moment.[676]
$[Co(NCS)_2(PEt_3)_2]$. XIV. Red, dec. 72 to 74°, mag. moment, spect., high-192 spin equilibria in soln.[1052,1285]
$[CoCl_3(PEt_3)_2]$. XII. Mag. moment 3.02 μB.[844]
trans-$[RhCl_3(PEt_3)_3]$. X. (MeOH). Orange-red, m. 114 to 117°, μ 7.0 D.[318]
cis-$[RhCl_3(PEt_3)_3]$. X. Yellow, dec. 169 to 174°.[318]
$[Rh_2Cl_6(PEt_3)_4]$. X. (EtOH). Red-purple, dec. 248 to 261°.[318]
$[Rh_2Cl_3(PEt_3)_3]$. X. (EtOH). Brown, dec. 260 to 270°.[318]
trans-$[IrCl_3(PEt_3)_3]$. X. (EtOH). Yellow, m. 116 to 117°, μ 6.85 D.[294]
cis-$[IrCl_3(PEt_3)_3]$. X. Colorless, m. 261 to 264°.[294]
$[Ir_2Cl_6(PEt_3)_4]$. X. Yellow-orange, dec. 235 to 275°.[294]
$[NiCl_2(PEt_3)_2]$. X. (EtOH). Red, m. 112 to 113°,[838] IR.[619]
$[NiBr_2(PEt_3)_2]$. X. Red, m. 106 to 107°,[838] IR,[619] x-ray.[600]
$[NiI_2(PEt_3)_2]$. X. Brown, m. 91 to 92°.[838]
$[Ni(NO_2)_2(PEt_3)_2]$. XIV. Orange, dec. 163 to 169°.[194]
$[Ni(NO_3)_2(PEt_3)_2]$. X. Green, m. 131 to 132°,[838] mag. moment,[67] IR,[598] spect.,[66] x-ray.[1182]
$[Ni(SCN)_2(PEt_3)_2]$. XIV. Yellow, dec. 141 to 142°, diamag.[457]
$[Ni(CNO)_2(PEt_3)_2]$. XI. Yellow, m. 158 to 161°, diamag.[126]
$[Ni(CN)_2(PEt_3)_2]$. XIV. Yellow, m. 156 to 157°.[126,841]
$[NiBr_3(PEt_3)_2]$. XII. Violet-black, m. 83 to 84°,[838] mag. moment, dipole.[843]
$[NiCl_3(PEt_3)_2]$. XII. Mag. moment 2 μB.[844]
$[PdCl_2(PEt_3)_2]$. X. Yellow, m. 139°,[954] IR.[619]
$[PdBr_2(PEt_3)_2]$. X. Orange, m. 131 to 132°,[757] IR.[619]
$[PdI_2(PEt_3)_2]$. XIV. Orange, m. 138°,[192] IR.[619]
$[Pd(CN)_2(PEt_3)_2]$. XIV. White, m. 174°, IR.[619]
$[Pd(NO_3)_2(PEt_3)_2]$. X. Yellow, dec. 182°.[192]
$[Pd(NCS)_2(PEt_3)_2]$. X.[115] IR.[619,1284]
$[PdCl_2PEt_3]_2$. XA. Orange-red, m. 230°,[955] IR.[622]
$[PdBr_2PEt_3]_2$. XA. Red, m. 212 to 213°.[12]
$[PdI_2PEt_3]_2$. XA. Red-brown, m. 191 to 192°.[12]
cis-$[PtCl_2(PEt_3)_2]$. X. White, m. 191 to 192°, μ 10.7 D,[836] IR.[619]
trans-$[PtCl_2(PEt_3)_2]$. X. Yellow, m. 142 to 143°,[836] IR,[619] x-ray.[567,999]
cis-$[PtBr\ (PEt_3)_2]$. X. White, m. 201 to 202°, μ 11.2 D,[836] IR.[619]
trans-$[PtBr_2(PEt_3)_2]$. X. Yellow, m. 134 to 135°,[836] IR,[619] x-ray.[567,999]
cis-$[PtI_2(PEt_3)_2]$. XIV. Pale yellow, m. → trans, μ 8.2 D.[836]
trans-$[PtI_2(PEt_3)_2]$. XIV. Yellow, m. 136 to 137°,[836] IR.[619]

cis-$[Pt(NO_2)_2(PEt_3)_2]$. XIV. White, dec. 192 to 193°.[836]
trans-$[Pt(NO_2)_2(PEt_3)_2]$. X. White, m. 201 to 202°.[836]
cis-$[(Pt(NCS)_2(PEt_3)]$. XIV. White, IR.[619]
trans-$[Pt(NCS)_2(PEt_3)_2]$. XIV. m. 147°,[266] IR.[619,1284]
trans-$[Pt(CN)_2(PEt_3)_2]$. XIV. White, m. 177°, IR.[619]
cis-$[Pt(NO_3)_2(PEt_3)_2]$. XIV. White, m. 182 to 183°.[836]
trans-$[Pt(NO_3)_2(PEt_3)_2]$. XIV. Yellow, m. 125 to 126°.[836]
$[PtSO_4(PEt_3)_2]$. XIV. White.[836]
$[Pt(PEt_3)_4]Cl_2$. X.[837]
$[PtCl_2PEt_3]_2$. XA. Orange, dec. 224 to 225°,[623] IR.[622]
$[PtBr_2PEt_3]_2$. XA. Red, m. 202°.[12]
$[PtI_2PEt_3]_2$. XA. Red, m. 209.5°.[12]
$[PtCl_4(PEt_3)_2]$. XII. X-ray,[65] reduction with thiosulfate.[1099]
$[CuI(PEt_3)_4]$. X. White, m. 236 to 240°, x-ray.[958]
$[AgIPEt_3]_4$. X. Colorless, m. 208 to 209°.[962]
$[Ag(SCN)PEt_3]$. X. m. 121°.[1283]
$[AuClPEt_3]$. XIII. (EtOH). Colorless, m. 78°, b. 210° (0.3 mm),[962] dipole.[244]
$[AuBrPEt_3]$. XIII. Colorless, m. 87°.[957]
$[AuIPEt_3]$. XIII. (EtOH). Colorless, m. 67°.[962]
$[Au(SCSOR)PEt_3]$. XIV.[889]
$[AuCl_3PEt_3]$. XII. Yellow, m. 121°.[957]
$[AuBr_3PEt_3]$. XII. Red, m. 129°.[957]
$[CdBr_2(PPr_3)_2]$. X. m. 75 to 77°.[554]
$[CdI_2(PPr_3)_2]$. X. m. 72 to 73°.[554]
$[(CdBr_2)_2(PPr_3)_2]$. X. m. 105 to 106°.[554]
$[Cd_2Br_4(PPr_3)_3]$. X. m. 126 to 128°.[554]
$[HgBr_2PPr_3]_2$. X. m. 133°.[554]
$[HgI_2PPr_3]_2$. X. α form, m. 114 to 115°; β form, yellow dec. 104 to 107°.[554]
$[(HgCl_2)_3(PPr_3)_2]$. X. m. 173 to 174°.[554]
$[(HgI_2)_2(PPr_3)_3]$. X. m. 124 to 125°.[554]
$[SnCl_4(PPr_3)_2]$. X. (EtOH). White, m. 157 to 159°.[34]
$[SnBr_4(PPr_3)_2]$. X. m. 103 to 106°.[34]
$[VCl_3(PPr_3)_2]$. X. Red.[801]
$[OsCl_3(PPr_3)_3]$. XIII. Red, dec. 165 to 168°, mag. moment 1.9 μB.[324]
$[OsBr_3(PPr_3)_3]$. XIII. Purple, m. 169 to 170°, mag. moment 2.0 μB.[324]
$[OsCl_4(PPr_3)_2]$. XII. Green-brown, m. 187 to 191°, mag. moment 1.6 μB.[324]
$[OsBr_4(PPr_3)_2]$. XII. Violet, m. 203 to 204°, mag. moment 1.6 μB.[324]
$[CoCl_2(PPr_3)_2]$. X. (EtOH). Blue, m. 80°.[839]
$[Co(NCS)_2(PPr_3)_3]$. Green-brown, dec. 91 to 93°, mag. moment 2.05 μB.[199]
trans-$[RhCl_3(PPr_3)_3]$. X. (EtOH). Orange, dec. 179 to 186°.[318]
trans-$[RhBr_3(PPr_3)_3]$. XIV. Red, m. 160 to 163°.[318]
trans-$[IrCl_3(PPr_3)_3]$. X. (EtOH). Yellow, m. 236 to 242°.[294]

$[IrCl_4(PPr_3)_2]$. XII. Violet, dec. 147 to 149°, mag. moment 1.9 μB.[324]
$[NiCl_2(PPr_3)_2]$. X. Red, m. 92 to 93°,[838] UV.[601]
$[Ni(NO_2)_2(PPr_3)_2]$. Yellow, m. 166 to 167°.[841]
$[Ni(NO_3)_2(PPr_3)_2]$. X. Green, m. 89 to 92°.[841]
$[PdCl_2(PPr_3)_2]$. X. Yellow, m. 96°.[954]
$[PdBr_2(PPr_3)_2]$. X. Orange-yellow, m. 87 to 88°.[34]
$[Pd(NO_2)_2(PPr_3)_2]$. White, dec. 167 to 168°.[954]
$[Pd(NCS)_2(PPr_3)_2]$. X-ray.[579]
$[PdCl_2PPr_3]_2$. XA. Orange-red, m. 189°,[955] IR.[622]
$[PdBr_2PPr_3]_2$. XA. Red, m. 178 to 179°.[354]
$[PdI_2PPr_3]_2$. XIV. Purple-black, m. 202 to 203°.[354]
cis-$[PtCl_2(PPr_3)_2]$. X. White, m. 149 to 150°, μ 11.5 D.[836]
trans-$[PtCl_2(PPr_3)_2]$. X. Yellow, m. 85 to 86°,[836] NMR.[644]
cis-$[PtBr_2(PPr_3)_2]$. X. White, m. 160 to 161°.[275]
trans-$[PtBr_2(PPr_3)_2]$. X. Yellow, m. 96 to 97°.[275]
trans-$[PtI_2(PPr_3)_2]$. XIV. Yellow, m. 118 to 119°.[360]
$[PtCl_2(PPr_3)]_2$. XA. Yellow-orange, m. 181 to 183°,[274] x-ray.[177]
cis-$[PtCl_4(PPr_3)_2]$. XII. Yellow, dec. 123 to 127°.[274]
trans-$[PtCl_4(PPr_3)_2]$. XII. Yellow, dec. 174 to 180°.[274]
$[CuIPPr_3]_4$. X. White, m. 206 to 207°.[958]
$[AgIPPr_3]_4$. X. Colorless, m. 258 to 265°.[962]
$[Ag(SCN)PPr_3]$. X. m. 66°, x-ray structure.[1283]
$[AuClPPr_3]$. XIII. Colorless, m. 40°.[962]
$[NiCl_2(PPr^i_3)_2]$. X. Red, m. 185 to 187°, UV.[601]
$[NiBr_2(PPr^i_3)_2]$. X. Red, m. 151 to 154°, UV.[601]
$[PdCl_2(PPr^i_3)_2]$. UV spect.,[1282] IR.[1097]
$[PdBr_2(PPr^i_3)_2]$. UV spect.[1282]
$[Pd(NCS)_2(PPr^i_3)_2]$. IR.[1097,1284]
$[Pt(NCS)_2(PPr^i_3)_2]$. IR.[1097,1284]
$[AgI(PPr^i_3)]$. X. m. 205°.[1283]
$[Ag(SCN)(PPr^i_3)]$. X. m. 99°.[1283]
$[Cd_2Br_4(PBu_3)_3]$. X. m. 93 to 94.5°.[554]
$[Cd_2I_4(PBu_3)_3]$. X. m. 100 to 101°.[554]
$[HgBr_2(PBu_3)_2]$. NMR.[869]
$[HgBr_2PBu_3]_2$. X. m. 116°.[554]
$[HgI_2PBu_3]_2$. X. m. 84 to 85°.[554]
$[(HgCl_2)_3(PBu_3)_2]$. X. m. 72 to 74°.[554]
$[(HgI_2)_2(PBu_3)_3]$. X. m. 162°.[554]
$[CrCl_3(PBu_3)_2]_2$. X. Green, m. 134°, μ 8.0 D.[806]
$[RuCl_3(PBu_3)_4]_2$. X. (EtOH). Red, m. 132 to 134°, mag. moment 1.93 μB.[1051]
$[Ru_2Cl_5(PBu_3)_4]$. X. Red, m. 95 to 96° (by-product from above complex),[1051] x-ray.[365]
$[OsCl_3(PBu_3)_3]$. XIII. Orange, dec. 132 to 134°.[324]
trans-$[RhCl_3(PBu_3)_3]$. X. Orange,red, dec. 139 to 142°.[318]
trans-$[IrCl_3(PBu_3)_3]$. X. (EtOH). Yellow, m. 155 to 158°.[294]
$[NiCl_2(PBu_3)_2]$. X. Red, m. 48 to 49°.[838]

$[NiBr_2(PBu_3)_2]$. X. Red, m. 51 to 53°.[841]
$[NiI_2(PBu_3)_2]$. Green-black, m. 32 to 33°.[841]
$[Ni(NO_2)_2(PBu_3)_2]$. XIV. Yellow, m. 137 to 138.5°.[562]
$[PdCl_2(PBu_3)_2]$. X. Yellow, m. 66°.[954]
$[PdBr_2(PBu_3)_2]$. X. Yellow, m. 73°.[954]
$[PdI_2(PBu_3)_2]$. X. Orange, m. 64 to 65°.[954]
$[Pd(NO_2)_2(PBu_3)_2]$. XIV. m. 141 to 142°, IR.[284]
$[Pd(NCS)_2(PBu_3)_2]$. XIV. White, m. 112°,[955] IR.[233]
$[PdCl_2PBu_3]_2$. XA. Orange-red, m. 145°, μ 2.34 D.[955]
cis-$[PtCl_2(PBu_3)_2]$. X. White, m. 144°, μ 11.5 D,[836] NMR.[644]
trans-$[PtCl_2(PBu_3)_2]$. X. Yellow, m. 65 to 66°,[836] NMR.[644]
trans-$[Pt(NO_2)_2PBu_3)_2]$. XIV. White, m. 148°, IR.[284]
cis-$[Pt(NO_2)_2(PBu_3)_2]$. XIV. Cream, m. 140 to 142°, IR.[284]
$[PtSO_4(PBu_3)_2]$. XIV. White, m. 175 to 176°.[284]
$[PtCl_2PBu_3]_2$. XA. Yellow-orange, m. 143 to 144°.[352]
$[CuIPBu_3]_4$. X. White, m. 75°,[958] x-ray.[1331]
$[AgIPBu_3]_4$. X. Colorless, m. 43°.[962]
$[AuClPBu_3]$. XIII. Colorless, b. 215 to 225° (0.03 mm).[962]
$[AuIPBu_3]$. XIII. b. 220 to 225° (0.2 mm).[962]
$[CrCl_3PCy_3]$. X. Violet, dec. 104°.[806]
$[FeCl_3PCy_3]$. X. Yellow, dec. 240°.[802]
$[CoBr_2(PCy_3)_2]$. X. Blue, dec. 205°,[802] mag. moment.[444]
$[CoI_2(PCy_3)_2]$. X. Green, dec. 221 to 223°,[802] mag. moment.[444]
$[NiCl_2(PCy_3)_2]$. X. Red, m. 227°,[641,1286] x-ray.[136]
$[NiBr_2(PCy_3)_2]$. X. Olive-green, dec. 204°.[802,1286]
$[NiI_2(PCy_3)_2]$. X. Green, m. 214°.[443]
$[Ni(SCN)_2(PCy_3)_2]$. Orange-yellow, m. 230°.[1286]
$[CuCl_2(PCy_3)_2]$. X. Green, dec. 214°.[802]
$[Ag(SCN)(PCy_3)_2]$. IR.[1283]
$[ZnCl_2(PMe_2Ph)_2]$. X. (EtOH). m. 118 to 119°.[398]
$[ZnBr_2(PMe_2Ph)_2]$. X. m. 119°.[398]
$[ZnI_2(PMe_2Ph)_2]$. X. (EtOH). m. 135°.[265]
$[CdCl_2(PMe_2Ph)_2]$. X. m. 302 to 304°.[265]
$[CdBr_2(PMe_2Ph)_2]$. X. m. 180°.[265]
$[CdI_2(PMe_2Ph)_2]$. X. m. 75°.[265]
$[HgCl_2(PMe_2Ph)_2]$. X. m. 168 to 169°.[265]
$[HgBr_2(PMe_2Ph)_2]$. X. m. 133 to 135°.[265]
$[HgI_2(PMe_2Ph)_2]$. X. m. 115 to 117°.[265]
$[TiCl_4(PMe_2Ph)]$. X. NMR suggests equilibrium with 2:1 adduct.[241]
$[WCl_4(PMe_2Ph)_2]$. XII. Orange,[1026] x-ray.[65]
$[ReCl_4(PMe_2Ph)_2]$. XII. (CCl_4). Purple,[323,324] far IR.[321]
mer-$[ReCl_3(PMe_2Ph)_3]$. XIII. Orange, m. 155 to 160°,[324] NMR.[1135,1208]
mer-$[ReBr_3(PMe_2Ph)_3]$. XIII. Red-brown, dec. 173 to 180°,[515] NMR.[1135]
$[RuCl_3(PMe_2Ph)_3]$. X. (EtOH). Brown, m. 175 to 177°, mag. moment 2.1 μB.[324]
$[RuBr_3(PMe_2Ph)_3]$. X. Purple, m. 168 to 171°.[324]

$[Ru_2Cl_3(PMe_2Ph)_6]Cl$. X. (MeOH). Yellow, m. 118 to 120°.[310]
$[OsCl_3(PMe_2Ph)_3]$. XIII. Red, dec. 200 to 204°, mag. moment 2.1 μB.[324]
$[OsBr_3(PMe_2Ph)_3]$. XIII. Purple, dec. 179 to 182°, mag. moment 2.0 μB.[324]
$[OsCl_4(PMe_2Ph)_2]$. XII. Brown, m. 206 to 207°, mag. moment 1.5 μB.[324]
mer-$[RhCl_3(PMe_2Ph)_3]$. X. (MeOH). Orange, dec. 218 to 224°,[318] NMR.[215]
fac-$[RhCl_3(PMe_2Ph)_3]$. X. (EtOH). Yellow, dec. 215 to 226°, IR.[215]
mer-$[RhBr_3(PMe_2Ph)_3]$. XIV. Orange, m. 195 to 201°, NMR.[215]
mer-$[RhI_3(PMe_2Ph)_3]$. XIV. Maroon, m. 179 to 182°, NMR.[215]
$[RhBrCl_2(PMe_2Ph)_3]$. XIV. Orange, m. 206 to 224°, NMR.[215]
mer-$[Rh(NCO)_3(PMe_2Ph)_3]$. XIV. Yellow, m. 159 to 161°, NMR.[215]
mer-$[Rh(SCN)_3(PMe_2Ph)_3]$. XIV. Yellow, m. 196 to 202°, NMR.[215]
$[IrCl_3(PMe_2Ph)_3]$. NMR.[831]
trans-$[IrCl_4(PMe_2Ph)_2]$. XII. Violet, dec. 152 to 154°,[324] NMR,[831] x-ray.[65]
$[NiCl_2(PMe_2Ph)_2]$. X. Red.[35,265]
$[NiBr_2(PMe_2Ph)_2]$. X. Red.[35]
$[NiI_2(PMe_2Ph)_3]$. X. Blue-green.[35]
$[NiBr_3(PMe_2Ph)_2 \cdot 0.5NiBr_2(PMe_2Ph)_2]$. X-ray structure.[991,1247]
$[Ni(CN)_2(PMe_2Ph)_3]$. Red, diamag.,[35] x-ray.[1246]
$[PdCl_2(PMe_2Ph)_2]$. X. (EtOH). Yellow, m. 192 to 194°, NMR.[833]
$[PdBr_2(PMe_2Ph)_2]$. XIV. Yellow, m. 180 to 182°, NMR.[833]
$[PdI_2(PMe_2Ph)_2]$. XIV. cis-, yellow; trans-, red, x-ray,[74,76] NMR.[831,833]
$[Pd(SCN)_2(PMe_2Ph)_2]$. XIV. Yellow, m. 165 to 167°, NMR.[833]
$[PdCl_2(PMe_2Ph)]_2$. XA. Orange, m. 226 to 228°, NMR.[833]
$[PtCl_2(PMe_2Ph)_2]$. X. (EtOH). Colorless, m. 199 to 200°, NMR.[833]
$[PtBr_2(PMe_2Ph)_2]$. XIV. Colorless, m. 201 to 204°, NMR.[833]
$[PtI_2(PMe_2Ph)_2]$. XIV. Yellow, m. 168 to 171°, NMR.[833]
$[Pt(SCN)_2(PMe_2Ph)_2]$. XIV. Colorless, m. 184 to 186°, NMR.[833]
$[PtCl_2(PMe_2Ph)]_2$. XA. Yellow, dec. 215 to 230°, NMR.[833]
$[PtCl_4(PMe_2Ph)_2]$. XII. Yellow, m. 176 to 178°.[1167]
$[CuI(PMe_2Ph)_2]$. X. m. 98 to 99°.[265]
$[AgI(PMe_2Ph)_2]$. X. m. 114 to 115°.[264,265]
$[AuI_3(PMe_2Ph)]$. X. (EtOH). Colorless, dec. 142°.[265]
$[ZnI_2(PEt_2Ph)_2]$. X. m. 172.5 to 174°.[265]
$[CdI_2(PEt_2Ph)_2]$. X. m. 102 to 103°.[265]
$[HgBr_2(PEt_2Ph)_2]$. X. m. 120 to 121°.[265]
$[HgI_2(PEt_2Ph)_2]$. X. m. 130 to 131°.[265]
$[ReCl_3(PEt_2Ph)]$. X. Purple, m. >350°,[335] x-ray structure,

trimeric cluster.[451]
$[ReCl_3(PEt_2Ph)_3]$. X. Orange, m. 163 to 166°, dipole,[335] NMR.[1135,1208]
$[ReBr_3(PEt_2Ph)_3]$. X. Brown, dec. 140°, μ 6.5 D.[297]
trans-$[ReCl_4(PEt_2Ph)_2]$. XII. Violet, dec. 161°, NMR, mag. moment 3.64 μB.[296,324]
$[ReOI(PEt_2Ph)_3]$. XI.[585]
trans-$[ReOCl_3(PEt_2Ph)_2]$. X. Green,[335] x-ray.[532]
$[ReBr_4(PEt_2Ph)_2]$. X. Purple, dec. 182°.[296]
$[FeCl_2(PEt_2Ph)_2]$. X. (C_6H_6). Cream, m. 70 to 73°, mag. moment 4.7 μB.[194]
$[RuCl_3(PEt_2Ph)_3]$. XI. (CO). Red-brown, dec. 178 to 184°.[347] X.[324]
$[RuBr_3(PEt_2Ph)_3]$. X. Purple, m. 173 to 176°.[324]
$[Ru_2Cl_3(PEt_2Ph)_6]Cl$. X. (MeOH). Yellow, m. 150 to 152°, diamag.[310,564]
$[Ru_2Cl_3(PEt_2Ph)_6][RuCl_3(PEt_2Ph)_3]$. X-ray structure.[1136]
$[Ru_2Cl_4(PEt_2Ph)_5]$. By heating above complex,[1131] x-ray structure (Cl bridge).[23]
$[OsCl_3(PEt_2Ph)_3]$. XIII. Red, dec. 191 to 195°, mag. moment 2.1 μB.[324]
$[OsBr_3(PEt_2Ph)_3]$. XIII. Purple, dec. 160 to 162°.[324]
$[CoCl_2(PEt_2Ph)_2]$. X. Blue, m. 71 to 73°.[342]
$[CoBr_2(PEt_2Ph)_2]$. X. Blue-green, m. 80 to 82°, mag. moment 4.5 μB.[342]
$[CoI_2(PEt_2Ph)_2]$. X. Green, m. 90 to 94°.[192]
$[Co(NCS)_2(PEt_2Ph)_3]$. Brown, dec. 92 to 94°, mag. moment 1.9 μB.[199]
$[Co(CN)_2(PEt_2Ph)_3]$. XIV.[1147]
trans-$[RhCl_3(PEt_2Ph)_3]$. X. (EtOH). Orange, dec. 183 to 196°.[318]
trans-$[IrCl_3(PEt_2Ph)_3]$. X. (EtOH). Yellow, dec. 235 to 241°,[294] mer- photochemically isom. to fac-.[216]
trans-$[IrBr_3(PEt_2Ph)_3]$. X. (MEK). Orange, dec. 235 to 250°.[294]
$[IrCl_4(PEt_2Ph)_2]$. XII. Violet, dec. 147 to 148°, mag. moment 1.9 μB.[324]
$[NiCl_2(PEt_2Ph)_2]$. X. Red, m. 112 to 113°.[838]
$[NiBr_2(PEt_2Ph)_2]$. X. Red, m. 114 to 116°.[337]
$[Ni(NCO)_2(PEt_2Ph)_2]$. X. (H_2O). Red-orange, m. 139 to 140°.[1150]
$[Ni(NO_3)_2(PEt_2Ph)_2]$. X. Green, m. 100 to 105°.[194]
$[Ni(CN)_2(PEt_2Ph)_2]$. From complex below, yellow. m. 169 to 170°.[1150]
$[Ni(CN)_2(PEt_2Ph)_3]$. X. (CH_2Cl_2). Red, m. 126 to 128°, visible spect.[1150]
$[PdCl_2(PEt_2Ph)_2]$. X. Yellow, m. 138°.[192]
$[PdBr_2(PEt_2Ph)_2]$. X. Yellow-orange.[409,757]
cis-$[PtCl_2(PEt_2Ph)_2]$. X. White, m. 202 to 203°,[836] NMR.[644]
trans-$[PtCl_2(PEt_2Ph)_2]$. X. Yellow, m. 123°.[836]
$[PtBr_2(PEt_2Ph)_2]$. X. Colorless, m. 193 to 194°.[409]

trans-$[PtI_2(PEt_2Ph)_2]$. X. Yellow, m. 137 to 138°.[836]
$[PtCl_2(PEt_2Ph)_3]$. X. Colorless, dec. 119 to 120°.[409]
$[PtBr_2(PEt_2Ph)_3]$. X. Yellow, dec. 121 to 122°.[409]
$[CuI(PEt_2Ph)]$. X. m. 153 to 155°.[265]
$[AgI(PEt_2Ph)]_4$. X. m. 138 to 139°.[264,265]
$[HgBr_2(PBu_2Ph)_2]$. NMR.[869]
$[ReCl_3(PBu_2Ph)_3]$. XIII. Yellow, mag. moment 2.1 μB,[324] NMR.[1135]
$[ReCl_4(PBu_2Ph)_2]$. XII. Purple, mag. moment 3.7 μB.[324]
$[RuCl_3(PBu_2Ph)_3]$. X. Brown, m. 136 to 139°, mag. moment 2.2 μB.[324]
$[OsCl_3(PBu_2Ph)_3]$. XIII. Red, dec. 160 to 164°, mag. moment 2.2 μB,[324] reduction.[320,326]
$[OsBr_3(PBu_2Ph)_3]$. XIII. Purple, dec. 158 to 160°.[324]
$[OsCl_4(PBu_2Ph)_2]$. XII. Yellow, m. 185 to 196°, mag. moment 1.5 μB.[324]
$[NiX_2(PBu_2Ph)_2]$. X.[457]
$[PdCl_2(PBu_2Ph)_2]$. X. Orange, m. 47°.[329]
$[PtCl_2(PBu_2Ph)_2]$. X. cis and trans, NMR.[644]
$[ReCl_3(PMePh_2)_3]$. XIII. Yellow.[324]
$[Ru_2Cl_3(PMePh_2)_6]Cl$. X. (EtOH). Orange, dec. 175 to 178°.[310]
$[Os_2Cl_3(PMePh_2)_6]Cl$. X. (EtOH). Yellow, dec. 185 to 186°.[310]
$[NiCl_2(PMePh_2)_2]$. X. Red-violet, dec. 148 to 150°, UV.[689]
$[NiBr_2(PMePh_2)_2]$. X. Green-brown, dec. 159 to 164°, UV, dipole.[689]
$[NiI_2(PMePh_2)_2]$. X. Brown-red, dec. 145 to 147°, UV, dipole.[689]
$[PtCl_2(PMePh_2)_2]$. XI. Yellow-white,[800] NMR.[644]
$[PtCl(PMePh_2)_3]Cl$. XI. Yellow, NMR.[644]
$[PtI_2(PMePh_2)_2]$. XI. Yellow.[800]
$[FeCl_2(PEtPh_2)_2]$. X. (C_6H_6). Cream, m. 185 to 187°, mag. moment, dipole.[194]
$[Ru_2Cl_3(PEtPh_2)_6]Cl$. X. (MeOH). Orange, dec. 135°.[310]
$[OsCl_3(PEtPh_2)_3]$. XIII. Red, m. 163 to 164°.[324]
$[Os_2Cl_3(PEtPh_2)_6]Cl$. X. (EtOH). Yellow, dec. 180 to 182°.[310]
$[CoCl_2(PEtPh_2)_2]$. X. Blue, m. 177 to 178°, mag. moment 4.5 μB.[199]
$[CoBr_2(PEtPh_2)_2]$. X. Green-blue, dec. 186 to 196°.[199,342]
$[Co(NCS)_2(PEtPh_2)_2]$. Green, m. 150 to 152°.[199]
$[Co(CN)_2(PEtPh_2)_3]$. XIV.[1147]
$[RhCl(PEtPh_2)_3]$. XIII. Orange, m. 160°.[1181]
trans-$[RhCl_3(PEtPh_2)_3]$. X. (EtOH). Orange, dec. >200°.[318,1181]
$[NiCl_2(PEtPh_2)_2]$. X. Red, dec. 146 to 151°, diamag., dipole.[687,689]
$[NiBr_2(PEtPh_2)_2]$. X. Green, m. 173 to 175°,[337] dipole, mag.[687,689]
$[NiI_2(PEtPh_2)_2]$. Brown-red, dec. 127 to 138°, dipole, mag.[687,689]

$[Ni(NO_3)_2(PEtPh_2)_2]$. X. Green, m. 155 to 175°.[194]
$[Ni(CN)_2(PEtPh_2)_2]$. X. (CH_2Cl_2). Yellow, dec. 215 to 217°.[1150]
$[AuCl(PEtPh_2)]$. X.[255]
$[Au_3Cl(PEtPh_2)_3H_2O]$. From above complex + $NaBH_4$.[255]
$[Au_3X(PEtPh_2)_3]$. From above complex, intermetallic bonds.[255]
$[HgBr_2(PBuPh_2)_2]$. NMR.[869]
$[CoBr_2(PBuPh_2)_2]$. X. Blue, m. 179 to 181°.[1112]
$[CoI_2(PBuPh_2)_2]$. X. Brown, dec. 180°.[1112]
$[NiX_2(PBuPh_2)_2]$. X.[457,689]
$[ZnCl_2(PPh_3)_2]$. X. m. 209°,[1144] m.,[295,398] IR.[481]
$[ZnBr_2(PPh_3)_2]$. X. m. 220 to 221°,[398,1144] IR.[481]
$[ZnI_2(PPh_3)_2]$. X. IR.[481]
$[CdCl_2(PPh_3)_2]$. X. IR.[481]
$[CdBr_2(PPh_3)_2]$. X. m. 225 to 226°,[554] IR.[481]
$[CdI_2(PPh_3)_2]$. X. m. 245°,[554] IR.[481]
$[HgCl_2(PPh_3)_2]$. X. White, dec. 273°,[554] IR.[481]
$[HgI_2(PPh_3)_2]$. X. (C_6H_6). Cream, m. 250°,[482,554] IR.[481]
$[HgCl_2PPh_3]_2$. X. White, m. 306 to 309°,[554] IR.[481]
$[HgBr_2PPh_3]_2$. X. Dec. 240 to 250°,[554] IR.[481]
$[HgI_2PPh_3]_2$. X. m. 242°.[482]
$[GaX_3PPh_3]$. X. IR.[87]
$[InCl_3(PPh_3)_2]$. X. White, m. 220 to 222°,[262] x-ray structure.[1319]
$[InBr_3(PPh_3)_2]$. X. (EtOAc). m. 182 to 184°.[262,1281]
$[InI_3(PPh_3)_2]$. X. (EtOAc). Yellow, dec. 135°; colorless, m. 195°.[262]
$[InI_3PPh_3]$. X. Colorless, from $InI_3(PPh_3)_2$.[262]
$[In(PPh_3)_4](ClO_4)_3$. X.[1281]
$[PbCl_4(PPh_3)_2]$. X. (CCl_4). Yellow, unstable.[388]
$[TiCl_4PPh_3]$. X. (Benzene). Red-violet, m. 147 to 149°, IR.[1341]
$[TiCl_4(PPh_3)_2]$. X. Red, diamag.,[312,578] m. 149°, IR,[1341] reaction with N_2.[1325]
$[VOCl_3PPh_3]$. X. Red-brown, dec. 210°.[1264]
$[CrCl_3PPh_3]$. X. (240°). Violet, m. 149°.[806]
$[(Mo_6Cl_8)Cl_4(PPh_3)_2]$. X. (THF). Yellow, UV spect.[566]
$[(Mo_6Cl_8)Cl_3(PPh_3)_3]Cl$. X. (THF). Yellow, Mo(II) cluster.[566]
$[MoOCl_3(PPh_3)_2]$. XI. Green.[531]
$[UCl_5PPh_3]$. XI. Green, m. >300°.[1201]
$[MnCl_2(PPh_3)_2]$. X. (THF). m. 232°, mag. moment 5.81 μB.[1036]
$[MnBr_2(PPh_3)_2]$. X. (THF). m. 229°, mag. moment 6.04 μB.[1036]
$[MnI_2(PPh_3)_2]$. X. (THF). m. 187°, mag. moment 5.92 μB.[1036]
$[ReX_3PPh_3]_n$. Polymers.[441]
$[ReCl_3PPh_3]$. X. (Acetone). Red-purple, dec. 230°.[423,591]
$[ReCl_4(PPh_3)_2]$. XII. Red-violet, dec. 226°,[296] mag. moment 3.84 μB.[945]
$[ReI_4(PPh_3)_2]$. X. Purple, paramag.[423]

$[ReCl_5PPh_3]^-$. XII.[594]
$[ReOX_3(PPh_3)_2]$. X.[335,450] Ligand displacement reactions.[296,851,1166]
$[ReO_2I(PPh_3)_2]$. Hydrolysis reaction.[584,585]
$[TcCl_4(PPh_3)_2]$. X. Green, mag. moment 3.92 μB.[565]
$Na_3[Fe(CN)_5PPh_3]$. XI. Diamag., IR.[1039]
$[FeCl_2(PPh_3)_2]$. X. (C_6H_6). Colorless.[194] X. (Fused). Yellow, m. 172°, mag. moment, dimer.[1034]
$[FeBr_2(PPh_3)_2]$. X. Yellow-orange, m. 158°, mag. moment,[1034] NMR.[1110]
$[FeI_2(PPh_3)_2]$. X. Yellow, m. 165 to 167°, mag. moment,[1034,1110] XI.[721]
$[FeCl_3(PPh_3)_2]$. X. (EtOAc). m. 113°, mag. moment 5.94 μB,[1035] reaction with N_2.[1325]
$[Fe(NO_3)_3(PPh_3)_2]$. X. (EtOAc). m. 175°, mag. moment 5.77 μB.[1035]
$[Fe(SCN)_3(PPh_3)_2]$. X. m. 120°, mag. moment 6.14 μB.[1035]
$[RuCl_2(PPh_3)_4]$. X. (MeOH). Brown, m. 130 to 132°.[1251]
$[RuBr_2(PPh_3)_4]$. X. (MeOH). Red-brown, m. 143 to 144°.[1251]
$[RuCl_2(PPh_3)_3]$. X. (MeOH). m. 132 to 134°,[1251] x-ray.[906]
$[RuBr_2(PPh_3)_3]$. X. (MeOH). m. 162°.[1251]
$[RuCl_3(PPh_3)_2]$. X. (MeOH). Brown, m. 160 to 170°,[324] MeOH.[1251]
$[RuBr_3(PPh_3)_2]$. X. (MeOH). m. 190°.[1251]
$[RuCl_4(PPh_3)_2]^-$. XII. Red.[1249]
$[OsBr_2(PPh_3)_3]$. X. $(MeOC_2H_4OH)$. Green, dec. 232°.[1305]
$[OsCl_3(NH_3)(PPh_3)_2]$. XVIII. Orange, x-ray.[213]
$[CoI(PF_3)_2(PPh_3)_2]$. XI.[893]
$[CoCl_2(PPh_3)_2]$. X. Blue, dec. 247 to 251°, mag. moment[221,342,444] reaction with butadiene.[1120]
$[CoBr_2(PPh_3)_2]$. X. Blue-green, m. 234 to 239°,[342] mag. moment.[444]
$[CoI_2(PPh_3)_2]$. X. Brown and green forms, dipole, mag. moment.[444,810]
$[Co(SCN)_2(PPh_3)_2]$. X. Green, m. 140°, mag. moment.[444,1175]
$[RhCl(PPh_3)_3]$. X. (EtOH). Red-purple, dec. 138°, absorbs H_2,[149,1373] x-ray.[758]
$[RhBr(PPh_3)_3]$. XIV. Orange, dec. 130°.[149,1072]
$[RhI(PPh_3)_3]$. X. Red, dec. 116°.[149,1072]
$[IrCl_4(PPh_3)_2]$. XII. Violet, dec. 161 to 163°.[151,324,1304]
$[IrCl_3(PPh_3)_2]$. XII. Orange.[151]
$[IrCl(PPh_3)_3]$. XI. Orange, dec. 212°.[151]
$[IrBr(PPh_3)_3]$. XI. Red-brown.[151]
$[NiCl_2(PPh_3)_2]$. X. (CH_3COOH). Blue, dec. 247 to 250°, dipole, mag.,[1320] x-ray.[597]
$[NiBr_2(PPh_3)_2]$. X. (BuOH). Green, dec. 222 to 225°, dipole,[1320] mag. moment 3.2 μB,[972] x-ray.[826]
$[NiI_2(PPh_3)_2]$. X. (BuOH). Brown, dec. 218 to 220°, dipole, mag.[1320]
$[Ni(NO_3)_2(PPh_3)_2]$. X. (CH_3COOH). Green, dec. 224 to 227°, dipole, mag.[1320]

$[Ni(SCN)_2(PPh_3)_2]$. X. (BuOH). Red, dec. 217 to 218°, diamag.[1320]
$[Ni_2(CN)_4(PPh_3)_2]$. XIV. Grey-green.[113]
$[NiX(PPh_3)_3]$. XI. Mag. moment ca. 1.9 μB.[1119]
$[PdCl_2(PPh_3)_2]$. X. Yellow, dec. 250 to 270°.[329]
$[PdBr_2(PPh_3)_2]$. X. Orange, dec. 250°.[380]
$[PdI_2(PPh_3)_2]$. X. Red, dec. 250°.[380]
$[PdCl_2PPh_3]_2$. XA. Brown-orange, dec. 250 to 270°,[354] IR.[622]
$[Pd(NO_2)_2(PPh_3)_2]$. X.[234]
$[Pd(NCO)_2(PPh_3)_2]$. XI[1056] and X.[123] From azide + CO.[123]
$[Pd(NCS)_2(PPh_3)_2]$. X.[115] IR.[233]
trans-$[Pd(N_3)_2(PPh_3)_2]$. X. Yellow.[125]
$[Pd(N_3)_2PPh_3]_2$. X. Red.[124]
$[Pd(CN)_2(PPh_3)_2]$. XII. (C_2N_2). m. >300°.[61]
$[Pd(CO_3)(PPh_3)_2]$. From $[Pd(PPh_3)_4]$ + O_2 + CO_2, yellow, dec. 91 to 93°.[1062]
$[Pd(OAc)_2(PPh_3)_2]$ and $[Pd(OAc)_2PPh_3]_2$. X.[1250,1252]
$[PtF_2(PPh_3)_2]$. XII. White.[981]
cis-$[PtCl_2(PPh_3)_2]$. X. White, dec. 310°,[836] imino deriv. by reaction with N_2H_4.[505]
trans-$[PtCl_2(PPh_3)_2]$. From hydride, yellow, m. 307 to 310°.[25]
cis-$[PtBr_2(PPh_3)_2]$. X and XII. Orange, dec. 300°.[944]
cis-$[PtI_2(PPh_3)_2]$. X. Orange-yellow, m. 285°.[944]
$[Pt(NO_2)_2(PPh_3)_2]$. X.[234]
$[Pt(NCO)_2(PPh_3)_2]$. XIV. IR.[1056]
$[Pt(NCS)_2(PPh_3)_2]$.[233]
$[Pt(N_3)_2(PPh_3)_2]$. X. Colorless, dec. 225°.[125]
$[Pt(CN)_2(PPh_3)_2]$. XII. (C_2N_2). m. >300°.[61] XIV.[73]
$[Pt(CO_3)(PPh_3)_2]$. XIV. Colorless, m. 202 to 205°, IR, also prepared from O_2 complex + CO_2,[1062] x-ray.[251]
$[Pt(C_2O_4)(PPh_3)_2]$. XIV. Photochem. dec. to $[Pt(PPh_3)_2]$.[179]
$[PtCl_2PPh_3]_2$. XA. Orange, dec. 270 to 280°,[623] IR.[622]
$[PtBr_2PPh_3]_2$. XA. Red-brown, dec. 260 to 280°.[12]
$[PtI_2PPh_3]_2$. XA. Red, dec. >300°.[944]
$[CuClPPh_3]_4$. X. (C_6H_6). White, m. 232°.[440]
$[CuCl(PPh_3)_2]$. X. (Melt). m. 173 to 175°.[110]
$[CuCl(PPh_3)_4]$. X. White, m. 166°.[440]
$[Cu_2Cl_2(PPh_3)_3]$. X. White, m. 236 to 237°.[440]
$[CuN_3(PPh_3)_2]$. X. Dec. 185°.[1381]
$[Cu(NCS)(PPh_3)_2]$. From azide + CS_2 (UV/$CHCl_3$), m. 210 to 213°.[1381]
$[Cu(NO_3)(PPh_3)_2]$. XIII. (MeOH). Colorless, m. 237°,[447] x-ray.[1001]
$[Cu(PPh_3)_4]ClO_4$. XIII. (EtOH). Colorless, dec. 275°.[447]
$[AgClPPh_3]_4$. X. (Pyridine). m. 294°.[253]
$[AgBrPPh_3]$. X. (Melt). m. 263 to 265°.[110]
$[AgIPPh_3]_4$. X. (Pyridine).[253]
$[AgCl(PPh_3)_3]$. X. $(CHCl_3)$. m. 133°.[253]
$[Ag(PPh_3)_4]ClO_4$. X. (EtOH). Colorless, m. 290°.[447]
$[Ag(PPh_3)_4]NO_3$. X. (EtOH). Colorless, m. 215°.[447]

$[Ag(OCOCH_3)PPh_3]$. XIV. (C_6H_6). White, m. 185 to 186°, IR.[1050]
$[AuClPPh_3]$. XIII.[889]
$[AuBrPPh_3]$. XIV. White, m. 250 to 251°, μ 8.6 D.[1340]
$[AuIPPh_3]$. XIV. m. 230 to 231°, μ 8.2 D.[1340]
$[AuCNPPh_3]$. XIV. White, m. 197 to 198°, IR.[1050]
$[AuN_3PPh_3]$. XIV. White, dec. 189 to 192°, IR.[1050]
$[Au(NCO)PPh_3]$. XIV. White, m. 200 to 202°, IR.[1050]
$[Au(NO_3)PPh_3]$. XIV. White, m. 165 to 168°, IR.[1050]
$[Au(PPh_3)_2]X$. X. IR, conductivity.[1002]
$Au_5Cl(PPh_3)_4$. Solvate from $AuClPPh_3$ + $NaBH_4$.[947]
$[Au_{11}(SCN)_3(PPh_3)_7]$. X-ray.[990]
$[AuBr_3PPh_3]$. XII. Scarlet, m. 155 to 157°, IR.[1050]
$[FeBr_2\{P(p\text{-}MeC_6H_4)_3\}_2]$. X. Yellow, PMR.[1110]
$[NiX_2\{P(p\text{-}MeC_6H_4)_3\}_2]$. X. Dipole, mag. moment.[221]
$[RhCl_2\{P(o\text{-}MeC_6H_4)_3\}_2]$. X. (EtOH). Blue-green, mag. moment 2.3 μB.[144,150]
trans-$[PdCl_2\{P(o\text{-}MeC_6H_4)_3\}_2]$. X. Yellow.[144,150]
$[PtI_2\{P(p\text{-}MeC_6H_4)_3\}_2]$. X. Yellow, m. 263°.[944]

D.3.4. Typical Carbonyl Halide Complexes

$[CrX(CO)_4PR_3]^-$. V.[26]
$[MoX(CO)_4PR_3]^-$. V.[26]
$[MoX(CO)_3PPh_3]$ and $[MoX(CO)_2(PPh_3)_2]$. V.[767]
$[MoX_2(CO)_2(PPh_3)_2]$. V.[425]
$[MoCl_2(CO)_2(PMe_2Ph)_3]$. V.[830]
$[MoX_2(CO)_3(PPh_3)_2]$. IXA.[424]
$[MoX_3(CO)_3PPh_3]^-$. V.[1277]
$[(\pi\text{-}C_5H_5)MoX(CO)_2PPh_3]$. V.[1273]
$[WX_2(CO)_2(PPh_3)_2]$. V.[424]
$[WX_2(CO)_3(PPh_3)_2]$. IXA.[424]
$[WX_3(CO)_3PPh_3]^-$. V.[1277]
$[MnX(CO)_4PPh_3]$. XIIA.[854]
$[Mn(SCN)(CO)_4PPh_3]$. V.[559]
$[MnX(CO)_3(PR_3)_2]$. V.[6,17,44,45,748,755] XIIA. Reactions.[1288]
$[Mn(NCS)(CO)_3(PPh_3)_2]$. V.[1355]
$[ReX(CO)_4PR_3]$. XIIA.[582,854,1019]
$[ReX(CO)_2(PPh_3)_2]$ and $[ReX(CO)_3(PPh_3)_2]$. V.[5,6] IXA.[591,593]
$[ReCl(CO)_2(PMe_2Ph)_3]$ and $[ReCl(CO)_3(PMePh)_2]$. IXA.[515]
$[ReX_2(CO)(PPh_3)_2]$. IXA.[589]
$[ReCl_3(CO)(PMePh)_3]$. IXA.[515]
$[ReCl_3(CO)_2(PMe_2Ph)_2]$. XIIA.[515]
$[Re(CO)_4(PPh_3)_2]Cl$. V.[749]
$[FeX_2(CO)_2(PR_3)_2]$. V.[752] IXA.[194] XIIA.[235,742]
$[FeX_2(CO)_3PR_3]$. XIIA.[742]
$[RuX_2(CO)(PR_3)_3]$. V.[665,829,927]
$[RuX_2(CO)_2PR_3)_2]$. V.[416,665,730,731,829,927]
$[RuX_2(CO)_3PPh_3]$ and $[RuX(CO)_2PPh_3]_2$. XIIA.[223,848]
$[RuBr_3(CO)(PPh_3)_2]$. IXA.[1305]
$[OsX_2(CO)_3PR_3]$. XIIA.[486]

$[OsX_2(CO)_2(PPh_3)_2]$. V.[724] IXA.[419]
$[OsX_3(CO)(PPh_3)_2]$. IXA.[1305]
$[CoX(CO)_3PPh_3]$. XIIA.[716,717]
$[CoX(CO)_2(PPh_3)_2]$. V.[718] XIIA.[1171]
$[CoX_2(CO)(PEt_3)_2]$. IXA.[194]
$[RhX(CO)_2PPh_3]$. V.[495]
$[RhX(CO)(PR_3)_2]$. V.[344,490,730,732,883,933,1297,1299]
IXA.[344,781,1139,1169]
$[RhX_3(CO)PPh_3]$. XIIA.[495]
$[RhX_3(CO)(PR_3)_2]$. IXA.[344] XIIA.[1139,1298]
$[RhCl_2Et(CO)(PMe_2Ph)_2]$. V.[1124]
$[IrX(CO)(PR_3)_2]$. V.[722] IXA.[151,319,421,1314]
$[IrCl(CO)_2(PPh_3)_2]$. X-ray.[1094]
$[IrI(CO)_2(PPh_3)_2]$ and $[IrI_2(CO)(PPh_3)_2]$. V.[52]
$[IrX_3(CO)(PR_3)_2]$. IXA.[316,1207] XIIA.[420,1313]
$[IrHX_2(CO)(PPh_3)_2]$. IXA.[1314] Dehydrohalogenation.[488]
$[Ir_2Cl_4(COMe)_2(CO)_2(PMe_2Ph)_2]$. V.[75]
$[PdX(CO)(PR_3)_2]^+$. IXA.[380,381]
$[PtX(CO)(PR_3)_2]^+$. IXA.[372,379,381] Reactions.[382]
$[PtX_2(CO)PR_3]$. IXA.[317]
$[PtF_2(CO)_2(PPh_3)_2]$. IXA.[981]

D.3.5. Mixcellaneous Tertiary Phosphines

(1) Forming carbonyl complexes:

$(CF_3)_2PF$, CF_3PF_2. Ni.[232]
$P(CF_3)_3$. Ni.[230,539]
CF_3PF_2, CCl_3PF_2, $(C_3F_7)_2PF$. Mo.[92]
$P(CH_2Cl)_3$. Mo.[835]
$P(CMe_3)_3$. Ni.[1196]
$P(MMe_3)_3$ (M = Si, Ge, Sn or Pb). Ni.[1195]
$P(C_2H_4CN)_3$. Ni.[993,995]
$[Ni_4(CO)_6\{P(C_2H_4CN)_3\}_4]$. X-ray.[154]
PPh_2Cl. Mo.[2,223]
$PPhCl_2$. Mo.[2]
$MePFNMe_2$ and $PhPFNEt_2$. Ni, Mo.[1189]
$C_6F_5PPh_2$. Fe, Mo.[762]
$P(p\text{-}FC_6H_4)_3$. Mo, W, Fe.[497]

(2) Forming halide complexes:

PMe_2CF_3. Ni,[130] Pt.[129]
$PMe(CF_3)_2$. Pt.[129]
$P(c\text{-}Pr)_3$. Ni.[1220]
PAm_3. Cu.[958]
PPe_3. Pt.[360]
$P(CH_2Cl)_3$. Ni,[457] Pd.[835]
$P(CH_2CH_2CN)_3$. Co, Ni.[518,1327]
PBz_3. Ni.[222]
PBz_2Ph. Ni,[222] Pd.[953]

$PBzPh_2$. Ni,[222] x-ray.[874]
$PPrPh_2$. Ni.[689]
PPr^iPh_2. Ni.[689]
PBu^iPh_2. Ni.[689]
$PAllPh_2$. Ni.[222]
$PAll_2Ph$. Pt.[1167]
$PMeBu^tPh$. Pt.[271]
$PEtCy_2$. Ni.[1150]
PPh_2CF_3. Pd, Pt.[1145]
$PPh(CF_3)_2$. Pt.[131]
$PPh(C_2H_4CN)_2$. Pd.[952]
$PPh_2(C_2H_4CN)$. Pd.[952]
$P(C_6F_5)_3$. Rh, Pd, Pt.[870]
$PMe_2(C_6F_5)$. Ni.[35]
$P(p\text{-}ClC_6H_4)_3$. Pt.[944]
$P(p\text{-}MeOC_6H_4)_3$. Pt.[944]
$PEtPh(p\text{-}BrC_6H_4)$. Pd.[462]
PPh(α-naphthyl)(4-biphenyl). Rh.[666]
$p\text{-}CF_3C_6H_4PEt_2$. Cd, Hg, Ni, Cu, Ag.[265]
$p\text{-}Me_2NC_6H_4PMe_2$. Zn, Cd, Hg, Co, Ni, Cu, Ag.[265]
$PPh(p\text{-}BrC_6H_4)(p\text{-}Me_2NC_6H_4)$. Pd.[462]
$PPh(NEt_2)_2$. Co, Ni, Pd, Pt, Cu.[902]
$PPh(OEt)_2$. Ni.[1245]
PCy_2(N-pip). Cu, Ag.[1222]

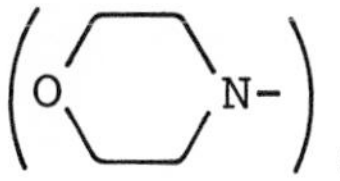
P. Co, Ni.[512]

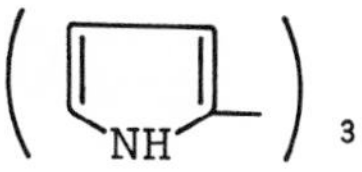
P and
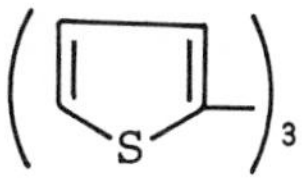
P. Cu.[803]

D.4. Cyclic Monodentate Tertiary Phosphines

D.4.1. Forming Carbonyl Complexes

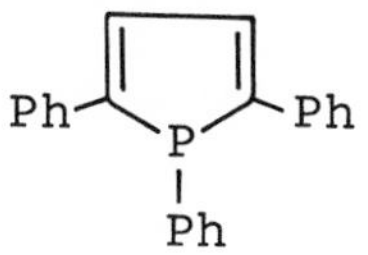
Cr, Mo, W, Fe, Ni.[434]

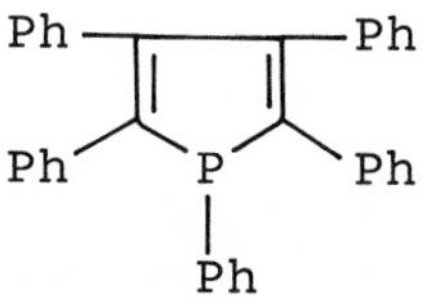
Fe.[205]

D.4.2. Forming Halide Complexes

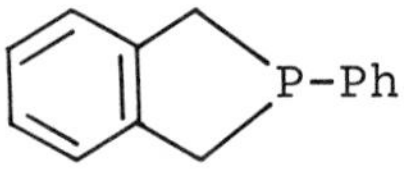

Co, Rh, Ir,[408] Ni, Pd, Pt.[409,953,960] Cu, Ag, Au.[407,953]

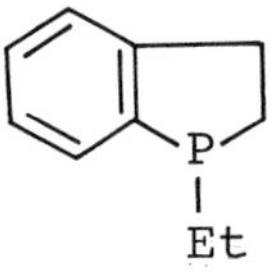

Pd.[952]

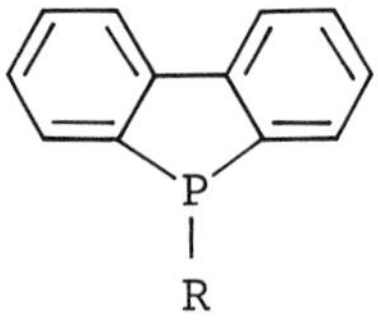

(R = Ph). Ni, Pd, Pt.[30]
(R = Me or Et). Co, Ni, Pd, Pt.[30,31]

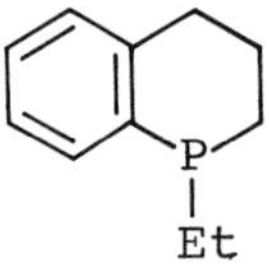

and

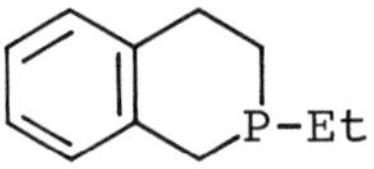

Pd.[127]

D.5. Complexes from Multidentate Phosphines

D.5.1. With No Other Ligands

$[V(Me_2PCH_2CH_2PMe_2)_3]$. IV. Black, dec. 253 to 254°, mag. moment 2.1 μB.[357]
$[Cr(Me_2PCH_2CH_2PMe_2)_3]$. IV. Yellow, dec. 293°.[357]
$[Mo(Me_2PCH_2CH_2PMe_2)_3]$. III. Yellow, dec. 140°.[356,357]
$[W(Me_2PCH_2CH_2PMe_2)_3]$. IV. Yellow, dec. 350°.[357]
$[Fe(Me_2PCH_2CH_2PMe_2)_2]$. IV. Brown, dec. 205°.[357]
$[Co(Me_2PCH_2CH_2PMe_2)_2]$. IV. Orange, m. 101 to 102°.[357]
$[Ni(Me_2PCH_2CH_2PMe_2)_2]$. IV. White, m. 120°, dipole.[307]
$[Pd(Me_2PCH_2CH_2PMe_2)_2]$. IV. White, m. 182 to 183°, dipole.[307]
$[Pt(Me_2PCH_2CH_2PMe_2)_2]$. IV. Colorless.[334]
$[Ni(Ph_2PCH_2PPh_2)_2]$. III.[134]
$[Pd(Ph_2PCH_2PPh_2)_2]$. IV. Scarlet, m. 195 to 210°.[307]
$[Pt(Ph_2PCH_2PPh_2)_2]$. II. Yellow, dec. >300°.[196]
$[Mo(Ph_2PCH_2CH_2PPh_2)_3]$. III. Red-orange, dec. 281.5°.[356]
$[Co(Ph_2PCH_2CH_2PPh_2)_2]$. IV. Red, dec. 280°.[306,1178]
$[Ni(Ph_2PCH_2CH_2PPh_2)_2]$. II. Orange, m. 253 to 256°, dipole,

IR.[301] III.[134,1302]

$[Pd(Ph_2PCH_2CH_2PPh_2)_2]$. IV. Yellow, m. 234°, dipole.[307]

$[Pt(Ph_2PCH_2CH_2PPh_2)_2]$. IV. Yellow, m. 252 to 255°.[334]

$[Ni(Ph_2PCH_2CH_2CH_2PPh_2)_2]$. III.[1302]

$[Mo\{o\text{-}C_6H_4(PEt_2)_2\}_3]$. III. Red-black, m. 236.5°.[356]

$[Co\{o\text{-}C_6H_4(PEt_2)_2\}_2]$. I. Red, sublimes 160° (0.01 mm).[306]

$[Ni\{o\text{-}C_6H_4(PEt_2)_2\}_2]$. I and II. Scarlet, m. 241 to 243°, dipole, IR.[301]

$[Pd\{o\text{-}C_6H_4(PEt_2)_2\}_2]$. IV. Yellow, m. 229 to 230°.[307] I. (200°).[306]

$[Pt\{o\text{-}C_6H_4(PEt_2)_2\}_2]$. II. Orange, dec. 248°.[196]

$[Ni\{o\text{-}C_6H_4(PPh_2)_2\}_2]$. I and II. Red, dec. 260 to 350°, dipole, IR.[301]

$[Ni\{MeC(CH_2PPh_2)_3\}_2]$. IV. α and β forms, orange.[307]

$[Pd\{MeC(CH_2PPh_2)_3\}_2]$. IV. α and β forms, yellow, dipole.[307]

D.5.2. Carbonyl Complexes Derived from $Ph_2PCH_2CH_2PPh_2$ (diph)

$[Co(CO)_4diph]$. V. Yellow, m. 211 to 212°.[358,1376]

cis-$[Cr(CO)_2diph_2]$. VI. Orange, m. 280°, dipole.[358]

trans-$[Cr(CO)_2diph_2]$. VI. Red, m. 279 to 280°, dipole.[358]

$[Cr_2(CO)_6diph_3]$. VI. Yellow, dec. 263°.[1376]

$[Mo(CO)_4diph]$. V. White, m. 193 to 194°.[358,1376]

cis$[Mo(CO)_2diph_2]$. VI. Yellow, m. 324 to 325°, dipole,[358] halogenation.[919]

$[Mo_2(CO)_6diph_3]$. VI. Colorless, dec. 210°.[1376]

$[W(CO)_4diph]$. V. Yellow, m. 208 to 209°.[358,1376]

cis-$[W(CO)_2diph_2]$. VI. Yellow, m. 320 to 323°, dipole,[358] halogenation.[919]

$[W_2(CO)_6diph_3]$. VI. Yellow, dec. 194°.[1376]

$[Mn(CO)_2diph_2]^+$ and $[Mn(CO)_2diph_2]^{2+}$, oxidation reaction.[1076,1237]

$[Mn(CO)_3diph]$. V. Mag. moment 1.7 μB.[1172]

$[MnX(CO)_3diph]$. V.[1074,1076]

$[Mn_2(CO)_8diph]$. V. Yellow, diamag.[727]

$[Re(CO)_3diph]$. V. cis and trans, monomer and dimer.[582]

$[Re(CO)_4diph]^+$. V.[5]

$[Fe(CO)_3diph]$. V. Yellow, dec. 135°,[965] IR,[1138] mass spect.[465]

$[Fe(CO)_4]_2diph$. V. (UV).[918] Mass spect.[465]

$[Co_2(CO)_4diph_2]$. VI. Purple-brown.[132]

$[Co_2(CO)_4diph_3]^{2+}$. V.[1172]

$[Ir(CO)diph_2]^+$. X-ray.[827]

$[Ni(CO)_2diph]$. V. Colorless, m. 138 to 140°, dipole,[301] NMR.[996]

D.5.3. Other Di-, Tri-, and Tetratertiary Phosphines Forming Carbonyl Complexes

Me_2PPMe_2. Mn,[692] Cr, Mo, W,[349,350,687] Fe, Ni.[686]

$[(CO)_4Fe(PMe_2)_2Fe(CO)_4]$. X-ray.[828]

Cy_2PPCy_2. Fe, Ni.[809]
Ph_2PPPh_2. Cr,[350] Rh, Ir,[738] Fe, Co, Ni,[112] Ni,[686] x-ray.[936]
$(CF_3)_2PP(CF_3)_2$. Ni.[230]
$Ph_2PCH_2PPh_2$. Cr, Mo, W,[358] Fe,[660] Ni.[1301]
$Ph_2P(CH_2)_3PPh_2$. Fe,[660] Ni.[1301]
$Me_2PCH_2CH_2PMe_2$. Mn,[692] Ni.[195]
$Et_2PCH_2CH_2PEt_2$. Cr, Mo, W,[358] Ni.[301]
$(CNC_2H_4)_2PCH_2CH_2P(C_2H_4CN)_2$. Ni.[36,629]
$(CF_3)_2PCF_2CF_2P(CF_3)_2$. Ni.[232]
$Ph_2PCH{=}CHPPh_2$. Cr, Mo, W, Fe, Ni.[878]
$Ph_2PC{\equiv}CPPh_2$. Ni.[261]
o-$C_6H_4(PMe_2)_2$. Ni.[301]
o-$C_6H_4(PEt_2)_2$. Cr, Mo, W,[358] Ni.[301]
o-$C_6H_4(PPh_2)_2$. Ni.[301]

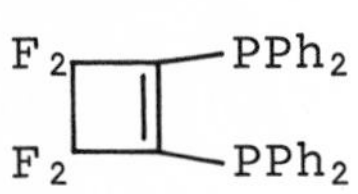

and
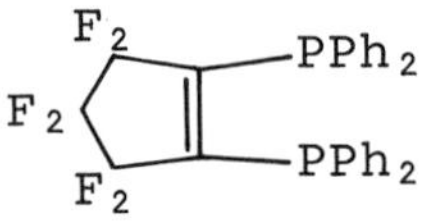

Cr, Mo, W,[463,464] Ni.[463]

$B_{10}H_{10}(CPPh_2)_2$. Ni.[1154,1374]
$(Ph_2PCH_2)_3CMe$. Cr, Mo, W,[358] Ni.[304]
(o-$Et_2PC_6H_4)_2PPh$. Mo,[358] Ni.[304]
(o-$Ph_2PC_6H_4)_2PPh$. Cr,[769] Mo.[358]
(o-$Ph_2PC_6H_4)_3P$. Cr,[769] Mo, W,[768] Mn.[366]
$C(CH_2PPh_2)_4$. Cr, Mo, W,[535] Ni.[534]

D.5.4. Halide Complexes Derived from $Ph_2PCH_2CH_2PPh_2$ (diph)

[$ZnCl_2$diph]. X. m. 290 to 291°.[398]
[$ZnBr_2$diph]. X. m. 287°.[398]
[$CdCl_2$diph]. X. m. 265 to 266°.[398]
[$CdBr_2$diph]. X. m. 285 to 286°.[398]
[$(HgCl_2)_2$diph]. X. m. 272 to 273°.[398]
[$HgBr_2$diph]. X.[398]
[$InBr_3$diph]. X. (EtOAc). Colorless, m. 223 to 224°.[262]
[$TiCl_4$diph]. X. Red, dec. 178°.[1341]
[$VOBr_2$diph]. X. Green.[1203]
[$(Mo_6Cl_8)Cl_3$diph]. X. (THF). Mo(II) cluster.[566]
[WCl_3diph]. XI. Brown, dimeric, unstable.[189]
[WCl_3(diph)$_2$]. XI. Yellow, air-stable, in equilibrium with [WCl_2(diph)$_2$]Cl.[189]
[WCl_4diph]. X. (220°). Yellow-brown, IR, mag. moment.[189]
[WCl_4diph]Cl. X. Green, mag. moment 0.9 μB.[190]
[$(UCl_4)_2$diph]. X. (THF).[1202]
[UCl_5diph]. XI. Green, m. >300°, mag. moment 2.7 μB.[1201]
[$ReCl_3$diph]$_n$. Purple.[441]
[Re_3Cl_9(diph)$_{1.5}$]. X. (Acetone). Red, UV.[454]
[ReO_2(diph)$_2$]X. X. Ionic, IR.[583]
[$TcCl_2$(diph)$_2$]Cl. X. Red-orange, mag. moment 2.68 μB.[565]
trans-[$RuCl_2$(diph)$_2$]. X and XI. Orange-yellow, m. 284 to

285°.[310]
trans-$[OsCl_2(diph)_2]$. X and XI. Yellow, dec. 293 to 296°.[310]
$[CoBr_2(diph)_2]$. X. Black, dec. 120 to 125°.[306]
$[CoX_2diph]$. X. (Acetone). Blue, spect., and mag. properties.[766]
$[CoX_2(diph)_2]$. X. (i-PrOH). Green, spect., and mag. properties.[766]
$[Co(NCS)_2(diph)_2]$. XIV.[1176]
$[Co(NO_3)_2(diph)_2]$. XIV.[1176]
$[Co(diph)_2]^+$. XI.[1179]
$[Rh(diph)_2]^+$. XI.[933,1180]
$[Ir(diph)_2]^+$. XI.[1179]
$[NiCl_2diph]$. X. (EtOH). Orange,[195] IR.[307]
$[NiBr_2diph]$. X. (EtOH). Red,[195] IR.[307]
$[NiI_2diph]$. X. Mauve.[772]
$[Ni(SCN)_2diph]$. X. Yellow, IR.[307]
$[NiBr_2(diph)_2]$. X. (CH_2Cl_2). Yellow.[195]
$[NiI_2(diph)_2]$. XIV. Yellow.[195]
$[Ni(diph)_2](NO_3)_2$. XIV. Orange, m. 217 to 230°.[307]
$[PdCl_2diph]$. X. (DMF). White, m. >360°.[193,1339]
$[PdCl_2(diph)_2]$. X. White, m. 283 to 306°.[1339]
$[PdBr_2diph]$. X. Pale yellow, dec. 333°.[193]
$[PdBr_2(diph)_2]$. XII. White, m. 240°.[307]
$[PdI_2(diph)_2]$. XIV. Yellow, m. 281 to 284°.[1339]
$[PtCl_2diph]$. X. White, m. >300°.[193,1339]
$[PtCl_2(diph)_2]$. X. m. 268 to 272°.[1339]
$[PtBr_2(diph)_2]$. XIV. m. 301 to 316°.[1339]
$[Pt(NO_3)_2(diph)_2]$. XIV. Dec. 310 to 311°.[1339]
[ClAu(diph)AuCl]. X.[1037] Forms cluster compounds (below) with $NaBH_4$.[1037]
$[Au_6(diph)_2X_2]$. Red-violet, polymeric cluster.[254]
$[Au_6(diph)_2X]Y$. Brown, ionic cluster.[254]
$[Au_6(diph)_3]X_2$. Green, ionic cluster.[254]

D.5.5. Other Di-, Tri-, and Tetratertiary Phosphines Forming Halide Complexes

Cy_2PPCy_2. Co, Ni, Pd, Cu.[813]
(pip)CyPPCy(pip). Ag, Cu.[1222]
Ph_2PPPh_2. Co, Ni, Pd, Cu.[812]
(pip)PhPPPh(pip), (pip)CyPP(Cy)$_2$. Cu, Ag.[1223]
$Ph_2PCH_2PPh_2$. (VO^{2+}),[1203] Ru, Os,[310] Ni,[195,576,1302] Pd.[307]
$Me_2PCH_2CH_2PMe_2$. Ti,[386] U,[963] Fe(II and III),[311] Ru, Os,[310] Rh,[276] Ni.[195]
$Et_2PCH_2CH_2PEt_2$. Ti(III and IV),[312] Cr(SCN),[805] Fe(II and III),[311] Ru, Os,[310] Zn, Cd, Hg, Co, Ni, Pd, Ag, Au.[1359]
$Cy_2P(CH_2)_nPCy_2$. (n = 3, 4, 5). Fe, Co, Ni, Cu.[808]
$MePhPCH_2CH_2PPhMe$. Ru,[161] x-ray.[863]
$Ph_2P(CH_2)_3PPh_2$. Co,[766,1147] Ni.[435,1148,1302]
$Ph_2P(CH_2)_4PPh_2$. Co, Ni,[1183,1186] Ni.[435,1148]

$Ph_2P(CH_2)_5PPh_2$.   Co, Ni.[1183]
$Ph_2P(CH_2)_2O(CH_2)_2PPh_2$.   Co, Ni.[1183]
MePhPCH=CHPPhMe.   Ru.[161]
$Ph_2PCH=CHPPh_2$.   Co, Ni,[1134] Ni.[980]
$Ph_2PC{\equiv}CPPh_2$.   Pd, Pt,[260] Cu, Au,[258,259] Ag, Hg.[40]
$o-C_6H_4(PMe_2)_2$.   Ti (8-coordinated).[386]
$o-C_6H_4(PEt_2)_2$.   Fe(II and III),[311] Ru, Os,[310] Ni, Pd,[307]
    Ni.[301]

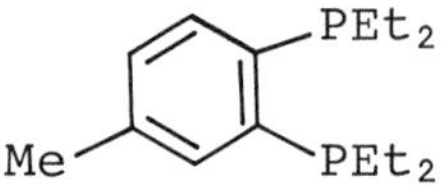

Pd, Au.[669]

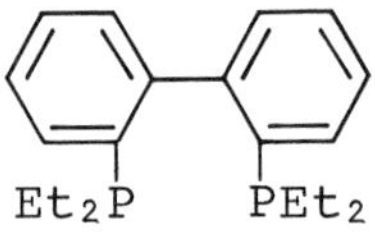

Ni, Pd, Au.[29,32]

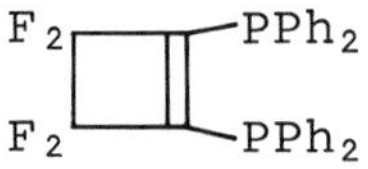

Hg, Pd, Pt.[463]

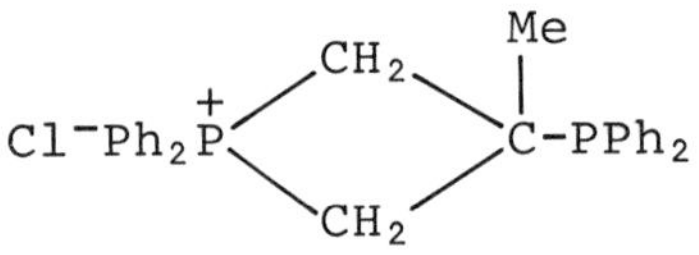

Co, Ni, Pd, Au.[162]

$(o-Ph_2PC_6H_4)_2PPh$.   Cr.[769]
$(o-Ph_2PC_6H_4)_3P$.   Cr,[769] Fe(II),[575,662] Co,[186,670] Ir,[477]
    Ni,[524] Pd, Pt.[671]
$C(CH_2PPh_2)_4$.   Co,[537] Ni.[653]

   D.5.6.   Cyclic Multidentate Phosphine Complexes (see
           also Chapter 2)

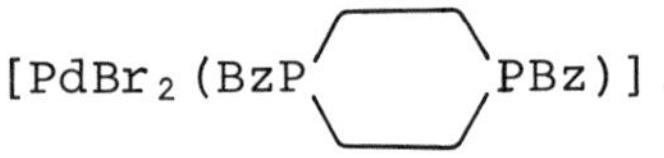

$[PdBr_2(BzP\qquad PBz)].$[756]

$[ClAuP\qquad PAuCl].$[756]

$[(EtP)_4W(CO)_4].$[466]
$[(PhP)_4M(CO)_4].$ Cr, Mo, W.[577]
$(PhP)_4[Fe(CO)_4]_2.$[42,43]
$[(PhP)_4Ni(CO)_3].$[42,43]
$[(EtP)_5Mo(CO)_4].$[466]   X-ray.[236]
$[(PhP)_5Mo(CO)_5].$[42]
$(CF_3P)_4[Ni(CO)_3]_n.$[230]
$[(EtP)_4PdCl_2].$[466]
$[(RP)_4CuX].$ (R = Et),[466] (R = Bu^i, Oct, Cy).[708]

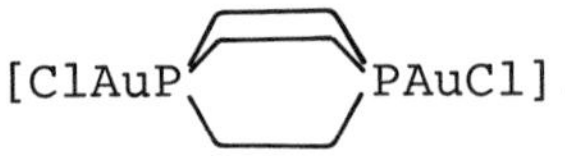

D.6. Unsymmetrical Multidentate Ligands, at Least One Donor Center Being a Phosphine

D.6.1. Forming Carbonyl Complexes

(a) P-N

$Ph_2PCH_2CH_2NEt_2$. Cr, Mo, W.[506]
$(CF_3)_2P(NMe)P(CF_3)_2$. Ni.[231]
$Ph_2PNEtPPh_2$. Mo.[1093]

(b) P-S

$Ph_2PCH_2CH_2SMe$. Cr, Mo, W, Mn.[1159]
$(CF_3)_2PSP(CF_3)_2$. Ni.[231]

(c) P-‖

$CH_2{=}CHCH_2CH_2PR_2$ (R = Et or Cy). Cr.[807]

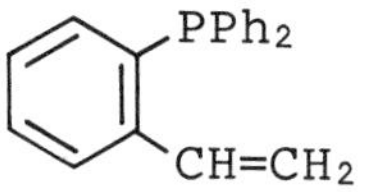

Cr, Mo, W,[152] Mn, Re.[799]

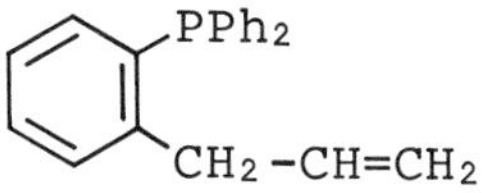

and

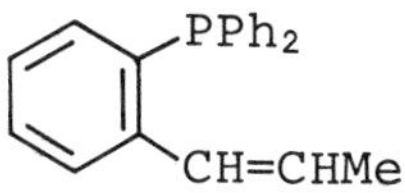

Cr, Mo, W,[146,798,928] Mn, Re.[799]

D.6.2. Forming Halide Complexes

(a) P-N

$N(CH_2CH_2PPh_2)_3$. Co, Ni.[469,1069,1182]
$Ph_2PNEtPPh_2$. Pd.[1093]
$R{\cdot}N(CH_2CH_2PPh_2)_2$. Co, Ni.[1068,1184]
$N(CH_2CH_2NEt_2)_2(CH_2CH_2PPh_2)$ and $N(CH_2CH_2NEt_2)(CH_2CH_2PPh_2)_2$. Co, Ni.[1185]
$(o\text{-}Me_2NC_6H_4)_3P$, $(o\text{-}Me_2NC_6H_4)_2PPh$, and $(o\text{-}Me_2NC_6H_4)PPh_2$. Co, Ni,[370] Pd, Pt.[595,1267]

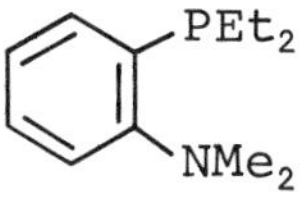

Pd.[959]

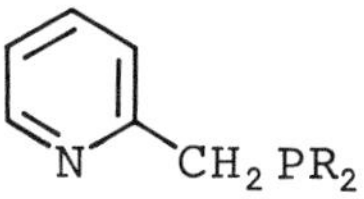

(R = Et or Ph) Co, Ni.[1292]

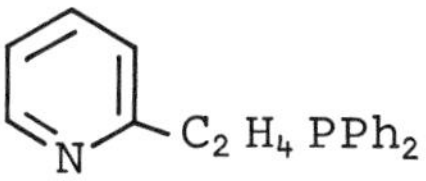

Zn, Co, Ni.[1291]

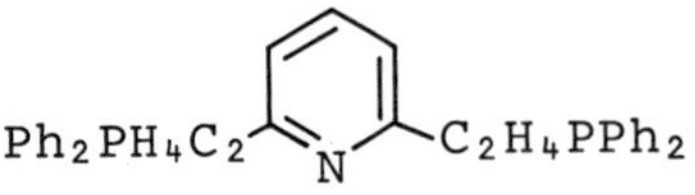

Ni.[1045]

(b) P-S and P-Se

$Et_2PCH_2CH_2SEt$. Ni.[1221]

$R_2PCH_2CH_2SH$, $RP(CH_2CH_2SH)_2$, $P(CH_2CH_2SH)_3$. Ni, Pd, Pt.[1198]

$S(CH_2CH_2PPh_2)_2$. Co, Ni,[1184] Ni, Pd.[496]

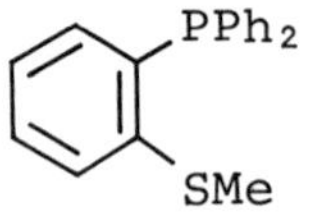 and 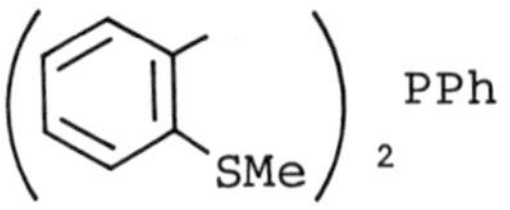

Co,[527] Ni,[924,992,1356] Pd.[528]

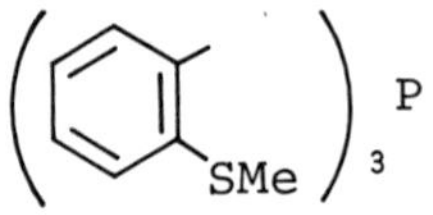

Ni,[525,677] Pd.[528]

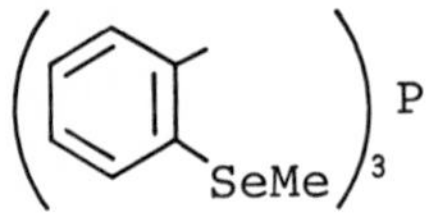

Ni.[526]

(c) P-As

$[Me_2As(CH_2)_3]_3P$. Ni.[1253]

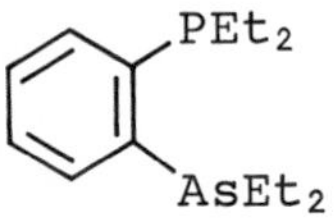

Pd,[856] Cu, Ag, Au.[399]

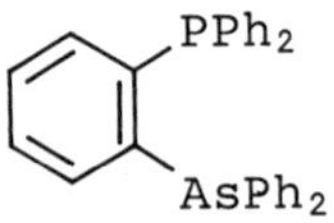

Ni,[519] Pd.[1053]

(d) P-ǁ

CH_2=$CHPPh_2$. Ag.[1358]

CH_2=$CHCH_2CH_2PR_2$ (R = Et or Cy). Ni, Pd, Cu, Ag.[807]

CH_2=$CH(CH_2)_3PR_2$. Pd, Pt.[148]

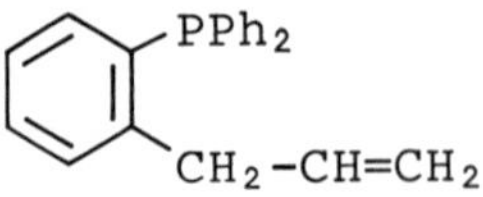

Pd, Pt, Cu, Ag.[146,147]

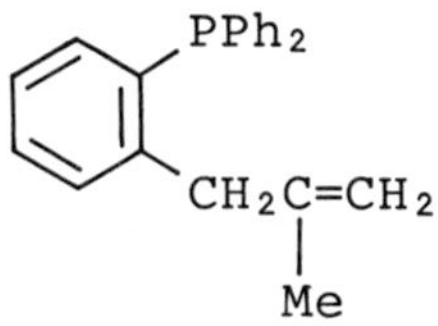

Cu.[147]

(e) Others

$Ph_2PCH_2CH_2COOH$. Ag.[401]

D.7. Typical Nitrosyl Complexes

$[V(CO)_4(NO)PPh_3]$. XVI.[1338]
$[CrI_2(NO)_2(PPh_3)_2]$. XVI.[135]
$[Cr(C_5H_5)(CO)(NO)PPh_3]$. XV.[226]
$[MoX_2(NO)_2(PPh_3)_2]$. XV.[449,846]
$[WX_2(NO)_2(PPh_3)_2]$. XV.[449,846]
$[Mn(NO)_3PPh_3]$. XV.[95,751] XVI.[713,901]
$[Mn(CO)_2(NO)(PPh_3)_2]$. XV.[751] X-ray.[541]
$[Mn(CO)_3(NO)PPh_3]$. XVI.[901] X-ray.[542]
$[MnBr(NO)_2(PPh_3)_2]$. XV.[751]
$[Re(NO_3)_2(NO)_2(PPh_3)_2]$. XVII.[587]
$[ReX_2(NO)_2(PPh_3)_2]$. XIV.[587]
$[Fe(NO)_2(PPh_3)_2]$. XV.[735,941]
$[Fe(NO)_2(Ph_2PPPh_2)_2]$. XV.[737]
$[Fe(NO)_2$diphosphine$]$. XV.[744,976,983]
$[Fe(CO)(NO)_2PPh_3]$. XV.[983]
$Hg[Fe(CO)_2(NO)PPh_3]_2$. XV.[876]
$[FeX(NO)_2PPh_3]$. XV.[735]
$[FeX(CO)_2(NO)(PPh_3)_2]$. XVIA.[460]
$[Fe(COR)(CO)(NO)(PPh_3)_2]$. XV.[734]
$[Ru(NO)_2(PPh_3)_2]$. XVI.[915]
$[RuX_3(NO)(PR_3)_2]$. XV.[345,555]
$[RuCl(NO)(PPh_3)_2]$. Zn reduction of $[RuCl_3(NO)(PPh_3)_2]$.[1255]
$[RuCl(NO)(CO)(PPh_3)_2]$. XVI.[894]
$[OsCl(NO)(CO)(PPh_3)_2]$. XVI.[895]
$[Co(NO)(PPh_3)_3]$. XV.[165,728]
$[Co(CO)(NO)(PR_3)_2]$. XV.[941,1301]
$[Co(CO)(NO)$diphosphine$]$. XV.[976,1301]
$[CoX(NO)_2PPh_3]$. XV.[165,728] XVI.[743]
$[CoX_2(NO)(PEt_3)_2]$. XVI.[194]
$[Co(NO)_2(PPh_3)_2]ClO_4$. XVI.[822]
$[Rh(NO)(PPh_3)_3]$. XV.[729] XVI.[411,915]
$[RhX_2(NO)(PPh_3)_2]$. XV.[729] XII.[411]
$[RhCl(NO_2)(NO)(PPh_3)_2]$. XVI and XVIA.[773]
$[Ir(NO)(PPh_3)_3]$. XVI.[54,940,1140]
$[Ir(NO)_2PPh_3]$. XVI.[54]
$[Ir(CO)(NO)(PPh_3)]$. XVI.[1140]
$[IrX_2(NO)(PPh_3)_2]$. XIV.[56,940] XII.[1140]
$[IrBr_3(NO)(PPh_3)_2]$. XII.[940]
$[Ir(NO)_2(PPh_3)_2]ClO_4$. XVI.[56,940]

$[IrCl(NO)(PPh_3)_2]ClO_4$. XIV.[1141]
$[Ir(NO)PPh_3]_2O$. X-ray.[263]
$[IrX(CO)(NO)(PPh_3)_2]^+$. X-ray.[759,760]
$[Ni(NO)_2(PPh_3)_2]$. XVI.[643]
$[NiX(NO)(PR_3)_2]$. XV.[16] XVI.[225,563] XVII.[194 562]
$[NiI(NO)PPh_3]_2$. XV.[711]
$[Pd(NO)(PPh_3)_2]_n$. XVI.[1115]
$[Pt(NO)(PPh_3)_2]_n$. XVI.[1115]

D.8. Typical Hydrido Complexes

$[GaH_3PR_3]$. XI.[642]
$[VH(CO)_5PPh_3]$. XXI.[747]
$[MoH(\pi\text{-}C_5H_5)(CO)_2PPh_3]$. XVIII.[964]
$[WH_6(PR_3)_3]$. XVIII.[1026]
$[MnH(CO)_3(PPh_3)_2]$. V.[191] XX.[1288]
$[MnH(CO)_4PR_3]$. XXI.[720,733]
$[ReH_3(PR_3)_4]$. $[ReH_5(PR_3)_2]$, and $[ReH_5(PR_3)_3]$. XVIII.[946]
$[ReH_3(PR_3)_2]$. XVIII.[592,945]
$[ReH_5(PR_3)_2]$ and $[ReH_7(PR_3)_2]$. XVIII.[277]
$[ReH_3(\text{diphosphine})_2]$ and $[ReH_3(\text{diphosphine})(PPh_3)_2]$. XI.[581]
$[ReH_8PR_3]^-$. XI.[608]
$[ReH_4X(PR_3)_3]$. XIV.[586]
$[FeH_2(PR_3)_3]$. XVIII.[1173]
$[FeH_2(\text{diphosphine})_2]$. XVIII.[305,311] XIX.[306]
$[FeHCl\ (\text{diphosphine})_2]$. XIV and XVIII.[305,311]
$[FeH(CO)_4PPh_3]^+$. XXI.[476]
$[FeH(C_6H_4PPhCH_2CH_2PPh_2)(Ph_2PCH_2CH_2PPh_2)]$. XXB.[673]
$[RuH_2(PR_3)_4]$. XVIII.[499,1363]
$[RuH_4(PPh_3)_3]$. XIX.[887]
$[RuH_2(CO)_2(PR_3)_2]$. XI.[455]
$[RuHX(PPh_3)_3]$. XVIII and XIX.[664] X-ray.[1232,1233]
$[RuHX(CO)(PR_3)_3]$. XXA.[340,347,829,1313]
$[RuH_2(\text{diphosphine})_2]$ and $[RuHX(\text{diphosphine})_2]$. XVIII.[308,313]
$[RuH(\beta\text{-}C_{10}H_7)(Me_2PCH_2CH_2PMe_2)_2]$. XXB.[280] X-ray.[776]
$[RuH(CH_2PMeCH_2CH_2PMe_2)(Me_2PCH_2CH_2PMe_2)]$. XXB.[280]
$[OsH_4(PR_3)_3]$. XVIII.[514,911]
$[OsH_6(PMe_2Ph)_2]$. XVIII.[514]
$[OsH_2(CO)_3PPh_3]$. XIX.[913] V.[543]
$[OsHCl_2(PR_3)_3]$ and $[OsHCl_3(PR_3)_2]$. XXA.[323] XVIII.[326]
$[OsHX(CO)(PR_3)_3]$. XXA.[341,1307,1313] X-ray.[1070]
$[OsH_2(\text{diphosphine})_2]$ and $[OsHX(\text{diphosphine})_2]$. XVIII.[305,308,313,778]
$[OsH(CO)_3(PPh_3)_2]^+$. XXI.[896]
$[CoH(PPh_3)_3]$ and $[CoH(PPh_3)_4]$. XIX.[1160]
$[CoH_2(PPh_3)_3]$. XIX.[1010,1239]
$[CoH_2(PR_3)_3]$. XVIII.[926,1177]
$[CoH(\text{diphosphine})_2]$. XI.[1177] XVIII.[1178,1378] XIX.[1178]
$[CoH(CO)(PR_3)_3]$. XXB.[1079] XI.[1006]
$[RhH(PPh_3)_3]$. XIX.[865]
$[RhH(PPh_3)_4]$. XVIII.[499,1363] XXB.[1262] X-ray.[86]

$[RhH(diphosphine)_2]$. XVIII.[1180]
$[RhHX_2(PR_3)_2]$. XIX.[82] XXA.[1181]
$[RhH_2Cl(PPh_3)_2]$. XIX.[1072]
$[RhH(CO)(PPh_3)_3]$. XVIII.[118] X-ray.[903,904]
$[RhH(CO)(PPh_3)_2]$ and $[RhH(CO)_2(PPh_3)_2]$. VIII.[552]
$[IrH_3(PR_3)_2]$ and $[IrH_3(PR_3)_3]$. XVIII.[279,943]
$[IrH_5(PEt_2Ph)_2]$. XVIII.[951]
$[IrH_3(PPh_3)_3]$. XVIII.[678,939,943]
$[IrH_3(PPh_3)_2]^+$. XIV.[59]
$[IrH_2(PPh_3)_3]^+$. XIV.[53]
$[IrH(CO)_2(PPh_3)_2]$. XVIII.[1361]
$[IrH(CO)(PPh_3)_2]$. XI.[943]
$[IrH(CO)(PPh_3)_3]$. XVIII.[118,943]
$[IrH_3(CO)(PPh_3)_2]$. XVIII.[943]
$[IrHX_2(PR_3)_3]$. XIV.[55,60,678,939] XVIII.[57] XXB.[340,1304,1316]
$[IrH_2X(PR_3)_3]$. XIV.[55,678,939] XXB.[341,1304]
$[IrHCl_2(CO)(PPh_3)_2]$. XIX.[1314,1315]
$[IrH_2X(CO)(PR_3)_2]$. XIX.[491,1315]
$[IrH_2(diphosphine)_2]Cl$. XIX.[1312]
$[NiHCl(PCy_3)_2]$. XVIII.[641] XIX.[855]
$[NiH(BH_4)(PCy_3)_2]$. XVIII.[638]
$[PdHX(PR_3)_2]$. XVIII.[218,638]
$[PtH_2(PPh_3)_2]$. XIX.[369,949]
$[PtHX(PR_3)_2]$. XVIII.[72,286,343] XIX.[257] XXA.[278,340,343] X-ray,[533,1080] IR,[7,287] Reactions.[1086]
$[PtHCl(diphosphine)]$.[390]
$[Pt_2H_2(PPh_2)_2(PEt_3)_2]$. XI.[281]
$[PtH(PEt_3)_2diph]Cl$. XI.[611]
$[CuH(PR_3)_3]$. XI.[929]
$[(CuH)_2PR_3]$. XIII.[503]
$[Cu(BH_4)(PPh_3)_2]$. XVIII.[252,470] X-ray.[921]
$[Cu(B_3H_8)(PPh_3)_2]$. XIV.[923] X-ray.[922]
$[Ag(BH_4)(PPh_3)_2]$. XVIII.[253]

D.9. Typical Hydrocarbyl Complexes

D.9.1. Alkyl, Aryl, and Acetylide Complexes

$[AlEt_3(Ph_2PPPh_2)]$. X.[389]
$[GaMe_3PMe_3]$. X.[392,498]
$[TlPhCl_2PPh_3]$. X.[471]
$[TiMeCl_3PPh_3]$. X.[128]
$[MnR(CO)_4PPh_3]$. XXIII.[58,720,1344] V.[853]
$[MnR(CO)_3(PR_3)_2]$. XXIII.[733]
$[ReAr_3(PR_3)_3]$ and $[ReAr_2(PR_3)_2]_n$. XXII.[299]
$[Fe(C_6Cl_5)_2(PEt_2Ph)_2]$. XXII.[342]
$[Fe(C_3F_7)I(CO)_3PPh_3]$. V.[1114]
$[RuR_2(diphosphine)_2]$ and $[RuXR(diphosphine)_2]$. XXII.[313,786]
$[OsR_2(diphosphine)_2]$ and $[OsXR(diphosphine)_2]$. XXII.[313,786]
$[OsMe_2(CO)_3PPh_3]$. V.[912]

$[CoAr_2(PR_3)_2]$. XXII.[342]
$[Co(mesityl)_2(PEt_2Ph)_2]$. X-ray,[1082] mag. properties.[159]
$[Co(C_6F_5)_2diphosphine]$. XXII.[1102]
$[CoHEt_2triphosphine]$. XXII.[536]
$[CoR(CO)_3PPh_3]$. XXII.[739,740,741,1344] (R = C_2HF_4; x-ray).[1345]
$[(MeC)Co_3(CO)_8PPh_3]$. V.[1152]
$[(\pi\text{-}C_5H_5)CoMe_2PPh_3]$. XXII.[877,1369,1370]
$[(\pi\text{-}C_5H_5)Co(C_2F_5)IPPh_3]$. XI.[1275]
$[RhMe(PPh_3)_3]$. XXII.[865]
$[RhPh(CO)(PPh_3)_2]$. VIII.[867]
$[RhMeIX(PPh_3)_2]$. XXIII.[908]
$[RhMeI_2(PPh_3)_2]$. X-ray.[1276]
$[RhBr(1\text{-naphthyl})_2(PR_3)_2]$. XXII.[351]
$[RhR(diene)PPh_3]$. XXII.[1262]
$[IrMe_3(PR_3)_3]$. XXII.[346]
$[IrMe_3PR_3]$, $[IrMe_2XPR_3]$, and $[IrMeX_2PR_3]$. XXII and XII.[1206]
$[IrMe(diene)(PPh_3)_2]$. XXII.
$[NiAr_2(PR_3)_2]$. XXII.[337,1102]
$[NiArX(PR_3)_2]$. XXII.[337,1102] XXII.[64]
$[Ni(C{\equiv}CR')_2(PR_3)_2]$. XXII.[337] X-ray.[473,1240]
$[Ni(C{\equiv}CPh)_2(PEt_2Ph)_3]$. XXII.[337]
$[Ni(CF{=}CF_2)_2(PEt_3)_2]$. XXII.[1146]
$[Ni(C{\equiv}CCF_3)I(PEt_3)_2]$. XXII.[224]
$[NiR_FI_2(PPh_3)_2]$. XXIII.[64]
$[NiR_FI$ diphosphine$]$. XXIII.[984]
$[(\pi\text{-}C_5H_5)NiR'(PR_3)]$. XXII.[1371] XXIII.[1137] XI.[982]
X-ray.[375-377]
$[PdR'_2(PR_3)_2]$. XXII.[243,393,1146]
$[PdRX(PR_3)_2]$. XXII.[243] XXIII.[572,574,1029,1158]
$[PdPhCl(PR_3)_2]$. XXV.[456]
$[Pd(C{\equiv}CCF_3)_2(PEt_3)_2]$. XXII.[224]
$[PtR'_2(PR_3)_2]$. XXII.[338,339,1146] XI.[516,884] IR.[14]
$[PtR'X(PR_3)_2]$. XXII.[338,339,1157] XI.[884] IR.[14]
$[Pt(C{\equiv}CR')_2(PR_3)_2]$. XXII.[224,339,406] Halogenation.[549]
$[PtRX(PEt_3)_2]$. XXIV.[278,383-385]
$[PtRX(PPh_3)_2]$. XXIII.[183,184,1158]
$[Pt(R_F)_2(PR_3)_2]$. XXII.[1157] XXIII.[1058]
$[PtR'_2X_2(PR_3)_2]$. XXIII.[338,339]
$[PtMe_3Br(PMe_2Ph)_2]$. XXIII.[1167] Pyrolysis.[1168]
$[PtMe_4(PMe_2Ph)_2]$. XXII.[1167]
$[CuMe(PPh_3)n]$. X.[439]
$[(CuPh)_nPPh_3]$. X.[438]
$[Cu(C{\equiv}CR')PR_3]$. X.[178,393,395]
$[Ag(C{\equiv}CPh)PR_3]$. X.[178,393] X-ray.[437]
$[AuMePR_3]$. XXII.[244]
$[AuMe_3PR_3]$. XI.[397]
$[AuCl(C_6F_5)_2PPh_3]$. XXIII.[1058] X-ray.[85]
$[Au(C{\equiv}CPh)PR_3]$. X and XXII.[396]
$[Au(\sigma\text{-}C_5H_5)PPh_3]$. XXII.[775]
$[Au(ferrocenyl)PPh_3]$. XXII.[1048]

D.9.2. π Complexes

$[V(\pi\text{-}C_5H_5)(CO)_3PR_3]$. V.[570] Reaction with CO.[1046]
$[Nb(\pi\text{-}C_5H_5)(CO)_2(PPh_3)_2]$. V.[1046]
$[Mo(\pi\text{-}C_5H_5)X(CO)_2PR_3]$. V.[661]
$[Mo_2(\pi\text{-}C_5H_5)_2(CO)_3\text{diphosphine}]$. V.[880]
$[W(\pi\text{-}C_5H_5)Cl(CO)_2PR_3]$. V.[569]
$[Mn(\pi\text{-}C_5H_5)(CO)_2PPh_3]$. V.[1059,1256]
$[Fe_2(\text{acetylene})_2(CO)_5PPh_3]$. V.[1330]
$[Fe(\text{diene})(CO)_2PPh_3]$. V.[362] XVIII.[475,1329]
$[Fe(\pi\text{-allyl})(CO)_2PR_3]$. V.[1032]
$[Fe(\pi\text{-allyl})I(CO)_2PPh_3]$. X-ray.[1004]
$[Fe(\pi\text{-}C_5H_5)X(CO)PPh_3]$. V. (UV).[1047,1118,1274]
$[Fe(\pi\text{-}C_5H_5)X(\text{diphosphine})]$. V.[879]
$[Fe(\pi\text{-}C_5H_5)(CO)Me_2PC_2H_4PMe_2]^+$. V.[882]
$[Fe_2(\pi\text{-}C_5H_5)_2(CO)_3PR_3]$. V.[659]
$[Ru(\text{diene})X_2(PPh_3)_2]$. XXVI.[1153]
$[Co(\pi\text{-}C_5H_5)(PPh_3)_{2\text{ and }3}]$. XXVII.[1151]
$[Co(\pi\text{-allyl})(CO)_2PPh_3]$. XI.[702]
$[Co(C_4Ph_4)(CO)BrPPh_3]$. XI.[937]
$[Rh(\text{olefin})Cl(PPh_3)_2]$. XXVI.[978]
$[Rh(\text{allene})IPPh_3]$. X-ray.[860]
$[Rh(\text{diene})XPR_3]$. X.[153,355,1163]
$[Rh(\pi\text{-}C_3H_5)(PPh_3)_2]$. XXII.[1143]
$[Ir(\text{olefin})X(CO)(PPh_3)_2]$. XXVI.[459,989,1027,1088]
$[Ir(\text{acetylene})Cl(CO)(PR_3)_2]$. XXVI.[413]
$[Ir(\text{diene})ClPPh_3]$. X.[1353]
$[Ir(\pi\text{-allyl})Cl_2(PR_3)_2]$. XXII.[1127]
$[Ni(\text{olefin})(PPh_3)_2]$. XXVI.[1348,1350] X-ray.[431,517]
$[Ni(CH_2{=}CHCN)_2(PPh_3)_{1\text{ and }2}]$. X.[1190]
$[Ni(\pi\text{-}C_3H_5)XPR_3]$. XXVII.[654,705]
$[Ni(\pi\text{-}C_3H_5)X(CO)PR_3]$.[654]
$[Ni(\pi\text{-2-methylallyl})Br\text{ diphosphine}]$. X.[374]
$[Ni(\pi\text{-}C_5H_5)PPh_3]Cl$. XXVII. Red, dec. 140° with formation of Ni powder.[548]
$[Ni(\pi\text{-}C_5H_5)P(CF_3)_2]_2$. XI.[504]
$[Pd(\text{olefin})(PPh_3)_2]$. XXVI.[1164]
$[Pd(\text{acetylene})(PPh_3)_2]$. XXVI.[632]
$[Pd(\pi\text{-}C_3H_5)ClPPh_3]$. XXVI.[1126]
$[Pd(\pi\text{-2-methylallyl})ClPPh_3]$. X.[1121] X-ray.[970]
$[Pd(\pi\text{-mesityloxide})ClPPh_3]$. X.[1090]
$[Pd(\pi\text{-2-methylallyl})(PPh_3)_2]^+$. X.[1126]
$[Pt(\pi\text{-}C_3H_5)X(PPh_3)_2]$. X.[1323]
$[Pt(\text{olefin})(PPh_3)_2]$. XXVI.[348,430,634,1164]
$[Pt(\text{tetracyanoethylene})(PPh_3)_2]$. XXVII.[70] X-ray.[1083]
$[Pt(\text{allene})(PPh_3)_2]$. XXVI.[1071,1078]
$[Pt(\text{acetylene})(PPh_3)_2]$. XXVI.[336,1162] XXVII.[200,348] Nature of bonding,[1044] x-ray.[614]
$[Cu(\pi\text{-}C_5H_5)PR_3]$. XXVI.[452,1342,1352]

D.10. Typical Nitrogen and Oxygen Complexes

[Mo(N_2)PPh_3]. XVIIIA.[709]
[Re(N_2)Cl(PR_3)$_4$]. XVIIIA.[282]
[Re(N_2)(CO)$_2$(NH_2)PPh_3]. XVIII.[1019]
[Fe(N_2)(PEt_3)$_4$]. XVIIIA.[245]
[Fe(N_2)H_2(PR_3)$_3$]. XVIIIA.[1173]
[Fe(N_2)H(diphosphine)$_2$]. XVIIIA.[89]
[Os(N_2)X_2(PR_3)$_3$]. XVIIIA.[320]
[Ru(N_2)H_2(PPh_3)$_2$]. XVIIIA.[887]
[Co(N_2)(PPh_3)$_3$]. XVIIIA.[926,1008,1010,1239,1364]
[Co(N_2)H(PPh_3)$_3$]. XVIIIA.[1007,1132,1243] X-ray.[474,540]
[Co(N_2)H(PR_3)$_3$]. XVIIIA.[1009]
[Ir(N_2)Cl(PPh_3)$_2$]. XXV.[412]
(N_2)[Ni(PCy_3)$_2$]$_2$. XVIIIA.[852,855]
[Ni(N_2)H(PR_3)$_2$]. XVIIIA.[1242]
[Ru(O_2)Cl(NO)(PPh_3)$_2$]. XXVI.[894] Reactions.[895]
[Ir(O_2)X(CO)(PR_3)$_2$]. XXVI.[491,1306] X-ray.[905,988]
[Ir(O_2)(diphosphine)$_2$]Cl. XXVI.[1312]
[Ni(O_2)(PPh_3)$_2$]. XXVI.[1350]
[Pt(O_2)(PPh_3)$_2$]. XXVI.[430,1261] X-ray,[427,861] reaction with acetone,[1290] SO_2.[428]

D.11. Typical Phosphidobridged Complexes

[V(PR_2)(CO)$_4$]$_2$. XXVIII.[754] XXIX.[736]
[Cr(PR_2)(CO)$_4$]$_2$. XXIX.[350]
[Mo(PR_2)(CO)$_4$]$_2$. XXIX.[350]
[Mo(PR_2)(CO)$_3$(PR'_3)]$_2$. V.[1271] (R = Me, R' = Et). X-ray.[935]
[Mo_2H(C_5H_5)$_2${PMe_2}(CO)$_4$]. XXX.[683] X-ray.[507]
[W(PR_2)(CO)$_4$]$_2$. XXIX.[350]
[Mn(PMe_2)(CO)$_4$]$_2$. XXX.[680]
[Mn(PPh_2)(CO)$_4$]$_2$. XXIX.[684] XXX.[4,637]
[Mn_2(CO)$_8$(PPh_2)H]. XXX.[637,681] X-ray.[509]
[Mn_2(CO)$_8$P(CF_3)$_2$X]. XXX.[538,650]
[Re(PPh_2)(CO)$_4$]$_2$. XXX.[745]
[Fe(PR_2)(CO)$_3$]$_2$. XXIX.[41,350,433]
[Fe(CO)$_3$P(CF_3)$_2$X]$_2$. XXX.[650,651]
[Fe(PPh_2)(NO)$_2$]$_2$. XXX.[744]
[Ru(PR_2)(CO)$_3$]$_2$. XXIX.[246,433]
[Co(PPh_2)(CO)$_3$]$_2$. XXIX.[684]
[Co_3(PMe_2)$_2$(CO)$_7$]. XXIX.[684]
[Co(PPh_2)(NO)$_2$]$_2$. XXX.[744]
[Co(PPh_2)(π-C_5H_5]$_2$ and [Ni(PPh_2)(π-C_5H_5)]$_2$. X-ray.[405]
[Ni(PR_2)(CO)$_2$]$_2$. XXIX.[686]
[PdX(PR'_2)(PR_3)$_2$]$_2$. XXVIII.[679]
[Pd_3Cl_2(PPh_2)$_2$(PPh_3)$_2$]. XXV.[456]
[PtX(PR'_2)(PR_3)$_2$]$_2$. XXVIII.[281]

(received December 7, 1970)

REFERENCES

1. Abel, E. W., A. M. Atkins, B. C. Crosse, and G. V. Hutson, J. Chem. Soc., A, 1968, 687.
2. Abel, E. W., M. A. Bennett, and G. Wilkinson, J. Chem. Soc., 1959, 2323.
3. Abel, E. W., R. A. N. McLean, and I. H. Sabherwal, J. Chem. Soc., A, 1969, 133.
4. Abel, E. W., and I. H. Sabherwal, J. Organometal. Chem., 10, 491 (1967).
5. Abel, E. W., and S. P. Tyfield, Chem. Commun., 1968, 465; Can. J. Chem., 47, 4627 (1969).
6. Abel, E. W., and G. Wilkinson, J. Chem. Soc., 1959, 1501.
7. Adams, D. M., Proc. Chem. Sec., 1961, 431.
8. Adams, D. M., J. Chem. Soc., 1962, 1220.
9. Adams, D. M., J. Chem. Soc., 1964, 1771
10. Adams, D. M., and G. Booth, J. Chem. Soc., 1962, 1112.
11. Adams, D. M., and P. J. Chandler, Chem. Commun., 1966, 69; J. Chem. Soc., A, 1967, 1009.
12. Adams, D. M., and P. J. Chandler, J. Chem. Soc., A, 1969, 588.
13. Adams, D. M., J. Chatt, J. Gerratt, and A. D. Westland, J. Chem. Soc., 1964, 734.
14. Adams, D. M., J. Chatt, and B. L. Shaw, J. Chem. Soc., 1960, 2047.
15. Adams, D. M., D. J. Cook, and R. D. W. Kemmitt, Nature, 205, 589 (1965); Chem. Commun., 1966, 103.
16. Addison, C. C., and B. F. G. Johnson, Proc. Chem. Soc., 1962, 305.
17. Addison, C. C., and M. Kilner, J. Chem. Soc., A, 1968, 1539.
18. Ahrland, S., J. Chatt, N. R. Davies, and A. A. Williams, Quart. Rev. (London), 12, 265 (1958); J. Chem. Soc., 1958, 276.
19. Albano, V., P. L. Bellon, and V. Scatturin, Chem. Commun., 1966, 507.
20. Albano, V., P. L. Bellon, and V. Scatturin, Chem. Commun., 1967, 730.
21. Albano, U. G., G. M. Basso-Ricci, and P. L. Bellon, Inorg. Chem., 8, 2109 (1969).
22. Albano, V. G., P. L. Bellon, and M. Sansoni, Chem. Commun., 1969, 899.
23. Alcock, N. W., and K. A. Raspin, J. Chem. Soc., A, 1968, 2108.
24. Alexander, J. J., and A. Wojcicki, J. Organometal. Chem., 15, P23 (1968).
25. Allen, A. D., and M. C. Baird, Chem. Ind. (London), 1965, 139.
26. Allen, A. D., and P. F. Barrett, Can. J. Chem., 46, 1649 (1968).

27. Allen, A. D., and C. D. Cook, Proc. Chem. Soc., 1962, 218.
28. Allen, A. D., and C. D. Cook, Can. J. Chem., 41, 1235 (1963); 42, 1063 (1964).
29. Allen, D. W., F. G. Mann, and I. T. Millar, Chem. Ind. (London), 1966, 196.
30. Allen, D. W., F. G. Mann, and I. T. Millar, Chem. Ind. (London), 1966, 2096; J. Chem. Soc., A, 1969. 1101.
31. Allen, D. W., F. G. Mann, I. T. Millar, H. M. Powell, and D. Watkin, Chem. Commun., 1969, 1004.
32. Allen, D. W., I. T. Millar, F. G. Mann, R. M. Canadine, and J. Walker, J. Chem. Soc., A, 1969, 1097.
33. Allen, F. H., and A. Pidcock, J. Chem. Soc., A, 1968, 2700.
34. Allison, J. A. C., and F. G. Mann, J. Chem. Soc., 1949, 2915.
35. Alyea, E. C., and D. W. Meek, J. Am. Chem. Soc., 91, 5761 (1969).
36. American Cyanamid Co., Brit. 886,501 (1958).
37. American Cyanamid Co., Brit. 898,324 (1960).
38. American Cyanamid Co., U. S. 3,294,870 (1964); U. S. 3,345,392 (1966).
39. American Cyanamid Co., L. S. Meriwether, and M. L. Fiene, U. S. 3,051,694 (1960), C. A., 58, 4423 (1963).
40. Anderson, W. A., A. J. Carty, and A. Efraty, Can. J. Chem., 47, 3361 (1969).
41. Ang, H. G., and J. M. Miller, Chem. Ind. (London), 1966, 944.
42. Ang, H. G., J. S. Shannon, and B. O. West, Chem. Commun., 1965, 10.
43. Ang, H. G., and B. O. West, Aust. J. Chem., 20, 1133 (1967).
44. Angelici, R. J., Inorg. Chem., 3, 1099 (1964).
45. Angelici, R. J., J. Inorg. Nucl. Chem., 28, 2627 (1966).
46. Angelici, R. J., and F. Basolo, J. Am. Chem. Soc., 84, 2495 (1962).
47. Angelici, R. J., and J. R. Graham, J. Am. Chem. Soc., 88, 3658 (1966).
48. Angelici, R. J., and C. M. Ingemanson, Inorg. Chem., 8, 83 (1969).
49. Angelici, R. J., and M. D. Malone, Inorg. Chem., 6, 1731 (1967).
50. Angelici, R. J., and E. E. Siefert, Inorg. Chem., 5, 1457 (1966).
51. Angelici, R. J., and E. E. Siefert, J. Organometal. Chem., 8, 374 (1967).
52. Angoletta, M., Gazz. Chim. Ital., 90, 1021 (1960).
53. Angoletta, M., Gazz. Chim. Ital., 92, 811 (1962).
54. Angoletta, M., Gazz. Chim. Ital., 93, 1591 (1963).

55. Angoletta, M., and A. Araneo, Gazz. Chim. Ital., 93, 1343 (1963).
56. Angoletta, M., and G. Caglio, Gazz. Chim. Ital., 93, 1584 (1963).
57. Angoletta, M., and G. Caglio, Gazz. Chim. Ital., 99, 46 (1969).
58. Anisimov, K. N., N. E. Kolobova, and A. A. Johannsson, Izv. Akad. Nauk SSSR, Ser. Khim., 1969, 1749.
59. Araneo, A., Gazz. Chim. Ital., 95, 1431 (1965).
60. Araneo, A., and S. Martinengo, Gazz. Chim. Ital., 95, 61 (1965).
61. Argento, B. J., P. Fitton, J. E. McKeon, and E. A. Rick, Chem. Commun., 1969, 1427.
62. Ashcroft, S. J., and C. T. Mortimer, J. Chem. Soc., A, 1967, 930.
63. Ashley-Smith, J., M. Green, N. Mayne, and F. G. A. Stone, Chem. Commun., 1969, 409.
64. Ashley-Smith, J., M. Green, and F. G. A. Stone, J. Chem. Soc., A, 1969, 3019.
65. Aslanov, L., R. Mason, A. G. Wheeler, and P. O. Whimp, Chem. Commun., 1970, 30.
66. Asmussen, R. W., and O. Bostrop, Acta Chem. Scand, 11, 1097 (1957).
67. Asmussen, R. W., A. Jensen, and H. Soling, Acta Chem. Scand, 9, 1391 (1955).
68. Atkins, P. W., J. C. Green, and M. L. H. Green, J. Chem. Soc., A, 1968, 2275.
69. Augustine, R. L., and J. F. VanPeppen, Chem. Commun., 1970, 495, 497.
70. Baddley, W. H., and L. M. Venanzi, Inorg. Chem., 5, 33 (1966).
71. Badley, E. M., J. Chatt, R. L. Richards, and G. A. Sim, Chem. Commun., 1969, 1322.
72. Bailar, J. C., and H. Itatani, Inorg. Chem., 4, 1618 (1965).
73. Bailar, J. C., and H. Itatani, J. Am. Chem. Soc., 89, 1592, 1600 (1967).
74. Bailey, N. A., J. M. Jenkins, R. Mason, and B. L. Shaw, Chem. Commun., 1965, 237.
75. Bailey, N. A., C. J. Jones, B. L. Shaw, and E. Singleton, Chem. Commun., 1967, 1051.
76. Bailey, N. A., and R. Mason, J. Chem. Soc., A, 1968, 2594.
77. Bainbridge, A., P. J. Craig, and M. Green, J. Chem. Soc., A, 1968, 2715.
78. Baird, M. C., J. Inorg. Nucl. Chem., 29, 367 (1967).
79. Baird, M. C., J. Organometal. Chem., 16, P16 (1969).
80. Baird, M., G. Hartwell, R. Mason, A. I. M. Rae, and G. Wilkinson, Chem. Commun., 1967, 92.
81. Baird, M. C., G. Hartwell, and G. Wilkinson, J. Chem. Soc., A, 1967, 2037.

82. Baird, M. C., D. N. Lawson, J. T. Mague, J. A. Osborn, and G. Wilkinson, Chem. Commun., 1966, 129; J. Chem. Soc., A, 1967, 1347.
83. Baird, M. C., and G. Wilkinson, Chem. Commun., 1966, 267, 514; J. Chem. Soc., A, 1967, 865.
84. Baird, M. C., C. J. Nyman, and G. Wilkinson, J. Chem. Soc., A, 1968, 348.
85. Baker, R. W., and P. Pauling, Chem. Commun., 1969, 745.
86. Baker, R. W., and P. Pauling, Chem. Commun., 1969, 1495.
87. Balls, A., N. N. Greenwood, and B. P. Straughan, J. Chem. Soc., A, 1968, 753.
88. Bamford, C. H., Trans. Faraday Soc., 61, 267 (1965).
89. Bancroft, G. M., M. J. Mays, and B. E. Prater, Chem. Commun., 1969, 585.
90. Bannister, W. D., B. L. Booth, M. Green, and R. N. Haszeldine, Chem. Commun., 1965, 154; J. Chem. Soc., A, 1969, 698.
91. Barlex, D. M., R. D. W. Kemmitt, and G. W. Littlecott, Chem. Commun., 1969, 613.
92. Barlow, C. G., J. F. Nixon, and M. Webster, J. Chem. Soc., A, 1968, 2216.
93. Barlow, C. G., J. F. Nixon, and J. R. Swain, J. Chem. Soc., A, 1969, 1082.
94. Barnett, K. W., and P. M. Treichel, Inorg. Chem., 6, 294 (1967).
95. Barraclough, C. G., and J. Lewis, J. Chem. Soc., 1960, 4842.
96. Barrett, P. F., and A. J. Poe, J. Chem. Soc., A, 1968, 429.
97. B.A.S.F., Brit. 714,202 (1954).
98. B.A.S.F., D.A.S. 1,026,959 (1955).
99. B.A.S.F., Brit. 1,070,207 (1963); Fr. 1,416,412 (1963).
100. B.A.S.F., D.A.S. 1,227,023 (1964).
101. B.A.S.F., Neth. 64.09121 (1964).
102. B.A.S.F., E. O. Fischer, and H. Werner, D.A.S. 1,181,708 (1962).
103. B.A.S.F., and W. Hieber, Fr. 1,196,549 (1957).
104. B.A.S.F., and H. Mueller, Brit. 923,470 (1960).
105. B.A.S.F., W. Reppe, and W. Schweckendiek, Ger. 805,641 (1951); C. A., 47, 602 (1953).
106. B.A.S.F., W. Reppe, and W. Schweckendiek, Ger. 876,094 (1953); C. A., 52, 10175 (1958).
107. B.A.S.F., W. Reppe, W. Schweckendiek, A. Magin, and K. Klager, Ger. 805,642 (1951); C. A., 47, 602 (1953).
108. B.A.S.F., and G. N. Schrauzer, D.A.S. 1,168,903 (1961).

109. B.A.S.F., and W. Schweckendiek, Ger. 834,991 (1952); C. A., 52, 9192 (1958).
110. B.A.S.F., and W. Schweckendiek, Ger. 836,647 (1952); C. A., 52, 14679 (1958).
111. B.A.S.F., and W. Schweckendiek, Ger. 841,590 (1952); C. A., 52, 9193 (1958).
112. B.A.S.F., and W. Schweckendiek, D.A.S. 1,072,244 (1959).
113. B.A.S.F., W. Schweckendiek, and K. Sepp, Ger. 888,849 (1953); C. A., 52, 11320 (1958).
114. B.A.S.F., and O. Weissbarth, Ger. 808,585 (1951); C. A., 47, 1736 (1953).
115. Basolo, F., W. H. Baddley, and J. L. Burmeister, Inorg. Chem., 3, 1212 (1964).
116. Basolo, F., A. T. Brault, and A. J. Poe, J. Chem. Soc., 1964, 676.
117. Basolo, F., J. Chatt, H. B. Gray, R. G. Pearson, and B. L. Shaw, J. Chem. Soc., 1961, 2207.
118. Bath, S. S., and L. Vaska, J. Am. Chem. Soc., 85, 3500 (1963).
119. Beattie, I. R., and R. Collis, J. Chem. Soc., A, 1969, 2960.
120. Beattie, I. R., and K. M. S. Livingston, J. Chem. Soc., A, 1969, 2201.
121. Beattie, I. R., and G. A. Ozin, J. Chem. Soc., A, 1968, 2373.
122. Becconsall, J. K., B. E. Job, and S. O'Brien, J. Chem. Soc., A, 1967, 423.
123. Beck, W., and W. P. Fehlhammer, Angew. Chem. Int. Ed., 6, 169 (1967).
124. Beck, W., W. P. Fehlhammer, P. Pollmann, and R. S. Tobias, Inorg. Chim. Acta, 2, 467 (1968).
125. Beck, W., K. Feldl, and E. Schuierer, Angew. Chem. Int. Ed., 4, 439 (1965); 5, 249 (1966).
126. Beck, W., and E. Schuierer, Chem. Ber., 98, 298 (1965).
127. Beeby, M. H., and F. G. Mann, J. Chem. Soc., 1951, 411.
128. Beermann, C., and H. Bestian, Angew. Chem., 71, 618 (1959).
129. Beg, M. A. A., and H. C. Clark, Can. J. Chem., 38, 119 (1960).
130. Beg, M. A. A., and H. C. Clark, Can. J. Chem., 39, 595 (1961).
131. Beg, M. A. A., and H. C. Clark, Can. J. Chem., 40, 283 (1962).
132. Behrens, H., and W. Aquila, Z. Anorg. Chem., 356, 8 (1967).
133. Behrens, H., and W. Klek, Z. Anorg. Chem., 292, 151 (1957).

134. Behrens, H., and A. Mueller, Z. Anorg. Chem., 341, 124 (1965).
135. Behrens, H., and H. Schindler, Z. Naturforsch., B, 23, 1109 (1968).
136. Bellon, P. L., V. Albano, V. D. Bianco, F. Pompa, and V. Scatturin, Ric. Sci., A3(8), 1213 (1963); C. A., 60, 14106a (1964).
137. Belluco, U., L. Cattalini, F. Basolo, R. G. Pearson, and A. Turco, J. Am. Chem. Soc., 87, 241 (1965).
138. Belluco, U., L. Cattalini, and A. Orio, Gazz. Chim. Ital., 93, 1422 (1963).
139. Belluco, U., L. Cattalini, and A. Turco, J. Am. Chem. Soc., 86, 226 (1964).
140. Belluco, U., U. Croatto, P. Uguagliati, and R. Pietropaolo, Inorg. Chem., 6, 718 (1967).
141. Belluco, U., M. Graziani, and P. Rigo, Inorg. Chem., 5, 1123 (1966).
142. Belluco, U., M. Martelli, and A. Orio, Inorg. Chem., 5, 582 (1966).
143. Belluco, U., P. Rigo, M. Graziani, and R. Ettore, Inorg. Chem., 5, 1125 (1966).
144. Bennett, M. A., R. Bramley, and P. A. Longstaff, Chem. Commun., 1966, 806.
145. Bennett, M. A., R. J. H. Clark, and D. L. Milner, Inorg. Chem., 6, 1647 (1967).
146. Bennett, M. A., L. V. Interrante, and R. S. Nyholm, Z. Naturforsch., B, 20, 633 (1965).
147. Bennett, M. A., W. R. Kneen, and R. S. Nyholm, Inorg. Chem., 7, 552,556 (1968).
148. Bennett, M. A., H. W. Kouwenhoven, J. Lewis, and R. S. Nyholm, J. Chem. Soc., 1964, 4570.
149. Bennett, M. A., and P. A. Longstaff, Chem. Ind. (London), 1965, 846.
150. Bennett, M. A., and P. A. Longstaff, J. Am. Chem. Soc., 91, 6266 (1969).
151. Bennett, M. A., and D. L. Milner, Chem. Commun., 1967, 581; J. Am. Chem. Soc., 91, 6983 (1969).
152. Bennett, M. A., R. S. Nyholm, and J. D. Saxby, J. Organometal. Chem., 10, 301 (1967).
153. Bennett, M. A., and G. Wilkinson, J. Chem. Soc., 1961, 1418.
154. Bennett, M. J., F. A. Cotton, and B. H. C. Winquist, J. Am. Chem. Soc., 89, 5366 (1967).
155. Bennett, M. J., and R. Mason, J. Chem. Soc., A, 1968, 75.
156. Benson, B. C., R. Jackson, K. K. Joshi, and D. T. Thompson, Chem. Commun., 1968, 1506.
157. Benson, R. E., and R. V. Lindsey, J. Am. Chem. Soc., 81, 4247 (1959).
158. Bentham, J. E., S. Cradock, and E. A. V. Ebsworth, Chem. Commun., 1969, 528.

159. Bentley, R. B., F. E. Mabbs, W. R. Smail, M. Gerloch, and J. Lewis, Chem. Commun., 1969, 119.
160. Benzoni, L., A. Andreetta, C. Zanzottera, and M. Camia, Chim. Ind. (Milan), 48, 1076 (1966).
161. Bercz, J. P., L. Horner, and C. V. Bercz, Ann. Chem., 703, 17 (1967).
162. Berglund, D., and D. W. Meek, Inorg. Chem., 8, 2602 (1969).
163. Berman, D. A., Dissertation Abstr., 18[4], 1240 (1958).
164. Bertrand, J. A., and D. L. Plymale, Inorg. Chem., 5, 879 (1966).
165. Bianco, T., M. Rossi, and L. Uva, Inorg. Chim. Acta, 3, 443 (1969).
166. Bibler, J. P., and A. Wojcicki, Inorg. Chem., 5, 889 (1966).
167. Biellmann, J. F., and H. Liesenfelt, Bull. Soc. Chim. Fr., 1966, 4029.
168. Bierling, B., K. Kirschke, H. Oberender, and M. Schulz, Z. Chem., 9, 105 (1969).
169. Bigorgne, M., C. R. Acad. Sci., Paris, 250, 3484 (1960).
170. Bigorgne, M., J. Inorg. Nucl. Chem., 26, 107 (1964).
171. Bigorgne, M., and C. Messier, J. Organometal. Chem., 2, 79 (1964).
172. Bigorgne, M., and A. Zelwer, Bull. Soc. Chim. Fr., 1960, 1986.
173. Birch, A. J., and K. A. M. Walker, J. Chem. Soc., C, 1966, 1894.
174. Birch, A. J., and G. S. R. Subbarao, Tetrahedron Lett., 1968, 3797.
175. Birk, J. P., J. Halpern, and A. L. Pickard, Inorg. Chem., 7, 2672 (1968); J. Am. Chem. Soc., 90, 4491 (1968).
176. Birk, J. P., J. Halpern, and A. L. Pickard, J. Am. Chem. Soc., 90, 4491 (1968).
177. Black, M., R. H. B. Mais, and P. G. Owston, Acta Crystallogr., 25, 1760 (1969).
178. Blake, D., G. Calvin, and G. E. Coates, Proc. Chem. Soc., 1959, 396.
179. Blake, D. M., and C. J. Nyman, Chem. Commun., 1969, 483.
180. Bland, W. J., J. Burgess, and R. D. W. Kemmitt, J. Organometal. Chem., 14, 201 (1968).
181. Bland, W. J., J. Burgess, and R. D. W. Kemmitt, J. Organometal. Chem., 15, 217 (1968).
182. Bland, W. J., J. Burgess, and R. D. W. Kemmitt, J. Organometal. Chem., 18, 199 (1969).
183. Bland, W. J., and R. D. W. Kemmitt, Nature, 211, 963 (1966), J. Chem. Soc., A, 1968, 1278.
184. Bland, W. J., and R. D. W. Kemmitt, J. Chem. Soc., A, 1969, 2062.

185. Bland, W. J., R. D. W. Kemmitt, I. W. Nowell, and D. R. Russell, Chem. Commun., 1968, 1065.
186. Blundell, T. L., H. M. Powell, and L. M. Venanzi, Chem. Commun., 1967, 763.
187. Bolton, E. S., R. Havlin, and G. R. Knox, J. Organometal. Chem., 18, 153 (1969).
188. Bonati, F., S. Cenini, D. Morelli, and R. Ugo, J. Chem. Soc., A, 1966, 1052; 1967, 932.
189. Boorman, P. M., N. N. Greenwood, and M. A. Hildon, J. Chem. Soc., A, 1968, 2466.
190. Boorman, P. M., N. N. Greenwood, M. A. Hildon, and R. V. Parish, J. Chem. Soc., A, 1968, 2002.
191. Booth, B. L., and R. N. Haszeldine, J. Chem. Soc., A, 1966, 157.
192. Booth, G., Advan. Inorg. Radiochem., 6, 1 (1964).
193. Booth, G., and J. Chatt, Proc. Chem. Soc., 1961, 67; J. Chem. Soc., A, 1966, 634.
194. Booth, G., and J. Chatt, J. Chem. Soc., 1962, 2099.
195. Booth, G., and J. Chatt, J. Chem. Soc., 1965, 3238.
196. Booth, G., and J. Chatt, J. Chem. Soc., A, 1969, 2131.
197. Booth, G. J. Chatt, and P. Chini, Chem. Commun., 1965, 639; J. Chem. Soc., A, 1970, 1538.
198. Bor, G., and L. Markó, Chem. Ind. (London), 1963, 912.
199. Boschi, T., M. Nicolini, and A. Turco, Coordination Chem. Rev., 1, 269 (1966).
200. Boston, J. L., S. O. Grim, and G. Wilkinson, J. Chem. Soc., 1963, 3468.
201. Bradford, C. W., and R. S. Nyholm, Chem. Commun., 1967, 384.
202. Bradford, C. W., and R. S. Nyholm, Chem. Commun., 1968, 867.
203. Braterman, P. A., and D. T. Thompson, J. Organometal. Chem., 10, P11 (1967); J. Chem. Soc., A, 1968, 1454.
204. Brault, A. T., E. M. Thorsteinson, and F. Basolo, Inorg. Chem., 3, 770 (1964).
205. Braye, E. H., and W. Hübel, Chem. Ind. (London), 1959, 1250.
206. Breitschaft, S., and F. Basolo, J. Am. Chem. Soc., 88, 2702 (1966).
207. Bresadola, S., P. Rigo, and A. Turco, Chem. Commun., 1968, 1205.
208. Brewis, S., and P. R. Hughes, Chem. Commun., 1965, 157, 489.
209. Brewis, S., and P. R. Hughes, Chem. Commun., 1966, 6.
210. Brewis, S., and P. R. Hughes, Chem. Commun., 1967, 71.
211. Bright, D., and J. A. Ibers, Inorg. Chem., 7, 1099 (1968).

212. Bright, D., and J. A. Ibers, Inorg. Chem., 8, 703 (1969).
213. Bright, D., and J. A. Ibers, Inorg. Chem., 8, 1078 (1969).
214. British Petroleum Co., Brit. 1,181,806 (1966).
215. Brookes, P. R., and B. L. Shaw, J. Chem. Soc., A, 1967, 1079.
216. Brookes, P. R., and B. L. Shaw, Chem. Commun., 1968, 919.
217. Brooks, E. H., and F. Glockling, Chem. Commun., 1965, 510; J. Chem. Soc., A, 1966, 1241.
218. Brooks, E. H., and F. Glockling, J. Chem. Soc., A, 1967, 1030.
219. Brown, C. K., and G. Wilkinson, Tetrahedron Lett., 1969, 1725.
220. Brown, T. H., and P. J. Green, J. Am. Chem. Soc., 91, 3378 (1969).
221. Browning, M. C., R. F. B. Davies, D. J. Morgan, L. E. Sutton, and L. M. Venanzi, J. Chem. Soc., 1961, 4816.
222. Browning, M. C., J. R. Mellor, D. J. Morgan, S. A. J. Pratt, L. E. Sutton, and L. M. Venanzi, J. Chem. Soc., 1962, 693.
223. Bruce, M. I., C. W. Gibbs, and F. G. A. Stone, Z. Naturforsch., B, 23, 1543 (1968).
224. Bruce, M. I., D. A. Harbourne, F. Waugh, and F. G. A. Stone, J. Chem. Soc., A, 1968, 356.
225. Brunner, H., Angew. Chem. Int. Ed., 6, 566 (1967); Chem. Ber., 101, 143 (1968).
226. Brunner, H., J. Organometal. Chem., 16, 119 (1969).
227. Bryan, R. F., Proc. Chem. Soc., 1964, 232; J. Chem. Soc., A, 1967, 172.
228. Bryan, R. F., and A. R. Manning, Chem. Commun., 1968, 1316.
229. Buckingham, A. D., and P. J. Stephens, J. Chem. Soc., 1964, 4583.
230. Burg, A. B., and W. Mahler, J. Am. Chem. Soc., 80, 2334 (1958).
231. Burg, A. B., and R. A. Sinclair, J. Am. Chem. Soc., 88, 5354 (1966).
232. Burg, A. B., and B. G. Street, Inorg. Chem., 5, 1532 (1966).
233. Burmeister, J. L., and F. Basolo, Inorg. Chem., 3, 1587 (1964).
234. Burmeister, J. L., and R. C. Timmer, J. Inorg. Nucl. Chem., 28, 1973 (1966).
235. Burt, R., M. Cooke, and M. Green, J. Chem. Soc., A, 1969, 2645.
236. Bush, M. A., and P. Woodward, J. Chem. Soc., A, 1968, 1221.
237. Butler, I. S., F. Basolo, and R. G. Pearson, Inorg. Chem., 6, 2074 (1967).

238. Cahours, A., and H. Gal, Ann. Chem., 155, 223, 355 (1870); C. R. Acad. Sci., Paris, 70, 897, 1380 (1870).
239. Cahours, A., and H. Gal, Ann. Chem., 156, 302 (1870); C. R. Acad. Sci., Paris, 71, 208 (1870).
240. Calderazzo, F., and F. A. Cotton, Chim. Ind. (Milan), 46, 1165 (1964).
241. Calderazzo, F., S. Losi, and P. B. Susz, Inorg. Chim. Acta, 3, 329 (1969).
242. California Research Corp., Brit. 1,031,144 (1963).
243. Calvin, G., and G. E. Coates, Chem. Ind. (London), 1958, 160; J. Chem. Soc., 1960, 2008.
244. Calvin, G., G. E. Coates, and P. S. Dixon, Chem. Ind. (London), 1959, 1628.
245. Campbell, C. H., A. R. Dias, M. L. H. Green, T. Saito, and M. G. Swanwick, J. Organometal. Chem., 14, 349 (1968).
246. Candlin, J. P., K. K. Joshi, and D. T. Thompson, Chem. Ind. (London), 1966, 1960.
247. Candlin, J. P., and A. C. Shortland, J. Organometal. Chem., 16, 289 (1969).
248. Capron-Cotigny, G., and R. Poilblanc, Bull. Sec. Chim. Fr., 1967, 1440.
249. Cardin, D. J., and M. F. Lappert, Chem. Commun., 1966, 506.
250. Carfagna, P. D., and E. L. Amma, A.C.S., 148th Meeting, Abstr. p. 31.
251. Cariati, F., R. Mason, G. B. Robertson, and R. Ugo, Chem. Commun., 1967, 408.
252. Cariati, F., and L. Naldini, Gazz. Chim. Ital., 95, 3 (1965); J. Inorg. Nucl. Chem., 28, 2243 (1966).
253. Cariati, F., and L. Naldini, Gazz. Chim. Ital., 95, 201 (1965).
254. Cariati, F., L. Naldini, G. Simonetta, and L. Malatesta, Inorg. Chim. Acta, 1, 315 (1967).
255. Cariati, F., L. Naldini, G. Simonetta, and L. Malatesta, Inorg. Chim. Acta, 1, 24 (1967).
256. Cariati, F., and R. Ugo, Chim. Ind. (Milan), 48, 1288 (1966).
257. Cariati, F., R. Ugo, and F. Bonati, Chem. Ind. (London), 1964, 1714; Inorg. Chem., 5, 1128 (1966).
258. Carty, A. J., and A. Efraty, Chem. Commun., 1968, 1559; Inorg. Chem., 8, 543 (1969).
259. Carty, A. J., and A. Efraty, Can. J. Chem., 46, 1598 (1968).
260. Carty, A. J., and A. Efraty, Can. J. Chem., 47, 2573 (1969).
261. Carty, A. J., A. Efraty, and T. W. Ng, Can. J. Chem., 47, 1429 (1969).
262. Carty, A. J., and D. G. Tuck, J. Chem. Soc., A, 1966, 1081.

263. Carty, P., A. Walker, M. Mathew, and G. J. Palenik, Chem. Commun., 1969, 1374.
264. Cass, R. C., G. E. Coates, and R. G. Hayter, Chem. Ind. (London), 1954, 1485.
265. Cass, R. C., G. E. Coates, and R. G. Hayter, J. Chem. Soc., 1955, 4007.
266. Cattalini, L., U. Belluco, M. Martelli, and R. Ettorre, Gazz. Chim. Ital., 95, 567 (1965).
267. Cattalini, L., A. Orio, and M. Nicolini, J. Am. Chem. Soc., 88, 5734 (1966).
268. Caulton, K. G., and F. A. Cotton, J. Am. Chem. Soc., 91, 6517 (1969).
269. Cetini, G., O. Gambino, R. Rossetti, and P. L. Stanghelleni, Inorg. Chem., 7, 609 (1968).
270. Chalk, A. J., and J. F. Harrod, J. Am. Chem. Soc., 87, 16 (1965).
271. Chan, T. H., Chem. Commun., 1968, 895.
272. Chapovskii, Y. A., V. A. Semion, V. G. Andrianov, and Y. T. Struchkov, Zh. Strukt. Khim., 9, 1100 (1968); 10, 664 (1969).
273. Chastain, B. B., D. W. Meek, E. Billig, J. E. Hix, and H. B. Gray, Inorg. Chem., 7, 2412 (1968).
274. Chatt, J., J. Chem. Soc., 1950, 2301.
275. Chatt, J., J. Chem. Soc., 1951, 652.
276. Chatt, J., and S. A. Butter, Chem. Commun., 1967, 501.
277. Chatt, J., and R. S. Coffey, Chem. Commun., 1966, 545; J. Chem. Soc., A, 1969, 1963.
278. Chatt, J., R. S. Coffey, A. Gough, and D. T. Thompson, J. Chem. Soc., A, 1968, 190.
279. Chatt, J., R. S. Coffey, and B. L. Shaw, J. Chem. Soc., 1965, 7391.
280. Chatt, J., and J. M. Davidson, Angew. Chem., 75, 1026 (1963); J. Chem. Soc., 1965, 843.
281. Chatt, J., and J. M. Davidson, J. Chem. Soc., 1964, 2433.
282. Chatt, J., J. R. Dilworth, and G. J. Leigh, Chem. Commun., 1969, 687.
283. Chatt, J., and L. A. Duncanson, Nature, 178, 997 (1956).
284. Chatt, J., L. A. Duncanson, B. M. Gatehouse, J. Lewis, R. S. Nyholm, M. L. Tobe, P. F. Todd, and L. M. Venanzi, J. Chem. Soc., 1959, 4073.
285. Chatt, J., L. A. Duncanson, F. A. Hart, and P. G. Owston, Nature, 181, 43 (1958).
286. Chatt, J., L. A. Duncanson, and B. L. Shaw, Proc. Chem. Soc., 1957, 343.
287. Chatt, J., L. A. Duncanson, and B. L. Shaw, Chem. Ind. (London), 1958, 859.
288. Chatt, J., L. A. Duncanson, and L. M. Venanzi, J. Chem. Soc., 1955, 4461.

289. Chatt, J., L. A. Duncanson, and L. M. Venanzi, J. Chem. Soc., 1958, 3203.
290. Chatt, J., L. A. Duncanson, and L. M. Venanzi, J. Inorg. Nucl. Chem., 8, 67 (1958).
291. Chatt, J., C. Eaborn, and S. Ibekwe, Chem. Commun., 1966, 700.
292. Chatt, J., C. Eaborn, S. Ibekwe, and P. N. Kapoor, Chem. Commun., 1967, 869; J. Chem. Soc., A, 1970, 881.
293. Chatt, J., C. D. Falk, G. J. Leigh, and R. J. Paske, J. Chem. Soc., A, 1969, 2288.
294. Chatt, J., A. E. Field, and B. L. Shaw, J. Chem. Soc., 1963, 3371.
295. Chatt, J., G. A. Gamlen, and L. E. Orgel, J. Chem. Soc., 1959, 1047.
296. Chatt, J., J. D. Garforth, N. P. Johnson, and G. A. Rowe, J. Chem. Soc., 1964, 601.
297. Chatt, J., J. D. Garforth, N. P. Johnson, and G. A. Rowe, J. Chem. Soc., 1964, 1012.
298. Chatt, J., J. D. Garforth, and G. A. Rowe, Chem. Ind. (London), 1963, 332.
299. Chatt, J., J. D. Garforth, and G. A. Rowe, J. Chem. Soc., A, 1966, 1834.
300. Chatt, J., and F. A. Hart, Nature, 169, 673 (1952); J. Chem. Soc., 1953, 2363.
301. Chatt, J., and F. A. Hart, Chem. Ind. (London), 1958, 1474; J. Chem. Soc., 1960, 1378.
302. Chatt, J., and F. A. Hart, J. Chem. Soc., 1960, 2807.
303. Chatt, J., and F. A. Hart, J. Chem. Soc., 1961, 1416.
304. Chatt, J., and F. A. Hart, J. Chem. Soc., 1965, 812.
305. Chatt, J., F. A. Hart, and R. G. Hayter, Nature, 187, 55 (1960).
306. Chatt, J., F. A. Hart, and D. T. Rosevear, J. Chem. Soc., 1961, 5504.
307. Chatt, J., F. A. Hart, and H. R. Watson, J. Chem. Soc., 1962, 2537.
308. Chatt, J., and R. G. Hayter, Proc. Chem. Soc., 1959, 153; J. Chem. Soc., 1961, 2605.
309. Chatt, J., and R. G. Hayter, J. Chem. Soc., 1961, 772.
310. Chatt, J., and R. G. Hayter, J. Chem. Soc., 1961, 896.
311. Chatt, J., and R. G. Hayter, J. Chem. Soc., 1961, 5507.
312. Chatt, J., and R. G. Hayter, J. Chem. Soc., 1963, 1343.
313. Chatt, J., and R. G. Hayter, J. Chem. Soc., 1963, 6017.
314. Chatt, J., and B. T. Heaton, Chem. Commun., 1968, 274.

315. Chatt, J., and B. T. Heaton, J. Chem. Soc., A, 1968, 2745.
316. Chatt, J., N. P. Johnson, and B. L. Shaw, J. Chem. Soc., 1964, 1625.
317. Chatt, J., N. P. Johnson, and B. L. Shaw, J. Chem. Soc., 1964, 1662.
318. Chatt, J., N. P. Johnson, and B. L. Shaw, J. Chem. Soc., 1964, 2508.
319. Chatt, J., N. P. Johnson, and B. L. Shaw, J. Chem. Soc., A, 1967, 604.
320. Chatt, J., G. J. Leigh, D. M. P. Mingos, and R. L. Richards, Chem. Ind. (London), 1969, 109; Chem. Commun., 1969, 515.
321. Chatt, J., G. J. Leigh, and D.M. P. Mingos, J. Chem. Soc., A, 1969, 1674.
322. Chatt, J., G. J. Leigh, and D. M. P. Mingos, J. Chem. Soc., A, 1969, 2972.
323. Chatt, J., G. J. Leigh, D. M. P. Mingos, and R. J. Paske, Chem. Ind. (London), 1967, 1324.
324. Chatt, J., G. J. Leigh, D. M. P. Mingos, and R. J. Paske, J. Chem. Soc., A, 1968, 2636.
325. Chatt, J., G. J. Leigh, D. M. P. Mingos, E. W. Randall, and D. Shaw, Chem. Commun., 1968, 419.
326. Chatt, J., G. J. Leigh, and R. J. Paske, Chem. Commun., 1967, 671.
327. Chatt, J., G. J. Leigh, and R. J. Paske, J. Chem. Soc., A, 1969, 854.
328. Chatt, J., and F. G. Mann, J. Chem. Soc., 1938, 1949.
329. Chatt, J., and F. G. Mann, J. Chem. Soc., 1939, 1622.
330. Chatt, J., F. G. Mann, and A. F. Wells, J. Chem. Soc., 1938, 2086.
331. Chatt, J., D. P. Melville, and R. L. Richards, J. Chem. Soc., A, 1969, 2841.
332. Chatt, J., and D. M. P. Mingos, J. Chem. Soc., A, 1969, 1770.
333. Chatt, J., R. L. Richards, J. E. Fergusson, and J. L. Love, Chem. Commun., 1968, 1522.
334. Chatt, J., and G. A. Rowe, Nature, 191, 1191 (1961).
335. Chatt, J., and G. A. Rowe, Chem. Ind. (London), 1962, 92; J. Chem. Soc., 1962, 4019.
336. Chatt, J., G. A. Rowe, and A. A. Williams, Proc. Chem. Soc., 1957, 208.
337. Chatt, J., and B. L. Shaw, Chem. Ind. (London), 1959, 675; J. Chem. Soc., 1960, 1718.
338. Chatt, J., and B. L. Shaw, J. Chem. Soc., 1959, 705.
339. Chatt, J., and B. L. Shaw, J. Chem. Soc., 1959, 4020.
340. Chatt, J., and B. L. Shaw, Chem. Ind. (London), 1960, 931.

341. Chatt, J., and B. L. Shaw, Chem. Ind. (London), 1961, 290.
342. Chatt, J., and B. L. Shaw, J. Chem. Soc., 1961, 285.
343. Chatt, J., and B. L. Shaw, J. Chem. Soc., 1962, 5075.
344. Chatt, J., and B. L. Shaw, J. Chem. Soc., A, 1966, 1437.
345. Chatt, J., and B. L. Shaw, J. Chem. Soc., A, 1966, 1811.
346. Chatt, J., and B. L. Shaw, J. Chem. Soc., A, 1966, 1836.
347. Chatt, J., B. L. Shaw, and A. E. Field, J. Chem. Soc., 1964, 3466.
348. Chatt, J., B. L. Shaw, and A. A. Williams, J. Chem. Soc., 1962, 3269.
349. Chatt, J., and D. T. Thompson, J. Chem. Soc., 1964, 2508.
350. Chatt, J., and D. A. Thornton, J. Chem. Soc., 1964, 1005.
351. Chatt, J., and A. E. Underhill, J. Chem. Soc., 1963, 2088.
352. Chatt, J., and L. M. Venanzi, J. Chem. Soc., 1955, 2787.
353. Chatt, J., and L. M. Venanzi, J. Chem. Soc., 1955, 3858; 1957, 2445.
354. Chatt, J., and L. M. Venanzi, J. Chem. Soc., 1957, 2351.
355. Chatt, J., and L. M. Venanzi, J. Chem. Soc., 1957, 4735.
356. Chatt, J., and H. R. Watson, Proc. Chem. Soc., 1960, 243.
357. Chatt, J., and H. R. Watson, Nature, 189, 1003 (1961); J. Chem. Soc., 1962, 2537.
358. Chatt, J., and H. R. Watson, J. Chem. Soc., 1961, 4980.
359. Chatt, J., and R. G. Wilkins, J. Chem. Soc., 1951, 2532.
360. Chatt, J., and R. G. Wilkins, J. Chem. Soc., 1952, 273, 4300; 1956, 525.
361. Chatt, J., and R. G. Wilkins, J. Chem. Soc., 1953, 70.
362. Chaudhari, F. M., and P. L. Pauson, J. Organometal. Chem., 5, 73 (1966).
363. Cheeseman, T. P., A. L. Odell, and H. A. Raethel, Chem. Commun., 1968, 1496.
364. Chini, P., A. Santambrogio, and N. Palladino, J. Chem. Soc., A, 1967, 830.
365. Chioccola, G., J. J. Daly, and J. K. Nicholson, Angew. Chem. Int. Ed., 7, 131 (1968); J. Chem. Soc., A, 1968, 1981.

366. Chiswell, B., and L. M. Venanzi, J. Chem. Soc., A, 1966, 417.
367. Chiswell, B., and L. M. Venanzi, J. Chem. Soc., A, 1966, 901.
368. Chock, P. B., and J. Halpern, J. Am. Chem. Soc., 88, 3511 (1966).
369. Chopoorian, J. A., J. Lewis, and R. S. Nyholm, Nature, 190, 528 (1961).
370. Christopher, R. E., I. R. Gordon, and L. M. Venanzi, J. Chem. Soc., A, 1968, 205.
371. Chuit, C., H. Felkin, C. Frajerman, G. Roussi, and G. Swierczewski, Chem. Commun., 1968, 1604.
372. Church, M. J., and M. J. Mays, Chem. Commun., 1968, 435; J. Chem. Soc., A, 1968, 3074.
373. Churchill, M. R., and J. P. Fennessey, Inorg. Chem., 7, 953 (1968).
374. Churchill, M. R., and T. A. O'Brien, Chem. Commun., 1968, 246.
375. Churchill, M. R., and T. A. O'Brien, J. Chem. Soc., A, 1969, 266.
376. Churchill, M. R., and T. A. O'Brien, J. Chem. Soc., A, 1968, 2970.
377. Churchill, M. R., T. A. O'Brien, M. D. Rausch, and Y. F. Chang, Chem. Commun., 1967, 992.
378. Cities Service Research and Development Co., Brit. 971,755 (1960); Brit. 979,553 (1960); Brit. 971,771 (1961).
379. Clark, H. C., P. W. R. Corfield, K. R. Dixon, and J. A. Ibers, J. Am. Chem. Soc., 89, 3360 (1967).
380. Clark, H. C., and K. R. Dixon, J. Am. Chem. Soc., 91, 596 (1969).
381. Clark, H. C., K. R. Dixon, and W. J. Jacobs, Chem. Commun., 1968, 93; J. Am. Chem. Soc., 90, 2259 (1968).
382. Clark, H. C., K. R. Dixon, and W. J. Jacobs, Chem. Commun., 1968, 548; J. Am. Chem. Soc., 91, 1346 (1969).
383. Clark, H. C., J. H. Tsai, and W. S. Tsang, Chem. Commun., 1965, 171.
384. Clark, H. C., and W. S. Tsang, Chem. Commun., 1966, 123.
385. Clark, H. C., and W. S. Tsang, J. Am. Chem. Soc., 89, 529, 533 (1967).
386. Clark, R. J. H., R. H. U. Negrotti, and R. S. Nyholm, Chem. Commun., 1966, 486.
387. Clarke, B., M. Green, R. B. L. Osborn, and F. G. A. Stone, J. Chem. Soc., A, 1968, 167.
388. Clees, H., and F. Huber, Z. Anorg. Chem., 352, 200 (1967).
389. Clemens, D. F., H. H. Sisler, and W. S. Brey, Inorg. Chem., 5, 527 (1966).

390. Clemmit, A. F., and F. Glockling, J. Chem. Soc., A, 1969, 2163.
391. Clifford, A. F., and A. K. Mukherjee, Inorg. Chem., 2, 151 (1963).
392. Coates, G. E., J. Chem. Soc., 1951, 2003.
393. Coates, G. E., Organometallic Compounds, Methuen, 1960.
394. Coates, G. E., C. Kowala, and J. M. Swan, Aust. J. Chem., 19, 539, 547, 555 (1966).
395. Coates, G. E., and C. Parkin, J. Inorg. Nucl. Chem., 22, 59 (1961).
396. Coates, G. E., and C. Parkin, J. Chem. Soc., 1962, 3220.
397. Coates, G. E., and C. Parkin, J. Chem. Soc., 1963, 421.
398. Coates, G. E., and D. Ridley, J. Chem. Soc., 1964, 166.
399. Cochran, W., F. A. Hart, and F. G. Mann, J. Chem. Soc., 1957, 2816.
400. Coffey, C. E., J. Lewis, and R. S. Nyholm, J. Chem. Soc., 1964, 1741.
401. Coffey, R. S., Tetrahedron Lett., No. 43, 3809 (1965).
402. Coffey, R. S., Chem. Commun., 1967, 923.
403. Coffey, R. S., Ann. Rept., 65B, 321 (1968), and references therein.
404. Cohen, I. A., and F. Basolo, J. Inorg. Nucl. Chem., 28, 511 (1966).
405. Coleman, J. M., and L. F. Dahl, J. Am. Chem. Soc., 89, 542 (1967).
406. Collamati, I., and A. Furlani, J. Organometal. Chem., 17, 457 (1969).
407. Collier, J. W., A. R. Fox, I. G. Hinton, and F. G. Mann, J. Chem. Soc., 1964, 1819.
408. Collier, J. W., and F. G. Mann, J. Chem. Soc., 1964, 1815.
409. Collier, J. W., F. G. Mann, D. G. Watson, and H. R. Watson, J. Chem. Soc., 1964, 1803.
410. Collman, J. P., J. N. Cawse, and J. W. Kang, Inorg. Chem., 8, 2574 (1969).
411. Collman, J. P., N. W. Hoffman, and D. E. Morris, J. Am. Chem. Soc., 91, 5659 (1969).
412. Collman, J. P., and J. W. Kang, J. Am. Chem. Soc., 88, 3459 (1966).
413. Collman, J. P., and J. W. Kang, J. Am. Chem. Soc., 89, 844 (1967).
414. Collman, J. P., J. W. Kang, W. F. Little, and M. F. Sullivan, Inorg. Chem., 7, 1298 (1968).
415. Collman, J. P., M. Kubota, J. Y. Sun, and F. Vastine, J. Am. Chem. Soc., 89, 169 (1967).

416. Collman, J. P., and W. R. Roper, J. Am. Chem. Soc., 87, 4008 (1965).
417. Collman, J. P., and W. R. Roper, J. Am. Chem. Soc., 1966, 180.
418. Collman, J. P., and W. R. Roper, Chem. Commun., 1966, 244.
419. Collman, J. P., and W. R. Roper, J. Am. Chem. Soc., 88, 3504 (1966).
420. Collman, J. P., and C. T. Sears, Inorg. Chem., 7, 27 (1968).
421. Collman, J. P., F. D. Vastine, and W. R. Roper, J. Am. Chem. Soc., 88, 5035 (1966).
422. Colton, R., R. Levitus, and G. Wilkinson, Nature, 186, 233 (1960).
423. Colton, R., R. Levitus, and G. Wilkinson, Chem. Ind. (London), 1959, 1314; J. Chem. Soc., 1960, 4121.
424. Colton, R., G. R. Scollary, and I. B. Tomkins, Aust. J. Chem., 21, 15, 1427 (1968).
425. Colton, R., and I. B. Tomkins, Aust. J. Chem., 19, 1143, 1519 (1966).
426. Columbrian Carbon, U. S. 3,187,062 (1962); U. S. 3,249,641 (1962).
427. Cook, C. D., P. T. Cheng, and S. C. Nyburg, J. Am. Chem. Soc., 91, 2123 (1969).
428. Cook, C. D., and G. S. Jauhal, J. Am. Chem. Soc., 89, 3066 (1967).
429. Cook, C. D., and G. S. Jauhal, Can. J. Chem., 45, 301 (1967).
430. Cook, C. D., and G. S. Jauhal, J. Am. Chem. Soc., 90, 1464 (1968).
431. Cook, C. D., C. H. Koo, S. C. Nyburg, and M. T. Shiomi, Chem. Commun., 1967, 426.
432. Cooke, J., W. R. Cullen, M. Green, and F. G. A. Stone, J. Chem. Soc., A, 1969, 1872.
433. Cooke, M., M. Green, and D. Kirkpatrick, J. Chem. Soc., A, 1968, 1507.
434. Cookson, R. C., G. W. A. Fowles, and D. K. Jenkins, J. Chem. Soc., 1965, 6406.
435. Corain, B., M. Bressan, P. Rigo, and A. Turco, Chem. Commun., 1968, 509.
436. Corfield, P. W. R., R. J. Doedens, and J. A. Ibers, Inorg. Chem., 6, 197 (1967).
437. Corfield, P. W. R., and H. M. M. Shearer, Acta Crystallogr., 20, 502 (1966).
438. Costa, G., A. Camus, N. Marsich, and L. Gatti, J. Organometal. Chem., 8, 339 (1967).
439. Costa, G., G. Pellizer, and F. Rubessa, J. Inorg. Nucl. Chem., 26, 961 (1964).
440. Costa, G., E. Reisenhofer, and L. Stefani, J. Inorg. Nucl. Chem., 27, 2581 (1965).

441. Cotton, F. A., N. F. Crutis, and W. B. Robinson, Inorg. Chem., 4, 1696 (1965).
442. Cotton, F. A., R. Eiss, and B. M. Foxman, Inorg. Chem., 8, 950 (1969).
443. Cotton, F. A., O. D. Faut, and D. M. L. Goodgame, J. Am. Chem. Soc., 83, 344 (1961).
444. Cotton, F. A., O. D. Faut, D. M. L. Goodgame, and R. H. Holm, J. Am. Chem. Soc., 83, 1780 (1961).
445. Cotton, F. A., and B. M. Foxman, Inorg. Chem., 7, 1784 (1968).
446. Cotton, F. A., and D. M. L. Goodgame, J. Am. Chem. Soc., 82, 2967 (1960).
447. Cotton, F. A., and D. M. L. Goodgame, J. Chem. Soc., 1960, 5267.
448. Cotton, F. A., D. M. L. Goodgame, M. Goodgame, and A. Sacco, J. Am. Chem. Soc., 83, 4157 (1961).
449. Cotton, F. A., and B. F. G. Johnson, Inorg. Chem., 3, 1609 (1964).
450. Cotton, F. A., and S. J. Lippard, Inorg. Chem., 5, 9 (1966).
451. Cotton, F. A., and J. T. Mague, Inorg. Chem., 3, 1094 (1964).
452. Cotton, F. A., and T. J. Marks, J. Am. Chem. Soc., 91, 7281 (1969).
453. Cotton, F. A., and R. V. Parish, J. Chem. Soc., 1960, 1440.
454. Cotton, F. A., and R. A. Walton, Inorg. Chem., 5, 1802 (1966).
455. Cotton, J. D., M. I. Bruce, and F. G. A. Stone, J. Chem. Soc., A, 1968, 2162.
456. Coulson, D. R., Chem. Commun., 1968, 1530.
457. Coussmaker, C. R. C., M. Hely-Hutchinson, J. R. Mellor, L. E. Sutton, and L. M. Venanzi, J. Chem. Soc., 1961, 2705.
458. Cowley, A. H., and J. L. Mills, J. Am. Chem. Soc., 91, 2915 (1969).
459. Cramer, R., and G. W. Parshall, J. Am. Chem. Soc., 87, 1392 (1965).
460. Crooks, G. R., and B. F. G. Johnson, J. Chem. Soc., A, 1968, 1238.
461. Cross, R. J., and F. Glockling, Proc. Chem. Soc., 1964, 143; J. Organometal. Chem., 3, 253 (1965).
462. Cule Davies, W., and F. G. Mann, J. Chem. Soc., 1944, 276.
463. Cullen, W. R., P. S. Dhaliwal, and C. J. Stewart, Inorg. Chem., 6, 2256 (1967).
464. Cullen, W. R., D. F. Dong, and J. A. J. Thompson, Can. J. Chem., 47, 4671 (1969).
465. Cullen, W. R., and D. A. Harbourne, Can. J. Chem., 47, 3371 (1969).

466. Cundy, C. S., M. Green, F. G. A. Stone, and A. Taunton-Rigby, Inorg. Nucl. Chem. Lett., 2, 233 (1966); J. Chem. Soc., A, 1968, 1776.
467. Dahm, D. J., and R. A. Jacobsen, Chem. Commun., 1966, 496; J. Am. Chem. Soc., 90, 5106 (1968).
468. Daniels, W. E., Inorg. Chem., 3, 1800 (1964).
469. Dapporto, P., and L. Sacconi, Chem. Commun., 1969, 1091.
470. Davidson, J. M., Chem. Ind. (London), 1964, 2021.
471. Davidson, J. M., and G. Dyer, J. Chem. Soc., A, 1968, 1616.
472. Davies, A. G., G. Wilkinson, and J. F. Young, J. Am. Chem. Soc., 85, 1692 (1963).
473. Davies, G. R., R. H. B. Mais, and P. G. Owston, J. Chem. Soc., A, 1967, 1750.
474. Davis, B. R., N. C. Payne, and J. A. Ibers, Inorg. Chem., 8, 2719 (1969); J. Am. Chem. Soc., 91, 1240 (1969).
475. Davison, A., M. L. H. Green, and G. Wilkinson, J. Chem. Soc., 1961, 3172.
476. Davison, A., W. McFarlane, L. Pratt, and G. Wilkinson, J. Chem. Soc., 1962, 3653.
477. Dawson, J. W., D. G. E. Kerfoot, C. Preti, and L. M. Venanzi, Chem. Commun., 1968, 1687.
478. Dawson, J. W., and L. M. Venanzi, J. Am. Chem. Soc., 90, 7229 (1968).
479. Day, J. P., and L. M. Venanzi, J. Chem. Soc., A, 1966, 1363.
480. Deacon, G. B., and J. H. S. Green, Chem. Ind. (London), 1965, 1031; Chem. Commun., 1966, 629.
481. Deacon, G. B., J. H. S. Green, and D. J. Harrison, Spectrochim. Acta, 24A, 845, 1921 (1968).
482. Deacon, G. B., and B. O. West, J. Chem. Soc., 1961, 5127.
483. Dean, R. R., and J. C. Green, J. Chem. Soc., A, 1968, 3047.
484. DeBoer, J. L., D. Rogers, A. C. Skapski, and P. G. H. Troughton, Chem. Commun., 1966, 756.
485. De Charentenay, F., J. A. Osborn, and G. Wilkinson, J. Chem. Soc., A, 1968, 787.
486. Deeming, A. J., B. F. G. Johnson, and J. Lewis, J. Organometal. Chem., 17, P40 (1969).
487. Deeming, A. J., and B. L. Shaw, Chem. Commun., 1968, 751; J. Chem. Soc., A, 1969, 1562.
488. Deeming, A. J., and B. L. Shaw, J. Chem. Soc., A, 1968, 1887.
489. Deeming, A. J., and B. L. Shaw, J. Chem. Soc., A, 1969, 443.
490. Deeming, A. J., and B. L. Shaw, J. Chem. Soc., A, 1969, 597.

491. Deeming, A. J., and B. L. Shaw, J. Chem. Soc., A, 1969, 1128.
492. Deeming, A. J., and B. L. Shaw, J. Chem. Soc., A, 1969, 1802.
493. Deganello, G., G. Carturan, and U. Belluco, J. Chem. Soc., A, 1968, 2873.
494. Deganello, G., G. Carturan, and P. Uguagliati, J. Organometal. Chem., 17, 179 (1969).
495. Deganello, G., P. Uguagliati, B. Crociani, and U. Belluco, J. Chem. Soc., A, 1969, 2726.
496. Degischer, G., and G. Schwarzenbach, Helv. Chim. Acta, 49, 1927 (1966).
497. Delbeke, F. T., G. P. Vanderkelen, and Z. Eeckhaut, J. Organometal. Chem., 16, 512 (1969).
498. Deroos, J. B., and J. P. Oliver, Inorg. Chem., 4, 1741 (1965).
499. Dewhirst, K. C., W. Keim, and C. A. Reilly, Inorg. Chem., 7, 546 (1968).
500. Dick, T. R., and S. M. Nelson, Inorg. Chem., 8, 1208 (1969).
501. Dieck, H. T., and H. Friedel, Chem. Commun., 1969, 411.
502. Dilts, J. A., and M. P. Johnson, Inorg. Chem., 5, 2079 (1966).
503. Dilts, J. A., and D. F. Shriver, J. Am. Chem. Soc., 91, 4088 (1969).
504. Dobbie, R. C., M. Green, and F. G. A. Stone, J. Chem. Soc., A, 1969, 1881.
505. Dobinson, G. C., R. Mason, G. B. Robertson, R. Ugo, F. Conti, D. Morelli, S. Cenini, and F. Bonati, Chem. Commun., 1967, 739.
506. Dobson, G. R., R. C. Taylor, and T. D. Walsh, Inorg. Chem., 6, 1929 (1967).
507. Doedens, R. J., and L. F. Dahl, J. Am. Chem. Soc., 87, 2576 (1965).
508. Doedens, R. J., and J. A. Ibers, Inorg. Chem., 6, 204 (1967).
509. Doedens, R. J., W. T. Robinson, and J. A. Ibers, J. Am. Chem. Soc., 89, 4323 (1967).
510. Dolcetti, G., M. Nicolini, M. Giustiniani, and U. Belluco, J. Chem. Soc., A, 1969, 1387.
511. Donda, A. F., and G. Moretti, J. Organometal. Chem., 31, 985 (1966).
512. Donoghue, J. T., J. A. McMillan, and D. A. Peters, J. Inorg. Nucl. Chem., 31, 3661 (1969).
513. Douek, I. C., and G. Wilkinson, J. Chem. Soc., A, 1969, 2604.
514. Douglas, P. G., and B. L. Shaw, Chem. Commun., 1969, 624.
515. Douglas, P. G., and B. L. Shaw, J. Chem. Soc., A, 1969, 1491.

516. Doyle, J. R., J. H. Hutchinson, N. C. Baenziger, and L. W. Tresselt, J. Am. Chem. Soc., 83, 2768 (1961).
517. Dreissig, W., and H. Dietrich, Acta Crystallogr., B24, 108 (1968).
518. Drew, M. G. B., D. F. Lewis, and R. A. Walton, Chem. Commun., 1969, 326.
519. Dubois, T. D., and D. W. Meek, Inorg. Chem., 6, 1395 (1967).
520. Duddell, D. A., P. L. Goggin, R. J. Goodfellow, and M. G. Norton, Chem. Commun., 1968, 879.
521. Duncanson, L. A., and L. M. Venanzi, J. Chem. Soc., 1960, 3841.
522. DuPont, Brit. 1,151,005 (1966).
523. DuPont, Brit. 1,181,137 (1966).
524. Dyer, G., J. G. Hartley, and L. M. Venanzi, J. Chem. Soc., 1965, 1293.
525. Dyer, G., and D. W. Meek, Inorg. Chem., 4, 1398 (1965).
526. Dyer, G., and D. W. Meek, Inorg. Chem., 6, 149 (1967).
527. Dyer, G., and D. W. Meek, J. Am. Chem. Soc., 89, 3983 (1967).
528. Dyer, G., M. O. Workman, and D. W. Meek, Inorg. Chem., 6, 1404 (1967).
529. Edgar, K., B. F. G. Johnson, J. Lewis, and S. B. Wild, J. Chem. Soc., A, 1968, 285.
530. Edgell, W. F., and M. P. Dunkle, Inorg. Chem., 4, 1629 (1965).
531. Edwards, D. A., J. Inorg. Nucl. Chem., 27, 303 (1965).
532. Ehrlich, H. W. W., and P. G. Owston, J. Chem. Soc., 1963, 4368.
533. Eisenberg, R., and J. A. Ibers, Inorg. Chem., 4, 773 (1965).
534. Ellermann, J., and K. Dorn, Angew. Chem. Int. Ed., 5, 516 (1966).
535. Ellermann, J., and K. Dorn, J. Organometal. Chem., 6, 157 (1966).
536. Ellermann, J., and W. H. Gruber, Angew. Chem. Int. Ed., 7, 129 (1968).
537. Ellermann, J., and W. H. Gruber, Z. Anorg. Chem., 364, 55 (1969).
538. Emeléus, H. J., and J. Grobe, Angew. Chem., 74, 467 (1962).
539. Emeléus, H. J., and J. D. Smith, J. Chem. Soc., 1958, 527.
540. Enemark, J. H., B. R. Davis, J. A. McGinnety, and J. A. Ibers, Chem. Commun., 1968, 96.
541. Enemark, J. H., and J. A. Ibers, Inorg. Chem., 6, 1575 (1967).
542. Enemark, J. H., and J. A. Ibers, Inorg. Chem., 7, 2339 (1968).

543. L'Eplattenier, F., and F. Calderazzo, Inorg. Chem., 6, 2092 (1967).
544. Esso, U. S. 3,310,576; Fr. 1,389,699 (1963).
545. Esso, Fr. 1,417,181 (1963).
546. Esso, W. L. Mertzweiller, and J. K. Bentley, U. S. 3,488,296 (1967).
547. Ethyl Corp., U. S. 3,256,260 (1961).
548. Ethyl Corp., and G. E. Schroll, U. S. 3,054,815 (1960); C. A., 58, 1494 (1963).
549. Ettorre, R., J. Organometal. Chem., 19, 247 (1969).
550. Evans, D., J. A. Osborn, F. H. Jardine, and G. Wilkinson, Nature, 208, 1203 (1965).
551. Evans, D., J. A. Osborn, and G. Wilkinson, J. Chem. Soc., A, 1968, 3133.
552. Evans, D., G. Yagupsky, and G. Wilkinson, J. Chem. Soc., A, 1968, 2660.
553. Evans, J. G., P. L. Goggin, R. J. Goodfellow, and J. G. Smith, J. Chem. Soc., A, 1968, 464.
554. Evans, R. C., F. G. Mann, H. S. Peiser, and D. Purdie, J. Chem. Soc., 1940, 1209.
555. Fairy, M. B., and R. J. Irving, J. Chem. Soc., A, 1966, 475.
556. Faller, J. W., A. S. Anderson, and C. C. Chen, J. Organometal. Chem., 17, P7 (1969).
557. Faraone, G., V. Ricevuto, R. Romeo, and M. Trozzi, Gazz. Chim. Ital., 96, 590 (1966); J. Inorg. Nucl. Chem., 28, 863 (1966).
558. Faraone, G., V. Ricevuto, R. Romeo, and M. Trozzi, Inorg. Chem., 8, 2207 (1969).
559. Farona, M. F., and A. Wojcicki, Inorg. Chem., 10, 1402 (1965).
560. Felkin, H., and G. Swierczewski, C. R. Acad. Sci., Paris, Ser. C, 266, 1611 (1968).
561. Fell, B., and W. Rupilius, Tetrahedron Lett., 32, 2721 (1969).
562. Feltham, R. D., J. Inorg. Nucl. Chem., 14, 307 (1960); Inorg. Chem., 3, 116 (1964).
563. Feltham, R. D., Inorg. Chem., 3, 119 (1964).
564. Feltham, R. D., and R. G. Hayter, J. Chem. Soc., 1964, 4587.
565. Fergusson, J. E., and J. H. Hickford, J. Inorg. Nucl. Chem., 28, 2293 (1966).
566. Fergusson, J. E., B. H. Robinson, and G. J. Wilkins, J. Chem. Soc., A, 1967, 486.
567. Ferrer, D. T., M. Praissman, and E. L. Amma, A.C.S., 142nd Meeting, Abstr., p. 35N (1962).
568. Fischer, E. O., E. Louis, W. Bathelt, E. Moser, and J. Muller, J. Organometal. Chem., 14, P9 (1968); 20, 147 (1969).
569. Fischer, E. O., and E. Moser, J. Organometal. Chem., 5, 63 (1966).

570. Fischer, E. O., and R. J. J. Schneider, Angew. Chem. Int. Ed., 6, 569 (1967).
571. Fischer, E. O., and H. Werner, Chem. Ber., 95, 703 (1962).
572. Fitton, P., M. P. Johnson, and J. E. McKeon, Chem. Commun., 1968, 6.
573. Fitton, P., and J. E. McKeon, Chem. Commun., 1968, 4.
574. Fitton, P., J. E. McKeon, and B. C. Ream, Chem. Commun., 1969, 370.
575. Fluck, E., and K. F. Brauch, Z. Anorg. Chem., 364, 107 (1969).
576. Foglemen, W. W., and H. B. Jonassen, J. Inorg. Nucl. Chem., 31, 1536 (1969).
577. Fowles, G. W. A., and D. K. Jenkins, Chem. Commun., 1965, 61.
578. Fowles, G. W. A., and R. A. Walton, J. Chem. Soc., 1964, 4330.
579. Frasson, E., C. Panattoni, and A. Turco, Nature, 199, 803 (1963).
580. Freni, M., Ann. Chim. (Rome), 48, 231 (1958); C. A., 52, 17150 (1958).
581. Freni, M., R. Demichelis, and D. Giusto, J. Inorg. Nucl. Chem., 29, 1433 (1967).
582. Freni, M., D. Giusto, and P. Romiti, J. Inorg. Nucl. Chem., 29, 761 (1967).
583. Freni, M., D. Giusto, and P. Romiti, Gazz. Chim. Ital., 97, 833 (1967).
584. Freni, M., D. Giusto, and P. Romiti, Gazz. Chim. Ital., 99, 641 (1969).
585. Freni, M., D. Giusto, P. Romiti, and G. Minghetti, Gazz. Chim. Ital., 99, 286 (1969).
586. Freni, M., D. Giusto, P. Romiti, and E. Zucca, J. Inorg. Nucl. Chem., 31, 3211 (1969).
587. Freni, M., D. Giusto, and V. Valenti, Gazz. Chim. Ital., 94, 797 (1964).
588. Freni, M., D. Giusto, and V. Valenti, J. Inorg. Nucl. Chem., 27, 755 (1965).
589. Freni, M., and V. Valenti, Gazz. Chim. Ital., 90, 1436 (1960).
590. Freni, M., and V. Valenti, Gazz. Chim. Ital., 90, 1445 (1960).
591. Freni, M., and V. Valenti, J. Inorg. Nucl. Chem., 16, 240 (1961).
592. Freni, M., and V. Valenti, Gazz. Chim. Ital., 91, 1357 (1961).
593. Freni, M., V. Valenti, and D. Giusto, J. Inorg. Nucl. Chem., 27, 2635 (1965).
594. Freni, M., V. Valenti, and R. Pomponi, Gazz. Chim. Ital., 94, 521 (1964).
595. Fritz, H. P., I. R. Gordon, K. E. Schwarzhans, and L. M. Venanzi, J. Chem. Soc., 1965, 5210.

596. Gans, P., and B. C. Smith, Chem. Ind. (London), 1963, 911.
597. Garton, G., D. E. Henn, H. M. Powell, and L. M. Venanzi, J. Chem. Soc., 1963, 3625.
598. Gatehouse, B. M., S. E. Livingstone, and R. S. Nyholm, J. Inorg. Nucl. Chem., 8, 75 (1958).
599. General Aniline and Film Corp., and R. F. Kleinschmidt, U. S. 2,542,417 (1949).
600. Giacometti, G., V. Scatturin, and A. Turco, Gazz. Chim. Ital., 88, 434 (1958).
601. Giacometti, G., and A. Turco, J. Inorg. Nucl. Chem., 15, 242 (1960).
602. Gibb, T. C., R. Greatrex, N. N. Greenwood, and D. T. Thompson, J. Chem. Soc., A, 1967, 1663.
603. Gibson, D., B. F. G. Johnson, J. Lewis, and C. Oldham, Chem. Ind. (London), 1966, 342.
604. Gilbert, J. D., and G. Wilkinson, J. Chem. Soc., A, 1969, 1749.
605. Gilchrist, T. L., F. J. Graveling, and C. W. Rees, Chem. Commun., 1968, 821.
606. Gillard, R. D., R. Ugo, F. Cariati, S. Cenini, and F. Bonati, Chem. Commun., 1966, 869.
607. Ginsberg, A. P., in Transition Metal Chemistry, R. L. Carlin, Ed., Arnold, 1965.
608. Ginsberg, A. P., Chem. Commun., 1968, 857.
609. Ginsberg, A. P., and E. Koubek, Inorg. Chem., 10, 1517 (1965).
610. Ginsberg, A. P., and W. E. Silverthorn, Chem. Commun., 1969, 823.
611. Giustiniani, M., G. Dolcetti, and U. Belluco, J. Chem. Soc., A, 1969, 2047.
612. Giustiniani, M., G. Dolcetti, M. Nicolini, and U. Belluco, J. Chem. Soc., A, 1969, 1961.
613. Giustiniani, M., G. Dolcetti, R. Pietropaolo, and U. Belluco, Inorg. Chem., 8, 1048 (1969).
614. Glanville, J. O., J. M. Stewart, and S. O. Grim, J. Organometal. Chem., 7, P9 (1967).
615. Glockling, F., and K. A. Hooton, J. Chem. Soc., 1962, 2658.
616. Glockling, F., and K. A. Hooton, Chem. Commun., 1966, 218; J. Chem. Soc., A, 1967, 1066.
617. Glockling, F., and K. A. Hooton, J. Chem. Soc., A, 1968, 826.
618. Glockling, F., and M. D. Wilbey, J. Chem. Soc., A, 1968, 2168.
619. Goggin, P. L., and R. J. Goodfellow, J. Chem. Soc., A, 1966, 1462.
620. Goodfellow, R. G., Chem. Commun., 1968, 114.
621. Goodfellow, R. J., J. G. Evans, P. L. Goggin, and D. A. Duddell, J. Chem. Soc., A, 1968, 1604.

622. Goodfellow, R. J., P. L. Goggin, and L. M. Venanzi, J. Chem. Soc., A, 1967, 1897.
623. Goodfellow, R. J., and L. M. Venanzi, J. Chem. Soc., 1965, 7533.
624. Goodgame, D. M. L., and M. Goodgame, Inorg. Chem., 4, 139 (1965).
625. Goodgame, D. M. L., M. Goodgame, and F. A. Cotton, J. Am. Chem. Soc., 83, 4161 (1961).
626. Gorsich, R. D., J. Am. Chem. Soc., 84, 2486 (1962).
627. Gray, H. B., E. Billig, A. Wojcicki, and M. Farona, Can. J. Chem., 41, 1281 (1963).
628. Gray, H. B., and A. Wojcicki, Proc. Chem. Soc., 1960, 358.
629. Grayson, M., P. T. Keough, and G. A. Johnson, J. Am. Chem. Soc., 81, 4803 (1959).
630. Graziani, M., F. Zingales, and U. Belluco, Inorg. Chem., 6, 1582 (1967).
631. Greaves, E. O., R. Bruce, and P. M. Maitlis, Chem. Commun., 1967, 860.
632. Greaves, E. O., and P. M. Maitlis, J. Organometal. Chem., 6, 104 (1966).
633. Green, M., N. R. Mayne, R. B. L. Osborn, and F. G. A. Stone, J. Chem. Soc., A, 1969, 1879.
634. Green, M., R. B. L. Osborn, A. J. Rest, and F. G. A. Stone, Chem. Commun., 1966, 502; J. Chem. Soc., A, 1968, 2525.
635. Green, M., R. B. L. Osborn, and F. G. A. Stone, J. Chem. Soc., A, 1968, 3083.
636. Green, M., A. Taunton-Rigby, and F. G. A. Stone, J. Chem. Soc., A, 1969, 1875.
637. Green, M. L. H., and J. T. Moelwyn-Hughes, Z. Naturforsch., B, 17, 783 (1962).
638. Green, M. L. H., H. Munakata, and T. Saito, Chem. Commun., 1969, 1287.
639. Green, M. L. H., M. Nehmé, and G. Wilkinson, Chem. Ind. (London), 1960, 1136.
640. Green, M. L. H., C. N. Street, and G. Wilkinson, Z. Naturforsch., B, 14, 738 (1959).
641. Green, M. L. H., and T. Saito, Chem. Commun., 1969, 208.
642. Greenwood, N. N., E. J. F. Ross, and A. Storr, J. Chem. Soc., 1965, 1400.
643. Griffith, W. P., J. Lewis, and G. Wilkinson, J. Chem. Soc., 1961, 2259.
644. Grim, S. O., R. L. Keiter, and W. McFarlane, Inorg. Chem., 6, 1133 (1967).
645. Grim, S. O., and D. A. Wheatland, Inorg. Chem., 8, 1716 (1969).
646. Grim, S. O., D. A. Wheatland, and P. R. McAllister, Inorg. Chem., 7, 161 (1968).

647. Grim, S. O., D. A. Wheatland, and W. McFarlane, J. Am. Chem. Soc., 89, 5573 (1967).
648. Grinberg, A. A., and Z. A. Razumova, Zh. Obshch. Khim., 18, 282 (1948); C. A., 42, 8689 (1948).
649. Grinberg, A. A., and Z. A. Razumova, Zh. Prik. Khim., 27, 105 (1954); C. A., 48, 6308 (1954).
650. Grobe, J., Angew. Chem., 75, 1113 (1963); Z. Anorg. Chem., 331, 63 (1964).
651. Grobe, J., Z. Anorg. Chem., 361, 32, 47 (1968).
652. Grobe, J., and H. Stierand, Z. Anorg. Chem., 371, 99 (1969).
653. Gruber, W. H., and J. Ellermann, Z. Naturforsch., B, 23, 1307 (1968).
654. Guerrieri, F., and G. P. Chiusoli, Chem. Commun., 1967, 781; J. Organometal. Chem., 15, 209 (1968).
655. Haake, P., and T. A. Hylton, J. Am. Chem. Soc., 84, 3774 (1962).
656. Haake, P., and M. Pfeiffer, Chem. Commun., 1969, 1330.
657. Haines, L. I. B., D. Hopgood, and A. J. Poe, J. Chem. Soc., A, 1968, 421.
658. Haines, L. I. B., and A. J. Poe, J. Chem. Soc., A, 1969, 2826.
659. Haines, R. J., and A. I. DuPreez, Chem. Commun., 1968, 1513.
660. Haines, R. J., A. L. DuPreez, and G. T. W. Wittmann, Chem. Commun., 1968, 611.
661. Haines, R. J., R. S. Nyholm, and M. H. B. Stiddard, J. Chem. Soc., A, 1967, 94.
662. Halfpenny, M. T., J. G. Hartley, and L. M. Venanzi, J. Chem. Soc., A, 1967, 627.
663. Hallman, P. S., D. Evans, J. A. Osborn, and G. Wilkinson, Chem. Commun., 1967, 305.
664. Hallman, P. S., B. R. McGarvey, and G. Wilkinson, J. Chem. Soc., A, 1968, 3143.
665. Halpern, J., B. R. James, and A. L. W. Kemp, J. Am. Chem. Soc., 88, 5142 (1966).
666. Harmon, R. E., J. L. Parsons, and S. K. Gupta, Chem. Commun., 1969, 1365.
667. Harrod, J. F., D. F. R. Gilson, and R. Charles, Can. J. Chem., 47, 1431 (1969).
668. Harrod, J. F., D. F. R. Gilson, and R. Charles, Can. J. Chem., 47, 2205 (1969).
669. Hart, F. A., and F. G. Mann, J. Chem. Soc., 1957, 3939.
670. Hartley, J. G., D. G. E. Kerfoot, and L. M. Venanzi, Inorg. Chim. Acta, 1, 145 (1967).
671. Hartley, J. G., L. M. Venanzi, and D. C. Goodall, J. Chem. Soc., 1963, 3930.
672. Haszeldine, R. N., R. V. Parish, and D. J. Parry, J. Organometal. Chem., 9, P13 (1967); J. Chem. Soc., A, 1969, 683.

673. Hata, G., H. Kondo, and A. Miyake, J. Am. Chem. Soc., 90, 2278 (1968).
674. Hata, G., and A. Miyake, Chem. Ind. (London), 1967, 921.
675. Hatfield, W. E., and J. T. Yoke, Inorg. Chem., 1, 470 (1962).
676. Hatfield, W. E., and J. T. Yoke, Inorg. Chem., 1, 475 (1962).
677. Haugen, L. P., and R. Eisenberg, Inorg. Chem., 8, 1072 (1969).
678. Hayter, R. G., J. Am. Chem. Soc., 83, 1259 (1961).
679. Hayter, R. G., Nature, 193, 872 (1962); J. Am. Chem. Soc., 84, 3046 (1962).
680. Hayter, R. G., Z. Naturforsch., B, 18, 581 (1963).
681. Hayter, R. G., J. Am. Chem. Soc., 85, 3120 (1963).
682. Hayter, R. G., Inorg. Chem., 2, 932 (1963).
683. Hayter, R. G., Inorg. Chem., 2, 1031 (1963).
684. Hayter, R. G., J. Am. Chem. Soc., 86, 823 (1964).
685. Hayter, R. G., Inorg. Chem., 3, 301 (1964).
686. Hayter, R. G., Inorg. Chem., 3, 711 (1964).
687. Hayter, R. G., and F. S. Humiec, J. Am. Chem. Soc., 84, 2004 (1962).
688. Hayter, R. G., and F. S. Humiec, Inorg. Chem., 2, 306 (1963).
689. Hayter, R. G., and F. S. Humiec, Inorg. Chem., 4, 1701 (1965).
690. Hayter, R. G., and L. F. Williams, Inorg. Chem., 3, 613 (1964).
691. Hayter, R. G., and L. F. Williams, Inorg. Chem., 3, 717 (1964).
692. Hayter, R. G., and L. F. Williams, J. Inorg. Nucl. Chem., 26, 1977 (1964).
693. Hayward, P. J., D. M. Blake, C. J. Nyman, and G. Wilkinson, Chem. Commun., 1969, 987.
694. Heathcock, C. H., and S. R. Poulter, Tetrahedron Lett., 32, 2755 (1969).
695. Heaton, B. T., and A. Pidcock, J. Organometal. Chem., 14, 235 (1968).
696. Heck, R. F., J. Am. Chem. Soc., 85, 657 (1963).
697. Heck, R. F., J. Am. Chem. Soc., 85, 1460, 3387 (1963).
698. Heck, R. F., J. Am. Chem. Soc., 86, 2796 (1964).
699. Heck, R. F., J. Am. Chem. Soc., 86, 2819 (1964).
700. Heck, R. F., J. Am. Chem. Soc., 86, 5138 (1964).
701. Heck, R. F., and D. S. Breslow, J. Am. Chem. Soc., 82, 4438 (1960).
702. Heck, R. F., and D. S. Breslow, J. Am. Chem. Soc., 83, 1097 (1961).
703. Heck, R. F., and D. S. Breslow, J. Am. Chem. Soc., 83, 4023 (1961).
704. Heck, R. F., and D. S. Breslow, J. Am. Chem. Soc., 84, 2499 (1962).

705. Heck, R. F., J. C. W. Chien, and D. S. Breslow, Chem. Ind. (London), 1961, 986.
706. Henrici-Olivé, G., and S. Olivé, Chem. Commun., 1969, 1482.
707. Hershman, A., and J. H. Craddock, Ind. Eng. Chem., Prod. Res. Develop., 7, 226 (1968).
708. Hicks, D. G., and J. A. Dean, Chem. Commun., 1965, 172.
709. Hidai, M., K. Tominari, Y. Uchida, and A. Misono, Chem. Commun., 1969, 814, 1392.
710. Hieber, W., W. Abeck, and H. K. Platzer, Z. Anorg. Chem., 280, 252 (1955).
711. Hieber, W., and I. Bauer, Z. Naturforsch., B, 16, 556 (1961).
712. Hieber, W., W. Beck, and E. Lindner, Z. Naturforsch., B, 16, 229 (1961).
713. Hieber, W., W. Beck, and H. Tengler, Z. Naturforsch., B, 15, 411 (1960).
714. Hieber, W., and R. Breu, Angew. Chem., 68, 679 (1956); Chem. Ber., 90, 1259, 1270 (1957).
715. Hieber, W., and H. Duchatsch, Z. Naturforsch., B, 18, 1132 (1963).
716. Hieber, W., and H. Duchatsch, Chem. Ber., 98, 1744 (1965).
717. Hieber, W., and H. Duchatsch, Chem. Ber., 98, 2530 (1965).
718. Hieber, W., and H. Duchatsch, Chem. Ber., 98, 2933 (1965).
719. Hieber, W., K. Englert, and K. Rieger, Z. Anorg. Chem., 300, 295 (1959).
720. Hieber, W., G. Faulhaber, and F. Theubert, Z. Naturforsch., B, 15, 326 (1960); Z. Anorg. Chem., 314, 125 (1962).
721. Hieber, W., and J. G. Floss, Z. Anorg. Chem., 291, 314 (1957).
722. Hieber, W., and V. Frey, Chem. Ber., 99, 2607 (1966).
723. Hieber, W., and V. Frey, Chem. Ber., 99, 2614 (1966).
724. Hieber, W., V. Frey, and P. John, Chem. Ber., 100, 1961 (1967).
725. Hieber, W., and W. Freyer, Chem. Ber., 91, 1230 (1958).
726. Hieber, W., and W. Freyer, Chem. Ber., 92, 1765 (1959).
727. Hieber, W., and W. Freyer, Chem. Ber., 93, 462 (1960).
728. Hieber, W., and K. Heinicke, Z. Naturforsch., B, 16, 553 (1961); Z. Anorg. Chem., 316, 305 (1962).
729. Hieber, W., and K. Heinicke, Z. Naturforsch., B, 16, 554 (1961); Z. Anorg. Chem., 316, 321 (1962).
730. Hieber, W., and H. Heusinger, Angew. Chem., 68, 678 (1956).

731. Hieber, W., and H. Heusinger, J. Inorg. Nucl. Chem., 4, 179 (1957).
732. Hieber, W., H. Heusinger, and O. Vohler, Chem. Ber., 90, 2425 (1957).
733. Hieber, W., M. Hofler, and J. Muschi, Chem. Ber., 98, 311 (1965).
734. Hieber, W., W. Klingshirn, and N. Beck, Chem. Ber., 98, 307 (1965).
735. Hieber, W., and R. Kramolowsky, Z. Naturforsch., B, 16, 555 (1961).
736. Hieber, W., and R. Kummer, Z. Naturforsch., B, 20, 271 (1965).
737. Hieber, W., and R. Kummer, Z. Anorg. Chem., 344, 292 (1966).
738. Hieber, W., and R. Kummer, Chem. Ber., 100, 148 (1967).
739. Hieber, W., and E. Lindner, Z. Naturforsch., B, 16, 137 (1961); Chem. Ber., 94, 1417 (1961).
740. Hieber, W., and E. Lindner, Chem. Ber., 95, 273 (1962).
741. Hieber, W., and E. Lindner, Chem. Ber., 95, 2042 (1962).
742. Hieber, W., and J. Muschi, Chem. Ber., 98, 3931 (1965).
743. Hieber, W., J. Muschi, and H. Duchatsch, Chem. Ber., 98, 3924 (1965).
744. Hieber, W., and G. Neumair, Z. Anorg. Chem., 342, 93 (1966).
745. Hieber, W., and W. Opavsky, Chem. Ber., 101, 2966 (1968).
746. Hieber, W., and J. Peterhans, Z. Naturforsch., B, 14, 462 (1959).
747. Hieber, W., J. Peterhans, and E. Winter, Chem. Ber., 94, 2572 (1961).
748. Hieber, W., and W. Schropp, Z. Naturforsch., B, 14, 460 (1959).
749. Hieber, W., and L. Schuster, Z. Anorg. Chem., 287, 214 (1956).
750. Hieber, W., and J. Sedlmeier, Jr., Chem. Ber., 87, 789 (1954).
751. Hieber, W., and H. Tengler, Z. Anorg. Chem., 318, 136 (1962).
752. Hieber, W., and A. Thalhofer, Angew. Chem., 68, 679 (1956).
753. Hieber, W., and G. Wagner, Z. Naturforsch., B, 12, 478 (1957).
754. Hieber, W., and E. Winter, Chem. Ber., 97, 1037 (1964).
755. Hieber, W., and K. Wollmann, Chem. Ber., 95, 1552 (1962).
756. Hinton, R. C., and F. G. Mann, J. Chem. Soc., 1959, 2835.

757. Hitchcock, C. H. S., and F. G. Mann, J. Chem. Soc., 1958, 2081.
758. Hitchcock, P. B., M. McPartlin, and R. Mason, Chem. Commun., 1969, 1367.
759. Hodgson, D. J., and J. A. Ibers, Inorg. Chem., 7, 2345 (1968).
760. Hodgson, D. J., and J. A. Ibers, Inorg. Chem., 8, 1282 (1969).
761. Hofmann, A. W., Ann. Chem., 103, 357 (1857).
762. Hogben, M. G., R. S. Gay, and W. A. G. Graham, J. Am. Chem. Soc., 88, 3457 (1966).
763. Hopgood, D., and A. J. Poë, Chem. Commun., 1966, 831.
764. Hopton, F. J., A. J. Rest, D. T. Rosevear, and F. G. A. Stone, J. Chem. Soc., A, 1966, 1326.
765. Horrocks, W. DeW., and L. H. Pignolet, J. Am. Chem. Soc., 88, 5929 (1966).
766. Horrocks, W. DeW., G. R. VanHecke, and D. DeW. Hall, Inorg. Chem., 6, 694 (1967).
767. Houk, L. W., and G. R. Dobson, J. Chem. Soc., A, 1966, 317; Inorg. Chem., 5, 2119 (1966).
768. Howell, I. V., and L. M. Venanzi, Inorg. Chim. Acta, 3, 121 (1969).
769. Howell, I. V., L. M. Venanzi, and D. C. Goodall, J. Chem. Soc., A, 1967, 395, 1007.
770. Huber, M., and R. Poilblanc, Bull. Soc. Chim. Fr., 1960, 1019.
771. Hudson, A., and M. J. Kennedy, J. Chem. Soc., A, 1969, 1116.
772. Hudson, M. J., R. S. Nyholm, and M. H. B. Stiddard, J. Chem. Soc., A, 1968, 40.
773. Hughes, W. B., Chem. Commun., 1969, 1126.
774. Hull, C. G., and M. H. B. Stiddard, J. Chem. Soc., A, 1968, 710.
775. Huttel, R., U. Raffay, and H. Reinheimer, Angew. Chem. Int. Ed., 6, 862 (1967).
776. Ibekwe, S. D., B. T. Kilbourn, U. A. Raeburn, and D. R. Russell, Chem. Commun., 1969, 433.
777. Ibekwe, S. D., and M. J. Newlands, Chem. Commun., 1965, 114.
778. Ibekwe, S. D., and U. K. Raeburn, J. Organometal. Chem., 19, 447 (1969).
779. Ibers, J. A., J. Organometal. Chem., 14, 423 (1968).
780. Imperial Chemical Industries Limited, Brit. 713,325 (1950).
781. Imperial Chemical Industries Limited, Belg. 600,192 (1960).
782. Imperial Chemical Industries Limited, Brit. 1,137,392 (1965).
783. Imperial Chemical Industries Limited, Brit. 1,121,642 (1965).

784. Imperial Chemical Industries Limited, Brit. 1,141,847; U. S. 3,488,400 (1966).
785. Imperial Chemical Industries Limited, J. Chatt, C. Eaborn, and S. Ibekwe, Brit. 1,177,702 (1967).
786. Imperial Chemical Industries Limited, J. Chatt, F. A. Hart, and R. G. Hayter, Brit. 928,441; 928,442 (1959).
787. Imperial Chemical Industries Limited, J. Chatt, F. A. Hart, and G. A. Rowe, Brit. 882,400 (1958).
788. Imperial Chemical Industries Limited, and R. S. Coffey, Brit. 1,135,979 (1965).
789. Imperial Chemical Industries Limited, and R. S. Coffey, Brit. 1,176,654 (1966).
790. Imperial Chemical Industries Limited, R. W. Dunning, K. A. Taylor, and J. Walker, Brit. 1,164,855 (1967).
791. Imperial Chemical Industries Limited, G. A. Gamlen, and A. Ibbotson, Brit. 1,129,551 (1965).
792. Imperial Chemical Industries Limited, K. A. Taylor, W. Hewertson, and J. A. Leonard, Brit. 1,131,146 (1966).
793. Imperial Chemical Industries Limited, and D. T. Thompson, Brit. 1,129,553 (1965).
794. I.G., Fr. 961,010 (1950); C. A., 46, 5617 (1952).
795. I.G., W. Schweckendiek, and O. Weissbarth, Ger. 824,047 (1951); C. A., 49, 6998 (1955).
796, Innorta, G., S. Pignataro, and A. Foffani, J. Organometal. Chem., 20, 284 (1969).
797. International Nickel, Brit. 1,116,128 (1966).
798. Interrante, L. V., M. A. Bennett, and R. S. Nyholm, Inorg. Chem., 5, 2212 (1966).
799. Interrante, L. V., and G. V. Nelson, Inorg. Chem., 7, 2059 (1968).
800. Irving, R. J., and E. A. Magnusson, J. Chem. Soc., 1957, 2018.
801. Issleib, K., and G. Bohn, Z. Anorg. Chem., 301, 188 (1959).
802. Issleib, K., and A. Brack, Z. Anorg. Chem., 277, 258 (1954).
803. Issleib, K., and A. Brack, Z. Anorg. Chem., 292, 245 (1957).
804. Issleib, K., and G. Döll, Z. Anorg. Chem., 305, 1 (1960).
805. Issleib, K., and H. Freitag, Z. Anorg. Chem., 332, 124 124 (1964).
806. Issleib, K., and H. O. Fröhlich, Z. Anorg. Chem., 298, 84 (1959); Chem. Ber., 95, 375 (1962).
807. Issleib, K., and M. Haftendorn, Z. Anorg. Chem., 351, 9 (1967).
808. Issleib, K., and G. Hohlfeld, Z. Anorg. Chem., 312, 169 (1961).

809. Issleib, K., and M. Keil, Z. Anorg. Chem., 333, 10 (1964).
810. Issleib, K., and B. Mitscherling, Z. Anorg. Chem., 304, 73 (1960).
811. Issleib, K., and H.-R. Roloff, Z. Anorg. Chem., 324, 250 (1963).
812. Issleib, K., and G. Schwager, Z. Anorg. Chem., 310, 43 (1961).
813. Issleib, K., and G. Schwager, Z. Anorg. Chem., 311, 83 (1961).
814. Issleib, K., and A. Tzschach, Z. Anorg. Chem., 297, 121 (1958).
815. Issleib, K., and A. Tzschach, Chem. Ber., 92, 705 (1959).
816. Issleib, K., and H. Weichmann, Z. Anorg. Chem., 362, 33 (1968).
817. Issleib, K., and E. Wenschuh, Z. Anorg. Chem., 305, 15 (1960).
818. Issleib, K., and E. Wenschuh, Z. Naturforsch., B, 19, 199 (1964).
819. Issleib, K., and G. Wilde, Z. Naturforsch., 312, 287 (1961).
820. Iwamoto, M., and S. Yuguchi, Chem. Commun., 1968, 28.
821. Iwashita, Y., and A. Hayata, J. Am. Chem. Soc., 91, 2525 (1969).
822. Jackson, T. B., M. J. Baker, J. O. Edwards, and D. Tutas, Inorg. Chem., 5, 2046 (1966).
823. Jardine, F. H., J. A. Osborn, and G. Wilkinson, J. Chem. Soc., A, 1967, 1574.
824. Jardine, F. H., J. A. Osborn, G. Wilkinson, and J. F. Young, Chem. Ind. (London), 1965, 560.
825. Jardine, F. H., and G. Wilkinson, J. Chem. Soc., C, 1967, 270.
826. Jarvis, J. A. J., R. H. B. Mais, and P. G. Owston, J. Chem. Soc., A, 1968, 1473.
827. Jarvis, J. A. J., R. H. B. Mais, P. G. Owston, and K. A. Taylor, Chem. Commun., 1966, 906.
828. Jarvis, J. A. J., R. H. B. Mais, P. G. Owston, and D. T. Thompson, J. Chem. Soc., A, 1968, 622.
829. Jenkins, J. M., M. S. Lupin, and B. L. Shaw, J. Chem. Soc., A, 1966, 1787.
830. Jenkins, J. M., J. R. Moss, and B. L. Shaw, J. Chem. Soc., A, 1969, 2796.
831. Jenkins, J. M., and B. L. Shaw, Proc. Chem. Soc., 1963, 279.
832. Jenkins, J. M., and B. L. Shaw, J. Chem. Soc., 1965, 6789.
833. Jenkins, J. M., and B. L. Shaw, J. Chem. Soc., A, 1966, 770.

834. Jenkins, J. M., and B. L. Shaw, J. Chem. Soc., A, 1966, 1407.
835. Jenkins, J. M., and J. G. Verkade, Inorg. Chem., 6, 2250 (1967).
836. Jensen, K. A., Z. Anorg. Chem., 229, 225 (1936).
837. Jensen, K. A., Z. Anorg. Chem., 229, 252 (1936).
838. Jensen, K. A., Z. Anorg. Chem., 229, 265 (1936).
839. Jensen, K. A., Z. Anorg. Chem., 229, 282 (1936).
840. Jensen, K. A., and O. Dahl, Acta Chem. Scand., 22, 1044 (1968); 23, 2342 (1969).
841. Jensen, K. A., P. Halfdan Nielsen, and C. T. Pedersen, Acta Chem. Scand., 17, 1115 (1963).
842. Jensen, K. A., and C. K. Jorgensen, Acta Chem. Scand., 19, 451 (1965).
843. Jensen, K. A., and B. Nygaard, Acta Chem. Scand., 3, 474 (1949).
844. Jensen, K. A., B. Nygaard, and C. T. Pedersen, Acta Chem. Scand., 17, 1126 (1963).
845. Job, B. E., R. A. N. McLean, D. T. Thompson, Chem. Commun., 1966, 895.
846. Johnson, B. F. G., J. Chem. Soc., A, 1967, 475.
847. Johnson, B. F. G., R. D. Johnston, P. L. Josty, J. Lewis, and I. G. Williams, Nature, 213, 901 (1967).
848. Johnson, B. F. G., R. D. Johnston, and J. Lewis, J. Chem. Soc., A, 1969, 792.
849. Johnson, B. F. G., J. Lewis, and M. S. Subramanian, Chem. Commun., 1966, 117.
850. Johnson, B. F. G., J. Lewis, J. M. Wilson, and D. T. Thompson, J. Chem. Soc., A, 1967, 1445.
851. Johnson, N. P., C. J. L. Lock, and G. Wilkinson, J. Chem. Soc., 1964, 1054.
852. Jolly, P. W., and K. Jonas, Angew. Chem. Int. Ed., 7, 731 (1968).
853. Jolly, P. W., and F. G. A. Stone, Chem. Commun., 1965, 85.
854. Jolly, P. W., and F. G. A. Stone, J. Chem. Soc., 1965, 5259.
855. Jonas, K., and G. Wilke, Angew. Chem. Int. Ed., 8, 519 (1969).
856. Jones, E. R. H., and F. G. Mann, J. Chem. Soc., 1955, 4472.
857. Kalck, P., and R. Poilblanc, J. Organometal. Chem., 19, 115 (1969).
858. Kasenally, A. S., J. Lewis, A. R. Manning, J. R. Miller, R. S. Nyholm, and M. H. B. Stiddard, J. Chem. Soc., 1965, 3407.
859. Kasenally, A. S., R. S. Nyholm, R. J. O'Brien, and M. H. B. Stiddard, Nature, 204, 871 (1964).
860. Kashiwagi, T., N. Yasuoka, N. Kasai, and M. Kukodo, Chem. Commun., 1969, 317.

861. Kashiwagi, T., N. Yasuoka, N. Kasai, M. Kakudo, S. Takahashi, and N. Hagihara, Chem. Commun., 1969, 743.
862. Kauffman, G. B., R. P. Pinnell, and L. T. Takahashi, Inorg. Chem., 1, 544 (1962).
863. Kawada, I., Tetrahedron Lett., 1969, 793.
864. Keen, I. M., J. Chem. Soc., 1965, 5751.
865. Keim, W., J. Organometal. Chem., 8, P25 (1967); 14, 179 (1968).
866. Keim, W., J. Organometal. Chem., 16, 191 (1969).
867. Keim, W., J. Organometal. Chem., 19, 161 (1969).
868. Keiter, R. L., Dissertation Abstr., 28B, 4455 (1968).
869. Keiter, R. L., and S. O. Grim, Chem. Commun., 1968, 521.
870. Kemmitt, R. D. W., D. I. Nichols, and R. D. Peacock, Chem. Commun., 1967, 599; J. Chem. Soc., A, 1968, 1898, 2149.
871. Kemmitt, R. D. W., R. D. Peacock, and J. Stocks, Chem. Commun., 1969, 554.
872. Kern, R. J., Chem. Commun., 1968, 706.
873. Kilbourn, B. T., and R. H. B. Mais, Chem. Commun., 1968, 1507.
874. Kilbourn, B. T., H. M. Powell, and J. A. C. Darbyshire, Proc. Chem. Soc., 1963, 207; J. Chem. Soc., A, 1970, 1688.
875. Kilbourn, B. T., U. A. Raeburn, and D. T. Thompson, J. Chem. Soc., A, 1969, 1906.
876. King, R. B., Inorg. Chem., 2, 1275 (1963).
877. King, R. B., Inorg. Chem., 5, 82 (1966).
878. King, R. B., and C. A. Eggers, Inorg. Chim. Acta, 2, 33 (1968).
879. King, R. B., L. W. Houk, and K. H. Pannell, Inorg. Chem., 8, 1042 (1969).
880. King, R. B., and P. N. Kapoor, J. Organometal. Chem., 18, 357 (1969).
881. King, R. B., and K. H. Pannell, Inorg. Chem., 7, 1510 (1968).
882. King, R. B., K. H. Pannell, C. A. Eggers, and L. W. Houk, Inorg. Chem., 7, 2353 (1968).
883. Kingston, J. V., and G. R. Scollary, J. Inorg. Nucl. Chem., 31, 2557 (1969).
884. Kistner, C. R., J. H. Hutchinson, J. R. Doyle, and J. C. Storlie, Inorg. Chem., 2, 1255 (1963).
885. Klason, P., and J. Wanselin, J. Prakt. Chem., 67, 41 (1903).
886. Kneise, W., H. J. Nienburg, and R. Fischer, J. Organometal. Chem., 17, 133 (1969).
887. Knoth, W. H., J. Am. Chem. Soc., 90, 7172 (1968).
888. Kobayashi, T., Y. Takahashi, S. Sakai, and Y. Ishii, Chem. Commun., 1968, 1373.
889. Kowala, C., and J. M. Swan, Aust. J. Chem., 19, 999 (1966).

890. Kraihanzel, C. S., and P. K. Maples, J. Am. Chem. Soc., 87, 5267 (1965); Inorg. Chem., 7, 1806 (1968).
891. Kraihanzel, C. S., and P. K. Maples, J. Organometal. Chem., 20, 269 (1969).
892. Kruck, T., and K. Baur, Z. Anorg. Chem., 364, 192 (1969).
893. Kruck, T., and W. Lang, Z. Anorg. Chem., 343, 181 (1966).
894. Laing, K. R., and W. R. Roper, Chem. Commun., 1968, 1556.
895. Laing, K. R., and W. R. Roper, Chem. Commun., 1968, 1568.
896. Laing, K. R., and W. R. Roper, J. Chem. Soc., A, 1969, 1889.
897. LaLancette, E. A., and D. R. Eaton, J. Am. Chem. Soc., 86, 5145 (1964).
898. LaMar, G. N., J. Chem. Phys., 43, 235 (1965).
899. LaMar, G. N., W. D. Horrocks, and L. C. Allen, J. Chem. Phys., 41, 2126 (1964).
900. LaMar, G. N., and E. O. Sherman, Chem. Commun., 1969, 809.
901. Lambert, R. F., and J. D. Johnstone, Chem. Ind. (London), 1960, 1267.
902. Lane, A. P., and D. S. Payne, J. Chem. Soc., 1963, 4004.
903. LaPlaca, S. J., and J. A. Ibers, J. Am. Chem. Soc., 85, 3501 (1963).
904. LaPlaca, S. J., and J. A. Ibers, Acta Crystallogr., 18, 511 (1965).
905. LaPlaca, S. J., and J. A. Ibers, Science, 145, 920 (1964); J. Am. Chem. Soc., 87, 2581 (1965).
906. LaPlaca, S. J., and J. A. Ibers, Inorg. Chem., 4, 778 (1965).
907. LaPlaca, S. J., and J. A. Ibers, Inorg. Chem., 5, 405 (1966).
908. Lawson, D. N., J. A. Osborn, and G. Wilkinson, J. Chem. Soc., A, 1966, 1733.
909. Layton, A. J., R. S. Nyholm, G. A. Pneumaticakis, and M. L. Tobe, Chem. Ind. (London), 1967, 465.
910. Layton, A. J., R. S. Nyholm, G. A. Pneumaticakis, and M. L. Tobe, Nature, 214, 1109 (1967); 218, 950 (1968).
911. Leigh, G. J., J. J. Levison, and S. D. Robinson, Chem. Commun., 1969, 705.
912. L'Eplattenier, F., Inorg. Chem., 8, 965 (1969).
913. L'Eplattenier, F., and F. Calderazzo, Inorg. Chem., 7, 1290 (1968).
914. Levison, J. J., and S. D. Robinson, Chem. Commun., 1967, 198.
915. Levison, J. J., and S. D. Robinson, Chem. Ind. (London), 1969, 1514.

916. Lewis, J., A. R. Manning, and J. R. Miller, J. Chem. Soc., A, 1966, 845.
917. Lewis, J., A. R. Manning, J. R. Miller, and J. M. Wilson, J. Chem. Soc., A, 1966, 1663.
918. Lewis, J., R. S. Nyholm, A. G. Osborne, S. S. Sandhu, and M. H. B. Stiddard, Chem. Ind. (London), 1963, 1398; J. Chem. Soc., 1964, 2825.
919. Lewis, J., and R. Whyman, Chem. Commun., 1965, 159; J. Chem. Soc., 1965, 5486, 6027.
920. Lewis, J., and R. Whyman, J. Chem. Soc., A, 1967, 77.
921. Lippard, S. J., and K. M. Melmed, J. Am. Chem. Soc., 89, 3929 (1967); Inorg. Chem., 6, 2223 (1967).
922. Lippard, S. J., and K. M. Melmed, Inorg. Chem., 8, 2755 (1969).
923. Lippard, S. J., and D. Ucko, Chem. Commun., 1967, 983; Inorg. Chem., 7, 1051 (1968).
924. Livingstone, S. E., and T. N. Lockyer, Inorg. Nucl. Chem. Lett., 3, 35 (1967).
925. Lock, C. J. L., and G. Wilkinson, Chem. Ind. (London), 1962, 40.
926. Lorberth, J., H. Nöth, and P. V. Rinze, J. Organometal. Chem., 16, Pl (1961).
927. Lupin, M. S., and B. L. Shaw, J. Chem. Soc., A, 1968, 741.
928. Luth, H., M. R. Truter, and A. Robson, Chem. Commun., 1967, 738; J. Chem. Soc., A, 1969, 28.
929. Lutsenko, I. F., M. A. Kazankova, and I. G. Malykhina, J. Gen. Chem. USSR, 37, 2364 (1967); C. A., 68, 83938 (1967).
930. Luttinger, L. B., Chem. Ind. (London), 1960, 1135.
931. Lyons, J. E., Chem. Commun., 1969, 564.
932. Magee, T. A., C. N. Matthews, T. S. Wang, and J. H. Wotiz, J. Am. Chem. Soc., 83, 3200 (1961).
933. Mague, J. T., and J. P. Mitchener, Inorg. Chem., 8, 119 (1969).
934. Maier, L., Angew. Chem., 71, 574 (1959).
935. Mais, R. H. B., P. G. Owston, and D. T. Thompson, J. Chem. Soc., A, 1967, 1735.
936. Mais, R. H. B., P. G. Owston, D. T. Thompson, and A. M. Wood, J. Chem. Soc., A, 1967, 1744.
937. Maitlis, P. M., and A. Efraty, J. Organometal. Chem., 4, 175 (1965).
938. Malatesta, L., and M. Angoletta, J. Chem. Soc., 1957, 1186.
939. Malatesta, L., M. Angoletta, A. Aràneo, and F. Canziani, Angew. Chem., 73, 273 (1961).
940. Malatesta, L., M. Angoletta, and G. Caglio, Angew. Chem., 75, 1103 (1963).
941. Malatesta, L., and A. Aràneo, J. Chem. Soc., 1957, 3803.

942. Malatesta, L., and G. Caglio, Chem. Commun., 1967, 420.
943. Malatesta, L., G. Caglio, and M. Angoletta, J. Chem. Soc., 1965, 6974.
944. Malatesta, L., and C. Cariello, J. Chem. Soc., 1958, 2323; J. Inorg. Nucl. Chem., 8, 561 (1958).
945. Malatesta, L., M. Freni, and V. Valenti, Angew. Chem., 73, 273 (1961).
946. Malatesta, L., M. Freni, and V. Valenti, Gazz. Chim. Ital., 94, 1228 (1964).
947. Malatesta, L., L. Naldini, G. Simonetta, and F. Cariati, Chem. Commun., 1965, 212; Coordination Chem. Rev., 1, 255 (1966).
948. Malatesta, L., and A. Sacco, Ann. Chim. (Rome), 44, 134 (1954); C. A., 48, 13516 (1954).
949. Malatesta, L., and R. Ugo, J. Chem. Soc., 1963, 2080.
950. Mango, F. D., and I. Dvoretsky, J. Am. Chem. Soc., 88, 1654 (1966).
951. Mann, B. E., C. Masters, and B. L. Shaw, Chem. Commun., 1970, 703.
952. Mann, F. G., and I. T. Millar, J. Chem. Soc., 1951, 2205; 1952, 4453.
953. Mann, F. G., I. T. Millar, and F. H. C. Stewart, J. Chem. Soc., 1954, 2832.
954. Mann, F. G., and D. Purdie, J. Chem. Soc., 1935, 1549.
955. Mann, F. G., and D. Purdie, J. Chem. Soc., 1936, 873.
956. Mann, F. G., and D. Purdie, J. Chem. Soc., 1940, 1230.
957. Mann, F. G., and D. Purdie, J. Chem. Soc., 1940, 1235.
958. Mann, F. G., D. Purdie, and A. F. Wells, J. Chem. Soc., 1936, 1503.
959. Mann, F. G., and H. R. Watson, J. Chem. Soc., 1957, 3950.
960. Mann, F. G., and H. R. Watson, Chem. Ind. (London), 1958, 1264.
961. Mann, F. G., and A. F. Wells, J. Chem. Soc., 1938, 702.
962. Mann, F. G., A. F. Wells, and D. Purdie, J. Chem. Soc., 1937, 1828.
963. Mannerskantz, H. C. E., G. W. Parshall, and G. Wilkinson, J. Chem. Soc., 1963, 3163.
964. Manning, A. R., J. Chem. Soc., A, 1968, 651.
965. Manuel, T. A., Inorg. Chem., 2, 854 (1963).
966. Manuel, T. A., Advan. Organometal. Chem., 3, 181 (1965).
967. Manuel, T. A., and F. G. A. Stone, J. Am. Chem. Soc., 82, 366 (1960).

968. Marvel, G., Progr. NMR Spectroscopy, 1, 296, 345 (1966).
969. Mason, R., G. B. Robertson, P. O. Whimp, and D. A. White, Chem. Commun., 1968, 1655; J. Chem. Soc., A, 1969, 2709.
970. Mason, R., and D. R. Russell, Chem. Commun., 1966, 26.
971. Matsumoto, K., Y. Odaira, and S. Tsutsumi, Chem. Commun., 1968, 832.
972. Matsuraga, Y., Can. J. Chem., 38, 621 (1960).
973. Matsuzaki, K., and T. Yasukawa, Chem. Commun., 1968, 1460.
974. Matthews, C. N., T. A. Magee, and J. H. Wotiz, J. Am. Chem. Soc., 81, 2273 (1959).
975. Mawby, R. J., F. Basolo, and R. G. Pearson, J. Am. Chem. Soc., 86, 3994, 5043 (1964).
976. Mawby, R. J., D. Morris, E. M. Thorsteinson, and F. Basolo, Inorg. Chem., 5, 27 (1966).
977. Mays, M. J., and S. M. Pearson, J. Chem. Soc., A, 1968, 2291.
978. Mays, M. J., and G. Wilkinson, J. Chem. Soc., 1965, 6629.
979. McArdle, P. A., and A. R. Manning, Chem. Commun., 1967, 417.
980. McAuliffe, C. A., and D. W. Meek, Inorg. Chem., 8, 904 (1969).
981. McAvoy, J., K. C. Moss, and D. W. A. Sharp, J. Chem. Soc., 1965, 1376.
982. McBride, D. W., E. Dudek, and F. G. A. Stone, J. Chem. Soc., 1964, 1752.
983. McBride, D. W., S. L. Stafford, and F. G. A. Stone, Inorg. Chem., 1, 386 (1962).
984. McBride, D. W., S. L. Stafford, and F. G. A. Stone, J. Chem. Soc., 1963, 723.
985. McCleverty, J. A., A. Davison, and G. Wilkinson, J. Chem. Soc., 1965, 3890.
986. McDonald, W. S., J. R. Moss, G. Raper, B. L. Shaw, R. Greatrex, and N. N. Greenwood, Chem. Commun., 1969, 1295.
987. McFarlane, W., Chem. Commun., 1967, 772; J. Chem. Soc., A, 1967, 1922.
988. McGinnety, J. A., R. J. Doedens, and J. A. Ibers, Inorg. Chem., 6, 2243 (1967); Science, 155, 709 (1967).
989. McGinnety, J. A., N. C. Payne, and J. A. Ibers, Chem. Commun., 1968, 235; J. Am. Chem. Soc., 91, 6301 (1969).
990. McPartlin, M., R. Mason, and L. Malatesta, Chem. Commun., 1969, 334.
991. Meek, D. W., E. C. Alyea, J. K. Stalick, and J. A. Ibers, J. Am. Chem. Soc., 91, 4920 (1969).

992. Meek, D. W., and J. A. Ibers, Inorg. Chem., 8, 1915 (1969).
993. Meriwether, L. S., E. C. Colthup, M. L. Fiene, and F. A. Cotton, J. Inorg. Nucl. Chem., 11, 181 (1959).
994. Meriwether, L. S., E. C. Colthup, G. W. Kennerly, and R. N. Reusch, J. Organometal. Chem., 26, 5155 (1961).
995. Meriwether, L. S., and M. L. Fiene, J. Am. Chem. Soc., 81, 4200 (1959).
996. Meriwether, L. S., and J. R. Leto, J. Am. Chem. Soc., 83, 3192 (1961).
997. Meriwether, L. S., M. F. Leto, E. C. Colthup, and G. W. Kennerly, J. Organometal. Chem., 27, 3930 (1962).
998. Mertweiller, J. K., and H. M. Tenney, U. S. 3,310,-576.
999. Messmer, G. G., and E. L. Amma, Inorg. Chem., 5, 1775 (1966).
1000. Messmer, G. G., E. L. Amma, and J. A. Ibers, Inorg. Chem., 6, 725 (1967).
1001. Messmer, G. G., and G. J. Palenik, Can. J. Chem., 47, 1440 (1969); Inorg. Chem., 8, 2750 (1969).
1002. Meyer, J. M., and A. L. Allred, J. Inorg. Nucl. Chem., 30, 1328 (1968).
1003. Mills, O. S., and A. D. Redhouse, Chem. Commun., 1966, 814; J. Chem. Soc., A, 1969, 1274.
1004. Minasyants, M. K., V. G. Andrianov, and Y. T. Struchkov, Zh. Strukt. Khim., 9, 1055 (1968).
1005. Misono, A., Y. Uchida, M. Hidai, and K. Kudo, J. Organometal. Chem., 20, P7 (1969).
1006. Misono, A., Y. Uchida, M. Hidai, and T. Kuse, Chem. Commun., 1968, 981.
1007. Misono, A., Y. Uchida, M. Hidai, and T. Kuse, Chem. Commun., 1969, 208.
1008. Misono, A., Y. Uchida, and T. Saito, Bull. Chem. Soc. Jap., 40, 700 (1967).
1009. Misono, A., Y. Uchida, T. Saito, M. Hidai, and M. Araki, Inorg. Chem., 8, 168 (1969).
1010. Misono, A., Y. Uchida, T. Saito, and K. M. Song, Chem. Commun., 1967, 419.
1011. Mitsui Chem. Ind. Co., Brit. 730,038 (1951).
1012. Mitsui Chem. Ind. Co., and M. Tanaka, Jap. 4918 (1953); C. A., 49, 6992 (1955).
1013. Mitsui Chem. Ind. Co., and M. Tanaka, Jap. 4919 (1953); C. A., 49, 6992 (1955).
1014. Mitsui Chem. Ind. Co., and M. Tanaka, Jap. 221 (1955); C. A., 50, 16852 (1956).
1015. Mitsui Chem. Ind. Co., M. Tanaka et al., Jap. 6619 (1953); C. A., 49, 9689 (1955); Jap. 2177 (1954); C. A., 49, 14804 (1955).

1016. Mitsui Chem. Ind. Co., M. Tanaka et al., Jap. 6625 (1953); C. A., 49, 9690 (1955).
1017. Mitsui Chem. Ind. Co., K. Yamamoto, and S. Kunizaki, Jap. 5087 (1954); C. A., 50, 6508 (1956).
1018. Mitsui Chem. Ind. Co., K. Yamamoto et al., Jap. 8268 (1954); C. A., 50, 13987 (1956).
1019. Moelwyn-Hughes, J. T., and A. W. B. Garner, Chem. Commun., 1969, 1309.
1020. Monsanto Research, U. S. 3,173,873 (1962).
1021. Montecatini-Edison, S.p.A., Fr. 1,514,495 (1966).
1022. Montecatini-Edison, S.p.A., Brit. 1,177,423 (1966).
1023. Montecatini-Edison, S.p.A., Brit. 1,182,353; Brit. 1,183,373.
1024. Montelatici, S., A. Van der Ent, J. A. Osborn, and G. Wilkinson, J. Chem. Soc., A, 1968, 1054.
1025. Moss, J. R., and B. L. Shaw, J. Chem. Soc., A, 1966, 1793.
1026. Moss, J. R., and B. L. Shaw, Chem. Commun., 1968, 632.
1027. Muir, K. W., and J. A. Ibers, J. Organometal. Chem., 18, 175 (1969).
1028. Muir, K. W., and J. A. Ibers, Inorg. Chem., 8, 1921 (1969).
1029. Mukhedkar, A. J., M. Green, and F. G. A. Stone, J. Chem. Soc., A, 1969, 3023.
1030. Muller, E., A. Segnitz, and E. Langer, Tetrahedron Lett., 1969, 1129.
1031. Muller, J., and K. Fenderl, J. Organometal. Chem., 19, 123 (1969).
1032. Murdoch, D., and E. A. Lucken, Helv. Chim. Acta, 47, 1517 (1964).
1033. Nagy-Magos, Z., G. Bor, and L. Markó, J. Organometal. Chem., 14, 205 (1968).
1034. Naldini, L., Gazz. Chim. Ital., 90, 391 (1960).
1035. Naldini, L., Gazz. Chim. Ital., 90, 1231 (1960).
1036. Naldini, L., Gazz. Chim. Ital., 90, 1337 (1960).
1037. Naldini, L., F. Cariati, G. Simonetta, and L. Malatesta, Chem. Commun., 1966, 647.
1038. Naldini, L., and A. Sacco, Gazz. Chim. Ital., 89, 2258 (1959).
1039. Nast, R., and K. W. Kruger, Z. Anorg. Chem., 341, 189 (1965).
1040. National Distillers, U. S. 3,251,893 (1962).
1041. National Distillers, J. Feldman, and B. A. Saffer, Fr. 1,388,444 (1963).
1042. Nauchno-Issledovatelsky Institute, USSR, Brit. 1,164,882 (1967).
1043. Nelson, J. H., H. B. Jonassen, and D. M. Roundhill, Inorg. Chem., 8, 2591 (1969).
1044. Nelson, J. H., K. S. Wheelock, L. C. Cusachs, and H. B. Jonassen, Chem. Commun., 1969, 1019.

1045. Nelson, S. M., and W. S. J. Kelly, Chem. Commun., 1968, 436.
1046. Nesmeyanov, A. N., K. N. Anisimov, N. E. Kolobova, and A. A. Pasynskii, Izv. Akad. Nauk, SSSR, Ser. Khim., 1968, 2814; C. A., 70, 78089.
1047. Newmeyanov, A. N., Yu. A. Chapousky, I. V. Polovyanyuk, and L. G. Makarova, J. Organometal. Chem., 7, 329 (1967).
1048. Newmeyanov, A. N., E. G. Perevalova, D. A. Lemenovsky, A. N. Kosina, and K. I. Grandberg, Izv. Akad. Nauk SSSR, Ser. Khim., 9, 2030, 2032 (1969).
1049. Nicholls, B., and M. C. Whiting, J. Chem. Soc., 1959, 551.
1050. Nichols, D. J., and A. S. Charleston, J. Chem. Soc., A, 1969, 2581.
1051. Nicholson, J. K., Angew. Chem. Int. Ed., 6, 264 (1967).
1052. Nicolini, M., C. Pecile, and A. Turco, J. Am. Chem. Soc., 87, 2379 (1965); Coordination Chem. Rev., 1, 133 (1966).
1053. Nicpon, P., and D. W. Meek, Inorg. Chem., 6, 145 (1967).
1054. Nixon, J. F., and M. D. Sexton, J. Chem. Soc., A, 1967, 1136; 1969, 1089.
1055. Noack, K., M. Ruch, and F. Calderazzo, Inorg. Chem., 7, 345 (1968).
1056. Norbury, A. H., and A. I. P. Sinha, J. Chem. Soc., A, 1968, 1598.
1057. Norris, A. R., and M. C. Baird, Can. J. Chem., 47, 3003 (1969).
1058. Nyholm, R. S., and P. Royo, Chem. Commun., 1969, 421.
1059. Nyholm, R. S., S. S. Sandhu, and M. H. B. Stiddard, J. Chem. Soc., 1963, 5916.
1060. Nyholm, R. S., J. F. Skinner, and M. H. B. Stiddard, J. Chem. Soc., A, 1968, 38.
1061. Nyholm, R. S., and K. Vrieze, Chem. Ind. (London), 1964, 318; J. Chem. Soc., 1965, 5337.
1062. Nyman, C. J., C. E. Wymore, and G. Wilkinson, Chem. Commun., 1967, 407; J. Chem. Soc., A, 1968, 561.
1063. Nyman, F., Chem. Ind. (London), 1965, 604.
1064. O'Connor, C., and G. Wilkinson, J. Chem. Soc., A, 1968, 2665.
1065. O'Connor, C., G. Yagupsky, D. Evans, and G. Wilkinson, Chem. Commun., 1968, 420.
1066. Odell, A. L., and H. A. Raethel, Chem. Commun., 1968, 1323.
1067. Ohno, K., and J. Tsuji, J. Am. Chem. Soc., 90, 99 (1968).
1068. Orioli, P. L., and L. Sacconi, Chem. Commun., 1968, 1310.

1069. Orioli, P. L., and L. Sacconi, Chem. Commun., 1969, 1012.
1070. Orioli, P. L., and L. Vaska, Proc. Chem. Soc., 1962, 333.
1071. Osborn, J. A., Chem. Commun., 1968, 1231.
1072. Osborn, J. A., F. H. Jardine, J. F. Young, and G. Wilkinson, J. Chem. Soc., A, 1966, 1711.
1073. Osborn, J. A., G. Wilkinson, and J. F. Young, Chem. Commun., 1965, 17.
1074. Osborne, A. G., and M. H. B. Stiddard, J. Chem. Soc., 1962, 4715.
1075. Osborne, A. G., and M. H. B. Stiddard, J. Chem. Soc., 1964, 634.
1076. Osborne, A. G., and M. H. B. Stiddard, J. Chem. Soc., 1965, 700.
1077. Osborne, A. G., and M. H. B. Stiddard, J. Organometal. Chem., 3, 340 (1965).
1078. Otsuka, S., A. Nakamura, and K. Tani, J. Organometal. Chem., 14, P30 (1968).
1079. Otsuka, S., and M. Rossi, J. Chem. Soc., A, 1969, 497.
1080. Owston, P. G., J. M. Partridge, and J. M. Rowe, Acta Crystallogr., 13, 246 (1960).
1081. Owston, P. G., and J. M. Rowe, Acta Crystallogr., 13, 253 (1960).
1082. Owston, P. G., and J. M. Rowe, J. Chem. Soc., 1963, 3411.
1083. Panattoni, C., G. Bombieri, U. Belluco, and W. H. Baddley, J. Am. Chem. Soc., 90, 798 (1968).
1084. Park, P. J. D., and P. J. Hendra, Spectrochim. Acta, 25A, 227, 909 (1969).
1085. Parker, D. J., and M. H. B. Stiddard, J. Chem. Soc., A, 1966, 695.
1086. Parshall, G. W., J. Am. Chem. Soc., 87, 2133 (1965).
1087. Parshall, G. W., J. Am. Chem. Soc., 88, 704 (1966).
1088. Parshall, G. W., and F. N. Jones, J. Am. Chem. Soc., 87, 5356 (1965).
1089. Parshall, G. W., W. H. Knoth, and R. A. Schunn, J. Am. Chem. Soc., 91, 4990 (1969).
1090. Parshall, G. W., and G. Wilkinson, Chem. Ind. (London), 1962, 261; Inorg. Chem., 1, 896 (1962).
1091. Patmore, D. J., and W. A. G. Graham, Inorg. Chem., 5, 1405 (1966).
1092. Paulik, F. E., and J. F. Roth, Chem. Commun., 1968, 1578.
1093. Payne, D. S., J. A. A. Mokuolu, and J. C. Speakman, Chem. Commun., 1965, 599.
1094. Payne, N. C., and J. A. Ibers, Inorg. Chem., 8, 2714 (1969).
1095. Pearson, R. G., and M. M. Muir, J. Am. Chem. Soc., 88, 2163 (1966).

1096. Pearson, R. G., M. M. Muir, and L. M. Venanzi, J. Chem. Soc., 1965, 5521.
1097. Pecile, C., Inorg. Chem., 5, 210 (1966).
1098. Pegot, C., and R. Poilblanc, C. R. Acad. Sci., Paris, 268, C955 (1969).
1099. Peloso, A., and G. Dolcetti, J. Chem. Soc., A, 1967, 1944.
1100. Perutz, M. F., and O. Weisz, J. Chem. Soc., 1946, 488.
1101. Pettit, L. D., and H. M. N. H. Irving, J. Chem. Soc., 1964, 5336
1102. Phillips, J. R., D. T. Rosevear, and F. G. A. Stone, J. Organometal. Chem., 2, 455 (1964).
1103. Phillips Petroleum, U. S. 3,427,365 (1966).
1104. Piacenti, F., M. Bianchi, E. Benedetti, and G. Braca, Inorg. Chem., 7, 1815 (1968).
1105. Piacenti, F., M. Bianchi, E. Benedetti, and G. Sbrana, J. Inorg. Nucl. Chem., 29, 1389 (1967).
1106. Pidcock, A., Chem. Commun., 1968, 92.
1107. Pidcock, A., R. E. Richards, and L. M. Venanzi, Proc. Chem. Soc., 1962, 184.
1108. Pidcock, A., R. E. Richards, and L. M. Venanzi, J. Chem. Soc., A, 1966, 1707.
1109. Pidcock, A., R. E. Richards, and L. M. Venanzi, J. Chem. Soc., A, 1968, 1970.
1110. Pignolet, L. H., D. Forster, and W. D. Horrocks, Inorg. Chem., 7, 828 (1968).
1111. Pignolet, L. H., and W. De W. Horrocks, Chem. Commun., 1968, 1012.
1112. Pignolet, L. H., and W. De W. Horrocks, J. Am. Chem. Soc., 90, 922 (1968).
1113. Plastas, H. J., J. M. Stewart, and S. O. Grim, J. Am. Chem. Soc., 91, 4326 (1969).
1114. Plowman, R. A., and F. G. A. Stone, Inorg. Chem., 1, 518 (1962).
1115. Pneumaticakis, G. A., Chem. Commun., 1968, 275.
1116. Poilblanc, R., and M. Bigorgne, C. R. Acad. Sci., Paris, 250, 1064 (1960); 252, 3054 (1961); Bull. Soc. Chim. Fr., 1962, 1301.
1117. Poilblanc, R., and M. Bigorgne, J. Organometal. Chem., 5, 93 (1966).
1118. Polovyanyuk, I. K., et al., Izv. Akad. Nauk SSSR, Ser. Khim., 1966, 387.
1119. Porri, L., M. C. Gallazzi, and G. Vitulli, Chem. Commun., 1967, 228.
1120. Porri, L., G. Vitulli, M. Zocchi, and G. Allegra, Chem. Commun., 1969, 276.
1121. Powell, J., S. D. Robinson, and B. L. Shaw, Chem. Commun., 1965, 78; J. Chem. Soc., A, 1967, 1839.
1122. Powell, J., and B. L. Shaw, J. Chem. Soc., 1965, 3879.

1123. Powell, J., and B. L. Shaw, J. Chem. Soc., A, 1968, 159.
1124. Powell, J., and B. L. Shaw, J. Chem. Soc., A, 1968, 211.
1125. Powell, J., and B. L. Shaw, J. Chem. Soc., A, 1968, 617.
1126. Powell, J., and B. L. Shaw, J. Chem. Soc., A, 1968, 774.
1127. Powell, J., and B. L. Shaw, J. Chem. Soc., A, 1968, 780.
1128. Powell, P., and H. Noth, Chem. Commun., 1966, 637.
1129. Pregaglia, G., A. Andreetta, G. Ferrari, and R. Ugo, Chem. Commun., 1969, 590.
1130. Prince, R. H., and K. A. Raspin, Chem. Commun., 1966, 156; J. Chem. Soc., A, 1969, 612.
1131. Prince, R. H., and K. A. Raspin, J. Inorg. Nucl. Chem., 31, 695 (1969).
1132. Pu, L. S., A. Yamamoto, and S. Ikeda, J. Am. Chem. Soc., 90, 3896 (1968).
1133. Quin, L. D., J. Am. Chem. Soc., 79, 3681 (1957).
1134. Ramaswamy, H. N., H. B. Jonassen, and A. M. Aguiar, Inorg. Chim. Acta, 1, 141 (1967).
1135. Randall, E. W., and D. Shaw, J. Chem. Soc., A, 1969, 2867.
1136. Raspin, K. A., J. Chem. Soc., A, 1969, 461.
1137. Rausch, M. D., Y. F. Chang, and H. B. Gordon, Inorg. Chem., 8, 1355 (1969).
1138. Reckziegel, A., and M. Bigorgne, J. Organometal. Chem., 3, 341 (1965).
1139. Reddy, G. K. N., and E. G. Leelamani, Current Sci., 34, 146 (1965).
1140. Reed, C. A., and W. R. Roper, Chem. Commun., 1969, 155.
1141. Reed, C. A., and W. R. Roper, Chem. Commun., 1969, 1459.
1142. Reed, H. W. B., J. Chem. Soc., 1954, 1931.
1143. Reilly, C. A., and H. Thyret, J. Am. Chem. Soc., 89, 5144 (1967).
1144. Reppe, W., and W. J. Schweckendiek, Ann. Chem., 560, 104 (1948).
1145. Rest, A. J., J. Chem. Soc., A, 1968, 2212.
1146. Rest, A. J., D. T. Rosevear, and F. G. A. Stone, J. Chem. Soc., A, 1967, 66.
1147. Rigo, P., M. Bressan, and A. Turco, Inorg. Chem., 7, 1460 (1968).
1148. Rigo, P., B. Corain, and A. Turco, Inorg. Chem., 7, 1623 (1968).
1149. Rigo, P., G. Guastalla, and A. Turco, Inorg. Chem., 8, 375 (1969).
1150. Rigo, P., C. Pecile, and A. Turco, Inorg. Chem., 6, 1636 (1967).

1151. Rinze, P. V., J. Lorberth, H. Nöth, and B. Stutte, J. Organometal. Chem., 19, 399 (1969).
1152. Robinson, B. H., and W. S. Tham, J. Organometal. Chem., 16, P45 (1969).
1153. Robinson, S. D., and G. Wilkinson, J. Chem. Soc., A, 1966, 300.
1154. Rohrscheid, F., and R. H. Holm, J. Organometal. Chem., 4, 335 (1965).
1155. Rose, D., J. D. Gilbert, R. P. Richardson, and G. Wilkinson, J. Chem. Soc., A, 1969, 2610.
1156. Rose, J. D., and F. S. Statham, J. Chem. Soc., 1950, 69.
1157. Rosevear, D. T., and F. G. A. Stone, J. Chem. Soc., 1965, 5275.
1158. Rosevear, D. T., and F. G. A. Stone, J. Chem. Soc., A, 1968, 164.
1159. Ross, E. P., and G. R. Dobson, J. Inorg. Nucl. Chem., 30, 2363 (1968).
1160. Rossi, M., and A. Sacco, Chem. Commun., 1969, 471.
1161. Roundhill, D. M., Chem. Commun., 1969, 567.
1162. Roundhill, D. M., and H. B. Jonassen, Chem. Commun., 1968, 1233.
1163. Roundhill, D. M., D. N. Lawson, and G. Wilkinson, J. Chem. Soc., A, 1968, 845.
1164. Roundhill, D. M., and G. Wilkinson, J. Chem. Soc., A, 1968, 506.
1165. Rouschias, G., and G. Wilkinson, J. Chem. Soc., A, 1966, 465.
1166. Rouschias, G., and G. Wilkinson, J. Chem. Soc., A, 1967, 993.
1167. Ruddick, J. D., and B. L. Shaw, Chem. Commun., 1967, 1135; J. Chem. Soc., A, 1969, 2801.
1168. Ruddick, J. D., and B. L. Shaw, J. Chem. Soc., A, 1969, 2969.
1169. Rusina, A., and A. A. Vlček, Nature, London, 206, 295 (1965).
1170. Sacco, A., Ann. Chim. (Rome), 43, 495 (1953); C. A., 48, 5012 (1954).
1171. Sacco, A., Gazz. Chim. Ital., 93, 542 (1963).
1172. Sacco, A., Gazz. Chim. Ital., 93, 698 (1963).
1173. Sacco, A., and M. Aresta, Chem. Commun., 1968, 1223.
1174. Sacco, A., and M. Freni, J. Inorg. Nucl. Chem., 8, 566 (1958); Ann. Chim. (Rome), 48, 218 (1958); C. A., 52, 19656 (1958).
1175. Sacco, A., and M. Freni, Int. Conf. Pure Appl. Chem., Munich, 1959; C. A., 55, 15386 (1961).
1176. Sacco, A., and F. Gorieri, Gazz. Chim. Ital., 93, 687 (1963)
1177. Sacco, A., and M. Rossi, Chem. Commun., 1967, 316; Inorg. Chim. Acta, 2, 127 (1968).
1178. Sacco, A., and M. Rossi, Chem. Commun., 1965, 602.

1179. Sacco, A., M. Rossi, and C. F. Nobile, Chem. Commun., 1966, 589.
1180. Sacco, A., and R. Ugo, J. Chem. Soc., 1964, 3274.
1181. Sacco, A., R. Ugo, and A. Moles, J. Chem. Soc., A, 1966, 1670; Coordination Chem. Rev., 1, 234 (1966).
1182. Sacconi, L., and I. Bertini, J. Am. Chem. Soc., 89, 2235 (1967); 90, 5443 (1968).
1183. Sacconi, L., and J. Gelsomini, Inorg. Chem., 7, 291 (1968).
1184. Sacconi, L., and R. Morassi, J. Chem. Soc., A, 1968, 2997.
1185. Sacconi, L., and R. Morassi, J. Chem. Soc., A, 1969, 2904.
1186. Sandhu, S. S., and M. Gupta, Chem. Ind. (London), 1967, 1876.
1187. Scatturin, V., and A. Turco, J. Inorg. Nucl. Chem., 8, 447 (1958).
1188. Schmid, G., W. Petz, W. Arloth, and N. Nöth, Angew. Chem. Int. Ed., 6, 696 (1967).
1189. Schmutzler, R., J. Chem. Soc., 1965, 5630.
1190. Schrauzer, G. N., J. Am. Chem. Soc., 81, 5310 (1959); 82, 1008 (1960); Chem. Ber., 94, 642 (1961).
1191. Schrauzer, G. N., Chem. Ber., 94, 1403 (1961).
1192. Schrauzer, G. N., and P. Glockner, Chem. Ber., 97, 2451 (1964).
1193. Schrauzer, G. N., V. P. Mayweg, H. W. Finck, and W. Heinrich, J. Am. Chem. Soc., 88, 4604 (1966).
1194. Schrauzer, G. N., V. P. Mayweg, and W. Heinrich, J. Am. Chem. Soc., 88, 5174 (1966).
1195. Schumann, H., and O. Stelzer, Angew. Chem. Int. Ed., 6, 701 (1967).
1196. Schumann, H., O. Stelzer, and U. Niederreuther, J. Organometal. Chem., 16, P64 (1969).
1197. Schuster-Wolden, H. G., and F. Basolo, J. Am. Chem. Soc., 88, 1657 (1966).
1198. Schwarzenbach, G., Chem. Zvest., 19, 200 (1965).
1199. Scott, R. N., D. F. Shriver, and L. Vaska, J. Am. Chem. Soc., 90, 1079 (1968).
1200. Seel, F., K. Ballreich, and R. Schmutzler, Chem. Ber., 94, 1173 (1961).
1201. Selbin, J., N. Ahmad, and M. J. Pribble, Chem. Commun., 1969, 759.
1202. Selbin, J., and J. D. Ortego, J. Inorg. Nucl. Chem., 29, 1449 (1967).
1203. Selbin, J., and G. Vigee, J. Inorg. Nucl. Chem., 30, 1644 (1968).
1204. Shapley, J. R., R. R. Schrock, and J. A. Osborn, J. Am. Chem. Soc., 91, 2816 (1969).
1205. Shaw, B. L., and E. Singleton, J. Chem. Soc., A, 1967, 1683.

1206. Shaw, B. L., and A. C. Smithies, J. Chem. Soc., A, 1967, 1047.
1207. Shaw, B. L., and A. C. Smithies, J. Chem. Soc., A, 1968, 2784.
1208. Shaw, D., and E. W. Randall, Chem. Commun., 1965, 82.
1209. Shell, U. S. 3,110,747; D.A.S., 1,273,532, 1,220,843 (1960).
1210. Shell, Fr. 1,370,507 (1962).
1211. Shell Development Co., A. J. Van Peski, and J. A. Van Melsen, U. S. 2,150,349 (1938).
1212. Shell Int. Res. Maat, Brit. 988,941; 988,942; 988,943 (1960).
1213. Shell Int. Res. Maat., Belg. 648,356 (1963).
1214. Shell Oil, Brit. 942,435, U. S. 3,130,237 (1961).
1215. Shell Oil, U. S. 3,168,553 (1961).
1216. Shell Oil, U. S. 3,168,507, 3,168,508 (1962).
1217. Shell Oil, U. S. 3,239,566 (1963).
1218. Shell Oil, U. S. 3,239,569 (1963).
1219. Shell Oil, U. S. 3,239,571 (1963).
1220. Shupack, S. I., J. Inorg. Nucl. Chem., 28, 2418 (1966).
1221. Sieckhaus, J. F., and T. Layloff, Inorg. Chem., 6, 2185 (1967).
1222. Siedel, W., Z. Anorg. Chem., 335, 316 (1965).
1223. Siedel, W., Z. Anorg. Chem., 341, 70 (1965); 342, 165 (1966).
1224. Siefert, E. E., and R. J. Angelici, J. Organometal. Chem., 8, 374 (1967).
1225. Siekman, R. W., and D. L. Weaver, Chem. Commun., 1968, 1021.
1226. Simo, C., and S. Holt, Inorg. Chem., 7, 2655 (1968).
1227. Simon, A., Z. Nagy-Magos, J. Palagyi, G. Palyi, G. Bor, and L. Marko, J. Organometal. Chem., 11, 634 (1968).
1228. Sims, J. J., V. K. Honwad, and L. H. Selman, Tetrahedron Lett., 1969, 87.
1229. Singer, H., and G. Wilkinson, J. Chem. Soc., A, 1968, 849.
1230. Singer, H., and G. Wilkinson, J. Chem. Soc., A, 1968, 2516.
1231. Singh, S., P. P. Singh, and R. Rivest, Inorg. Chem., 7, 1236 (1968).
1232. Skapski, A. C., and F. A. Stephens, Chem. Commun., 1969, 1008.
1233. Skapski, A. C., and P. G. H. Troughton, Chem. Commun., 1968, 1230.
1234. Skapski, A. C., and P. G. H. Troughton, Chem. Commun., 1969, 170; J. Chem. Soc., A, 1969, 2772.
1235. Smith, B. C., and M. A. Wassef, J. Chem. Soc., A, 1968, 1817.

1236. Smith, J. G., and D. T. Thompson, J. Chem. Soc., A, 1967, 1694.
1237. Snow, M. R., and M. H. B. Stiddard, J. Chem. Soc., A, 1966, 777.
1238. Socrates, G., J. Inorg. Nucl. Chem., 31, 1667 (1969).
1239. Spier, G., and L. Marko, Inorg. Chim. Acta, 3, 126 (1969).
1240. Spofford, W. A., P. D. Carfagna, and E. L. Amma, Inorg. Chem., 6, 1553 (1967).
1241. Spooncer, W. W., A. C. Jones, and L. H. Slaugh, J. Organometal. Chem., 18, 327 (1969).
1242. Srivastava, S. C., and M. Bigorgne, J. Organometal. Chem., 18, P30 (1969).
1243. Srivastava, S. C., and M. Bigorgne, J. Organometal. Chem., 19, 241 (1969).
1244. Stalick, J. K., and J. A. Ibers, Inorg. Chem., 8, 419 (1969).
1245. Stalick, J. K., and J. A. Ibers, Inorg. Chem., 8, 1084 (1969).
1246. Stalick, J. K., and J. A. Ibers, Inorg. Chem., 8, 1090 (1969).
1247. Stalick, J. K., and J. A. Ibers, Inorg. Chem., 9, 453 (1970).
1248. Staples, P. J., and A. Thompson, J. Chem. Soc., A, 1969, 1058.
1249. Stephenson, T. A., Inorg. Nucl. Chem. Lett., 4, 687 (1968).
1250. Stephenson, T. A., S. M. Morehouse, A. R. Powell, J. P. Heffer, and G. Wilkinson, J. Chem. Soc., 1965, 3632.
1251. Stephenson, T. A., and G. Wilkinson, J. Inorg. Nucl. Chem., 28, 945 (1966).
1252. Stephenson, T. A., and G. Wilkinson, J. Inorg. Nucl. Chem., 29, 2122 (1967).
1253. Stevenson, D. L., and L. F. Dahl, J. Am. Chem. Soc., 89, 3424 (1967).
1254. Stiddard, M. H. B., J. Chem. Soc., 1963, 756.
1255. Stiddard, M. H. B., and R. E. Townsend, Chem. Commun., 1969, 1372.
1256. Strohmeier, W., and C. Barbeau, Z. Naturforsch., B, 17, 848 (1962).
1257. Strohmeier, W., and F. J. Müller, Z. Naturforsch., B, 22, 451 (1967).
1258. Strohmeier, W., and W. Rehder-Stirnweiss, J. Organometal. Chem., 18, P28 (1969).
1259. Takahashi, S., and N. Hagihara, J. Chem. Soc., Jap., Ind. Chem. Sect., 72, 1637 (1969).
1260. Takahashi, S., T. Shibano, and N. Hagihara, Tetrahedron Lett., 26, 2451 (1967).

1261. Takahashi, S., K. Sonogashira, and N. Hagihara, J. Chem. Soc. Jap., 87, 610 (1966); C. A., 65, 14485 (1966).
1262. Takesada, M., H. Yamazaki, and N. Hagihara, J. Chem. Soc. Jap., 89, 1121 (1968); C. A., 70, 87940.
1263. Takesada, M., H. Yamazaki, and N. Hagihara, J. Chem. Soc. Jap., 89, 1126 (1968); C. A., 70, 47912.
1264. Tarama, K., S. Yoshida, H. Kanai, and S. Osaka, Bull. Chem. Soc. Jap., 41, 1271 (1968).
1265. Tayim, H. A., and J. C. Bailar, J. Am. Chem. Soc., 89, 3420 (1967).
1266. Tayim, H. A., and J. C. Bailar, J. Am. Chem. Soc., 87, 4330 (1967).
1267. Taylor, R. C. G. R. Dobson, and R. A. Kolodny, Inorg. Chem., 7, 1886 (1968).
1268. Taylor, R. C., J. F. Young, and G. Wilkinson, Inorg. Chem., 5, 20 (1966).
1269. Taylor, T., and L. R. Hathaway, Inorg. Chem., 8, 2135 (1969).
1270. Thomas, G., Z. Anorg. Chem., 360, 15 (1968).
1271. Thompson, D. T., J. Organometal. Chem., 4, 74 (1965).
1272. Thorsteinson, E. M., and F. Basolo, Inorg. Chem., 5, 1691 (1966); J. Am. Chem. Soc., 88, 3929 (1966).
1273. Treichel, P. M., K. W. Barnett, and R. L. Shobkin, J. Organometal. Chem., 7, 449 (1967).
1274. Treichel, P. M., R. L. Shubkin, K. W. Barnett, and D. Reichard, Inorg. Chem., 5, 1177 (1966).
1275. Treichel, P. M., and G. Werber, Inorg. Chem., 4, 1098 (1965).
1276. Troughton, P. G. H., and A. C. Skapski, Chem. Commun., 1968, 575.
1277. Tsang, W. S., D. W. Meek, and A. Wojcicki, Inorg. Chem., 7, 1263 (1968).
1278. Tsuji, J., and K. Ohno, Tetrahedron Lett., 1965, 3969.
1279. Tsuji, J., and K. Ohno, J. Am. Chem. Soc., 88, 3452 (1966).
1280. Tucci, E. R., Ind. Eng. Chem. Prod. Res. Develop., 7, 32 (1968); 7, 227 (1968).
1281. Tuck, D. G., Coordination Chem. Rev., 1, 286 (1966).
1282. Turco, A., and G. Giacometti, Ric. Sci., 30, 1051 (1960).
1283. Turco, A., C. Panattoni, and E. Frasson, Nature, 187, 772 (1960).
1284. Turco, A., and C. Pecile, Nature, 191, 66 (1961).
1285. Turco, A., C. Pecile, M. Nicolini, and M. Martelli, J. Am. Chem. Soc., 85, 3510 (1963).
1286. Turco, A., V. Scatturin, and G. Giacometti, Nature, 183, 601 (1959); Gazz. Chim. Ital., 89, 2005 (1959).
1287. Ugo, R., Coordination Chem. Rev., 3, 319 (1968).

1288. Ugo, R., and F. Bonati, J. Organometal. Chem., 8, 189 (1967).
1289. Ugo, R., F. Cariati, and G. La Monica, Chem. Commun., 1966, 868.
1290. Ugo, R., F. Conti, S. Cenini, R. Mason, and G. B. Robertson, Chem. Commun., 1968, 1498.
1291. Uhlig, E., and M. Maaser, Z. Anorg. Chem., 343, 205 (1966).
1292. Uhlig, E., and M. Schafer, Z. Chem., 8, 470 (1968); Z. Anorg. Chem., 359, 67 (1968).
1293. Union Carbide Corp., U. S. 3,499,932 (1967).
1294. U. S. Rubber, Brit. 981,840 (1962).
1295. U. S. Rubber, U. S. 3,267,087 (1963).
1296. U. S. Rubber, Brit. 1,036,984 (1964).
1297. Vallarino, L., J. Chem. Soc., 1957, 2287.
1298. Vallarino, L., J. Inorg. Nucl. Chem., 8, 288 (1958).
1299. Vallarino, L. M., Inorg. Chem., 4, 161 (1965).
1300. Van Den Akker, M., and F. Jellinek, J. Organometal. Chem., 10, P37 (1967); Rec. Trav. Chim., 86, 897 (1967).
1301. Van Hecke, G. R., and W. DeW. Horrocks, Inorg. Chem., 5, 1960 (1966).
1302. Van Hecke, G. R., and W. DeW. Horrocks, Inorg. Chem., 5, 1968 (1966).
1303. Vaska, L., Z. Naturforsch., B, 15, 56 (1960).
1304. Vaska, L., J. Am. Chem. Soc., 83, 756 (1961).
1305. Vaska, L., Chem. Ind. (London), 1961, 1402.
1306. Vaska, L., Science, 140, 808 (1963).
1307. Vaska, L., J. Am. Chem. Soc., 86, 1943 (1964).
1308. Vaska, L., Chem. Commun., 1966, 614.
1309. Vaska, L., J. Am. Chem. Soc., 88, 4100 (1966).
1310. Vaska, L., J. Am. Chem. Soc., 88, 5325 (1966).
1311. Vaska, L., and S. S. Bath, J. Am. Chem. Soc., 88, 1333 (1966).
1312. Vaska, L., and D. L. Catone, J. Am. Chem. Soc., 88, 5324 (1966).
1313. Vaska, L., and J. W. Diluzio, J. Am. Chem. Soc., 83, 1262 (1961).
1314. Vaska, L., and J. W. Diluzio, J. Am. Chem. Soc., 83, 2784 (1961).
1315. Vaska, L., and J. W. Diluzio, J. Am. Chem. Soc., 84, 679 (1962).
1316. Vaska, L., and J. W. Diluzio, J. Am. Chem. Soc., 84, 4989 (1962).
1317. Vaska, L., and R. E. Rhodes, J. Am. Chem. Soc., 87, 4970 (1965).
1318. Vaska, L., and E. M. Sloane, J. Am. Chem. Soc., 82, 1263 (1960).
1319. Veidis, M. V., and G. J. Palenik, Chem. Commun., 1969, 586.

1320. Venanzi, L. M., J. Chem. Soc., 1958, 719; J. Inorg. Nucl. Chem., 8, 137 (1958).
1321. Venanzi, L. M., Chem. Brit. 4, 162 (1968).
1322. Vohler, O., Chem. Ber., 91, 1235 (1958).
1323. Volger, H. C., and K. Vrieze, J. Organometal. Chem., 6, 297 (1966); 9, 527, 537 (1967).
1324. Volger, H. C., K. Vrieze, and A. P. Praat, J. Organometal. Chem., 14, 185, 429 (1968).
1325. Volpin, M. E., N. K. Chapovskaya, and V. B. Shur, Izv. Akad. Nauk SSSR, Ser. Khim., 1966, 1083.
1326. Vranka, R. G., L. F. Dahl, P. Chini, and J. Chatt, J. Am. Chem. Soc., 91, 1574 (1969).
1327. Walton, R. A., and R. Whyman, J. Chem. Soc., A, 1968, 1394.
1328. Wawersik, H., and F. Basolo, Chem. Commun., 1966, 366.
1329. Weiss, E., and W. Hübel, J. Inorg. Nucl. Chem., 11, 42 (1959).
1330. Weiss, E., W. Hübel, and R. Merényi, Chem. Ber., 95, 1155 (1962).
1331. Wells, A. F., Z. Kristallogr., 94, 447 (1936); C. A., 31, 590 (1937).
1332. Werner, H., J. Organometal. Chem., 5, 100 (1966).
1333. Werner, H., and R. Prinz, J. Organometal. Chem., 5, 79, 100 (1966).
1334. Werner, H., and R. Prinz, Chem. Ber., 99, 3582 (1966).
1335. Werner, H., R. Prinz, E. Bundschuh, and K. Deckelmann, Angew. Chem., 78, 646 (1966).
1336. Werner, H., and H. Rascher, Inorg. Chim. Acta, 2, 181 (1968).
1337. Werner, R. P. M., Z. Naturforsch., B, 16, 477 (1961).
1338. Werner, R. P. M., Z. Naturforsch., B, 16, 478 (1961).
1339. Westland, A. D., J. Chem. Soc., 1965, 3060.
1340. Westland, A. D., Can. J. Chem., 47, 4135 (1969).
1341. Westland, A. D., and L. Westland, Can. J. Chem., 43, 426 (1965).
1342. Whitesides, G. M., and J. S. Fleming, J. Am. Chem. Soc., 89, 2855 (1967).
1343. Whitesides, G. M., and G. Maglio, J. Am. Chem. Soc., 91, 4980 (1969).
1344. Wilford, J. B., A. Forster, and F. G. A. Stone, J. Chem. Soc., 1965, 6519.
1345. Wilford, J. B., and H. M. Powell, J. Chem. Soc., A, 1967, 2092.
1346. Wilford, J. B., and H. M. Powell, J. Chem. Soc., A, 1969, 8.
1347. Wilke, G., and B. Bogdanovie, Angew. Chem., 73, 756 (1961).

1348. Wilke, G., and G. Herrmann, Angew. Chem., 74, 693 (1962).
1349. Wilke, G., E. W. Muller, and M. Kroner, Angew. Chem., 73, 33 (1961).
1350. Wilke, G., H. Schott, and P. Heimbach, Angew. Chem. Int. Ed., 6, 92 (1967).
1351. Wilkinson, G., Brit. 1,130,749 (1965).
1352. Wilkinson, G., and T. S. Piper, J. Inorg. Nucl. Chem., 2, 33 (1956).
1353. Winkhaus, G., and H. Singer, Chem. Ber., 99, 3610 (1966).
1354. Wittle, J. K., and G. Urry, Inorg. Chem., 7, 560 (1968).
1355. Wojcicki, A., and M. F. Farona, Inorg. Chem., 3, 151 (1964).
1356. Workman, M. O., G. Dyer, and D. W. Meek, Inorg. Chem., 6, 1543 (1967).
1357. Wright, D., Chem. Commun., 1966, 197.
1358. Wu, Chisung, and F. J. Welch, J. Organometal. Chem., 30, 1229 (1965).
1359. Wymore, C. E., and J. C. Bailar, J. Inorg. Nucl. Chem., 14, 42 (1960).
1360. Yagupsky, G., C. K. Brown, and G. Wilkinson, Chem. Commun., 1969, 1244.
1361. Yagupsky, G., and G. Wilkinson, J. Chem. Soc., A, 1969, 725.
1362. Yagupsky, M., and G. Wilkinson, J. Chem. Soc., A, 1968, 2813.
1363. Yamamoto, A., S. Kitazume, and S. Ikeda, J. Am. Chem. Soc., 90, 1089 (1968).
1364. Yamamoto, A., S. Kitazume, L. S. Pu, and S. Ikeda, Chem. Commun., 1967, 79.
1365. Yamamoto, K., Bull. Chem. Soc. Jap., 27, 501, 516 (1954).
1366. Yamamoto, K., and M. Oku, Bull. Chem. Soc., Jap., 27, 382 (1954).
1367. Yamamoto, K., and M. Oku, Bull. Chem. Soc. Jap., 27, 505 (1954).
1368. Yamamoto, K., and M. Oku, Bull. Chem. Soc. Jap., 27, 509 (1954).
1369. Yamazaki, H., and N. Hagihara, Bull. Chem. Soc. Jap., 38, 2212 (1965).
1370. Yamazaki, H., and N. Hagihara, J. Organometal. Chem., 7, P22 (1967).
1371. Yamazaki, H., T. Nishido, Y. Matsumoto, S. Sumida, and Hagihara, J. Organometal. Chem., 6, 86 (1966).
1372. Yamazaki, H., M. Takesada, and N. Hagihara, Bull. Chem. Soc. Jap., 42, 275 (1969).
1373. Young, J. F., J. A. Osborn, F. H. Jardine, and G. Wilkinson, Chem. Commun., 1965, 131.

1374. Zakharkin, L. I., and G. G. Zhigareua, Zh. Obshch. Khim., 37, 1791 (1967).
1375. Ziegler, M. L., H. Haas, and R. K. Sheline, Chem. Ber., 98, 2454 (1965).
1376. Zingales, F., and F. Canziani, Gazz. Chim. Ital., 92, 343 (1962).
1377. Zingales, F., F. Canziani, and F. Basolo, J. Organometal. Chem., 7, 461 (1967).
1378. Zingales, F., F. Canziani, and A. Chiesa, Inorg. Chem., 2, 1303 (1963).
1379. Zingales, F., A. Chiesa, and F. Basolo, J. Am. Chem. Soc., 88, 2707 (1966).
1380. Zingales, F., U. Sartorelli, and A. Trovati, Inorg. Chem., 6, 1246 (1967).
1381. Ziolo, R. G., and Z. Dori, J. Am. Chem. Soc., 90, 6560 (1968).